FORCE

$$1\ N = 1\ kg\ m/s^2$$
$$= 0.224\ 809\ lbf$$

$$1\ lbf = 4.448\ 222\ N$$
$$1\ dyne = 1 \times 10^{-5}\ N$$

ENERGY

$$1\ Btu = 778.169\ ft\ lbf$$
$$1\ J = 9.478 \times 10^{-4}\ Btu$$
$$1\ cal = 4.1840\ J$$

$$1\ Btu = 1.055\ 056\ kJ$$
$$1\ ft\ lbf = 1.3558\ J$$
$$1\ IT\ cal = 4.1868\ J$$

SPECIFIC ENERGY

$$1\ kJ/kg = 0.42992\ Btu/lbm$$
$$1\ kJ/kg\ mol = 0.4299\ Btu/lbmol$$

$$1\ Btu/lbm = 2.326\ kJ/kg$$
$$1\ Btu/lbmol = 2.326\ kJ/kg\ mol$$

SPECIFIC ENTROPY, SPECIFIC HEAT, GAS CONSTANT

$$1\ kJ/kg\ K = 0.2388\ Btu/lbm\ R$$
$$1\ kJ/kg\ mol\ K = 0.2388\ Btu/lbmol\ R$$

$$1\ Btu/lbm\ R = 4.1868\ kJ/kg\ K$$
$$1\ Btu/lbmol\ R = 4.1868\ kJ/kg\ mol\ K$$

DENSITY

$$1\ kg/m^3 = 0.062\ 428\ lbm/ft^3$$

$$1\ lbm/ft^3 = 16.0185\ kg/m^3$$

SPECIFIC VOLUME

$$1\ m^3/kg = 16.018\ ft^3/lbm$$

$$1\ ft^3/lbm = 0.062\ 428\ m^3/kg$$

POWER

$$1\ W = 1\ J/s$$
$$1\ kW = 1.3410\ hp = 3412\ Btu/h$$

$$1\ Btu = 1.055\ 056\ kJ/s$$
$$1\ hp = 550\ ft\ lbf/s$$
$$= 2545\ Btu = 745.7\ W$$

VELOCITY

$$1\ m/s = 3.281\ ft/s = 2.237\ mph$$

$$1\ mph = 1.467\ ft/s = 0.4470\ m/s$$
$$1\ ft/s = 0.3048\ m/s$$

TEMPERATURE

$$T[°C] = \tfrac{5}{9}\left(T[°F] - 32\right)$$
$$T[°C] = T[K] - 273.15$$
$$T[K] = \tfrac{5}{9}T[R]$$

$$T[°F] = \tfrac{9}{5}T[°C] + 32$$
$$T[°F] = T[R] - 459.67$$

$$\Delta T[K] = 1.8\Delta T[R]$$
$$\Delta T[K] = \Delta T[°C]$$
$$\Delta T[R] = \Delta T[°F]$$

D1313746

F. W. Olin College Library

Thermodynamics
Principles and Practice

Michel Saad
Santa Clara University
Santa Clara, California

Prentice Hall, Upper Saddle River, New Jersey 07458

F.W. Olin College Library

Library of Congress Cataloging-in-Publication Data

Saad, Michel A.
    Thermodynamics : principles & practice / Michel A. Saad.
      p. cm.
    Includes bibliographical references and index.
    ISBN 0-13-490525-3
    1. Thermodynamics.   2. Statistical thermodynamics.
    3. Engineering.   I. Title.
    QC311.S153   1997
    536'.7—DC20   96-33305
                                  CIP

Acquisitions editor: William Stenquist

Production services: Thompson Steele Production Services

Editorial/production supervision: Sharyn Vitrano

Editor-in-chief: Marcia Horton

Managing editor: Bayani Mendoza de Leon

Copy editing: Thompson Steele Production Services

Art director and cover designer: Jayne Conte

Director of production and manufacturing: David W. Riccardi

Manufacturing buyer: Julia Meehan

Editorial assistant: Meg Weist

 © 1997 by Prentice-Hall, Inc.
Simon & Schuster / A Viacom Company
Upper Saddle River, NJ 07458

All rights reserved. No part of this book may be reproduced, in any form or by any means, without permission in writing from the publisher.

The author and publisher of this book have used their best efforts in preparing this book. These efforts include the development, research, and testing of the theories and programs to determine their effectiveness. The author and publisher make no warranty of any kind, expressed or implied, with regard to these programs or the documentation contained in this book. The author and publisher shall not be liable in any event for incidental or consequential damages in connection with, or arising out of, the furnishing, performance, or use of these programs.

Printed in the United States of America

10 9 8 7 6 5 4 3 2 1

ISBN 0-13-490525-3

Prentice-Hall International (UK) Limited, *London*
Prentice-Hall of Australia Pty. Limited, *Sydney*
Prentice-Hall Canada Inc., *Toronto*
Prentice-Hall Hispanoamericana, S.A., *Mexico*
Prentice-Hall of India Private Limited, *New Delhi*
Prentice-Hall of Japan, Inc., *Tokyo*
Simon & Schuster Asia Pte. Ltd., *Singapore*
Editora Prentice-Hall do Brasil, Ltda., *Rio de Janeiro*

Contents

Preface

This book is intended primarily for use in undergraduate and senior-level engineering courses. Its main objective is to increase the depth and breadth of the undergraduate understanding of thermodynamics. The writing follows the general lines of my previous book on the same subject with some additions and reorganization of the material. In the first six chapters, the language and the fundamental principles of thermodynamics are presented in a comprehensive and coherent context. The material is supplemented by illustrative examples that are found to be helpful to the student especially during the early part of the thermodynamic course sequence. All dimensional quantities are given solely in SI units. In Chapter 4, thermodynamic properties of pure substances are presented and subsequently used in Chapter 7, which is devoted entirely to cycle applications. Thermodynamic relations and equations of state are presented in Chapter 8. Nonreacting and reacting gas mixtures are presented in Chapters 9 and 10. Chapter 11 discusses advanced energy systems and innovative methods of energy utilization.

As the student acquires a sufficient understanding of the subject, the importance of the fundamentals becomes obvious, and the underlying principles of thermodynamics unveil their logic and simplicity. To enforce this issue, and because of the intimate connection between kinetic theory and thermodynamics, attempts are made at the outset of the book to evoke the student's insight into microscopic matter. This serves to relate macroscopic behavior to its microscopic counterpart.

The subject of statistical thermodynamics is important in addressing a variety of current and future engineering applications. Interpretation of classical phenomena in terms of the behavior of molecules not only agrees with intuition but also provides physical insight and understanding of macroscopic behavior. Chapters 12 and 13 are devoted to statistical thermodynamics and provide the methodology used in that discipline. The book concludes by presenting an introduction to irreversible thermodynamics in Chapter 14.

I would like to thank my students for their suggestions and indirect contributions to this book. In particular, I am grateful to Miguel Mateos for his invaluable assistance during the different phases of bringing this text to completion. I would also like to express my gratitude to Eileen Tan for her expert typing of the many drafts and for preparing the final manuscript.

Michel A. Saad

Nomenclature

Symbol	Meaning
A	Area
a	Acceleration, activity
B	Magnetic field strength
A, B, C, D, K	Constants
a, b, c	Constants, distance
AF	Air-fuel ratio
C	Concentration, charge, quantity of electricity, number of components, number of molecules striking an area
c	Velocity of light, specific heat, wave velocity
c_p	Specific heat at constant pressure
c_v	Specific heat at constant volume
D	Thermal diffusion coefficient
E	Bulk modulus of elasticity, total internal energy, electric field strength
e	Total internal energy per unit mass, effectiveness
F	Force, thrust, Helmholtz function, number of degrees of freedom
F, f	Function
f	Number of degrees of freedom, fugacity
$f(v)$	Speed distribution function
G	Gibbs function
g	Acceleration due to gravity, Gibbs function per unit mass
g_c	Constant of proportionality in Newton's second law
g_i	Degeneracy of the ith level
H	Total head, magnetic intensity, enthalpy
HHV	Higher heating value
ΔH°	Enthalpy of reaction at the standard state
ΔH	Enthalpy of reaction
h	Height, enthalpy per unit mass, Planck's constant
h_f°	Enthalpy of formation
h_{fg}	Enthalpy of vaporization per unit mass
I	Magnetization, irreversibility, moment of inertia

Symbol	Meaning
I, i	Electric current
J	Mechanical equivalent of heat, flux
j	Current density
K_p	Equilibrium constant
K_e	Electrical conductivity
K_t	Thermal conductivity
KE	Kinetic energy
k	Boltzmann's constant
L, l	Length
L_{ij}	Onsager coefficients
LHV	Lower heating value
M	Molar mass, moment
MEP	Mean effective pressure
m	Mass, molecular mass
m, n, j	Integers
m_f	Mass fraction
N	Number of molecules
N_a	Avogadro's number
n	Number of moles, number of molecules per unit volume, exponent
P	Property, power, electric polarization, number of phases, Legendre's polynomial, collision probability
PE	Potential energy
p	Pressure
p_r	Relative pressure
p_R	Reduced pressure
Q	Heat transfer
q	Heat transfer per unit mass
R	Gas constant, electric resistance
$\overline{R}$	Universal gas constant
R_e	Electrical resistance
r	Radius, distance
r_c	Cut-off ratio
r_v	Compression ratio
r_p	Pressure ratio
S	Entropy, Seebeck coefficient
s	Entropy per unit mass
T	Absolute temperature, torque, temperature
T_r	Relative temperature
T_R	Reduced temperature
T_0	Surroundings temperature
t	Time
U	Internal energy
ΔU^o	Internal energy of reaction at the standard state
u	Internal energy per unit mass

Symbol	Meaning
V	Volume, velocity, voltage
v	Specific volume or velocity
v_r	Relative volume
v_R	Pseudo-reduced specific volume
v_{rms}	Root mean square speed
v_{avg}	Average speed
v_{mp}	Most probable speed
W	Weight, work transfer, probability
w	Work transfer per unit mass
X	Property, thermometric property, generalized intensive property, driving force
x	Quality, mole fraction
X, Y, Z	Functions of x, y, and z
x, y, z	Cartesian coordinates
Z	Compressibility factor, collision frequency between molecules, molecular partition function, figure of merit
z	Affinity
$\mathcal{Z}$	Partition function

Greek Letters

Γ	Gamma function
γ	Specific weight, ratio of specific heats
β	Coefficient of volumetric expansion, coefficient of performance
κ	Coefficient of compressibility
ρ	Density, density per unit length, electrical resistivity
θ, φ	Angles
Θ, Φ	Functions
α	Angle, constant, molar fraction dissociated
a, β	Lagrange's multipliers
$\mathcal{E}$	Electric potential
ω	Angular velocity, solid angle, circular frequency, specific humidity
σ	Surface tension, collision cross section, symmetry number, rate of entropy production per unit volume, Stefan–Boltzmann constant
μ	Chemical potential
μ_h	Joule–Thomson coefficient
μ_T	Isothermal coefficient
v	Frequency, stoichiometric coefficient
Φ	Availability function for a closed system
φ	Availability function per unit mass, relative humidity, potential function, work function
π	Peltier coefficient

Symbol	*Meaning*
Ψ	Availability function for steady-flow processes, wave function
ψ	Availability function per unit mass, saturation ratio
ψ, φ	Functions
Λ	Molecular free path
λ	Integrating factor, Lagrange's multiplier, degree of advance of reaction, wave length
ε	Internal energy per molecule, energy per molecule
η	Efficiency
$\displaystyle\sum_{i}$	Summation over indices i
$\displaystyle\prod_{i}$	Product over indices i
τ	Thomson coefficient
$\theta_E = \dfrac{h\nu_E}{k}$	Einstein characteristic temperature
$\theta_D = \dfrac{h\nu_D}{k}$	Debye characteristic temperature
$\theta_r = \dfrac{h^2}{8\pi^2 Ik}$	Rotational characteristic temperature
$\theta_v = \dfrac{h\nu}{k}$	Vibrational characteristic temperature

Subscripts

a	Air
c	Critical point property
f	Fuel
f, g	Refer to fluid and gas
i, j	Refer to components i and j
j	Refers to all i components except one
P	Products
R	Refers to reduced properties, reactants
r	Refers to relative properties
(o)	Refers to environment conditions
i, r	Refer to irreversible and reversible processes

Superscripts

$*$	Refers to ideal gas
$(-)$	Refers to property per mole
$(°)$	Denotes standard reference state
e	Denotes equilibrium
$(\cdot)$	Denotes quantity per unit time

Scope of Thermodynamics

Thermodynamics plays an important role in the analysis of systems and devices in which energy transfer and energy transformation take place. Thermodynamics' implications are far-reaching, and its applications span the whole range of the human enterprise. All along our technological history, the development of science has enhanced our ability to harness energy and use it for society's needs. The Industrial Revolution is a result of the discovery of how to exploit energy and how to convert heat into work. Nature allows the conversion of work completely into heat, but heat is taxed when converted into work in a cyclic engine. For this reason, the return on our investment of heat transfer is compared with the output work transfer and attempts are made to maximize this return.

Most of our daily activities involve energy transfer and energy change. The human body is a familiar example of a biological system in which the chemical energy of the food or body fat is transformed into other forms of energy such as heat transfer and work transfer. Our encounter with the environment also reveals a wide area of engineering applications. These include power plants to generate electricity, engines to propel automobiles and aircraft, refrigeration and air conditioning systems, and so on. This section describes briefly some of the applications shown schematically in Figure 1.

In the hydroelectric power system (a) the potential energy of the water is converted into mechanical energy through the use of a hydraulic turbine. The mechanical energy is then converted into electric energy by an electric generator coupled to the shaft of the turbine. In the steam power generating plant (b), chemical or nuclear energy is converted into thermal energy in a boiler or a reactor. The energy is imparted to water, which vaporizes into steam. The energy of the steam is used to drive a steam turbine, and the resulting mechanical energy is used to run a generator to produce electric power. The steam leaving the turbine is then condensed, and the condensate is pumped back to the boiler to complete the cycle. Breeder reactors (c) use uranium-235 as a fuel source and can produce some more fuel in the process. A solar power plant (d) uses solar concentrators (parabolic or flat mirrors) to heat a working fluid in a receiver located on a tower. The heated fluid then expands in a turbogenerator in a similar manner as in a conventional power plant.

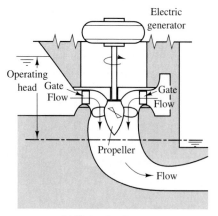

(a) Hydroelectric system

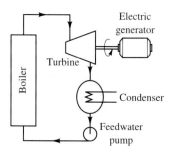

(b) Conventional steam power plant

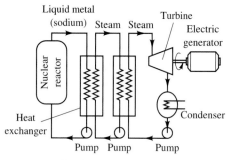

(c) Breeder reactor power plant

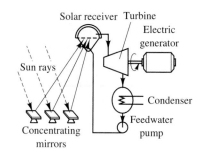

(d) Solar power plant

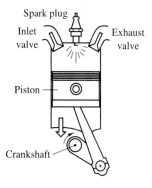

(e) Reciprocating internal combustion engine

(f) Aircraft gas turbine

Figure 1   Schematics of some of the applications of thermodynamics.

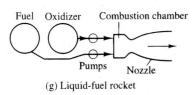

(g) Liquid-fuel rocket

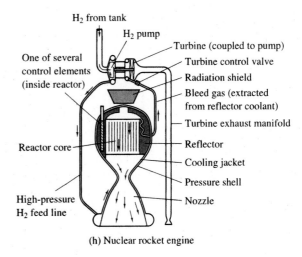

(h) Nuclear rocket engine

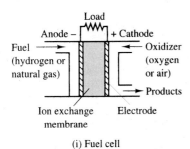

(i) Fuel cell

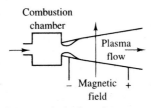

(j) Magnetohydrodynamic generator

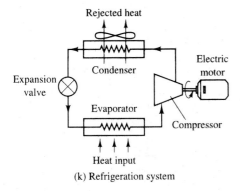

(k) Refrigeration system

Figure 1   *(continued)*

In a spark-ignition internal combustion engine (e), the chemical energy of the fuel is converted into mechanical work. An air-fuel mixture is compressed and combustion is initiated by a spark device. The expansion of the combustion gases pushes against the piston, which results in the rotation of the crankshaft. Gas turbine engines (f), commonly used for aircraft propulsion, convert the chemical energy of the fuel into thermal energy that is used to run a gas turbine. The turbine is directly coupled to a compressor that supplies the air required for combustion. The exhaust gases, upon expanding in a nozzle, create the necessary thrust. For power generation, the turbine is coupled to an electric generator and drives both the compressor and the generator. In a liquid-fuel rocket (g), a fuel and an oxidizer are combined, and the combustion gases expand in a nozzle creating a propulsive force (thrust) to propel the rocket. A typical nuclear rocket propulsion engine (h) offers a higher specific impulse when compared to chemical rockets.

The fuel cell (i) converts chemical energy into electric energy directly making use of an ion-exchange membrane. When a fuel such as hydrogen is ionized, it flows from the anode through the membrane toward the cathode. The released electrons at the anode flow through an external load. In a magnetohydrodynamic (MHD) generator (j), electricity is produced by moving a high-temperature gas (plasma) through a magnetic field.

The refrigeration system (k) utilizes work supplied by the electric motor to transfer heat from a refrigerated space. Low-temperature boiling fluids such as ammonia and refrigerant-12 absorb energy in the form of heat transfer, as they vaporize in the evaporator causing a cooling effect in the region being cooled.

The actual hardware used in the preceding applications is much more complicated, as shown, for example, in the cut-away section of the jet engine in Figure 2. Air is drawn through several stage compressors and enters the combustion chamber where fuel is added and burned. The high-temperature combustion products expand through a turbine section and produce power. A portion of that power drives the compressors, and the balance either drives an electric generator in the case of a stationary power plant or is used to discharge the gases at high velocity to generate a forward thrust to propel an aircraft.

These are only a meager number of engineering applications, and the study of thermodynamics is relevant to the analysis of a much wider range of processes and applications not only in engineering, but also in other fields of science.

In academic preparation, a rigorous presentation of the subject provides the scientific basis required to improve the design and performance of energy-transfer systems. Proficiency in teaching and learning thermodynamics requires not only higher-order reasoning skills but also a body of knowledge with a certain context about their implication and use. This book has been written with these goals in mind.

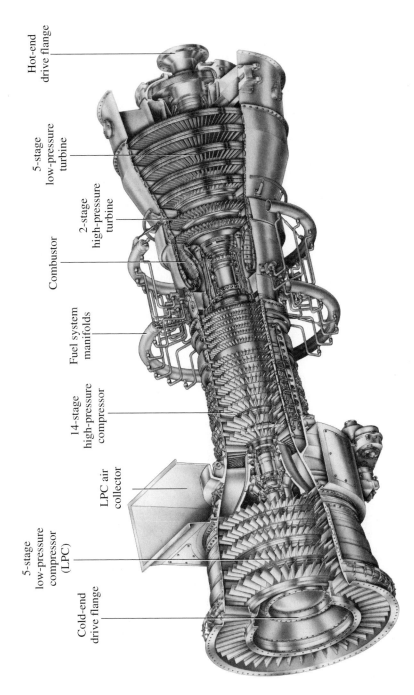

Figure 2   A cut-away section of a jet engine. (Courtesy of General Electric Aircraft Engines)

Hot-end drive flange

5-stage low-pressure turbine

2-stage high-pressure turbine

Combustor

Fuel system manifolds

14-stage high-pressure compressor

LPC air collector

5-stage low-pressure compressor (LPC)

Cold-end drive flange

Chapter 1

Introductory Concepts and Definitions

1.1 Introduction

Thermodynamics[1] is an axiomatic science concerned with the transformation of energy from one form to another. But energy and matter are closely related, in as much as the transfer of energy results in a change of the state of matter. Hence thermodynamics becomes involved in describing how energy interacts with matter, and in doing so, it becomes concerned with certain physical variables and their interrelationships. Classical thermodynamics was formalized mainly by Carnot, Joule, Kelvin, and Clausius. Boltzmann bridged the gap between the classical and the microscopic points of view of thermodynamics. Through his work, it was possible to explain macroscopic behavior of matter in terms of the probabilistic behavior of its microscopic particles. In 1876, J.W. Gibbs extended the classical approach of thermodynamics to substances undergoing physical and chemical changes. He laid the foundation for developing thermodynamics into a science of such a broad scope that it can be applied to almost all physical and chemical phenomena.

The principles of thermodynamics may be summarized as four laws or axioms known as the *Zeroth, First, Second,* and *Third* laws of Thermodynamics.

---

[1] The word *thermodynamics* combines the Greek words *therme* (heat) and *dynamics* (power) for the field that involves "heat" and "work" interactions.

Although the formulation of these laws is simple, their implications are remarkably extensive, and through the years they have formed the theoretical and practical foundation for virtually all thermal energy conversion systems. The *Zeroth Law,* formulated as a logical afterthought by about 1931, deals with thermal equilibrium and the possibility of defining the concept of temperature. The *First Law* introduces the concept of internal energy and establishes the principle of conservation of energy. It ascertains the equivalence of work transfer and heat transfer as possible forms of energy interactions. There is however a penalty attached to the energy transformation process resulting in a loss in the availability of energy for future performance of work. The *Second Law* indicates the natural direction of change of the distribution of energy and introduces the principle of increase of entropy.[2] The concept of entropy describes quantitatively the loss in available energy in all naturally occurring transformations. Finally, the *Third Law* defines the absolute zero of entropy. These laws were deduced from experimental observations, and there is no mathematical proof for them but, like all physical laws, thermodynamic laws are based on logical reasoning; evidence that justifies their continued use is obtained from experiments that verify their consequences.

The first 11 chapters of this text are based on the classical, which is the macroscopic view of matter. In this approach, matter is considered a continuum without any concern to its atomistic structure. Chapters 12 and 13 outline the statistical approach of thermodynamics, and Chapter 14 is an introduction to Irreversible (Non-Equilibrium) Thermodynamics.

When matter is considered from the microscopic viewpoint, the subject is called *Statistical Thermodynamics,* which may be regarded as a branch of statistical mechanics. The microscopic approach focuses on the statistical behavior of a mass consisting of numerous individual molecules, and correlates macroscopic properties of matter with molecular configuration and with intermolecular forces. In this sense, the thermodynamic behavior of a system represents its time-average behavior. The difference between the two approaches may be illustrated by considering the pressure exerted by a gas confined in a container. The gas is composed of a large number of molecules, each molecule having at a given instant certain characteristics such as velocity, momentum, and position. Statistical and quantum mechanics describe the behavior of the gas by first describing the behavior of the individual molecules and then by averaging their individual properties. From a microscopic point of view, the pressure exerted by the gas at a given point and at a certain instant depends on the momentary behavior of the molecules in the neighborhood of that particular point. It is clear that pressure does fluctuate with time owing to the random motion of the molecules. Statistical methods employ the concept of probability, which predicts that the average behavior of the molecules remains uniform although the behavior of the individual molecules does not. In this sense, pressure has a meaning only if averaged

---

[2] The terms *internal energy* and *entropy* are defined in Chapters 3 and 5.

over a large number of molecules. On the other hand, the macroscopic point of view concerns itself with the overall force per unit area, regardless of the atomic or microscopic origins of the force.

Classical thermodynamics, because of its generality, does not explain certain phenomena adequately. These phenomena include, for example, the kinetics involved in the approach to equilibrium states, the specific heats of certain molecules, and entropy from the physical, rather than the mathematical, standpoint. The desire to promote understanding of these problems led to the development of the microscopic approach, which provides a deeper insight into the principles involved. This approach employs simplified models of matter and interprets macroscopic behavior in terms of molecular properties. The microscopic treatment of matter is particularly helpful when dealing with systems in which the mean free path[3] of the molecules is large compared with the dimensions of the system.

Classical and statistical thermodynamics tend to complement and reinforce each other so that the two disciplines provide more insight into the behavior of matter than either of them alone can offer. If, however, both approaches can be integrated, classical thermodynamics becomes dependent on the laws and assumptions governing the behavior of individual particles and, therefore, loses its primary quality of generality. Evidently, if a problem can be solved by the two approaches, the final result should be the same. This fact allows classical thermodynamics to be used as a check on molecular theories.

1.2 Dimensions and Units

Several systems of dimensions may be used to express physical quantities in terms of primary or fundamental dimensions. Primary dimensions may be chosen as:

Force	F	Mass	M
Length	L	Time	t
Temperature	T	Electric current	I

Other physical quantities of interest can be expressed in terms of these primary dimensions. For example, the dimensions of velocity, density, and area are[4]

$$[V] = \frac{L}{t} \qquad [\rho] = \frac{M}{L^3} \qquad [A] = L^2$$

---

[3] The mean free path is the average distance traveled by a molecule between collisions.
[4] The square bracket [ ] indicates "dimensions of."

Any consistent set of units can be used to measure the primary dimensions of an adopted dimensional system. For example, in the English Engineering System, there are four primary dimensions: force, mass, length, and time (FMLt). The unit of force is the pound force (lbf), the unit of mass is the pound mass (lbm), the unit of length is the foot (ft), and the unit of time is the second (s). The pound force (lbf) is defined as that force necessary to accelerate 1 pound mass (lbm) at the rate of 32.174 ft/s². Newton's second law may be used to establish the equivalence between dimensions. Newton's law is written as

$$F = \frac{1}{g_c} ma \tag{1.1a}$$

Substitution leads to

$$1 \text{ lbf} = \frac{1}{g_c}(1 \text{ lbm})(32.174 \text{ ft/s}^2)$$

The proportionality constant[5] g_c is determined by experiment or by definition of units and is equal to

$$g_c = 32.174 \frac{\text{lbm ft}}{\text{lbf s}^2}$$

Alternatively, if the unit of force is the kilogram force (kgf), the unit of mass is the kilogram (kg), the unit of length is the meter (m), and the unit of time is the second (s), then the kilogram force (kgf) is defined as that force necessary to accelerate 1 kg mass at the rate of 9.80665 m/s² so that

$$1 \text{ kgf} = \frac{1}{g_c}(1 \text{ kg})(9.80665 \text{ m/s}^2)$$

and

$$g_c = 9.80665 \frac{\text{kg m}}{\text{kgf s}^2}$$

The force that imparts an acceleration of 1 cm/s² to a mass of 1 gram is defined as a *dyne* so that

$$1 \text{ dyne} = 1 \frac{\text{gm cm}}{\text{s}^2}$$

---

[5] Note that g_c is completely different from the gravitational constant g, which has a standard value of $g = 32.174 \text{ ft/s}^2 = 9.80665 \text{ m/s}^2$.

and in this case (*FMLt* system)

$$g_c = 1.0 \frac{\text{gm cm}}{\text{dyne s}^2}$$

The constant g_c can also be defined to be a nondimensional conversion factor of absolute magnitude equal to unity so that Newton's law is written as

$$F = ma \qquad\qquad (1.1\text{b})$$

In this case, either the force (*F*) or the mass (*M*) can be eliminated from the list of primary dimensions, resulting in the *MLt* or the *FLt* systems of dimensions.

In the *MLt* system, the three primary dimensions are mass, length, and time, whereas force is a derived dimension (ML/t^2). For example, if the centimeter is the unit of length, the gram is the unit of mass, and the second is the unit of time (cgs units), the units of force are gm cm/s^2, which has been previously defined as a dyne. In this case, the force is a derived dimension whose units are determined by setting $g_c = 1.0$.

Similarly, in the *FLt* system of dimensions, force, length, and time are the primary dimensions, whereas mass is a derived dimension (Ft^2/L). The units of mass result from arbitrarily setting $g_c = 1.0$.

The international system of units (SI), which has been adopted by many countries, is based on a coherent form of the *MLt* system of dimensions. It is an absolute system because it is independent of the gravitational acceleration *g*. The meter (m) is the unit of length, the kilogram (kg) is the unit of mass, and the second (s) is the unit of time. The force that imparts an acceleration of 1 m/s^2 to a mass of 1 kg is a derived dimension. The SI unit of force is called the newton (N); (1 N = 1 kg m/s^2). The joule (J) is the unit of energy and is equal to the energy expended in moving against a force of 1 newton for a distance of 1 m; (1 J = 1 N m = 1 kg m^2/s^2). Power is energy transfer per unit time, and its unit is the watt (W); (1 W = 1 J/s = 1 N m/s = 1 kg m^2/s^3). The SI system provides the following set of base units:

Physical Quantity	Unit	Symbol
Length	meter	(m)
Mass	kilogram	(kg)
Time	second	(s)
Electric current	ampere	(A)
Thermodynamic temperature	kelvin	(K)
Luminous intensity	candela	(cd)
Amount of substance	mole	(mol)

Since it is frequently necessary to work with large or small numbers, multiples and submultiples of these units, based on powers of 10, have been introduced as prefixes:

Multiples			*Submultiples*		
Name	*Symbol*	*Multiplier*	*Name*	*Symbol*	*Multiplier*
deca	da	10	deci	d	10^{-1}
hecto	h	10^2	centi	c	10^{-2}
kilo	k	10^3	milli	m	10^{-3}
mega	M	10^6	micro	μ	10^{-6}
giga	G	10^9	nano	n	10^{-9}
tera	T	10^{12}	pico	p	10^{-12}

The mole has been recommended as a unit of matter for the SI system. However, it is more precise to use the kilogram-mole (kg-mol), which is defined as the amount of substance that has as many atoms or molecules as there are atoms in 12 kg of carbon-12. One kg-mol of a substance is the amount of that substance in kilograms numerically equal to its molar mass. For monatomic oxygen, as an example, one kg-mol has a mass of 16 kg, and one gm-mol has a mass of 16 gm. The number of kg-moles of a substance, n, is equal to the mass of the substance, m, in kilograms divided by its molar mass, M, in kg/kg-mol.

Table 1.1 presents some values of g_c for several systems of dimensions in consistent sets of units.

Table 1.1   Values of g_c for Several Systems of Dimensions in Consistent Sets of Units

					g_c	
Force	*Mass*	*Length*	*Time*	*FMLt System*	*MLt System**	*FLt System**
(lbf)	lbm	ft	s	$32.174 \dfrac{\text{lbm ft}}{\text{lbf s}^2}$	1.0	—
lbf	(slug)	ft	s	$1.0 \dfrac{\text{slug ft}}{\text{lbf s}^2}$	—	1.0
(kgf)	kg	m	s	$9.80665 \dfrac{\text{kg m}}{\text{kgf s}^2}$	1.0	—
(dyne)	gm	cm	s	$1.0 \dfrac{\text{gm cm}}{\text{dyne s}^2}$	1.0	—
(newton)	kg	m	s	$1.0 \dfrac{\text{kg m}}{\text{N s}^2}$	1.0	—

*Whenever the *MLt* or the *FLt* system of dimensions is used, the value of g_c = 1.0, and the units shown in parentheses are derived units.

The SI system of units will be adopted throughout this text because it is operationally more convenient than the English system of units. Conversion factors between the two systems are given in the Appendix.

Example 1.1

A body has a mass of 5 kg and is attracted to the earth at a location where the gravitational acceleration is 8 m/s^2. What is its weight in newtons? If the body moves at a velocity of 10 m/s, what is its kinetic energy?

SOLUTION

Referring to Figure 1.1, the force exerted on the body to hold it in equilibrium against the action of the field of gravity is equal to its weight. Therefore,

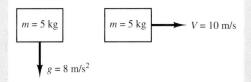

$m = 5$ kg

$g = 8$ m/s^2

$m = 5$ kg $\longrightarrow$ $V = 10$ m/s

Figure 1.1

$$W = F = mg = (5 \text{ kg})(8 \text{ m/s}^2) = 40 \, \frac{\text{kg m}}{\text{s}^2} = 40 \text{ N}$$

Note that the weight of a body can vary depending on any variation of the local gravitational acceleration but its mass remains constant. The kinetic energy is

$$\frac{1}{2}mV^2 = \frac{1}{2}(5 \text{ kg})(10 \text{ m/s})^2 = 250 \, \frac{\text{kg m}^2}{\text{s}^2} = 250 \text{ J}$$

1.3 Thermodynamic Systems

In thermodynamic analysis, it is important to identify clearly the entity or the system under consideration. There are two main types of thermodynamic systems: the closed system and the control volume. In a *closed* system (simply called *system*), the analysis is focused on a quantity of matter of fixed mass and identity. The system is surrounded by a boundary that may change position, size, or shape but is impervious to the flow of matter. Heat and work,[6] which are means of energy transfer, can cross the boundary of the system. The region out-

[6] The terms *heat* and *work* are defined in Chapter 2.

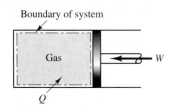

Figure 1.2   Fixed-mass system, mass does not cross the boundary of the system.

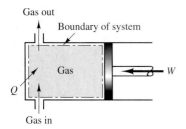

Figure 1.3   Control volume, mass crosses the boundary of the control volume.

side the boundaries of a system and contiguous to it is called the *surroundings.* A system that exchanges neither energy nor matter with its surroundings is called an *isolated system.* Consider the system consisting of the gas confined between the cylinder and the piston, as shown in Figure 1.2. Although heat and work may cross the boundary, and the volume of the system may change owing to the motion of the piston, the system is a closed system since no mass crosses its boundary. Other examples of closed systems include a free body and a point mass as used in mechanics.

In the *control volume* shown in Figure 1.3, the analysis centers about a region in space through which matter and energy flow. The surface of the control volume is called the control surface and always consists of a closed surface. The control volume comprises our region of interest in applying various thermodynamic principles and is a useful concept in the analysis of devices such as turbines and pumps through which mass flows. The control volume may be either stationary or moving at a constant or a variable velocity relative to a coordinate system. If no mass transfer occurs across the control surface, the control volume becomes identical with the closed system.

Thermodynamic analyses employ systems, referred to as thermodynamic systems, which are idealized versions of the complex real systems. Regardless of the type of system considered, the boundaries of the system must be well defined as a first and important step before proceeding with any thermodynamic analysis. It is also important to define the boundary of the system or control volume in such a way to allow better understanding of the problem.

Thermodynamics is concerned with systems in *thermodynamic equilibrium.* A system is said to have attained a state of thermodynamic equilibrium when a spontaneous change of its state is impossible. The properties of a system, which will be discussed in the next section, are meaningful, ordinarily, only when a system is in thermodynamic equilibrium since they apply to the system as a whole. Any change of state of the system, however, leads to a departure from thermodynamic equilibrium, which results in difficulty in specifying the interior state of the system. This difficulty is surmounted by defining a *quasi-equilibrium* state, which deviates infinitesimally from thermodynamic equilibrium. When a system undergoes a quasi-equilibrium process, each successive state through which the system passes is assumed to be in equilibrium, and the thermodynamic potentials of the system and its surroundings are equal.

1.4 Properties of Systems

The equilibrium state of a thermodynamic system at a particular time is described by a set of state functions called *properties.* Properties are functions of the state of the system only; hence they do not depend on the history of the system or the process by which the state was attained. The change in the value of a property is thus fixed only by the initial and final state of the system.

Properties may be divided into two categories: *intensive* and *extensive*. Intensive properties such as temperature, pressure, and density do not depend on the size, mass, or configuration of the system. Intensive properties define the *intensive state* of a system and have meaning only for systems in equilibrium states. An intensive property may also be defined at a point when the size of the system surrounding the point approaches zero. Properties that depend on the size of the system, such as length, volume, mass, and energy, are *extensive properties*. Some intensive and extensive properties are shown in Figure 1.4. Any extensive property of the whole system is equal to the sum of the respective partial properties of the components of the system. A property indicating the extent or mass of the system, in addition to intensive properties, is needed to define the extensive state of the system. Three properties,[7] one of which may be mass, are required to determine uniquely an extensive property of a single-component, single-phase substance.[8] For example, the volume of such a substance can be determined by its mass, temperature, and pressure. Further, the ratio of two extensive properties of a homogeneous system is an intensive property. Mass per unit volume, for example, is an intensive property.

The number of properties necessary to define a system depends on the complexity of the system. In a simple system, the intensive state has two degrees of freedom. If such a system is in equilibrium, the intensive state is specified by two independent intensive properties. Systems that contain more than one component or more than one phase require more than two independent properties to specify their state.

Those properties that define the state of a system are called *independent* properties. Those properties that become fixed when the state of the system is defined by the independent properties are called *dependent* properties. As an example, if pressure is selected as the independent property, and a value of 1 atmosphere is chosen, then the temperature (dependent property) at which water boils is 100°C.

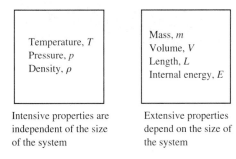

Temperature, T Pressure, p Density, ρ	Mass, m Volume, V Length, L Internal energy, E
Intensive properties are independent of the size of the system	Extensive properties depend on the size of the system

Figure 1.4   Intensive and extensive properties.

[7] This requires the absence of kinetic, electrical, surface, magnetic, and gravitational effects with no changes in the components of the system.

[8] A single component substance is a substance that has a homogeneous and invariable chemical composition. The phase of a substance is the homogeneous, chemical, and physical state of its molecules.

It must be realized that macroscopic properties are manifestations of the microscopic behavior of the particles of a system. The microscopic property continuously fluctuates about a time-average value so that the macroscopic property represents the time-average of microscopic properties. The microscopic property of a system at a certain instant is thus a characteristic of the system at that instant. The instantaneous state of each particle is necessary to define the microscopic state of the system.

Properties may be expressed in a functional relationship known as the *equation of state* of the system. As an example, consider an equation of state relating three properties of a system such as x, y, z,

$$f(x, y, z) = 0 \tag{1.2}$$

Suppose it is possible to solve Eq. (1.2) explicitly for each of the variables x, y, z, as

$$\begin{aligned} x &= x(y, z) \\ y &= y(z, x) \\ z &= z(x, y) \end{aligned} \tag{1.3}$$

In the last expression, for example, z is the dependent variable whereas x and y are the independent variables. The differential of the dependent property can be written in terms of its partial derivatives and the differentials of the independent properties:

$$dz = \left(\frac{\partial z}{\partial x}\right)_y dx + \left(\frac{\partial z}{\partial y}\right)_x dy$$

If

$$M = \left(\frac{\partial z}{\partial x}\right)_y \qquad \text{and} \qquad N = \left(\frac{\partial z}{\partial y}\right)_x$$

then

$$dz = M\, dx + N\, dy \tag{1.4}$$

The subscripts on the partial derivatives indicate the independent variables. If Eq. (1.4) has a solution, then dz is called a *perfect* or *an exact differential*, and Eq. (1.4) is integrable. Otherwise, it is an *inexact differential* and is written as δz.

If variable z is to be adequately defined by the thermodynamic coordinates x, y, its value must be the same regardless of the order taken in approaching it. In evaluating dz, it should be possible to make the changes dx and dy in either sequence, without changing the result. If the function $z = z(x, y)$ and its partial derivatives are continuous, then the second derivative of z, with respect to x and y, is independent of the order of the successive differentiation. Thus

$$\frac{\partial}{\partial y}\left[\left(\frac{\partial z}{\partial x}\right)_y\right]_x = \frac{\partial}{\partial x}\left[\left(\frac{\partial z}{\partial y}\right)_x\right]_y \qquad (1.5)$$

or

$$\left(\frac{\partial M}{\partial y}\right)_x = \left(\frac{\partial N}{\partial x}\right)_y \qquad (1.6)$$

Equation (1.6) is a necessary condition for the existence of a function of x and y satisfying Eq. (1.4). Equation (1.6) is also a sufficient condition, for when Eq. (1.6) is integrated twice, the expression of z is obtained. A function that satisfies Eq. (1.6) is called a *point function* or a *property* of the system.

It may be shown that the sum, or the difference, of two point functions is also a point function, provided that the functions and their derivatives are continuous. Similarly, the product of two point functions is a point function; on the other hand, the product of a point function and an inexact differential is an inexact differential. Proofs of these statements are left for the reader as exercises.

Example 1.2

The heat interaction with a system is given in terms of two independent functions T and v by the equation

$$dq = f(T)\,dT + \frac{RT}{v}\,dv$$

where R is a constant and T and v are the temperature and the specific volume of the system. Is dq an exact differential?

SOLUTION

The test for exactness requires the equality of

$$\frac{\partial f(T)}{\partial v} = 0 \qquad \text{and of} \qquad \frac{\partial(RT/v)}{\partial T} = \frac{R}{v}$$

Since $0 \neq R/v$, dq is not an exact differential, hence no state function exists that has a differential equal to δq.

Example 1.3

The p-v-T relation for a gas is given by $p(v - b) = RT$, where b and R are constants. Show that the pressure p is a property (an exact differential).

SOLUTION

From $p = p(T, v)$,

$$dp = \left(\frac{\partial p}{\partial T}\right)_v dT + \left(\frac{\partial p}{\partial v}\right)_T dv$$

but $p = RT/(v - b)$ so that

$$\left(\frac{\partial p}{\partial T}\right)_v = \frac{R}{v - b} \qquad \text{and} \qquad \frac{\partial^2 p}{\partial v\, \partial T} = -\frac{R}{(v - b)^2}$$

Also

$$\left(\frac{\partial p}{\partial v}\right)_T = -\frac{RT}{(v - b)^2} \qquad \text{and} \qquad \frac{\partial^2 p}{\partial T\, \partial v} = -\frac{R}{(v - b)^2}$$

Since $\partial^2 p/\partial v\, \partial T = \partial^2 p/\partial T\, \partial v$, that is, the order of differentiation is immaterial (the path along which the pressure changed is immaterial), the condition of exactness has been satisfied and hence p is a property.

1.5 Density, Specific Weight, Specific Volume, Specific Gravity

The *average density* of a system is its total mass divided by its total volume. When matter is treated as a continuum, that is, enough particles exist for statistical averages to be meaningful, it is possible to define intensive properties at a point. Density is an intensive property and can be described as a continuous function that may vary from point to point within a system. In determining the density at a point, a small volume ΔV is chosen so as to include the point. The corresponding mass is Δm, and the density ρ at a point is, therefore,

$$\rho \equiv \lim_{\Delta V \to \Delta V'} \frac{\Delta m}{\Delta V} \qquad (1.7)^9$$

[9] The symbol $\equiv$ means "defined as."

The volume $\Delta V'$ must contain enough particles to comply with the continuum hypothesis, and yet it must be small compared to the dimensions of the system. The volume $\Delta V'$ is the smallest volume for which a continuous distribution of matter is allowed. Density is expressed in kg/m^3.

The *specific volume* v is the reciprocal of the density, $v = 1/\rho$. It is the volume per unit mass and may be expressed in m^3/kg (Figure 1.5). The specific volume is an intensive property and may vary from point to point within a system.

Specific weight, which is also known as weight density, is weight per unit volume. It is expressed in N/m^3 and depends on both the density of the substance and the value of the gravitational acceleration g. Let w represent the weight corresponding to a certain mass m; then,

$$w = mg$$

If both sides of this equation are divided by the volume, then

$$\gamma = \rho g \qquad\qquad (1.8)$$

where γ is weight density and ρ is mass density. For example, if the density of water at 20°C is 998 kg/m^3, its specific weight is

$$\gamma = (998 \text{ kg/m}^3)(9.80665 \text{ m/s}^2) = 9787 \text{ N/m}^3$$

Note that γ is force per unit volume, whereas ρ is mass per unit volume.

The *specific gravity* of a substance is the ratio of the density of the substance to the density of a reference substance. Specific gravity can also be expressed as the ratio of specific weights, rather than densities, provided that the specific weights are evaluated in regions having the same gravitational acceleration. For liquids and solids, the reference substance is pure water at atmospheric pressure; for gases, the reference substance is air. Common reference temperatures for liquids and solids are 4°C (39.2°F), 15°C (59.0°F), or 20°C (68°F). At these temperatures, the density of water is respectively 1000, 999, and 998 kg/m^3. In the case of a gas, specific gravity is the ratio of the density of the gas to the density of air, at the same temperature and pressure.

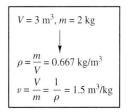

Figure 1.5 Density ρ is mass per unit volume, and specific volume v is volume per unit mass.

1.6 Pressure

The *pressure* exerted by a system is the force exerted normal to a unit area of the boundary. When a fluid is contained within a vessel, the pressure exerted on the vessel is equal to the mean change of momentum of the molecules exerted perpendicular to the confining boundary per unit area and per unit time. Pressure of fluids is analogous to normal stress in solids.

For a pressure-continuum model, the pressure at a point is defined as the force per unit area passing through the point in the limit when the area approaches a small area $\Delta A'$. If ΔF_n is the force normal to the area ΔA, then the pressure at the specified point is defined as

$$p \equiv \lim_{\Delta A \rightarrow \Delta A'} \frac{\Delta F_n}{\Delta A} \qquad (1.9)$$

where the area $\Delta A'$ is the smallest possible area capable of maintaining the continuum model.

A *fluid* is defined as a substance in which the shear stresses are zero whenever it is at rest relative to its container. When a fluid is at rest, only normal stresses exist. In this case, the pressure at a point becomes a scalar point function (independent of direction) and is called static pressure. When a fluid is in motion, the pressure force exerted on an area passing through a point can be resolved into three mutually perpendicular components: one normal to the area and two in the plane of the area. This results in a normal stress perpendicular to the area and two tangential stresses in the plane of the area. The magnitude of these stresses generally varies according to direction except for the case of the ideal or inviscid fluid. In an ideal fluid, shear stresses are absent even if there is a relative motion within the fluid. Pressure, in this case, is independent of direction whether the ideal fluid is in motion or at rest. Consider a homogeneous fluid of density ρ in static equilibrium as shown in Figure 1.6. A pressure difference exists between two points that are separated by distance h in the vertical direction. The weight of a cylinder of fluid can be equated to the difference between forces due to pressure at the two ends of the cylinder so that

$$A\Delta p = \gamma hA$$

or

$$\Delta p = \gamma h = \rho gh \qquad (1.10)$$

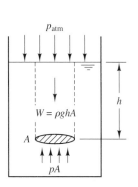

Figure 1.6   The weight of a fluid cylinder is equated to forces due to pressure.

According to this equation, the hydrostatic pressure is the same at all points in a horizontal plane and varies only with depth.

The unit of pressure in the SI system is the pascal (Pa); (1 Pa = 1 N/m^2). A larger unit that is slightly less than the standard atmospheric pressure is the bar (1 bar = 100 kPa). The standard atmospheric pressure is defined as the pressure produced by a column of mercury 760 mm high, the mercury density being 13.5951 gm/cm^3 and the acceleration due to gravity being its standard value of 9.80665 m/s^2. The standard atmospheric pressure is 14.6959 lbf/in^2 abs (psia), 29.92 in. Hg abs or 101.325 kPa (k N/m^2).

Thermodynamic analysis is concerned with values of absolute pressure. However, most pressure-measuring devices indicate a gauge pressure that is the difference between the absolute pressure of a system and the absolute pressure of the atmosphere. A Bourdon-tube pressure gauge, for example, measures the pressure relative to the pressure of the surrounding atmosphere. Conversion from gauge pressure to absolute pressure is accomplished according to the relation

$$p_{abs} = p_{gauge} + p_{atm} \tag{1.11}$$

Figure 1.7 shows this relationship. Note that the datum of absolute pressure is perfect vacuum, whereas the datum of the gauge scale is atmospheric pressure.

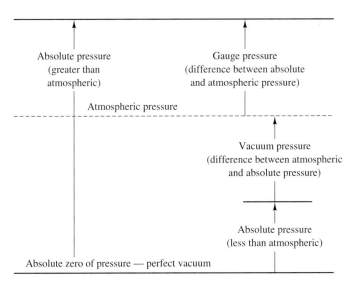

Figure 1.7   Relationships among the absolute, atmospheric, gauge, and vacuum pressures.

For pressures below atmospheric, the gauge pressure is negative, and the term *vacuum* indicates the magnitude of the difference between the atmospheric pressure and the absolute pressure so that

$$p_{abs} = p_{atm} - p_{vac} \tag{1.12}$$

Table 1.2 gives some conversion factors for pressure.

In dealing with fluid flow problems, various types of pressure are commonly used. *Static pressure* is the pressure sensed by a measuring device if it were moving with the same velocity as the fluid stream. *Stagnation* or *total pressure*

Table 1.2   Conversion Factors for Pressure

		Pa	bar	lbf/in²	dyne/cm²	kgf/cm²	mm Hg (1 torr) (at 25°C)	atm
1 Pa	=	1	10^{-5}	1.4504×10^{-4}	10	10.197×10^{-6}	7.501×10^{-3}	9.876×10^{-6}
1 bar	=	10^5	1	14.504	10^6	1.0197	750.1	0.9876
1 lbf/in²	=	6894.76	6894.76×10^{-5}	1	689.476×10^2	0.07031	51.715	0.06805
1 dyne/cm²	=	0.1	1.0×10^{-6}	145.0383×10^{-7}	1	101.972×10^{-8}	750.062×10^{-6}	986.923×10^{-9}
1 kgf/cm²	=	98.067×10^3	0.98067	12.2234	980.665×10^3	1	735.559	967.841×10^{-3}
1 mm Hg (at 25°C) (1 torr)	=	133.3	133.3×10^{-5}	0.01934	1333.223	1.359×10^{-3}	1	1.315×10^{-3}
1 atm	=	1.01325×10^5	1.01325	14.6959	101.325×10^4	1.03323	760.0	1

is the force per unit area perpendicular to the direction of flow when the fluid is brought reversibly[10] to rest. For a constant density fluid, the stagnation pressure is given by

$$p_{stagnation} = \rho \frac{V^2}{2} + p_{static} \tag{1.13}$$

where ρ is the fluid density and V is the velocity. The difference between the total pressure and the static pressure is due to the velocity. This difference, $\rho V^2 / 2$, is called *dynamic* or *velocity pressure*. Figure 1.8 shows the different types of pressures.

In fluid flow through a duct, static conditions prevail at the walls of the duct. The velocity at the wall is zero, and therefore the pressure measured at the wall is the static pressure. If the fluid particles move parallel to the center line of the duct, the static pressure is uniform across any section of the duct.

To measure pressures slightly different from atmospheric, a manometer is normally used, and the pressure is determined according to the hydrostatic for-

---

[10] See Section 1.15 for a definition of reversibility.

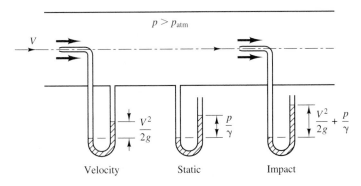

Figure 1.8 Types of pressures.

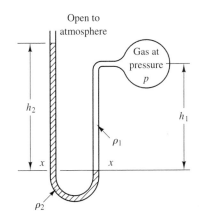

Figure 1.9 Pressures measurement by a manometer.

mula given by Eq. (1.10). The manometer liquid may be mercury, water, alcohol, and so on. A typical manometer is shown in Figure 1.9. Since the manometer fluid is in equilibrium, the pressure along a horizontal line xx is the same for either branch of the manometer. Then,

$$p + \rho_1 gh_1 = p_a + \rho_2 gh_2 \qquad (1.14)$$

where p is the absolute pressure in the bulb, p_a is the atmospheric pressure exerted on the liquid-free surface, and ρ_1 and ρ_2 are the densities of the fluid in the bulb and the manometer, respectively.

If $\rho_1 << \rho_2$, then

$$p - p_a = \rho_2 gh_2 \qquad (1.15)$$

1.7 Molecular Interpretation of Pressure

When an ideal gas[11] at normal temperature and pressure is examined from a macroscopic viewpoint, the gas appears as though it were a homogeneous system, and it is described with such parameters as pressure, volume, and temperature. However, when the gas is considered from the microscopic viewpoint, it consists of a multitude of small particles that are separated from each other by considerable space and that are in constant random motion. The properties of a gas that we observe on the macroscopic scale are determined by the behavior of the particles on a microscopic scale. It will be shown that such properties of molecules as mass and velocity are related to the macroscopic properties of pressure and temperature. Furthermore, information deduced about a system from observations made

[11] See Section 1.16 for a definition of an ideal gas.

on one scale should be consistent with information deduced from observations made on the other scale.

On the microscopic scale, pressure may be considered the result of the force exerted by molecules as they bombard a surface. The magnitude of the force depends on the momentum of the molecules and the frequency with which they collide with the walls of the system. An expression for the pressure of an ideal gas can be derived by employing a model that arises from the kinetic theory of gases.

Consider a gas occupying a spherical volume as shown in Figure 1.10. Molecules of the gas move continuously in all directions at velocities that encompass a wide range of values. The size of molecules is assumed to be small compared with the distance between molecules. The pressure exerted by the gas arises from the collisions of the molecules with the wall of the vessel. Assume that there are N molecules of masses $m_1, m_2, m_3, \ldots, m_N$ and velocities $v_1, v_2, v_3, \ldots, v_N$; also assume, for the present, that the molecules do not collide with each other. When a particle of mass m_i collides with the wall of the vessel, it rebounds at an angle equal to the angle of incidence. The radial components of momentum of the particle before and after collision are $-m_i v_i \cos \theta_i$ and $+m_i v_i \cos \theta_i$. The rate of change of momentum in the radial direction is

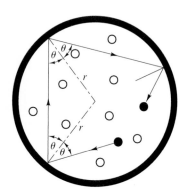

Figure 1.10 Molecules in a spherical vessel.

$$\frac{m_i v_i \cos \theta_i - (-m_i v_i \cos \theta_i)}{\Delta t_i} = \frac{2m_i v_i \cos \theta_i}{\Delta t_i}$$

where Δt_i is the mean time of impact with the wall. The total rate of change of momentum of all the molecules is the sum of the rate of change of momentum of the individual molecules:

$$\sum_{i=1}^{N} \frac{2m_i v_i \cos \theta_i}{\Delta t_i}$$

where N is the number of molecules. The pressure on the wall of the vessel is equal to the total rate of change of momentum of the molecules per unit area or

$$p = \frac{1}{4\pi r^2} \sum_{i=1}^{N} \frac{2m_i v_i \cos \theta_i}{\Delta t_i}$$

For the pressure to be single-valued, the average time of impact must be equal to the average time between collisions with the wall of the vessel. Since the distance between collisions with the wall is $2r \cos \theta_i$, the time between collisions is

$$\Delta t_i = \frac{2r \cos \theta_i}{v_i}$$

Substituting for Δt_i, the expression for pressure becomes

$$p = \frac{1}{4\pi r^3} \sum_{i=1}^{N} m_i v_i^2$$

But the volume of the sphere $V = \frac{4}{3}\pi r^3$ so that

$$pV = \frac{1}{3} \sum_{i=1}^{N} m_i v_i^2 \qquad (1.16)$$

Effective velocity, or *root-mean-square* (rms) velocity, is defined by

$$v_{rms}^2 \equiv \frac{\sum_i m_i v_i^2}{\sum_i m_i} \qquad (1.17)$$

When only one chemical species is present, the rms velocity is defined by

$$v_{rms}^2 = \frac{\sum_i v_i^2}{N}$$

Substitution into Eq. (1.16) gives

$$pV = \frac{1}{3} N m v_{rms}^2 \qquad (1.18)$$

where m is the mass of one molecule. This equation shows the relationship between the macroscopic properties pressure and volume with the microscopic properties molecular mass and molecular velocity. Note that the root-mean-square velocity is different from the *average* velocity of the molecule. Since a molecule shows equal probability of moving in any direction, the average velocity of a molecule is zero.

In the derivation of Eq. (1.18), collisions between the molecules were not considered. Such collisions can occur when the molecules rebound between the walls of the vessel. However, intermolecular collisions tend to be elastic. Momentum is therefore conserved despite the collisions, and so Eq. (1.18) is still valid. Pressure does not depend upon orientation of the areas on which it acts, and this can be attributed to the random motion of the particles.

Although the derivation of the expression of pressure was based on a spherical volume, it can be shown that Eq. (1.18) is valid irrespective of the shape of the volume.

1.8 Temperature and the Zeroth Law of Thermodynamics

In the macroscopic analysis of thermodynamic systems, temperature is considered a primary concept like length and mass. However, it will be shown in Chapter 5 that temperature can be derived in terms of primary properties and therefore need not be considered a primary concept.

A satisfactory definition of temperature[12] cannot be stated at this point. In fact, absolute temperature cannot be rigorously defined unless the second law of thermodynamics is applied, employing a temperature unit that does not depend on the properties of a particular substance. In this section, some important characteristics of temperature based on experimental observation are discussed.

The concept of temperature originates from the sensory perception of *hotness* or *coldness*. It is evident, however, that such a physiological sensation is insufficient for precise evaluation of temperature. Maxwell defined the temperature of a body as "its thermal state considered with reference to its ability to communicate heat to other bodies."

To establish a method of defining and measuring temperature, consider a body *A* that is brought into thermal contact with a body *B*, both bodies being isolated from their surroundings. Energy in the form of heat will be transferred from the body at the higher temperature to the body at the lower temperature. If sufficient time is allowed, bodies *A* and *B* approach a state at which no further change is observed so that the two bodies are in a state of *thermal equilibrium*. When this state is reached, we may postulate that the two bodies are at the *same temperature*. As a corollary to this observation, it may be noted that *if two systems are each in thermal equilibrium with a third system, they are also in thermal equilibrium with each other*. This statement is known as the *Zeroth Law of Thermodynamics*, and its meaning is illustrated in Figure 1.11. When system A and B are each in thermal equilibrium with system C, then system A and B are also in mutual equilibrium. The zeroth law is the basis of the concept of temperature and enables us to compare the temperatures of two bodies *A* and *B* with the help of a third body *C*, and say that the temperature of *A* is the same as the temperature of *B* without actually bringing *A* and *B* in contact with each other. The test body *C* is called a *thermometer*. It must be noted that thermal equilibrium does not describe temperature in a physical sense. It merely defines *equality of temperature*. In other words, equality or inequality of temperature of two systems is the property of being or not being in thermal equilibrium when the two systems are brought into contact. It is important to realize that the thermometer *C* indicates its own temperature, which is the same as the temperature of a system in thermal equilibrium with the thermometer. Note that the concept of temperature just described applies to equilibrium states only.

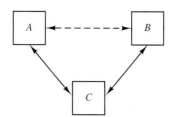

Figure 1.11   Two systems *A* and *B* in thermal equilibrium with system *C* are also in thermal equilibrium between themselves.

---

[12] The word *temperature* comes from the Latin word *temperatura* meaning "proper mixing or tempering," implying attainment of thermal equilibrium.

From the microscopic point of view, temperature is a manifestation of the activity of the molecules. An increase in temperature is accompanied by a simultaneous increase in the kinetic energy of the molecules. When two systems of ideal gases are in thermal equilibrium, the average kinetic energy of the molecules is the same for the two systems.

Temperature can be measured only by indirect methods. Generally, heat is transferred to an instrument such as body C, and the change due to temperature in some property or response of C is measured. A property that changes in value as a function of temperature is called a *thermometric property*. Examples of thermometric properties include the length of the column of liquid in a capillary connected to a bulb, the pressure of a fixed mass of gas kept at constant volume, the volume of a fixed mass of gas kept at constant pressure, the electrical resistance of a metallic wire at atmospheric pressure, and the emf of a thermocouple.

In establishing a temperature scale, an arbitrary number is assigned to represent the temperature of one *fixed point,* and other temperatures are specified with this fixed point as a reference. In 1854, Kelvin pointed out that a single fixed point, such as the triple point of water (ice, liquid water, and steam coexisting in equilibrium), is sufficient to define the datum of an absolute temperature scale. This fixed point was adopted in 1954 by the Tenth International Conference on Weights and Measures owing to the great accuracy with which it can be determined, and its value was set at 273.16 K. Previously, the two fixed points that defined the temperature scale were the ice point, which is the temperature at which ice melts under standard atmospheric pressure and is 0.01°C below the triple point, and the steam point, which is the temperature at which water boils at standard atmospheric pressure.

1.9 Comparison of Thermometric Substances

In establishing a temperature scale, it is necessary to clarify the relationship between temperature and the thermometric property in order to make interpolation and extrapolation possible. Consider the thermometric property X, such that the temperature T is a linear function of this property:

$$T = a + bX \tag{1.19}$$

where a and b are arbitrary constants. Note that equal temperature intervals are defined as those that produce equal changes in the property X. To determine the constants a and b, first assign numerical values to any two temperatures, for example, to the steam point and the ice point of water. Let X_i represent the value of the thermometric property of a substance in thermal equilibrium with ice melting under atmospheric pressure, and let X_s represent the value of the thermometric property of this substance in thermal equilibrium with steam at atmospheric pressure. The numerical change in the thermometric property accompanying the change in temperature over this standard interval is $X_s - X_i$, which is called the *standard property change.*

The *Celsius* scale, for example, assigns 0 to the *ice point,* and 100 to the *steam point.* Then $(X_s - X_i)$ is 100 degrees or intervals. By substituting these values for T, Eq. (1.19) gives

$$a = -\frac{100\, X_i}{X_s - X_i} \qquad b = \frac{100}{X_s - X_i}$$

Thus the temperature T is given by

$$T = -\frac{100\, X_i}{X_s - X_i} + \frac{100}{X_s - X_i}X = 100\frac{X - X_i}{X_s - X_i} \tag{1.20}$$

The *Fahrenheit* scale, on the other hand, assigns the number 32 to the ice point temperature and the number 212 to the steam point temperature. There are then $212 - 32 = 180$ intervals involved, and after the constants a and b are evaluated, Eq. (1.19) becomes

$$T = 180\frac{X - X_i}{X_s - X_i} + 32 \tag{1.21}$$

To measure a wide range of temperature, several thermometers, each having different thermometric properties, are used. Discontinuities occur at the transition points, and the scale used in these thermometers depends on the manner in which the thermometric property changes with temperature.

1.10 Ideal-Gas Thermometer

In an ideal-gas thermometer, the expansion and contraction of a gas as a function of temperature serves as the thermometric property. Consider a fixed mass of an ideal gas that undergoes a change of temperature. The ratio of pressures associated with the end states, if the volume is maintained constant, is the same as the ratio of volumes associated with the end states if the pressure is maintained constant:

$$\left(\frac{p_1}{p_2}\right)_V = \left(\frac{V_1}{V_2}\right)_p \tag{1.22}$$

A temperature scale may therefore be established in which the ratio between any two temperatures, T_1 and T_2, is indicated by the pressure ratio or volume ratio of the enclosed ideal gas:

$$\frac{T_1}{T_2} = \left(\frac{p_1}{p_2}\right)_V = \left(\frac{V_1}{V_2}\right)_p \qquad (1.23)$$

In this thermometer, the gas may undergo changes in two successive steps. In the constant-volume process, following path 1-*a* in Figure 1.12, the temperature changes from T_1 to T_a:

$$\frac{T_a}{T_1} = \frac{p_a}{p_1} = \frac{p_2}{p_1} \qquad (1.24)$$

Then in the constant-pressure process, along path *a*-2, the temperature changes from T_a to T_2:

$$\frac{T_2}{T_a} = \frac{V_2}{V_a} = \frac{V_2}{V_1} \qquad (1.25)$$

When T_a is eliminated, then

$$\frac{T_2}{T_1} = \frac{p_2 V_2}{p_1 V_1} \qquad (1.26)$$

or pV/T is a constant (ideal gas relation). At each temperature, the product (pV) is a constant so that families of curves are generated when pressure-volume relationships at constant temperature are plotted on a p-V diagram, as shown in Figure 1.13. It is important to note that the products $(pV)_{T_1}$, $(pV)_{T_2}$, and so on are functions of temperature only and can therefore be used as a thermometric property to measure temperature.

The *constant-volume* and the *constant-pressure* gas thermometers operate according to the foregoing principle. In the constant-volume (or constant-density) thermometer shown in Figure 1.14, a real gas (normally hydrogen or helium) in bulb *A* must exist at a specified pressure, which is characteristic of the gas, if the bulb is maintained at the triple point of water. The volume of the gas in the calibrated bulb is kept constant by raising or lowering the mercury column until the mercury level is at the mark indicated in the figure. Thus the height h_{tp} of the mercury column indicates the gauge pressure of the gas in the bulb when the gas is at the triple point of water. This corresponds to an absolute pressure of p_{tp} at the triple point. When the gas is maintained at a different temperature, the mercury will be at some other level that corresponds to an absolute pressure p. The ratio p/p_{tp} can now be calculated. Note that this thermometer is an idealized model, for it assumes that a uniform temperature can be easily maintained and that the bulb's volume does not change with temperature. However, the values of p and p_{tp} depend in part on the amount of the gas in the bulb. This difficulty is surmounted by removing discrete portions of the gas from the bulb and repeating the previous measurement each time. This results in a new set of values of

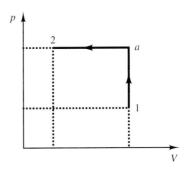

Figure 1.12   Change of state from 1 to 2 along 1-*a* and *a*-2.

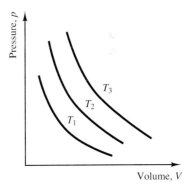

Figure 1.13   Constant-temperature curves on a p-V diagram.

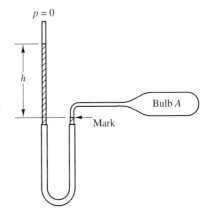

Figure 1.14   Constant-volume gas thermometer.

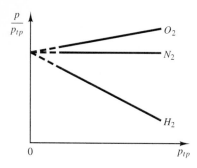

Figure 1.15   At zero pressure, different constant-volume gas thermometers indicate the same ratio of p/p_{tp}.

p_{tp}, p, and their ratio p/p_{tp} for each measurement. Another variation of this experiment is to use different gases. This again yields different ratios of p/p_{tp}.

As shown in Figure 1.15, when the pressure ratios p/p_{tp} are plotted against the reference pressure at the triple point, several curves result. If these curves are extrapolated to the ordinate where $p_{tp} = 0$, it is found that they all meet at a unique value irrespective of the gas in the bulb. Assigning a value of 273.16 to the temperature at the triple point of water, an absolute temperature T can be defined as

$$T = (273.16) \lim_{p \to 0} \left(\frac{p}{p_{tp}} \right)_V \qquad (1.27)$$

1.11 Thermodynamic Temperature Scale

If temperatures are measured with several thermometers, and if each thermometer employs a different thermometric substance, the readings of the various thermometers are likely to be *identical* only at one fixed point. This is because in calibrating a thermometer it is assumed that the thermometric property varies with temperature in a linear fashion. Such a linear relation is valid only as a first approximation; furthermore, the thermometric property of each substance varies with temperature in its own characteristic and complex manner. For this reason, the temperature measured by the different thermometers cannot be expected to exactly coincide except at the fixed point. The *thermodynamic temperature* (or the absolute temperature) must be independent of the properties of a particular system and is based on the second law of thermodynamics. An *ideal gas* scale indicates temperatures that coincide with the thermodynamic temperatures. Discussion of this temperature and insight into its elusive nature will be reconsidered in more detail in Chapter 5.

Two absolute scales commonly used are the *Kelvin* and the *Rankine* scales. In the Kelvin scale, a value of 273.16 is assigned to the triple point of water. The steam point on this scale is found to be 100.00°C. Conversions between the Kelvin and Celsius temperature scales are accomplished through the following relationship:

$$T(K) = T(°C) + 273.15 \qquad (1.28)$$

Note that temperature *differences* are equivalent for the Celsius and the Kelvin scales so that both scales have the same numerical intervals. In the Rankine scale, the triple point of water is at 491.69R, whereas the ice point is 491.67R. Since the Rankine scale, like the Kelvin scale, is an absolute scale, the numerical value of a temperature on the Rankine scale is related to the value on the Kelvin scale by the following ratio:

$$\frac{T(R)}{T(K)} = \frac{491.69}{273.16} = \frac{180}{100} = 1.8 \tag{1.29}$$

The Rankine scale has the same unit temperature intervals as the Fahrenheit scale, but its zero value is 459.67°F below the zero of the Fahrenheit scale. The two scales are related by the following equation:

$$T(R) = T(°F) + 459.67 \tag{1.30}$$

Substituting Eqs. (1.28) and (1.29) into Eq. (1.30) gives

$$T(°F) = 1.8T(°C) + 32 \tag{1.31}$$

Table 1.3 and Figure 1.16 compare the several temperature scales.

Table 1.3 Comparison of Temperature Scales at Some Standard and Fixed Points

Fixed point	Celsius	Kelvin	Fahrenheit	Rankine
Absolute zero	−273.15	0.00	−459.67	0.00
Ice point	0.00	273.15	32.00	491.67
Triple point of water	0.01	273.16	32.02	491.69
Steam point	100.00	373.15	212.00	671.67

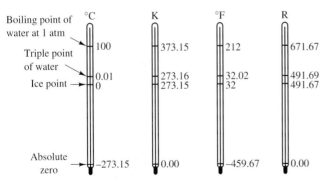

Figure 1.16 Relationships among Celsius, Kelvin, Fahrenheit, and Rankine scales.

1.12 The International Temperature Scale

In order to practically measure temperatures, the International Committee of Weights and Measurements specified the procedure to be followed. The procedure yields values that are reproducible and universally comparable over the range of temperature from 13.81 K to 1377.58 K. A revised International Practical Temperature Scale (IPTS-68) was adopted by the International Committee of Weights and Measurements in 1968. The scale conforms very closely to the thermodynamic temperature scale. In establishing the International Scale, the Committee specified (a) the temperature values at certain selected fixed points; (b) the fixed points at which instruments (thermocouple, resistance thermometer, and optical pyrometer) are to be calibrated; (c) the forms of equations for interpolating temperatures between these fixed points; and (d) the experimental procedure for calibration and measurements. The primary fixed-point temperatures in degree Kelvin are given in Table 1.4.

For temperature interpolation, the International Scale is divided into four ranges. Each range uses an interpolation equation chosen to give the best agreement with gas thermometer measurements. From 13.81 K to 273.15 K, the temperature is measured by a platinum resistance thermometer; from 273.15 K to 1003.89 K, a platinum resistance thermometer is also used, but the temperature is defined by a different formula; from 100.389 K to 1377.58 K, a platinum/platinum-rhodium thermocouple is used; above 1377.58 K, an optical pyrometer is used.

Table 1.4   Basic Fixed Points (K)

Triple point of equilibrium—hydrogen	13.81
Boiling point of equilibrium—hydrogen at 33.33 kPa	17.042
Boiling point at 1 atm pressure of equilibrium—hydrogen	20.28
Boiling point of neon at 1 atm	27.102
Triple point of oxygen	54.361
Boiling point of oxygen at 1 atm	90.188
Triple point of water	273.16
Boiling point of water at 1 atm	373.15
Freezing point of zinc at 1 atm	692.73
Freezing point of silver at 1 atm	1235.08
Freezing point of gold at 1 atm	1377.58

1.13 Thermodynamic Processes and Cycles

A *process* is the path followed by a system as it undergoes a change of state. It is described by the successive thermodynamic states through which the system passes. Figure 1.17 shows a *p-V* diagram of an expansion process from state 1 to state 2. When a system undergoes a practical process, there is a departure from thermodynamic equilibrium and some or all of the intervening states may be nonequilibrium states. To assess this departure from equilibrium, an idealized process is considered. An ideal or *quasi-equilibrium* process is one in which the state of the system deviates from thermodynamic equilibrium by only infinitesimal amounts throughout the entire process. Consider, for example, a frictionless piston subjected to a gas pressure p_1 on one face and a pressure p_2 on the opposite face. If the two pressures are equal, the piston is in equilibrium. If p_1 is made infinitesimally larger than p_2, the gas on side 1 will undergo a small expansion as a result of the imbalance of forces on the two faces of the piston. But because of the infinitesimal imbalance of forces, the gas undergoes a process without ever departing significantly from thermodynamic equilibrium. This is called a *quasi-equilibrium expansion.*

During the course of a process, energy transfer may take place at the system's boundary by heat or by work, and changes in properties of the system may occur. Those properties or phenomena that do not change during a process serve in describing the process. For example, an *isochoric* process describes a process in which the volume of a system remains constant. Similarly, an *isobaric* process is one in which the pressure remains constant, whereas an *isothermal* process is one in which the temperature remains constant. A process that involves no work interaction is called a *no-work* process, whereas one that involves no heat interaction is called an *adiabatic* process.

When a system at a given initial state undergoes a sequence of processes and then returns to its original state, the system has completed a thermodynamic *cycle.* Figure 1.18 shows a *p-V* diagram of a thermodynamic cycle. The properties of the system, at the completion of the cycle, have the same values as those of the initial state.

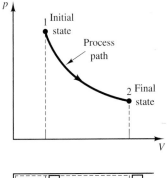

Figure 1.17   *p-V* diagram of an expansion process.

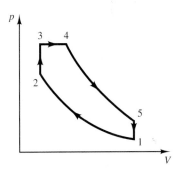

Figure 1.18   *p-V* diagram of a thermodynamic cycle.

1.14   Reversible and Irreversible Processes

If a process can occur in a reverse order, and if the initial state and all energies transferred or transformed during the process can be completely restored in both the system and its surroundings, then the process is called *reversible.* If a process is totally reversible, then no aftereffects or changes are evident in the system or in the surroundings when the process occurs in the forward and then in the reverse direction. A process is *irreversible* if either the system or its surroundings or both the system and its surroundings cannot be restored to their respective initial states after the process has occurred. A quasi-equilibrium process is an *internally reversible* process that proceeds at an infinitely slow rate. Such a process may be thought of as an infinite succession of equilibrium states. It may be stopped at any time and made to proceed in the opposite direction, thereby reversing the original process in every detail and restoring the system to its original state without placing any requirements on the surroundings. On the other hand, if a process is not in equilibrium, it cannot be reversed along its original path without causing a change in the surroundings. Therefore, quasi-equilibrium conditions are necessary but not sufficient for a process to be totally reversible. Conversely, if a process occurs at a finite rate, it is an irreversible process.

In reversible processes, any energy transfer, such as mechanical, thermal, and chemical, would have to be transferred reversibly. Thus processes involving dissipative effects, such as fluid flow, in which frictional losses occur, unrestrained expansion of a gas to a lower pressure, electrical resistance, and convective heat transfer, cannot be reversible. For heat transfer to occur, there must be a temperature difference. If heat is transferred from system A to system B during a process, then system A must be at a higher temperature than system B. In order for the reverse process to occur, heat would have to flow in the opposite direction, from system B to system A. This is impossible since system B is at a lower temperature than system A and heat never flows in that direction without creating another change in the surroundings. Thus it can be stated that actual processes involving heat interaction are irreversible. For a heat transfer process to be reversible, the difference in temperature must be infinitesimal in magnitude and the time allowed to reach equilibrium must be infinite in length.

Another factor that causes irreversibility is friction. When a piston moves in a cylinder as a gas expands to lower pressure, some energy in the form of work is expended in overcoming the friction between the piston and the cylinder. This friction produces heat. If the gas is compressed to its original volume, more heat is produced, but the work energy expended in overcoming friction is not recovered. This causes a permanent change in the surroundings. Therefore, the flow of real fluids is irreversible since it involves inherently irreversible effects. Additional examples of irreversible processes are given in Figure 1.19.

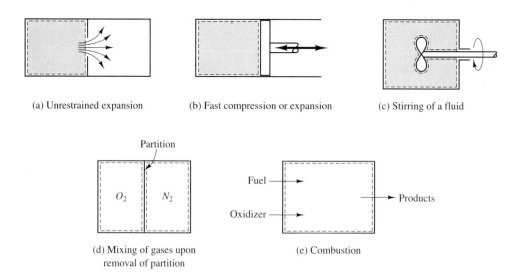

(a) Unrestrained expansion   (b) Fast compression or expansion   (c) Stirring of a fluid

(d) Mixing of gases upon
removal of partition

(e) Combustion

Figure 1.19   Examples of irreversible processes.

To visualize a reversible process, consider the gas enclosed in the vertical piston-cylinder arrangement shown in Figure 1.20a. The gas pressure supports the weight of the piston at level z_1. When heat is transferred to the system at a slow rate, the gas expands and the piston moves up in the cylinder maintaining the pressure constant. In the absence of mechanical and fluid friction, the piston will reach level z_2 and the gas undergoes a quasi-equilibrium expansion. Now suppose the reverse process takes place. When the same amount of heat is transferred from the system, the piston moves down to its original position. This is a reversible process because both the system and the surroundings regained their original states. If the piston were not frictionless, the piston would reach a lower level z_3 for the same amount of heat transfer as shown in Figure 1.20b. Upon the reverse process the piston would reach level z_4. We conclude that case (b) is an irreversible process because the system did not regain its original state.

Another example is a suspended pendulum in which kinetic and potential energies are successively exchanged. In the absence of friction, the process is reversible and the pendulum would swing forever; otherwise, it is irreversible because the pendulum and its surroundings do not regain their original states.

All actual or natural processes are irreversible since they take place with finite differences of potential between parts of the system or between the system and its surroundings. Reversible processes do not occur in nature, and this reduces the possibility of obtaining work energy from a system. Nevertheless, certain processes can be made almost reversible. Friction can be reduced, processes can be performed slowly, and temperature differences can be made small so that conditions can be made virtually reversible. All these conditions imply that reversible processes are carried out with an infinitesimal departure from equilibrium. The concept of reversibility is, therefore, useful because it indicates the limits of possible change in a real system and, as will be shown in Chapter 6, is a consequence of the second law of thermodynamics.

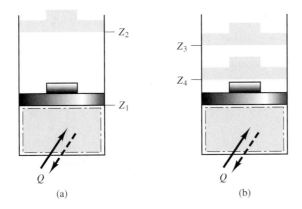

(a)   (b)

Figure 1.20   Reversible and irreversible processes.

1.15 Kinetic Theory and the Ideal-Gas Model

From a microscopic point of view, matter is not continuous but consists, rather, of discrete invisible particles. Although this concept was hypothesized by the ancient philosophers, it was only at the beginning of the nineteenth century that this hypothesis acquired a scientific form and served as the basis on which the atomic theory was built. The kinetic theory of gases attempts to explain the macroscopic properties of a gas in terms of the motion of its molecules. Kinetic theory is applicable to all matter, but it has been applied more extensively, so far, to gases than to solids or liquids.

The kinetic theory of gases was developed in the middle of the nineteenth century when the dynamic theory of heat was beginning to receive acceptance. The historical development of both theories reflects the struggle of primary hypotheses against established mathematical theories. In his book, *Hydrodynamics* (1738), Daniel Bernoulli presented an account of the kinetic theory of gases, but his work passed unnoticed and had, at the time, little or no influence on investigation in this area. Scientists during that period favored the static theory of gases, in which atoms are considered to repel each other, thus exerting pressure on the walls of the confining vessel. More than a century later, in 1859, Bernoulli's theory was revived, and the kinetic theory developed rapidly, mainly because of the work of Maxwell, Clausius, and Boltzmann. Planck, in 1901, formulated the quantum theory, which helped explain certain aspects of kinetic theory. According to quantum theory, the energy (and momentum) of very small particles, such as molecules and electrons, is subject to quantum restrictions so that the particles can have only certain discrete values of energy.

The movement of the molecules of a gas is extremely complex, and it is pointless to describe in detail the motion of each of the molecules. Through statistical methods, however, it is possible to describe the motion of numerous molecules in terms of the average motion of these molecules in space or time. Molecular motion is studied because it is closely associated with the macroscopic characteristics of a gas, such as pressure and temperature. The application of statistical techniques to relate the behavior of individual molecules to the observable behavior of material systems is called *statistical thermodynamics*.

The main assumption of the kinetic theory of gases lies in its definition of the gaseous state. The gas of a single-component substance is assumed to consist of a large number of identical, discrete particles called *molecules,* a molecule being the smallest unit with the chemical properties of the substance. The atoms of molecules are bound together by complex intermolecular forces as a result of the orbiting of electrons about the nucleus of each atom. The number of molecules is usually extremely large. For example, in 1 cubic cm of a gas at standard conditions (273 K and standard atmospheric pressure), there are 2.69×10^{19} molecules; at a pressure of 10^{-10} atm, 1 cubic mm contains 2.69 million molecules.

The volume occupied by the molecules is negligibly small compared to the total volume of the gas. At atmospheric temperature and pressure, for example, the molecules occupy one thousandth of the volume of the gas. The distance between molecules is therefore much larger than the molecules themselves so that on a molecular scale a gas is neither continuous nor homogeneous. Matter appears to be continuous on a macroscopic scale because of the large number of molecules in a small volume and because properties of matter are continuously subjected to an averaging effect owing to the contributions of the numerous molecules. The molecules are constantly in chaotic motion and may be visualized as point masses or small elastic spheres that move in straight lines until they collide with each other or with the walls of the confining vessel. Kinetic theory postulates that the molecules exert no force on each other except when they collide. The collisions are assumed perfectly elastic, and each collision changes both the magnitude and direction of movement of the two colliding molecules. The relative velocities of the molecules of a gas vary from small values at the instant of collision to very large values. No attractive forces are considered to be exerted by the molecules, whereas repulsive forces are assumed to act only during collision contact.

Experimental proof that molecules are in continuous motion was given when Brownian motion was discovered in liquids by Robert Brown in 1827 and in gases by De Broglie in 1908. Microscopic observations reveal that macroscopic masses (large by molecular standards) are in continuous and irregular motion because of the collisions between molecules.

Kinetic theory first assumed that motion of molecules was governed by the classical laws of mechanics. It was found later that this assumption is invalid for molecular-scale systems. When the quantum theory was developed, it became possible to explain the detailed motion of microscopic particles.

Although the position and velocity of molecules constantly change because of the large number of molecules, the number of molecules per unit volume of the gas remains essentially constant. Similarly, motion of the molecules occurs uniformly in all directions so that all magnitudes and directions of molecular velocities are equally probable. By statistical laws, the number of molecules having velocity between any two limits can be predicted with a known accuracy. It is thus seen that the concept of mechanical equilibrium of a gas is replaced by the concept of *statistical equilibrium* based on use of statistical laws. Properties evaluated by means of statistical equilibrium represent average properties; properties measured experimentally on a macroscopic scale are also average properties.

1.16   Equation of State of an Ideal Gas

Thermodynamic properties of simple systems are established by any two independent properties. A relationship between properties x and y,

$$f(x, y, z) = 0 \tag{1.32}$$

is called an *equation of state*. If values of any two of the properties are known, then the value of the third property is established. In the case of an ideal gas, the equation of state relating pressure, volume, and temperature is

$$pv = RT \qquad \text{or} \qquad p = \rho RT \qquad (1.33)$$

where

p = absolute pressure
v = specific volume
R = gas constant
T = absolute temperature
ρ = density

Properties may be expressed on a mole basis (that is, per kg-mol or pound-mol). A kg-mol is defined as the quantity of a substance whose mass in kg is numerically equal to the molar mass M of the substance. The equation of state for an ideal gas, on a mole basis, can thus be written

$$pV = mRT = n\overline{R}T \qquad (1.34)$$

where

V = volume of m kg or n moles
$\overline{R} = MR$ = molar (universal) gas constant approximately independent of the nature of the gas and equals 8314.4 J/kg-mol K.

Values of $\overline{R}$ in other units are given in the Appendix. Noting that $n = m/M$ and $v = V/m = V/nM$, it is clear that Eqs. (1.33) and (1.34) are equivalent.

A number of real gases, such as hydrogen, nitrogen, oxygen, and helium, follow the ideal-gas law at room temperatures so closely that they can be treated as ideal gases. Equations of state for ideal and real gases will be discussed in more detail in Chapter 8.

Example 1.4

Air is at 25°C and 101.325 kPa. If the gas constant $R = 287$ J/kg K, find the specific volume and the molar mass of this gas, assuming it behaves as an ideal gas.

SOLUTION

A sketch of the system is given in Figure 1.21.

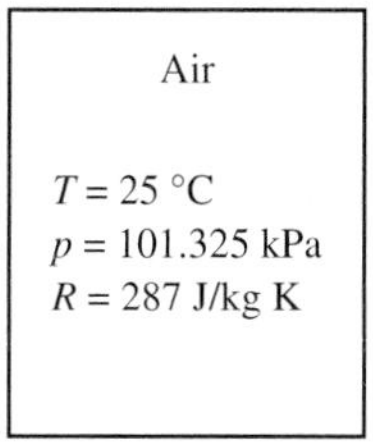

Figure 1.21

$$pv = RT$$
$$(101.325 \text{ kPa})v = (0.287 \text{ kJ/kg K})[(25 + 273.15) \text{ K}]$$

From which

$$v = 0.8445 \text{ m}^3/\text{kg}$$

The molar mass is

$$M = \frac{\overline{R}}{R} = \frac{8314.4 \text{ J/kg-mol K}}{287 \text{ J/kg K}} = 28.97 \text{ kg/kg-mol}$$

1.17 Avogadro's Number and Boltzmann's Constant

According to Avogadro's law, equal volumes of ideal gases at the same temperature and pressure contain an equal number of molecules. Experimental measurements indicate that the number of molecules in 1 kg-mol of an ideal gas is equal to $(6.0248 \pm 0.0003) \times 10^{26}$ (*Avogadro's number*, N_a). Note that regardless of temperature and pressure, 1 kg-mol of a gas always contains this number of molecules.

One kg-mol of any ideal gas at standard atmospheric pressure and at $0°C$ occupies a volume of 22.41 m^3. The specific volume of a gas (the volume occupied by a unit mass of gas) is equal to its molar volume (the volume occupied by 1 mole) divided by the molar mass of the gas. From this it follows that the density of a gas at standard conditions of temperature and pressure is proportional to its molar mass.

The ratio of the universal gas constant $\overline{R}$ to Avogadro's number N_a is called *Boltzmann's constant*. It is denoted by k and may be regarded as the universal gas constant per molecule. The value of k is 1.38066×10^{-23} J/molecule K. The equation of state of an ideal gas in terms of Boltzmann's constant takes the form

$$pV = nN_akT = NkT \tag{1.35}$$

where N is the number of molecules in the volume V.

1.18 Coefficients of Volumetric Expansion and Compressibility

In a single-component, single-phase system, changes in volume may occur owing to change in pressure or temperature, and can by expressed as

$$dv = \left(\frac{\partial v}{\partial T}\right)_p dT + \left(\frac{\partial v}{\partial p}\right)_T dp \tag{1.36}$$

Dividing by v, this becomes

$$\frac{dv}{v} = \frac{1}{v}\left(\frac{\partial v}{\partial T}\right)_p dT + \frac{1}{v}\left(\frac{\partial v}{\partial p}\right)_T dp$$

The expression $1/v\,(\partial v/\partial T)_p$ is called the *coefficient of volumetric expansion, β,* and the expression $-1/v\,(\partial v/\partial p)_T$ is called the *coefficient of isothermal compressibility, κ.* Both β and κ are intensive properties independent of the volume of the system.

Volume change can be expressed in terms of β and κ as

$$\frac{dv}{v} = \beta\, dT - \kappa\, dp \tag{1.37}$$

Density change can be expressed in a similar way:

$$\frac{d\rho}{\rho} = \kappa\, dp - \beta\, dT \tag{1.38}$$

The coefficient of volumetric expansion, β, is defined as the fractional change in volume at constant pressure, per unit change of temperature when the change in temperature and the corresponding change in volume become infinitesimal:

$$\beta \equiv \frac{1}{v}\left(\frac{\partial v}{\partial T}\right)_p = -\frac{1}{\rho}\left(\frac{\partial \rho}{\partial T}\right)_p \tag{1.39}$$

In the case of an ideal gas, the coefficient of volumetric expansion is

$$\beta = \frac{1}{v}\left(\frac{\partial v}{\partial T}\right)_p = \frac{p}{RT}\left(\frac{R}{p}\right) = \frac{1}{T} \tag{1.40}$$

Thus, the coefficient of volumetric expansion of an ideal gas varies inversely with absolute temperature and is independent of both pressure and volume. The coefficient of volumetric expansion of a real gas, however, depends on both pressure and temperature.

The coefficient of compressibility, κ, is defined as the fractional change in volume at constant temperature per unit change of pressure when the change in pressure and the corresponding change in volume become infinitesimal:

$$\kappa \equiv -\frac{1}{v}\left(\frac{\partial v}{\partial p}\right)_T = \frac{1}{\rho}\left(\frac{\partial \rho}{\partial p}\right)_T \qquad (1.41)$$

The negative sign is included in the preceding equation because an increase in pressure generally results in a decrease in volume. When the system consists of an ideal gas, the coefficient of compressibility becomes

$$\kappa = -\frac{1}{v}\left(\frac{\partial v}{\partial p}\right)_T = -\frac{1}{v}\left(-\frac{RT}{p^2}\right) = \frac{1}{p} \qquad (1.42)$$

The compressibility of a substance is measured by its change of volume when the external pressure changes. Let the pressure on a certain substance of specific volume v be increased by dp. As a result, the specific volume is diminished by dv. The *bulk modulus of elasticity, E,* is given by

$$E = -\frac{dp}{dv/v} = -v\frac{dp}{dv} \qquad (1.43)$$

where dv/v is the volumetric strain. Note that E is not a property since the derivative dp/dv depends on the process. Substituting from Eq. (1.37), the bulk modulus can be expressed as

$$E = \frac{1}{\kappa - \beta(dT/dp)}$$

In the case of solids and liquids, changes in pressure produce negligible changes in temperature so that the term dT/dp is essentially zero. Thus the reciprocal of the coefficient of compressibility of solids and liquids is the bulk modulus of elasticity. In the case of gases, the term dT/dp cannot be neglected and therefore the bulk modulus of elasticity, E, is not a property.

Classification of a substance as *compressible* or *incompressible* depends on the magnitude of the coefficients of volumetric expansion and compressibility. The degree of compressibility, however, depends on the process itself. At atmospheric pressure, for example, the volume change, ΔV, due to pressure change, of a liquid such as water ($E = 20,700$ N/cm^2) is small, but at large pressures its compressibility becomes evident. Similarly, the volume change of a flowing gas as the pressure changes may be quite large. But when a gas flows at low velocity at constant temperature, it behaves as though it were incompressible.

Example 1.5

The isothermal coefficient of compressibility of water at 10°C and atmospheric pressure is 50×10^{-6} atm^{-1}. What absolute pressure must be exerted on a certain mass of water to decrease its volume 5 percent at the same temperature?

SOLUTION

The coefficient of compressibility is

$$\kappa = -\frac{1}{v}\left(\frac{\partial v}{\partial p}\right)_T = 50 \times 10^{-6} \text{ atm}^{-1}$$

Separating variables and integrating gives

$$\int_{p=1 \text{ atm}}^{p} dp = -(2 \times 10^4 \text{ atm})\int_{v}^{0.95v} \frac{dv}{v} = -(2 \times 10^4) \ln 0.95 = 1026 \text{ atm}$$

and the absolute pressure that must be exerted is $= 1026 + 1 = 1027$ atm.

Example 1.6

The change of the specific volume of water as a function of temperature at atmospheric pressure in the range of 0°C to 4°C is as shown in Figure 1.22. Determine the coefficient of volumetric expansion of water at 2°C.

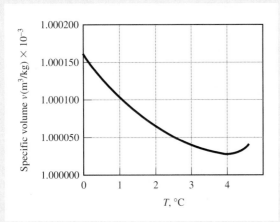

Figure 1.22

SOLUTION

The coefficient of volumetric expansion is

$$\beta = \frac{1}{v}\left(\frac{\partial v}{\partial T}\right)_p$$

Using two neighboring points at 1°C and 3°C, an average value of β can be written as

$$\beta = \frac{1}{v}\left(\frac{\Delta v}{\Delta T}\right)$$

Substituting values from Figure 1.22 gives

$$\beta = \frac{1}{1.000062 \text{ m}^3/\text{kg}}\left[\frac{1.000038 \text{ m}^3/\text{kg} - 1.000104 \text{ m}^3/\text{kg}}{(3-1)\text{ K}}\right]$$
$$= -3.3 \times 10^{-5}(\text{K})^{-1}$$

From 0°C to 4°C, water contracts as the temperature rises. At 4°C there is a point of maximum density, and above this temperature it expands with rising temperature.

1.19 Guidelines for Solving Thermodynamic Problems

The analysis of thermodynamic problems is governed by the first law of thermodynamics, which relates to energy balance, and by the second law of thermodynamics, which relates heat transfer and irreversibility to entropy. The analysis must also fulfill the requirements of the conservation of mass. In problems that involve kinetic and dynamic effects, Newton's laws of motions are applied.

A systematic approach for solving thermodynamic problems is as follows:

1. Read the problem statement carefully in order to define the process, the assumptions involved, and the answers needed. Note the input data, the interaction with the surroundings, and state what unknowns are required.
2. Draw a schematic diagram of the physical system, and decide whether the system approach or the control volume approach is more appropriate for solving the problem. Accordingly, define clearly the boundary of the system (or control volume).
3. Identify the substance involved in the problem, and list the given properties with the proper symbols on the schematic. Identify the source of other relevant property values necessary for the solution such as thermodynamic relations, tables, or graphs.

4. Indicate on your schematic all energy transfers with the surroundings, and identify each transfer whether it is heat, work, or energy associated with mass transfer.
5. Additional insight into solving the problem can be achieved by sketching the process on a relevant property diagram such as a *p-v* or a *T-v* diagram, indicating for each process to be analyzed the initial and final states of the system.
6. State the simplifying assumptions that you would use to model the problem.
7. Apply the thermodynamic laws and relationships pertinent to the solution, and simplify the equations using the applicable assumptions.
8. Substitute numerical values into the equations using a consistent set of units, and make sure that the terms in each equation are dimensionally consistent. It is important to include the appropriate sign for each energy transfer. Energy transfer to the system is considered positive, whereas energy transfer from the system is considered negative.
9. Proceed with the calculations to obtain the answer to the problem, and review the solution for errors. Do your results seem reasonable?

1.20 Summary

Thermodynamics is concerned with the study of energy transformation and the relationships among properties of substances involved in these transformations. Thermodynamic analysis is based on a fixed-mass system or a control volume. The region outside the boundary of the system or control volume is called the surroundings. An isolated system excludes all interactions with the surroundings.

A property is any characteristic of a system. Intensive properties (T, p, ρ) are independent of the mass of the system, whereas extensive properties (V, system energy) depend on the mass. Differentials of properties are mathematically described as exact differentials. A quasi-static process is a process proceeding through a sequence of equilibrium states. When a process can be reversed so that the initial states of both system and surroundings are completely restored, the process is called reversible.

Pressure is the normal force per unit area; it is attributed to the change of momentum of the molecules. The absolute pressure is related to the gauge pressure by the relation

$$p_{abs} = p_{gauge} + p_{atm}$$

Density is the mass per unit volume; it is the reciprocal of specific volume. Temperature is based on the zeroth law of thermodynamics, which defines temperature in terms of thermal equilibrium. The units of temperature in SI system are °C or K, and in the English system, °F or R. The Kelvin and the Rankine temperature scales are the absolute temperature scales and are given by

$$T(K) = T(°C) + 273.15°$$
$$T(R) = T(°F) + 459.67°$$

and

$$\frac{T(R)}{T(K)} = 1.8$$

The behavior of gases at low pressure and moderate or high temperature can be approximated by the ideal-gas equation $pV = mRT = n\overline{R}T = NkT$, where R, $\overline{R}$, and k are the gas constant, universal gas constant, and Boltzmann constant, and m, n, and N are the mass, number of moles, and number of molecules, respectively. The ideal-gas temperature scale is equivalent to the absolute temperature scale.

The coefficient of volumetric expansion and the isothermal compressibility factor are defined as

$$\beta = \frac{1}{v}\left(\frac{\partial v}{\partial T}\right)_p \quad \text{and} \quad \kappa = -\frac{1}{v}\left(\frac{\partial v}{\partial p}\right)_T$$

Problems

1.1 Calculate the force exerted on a mass of 2 kg moving at an acceleration of 3 m/s^2 in (a) newtons, (b) lbf, (c) dynes.

1.2 The weight of 1 m^3 of air at 25°C is 11.6 N at a location where $g = 9.80665$ m/s^2. What is the density of air in kg/m^3 and lbm/ft^3?

1.3 A force of 300 N is applied to a mass of 30 kg along an incline at a location where $g = 9$ m/s^2. If the force and the incline make a 30° angle with the vertical, determine the acceleration of the mass if the force is acting (a) upward, (b) downward.

1.4 Prove that the weight of a body at an elevation z above sea level is given by

$$w = mg\left(\frac{r}{r+z}\right)^2$$

where r is the radius of the earth at the equator.

1.5 The weight of an automobile is 12,000 N at a location where the gravitational acceleration $g = 9.81$ m/s^2. What is the net force necessary to accelerate the car at a constant rate of 1.2 m/s^2?

1.6 A force of 800 N is applied to an air-streamed object of mass of 25 kg in an upward direction. Neglecting air drag, determine the acceleration of the object assuming the average gravitational acceleration is $g = 9.8$ m/s^2.

1.7 The weight of an apple is 1 N at a location where the local acceleration of gravity is 9.8 m/s^2. What is the mass of the apple in kilograms? What is its weight on the surface of the moon, where $g = 1.67$ m/s^2?

1.8 If dU is a function of three properties x, y, and z such that

$$dU = M\ dx + N\ dy + P\ dz$$

where M, N, and P are functions of the properties x, y, and z, prove that the following conditions are necessary in order for dU to be an exact differential:

$$\frac{\partial P}{\partial y} = \frac{\partial N}{\partial z} \qquad \frac{\partial M}{\partial z} = \frac{\partial P}{\partial x} \qquad \frac{\partial N}{\partial x} = \frac{\partial M}{\partial y}$$

1.9   Ascertain the exactness of the following differentials:
(a) $df = (x^2 + y^2)dx + 2xy\ dy$
(b) $df = dx/x + dy/y$
(c) $df = [y/(x^2 + y^2)]dx - [x/(x^2 + y^2)]dy$
and find an expression for the function $f(x, y)$.

1.10   The differential of the pressure of a certain gas is given by one of the following equations:

$$dp = \frac{2RT}{(v - b)^2}dv + \frac{R}{v - b}dT$$

or

$$dp = -\frac{RT}{(v - b)^2}dv + \frac{R}{v - b}dT$$

Identify the correct equation, and find the equation of state of that gas.

1.11   The differential of a function $f(x, y)$ is given by

$$df = (x^2 + y^2)dx + (2xy + \cos y)dy$$

Prove that df is an exact differential, and find the expression of f in terms of x and y.

1.12   For the van der Waals equation of state,

$$p = \frac{RT}{v - b} - \frac{a}{v^2}$$

where a and b are constants, determine

$$\left.\frac{\partial p}{\partial v}\right|_T \qquad \left.\frac{\partial v}{\partial T}\right|_p \qquad \text{and} \qquad \left.\frac{\partial T}{\partial p}\right|_v$$

1.13   A vessel containing 0.3 kg of a gas on the surface of the moon has a volume of 0.4 m³. If the local gravity is 1.67 m/s², determine (a) the density, (b) the specific volume, and (c) the specific weight of the gas. What will these values be at sea level ($g = 9.81$ m/s²)?

1.14   The pressure in a tank is 80 kPa. Determine the pressure equivalent in terms of meters of a column of liquid at room temperature if the liquid is (a) water, (b) mercury (s.g. = 13.59), (c) alcohol (s.g. = 0.79). The local value of $g = 9.81$ m/s².

1.15   A vacuum pressure gauge reads 620 mm of mercury. If the barometric pressure is 760 mm Hg, what is the absolute pressure in lbf/in² (psia) and in SI units?

1.16   Complete the following table if atmospheric pressure is 101.325 kPa:

kPa, abs	psia	psig	cm Hg abs
101.325	____	____	_____
_____	100	____	_____
_____	____	25	_____
_____	____	____	200

1.17   A pressure gauge reads 150 kPa, and the barometer reads 95 kPa. Find the absolute pressure in kPa and in psia.

1.18   Calculate the height of a column of water equivalent to atmospheric pressure of 101.325 kPa if the water is at 15°C. What is the height if the water is replaced by mercury?

1.19   If all atmospheric air is liquified, what will be the equivalent height of this liquid air in meters of water at 15°C?

1.20   Find the weight of air in the atmosphere surrounding the earth if the pressure is 101.325 kPa everywhere on the surface. Assume that the earth is a sphere of diameter 13,000 km.

1.21   The pressure in the bulb *A* of Figure 1.23 is 50 cm Hg vacuum. What is the height *h* of the mercury in the tube? (s.g. of Hg = 13.58, and barometric pressure = 76 cm Hg.)

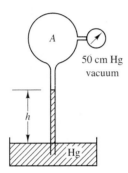

50 cm Hg vacuum

A

h

Hg

Figure 1.23

1.22   For the conditions shown in Figure 1.24, determine $(p_a - p_b)$ in kPa if the prevailing temperature is 20°C.

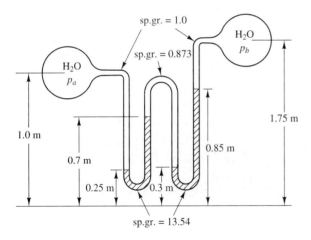

sp.gr. = 1.0

H_2O
p_b

sp.gr. = 0.873

H_2O
p_a

1.0 m

1.75 m

0.85 m

0.7 m

0.25 m   0.3 m

sp.gr. = 13.54

Figure 1.24

1.23   Plot the pressure difference across the hull of a submarine as a function of depth in seawater with a specific gravity of 1.03. The pressure inside the submarine is atmospheric. Assume an average local acceleration of gravity to be 9.8 m/s².

1.24   The absolute pressure of a gas in a closed vessel is 70 kPa. If the atmospheric pressure is 750 mm of mercury, determine the vacuum pressure in the vessel, in kPa. The density of mercury is 13.59 gm/cm³, and the local gravitational acceleration is 9.81 m/s².

1.25   The temperature variation in the atmosphere near the surface of the earth is given by

$$T(z) = 15 - 0.00651z$$

where *T* is in °C, and *z* is the elevation above the earth's surface in meters. If the atmospheric pressure at *z* = 0 is 101.325 kPa, determine the pressure at an elevation of 2 kilometers above the earth.

1.26   Derive Eq. (1.18) for a gas occupying a cubical volume.

1.27   Complete the following table:

°F	R	°C	K
60	___	___	___
___	600	___	___
___	___	-5	___
___	___	___	400

1.28   A Celsius and a Fahrenheit thermometer are immersed in a fluid. If the reading of the Fahrenheit thermometer is numerically twice that of the Celsius thermometer, what is the temperature of the fluid in R and K? Find the temperature at which the Fahrenheit and Celsius scale coincide.

1.29   A temperature scale is graduated according to the equation $T = 100 + 3T_c$, where *T* is the temperature reading on the scale, and T_c is the Celsius temperature. Find the freezing and boiling points of the thermometric substance. What is the absolute temperature corresponding to 20° temperature reading on this scale?

1.30   A thermometric property *x* (length of a mercury column in a mercury-in-glass thermometer) is equal to 8 cm and 50 cm when the thermometer is at the ice and steam points. The temperature T varies linearly with x. Now assume that a temperature $T*$ on a certain Celsius scale is defined by the equation $T* = a + bx^2$, where $T = 0°$ and $100°$ at the ice and steam points, and a and b are constants. Find the temperature $T*$ on this scale when the temperature T is $40°C$.

1.31   The temperature scale of a thermometer is given by $T* = ae^x + b$, where a and b are constants, and x is the same thermometric property as defined in Problem 1.30. If 0 and 200 denote the ice and steam points on the new scale, plot a graph of T versus $T*$, and find the temperature reading corresponding to $40°C$ on this scale. Assume a linear variation of T with x.

1.32   A balloon containing an ideal gas has a volume of 0.2 m^3. The temperature and pressure of the gas are $15°C$ and 101.325 kPa. If the gas is heated to $60°C$, what must the applied pressure be in order that the volume remains constant?

1.33   Find the mass of air enclosed in a cylindrical tank 0.15 m in diameter and 1 m long (inside dimensions). The temperature of the air is $20°C$, and the pressure is 600 kPa. If a valve mounted on the tank is opened until the cylinder pressure becomes atmospheric, find the mass of air that escaped from the tank if the final temperature is $20°C$.

1.34   Air at 1 atm and $300°C$ is compressed isothermally from a volume of 100 m^3 to 5 m^3. Determine the mass of the air and the final pressure.

1.35   Two tanks are connected by a valve. One tank contains 1 kg of nitrogen gas at $60°C$ and 60 kPa. The other tank contains 0.4 kg of the same gas at $35°C$ and 200 kPa. The valve is opened, and the gases are allowed to mix. If the equilibrium temperature is $50°C$, determine the final equilibrium pressure.

1.36   Two kg of an ideal gas at 100 kPa and $25°C$ are compressed to 350 kPa according to the relation $pv^n = $ constant, where $n = 1.4$. If the initial volume is 1.2 m^3, determine the final volume and temperature of the gas.

1.37   Determine the value of the universal gas constant if one kg-mol of gas occupies 22.41 m^3 at $0°C$ and standard atmospheric pressure.

1.38   The average temperature of the atmosphere as a function of altitude above sea level is given by the relation

$$T_{atm} = 288.15 - 0.00651z$$

Prove that the pressure is given by

$$p_{atm} = 101.325(1 - 0.02259 \times 10^{-3}z)^{5.249}$$

where T_{atm} is in K, p_{atm} in kPa, and z in m. What is the density of air at an altitude of 10 km above sea level?

1.39   Prove that

$$\left(\frac{\partial \beta}{\partial p}\right)_T = -\left(\frac{\partial \kappa}{\partial T}\right)_p$$

1.40   The coefficient of volumetric expansion and the compressibility are defined as

$$\beta = \frac{1}{v}\left(\frac{\partial v}{\partial T}\right)_p \qquad \kappa = -\frac{1}{v}\left(\frac{\partial v}{\partial p}\right)_T$$

Calculate $(\partial p/\partial T)_v$ for an ideal gas in terms of β and κ.

1.41   The coefficient of volumetric expansion and the coefficient of compressibility for a certain substance are given by

$$\beta = \frac{2bT}{v} \qquad \text{and} \qquad \kappa = \frac{a}{v}$$

where a and b are constants. Determine the equation of state of this substance.

1.42   Determine an expression for the coefficient of volumetric expansion β and the compressibility κ for a gas obeying the Clausius equation of state: $p(v - b) = RT$.

1.43   The equation of state for 1 mole of a "van der Waals" gas is

$$\left(p + \frac{a}{v^2}\right)(v - b) = RT$$

where a and b are constants. Prove that

$$\beta = \frac{Rv^2(v - b)}{RTv^3 - 2a(v - b)^2}$$

and

$$\kappa = \frac{v^2(v - b)^2}{RTv^3 - 2a(v - b)^2}$$

where β is the coefficient of volumetric expansion, and κ is the coefficient of compressibility.

C1.1   Write an interactive computer program to express absolute pressure in kPa and psi.

C1.2   Write an interactive computer program to express temperature in K, °C, °F, and R.

Chapter 2

Energy Types and Conservation Laws

2.1 Introduction

Physical laws are relationships that model patterns of behavior in physical systems. Conservation laws comprise a particular class of physical laws and are characterized by their simplicity, generality, and utility. They describe the behavior of a system when certain physical quantities are maintained constant.

The principle of conservation of energy will be discussed in the context of the first law of thermodynamics (Chapter 3). The present chapter discusses the law of conservation of mass and Newton's second law of motion. The law of conservation of mass simply states that mass can neither be created nor destroyed. The laws of conservation of mass and energy are utilized in dealing with energy relationships and will be introduced with emphasis on fluid flow. Several forms of energy and methods of energy transfer are also introduced in this chapter.

2.2 Work

Work is the name of a method of energy interaction between a system and its surroundings. It is identified at the interaction boundary if *the sole effect external to the system can be reduced to a change in the level of a weight.* This definition does not say that a weight is actually raised or lowered; rather, it indicates that the sole effect external to the system can be shown to be equivalent to the raising or lowering of a weight. Work done on the system by the surroundings is considered positive; work done by the system on the surroundings is negative.[1]

Consider mechanical work as defined in mechanics. Mechanical work is the product of an external force acting at the boundaries of a system and the distance through which the force moves along its line of action. Considering motion in only one dimension, the work done is

$$W_{1\text{-}2} = \int_{x_1}^{x_2} F_x \, dx \tag{2.1}$$

where F_x is the component of force in the direction of the displacement dx. But from the standpoint of thermodynamics, work can also arise from other effects, not only from mechanical motion. It is therefore necessary to adopt a broader interpretation of work and express it in terms of the concept of systems and processes.

Work that is not of a mechanical nature and cannot be expressed by Eq. (2.1) can still be recognized as work by the following criterion: Imagine a device that utilizes this nonmechanical energy to produce mechanical work, resulting then in the change of level of a weight. If this is feasible, then this nonmechanical energy is work, provided that no other effects are produced. As an example, electrical energy from a storage battery may cross the boundaries of a system, as shown in Figure 2.1. Does work cross the boundaries? This question is

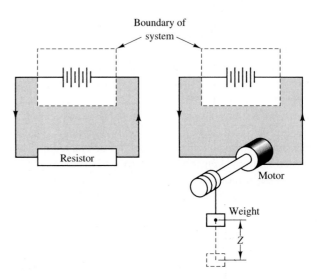

Figure 2.1   Equivalence of electrical energy to work.

[1] This sign convention differs from that in many texts, but it is less confusing to consider energies entering the system as positive and those leaving the system as negative. This convention is in accordance with mechanics that the student has usually studied before thermodynamics.

answered by imagining that the electrical energy drives an ideal motor, which rotates a frictionless pulley, which, in turn, raises a weight. Since the only effect on the surroundings is reduced to the raising of a weight, electrical energy is work.

Now consider the frictionless piston-cylinder assembly shown in Figure 2.2. Let the system be the fluid enclosed between the piston and the cylinder. The system exerts a pressure p on the piston face of area A. If the piston is allowed to move a distance dx, the force F_e acting on the piston may vary but at any instant is given by

$$F_e = p_e A \tag{2.2}$$

where p_e is the external pressure exerted on the system at the moving boundaries.

Suppose the piston moves to the left a distance dx in the direction of the force F_e. The work done is $\delta W = F_e \, dx$, and the surroundings has done work on the system. If, on the other hand, the piston moves to the right, work is done by the system on the surroundings, which is the same, algebraically, as saying that negative work is done on the system.

In order to express the external force in terms of the pressure of the system rather than that of the surroundings, the piston must be maintained in equilibrium at all instants. This requires that the external force F_e be opposite and differs

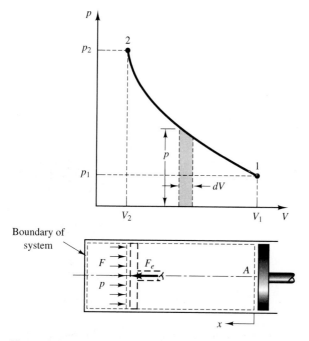

Figure 2.2 Work done on a system in a quasi-equilibrium process is $-\int p \, dV$.

infinitesimally from the force F exerted by the system. This restriction implies that the process is reversible, and the work done on the system is

$$\delta W = F dx$$

But since

$$F = pA$$

therefore,

$$\delta W = pA\ dx$$

Let V be the initial volume of the system, and $V + dV$ the final volume after the motion of the piston. Since compression is involved, dV is a negative quantity and equal to $-A dx$. Therefore the work done on the system is

$$\delta W = -p\ dV \tag{2.3}$$

In the opposite case, when the piston moves to the right so that expansion occurs, dV is positive. Work is done by the system (or negative work is done on the system), and therefore $\delta W = -p\ dV$. Hence, Eq. (2.3) indicates the work done on the system when either expansion or compression occurs. In a finite process in which the volume changes from V_1 to V_2, the total work interaction is obtained by integrating Eq. (2.3):

$$W_{1\text{-}2} = \int_1^2 \delta W = -\int_{V_1}^{V_2} p\ dV \tag{2.4}$$

To integrate this equation, a functional relation between p and V must be known. As an example, an analytical form for the pressure-volume relationship is $pV^n = C$, where the value of n is a constant for a particular process. For the case where the pressure remains constant during a process, the work is given by

$$W_{1\text{-}2} = -p\int_{V_1}^{V_2} dV = p(V_1 - V_2)$$

Equation (2.4) also applies to any process as long as p is the pressure at the face of the piston (or at the boundaries of the system that move); the pressure at other points in the system need not be the same. The pressure on the face of the moving piston is the same as the pressure throughout the system only if the process is a quasi-equilibrium process proceeding through a sequence of equilibrium states. Recall that one aspect of the quasi-equilibrium process is that the values of the intensive properties are uniform throughout the system. A process

involving rapid expansion or compression is usually accompanied by fluid friction or mechanical friction, and in such nonquasi-equilibrium processes, the
fluid pressure at the boundary differs from that in the interior of the system. It is
then difficult to determine actual pressures, and consequently the work done is
uncertain. Note that the work done on the system, whether the compression is
quasi-equilibrium or not, is the work of the external force F_e. The value of the
force F_e as a function of the position of the piston is then required in order to calculate work interaction.

Work interaction in a quasi-equilibrium process can be represented graphically on a p-V diagram, appearing as the area under the curve 1-2 of Figure 2.2.

In an expansion process, in which the system changes from state 1 to state 2,
the work interaction is negative since the volume increases. In this case, work is
done by the system on the surroundings. Conversely, if the process proceeds
from state 2 to state 1, as in compression, the work interaction is positive, and
work is done on the system by the surroundings.

A cyclic process can be represented on a p-V diagram by the closed curve
1-A-2-B-1 of Figure 2.3. The work done on the system during the expansion process 1-A-2 is

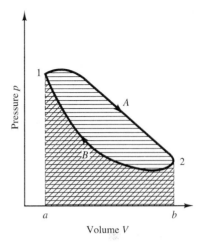

Figure 2.3 Work done in a cycle $-\oint p\,dV$.

$$-\int_{1(A)}^{2} p\,dV = \text{area } a\text{-}1\text{-}A\text{-}2\text{-}b\text{-}a$$

and the work done on the system during the compression process 2-B-1 is

$$-\int_{2(B)}^{1} p\,dV = \text{area } b\text{-}2\text{-}B\text{-}1\text{-}a\text{-}b$$

Hence, the net work done is

$$W = -\int_{1(A)}^{2} p\,dV - \int_{2(B)}^{1} p\,dV = -\oint p\,dV = -(\text{area } 1\text{-}A\text{-}2\text{-}B\text{-}1)$$

where the symbol $\oint$ denotes a cyclic integral. Note that the work interaction is
negative. If the cycle of Figure 2.3 were performed in the opposite direction
1-B-2-A-1, the area of the closed curve would represent work done on the system.

In representing work done in a cycle by means of a p-V diagram, the following rule can be applied to distinguish positive work from negative work: If an
observer is moving along the boundary of the $\int p\,dV$ area in the direction of the
process, and if the area lies to the right of the observer, then work is done by the
system and the work interaction is negative.

Indicator diagrams or cards provide a plot of gauge pressure versus crank
angle, as shown in Figure 2.4, or versus piston travel, as shown in Figure 2.5. The
area of an indicator diagram may be used to determine actual work done on the
piston during a complete cycle. The area can be determined by counting squares

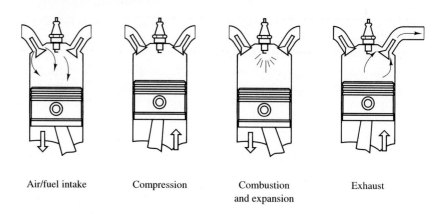

Air/fuel intake　　　Compression　　　Combustion　　　Exhaust
and expansion

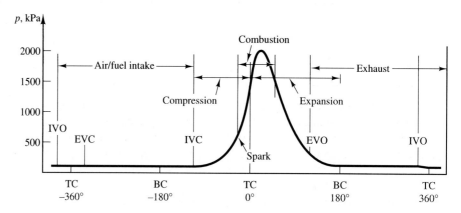

Figure 2.4　Indicator diagram (pressure versus crank angle) of a four-stroke spark ignition cycle. (TC = top center, BC = bottom center, IVO and IVC are inlet valve opens and closes, EVO and EVC are exhaust valve opens and closes)

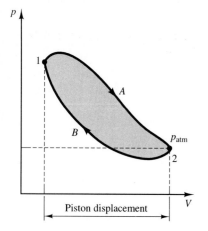

Figure 2.5　Indicator diagram.

on graph paper, by measuring directly with a planimeter, or by using a numerical integration method (such as Simpson's rule). The work done is given by

$$W/cycle = -\oint p\,dV = -\oint pA\,dL = -p_m\,AL$$

where

A = area of the piston face
L = stroke
p_m = indicated mean effective pressure

$$= \frac{-\oint p\,dV}{V_2 - V_1} = \frac{\text{area } 1\text{-}A\text{-}2\text{-}B\text{-}1}{V_2 - V_1}$$

$V_2 - V_1$ = volume swept by the piston (the piston displacement)

Example 2.1

Find the work done per cycle by a 10 cm bore, 12.5 cm stroke engine if the area of the indicator card is 11 cm², the card length is 5 cm, and the pressure scale is 200 kPa for every cm.

SOLUTION

$$\text{Piston area A} = \frac{\pi}{4}(0.1^2 \text{ m}^2) = 0.00785 \text{ m}^2$$

$$\text{W/cycle} = -p_m AL$$

$$= -[(11/5) \text{ cm}](200 \times 10^3 \text{ Pa/cm})(0.00785 \text{ m}^2)(0.125 \text{ m}) = -4.32 \text{ J}$$

2.3 Work as a Path Function

Several reversible processes leading from state 1 to state 2 are shown in Figure 2.6. The work interaction in each case is represented by the area underneath the corresponding path. The work done is obviously different in each case because the path depends on the nature of the process. This means that work is not a property or a state function; rather, it is a *path function,* and an infinitesimal increment of work is an inexact differential. Therefore, a system does not possess work; instead, work is a mode of transfer of energy. This transfer occurs only at the boundaries of the system as the system undergoes a change of state. For this reason, work is expressed by $\int_1^2 \delta W = W_{1\text{-}2}$, and not by $W_2 - W_1$.

Hereafter, the symbols d and δ will be used to distinguish between differentials of properties, such as pressure, which are fixed by the end states of the system, and nonproperties, such as work, which are functions of the path. Expressions of work interaction during various processes are discussed in the following sections.

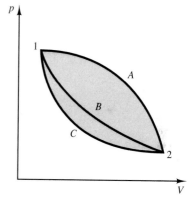

Figure 2.6   Work transfer depends on the path.

2.4 Units of Energy Transfer by Work

Work is a scalar quantity having the dimensions of energy. The unit of work in SI units is the joule (J), $(1 \text{ J} = 1 \text{ Nm} = 1 \text{ kg-m}^2/\text{s}^2)$. In the English engineering system, the unit of work is the *foot-pound force,* which is equivalent to a force of 1 lbf acting through a distance of 1 ft in the direction of the displacement. In the CGS system, the unit of work is the *erg,* defined as the amount of work done by a force of 1 dyne when it acts through a distance of 1 cm in the direction of the displacement. The joule is 10^7 ergs.

The time rate of energy transfer by work is called power and is denoted by $\dot{W}$. In the SI system of units, the unit of power is J/s, called the watt (W). The W-s (joule) is the energy associated with an electromotive force of 1 volt and the passage of 1 coulomb of electricity.

Example 2.2

A reversible cycle of a work-producing machine is represented by a circle of 5 cm diameter on a p-V diagram, as shown in Figure 2.7. The pressure and specific volume scales are

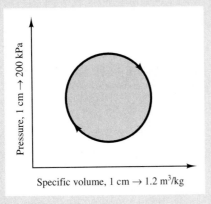

Figure 2.7

$$p\text{-scale:} \quad 1 \text{ cm} = 200 \text{ kPa}$$
$$v\text{-scale:} \quad 1 \text{ cm} = 1.2 \text{ m}^3/\text{kg}$$

Compute the work done on 1 kg of fluid.

SOLUTION

$$w = -\oint p\,dv$$
$$= -(\text{area of circle shown in Figure 2.7})$$
$$= -\frac{\pi}{4}(5 \text{ cm})^2(200 \text{ kPa/cm})[1.2 \ (\text{m}^3/\text{kg})/\text{cm}] = -0.471 \text{ kJ/kg}$$

Example 2.3

Helium contained in a cylinder fitted with a piston expands reversibly according to the relation $pV^{1.5} = $ constant. The initial volume is 0.1 m^3, the initial pressure is 450 kPa, and the initial temperature is 250 K. After expansion, the pressure is 200 kPa. Calculate the work done during the expansion process.

SOLUTION

A sketch of the system and a p-V diagram of the process is shown in Figure 2.8. Using the relation $pV^n = C$, the work interaction can be expressed as

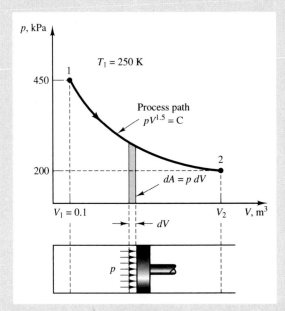

Figure 2.8

$$W_{1\text{-}2} = -\int_{V_1}^{V_2} p \, dV$$

$$= -\int_{V_1}^{V_2} \frac{C}{V^n} dV = C \left[\frac{V^{-n+1}}{n-1} \right]_{V_1}^{V_2} = C \frac{\left(V_2^{-n+1} - V_1^{-n+1} \right)}{n-1}$$

But

$$C = p_1 V_1^n = p_2 V_2^n$$

so that

$$W_{1\text{-}2} = \frac{p_2 V_2 - p_1 V_1}{n-1}$$

To determine V_2,

$$p_1 V_1^n = p_2 V_2^n = C$$

or

$$\frac{V_2}{V_1} = \left(\frac{p_1}{p_2}\right)^{1/n} \qquad V_2 = 0.1\left(\frac{450}{200}\right)^{0.667} = 0.1717 \text{ m}^3$$

Then

$$W_{1\text{-}2} = \frac{(200 \text{ kPa})(0.1717 \text{ m}^3) - (450 \text{ kPa})(0.1 \text{ m}^3)}{0.5} = -21.32 \text{ kJ}$$

The negative result of $W_{1\text{-}2}$ means work is done by the system on the surroundings. Another way of solving this problem is to introduce the ideal-gas law ($pV = mRT$); thus, the expression for the work interaction becomes

$$W_{1\text{-}2} = \frac{mRT_2 - mRT_1}{n - 1} = \frac{mR(T_2 - T_1)}{n - 1}$$

To determine m,

$$p_1V_1 = mRT_1$$

where

$$R = \frac{\overline{R}}{M} = \frac{8.3144 \text{ kJ/kg-mol K}}{4.003 \text{ kg/kg-mol}} = 2.077 \text{ kJ/kg K}$$

Therefore,

$$m = \frac{(450 \text{ kPa})(0.1 \text{ m}^3)}{(2.077 \text{ kJ/kg K})(250 \text{ K})} = 0.0867 \text{ kg}$$

To determine T_2,

$$\frac{p_2V_2}{p_1V_1} = \frac{T_2}{T_1}$$

But

$$\frac{V_2}{V_1} = \left(\frac{p_1}{p_2}\right)^{1/n}$$

which, when substituted in the previous equation, gives

$$\frac{p_2}{p_1}\left(\frac{p_1}{p_2}\right)^{1/n} = \frac{T_2}{T_1} \qquad \text{or} \qquad \frac{T_2}{T_1} = \left(\frac{p_2}{p_1}\right)^{(n-1)/n}$$

Hence,

$$T_2 = (250 \text{ K})\left(\frac{200}{450}\right)^{0.333} = 250 \times 0.763 = 191 \text{ K}$$

and

$$W_{1\text{-}2} = \frac{(0.0867 \text{ kg})(2.077 \text{ kJ/kg K})[(191 - 250) \text{ K}]}{0.5} = -21.32 \text{ kJ}$$

Example 2.4

Calculate the work done on a mass of 2 kg of air when it expands reversibly and isothermally at 300 K from a volume of 2 m³ to a volume of 4 m³.

SOLUTION

A sketch of the system and a p-V diagram of the process is shown in Figure 2.9.

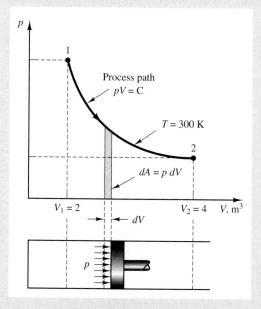

Figure 2.9

$$W_{1\text{-}2} = -\int_{V_1}^{V_2} p \, dV$$

and, considering air as an ideal gas,

$$p = \frac{mRT}{V}$$

Then,

$$W_{1\text{-}2} = -mRT \int_{V_1}^{V_2} \frac{dV}{V} = -mRT \ln \frac{V_2}{V_1}$$

$$= -(2 \text{ kg})(0.287 \text{ kJ/kg K})(300 \text{ K}) \ln \left(\frac{4}{2}\right) = -119.36 \text{ kJ}$$

The work done in the preceding example may be calculated in terms of pressures instead of volumes since, for an isothermal process,

$$p_1 V_1 = p_2 V_2 = mRT = \text{constant} \qquad \text{or} \qquad \frac{V_2}{V_1} = \frac{p_1}{p_2}$$

and

$$W_{1\text{-}2} = -mRT \ln \frac{V_2}{V_1} = -mRT \ln \frac{p_1}{p_2}$$

2.5 Flow Energy

Energy transfer in the form of work is associated with flow phenomena and represents the work necessary to advance a fluid against the existing pressure. Consider an element of a fluid in a duct of mass dm and volume dV. As shown in Figure 2.10(a), the mass dm is at the threshold of a control volume, and at this point the pressure is p. It is required to calculate the work done by the matter behind this element of fluid in pushing it into the control volume. Let an imaginary piston be placed behind dm, as shown in Figure 2.10(b), and calculate the amount of work done by the piston in moving dm past line a-a.

$$\text{Flow energy per unit mass} = \frac{(pA)dx}{dm} = \frac{pA(dV/A)}{dm} = p\frac{dV}{dm} = pv$$

Thus the flow energy per unit mass is simply the product of the absolute pressure and the specific volume of the fluid. It is sometimes called *flow work* since it rep-

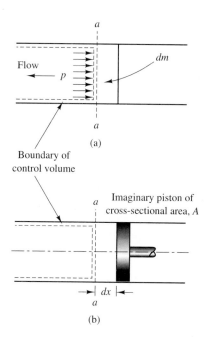

a

Flow ← p dm

(a)

Boundary of
control volume

Imaginary piston of
cross-sectional area, A

dx

a

(b)

Figure 2.10   Flow energy per unit
mass = pv.

resents the amount of work that must be done on a control volume to introduce a
unit mass of material into it. When mass leaves the boundary of a control volume, flow work must be done by the control volume on the surroundings.

2.6 Expressions of Work for Some Thermodynamic Systems

Most engineering thermodynamics involves systems whose equilibrium states
are described by the coordinates p, V, and T. For such systems, the work of compression or expansion during a quasi-static process is equal to $-\int p\, dV$. In
determining work for other types of systems, an analogy may be made between
such systems and the p-V-T system. Some examples are now cited.

(a) *Electrical work:* Energy transfer due to an electric potential is called *electrical work.* It is defined as the work that must be done by an electric field in moving a charge against an electric potential. If a quantity of electricity or a charge dC
of a condenser is flowing across an electric potential $\mathcal{E}$, then the electrical work
done on the condenser (system) expressed in a differential form is given by

$$\delta W_e = \mathcal{E}\, dC$$

The current is the time rate of change of the electrical charge, or

$$i = \frac{dC}{dt}$$

Then,

$$W_e = \int_{t_1}^{t_2} \mathcal{E}i \, dt \tag{2.5}$$

and the power is

$$\dot{W}_e = \frac{\delta W_e}{dt} = \mathcal{E}i \tag{2.6}$$

where

W_e = electrical work done on the system (J)
C = quantity of electricity flowing (coulombs)
$\mathcal{E}$ = electrical potential (volts)
i = rate of flow of electricity or current (amperes)
t = time (s)
$\dot{W}_e$ = electric power (W)

Figure 2.11 gives the relation between $\dot{W}_e$ and $\mathcal{E}$, i and the resistance R. The usual unit of electrical work is the joule.

1 J = 1 volt × 1 coulomb   (1 coulomb = 1 ampere-s)

also

1 J = 1 W-s = 0.7376 ft-lbf

The watt is the unit of electric power defined as the energy transferred at the rate of 1 joule per second or the power developed by a current of 1 ampere flowing through a potential of 1 volt.

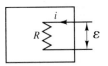

Figure 2.11   Electric power:
$W_e = \mathcal{E}i = i^2 R$.

Example 2.5

A current of 0.3 amp is flowing through an electric resistance. If the voltage across the resistance is 110 volts, find the power consumption.

SOLUTION

$$P = \mathcal{E}i = 110 \times 0.3 = 33 \text{ W}$$

(b) *Work done in stretching a wire:* An expression for the differential work done in stretching a wire (Figure 2.12) from a length L to $L + \Delta L$ under a force F is

$$\delta W = F \, dL$$

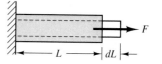

Figure 2.12   Stretching a wire.

which when integrated gives

$$W_{1\text{-}2} = \int_{L}^{L+\Delta L} F \, dL \qquad (2.7)$$

The work in this case is positive since work must be done on the system (the wire) in order to increase the length of the wire.

(c) *Work done in changing the area of a surface film:* The work done on a homogeneous liquid film (Figure 2.13) in changing its surface area by an infinitesimal amount dA is

$$\delta W = \sigma \, dA$$

and the work done for an increase in surface area from A_1 to A_2 is

$$W_{1\text{-}2} = \int_{A_1}^{A_2} \sigma \, dA \qquad (2.8)$$

where σ is the interfacial tension per unit length.

(d) *Magnetization of a paramagnetic solid:* A differential expression of the work done per unit volume on a magnetic material through which the magnetic and magnetization fields are uniform is

$$\delta W = H \, dI$$

and

$$W_{1\text{-}2} = \int_{I_1}^{I_2} H \, dI \qquad (2.9)$$

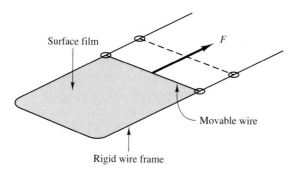

Figure 2.13   Stretching a liquid film with a movable wire.

where H is the magnetic field strength, and I is the component of the magnetization field in the direction of the field. Work must be done by the system to increase the magnetization (positive dI).

(e) *Polarization of a dielectric:* A differential expression of the work done per unit volume on a dielectric material through which the electric and polarization fields are uniform is

$$\delta W = E \, dP$$

and

$$W_{1\text{-}2} = \int_{P_1}^{P_2} E \, dP \tag{2.10}$$

where E is the electric field intensity within the dielectric, and P is the component of the polarization field in the direction of the electric field.

The similarity of the foregoing equations for work in quasi-equilibrium processes indicates that work can be expressed as a function of a generalized force X and a differential of a generalized displacement dx. If X is an intensive property and dx is a differential of an extensive property, then the general expression for work can be written as

$$\delta W = X \, dx \tag{2.11}$$

If the intensive property is plotted against the extensive property, the area underneath the curve is a measure of work.

2.7 Work Done in Two Irreversible Processes

It was previously mentioned that work is equal to $-\int p \, dV$ only if the process proceeds in a quasi-equilibrium manner. Two examples will be considered in which the processes are not in quasi-equilibrium. The first is the free expansion of a gas to a larger volume.

Consider an insulated vessel divided by a diaphragm into two compartments, as shown in Figure 2.14. One compartment contains gas in an equilibrium state, whereas the other compartment is evacuated. If the diaphragm is removed, the gas will expand suddenly into the evacuated compartment. No work or heat interaction takes place between the system and its surroundings. This nonequilibrium process is called *free expansion.* According to the conditions of reversibility, the initial state of the system and the surroundings cannot be restored unless some energy is supplied from an external source. Therefore, the free expansion process is irreversible, and for the system defined in Figure 2.14, no work is done at the boundaries.

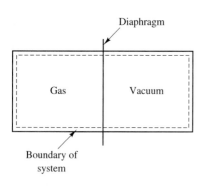

Figure 2.14   Work done in a free expansion process is zero.

The second irreversible process involves a system in which a paddle wheel stirs a fluid of fixed volume, as shown in Figure 2.15. The work done increases the energy of the system, and after paddling, the system settles down to equilibrium with a higher temperature. The turbulence in the fluid due to the stirring action is the cause of the irreversibility of the process, and the amount of work done is not equal to $-\int p\,dV$. Note that work is done in this case while the system is maintained at a constant volume. This is in contrast to the free expansion process in which no work is done, but there is a change in the volume of the system. During both processes, the pressure and specific volume are not uniform throughout the system and therefore the state of the system cannot be adequately described.

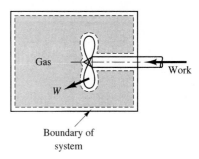

Figure 2.15   Work done on a system at constant volume.

2.8 Energy

The concept of energy was first introduced in mechanics by Newton when he hypothesized about kinetic and potential energies. But the emergence of energy as a unifying concept in physics was not adopted until the middle of the nineteenth century and was considered a major scientific achievement of that century. The concept of energy is so familiar to us today that it is intuitively obvious, yet we have difficulty in defining it exactly. Energy is a *scalar* quantity that cannot be observed directly but can be recorded and evaluated by indirect measurements. The absolute value of the energy of a system is difficult to measure, whereas its energy change is rather easy to calculate.

The sun is the major source of the earth's energy. It emits a spectrum of energy that travels across space as electromagnetic radiation. Energy is also associated with the structure of matter and can be released by chemical and atomic reactions. Throughout history, the emergence of civilizations has been characterized by the discovery and effective application of energy to society's needs.

Energy manifests itself in many forms, which are either internal or transient, and any one form of energy can be converted to any other form. The extent of conversion, however, may be complete or partial. Mechanical, electrical, chemical, or other forms of energy can be converted completely to heat. The conversion of heat to mechanical energy (in cyclic operation), on the other hand, is only partial.

Energy may be classified as:

(**1**) *Energy in transition due to a potential difference:* Gradients of force, temperature, and electrical potential[2] result in the transfer of mechanical work, heat, and electrical energy.

There are numerous modes of energy transmission; some are more familiar than others. One form of mechanical energy consists of sound waves, where the energy is transmitted by adjacent masses of air that alternately expand and contract.

---

[2] Potentials can also include chemical and magnetic.

Although the energy is transmitted by local movement of the molecules of the medium, as an overall effect, the medium remains at one general location. Another mode of energy transmission is electromagnetic radiation. Visible light, X-rays, gamma rays, radio waves, and radiant heat are examples of energy transmitted by this means. Electromagnetic waves are emitted by vibrating molecules, and the waves may be transmitted to other molecules with or without the aid of an intervening medium.

(2) *System energy:* Examples of this type are potential, kinetic, chemical, and atomic energy. Internal energy is present in a system by virtue of its orientation in a force field, its motion, its chemical composition, or its atomic structure.

From the microscopic point of view, particles or molecules of a system that move translationally possess kinetic translational energy; polyatomic molecules that rotate about their centers of mass possess rotational energy; and molecules that vibrate along their centers of mass possess vibrational energy. The magnitude of the translational, rotational, and vibrational energies is a function of temperature. Translational and rotational energies comprise most of the molecular kinetic energy at room temperature, whereas vibrational energy becomes significant only at high temperatures. Additional energies attributed to the molecular structure are bonding, electronic, ionic, nuclear, and so forth.

Thermodynamics is concerned with the *change* of energy rather than with the absolute value of energy. In most cases, there are no means of measuring absolute energies; further, since only knowledge of changes of energy is needed for solving most problems, the zero level of energy may be set at an arbitrary state.

2.9 Potential Energy

The energy that a system possesses by virtue of its position in a force field is called *potential energy.* For example, gravitational potential energy is associated with the gravitational force field. Referring to Figure 2.16, the amount of work done on a system of mass m when it is lifted with negligible acceleration through a distance dz in a gravitational field is equal to the product of its weight, mg, and the height, dz. Since the force required to lift the system is equal and opposite to its weight,

$$\delta W = mg\, dz$$

The increase in potential energy is equal to work done on a system (in the absence of friction) in changing its position. Therefore, the change in potential energy of a system that has been moved through a uniform gravitational field is

$$\Delta(PE) = \int_{z_1}^{z_2} F_z\, dz = \int_{z_1}^{z_2} mg\, dz$$
$$= mg(z_2 - z_1) \tag{2.12}$$

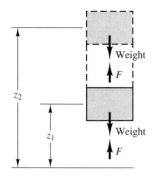

Figure 2.16   Work is done on a system to lift it in a gravitational field from elevation z_1 to elevation z_2.

Similar expressions can be obtained for the potential energy in any conservative force field, such as magnetic or electric fields. Note that the potential energy is an extensive scalar property and depends on the initial and final states of the system. It is not possible to ascribe an absolute value to potential energy. Only changes in potential energies between two states are required.

2.10 Kinetic Energy

The energy a system possesses owing to its motion is *kinetic energy*. According to Newton's second law of motion, the force F, acting on a particle, is equal to the product of its mass m and its acceleration a, or

$$F = ma$$

Acceleration along the x-direction can be expressed as

$$a = \frac{dV}{dt} = \frac{dV}{dx}\frac{dx}{dt} = V\frac{dV}{dx}$$

Therefore,

$$F_x = mV\frac{dV}{dx}$$

Referring to Figure 2.17, the infinitesimal amount of work done on the particle in moving it a distance dx is

$$\delta W = F_x\,dx$$

This work is equal to the change in kinetic energy of the system (in the absence of friction). Therefore,

$$\Delta(KE) = \int_{x_1}^{x_2} F_x\,dx = m\int_{V_1}^{V_2} V\,dV$$
$$= \frac{m(V_2^2 - V_1^2)}{2}$$

(2.13)

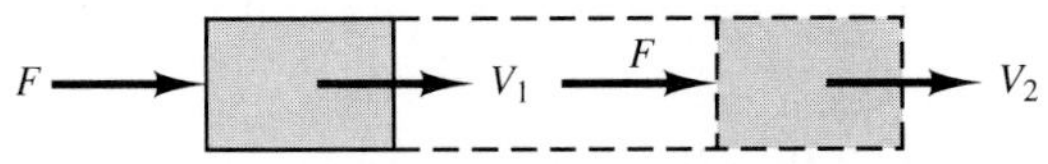

Figure 2.17   Work is done on a system to accelerate it from velocity V_1 to velocity V_2.

Note that the energy transfer to the system in the form of work is stored in the system as kinetic energy. When rotational motion about a fixed axis occurs, the change in the kinetic energy of a rigid body is given by

$$\Delta(KE) = \frac{mr^2\left(\omega_2^2 - \omega_1^2\right)}{2} \tag{2.14}$$

where r is the radius of gyration, and ω is the angular velocity of the system.

Kinetic energy, like potential energy, is a scalar extensive property of the system. The kinetic energy of a system with zero velocity relative to the earth is taken to be zero.

The combined potential and kinetic energies of a system constitute the *mechanical energy* of the system. Any work done on a system appears as an increase in the mechanical energy of the system if no friction occurs and if only mechanical energy is involved. If the work is positive, the total mechanical energy of the system increases. Similarly, a decrease in the mechanical energy of a system occurs when work is done by the system. Since the change in mechanical energy of a system is entirely defined by its end states,[3] any work done is also a function of the end states. Two cases arise in which the mechanical energy of a system remains constant. The first occurs when the system has identical initial and final states. The second case is that of an isolated system in which no energy crosses its boundary. In both cases, no work is done. Consequently, if there is an increase in kinetic energy, there is an equal decrease in potential energy. This has led to the concept of conservation of energy, which was first applied to freely falling bodies. As an example, consider a pendulum as an isolated system. Referring to Figure 2.18, the vertical distance z represents the height of the center of mass of the pendulum at position M above its middle position B. The total mechanical energy of the system is given by

$$mgz + \frac{1}{2}mV^2 = \text{constant}$$

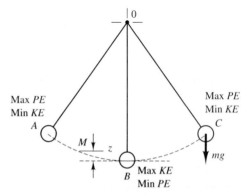

Figure 2.18   Conservation of mechanical energy.

[3] Provided that the system is subjected to conservative forces only.

where m is the mass of the pendulum, and V is its velocity at position M. The mechanical energy remains constant as the pendulum swings between the extreme positions A and C. At points A and C, the potential energy is maximum, and the kinetic energy is zero. As the pendulum moves from point A to point B, potential energy is transformed into kinetic energy; as the pendulum moves from point B to point C, kinetic energy is transferred back to potential energy.

2.11 Internal Energy

The internal energy, U, of a substance is energy associated with the configuration and motion of its molecules, atoms, and subatomic particles relative to its center of mass. Internal energy U is a property consisting of the combined molecular kinetic energy and molecular potential energy, and is determined by properties, such as pressure and temperature. It is part of the total energy E, which comprises all forms of internal energy, such as chemical energy in addition to the macroscopic kinetic and potential energies of the system:

$$E = (U + ..., etc.) + KE + PE \qquad (2.15)$$

The energy per unit mass (specific energy) at a point is defined by the equation

$$e = \lim_{\Delta m \to 0} \frac{\Delta E}{\Delta m} \qquad (2.16)$$

where Δm is an elementary mass surrounding the point. Note that the ratio $\Delta E / \Delta m$ is the average energy per unit mass. The total energy E in terms of the specific energy e is

$$E = \int e \, dm \qquad (2.17)$$

A more detailed discussion of internal energy will be given in Chapter 3.

2.12 Energy Transfer by Heat

In the eighteenth century, the accepted theory of the nature of heat was the caloric theory. It assumed that heat was a fluidlike substance without mass that was evolved from a body owing to the propulsion of its caloric particles. It was further assumed that the amount of caloric particles in a system is a property of the system and their number increases as the temperature is raised. The caloric theory was replaced in the latter part of the eighteenth century by the dynamic theory of heat, now generally accepted, which considers heat as a mode of energy transfer.

In previous sections, it was shown that adiabatic work interaction between the surroundings and a closed system changes the energy of the system. The same energy change may also be obtained by heat interaction. Hence, heat and work are simply modes of energy transfer between a system and its surroundings.

Let two systems at different temperatures be brought into contact with each other so that they have a common boundary. Energy in the form of heat is transferred between the two systems owing to the temperature difference, the high-temperature system losing a certain amount of energy and the low-temperature system gaining that same amount. This process continues until thermal equilibrium is attained. Note that heat is not observed; instead, heat transfer is inferred through temperature changes or through other similar effects. Energy transfer in the form of heat may be regarded as the only energy interaction between two systems in the absence of work interaction. Hence, heat may be defined as the name of *a method of energy transfer (independent of mass transfer) between a system and its surroundings due to temperature difference*. Regardless of the mechanism of heat transfer, a temperature difference is necessary for effecting heat interaction. Heat transferred to a system is considered positive, and heat transferred from a system is negative.

Energy may be transferred from one system to another by either or both heat transfer mechanisms: *conduction* and *radiation*. Energy is transferred by conduction when two systems at different temperatures are brought into direct or close contact with each other. In conduction heat flow, energy is transmitted from the more energetic to the less energetic particles of a substance by direct molecular collisions without appreciable macroscopic displacement of the molecules. Thermal radiation is energy emitted by matter due to changes in the electron configurations of the constituent atoms or molecules. The energy is transported by electromagnetic waves. These waves, assuming nonrelativistic conditions and no diffraction, travel in straight lines in homogeneous media or vacuum until they are reflected or absorbed. The velocity of propagation of these waves in vacuum is equal to the velocity of light. The mechanism of heat transfer by radiation is basically the same as that of light radiation.

Convection is a process that involves both mass transport and heat transfer in which the heat is transferred by both conduction and radiation. Convection is generally classified as natural or forced, depending on the forces acting on the fluid. When heat is transferred from a solid to a fluid, a density field is created in the fluid, and this nonuniform density field results in fluid motion. This process is called *natural* convection or *free* convection. If the fluid is caused to flow by an external means such as a pump or a fan, the process is called *forced* convection. Figure 2.19 shows the different modes of heat transfer.

Heat is a path function and, therefore, like work, is an inexact differential. Heat depends on the details of the process connecting two states and may be identified at the boundaries of a system, but it is not really possessed by the system. When

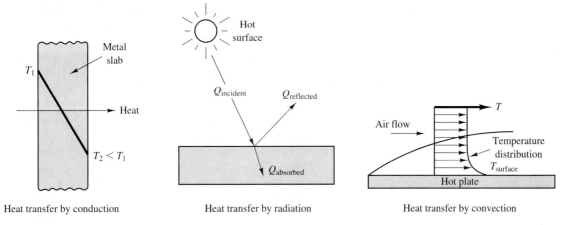

Heat transfer by conduction   Heat transfer by radiation   Heat transfer by convection

Figure 2.19   Modes of heat transfer.

energy enters a system, a change in properties occurs, and it then becomes impossible to identify the energy transfer mechanism as work or heat. This means that at the completion of a process, the terms *work* and *heat* have no meaning as far as the system is concerned. Only by investigating phenomena occurring at the boundaries of a system during the interaction can energy transfer as heat be distinguished from energy transfer as work.

An *adiabatic* process is defined as one that involves no heat interaction. Adiabatic conditions exist if the system is surrounded by an enclosure impermeable to heat or if the system is subjected to very rapid compression or expansion. In the latter case, the rate of heat interaction is relatively small, and therefore only a negligible amount of heat is transferred during the short duration of compression or expansion.

2.13 Units of Energy Transfer by Heat

The unit of heat[4] in the SI system of units is the *joule* (J). In the English system, the *British thermal unit* (Btu) is the unit of heat. The quantity of heat required to raise the temperature of a 1 lbm of water from 59.5°F to 60.5°F under atmospheric pressure is called the 60°F Btu. Similarly, in the CGS system, the quantity of heat required to increase the temperature of 1 gram of water from 14.5°C to 15.5°C under atmospheric pressure is called the 15°C calorie.

After the first law of thermodynamics was established, it was realized that heat and work, even though they are different means of energy transfer, were expressed

---

[4] The words *heat* and *work* are used in this text instead of *the transfer of heat energy* and *the transfer of work energy*. The reader is reminded that heat and work are *methods* of energy transfer.

in different energy units. In order to make it possible to express heat in units of work and vice versa, electrical measurements were made that established a unit of heat. In 1929, the International Steam Table Conference adopted the *International Calorie* (IT) as

$$1 \text{ IT calorie} = 4.1868 \text{ kg m}^2/\text{s}^2 = 4.1868 \text{ J}$$

The British thermal unit of heat could be expressed in terms of the IT calorie:

$$1 \text{ Btu} = 251.996 \text{ IT calories} = 1055 \text{ J}$$

Note also that 1 Btu/lbm°F = 1 cal/gm°C = 4.1868 kJ/kg°C. The heat flux, $(\dot{Q}/A)$, is the rate of heat transfer per unit of system surface area. The units for the heat flux are W/m^2 or Btu/h.ft^2.

2.14 Specific Heat[5]

When energy is transferred between a system and its surroundings, the energy interaction causes a change in the internal energy of the system. Two methods are used to calculate the change of internal energy. In one method, the temperature of the system changes, but the physical and chemical states are unaffected by the energy transfer. In the second method, the physical or chemical states change, but the temperature remains constant. This section deals with the first case.

The *specific heat* of a substance is equal to the amount of energy required to raise the temperature of a unit mass of the substance 1 degree. The *average* specific heat is equal to the amount of energy transferred per unit mass divided by the accompanying increase of its temperature or

$$c_{avg} = \frac{Q}{m\Delta T} = \frac{q}{\Delta T} \tag{2.18}$$

where

c_{avg} = average specific heat (J/kg K)
Q = heat interaction (J)
ΔT = temperature difference (K or °C)
m = mass (kg)
q = heat per unit mass (J/kg)

---

[5] The names *specific heat* and *heat capacity* are inherited from the caloric theory, which considered heat as a fluidlike substance. These names are unfortunate since they erroneously imply that heat can be stored in a system. See Sections 3.3 and 3.5 for better definitions of specific heats.

When q and ΔT are very small, then the ratio $q/\Delta T$ tends to a limit that indicates the specific heat at a temperature T. Thus,

$$c = \lim_{\Delta T \to 0} \frac{q}{\Delta T} = \frac{\delta q}{dT} \qquad (2.19)$$

To obtain a mean value of specific heat over a wide temperature range, the specific heat is integrated, by direct, numerical, or graphical methods, according to the equation

$$c_{\text{avg}} = \frac{\displaystyle\int_{T_1}^{T_2} c \, dT}{\displaystyle\int_{T_1}^{T_2} dT} = \frac{\displaystyle\int_{T_1}^{T_2} c \, dT}{T_2 - T_1} \qquad (2.20)$$

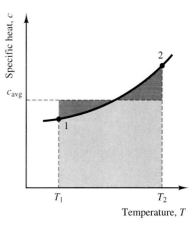

Figure 2.20   Average specific heat.

In evaluating this integral graphically, values of specific heat c are first plotted against temperature, as shown in Figure 2.20. The average value of specific heat divides the figure into two portions, shown as hatched areas, that are equal.

The specific heat of solids and liquids depends upon temperature, but, unlike gases, is not sensitive to the type of process involved during heat interaction. The law of Dulong and Petit indicates that the atomic specific heat (that is, heat transfer per unit mass per degree) of an element in the solid state at room temperature is equal to approximately 6.4 cal/gm-atom°C. If a substance undergoes a change of phase when heated, its specific heat appears to be infinitely large since heating produces no change in temperature.

The specific heat of a gas depends on whether the gas is at constant pressure or at constant volume when the heat interaction takes place:

$$c_p = \left(\frac{\delta q}{dT}\right)_p \qquad (2.21)$$

$$c_v = \left(\frac{\delta q}{dT}\right)_v \qquad (2.22)$$

where c_p is the constant pressure specific heat, and c_v is the constant volume specific heat. Both c_p and c_v are properties of the substance. Molar specific heat is given by

$$\bar{c}_p = \frac{mc_p}{n} \qquad (2.23)$$

$$\bar{c}_v = \frac{mc_v}{n} \qquad (2.24)$$

where $\bar{c}_p$ and $\bar{c}_v$ are in J/kg-mol K, m is mass in kg, and n is number of moles.

The specific heat of a gas changes with temperature in a complex manner, and various equations have been developed that indicate values over wide temperature ranges (see the Appendix).

The unit of specific heat in the SI system is J/kg K. In the English system, the unit of specific heat is Btu/lbm°F, and in the CGS system is cal/gm°C. The numerical value of the specific heat is the same when expressed in the English or the CGS system.

2.15 Latent Heat[6]

The amount of energy necessary to change the physical structure (phase) of a unit mass of a substance, without changing the temperature and without doing work other than flow energy pv, is called the *latent heat.* Latent heat that is associated with the melting and solidification of a solid is called *heat of fusion,* that related to the vaporization and condensation of a liquid is called *heat of vaporization,* and that related to the sublimation of a solid is called the *heat of sublimation.*

Latent heats are expressed in terms of particular temperature and pressure conditions. The latent heats of fusion and sublimation are essentially constant over wide ranges in temperature; the latent heats of vaporization vary markedly with pressure.

Example 2.6

Find the amount of heat required for each change of phase during the heating of 1 kg of ice initially at $-7°C$ till it becomes superheated steam at 150°C. The pressure is atmospheric.

Assume	c of ice	=	2.104 kJ/kg K
	c of water	=	4.207 kJ/kg K
	c_p of steam	=	1.872 kJ/kg K
	Latent heat of fusion of ice	=	333.3 kJ/kg
	Latent heat of vaporization of water	=	2256.2 kJ/kg

SOLUTION

Heating of ice:	$Q_{1\text{-}2} = (1 \text{ kg})(2.104 \text{ kJ/kg K})[0\text{K} - (-7\text{K})]$	=	14.7 kJ
Melting of ice:	$Q_{2\text{-}3} = (1 \text{ kg})(333.3 \text{ kJ/kg})$	=	333.3 kJ
Heating of water:	$Q_{3\text{-}4} = (1 \text{ kg})(4.207 \text{ kJ/kg K})[(100 - 0) \text{ K}]$	=	4.207 kJ
Vaporization of water:	$Q_{4\text{-}5} = (1 \text{ kg})(2256.2 \text{ kJ/kg})$	=	2256.2 kJ
Heating of steam:	$Q_{5\text{-}6} = (1 \text{ kg})(1.872 \text{ kJ/kg K})[(150 - 100) \text{ K}]$	=	93.6 kJ
		Total Q =	3118.5 kJ

[6] The word *heat* is wrong because the transfer of energy could be by heat or work. See Section 3.3.

Example 2.7

A 2.5 kg system undergoes a process wherein its temperature changes from 280 K to 340 K. Determine the heat interaction if the specific heat of the system for the process in question is given by

$$c = \left(0.84 + \frac{35}{T + 320}\right) \text{ kJ/kg K}$$

What is the average specific heat?

SOLUTION

$$Q = m \int_{T=280}^{T=340} c \, dT = m \int_{T=280}^{T=340} \left(0.84 + \frac{35}{T + 320}\right) dT$$

$$= m[0.84T + 35 \ln (T + 320)]_{280}^{340}$$

$$= (2.5 \text{ kg})\left[(0.84 \text{ kJ/kg K})[(340 - 280) \text{ K}] + (35 \text{ kJ/kg}) \ln \frac{660}{600}\right]$$

$$= 134.34 \text{ kJ}$$

The average specific is

$$c_{avg} = \frac{Q}{m\Delta T} = \frac{134.34 \text{ kJ}}{(2.5 \text{ kg})(60 \text{ K})} = 0.896 \text{ kJ/kg K}$$

2.16 Relation Between Kinetic Theory and Thermodynamics

The relationship between kinetic theory and thermodynamics arises from the molecular velocity distribution of a gas at the equilibrium state. The law governing the velocity distribution will be derived in Chapter 12. In this section, some preliminary kinetic interpretations of energy and thermodynamic functions are discussed.

Consider a gaseous system in which each molecule of mass m travels at a velocity of v and possesses a kinetic energy $mv^2/2$. The internal energy of translation of the gas is equal to the sum of the translational kinetic energy of its individual molecules, or

$$U = \sum_{i=1}^{N} \frac{m_i v_i^2}{2}$$

$$= Nm \frac{v_{rms}^2}{2} \qquad \text{(for identical molecules)}$$

(2.25)

where v_{rms} is the root mean square speed, $mv_{rms}^2/2$ is the average kinetic energy of a single molecule, and N is the number of molecules.

The root mean square speed of molecules of an ideal gas is related to pressure. This was shown in Chapter 1 with a derivation based on the laws of mechanics. Pressure was interpreted as the momentum transmitted by the molecules to a unit surface area in a unit time. The expression was given by Eq. (1.18) as

$$pV = \frac{1}{3} Nmv_{rms}^2 \qquad (2.26)$$

From this equation and the ideal-gas law, an expression of the root mean square speed is

$$v_{rms} = \sqrt{\frac{3pV}{Nm}} = \sqrt{\frac{3\overline{R}T}{M}} = \sqrt{\frac{3kT}{m}} \qquad (2.27a)$$

Note that v_{rms} is proportional to the square root of the absolute temperature of the gas and is inversely proportional to the square root of the molar mass. Equation (2.27) serves also as a definition of absolute temperature in terms of average microscopic properties, for it is a measure of the average translational kinetic energy of the molecules:

$$T = \frac{m}{3k} v_{rms}^2 \qquad (2.27b)$$

It will be shown in Chapter 12 that the average kinetic energy of an ideal gas is independent of pressure, depending only on the temperature of the gas. The average kinetic energy of translation per molecule of different gases at a given temperature is the same for all the gases irrespective of molar mass as expressed by

$$(KE)_{avg} = \frac{mv_{rms}^2}{2} = \frac{3}{2} \frac{\overline{R}}{N_a} T = \frac{3}{2} kT \qquad (2.28)$$

This result, when generalized, will lead to the *principle of equipartition of energy* discussed in the following section.

When heat is transferred to a gas at constant volume, the kinetic energy of its molecules is increased and the temperature rises in accordance with Eq. (2.27). Conversely, if heat is transferred from the gas, the temperature decreases because the kinetic energy of the molecules is diminished. When two gases at different temperatures (different average kinetic energies) come in contact, the collisions between their molecules tend to equalize their momenta. When thermal equilibrium is attained, the molecules of both gases have the same average kinetic energy, and the decrease in kinetic energy of the molecules of the hot gas is equal to the increase in kinetic energy of the molecules of the cold gas. A similar situation occurs when heat is transferred between two gases separated by a heat-conducting wall. The molecules on the hot side, when colliding with the walls, impart this energy to molecules of the cold gas. The process continues until the average kinetic energies of the molecules on both sides are equal.

Energy interaction as just described excludes heat as a distinct means of energy transfer. It may be considered a form of microscopic work that eventually appears as kinetic energy. This form of kinetic energy must, however, be distinguished from macroscopic mechanical kinetic energy in the sense that it is *disordered* kinetic energy exhibited by the agitation of the molecules in all directions and at different speeds.

Consider now the adiabatic compression of a gas in an insulated piston-cylinder assembly. As the piston does work on the gas, it imparts additional kinetic energy to the molecules that collide with its face. This additional energy is subsequently transmitted and distributed by molecular collisions to the other molecules, resulting in an increase in the average kinetic energy of the molecules. The increase in temperature of the gas due to compression is thus a consequence of the increase in the average kinetic energy of the molecules. The temperature of the piston and cylinder walls also increases as a result of the collisions between the molecules of the gas and the atoms of the piston and cylinder walls. Work done on a gas is therefore accomplished by doing work on the molecules and may be interpreted as an increase in the kinetic energy of the molecules in a *predetermined* or fixed direction. On the other hand, heat transfer to a system results in molecular motion in an entirely disordered pattern. The reverse process, of recoordination of molecular collisions, can also be accomplished, but some loss of efficiency occurs, as will be described by the second law of thermodynamics.

Example 2.8

Calculate the root mean square speed of oxygen and hydrogen molecules at 70°C. What is the kinetic energy per molecule?

SOLUTION

Assume O_2 and H_2 may be treated as ideal gases, then

$$(v_{rms})_{O_2} = \sqrt{\frac{3\bar{R}T}{M}} = \sqrt{\frac{3(8314.4 \text{ J/kg-mol K})(343.15 \text{ K})}{32.0 \text{ kg/kg-mol}}} = 517 \text{ m/s}$$

and

$$(v_{rms})_{H_2} = \sqrt{\frac{3(8314.4 \text{ J/kg-mol K})(343.15 \text{ K})}{2.0160 \text{ kg/kg-mol}}} = 2060 \text{ m/s}$$

From Eq. (2.28), the kinetic energy of either molecule is

$$KE = \frac{3}{2}kT = \frac{3}{2}(1.38065 \times 10^{-23} \text{ J/K})(343.15 \text{ K}) = 7.1 \times 10^{-21} \text{ J}$$

2.17 Equipartition of Energy

The molecules of a gas possess four types of energy: translational, rotational, vibrational, and electronic. The translational energy is given by

$$U = \sum_i N_i(\varepsilon_{\text{trans}})_i = N\frac{mv_{\text{rms}}^2}{2} = N(\varepsilon_{\text{trans}})_{\text{avg}} \tag{2.29}$$

and the average translational energy per molecule is

$$(\varepsilon_{\text{trans}})_{\text{avg}} = \frac{U}{N} = \frac{3}{2}kT \tag{2.30}$$

The translational energy of a molecule, which is a form of kinetic energy, may be described in terms of its velocity components in the three perpendicular directions x, y, and z. It is equal to the sum of its translational energies in the three directions so that

$$\varepsilon_{\text{trans}} = \frac{mv^2}{2} = \frac{mv_x^2}{2} + \frac{mv_y^2}{2} + \frac{mv_z^2}{2}$$

Since all three directions are equally probable, the average kinetic energy is the same for each of the three directions. Consequently, the average kinetic energy of a molecule associated with each direction is equal to one-third of the average translational energy, or

$$\varepsilon_{x, y, \text{ or } z} = \frac{1}{2}kT \tag{2.31}$$

This equipartition of translational energy means that the kinetic energy is equally divided among three degrees of freedom each contributing an average value of $\frac{1}{2}kT$ to the internal energy of the molecule. This is called the *principle of equipartition of energy*. The magnitude of the average translational energy of a molecule at 25°C in any one direction is

$$\frac{1}{2}kT = \frac{1}{2}(1.38065 \times 10^{-23} \text{ J/K})(298.15 \text{ K}) = 2.06 \times 10^{-21} \text{ J}$$

and its total translational kinetic energy is

$$\frac{3}{2}kT = 6.18 \times 10^{-21} \text{ J}$$

Since $\overline{R} = N_a k$, the average kinetic energy due to translational motion on a mole basis is

$$\frac{3}{2}\overline{R}T = \frac{3}{2}(8.3144 \text{ kJ/kg-mol K})(298.15 \text{ K}) = 3718.4 \text{ kJ/kg-mol}$$

The principle of equipartition of energy applies to other manifestations of kinetic energy. Considering rotational energy, a molecule may rotate about three perpendicular axes through its center of mass and thus possesses three rotational degrees of freedom. In the case of a *linear* molecule, the atoms lie on the same axis so that contribution to rotational energy is a result of rotation about two axes that are perpendicular to each other and to the line connecting the centers of the atoms. Diatomic molecules are in this category and have only two rotational degrees of freedom. Since each degree of freedom of rotational energy contributes $\frac{1}{2}kT$ to the internal energy of the molecule, the translational and rotational energies of a diatomic molecule are specified by five degrees of freedom, whereas the kinetic energy of a monatomic molecule is specified by only three degrees.

The third type of molecular energy is vibrational energy. The atoms comprising a molecule tend to vibrate about their equilibrium positions. If a molecule has N atoms, each atom may independently vibrate in three directions and the molecule may have $3N$ degrees of freedom. Some of this vibrational motion results in translational and rotational motion of the entire molecule. Consequently, vibrational motion of a nonlinear molecule shows $(3N - 6)$ degrees of freedom, whereas vibrational motion of a linear molecule shows $(3N - 5)$ degrees of freedom. A vibrating molecule possesses both kinetic and potential energies, and each contributes $\frac{1}{2}kT$. For a diatomic molecule, the total average energy should be $\frac{7}{2}kT$, $\left(\frac{5}{2}kT + kT\right)$. This result contradicts experimental observations because the vibration of the molecule does not occur until high temperatures are reached. As will be shown in Chapter 12, the restrictions imposed by the quantum theory limit the conformity of the vibrational motion with the principle of equipartition of energy except at very high temperatures. Finally, the energy of the molecule may be in the form of electronic energy. This is associated with the structural arrangement of the electrons in the molecule. A change in the electronic energy of a molecule arises from a change in the kinetic and potential energy of one or more of its electrons. No degrees of freedom are assigned to electronic energy; rather, the total change of electronic energy is computed. Information about the electronic energy inherent in a molecule is obtained by means of spectroscopy. Through measurements of the radiation absorbed or emitted by a molecule, the allowed energy states of the molecule can be determined, and this is related to the molecular structure. Figure 2.21 shows the different modes of energy of a diatomic molecule.

The application of the principle of equipartition of energy to calculate specific heat values will be discussed in Chapter 13.

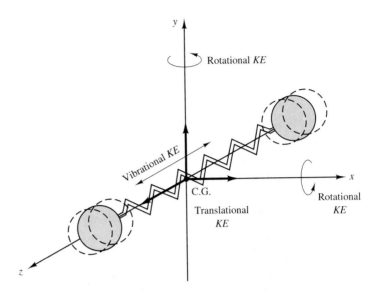

Figure 2.21   Energy modes of a diatomic molecule.

2.18 Law of Conservation of Mass for a Control Volume

In order to interpret the phenomena that occur in flow systems, the principles of conservation of mass, energy, and momentum are usually applied. A general form of the law of conservation of mass for a control volume will be developed in this section.

Consider a control volume fixed in space in a fluid flow field. At an initial time t_1, let a closed system coincide with the control volume, and, therefore, the mass in the system is identical to that within the control volume. Figure 2.22 shows the control volume and the system at time t_1 and also at subsequent time t_2 after an infinitesimally small time interval. In the interval $\Delta t = (t_2 - t_1)$, some matter leaves the control volume while new matter enters the control volume. The dotted boundary represents the control surface at any time, and the solid boundaries represent the system at times t_1 and t_2. Let primes denote values associated with the control volume. Since the mass of the system is equal to the mass within the control surface at time t_1,

$$m_{t_1} = m'_{t_1}$$

At time t_2, the mass of the system is equal to the mass within the control surface plus the net mass leaving the control volume, or,

$$m_{t_2} = m'_{t_2} + m_e - m_i$$

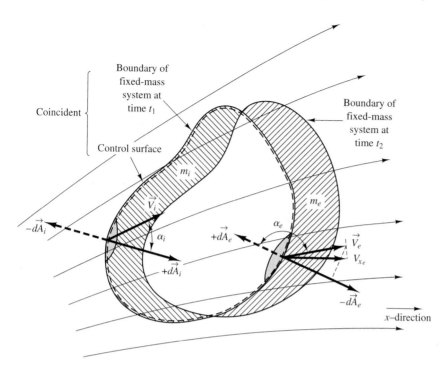

Figure 2.22   Notation for principles of conservation of mass and momentum.

Subtracting the two preceding equations gives

$$m_{t_2} - m_{t_1} = m'_{t_2} - m'_{t_1} + m_e - m_i \tag{2.32}$$

But the mass within the system is conserved ($m_{t_2} = m_{t_1}$), therefore,

$$m'_{t_2} - m'_{t_1} = m_i - m_e$$

When Δt approaches 0, the foregoing equation becomes

$$dm' = \int_{cs} dm_i - \int_{cs} dm_e \tag{2.33}$$

where $\int_{cs}$ designates the integration over the surface area of the control volume to account for all the locations on the boundary through which mass enters or exits. Equation (2.33) may also be expressed on a time rate basis:

$$\frac{dm'}{dt} = \int_{cs} d\dot{m}_i - \int_{cs} d\dot{m}_e$$

which states that the rate of change of mass within the control volume at time t is equal to the net rate of mass entering the control surface. In general,

$$\int_{cv} d\dot{m}' = \int_{cs} d\dot{m} \qquad (2.34)$$

Expressing the mass flow rate in terms of density, velocity, and area, the preceding equation becomes

$$\underbrace{\int_{cv} \frac{dm'}{dt}}_{\substack{\text{rate of change} \\ \text{of mass within} \\ \text{control volume}}} = \underbrace{\int_{cs} \rho_i V_i \cos a_i \, d A_i - \int_{cs} \rho_e V_e \cos a_e \, d A_e}_{\substack{\text{rate of mass transfer} \\ \text{across control surface}}} \qquad (2.35)$$

where a is the angle between the velocity vector $\vec{V}$ and the area vector $\vec{dA}$ as shown in Figure 2.22. Note that the velocities in Eq. (2.35) are velocities relative to the control surface. Also the product $\rho V \cos a$ is the time rate of mass flow per unit area, which is called the mass flux.

Under steady-state conditions, there is no change of mass within the control surface so that the mass flow rates entering and leaving the control volume are equal. Hence,

$$\int_{cs} \rho_i V_i \cos a_i \, dA_i = \int_{cs} \rho_e V_e \cos a_e \, dA_e \qquad (2.36)$$

For one-dimensional flow in which all intensive fluid properties are uniform over the cross-sectional area of each inlet or exit area through which matter flows, integration of the preceding equation gives the mass rate of flow $\dot{m}$ as

$$\dot{m} = \rho \, AV = \text{constant} \qquad (2.37)$$

where A is the area of flow perpendicular to the velocity V. Taking the logarithm and differentiating Eq. (2.37) leads to the following:

$$d(\ln \rho) + d(\ln A) + d(\ln V) = 0$$

or

$$\frac{d\rho}{\rho} + \frac{dA}{A} + \frac{dV}{V} = 0 \qquad (2.38)$$

Under steady-state conditions, the mass rate of flow across any two sections of areas A_i and A_e, as shown in Figure 2.23, is

$$\dot{m} = \rho_i A_i V_i = \rho_e A_e V_e$$

or

$$\frac{A_i V_i}{v_i} = \frac{A_e V_e}{v_e}$$

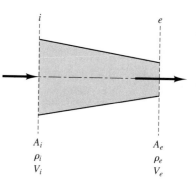

where v_i and v_e are the specific volumes at sections i and e. It must be pointed out that the preceding equation is applicable at sections i and e as long as the flow properties are uniform at these two sections even though the flow properties may not be uniform between the two sections. Note that the assumption of steady-state implies that *every* property is independent of time. Note also that when only the mass within a control volume remains constant, it does not necessarily mean that steady-state conditions prevail.

Figure 2.23 Flow in one dimension.

2.19 Momentum Principle

When no net external force acts on a system, the linear momentum along each of three mutually perpendicular directions is conserved both in magnitude and direction. A familiar example from mechanics is the collision of two spheres. Conservation of momentum occurs whether the collision is elastic or not, such that the sums of the momenta of the two spheres before and after collision are equal in magnitude.

When a net external force acts on a system, the linear momentum is no longer conserved. Under this condition, the phenomenon is described by a principle that is more general than the momentum principle, namely, *Newton's second law of motion.*

According to Newton's second law, the sum of all the forces acting in a certain direction on a particle at rest or in motion is equal to the time rate of change of its momentum in the same direction. The motion of the particle must be described relative to an inertial coordinate system, that is, a coordinate system moving at constant velocity in one direction. Considering the x-direction, for example, Newton's law for a particle yields

$$\sum F_x = \frac{d}{dt}(mV_x) \tag{2.39}$$

where $\sum F_x$ is the net force acting on the particle in the x-direction. The net force includes body forces (such as gravitational, magnetic, and electrical forces that act on the mass of the particle) and surface forces (such as pressure, shear, and surface tension forces that act on the surface of the particle).

In a system in which there are many particles each of a different acceleration, internal forces between particles balance according to *Newton's third law.* Hence Newton's second law can be applied to a system containing numerous particles by considering the momentum of the system to be the vectorial sum of momenta of the particles in the system.

Referring to Figure 2.22, the rate of change of the x-momentum of a system is equal to the rate of change of the x-momentum of the control volume plus the net rate of change of momentum leaving the control volume in the same direction, or

$$
\frac{\Delta(mV_x)}{\Delta t} = \frac{(mV_x)'_{t_2} + (mV_x)_e - (mV_x)'_{t_1} - (mV_x)_i}{\Delta t}
$$

$$
= \frac{\Delta(mV_x)'}{\Delta t} + \frac{(mV_x)_e - (mV_x)_i}{\Delta t}
\tag{2.40}
$$

where $(mV_x)_i$ and $(mV_x)_e$ are the momenta of the fluid entering and leaving the control volume in time Δt.

The first term on the right-hand side of Eq. (2.40) represents the time rate of change of the x-momentum of the fluid within the control volume. The second term represents the net rate of momentum leaving (outflow − inflow) the control volume. When very small time intervals are considered, Newton's second law can be written as

$$
\sum F_x = \frac{d}{dt}(mV_x) = \frac{d}{dt}(mV_x)' + \int_e V_x \, d\dot{m} - \int_i V_x \, d\dot{m}
$$

or

$$
\sum F_x + \int_{net\ in} V_x \, d\dot{m} = \frac{d}{dt}(mV_x)'
\tag{2.41}
$$

where $\dot{m}$ is the mass rate of flow. Equation (2.41) is the momentum equation applied to a control volume in the x-direction. It states that the sum of the external forces acting on the fluid within a control volume in a certain direction plus the net rate of inflow of momentum across the control surface in that direction is equal to the time rate of change of momentum within the control volume in the same direction. By expressing mass flow in terms of density, velocity, and area, Eq. (2.41) becomes

$$
\sum F_x + \int_{net\ in} \rho V \cos a V_x \, dA = \frac{d}{dt}(mV_x)'
\tag{2.42}
$$

where a is the angle between the velocity vector $\vec{V}$ and the area vector $\vec{dA}$. Equations similar to Eq. (2.42) can be derived for the y- and z-directions.

Under steady-state conditions, the rate of change of momentum within the control surface is zero and Eq. (2.42) reduces to

$$\sum F_x + \int_{\text{net in}} \rho V \cos \alpha V_x \, dA = 0 \qquad (2.43)$$

or, in general,

$$\sum \vec{F} + \int_{\text{net entering cs}} \vec{V} d\dot{m} = 0 \qquad (2.44)$$

A positive value of the second term of Eq. (2.44) means an increase of momentum of the control volume.

Even if there are frictional forces or nonequilibrium regions within the control surface, the momentum equation still applies. The momentum principle has wide applications in the field of propulsion and is used, in particular, in evaluating the thrust forces developed by propulsion systems.

In applying the momentum equation, the following procedure is recommended: (a) select a control volume and define a positive direction; (b) identify the body forces and the surface forces; (c) investigate the control volume to determine whether the flow is steady, one-dimensional, incompressible, and so on; (d) investigate the control surface to determine the momentum flux terms. Note that all velocities must be expressed in the same coordinate system as the control surface.

Example 2.9

What is the propulsive force developed by a jet engine propelling an airplane at a constant speed of 150 m/s? The combustion products are discharged at 1000 m/s relative to the airplane, and the engine exhausts 8 kg/s. Neglect the mass and momentum of the fuel, and assume horizontal flight.

SOLUTION

Considering the jet engine as a control volume as shown in Figure 2.24, the air enters the engine with a velocity of 150 m/s. Assuming the inlet and exit pressures are atmospheric, the force F opposite to the direction of motion of the airplane is given by the momentum equation as

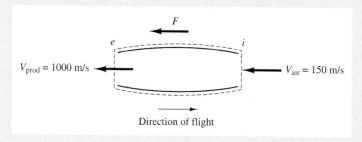

Figure 2.24

$$F = \dot{m}(V_e - V_i) = (8 \text{ kg/s})[(1000 - 150) \text{ m/s}] = 6800 \text{ N}$$

Example 2.10

Water flows in a 5 cm inside diameter pipe at 40°C. The water flows through a 90° bend in the pipe and then discharges through a convergent nozzle to the atmosphere, as shown in Figure 2.25. At the entrance to the bend, the fluid pressure is 600 kPa gauge, and the flow velocity is 600 m/min. At the nozzle discharge, the velocity is 2400 m/min. Calculate the magnitude and direction of the net force that the bend exerts on the remaining pipe. Neglect the weight of the bend.

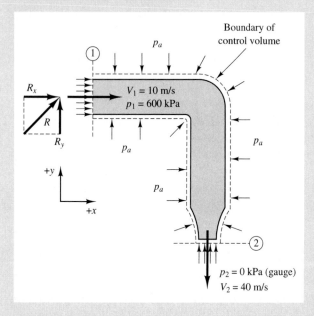

Figure 2.25

SOLUTION

Let R_x and R_y be the forces acting on the bend from the remaining pipe. Assuming steady state and one-dimensional flow, the mass rate of flow is given by

$$\dot{m} = \rho \, AV$$

$$= (1000 \text{ kg/m}^3)\left(\frac{\pi}{4} \times 0.05^2 \text{ m}^2\right)\left(\frac{600 \text{ m/min}}{60 \text{ s/min}}\right) = 19.6 \text{ kg/s}$$

Assuming that atmospheric pressure surrounds the pipe, gauge pressures are considered in the momentum equation. For the x-direction,

$$\sum F_x + \int_{\text{net in}} V_x \, d\dot{m} = \frac{d}{dt}(mV_x)'$$

For steady state,

$$\frac{d}{dt}(mV_x)' = 0$$

Therefore,

$$F_x = (p_1 A_1)_x + R_x = \dot{m}(V_2 - V_1)_x$$

or

$$R_x = (19.6 \text{ kg/s})[(0 - 10) \text{ m/s}] - (600 \times 10^3 \text{ Pa})\frac{\pi}{4}(0.05^2 \text{ m}^2)$$

$$= -1374.1 \text{ N}$$

Noting that $p_2 = 0$ kPa (gauge), the momentum equation for the y-direction is

$$F_y = (p_2 A_2)_y + R_y = \dot{m}(V_2 - V_1)_y$$

therefore,

$$R_y = (19.6 \text{ kg/s})[(-40 - 0) \text{ m/s}] = -784.0 \text{ N}$$

The negative signs mean that R_x and R_y are in opposite directions from those shown in the figure. The total reaction is

$$R = \sqrt{R_x^2 + R_y^2} = \sqrt{(-1374.1 \text{ N})^2 + (-784.0 \text{ N})^2}$$

$$= 13763.55 \text{ N}$$

The line of action of this force is inclined at an angle $a = \tan^{-1}(784.0/1374.1) = 29.7°$ with the horizontal.

Example 2.11

A 5 cm diameter water jet of constant area has a velocity of 25 m/s. It strikes a blade, as shown in Figure 2.26a. If the deflection angle of the jet is 150°, calculate the x- and y-components of the force exerted by the jet on the blade for the following cases: (a) the blade is stationary; (b) the blade is moving at 10 m/s in the same direction as the incoming jet; (c) the blade is moving at 10 m/s in the opposite direction of the incoming jet. Assume there is no friction, and neglect gravitational effects.

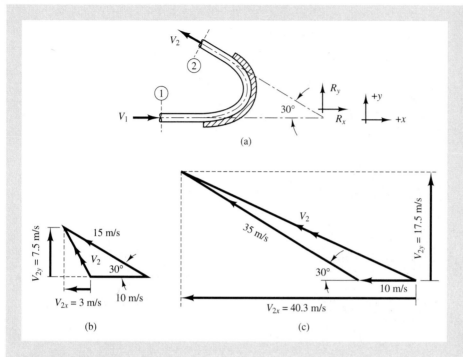

Figure 2.26

SOLUTION

(a) The mass rate of flow is

$$\dot{m} = \rho \, AV = (1000 \text{ kg/m}^3) \frac{\pi}{4} (0.05^2 \text{ m}^2)(25 \text{ m/s}) = 49.1 \text{ kg/s}$$

Applying the momentum principle in the x-direction gives

$$\sum F_x + \int_{\text{net in}} V_x \, d\dot{m} = \frac{d}{dt} (mV_x)'$$

For steady state,

$$\frac{d}{dt} (mV_x)' = 0$$

Therefore,

$$
\begin{aligned}
R_x &= \dot{m}(V_{2x} - V_{1x}) \\
&= \dot{m}(-V_2 \cos 30 - V_1) \\
&= (49.1 \text{ kg/s})[(-25 \cos 30 - 25) \text{ m/s}] = -2290.5 \text{ N}
\end{aligned}
$$

and for the y-direction,

$$R_y = \dot{m}(V_{2y} - V_{1y}) = \dot{m}(V_2 \sin 30 - 0)$$
$$= (49.1 \text{ kg/s})(12.5 \text{ m/s}) = 613.8 \text{ N}$$

(b) When the blade moves to the right, less mass strikes the blade per unit time, and therefore, less mass has its momentum changed. The relative velocity at section 1 = 25 − 10 = 15 m/s. The mass striking the blade per unit time is

$$\dot{m} = (1000 \text{ kg/m}^3)\left[\frac{\pi}{4}(0.05)^2 \text{ m}^2\right](15 \text{ m/s}) = 29.5 \text{ kg/s}$$

The absolute velocity of water at section 2 is the vectorial sum of the blade velocity and the relative velocity of the jet with respect to the blade. From Figure 2.26(b),

$$V_{2x} = (3.0 \text{ m/s}) \text{ to the left}$$
$$V_{2y} = (7.5 \text{ m/s}) \text{ upward}$$
$$R_x = \dot{m}(V_{2x} - V_{1x})$$
$$= (29.5 \text{ kg/s})[(-3 - 25) \text{ m/s}] = -826 \text{ N}$$

and

$$R_y = \dot{m}(V_{2y} - V_{1y})$$
$$= (29.5 \text{ kg/s})[(7.5 - 0) \text{ m/s}] = 221 \text{ N}$$

(c) When the blade moves to the left, the relative velocity at section 1 = 25 + 10 = 35 m/s, and the mass striking the blade per unit time is

$$\dot{m} = (1000 \text{ kg/m}^3)\frac{\pi}{4}\left[(0.05)^2 \text{ m}^2\right](35 \text{ m/s}) = 68.7 \text{ kg/s}$$

From Figure 2.26(c),

$$V_{2x} = (40.3 \text{ m/s}) \text{ to the left}$$
$$V_{2y} = (17.5 \text{ m/s}) \text{ upward}$$
$$R_x = (68.7 \text{ kg/s})[(-40.3 - 25) \text{ m/s}] = -4486 \text{ N}$$
$$R_y = (68.7 \text{ kg/s})[(17.5 - 0) \text{ m/s}] = 1202 \text{ N}$$

This example can also be solved using relative velocities rather than absolute velocities.

In cases when a series of blades is involved, as in a turbine wheel, the impinging fluid strikes one or more blades as the turbine rotates, and in this case, the rate of mass striking the blades is constant.

2.20 Dynamic Analysis and Euler's Equation

Euler's equation is one of the basic equations in fluid dynamics and can be derived from Newton's second law of motion. The product of the mass and the acceleration of a fluid particle can be equated vectorially, with the external forces acting on the particle. Consider the frictionless steady flow of a fluid in a streamtube[7] of infinitesimal length. As shown in Figure 2.27, the center streamline is inclined at an angle θ with the vertical, and ds is the distance along the center between two adjacent sections. The cross-sectional areas at the two sections are A and $A + dA$, and the properties at the two sections are assumed to differ from each other by infinitesimal quantities.

The external forces acting on the fluid element are the pressure forces and the weight of the fluid element in the streamtube. The weight is equal to the volume multiplied by the specific weight of the fluid or

$$\gamma\left(A + \frac{dA}{2}\right) ds$$

where $[A + (dA/2)]$ is the average cross-sectional area between the two sections. The component of weight in the s-direction is

$$-\gamma\left(A + \frac{dA}{2}\right) ds \cos\theta = -\gamma\left(A + \frac{dA}{2}\right) ds \frac{dz}{ds} = -\gamma\left(A + \frac{dA}{2}\right) dz$$

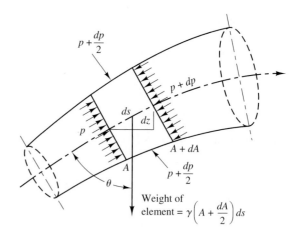

Figure 2.27   Flow in a streamtube.

[7] A streamline is a line whose tangent at each instant and at any point gives the direction of the velocity of the fluid at that point. A streamtube is formed by a family of streamlines passing through each point of a closed curve. The mass flow across each section of the streamtube is constant in steady flow.

where $\cos \theta = dz/ds$. Applying the momentum equation in the *s*-direction gives

$$pA + \left(p + \frac{dp}{2}\right) dA - (p + dp)(A + dA) - \gamma\left(A + \frac{dA}{2}\right) dz = \rho A V \, dV$$

The second term in the foregoing equation is the component of the force due to pressure that is exerted on the sides of the fluid element in the direction of motion. By expanding terms and by neglecting second-order differentials, this equation becomes

$$A \, dp + \gamma A \, dz + \rho A V \, dV = 0$$

Substituting ρg for γ and dividing by $A\rho$ yields

$$\frac{dp}{\rho} + V \, dV + g \, dz = 0 \qquad (2.45)$$

Equation (2.45) is Euler's equation for steady flow along a streamline. Note that the fluid was assumed frictionless and that gravity was assumed to be the only field force. Euler's equation applies to both compressible and incompressible flows.

In treating one-dimensional incompressible flow, since density is constant, Euler's equation can be integrated directly to give

$$\frac{1}{\rho}(p_2 - p_1) + \frac{V_2^2 - V_1^2}{2} + g(z_2 - z_1) = 0$$

or

$$\frac{p_1}{\rho} + \frac{V_1^2}{2} + gz_1 = \frac{p_2}{\rho} + \frac{V_2^2}{2} + gz_2 \qquad (2.46)$$

Each term in Eq. (2.46) represents energy per unit mass. This equation can also be expressed as

$$\frac{p_1}{\gamma} + \frac{V_1^2}{2g} + z_1 = \frac{p_2}{\gamma} + \frac{V_2^2}{2g} + z_2 = H \qquad (2.47)$$

Each term now has the dimensions of length. In Eq. (2.47), which is *Bernoulli's equation*, H is called the *total head* and is equal to the sum of the *pressure head* p/γ, the *velocity head* $V^2/2g$, and the *potential head z*. According to Bernoulli's equation, the total head H, which represents the sum of the pressure, velocity, and potential heads, is a constant along a streamline if the flow is steady and the fluid is frictionless and incompressible.

Example 2.12

Water flows through the reducer shown in Figure 2.28. If the velocity at section 2 is 1 m/s, what is the pressure at that section? Assume frictionless flow.

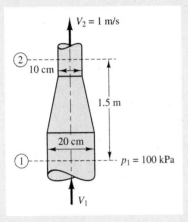

Figure 2.28

SOLUTION

The velocity at section 1 is

$$V_1 = \left(\frac{A_2}{A_1}\right)V_2 = \left(\frac{10}{20}\right)^2 (1 \text{ m/s}) = 0.25 \text{ m/s}$$

Applying Eq. (2.46) between sections 1 and 2 gives

$$\frac{(100 \times 10^3 \text{ Pa})}{(1000 \text{ kg/m}^3)} + \frac{(0.25^2 \text{ m}^2/\text{s}^2)}{2} + 0 = \frac{p_2}{(1000 \text{ kg/m}^3)} + \frac{(1^2 \text{ m}^2/\text{s}^2)}{2} + (9.81 \text{ m/s}^2)(1.5 \text{ m})$$

from which

$$p_2 = 54.81 \text{ kPa}$$

2.21 Summary

Work is the name of a method of transferring energy between a system and its surroundings. It is identified at the interaction boundary if the sole effect external to the system can be shown to be equivalent to the raising or lowering of a weight. Work associated with a moving boundary during a quasi-static process is equal to $-\int p\, dV$. A functional relation between p and V is required to evaluate the integral, which is also equal to the area under the process curve on a *p-V* diagram. Work is not a property; it depends on the path followed to change the state of a system (path function). Work done on a system is considered positive.

Energy can be classified as (1) energy in transit such as heat energy or work energy and (2) internal energy such as chemical or atomic energy associated with the microscopic structure and activity of the particles of matter. Macroscopic changes in potential and kinetic energies are defined by

$$\Delta(PE) = mg(z_2 - z_1) \quad \text{and} \quad \Delta(KE) = \frac{m(V_2^2 - V_1^2)}{2}$$

The energy *E* is the sum of all forms of energy associated with a system.

$$E = (U + \ ...etc.) + KE + PE$$

Heat is the name of a means of transferring energy (other than energy carried by mass transfer) due to temperature difference. Heat and work are two terms relating to the transfer of energy between a system and its surroundings. For a control volume, energy can also be carried across the boundary by mass transfer. Heat transfer to a system or a control volume is considered positive.

For steady state the momentum equation indicates that the sum of all external forces acting in a certain direction on a control volume is equal to the net time rate of outflow of momentum in that direction.

Problems

2.1 Determine the work required to move an object horizontally a distance of 12 m with a force inclined at an angle $30°$ from the horizontal. The force-displacement relationship is given by

$$F = 300x + 0.2x^2$$

where *F* is in newtons and *x* in meters.

2.2 The force-displacement relationship of a spring is given by

$$F = 15x^2$$

where *F* is the force in newtons and *x* is the displacement in cm. Calculate the amount of work required to compress the spring 2 cm from its uncompressed state.

2.3 A spherical balloon has a diameter of 25 cm and contains air at a pressure of 150 kPa. The diameter of the balloon increases to 30 cm because of heating, and during this process the pressure is proportional to the diameter. Calculate the work done on the gas assuming reversible work interaction.

2.4 If the initial state of an ideal gas is 100 kPa and 0.3 m³/kg, calculate the reversible work interaction per kg for the following processes:
(a) Constant pressure process if the final volume is 1.5 m³/kg.
(b) Isothermal process according to $pV =$ constant and the final volume 0.5 m³/kg.
(c) Constant volume process till the final pressure is 400 kPa. Sketch the *p-V* diagram for each case.

2.5 A mass of 5 kg of air is compressed reversibly from an initial pressure of 100 kPa to 700 kPa according to the relation $pV = C$. If the initial density of the air is 1.2 kg/m^3, find the work necessary for compression.

2.6 One kg of hydrogen at 30°C expands reversibly till it doubles its volume according to the relation $pV^n = C$. Determine the value of n if the final temperature is $-20°$ C. What is the work done?

2.7 Find the minimum work done per kg of air in a cycle according to the following processes:
(a) Isothermal expansion from state 1 to state 2.
(b) Constant volume compression from state 2 to state 3.
(c) Constant pressure compression from state 3 to the initial state 1.
Data: $p_1 = 350$ kPa
            $v_1 = 1$ m^3/kg
            $T_3 = 1650°C$
Represent the work on a p-V diagram.

2.8 The relationship between the pressure and volume of a system during an expansion process is $p = 250 - 300\,V$, where p is in kPa and V in m^3. Find the maximum work done by the system in expanding from an initial volume of 0.2 m^3 to a final volume of 0.4 m^3.

2.9 An ideal gas changes from an initial state 1 to a final state 3 first at a constant temperature following the path 1-2 and then at constant pressure following the path 2-3, as shown in Figure 2.29. Find the work interactions during each of the processes 1-2 and 2-3. All processes are reversible.

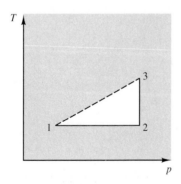

Figure 2.29

2.10 Air, which may be treated as an ideal gas, is compressed from a pressure of 100 kPa and a temperature 300 K to a pressure of 1500 kPa in a quasi-static manner so that it obeys the relation

$$pv^{1.3} = \text{constant}$$

Determine the amount of work necessary to compress 1 kg of air. Compare this work with the work of an isothermal compression.

2.11 An ideal gas has an initial volume of 0.06 m^3 expands reversibly in a piston-cylinder assembly from 3 MPa and 700 K to a final volume of 0.2 m^3 and a final temperature of 400 K according to the law $pV^n = C$. Find (a) the final pressure, and (b) the work done.

2.12 Derive an expression for the reversible work done when a gas expands from a volume V_1 to a volume V_2. The p-V relation for the process is

$$p(V - a) = \text{constant}$$

where a is a constant.

2.13 The compression of a liquid is expressed by the relation

$$V = V_0 e^{a(p-p_0)}$$

where V is volume, p is pressure, a is a dimensional constant, and subscript 0 indicates initial conditions. If the compression is quasi-static with no change in temperature, derive an expression for the work done.

2.14 The relationship between the pressure and volume of a gas during a compression process is $pV^2 = $ constant. If the initial pressure and volume of the gas are 100 kPa and 0.5 m^3 and the final pressure is 350 kPa, determine the reversible work interaction.

2.15 An indicator card of a 10 cm bore, 12 cm stroke cylinder has an area of 20 cm^2 and is 8 cm long. The indicator spring constant is 350 kPa per cm displacement.
(a) Find the mean effective pressure.

(b) If the cycle is repeated 80 times per minute, what is the indicated power?

2.16 The indicator card of an 8 cm bore, 10 cm stroke water pump is in the shape of a rectangle of dimension 2 × 10 cm. The indicator spring constant is 22 MPa/m.
(a) Find the mean effective pressure.
(b) If the cycle is repeated once every second, what is the power required by the pump?

2.17 A piston-cylinder device contains 0.2 kg of air initially at 100 kPa and 300 K. The air is compressed reversibly and isothermally until the volume is reduced by one-half. Determine the work transfer during this process. Sketch the *p-V* diagram of the process indicating the temperature, processes, and volume at the end points.

2.18 If 7 kJ of work are used to charge a 6-volt storage battery at 5 amp with 100 percent efficiency, how long would the charging continue?

2.19 An internal combustion engine uses fuel of calorific value 45000 kJ/kg and consumes 20 kg/hour when developing 75 kW. What is the ratio of work output to energy supplied?

2.20 Determine the power delivered by a motor rotating at 3400 revolutions per min against a torque of 40 Nm.

2.21 Water is delivered at a velocity of 20 m/s from a nozzle in a vertical direction. For 1 kg of water, compute the change in potential, kinetic, and mechanical energies at elevations zero, 10, and 20 m. What is the maximum elevation the water will reach? Neglect friction of the air.

2.22 A stone is directed vertically upward from ground level at a velocity of 30 m/s. If 15 percent of the initial kinetic energy is converted to frictional effects on ascent and 15 percent of the maximum potential energy is converted on descent, (a) determine the peak elevation, and (b) determine the final velocity just before the stone strikes the ground.

2.23 Water flows in a river at the rate of 2×10^3 kg/s. The water reaches the top of a waterfall 30 m high at a speed of 10 m/s. It is proposed to use the energy of the water to generate electric power in a hydroelectric plant. If the water leaves the plant at a speed of 6 m/s, determine the maximum power output assuming no losses.

2.24 The molar specific heat of an ideal gas (molar mass 16) at constant pressure is given by

$$\bar{c}_p = -673 + 140T^{\,0.25} - 0.79T^{\,0.75} + 3240T^{\,-0.5} \text{ kJ/kg-mol K}$$

where T is in K.

Find the heat transfer per kg during a constant pressure expansion from an initial volume of 0.5 m³/kg to a final volume of 3 m³/kg if the pressure is maintained at 140 kPa.

2.25 One kg of ice at 0°C is placed in a container with 2 kg of water at 65°C. The container is in an atmosphere that remains at 15°C. There is free exchange of energy between all items so that a final temperature of 15°C is reached. Calculate the amount of heat transferred to (a) the original 1 kg of ice, (b) the water, (c) the atmosphere. Indicate the direction of heat interaction in each case.

2.26 The specific heat at constant pressure of N_2 (in kJ/kg K) varies with temperature according to the relation

$$c_p = 1.415 - 287.9T^{\,-1} + 5.35 \times 10^4 T^{\,-2}$$

where T is in K.

Determine the average specific heat of N_2 in the temperature range of 1000 K to 2000 K.

2.27 Water flows at the rate of 50 kg/min through a heat exchanger operating at steady state. The water is heated from 30°C to 50°C. If the rate of heat transfer to the water is 69.617 kW, determine the average specific heat of water.

2.28   The magnitude of molecular velocities of an ideal gas is given by the following distribution:

Number of Molecules	Speed in m/s
10	20
20	50
40	100
60	125
80	150
100	200
60	300
40	350

Draw a graph showing the speed distribution among the molecules, and compute the average and the root mean square speeds.

2.29   Calculate the density and root mean square speed for the following gases at 0°C and 100°C. Avogadro's number $= 6.0221 \times 10^{26}$ molecules/kg-mol and $k = 1.38065 \times 10^{-23}$ J/molecule K. Assume atmospheric pressure.

Gas	Molar Mass (kg/kg-mol)
A	39.94
H_2	2.02
He	4.00
N_2	28.02
CO_2	44.00

2.30   A conical diverging tube is attached at the end of a pipeline as shown in Figure 2.30. If the water discharges into the air at atmospheric pressure, find the magnitude and sense of the force exerted by the conical tube on the pipe line. Assume no frictional losses in the system.

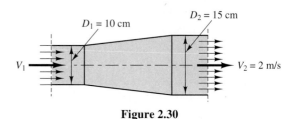

Figure 2.30

2.31   Determine the magnitude and sense of the x- and the y-components of the force necessary to support the water hose shown in Figure 2.31. The wall of the hose at section 1 cannot transmit any force. Neglect the weight of the hose and assume no losses.

2.32   A 5 cm diameter horizontal jet of water strikes a vertical plate. If the horizontal force needed to support the plate is 350 N, what is the velocity of the jet?

2.33   Water flows in a 15 cm diameter horizontal pipe. The pipe reduces to 10 cm diameter by means of a reducer. If the gauge pressure and velocity in the pipeline are 300 kPa and 6 m/s, find the force exerted by the water on the reducer.

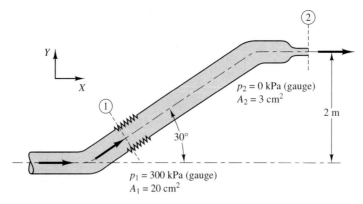

Figure 2.31

2.34 The deflector shown in Figure 2.32 divides the flow of an impinging jet 5 cm wide such that two-thirds of the flow is directed upward and one-third downward. If the velocity of the water jet is 3 m/s, find the magnitude and sense of the horizontal and vertical components of the force necessary to keep the deflector in equilibrium. (Neglect friction and gravitational effects.)

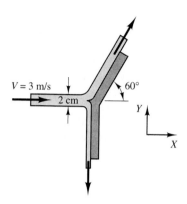

Figure 2.32

2.35 Solve the previous problem if the deflector is moving upstream at a velocity of 1 m/s.

2.36 The velocity distribution of a fluid flowing steadily in a circular pipe is given by

$$V = V_{max}\left[1 - \left(\frac{r}{r_0}\right)^2\right]$$

where V_{max} is the maximum velocity, and r_0 is the pipe radius. Determine the ratio of V_{av}/V_{max}, where V_{av} is the average velocity in the pipe.

2.37 An ideal venturi-meter is inclined at an angle 30° to the horizontal. From the data and dimensions shown in Figure 2.33, determine the flow rate of the water.

C2.1 The data file Water.dat, provided on the diskette, contains four columns of data for properties of water. The first and second columns represent temperature (°C) and pressure (kPa) at saturation, and the third and fourth columns represent the corresponding enthalpy (kJ/kg) and entropy (kJ/kg-K) for saturated liquid water. Using plotting software, produce a plot of enthalpy and entropy versus temperature. The graph should show the enthalpy scale as the left side ordinate and the entropy scale as the right side ordinate. The data file is in ASCII format and can be imported to any software.

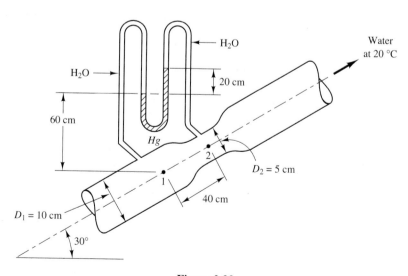

Figure 2.33

C2.2 Using plotting software and the data file Water.dat, produce a plot of temperature versus pressure with appropriate labels, captions, and so on. Using the same software, determine an equation that best fits the curve plotted (or the data), and plot the equation on the same graph for comparison with the actual data.

C2.3 Oxygen in a piston-cylinder arrangement expands reversibly, against the atmosphere, according to the relation $pV^{1.3} = $ constant. The initial volume, pressure, and temperature of the gas are 0.1 m^3, 450 kPa, and 320 K, respectively. Write a computer program to determine the work done during the expansion process. Tabulate your results for a pressure drop of 20 kPa increments, and plot the work, volume, and temperature versus the tank pressure. What is the maximum work that can be obtained from this experiment?

C2.4 Determine the work required to compress 10 kg of helium initially at 300 K from 5 m^3 to 1.5 m^3. During the compression process, assume $pV = $ constant. You should use either trapezoidal or Simpson's formula to numerically integrate for work. What is the final temperature of the system?

C2.5 A 1 kg block of steel at 100°C is placed on a 3 kg block of ice. The initial temperature of the ice is -20°C. Write a computer program to calculate the steel temperature for every two-degree temperature rise of the ice block. What will be the final temperature of the steel block? Will the ice completely melt? Plot the steel temperature versus ice temperature. Specific heat of ice and steel are 2.104 kJ/kg K and 0.4 kJ/kg K, respectively. Assume the specific heat of ice remains constant until it completely melts. Specific heat of water is 4.207 kJ/kg K.

Chapter 3

The First Law
of Thermodynamics

3.1 Joule's Experiments

Between the years 1843 and 1848, Joule conducted experiments that were the first steps in the quantitative analysis of thermodynamic systems, and that led to the first law of thermodynamics. Joule's experiments were performed with equipment similar to that shown in Figure 3.1(a). In the system that he studied, energy in the form of work was transferred to a fluid by means of a paddle wheel. This work transfer caused a rise in the temperature of the fluid. The amount of work done was measured by the change in potential energy as the weight W fell through distance z. The system was then placed in contact with a water bath so that energy in the form of heat was transferred from the fluid until the original state of the fluid was reestablished, as indicated by temperature and pressure. The system had, in this way, gone through a complete cycle. The amount of heat transfer from the system was equal to the increase of energy of the water bath and therefore could easily be determined by measuring the rise in temperature of the water bath. Joule carried out many similar experiments involving different work interactions. For example, Joule used work transfer in the form of electrical energy to heat the system. He measured the electrical energy and the amount of heat required to complete the cycle and restore the system to its initial state. By repeating the cycle for different amounts of work interaction and measuring the heat transfer to the bath in every case, Joule found that the net work input W was always proportional to the net heat Q transferred from the system, regardless

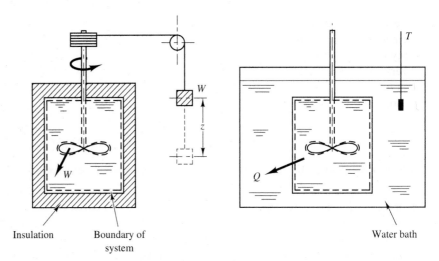

Insulation   Boundary of   Water bath
                      system

Figure 3.1   Joule's experiment.

of the kind of work done, the rate at which the work was done, and the scheme used for transforming it into heat. Since the system at the end of a cycle experiences no net change, the cyclic integral of the heat transfer is proportional to the cyclic integral of the work transfer, and the algebraic sum of heat and work during the cycle is zero. If the same units are used for both heat and work, this relation may symbolically be written as

$$\oint \delta Q + \oint \delta W = 0 \tag{3.1}$$

where δQ and δW are infinitesimal amounts of heat and work, and the symbol $\oint$ denotes integration along a closed path (cyclic integral). The symbol δ has been used to indicate that both heat and work are inexact differentials depending on the path of the cycle. In Eq. (3.1), the same units (J) are used for both heat and work. If different units are used, a proportionality factor relating work and heat, called the *mechanical equivalent of heat,* is required. It is equal to the ratio of the work done to the heat transferred from a system during a cycle or number of cycles. Thus there is an exact numerical equivalence between heat and work in the sense that each is a manner of transferring energy. This, however, does not mean that heat and work are equivalent in the manner in which they may interchange, nor does it mean that their basic definitions no longer apply. Equation (3.1) is a statement of the first law of thermodynamics for a system undergoing one or more cycles. According to the first law, when heat and work interactions take place between a system and its surroundings, the algebraic sum of the work and heat interactions during a complete cycle is zero.

The first law of thermodynamics cannot be proved analytically, but experimental evidence has repeatedly confirmed its validity, and since no phenomenon has been shown to contradict it, the first law is accepted as a law of nature. It may be remarked that no restriction was imposed that limited the first law to reversible energy transformations. Hence, the first law applies to reversible as well as to irreversible transformations.

For noncyclic processes, a more general formulation of the first law is required. The notion or the concept of *energy* fulfills this need.

3.2 First Law of Thermodynamics for a System

One important consequence of the first law of thermodynamics is that the energy of a system is a property. To prove this, consider a cyclic change of a system whose state is specified by two thermodynamic variables, such as p and V. As shown in Figure 3.2, the system may proceed from state 1 to state 2 along path A, and then back to the original state 1 along path B. Applying Eq. (3.1) to cycle 1-A-2-B-1, gives the following energy relationship:

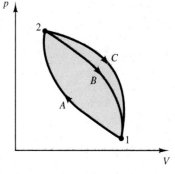

$$\left(\int_{1(A)}^{2} \delta Q + \int_{2(B)}^{1} \delta Q \right) + \left(\int_{1(A)}^{2} \delta W + \int_{2(B)}^{1} \delta W \right) = 0$$

Figure 3.2   The $\int_{1}^{2} (\delta Q + \delta W)$ is a function of states 1 and 2.

If the system changes from state 2 to state 1 along another arbitrary path C, rather than path B, so that the cycle is 1-A-2-C-1, the energy relationship is

$$\left(\int_{1(A)}^{2} \delta Q + \int_{2(C)}^{1} \delta Q \right) + \left(\int_{1(A)}^{2} \delta W + \int_{2(C)}^{1} \delta W \right) = 0$$

Subtracting and rearranging,

$$\int_{2(B)}^{1} \delta Q + \int_{2(B)}^{1} \delta W = \int_{2(C)}^{1} \delta Q + \int_{2(C)}^{1} \delta W$$

or

$$\int_{2(B)}^{1} (\delta Q + \delta W) = \int_{2(C)}^{1} (\delta Q + \delta W) \tag{3.2}$$

This means that the quantity $\int (\delta Q + \delta W)$ is a constant whether the system changes state along path B or path C. Thus, $\int (\delta Q + \delta W)$ is a function of the initial and final states of the system and does not depend on the process followed between the two states.

The quantity $(\delta Q + \delta W)$ is therefore a differential of a property of the system. This property is the *total energy* of the system and is given the symbol E. Hence, the change in the energy of the system is equal to the algebraic sum of the heat interaction and the work interaction with the surroundings. Thus the first law applied to a system undergoing a process is

$$\delta Q + \delta W = dE \tag{3.3}$$

where δQ is the net heat transfer to the system, δW is the net work done on the system, and dE is the difference between the final and the initial energy of the system. Equation (3.3) is simply an energy balance stating that the change of energy of the system is equal to the net amount of energy transfer across its boundary. According to the sign convention indicated in Chapter 2, heat or work transfer to a system is positive, whereas heat or work transfer from a system is negative. The work transfer interaction, W, may consist of the work due to expansion or compression of the system in addition to other types of work, such as mechanical, electrical, magnetic, potential work, so that if $\delta W'$ is work other than p-V work, then

$$\delta W = -pdV + \delta W' \tag{3.4}$$

On a unit mass basis, Eq. (3.3) becomes

$$\delta q + \delta w = de \tag{3.5}$$

When Eq. (3.3) is expressed as a time rate equation,

$$\frac{\delta Q}{dt} + \frac{\delta W}{dt} = \frac{dE}{dt} \tag{3.6}$$

where t represents time.

Note that a system can interact with its surroundings by energy transfer in the form of heat and work, and although δQ and δW are not properties, their algebraic sum is a property. This means that an infinite number of combinations of δQ and δW can accomplish a certain change of state dE; their sum, however, must be exactly equal to dE.

In an isolated system, $Q = 0$ and $W = 0$, and therefore, from Eq. (3.3), $E_1 = E_2$. It is therefore evident that the energy of the isolated system remains constant, which is a statement of the principle of conservation of energy. Note also that the general definition of energy was introduced through the first law,[1] and, therefore, the principle of the conservation of energy may be considered a consequence of the first law.

Any change in the total energy of a system is reflected in changes in the various forms of energy that comprise the total internal energy:

$$dE = (dU + \cdots, etc) + d(KE) + d(PE) \tag{3.7}$$

---

[1] Although the preceding form of the first law, Eq. (3.3), applies only to closed systems, other forms, as will be shown later in this chapter, are applicable to control volumes.

where U is internal energy due to the molecular kinetic and potential energies. It is independent of the macroscopic motion of the system as well as gravity, electricity, magnetism, and surface tension.

The integrated form of Eq. (3.3) is

$$Q_{1\text{-}2} + W_{1\text{-}2} = (E_2 - E_1) \qquad (3.8)$$

When the total energy consists only of internal energy U, kinetic energy, and potential energy, this equation becomes

$$\underbrace{Q_{1\text{-}2} + W_{1\text{-}2}}_{\text{energy transfer}} = \underbrace{(U_2 - U_1) + \frac{1}{2}m(V_2^2 - V_1^2) + mg(z_2 - z_1)}_{\text{energy change}} \qquad (3.9)$$

A device that creates energy or continuously produces work without any other interaction with the surroundings is called a *perpetual motion machine of the first kind.* Such a machine violates the first law of thermodynamics. All attempts to achieve perpetual motion have failed, thus providing experimental proof (by contradiction) of the validity of the first law.

Example 3.1

Utilizing Joule's experiments as a guide and assuming that energy can be measured only when it is in the form of work, evaluate the heat interaction between a system and its surroundings when the system changes its state from 1 to 2 along path B as shown in Figure 3.3. Assume that the states of the system are defined by p and V and that state 2 can be reached by an adiabatic path (1-A-2).

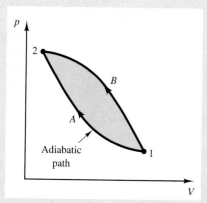

Figure 3.3

SOLUTION

The work in the actual process 1-*B*-2 is given per unit mass by

$$w_{1\text{-}2 \text{ actual}} = \Delta e - q_{1\text{-}2}$$

Measure the work done in the actual process 1-*B*-2 and in the adiabatic process 1-*A*-2. Since the internal energy change in process 1-*B*-2 is the same as that in process 1-*A*-2, the first law gives

$$w_{1\text{-}2 \text{ adia}} = \Delta e$$

Subtraction of these two equations yields

$$q_{1\text{-}2} = w_{1\text{-}2 \text{ adia}} - w_{1\text{-}2 \text{ actual}}$$

This result indicates that the heat interaction with a system is equal to the difference between the work done if the process proceeded adiabatically and the work actually done. Note also that, for an adiabatic process connecting two end states, the work done depends on the end states only.

3.3 Internal Energy

The internal energy, U, is an extensive property. It is a function of the state of the system, and its integral around a complete cycle is therefore zero. The cyclic integral of the function U is then *conserved*[2]

$$\oint dU = 0 \qquad\qquad (3.10)$$

If the kinetic and potential energies do not change, heat and work or both provide the only means by which the internal energy of a system can change, according to Eq. (3.9). Note also that the internal energy of an isolated system is constant. Internal energy is expressed in the same units as heat or work, the units usually being J, whereas specific internal energy is expressed in J/kg.

If the state of a system is determined by any two of the properties p, T, and v, and if electrical, magnetic, surface tension effects are negligible, then the work done per unit mass in a quasi-static process is

$$\delta w = -p \, dv$$

If only internal energy u changes, then Eq. (3.5) becomes

[2] A function X is said to be conversed if its integral around a cycle is zero or $\oint dX = o$.

$$\delta q - p \, dv = du \tag{3.11}$$

When u is expressed in terms of T and v, then $u = u(T, v)$, and the change of u is given by

$$du = \left(\frac{\partial u}{\partial T}\right)_v dT + \left(\frac{\partial u}{\partial v}\right)_T dv \tag{3.12}$$

Combining Eq. (3.11) with Eq. (3.12) yields

$$\delta q - p \, dv = \left(\frac{\partial u}{\partial T}\right)_v dT + \left(\frac{\partial u}{\partial v}\right)_T dv$$

This may be rearranged as

$$\delta q = \left(\frac{\partial u}{\partial T}\right)_v dT + \left[p + \left(\frac{\partial u}{\partial v}\right)_T\right] dv \tag{3.13}$$

In a constant-volume process, no volume change occurs so that

$$\delta q = \left(\frac{\partial u}{\partial T}\right)_v dT$$

But at constant volume, $\delta q = c_v \, dT$. Therefore c_v may be defined as

$$c_v \equiv \left(\frac{\partial u}{\partial T}\right)_v \tag{3.14}$$

Note that a change of temperature does not necessarily result from heat transfer since work can also cause a change of temperature. Thus the name *specific heat* is a misnomer since it implies that c_v is associated with a heat quantity when it is really associated with a property. This means that c_v may replace $(\partial u/\partial T)_v$ in any process, quasi-static or not, even when the process does not take place at constant volume. Note also that c_v is an intensive property and is equal to $(\delta q/dT)_v$ only in a constant-volume process that involves no work interaction. Therefore a better name for c_v would be specific internal energy per degree at constant volume.

3.4 Internal Energy of an Ideal Gas

In experiments performed by Gay-Lussac and by Joule, the free expansion of gases was studied. The apparatus used is essentially that shown in Figure 3.4. Chamber A contains a gas in thermodynamic equilibrium, while chamber B is

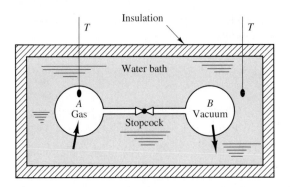

Figure 3.4   Experiment to show that the internal energy of an ideal gas is a function of temperature only.

evacuated. The two chambers constitute a closed system. Initially the system is in thermal equilibrium with its surroundings (water bath) at a uniform temperature T_1. When the stopcock between chambers A and B is opened, the gas in A expands freely into B, occupying both chambers. During the expansion process, the temperature of the gas in A decreases below its initial value while the temperature in B increases. These temperature changes develop temperature gradients so that heat flows from the water bath to chamber A and from chamber B to the water bath. This causes an additional flow of gas from A to B. After sufficient time has passed, the system and its surroundings reach thermodynamic equilibrium. The heat transferred to the system from the surroundings as a result of this process is

$$Q_{sys} = -Q_{surr} = mc(T_2 - T_1) *$$

where m and c are the mass and average specific heat of the system, and $(T_2 - T_1)$ is the temperature change of the system. Since the boundaries of the system do not move, no external work is done during the free expansion of the gas. When the first law of thermodynamics is applied to this system, it becomes

$$\Delta U = Q_{sys}$$

By substituting for Q_{sys}, this becomes

$$\Delta U = mc(T_2 - T_1)$$

Measurements show that the temperature of the water bath remains practically constant. This means that chamber A loses as much energy as chamber B gains so that there is no net heat interaction between the system and the surroundings. The absence of both heat interaction and work interaction indicates that the internal energy of this system remains constant even though a volume change has taken place. Therefore, in gases that tend to behave like ideal gases, the inter-

---

* The subscripts "sys" and "surr" refer to interaction with system and surroundings, respectively.

nal energy is independent of volume and depends only on temperature. This statement is called *Joule's law* and is expressed as

$$\left(\frac{\partial u}{\partial v}\right)_T = 0 \qquad (3.15)$$

For most substances (other than ideal gases), the internal energy, when no phase change is involved, depends strongly on temperature and rather weakly on pressure or volume. In Joule's experiment, any temperature changes that might have occurred were very small because they were obscured as a result of the large thermal capacity of chamber A, chamber B, and the water bath compared with the thermal capacity of the gas. In the Joule–Thomson experiment, explained later in this chapter, results of greater accuracy were obtained.

The change in internal energy of an ideal gas per unit mass, as the gas changes from state 1 to state 2, is

$$\int_1^2 du = \int_{T_1}^{T_2} c_v \, dT \qquad (3.16)$$

If c_v is a constant, then

$$u_2 - u_1 = c_v(T_2 - T_1)$$

Example 3.2

An insulated rigid tank contains 0.2 kg of air at a temperature of 300 K and a pressure of 100 kPa. A paddle wheel inside the tank transfers 5 kJ of energy to the air. Determine the change of the internal energy and the final temperature and pressure of the air.

SOLUTION

Referring to Figure 3.5, the first law, in the absence of kinetic and potential energy changes, is

$$Q + W = \Delta U \qquad (Q = 0)$$

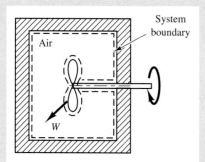

Figure 3.5

Therefore,

$$\Delta U = U_2 - U_1 = W = 5 \text{ kJ}$$

Assuming air to be an ideal gas,

$$\Delta U = mc_v(T_2 - T_1)$$
$$5 \text{ kJ} = (0.2 \text{ kg})(0.7165 \text{ kJ/kg K})[(T_2 - 300) \text{ K}]$$

from which $T_2 = 334.89 \text{ K}$
At constant volume,

$$p_2 = p_1\left(\frac{T_2}{T_1}\right) = (100 \text{ kPa})\left(\frac{334.89}{300}\right) = 111.63 \text{ kPa}$$

3.5 Enthalpy

Equation (3.3) can be rewritten as

$$\delta Q - p\,dV + \delta W' = dE \tag{3.17}$$

where $\delta W'$ represents all work done on the system other than p-V work, such as shaft, electrical, and magnetic. When $d(pV)$ is added to both sides, Eq. (3.17) becomes

$$\delta Q + V\,dp + \delta W' = d(E + pV)$$

If E comprises only internal energy, the foregoing equation becomes

$$\delta Q + V\,dp + \delta W' = d(U + pV) \tag{3.18}$$

The combination $(U + pV)$ often plays an important role in calculations where matter crosses the boundaries of a control volume. For this reason, it is convenient to combine $(U + pV)$ under the name of a new property called *enthalpy H*. As will be seen in the following section, enthalpy is directly associated with the energy transfer across a control surface. Enthalpy is defined by

$$H \equiv U + pV \tag{3.19}$$

Enthalpy is an extensive property, and specific enthalpy is given by

$$h \equiv u + pv \tag{3.20}$$

The unit of H is J and that of h is J/kg. Since enthalpy is a combination of functions of state (u, p, and v), it, too, is a function of state. Note that the difference between the enthalpy h and the internal energy u is the product pv. The term pv accounts for the work done per unit mass of a fluid as it flows across the boundaries of a control volume. Note that, although one can calculate the enthalpy of a system, the characteristic energy of a system is the internal energy u. If a system undergoes a constant pressure process, and if the only work involved is p-V work, then according to the first law

$$q_{1\text{-}2} - p(v_2 - v_1) = u_2 - u_1$$

or

$$q_{1\text{-}2} = h_2 - h_1 \tag{3.21}$$

The heat transfer to a system is thus equal to the change of enthalpy under the two constraints just mentioned.

When enthalpy is substituted into Eq. (3.18), it becomes, for a unit mass,

$$\delta q = dh - v\,dp - \delta w' \tag{3.22}$$

Since $\delta q = c_p\,dT$, under conditions of constant pressure and when p-V work is the only work, then c_p may be defined as

$$c_p \equiv \left(\frac{\partial h}{\partial T}\right)_p \tag{3.23}$$

The constant-pressure specific heat c_p is an intensive property depending on the state of the system. This means that c_p may replace $(\partial h/\partial T)_p$ in any process, quasi-static or not, even when the process is not at constant pressure. The constant-pressure specific heat is equal to $(\delta q/dT)_p$ only in a constant-pressure process in which $\delta w' = 0$. As stated in the case of c_v, c_p must be thought of as a change of property with respect to temperature. It is the specific enthalpy per degree at constant pressure.

If enthalpy is a function of two independent parameters, such as T and p, then the change in enthalpy for the process is

$$\begin{aligned}
dh &= \left(\frac{\partial h}{\partial T}\right)_p dT + \left(\frac{\partial h}{\partial p}\right)_T dp \\
&= c_p\,dT + \left(\frac{\partial h}{\partial p}\right)_T dp
\end{aligned} \tag{3.24}$$

In the case of an ideal gas, Eq. (3.20) becomes

$$dh = du + d(pv) = c_v\,dT + R\,dT = (c_v + R)dT$$

In this case, enthalpy depends only on temperature so that $(\partial h/\partial p)_T = 0$. Therefore, Eq. (3.24) becomes, for an ideal gas,

$$dh = c_p\,dT$$

Hence,

$$c_p\,dT = (c_v + R)dT$$

or

$$c_p = c_v + R \qquad (3.25)$$

Note from the preceding equations that, for an ideal gas, c_p and c_v are functions solely of temperature.

Example 3.3

One kilogram of air initially at 300 K and 100 kPa undergoes a quasi-equilibrium constant-pressure process in a piston-cylinder assembly. If the final temperature is 450 K, determine the work and heat interactions and the change in enthalpy.

SOLUTION

Referring to Figure 3.6, the first law in the absence of kinetic and potential energy changes is

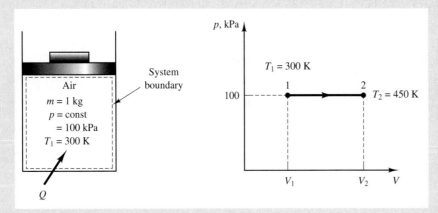

Figure 3.6

$$Q + W = \Delta U$$

$$W_{1\text{-}2} = -\int_1^2 p\,dV = -p(V_2 - V_1)$$

Assuming air to be an ideal gas,

$$V_1 = \frac{mRT_1}{p_1} = \frac{(1 \text{ kg})(0.287 \text{ kJ/kg K})(300 \text{ K})}{100 \text{ kPa}} = 0.861 \text{ m}^3$$

and

$$V_2 = \frac{mRT_2}{p_2} = \frac{(1 \text{ kg})(0.287 \text{ kJ/kg K})(450 \text{ K})}{100 \text{ kPa}} = 1.2915 \text{ m}^3$$

Therefore,

$$W_{1\text{-}2} = -(100 \text{ kPa})[(1.2915 - 0.861) \text{ m}^3] = -43.05 \text{ kJ}$$
$$Q_{1\text{-}2} = (U_2 - U_1) - W_{1\text{-}2} = mc_v(T_2 - T_1) - W_{1\text{-}2}$$
$$= (1 \text{ kg})(0.7165 \text{ kJ/kg K})[(450 - 300) \text{ K}] - (-43.05 \text{ kJ})$$
$$= 150.525 \text{ kJ}$$

The change in enthalpy is

$$H_2 - H_1 = mc_p(T_2 - T_1) = (1 \text{ kg})(1.0035 \text{ kJ/kg K})[(450 - 300) \text{ K}]$$
$$= 150.525 \text{ kJ}$$

Note that for the quasi-equilibrium constant-pressure process, the heat interaction is equal to the change in enthalpy of the system.

In the following sections, the first law of thermodynamics will be applied first to closed systems and then to control volumes.

3.6 Application of the First Law to a Closed System

The first law of thermodynamics applies to all energy interactions between a system and its surroundings. A change in the energy of a system is accompanied by an equal but opposite change in the energy of its surroundings. According to the first law, any heat or work that crosses the boundaries of a system represents energy that is equal in magnitude to the change in internal energy of the system. Remember the sign convention adopted for the transition energies: Energy entering the system is considered positive, and energy leaving the system is negative.

When applying the first law to typical processes, a sketch that shows energy flow across well-defined boundaries of a system will prove particularly helpful.

3.6.1 The Constant-Volume (Isochoric) Process

Consider the transfer of heat to a system consisting of a fluid in a rigid vessel, as shown in Figure 3.7.

The process is represented by a vertical line on the p-V diagram. Since $dV = 0$, the displacement work is zero. Assuming that only internal energy comprises e, the first law for this system, based on a unit mass, becomes

$$\delta q = du$$

Thus, the heat transfer to the system is accounted for as an increase in the internal energy of the system. Similarly, any heat transfer from the system represents an equivalent decrease in the internal energy of the system. Note that the same change of internal energy could be achieved using work interaction such as shaft or electric work. A combination of heat and work can also result in the same change of state.

When a solid or liquid is heated, the resulting change in volume is relatively small. Heat transfer to solids and liquids may therefore be considered to take place at constant volume, and the energy supplied is essentially equal to the increase in internal energy of the system.

Quasi-static work, $-\int p\, dV$, is represented by the area, in the p-V plane, below the process path. In the constant-volume process, this area is clearly zero so that it is impossible to have quasi-static work done in a constant-volume process. Work, however, still can be done on a constant-volume system (as demonstrated by Joule's paddle-wheel experiments), but in this case the process of work transfer is irreversible.

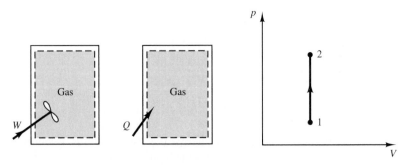

Figure 3.7   Constant-volume process.

Example 3.4

One kg of air (assumed ideal gas, $R = 0.287$ kJ/kg K) is confined to a constant-volume vessel. The volume and the initial pressure of the air are 0.2 m^3 and 350 kPa, respectively. If 120 kJ of heat are supplied to the gas, the temperature increases to 411.5 K. Find (a) the work done; (b) the change of internal energy; (c) the specific heat of the gas at constant volume.

SOLUTION

A sketch of the system and a *p-V* diagram of the process is shown in Figure 3.8.

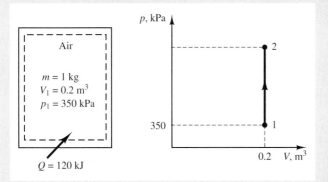

Figure 3.8

(a) The displacement work is zero since there is no change in volume.
(b) First law: $q = \Delta u = 120$ kJ/kg.
(c) The initial temperature may be determined from the ideal-gas equation of state,

$$pv = RT$$

from which

$$T = \frac{(350 \text{ kPa})(0.2 \text{ m}^3)}{0.287 \text{ kJ/kg K}} = 243.9 \text{ K}$$

Therefore

$$\Delta T = 411.5 - 243.9 = 167.6 \text{ K}$$

The average specific heat at constant volume is

$$c_v = \left(\frac{\Delta u}{\Delta T}\right)_v = \frac{120 \text{ kJ/kg}}{167.6 \text{ K}} = 0.716 \text{ kJ/kg K}$$

3.6.2 The Constant-Pressure (Isobaric) Process

A fluid maintained at constant pressure by means of a frictionless piston-cylinder assembly is shown in Figure 3.9. For a system of unit mass, work $w_{1\text{-}2} = -\int_{v_1}^{v_2} p\,dv = -p(v_2 - v_1)$ and is represented by the hatched area in Figure 3.9. The first law, as indicated by Eq. (3.11), becomes, after integration,

$$q_{1\text{-}2} - p(v_2 - v_1) = u_2 - u_1$$

or

$$q_{1\text{-}2} = (u_2 + pv_2) - (u_1 + pv_1) = h_2 - h_1$$

which is the same as Eq. (3.21). This relationship is valid for constant-pressure processes if only p-V work is done. Note that the heat interaction in this case depends only on the end states of the system.

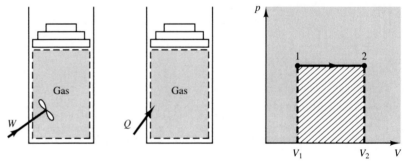

Figure 3.9   Constant-pressure process.

Example 3.5

Air at a temperature of 500°C is compressed at a constant pressure of 1.2 MPa from a volume of 2 m³ to a volume 0.4 m³. If the internal energy decrease is 4820 kJ, find (a) the work done during the reversible compression; (b) the heat transferred; (c) the change of enthalpy; (d) the average specific heat at constant pressure.

SOLUTION

(a) A sketch of the system and a p-V diagram of the process are shown in Figure 3.10.

$$W_{1\text{-}2} = -p(V_2 - V_1) = -(1.2 \times 10^3 \text{ kPa})[(0.4 - 2)\text{ m}^3] = 1920 \text{ kJ}$$

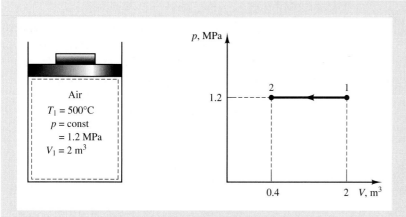

Figure 3.10

(b)

$$Q_{1\text{-}2} + W_{1\text{-}2} = (U_2 - U_1)$$

Hence,

$$Q_{1\text{-}2} = -1920 \text{ kJ} - 4820 \text{ kJ} = -6740 \text{ kJ}$$

(c) Since the process is at constant pressure, and $W' = 0$, then

$$H_2 - H_1 = Q_{1\text{-}2} = -6740 \text{ kJ}$$

(d)

$$T_2 = \frac{T_1 V_2}{V_1} = \frac{(773.15 \text{ K})(0.4 \text{ m}^3)}{2 \text{ m}^3} = 154.3 \text{ K}$$

$$m = \frac{p_1 V_1}{R T_1} = \frac{(1200 \text{ kPa})(2 \text{ m}^3)}{(0.287 \text{ kJ/kg K})(773.15 \text{ K})} = 10.816 \text{ kg}$$

and

$$c_p = \left(\frac{\Delta H}{m \Delta T}\right)_p = \frac{-6740 \text{ kJ}}{(10.816 \text{ kg})[(154.3 - 773.15) \text{ K}]}$$
$$= 1.007 \text{ kJ/kg K}$$

3.6.3 Process at Constant Internal Energy

When the internal energy of a system remains constant, the first law becomes

$$\delta q + \delta w = 0$$

If a system undergoes a process in which the internal energy does not change, the heat interaction and the work interaction must be equal to each other in magnitude but of opposite sign. This means that the heat transfer to the system produces an equivalent amount of work.

    If two dissimilar metals are joined, and if two such junctions are held at different temperatures, an electric potential is developed between the junctions. When the circuit is closed, an electric current flows through the metals. The thermoelectric generator utilizes this phenomenon to convert heat into electric energy. Note that the thermoelectric generator has a temperature gradient but is in thermal equilibrium ($\Delta U = 0$). The components of a thermoelectric generator, as shown in Figure 3.11, consist of a series of semiconductors, A and B, connected to the hot and cold conductors, C and D. The electrical terminals connected to the conductors D carry the output electrical energy to the load G.

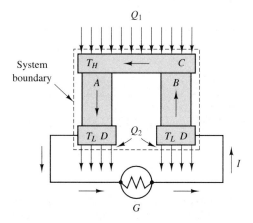

Figure 3.11   Thermoelectric generator.

Example 3.6

It is required to make a first law analysis of a thermoelectric device under the following conditions:

rate of heat transfer = 20 W
    emf generated = 2.5 volts
    current flow = 0.6 amp

Assume steady state.

SOLUTION

Since the current flow through the device is constant, the electric power output is equal to the voltage drop multiplied by the current, or

$$\dot{W} = VI = 2.5 \times 0.6 = 1.5 \text{ W}$$

The first law applied to the system shown is

$$\dot{W} + \sum \dot{Q} = \Delta \dot{E}$$

At steady state, there is no change in internal energy of the system so that $\Delta \dot{E} = 0$. Hence,

$$-1.5 \text{ W} + (20 + \dot{Q}_2) \text{ W} = 0 \qquad \text{or} \qquad \dot{Q}_2 = -18.5 \text{ W}$$

The efficiency of the energy conversion is

$$\eta = \frac{\text{electric power output}}{\text{rate of heat input}} = \frac{1.5 \text{ W}}{20 \text{ W}} = 7.5 \text{ percent}$$

3.6.4   The Constant Temperature (Isothermal) Process

During this process, the temperature of a system is maintained constant. If an ideal gas undergoes an isothermal process, its internal energy, being a function of temperature alone, remains constant and Eq. (3.11) becomes

$$\delta q + \delta w = 0$$

Thus, in an isothermal compression or expansion process that involves only an ideal gas, the sum of the heat and work input is zero, just as in a constant-internal energy process. Figure 3.12 shows a p-V diagram for an isothermal process of an ideal gas.

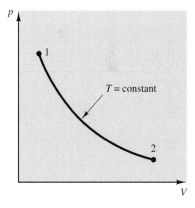

Figure 3.12   Isothermal process.

Example 3.7

An ideal gas occupying a volume of 0.2 m^3 at a pressure of 1.5 MPa expands isothermally in a quasi-equilibrium process to a volume of 0.5 m^3. Find the final pressure, W, Q, and ΔU.

SOLUTION

A sketch of the system and a p-V diagram of the process is shown in Figure 3.13.

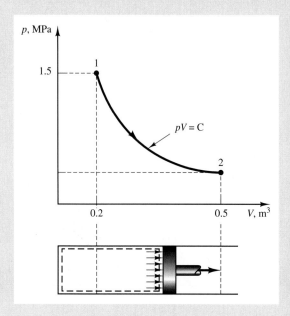

Figure 3.13

$$p_1 V_1 = mRT \qquad \text{and} \qquad p_2 V_2 = mRT$$

Then,

$$p_2 = p_1\left(\frac{V_1}{V_2}\right) = (1.5 \text{ MPa})\left(\frac{0.2}{0.5}\right) = 0.6 \text{ MPa}$$

Since the temperature remains constant, $\Delta U = 0$, and the first law gives

$$Q_{1\text{-}2} = -W_{1\text{-}2}$$

But

$$W_{1\text{-}2} = \int_1^2 \delta W = -\int_{V_1}^{V_2} p \, dV$$

$$= -\int_{V_1}^{V_2} \frac{mRT}{V} \, dV = -mRT \ln \frac{V_2}{V_1} = -p_1 V_1 \ln \frac{V_2}{V_1}$$

$$= -(1.5 \times 10^3 \text{ kPa})(0.2 \text{ m}^3) \ln 2.5 = -274.89 \text{ kJ}$$

3.6.5 The Adiabatic Process

In an adiabatic process, no heat interaction occurs between the system and the surroundings. The first law then becomes

$$\delta W = dU$$

which, when integrated, is

$$W_{1\text{-}2} = U_2 - U_1$$

Therefore, the work done on the system is equal to the change in internal energy of the system. When an ideal gas undergoes a reversible adiabatic process, the preceding equation per unit mass can be written in differential form as

$$c_v \, dT = -p \, dv = -\frac{RT}{v} dv$$

or

$$c_v \frac{dT}{T} = -R \frac{dv}{v}$$

which may be integrated to give

$$\ln \frac{T_2}{T_1} = \frac{R}{c_v} \ln \frac{v_1}{v_2}$$

or

$$\frac{T_2}{T_1} = \left(\frac{v_1}{v_2}\right)^{R/c_v} \tag{3.26}$$

The temperature ratio may also be expressed in terms of the pressure ratio. From the ideal-gas equation, the volume ratio can be replaced so that

$$\frac{T_2}{T_1} = \left(\frac{p_2 T_1}{p_1 T_2}\right)^{R/c_v} = \left(\frac{p_2}{p_1}\right)^{R/(R+c_v)} \tag{3.27}$$

In the case of an ideal gas, $c_p = c_v + R$. Combining Eqs. (3.26) and (3.27), the pressure-volume relationship in an adiabatic process is

$$\frac{v_2}{v_1} = \left(\frac{p_1}{p_2}\right)^{c_v/c_p} \tag{3.28}$$

or

$$p_2 v_2^\gamma = p_1 v_1^\gamma = \text{constant} \qquad (3.29)$$

where γ is the ratio of specific heats:

$$\gamma = \frac{c_p}{c_v} \qquad (3.30)$$

3.6.6 The Polytropic Process

A polytropic process is a real expansion or compression process in which the relationship betweeen p and V is given by

$$pV^n = C \qquad (3.31)$$

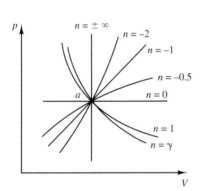

Figure 3.14   Processes described by $pV^n = C$.

where n is the "index" of the process. In general, n may have any value from $-\infty$ to $+\infty$. In a constant-pressure process, $n = 0$; in an isothermal process (for an ideal gas), $n = 1$; in an adiabatic process, $n = \gamma$; and in a constant-volume process, $n \rightarrow \infty$.

Several reversible polytropic processes corresponding to different values of n are plotted in Figure 3.14. Starting at point a, expansion and compression processes are curves located in the lower right and upper left quadrants. Processes that exhibit negative values of n are not commonly encountered in practice. Although these processes are possible, they imply a simultaneous decrease in volume accompanying a decrease in pressure. If the polytropic processes shown in Figure 3.14 were reversible, the area under the curve or line would represent the reversible work done in each case. For a reversible polytropic process, the work done on a system, as outlined in Chapter 2, is

$$W_{1\text{-}2} = \frac{p_2 V_2 - p_1 V_1}{n - 1} \qquad (3.32)$$

Note that Eq. (3.32) does not apply if $n = 1$. A summary of process relations of an ideal gas with constant specific heat is presented in Table 3.1.

Example 3.8

One kg of an ideal gas expands in a reversible polytropic process according to the law $pV^n = C$ (where $n = 1.3$). The initial pressure and volume are 620 kPa and 0.15 m^3. The final volume is 1.0 m^3. Find (a) the final

Table 3.1 Process-Relations for Ideal Gas

Process	Q	$-\int_1^2 p\,dV$	$\int_1^2 V\,dp$	ΔU	p-V-T Relations	Exponent in $pV_n = C$
Constant volume	$mc_v(T_2 - T_1)$	0	$V(p_2 - p_1)$	$mc_v(T_2 - T_1)$	$\dfrac{T_2}{T_1} = \dfrac{p_2}{p_1}$	$n = \infty$
Constant pressure	$mc_p(T_2 - T_1)$	$p(V_1 - V_2)$	0	$mc_v(T_2 - T_1)$	$\dfrac{T_2}{T_1} = \dfrac{V_2}{V_1}$	$n = 0$
Constant temperature	$p_1 V_1 \ln \dfrac{V_2}{V_1}$	$p_1 V_1 \ln \dfrac{V_1}{V_2}$	$p_1 V_1 \ln \dfrac{V_1}{V_2}$	0	$p_1 V_1 = p_2 V_2$	$n = 1$
Reversible adiabatic	0	$\dfrac{p_2 V_2 - p_1 V_1}{\gamma - 1}$	$\dfrac{\gamma}{\gamma - 1}(p_2 V_2 - p_1 V_1)$	$mc_v(T_2 - T_1)$	$p_1 V_1^\gamma = p_2 V_2^\gamma,$ $\dfrac{T_2}{T_1} = \left(\dfrac{V_1}{V_2}\right)^{\gamma-1} = \left(\dfrac{p_2}{p_1}\right)^{(\gamma-1)/\gamma}$	$n = \gamma$
Reversible polytropic	$mc_v\left(\dfrac{\gamma - n}{1 - n}\right)(T_2 - T_1)$	$\dfrac{p_2 V_2 - p_1 V_1}{n - 1}$	$\dfrac{n}{n - 1}(p_2 V_2 - p_1 V_1)$	$mc_v(T_2 - T_1)$	$p_1 V_1^n = p_2 V_2^n,$ $\dfrac{T_2}{T_1} = \left(\dfrac{V_1}{V_2}\right)^{n-1} = \left(\dfrac{p_2}{p_1}\right)^{(n-1)/n}$	$n = n$

Constant volume Constant pressure Constant temperature Reversible adiabatic Reversible polytropic

temperature; (b) the work done; (c) the change of internal energy; (d) the heat interaction. (R for the gas is 0.130 kJ/kg K, and $c_v = 0.511$ kJ/kg K.)

SOLUTION

(a) A sketch of the system and a p-V diagram of the process is shown in Figure 3.15.

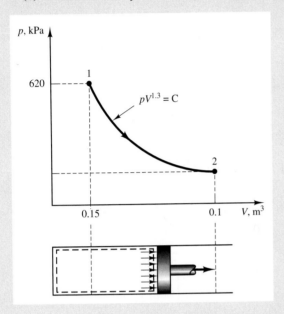

Figure 3.15

$$p_1 v_1^{1.3} = p_2 v_2^{1.3}$$

then

$$p_2 = (620 \text{ kPa})\left(\frac{0.15}{1.0}\right)^{1.3} = 52.6 \text{ kPa}$$

The temperatures T_1 and T_2 are determined from the ideal-gas law as

$$T_1 = \frac{p_1 v_1}{R} = \frac{(620 \text{ kPa})(0.15 \text{ m}^3/\text{kg})}{0.13 \text{ kJ/kg K}} = 715.4 \text{ K}$$

and

$$T_2 = \frac{p_2 v_2}{R} = \frac{(52.6 \text{ kPa})(1.0 \text{ m}^3/\text{kg})}{0.13 \text{ kJ/kg K}} = 404.6 \text{ K}$$

(b)

$$w_{1\text{-}2} = -\int_{v_1}^{v_2} p\, dv = \frac{p_2 v_2 - p_1 v_1}{n - 1}$$

$$= \frac{(52.6\ \text{kPa})(1.0\ \text{m}^3/\text{kg}) - (620\ \text{kPa})(0.15\ \text{m}^3/\text{kg})}{1.3 - 1}$$

$$= -134.67\ \text{kJ/kg}$$

(c)

$$\Delta u = c_v(T_2 - T_1) = (0.511\ \text{kJ/kg K})[(404.6 - 715.4)\ \text{K}]$$

$$= -158.8\ \text{kJ/kg}$$

(d)

$$q_{1\text{-}2} = \Delta u - w_{1\text{-}2} = -158.8\ \text{kJ/kg} + 134.67\ \text{kJ/kg}$$

$$= -24.13\ \text{kJ/kg}$$

3.7 The First Law of Thermodynamics for a Control Volume

Energy may be conveyed across the boundaries of a control volume not only by heat and work but also by the matter that crosses the boundaries. Matter may convey internal, kinetic, potential, chemical, and magnetic energies across boundaries. The following analysis is limited to the first three forms of energy, but the same procedure applies to systems in which any form of internal energy is involved. The first law indicates that an energy balance exists between a system and its surroundings. When energy in the form of heat and work is transferred to a system, the decrease in energy of the surroundings is equal to the increase in the internal energy of the system. As shown in Figure 3.16, matter enters a control volume at section i and leaves at section e. Also heat and work are exchanged between the control volume and surroundings. According to the first law of thermodynamics, the energy entering the control volume must be equal to the energy leaving the control volume plus any change of energy within the control volume:

	heat and work interaction	+	energy entering the control volume at section i
=	energy leaving the control volume through section e	+	change of energy within the control volume

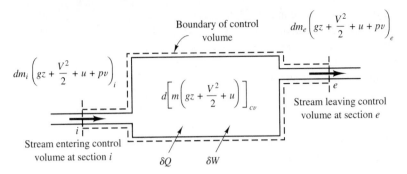

Figure 3.16   The first law of thermodynamics applied to a control volume.

Expressed mathematically, this becomes

$$\delta Q + \delta W + dm_i \left(u + pv + \frac{V^2}{2} + gz \right)_i$$
$$= dm_e \left(u + pv + \frac{V^2}{2} + gz \right)_e + d\left[m\left(u + \frac{V^2}{2} + gz \right) \right]_{cv} \tag{33.33a}$$

where dm_i and dm_e are the masses entering and leaving the control volume during the short time interval dt; δW is work other than flow work. A more general form of this statement is

$$\underbrace{ Q + W + \sum_{net=in-out} m\left(h + \frac{V^2}{2} + gz \right) }_{\text{energy transfer}} = \underbrace{ \Delta E_{cv} }_{\text{energy change}} \tag{3.33b}$$

where the specific enthalpy h is substituted for $u + pv$. The summation sign is introduced to account for several locations at the boundary through which mass enters and leaves the control volume.

As a rate equation, this becomes

$$\dot{Q} + \dot{W} + \sum_{net=in-out} \dot{m}\left(h + \frac{V^2}{2} + gz \right) = \frac{dE_{cv}}{dt} \tag{3.34}$$

where $\cdot$ indicates time rate. Equation (3.33b) is an energy accounting balance: it states that the net energy transfer to a control volume is accounted for as an energy change within the control volume. Although Eq. (3.33b) is the general energy equation for a control volume, it still serves as the energy equation for a closed system provided that the summation terms drop out since $dm_i = dm_e = 0$.

The procedure outlined in Section 2.21 may also be applied and leads to the preceding equations. For a system that coincides initially with the control volume, the change of internal energy is

$$dE = dE_{cv} + dE_e - dE_i$$

where E_{cv} is the internal energy of the control volume, E_e is the internal energy of the fluid leaving the control volume, and E_i is the internal energy of the fluid entering the control volume. Substituting for dE from the first law gives

$$\delta Q_{sys} + \delta W_{sys} = dE_{cv} + dE_e - dE_i$$

The energies δQ_{sys} and δW_{sys} are interactions with the system. But the heat transfer δQ_{sys} to (or from) the system is independent of the flow of fluid into (and out of) the control volume. Therefore, the amount of heat is the same for the control volume as for the system. On the other hand, the work δW_{sys} done on the system is not necessarily the same as that done on the control volume. Flow work[3] is necessary to introduce mass into, and to remove it from, the control volume. Separating the work done on a control volume into two terms, the foregoing equation becomes

$$\delta Q + \delta W + d(pV) = dE_{cv} - dE_{\text{net in}}$$

where δW in the preceding equation represents the differential of energy transfer as work across the boundary of the control volume other than flow work. If the properties are uniform at the points where matter enters and leaves the control volume, then flow work and internal energy can be grouped together:

$$\delta Q + \delta W + \sum (e + pv)dm = dE_{cv} \qquad (3.35)$$

where the summation sign accounts for several locations at the boundary through which masses cross the control surface. Integration of Eq. (3.35) gives

$$Q + W + \sum (e + pv)m = \Delta E_{cv}$$

When this equation is expressed on a rate basis,

$$\dot{Q} + \dot{W} + \sum (e + pv)\dot{m} = \frac{dE_{cv}}{dt} \qquad (3.36)$$

it becomes the same as Eq. (3.34).

---

[3] This neglects shear work on the fluid at the boundaries as well as magnetic, electric, and capillary effects.

Two cases will now be considered. First is the steady-state flow process, and second is the unsteady or transient flow process.

3.8 The Steady-State Flow Process

For a control volume at steady state, the following conditions apply:

1. The mass entering the control volume flows at a constant rate, and at any time, the mass flow at the inlet is the same as the mass flow at the exit. This implies that the mass within the control volume neither increases nor diminishes at any time.
2. The state and energy of the fluid at the inlet, at the exit, and at every point within the control volume are time independent.
3. The rate at which any energy, in the form of heat or work, crosses the control surface is constant.

When steady state exists, conservation of mass and energy require that

$$\sum dm = 0 \qquad \text{or} \qquad dm_i = dm_e$$

and

$$dE_{cv} = d\left[m\left(u + \frac{V^2}{2} + gz \right) \right]_{cv} = 0$$

Thus Eq. (3.33b) becomes, under steady-state conditions,

$$\delta Q + \delta W + \sum dm\left(h + \frac{V^2}{2} + gz \right) = 0$$

where h is the enthalpy function. Integration leads to

$$Q + W + \sum m\left(h + \frac{V^2}{2} + gz \right) = 0 \qquad (3.37)$$

On a unit mass basis, this is

$$q_{i-e} + w_{i-e} + (h_i - h_e) + \frac{V_i^2 - V_e^2}{2} + g(z_i - z_e) = 0 \qquad (3.38)$$

On a rate basis, the energy equation for steady state is

$$\frac{\delta Q}{dt} + \frac{\delta W}{dt} + \sum \dot{m}\left(h + \frac{V^2}{2} + gz\right) = 0$$

or

$$\dot{Q} + \dot{W} + \sum \dot{m}\left(h + \frac{V^2}{2} + gz\right) = 0 \qquad (3.39)$$

where $\dot{m}$ is the mass flow per unit time.

When more than one fluid enters and leaves the control volume in steady flow, as shown in Figure 3.17, the continuity equation becomes

$$m_a + m_b + \cdots = m_c + m_d + \cdots$$

where m_a, m_b, m_c, and m_d represent masses of the different constituents entering and leaving the control volume in a given time interval. The steady-flow energy equation is

$$Q + W + m_a\left(h + \frac{V^2}{2} + gz\right)_a + m_b\left(h + \frac{V^2}{2} + gz\right)_b + \cdots$$
$$= m_c\left(h + \frac{V^2}{2} + gz\right)_c + m_d\left(h + \frac{V^2}{2} + gz\right)_d + \cdots$$

where subscripts a, b, and so on refer to flows entering, and c, d, and so on refer to flows leaving the control volume.

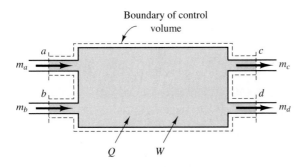

Figure 3.17   Control volume with multiple streams.

Example 3.9

Saturated water at 600 kPa ($u = 820$ kJ/kg, $h = 830$ kJ/kg) is injected into saturated steam at a pressure of 1400 kPa ($u = 2590$ kJ/kg, $h = 2790$ kJ/kg). If the mixing process is accomplished at constant pressure and the mixing ratio of water to steam is 1:10 by mass, find the enthalpy of the mixture. Assume steady-state and adiabatic mixing.

SOLUTION

A sketch of the mixing chamber is shown in Figure 3.18. The first law for steady flow is

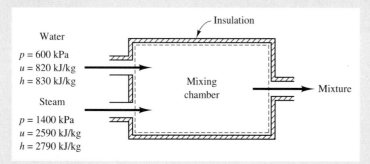

Figure 3.18

$$Q + W + \sum m\left(h + \frac{V^2}{2} + gz\right) = 0$$

There is no heat or work transfer ($Q = 0$, $W = 0$), and changes in kinetic and potential energies are considered negligible. The first law then reduces to

$$\sum m(h) = 0$$

or

$$m_s h_s + m_w h_w = m_{\text{mixt}} h_{\text{mixt}}$$

But

$$m_{\text{mixt}} = m_s + m_w$$

therefore,

$$m_s h_s + m_w h_w = (m_s + m_w) h_{\text{mixt}}$$

Dividing by m_s and rearranging gives

$$h_{\text{mixt}} = \frac{h_s + (m_w / m_s) h_w}{1 + (m_w / m_s)}$$

Substituting values,

$$h_{\text{mixt}} = \frac{(2790 \text{ kJ/kg}) + 0.1(830 \text{ kJ/kg})}{1 + 0.1} = 2611.8 \text{ kJ/kg}$$

3.9 Work Interaction in a Reversible Steady-State Flow Process

The work interaction per unit mass between a control volume and its surroundings is expressed, under steady-state flow conditions, by

$$\delta w = dh + V \, dV + g \, dz - \delta q \tag{3.40}$$

In a system, the reversible heat interaction δq is given by

$$\delta q = du + p \, dv = dh - v \, dp$$

But δq also indicates the heat interaction for a control volume. Therefore, when δq is replaced in Eq. (3.40), the work interaction for a control volume is

$$w_{\text{rev}} = \int_i^e v \, dp + \frac{V_e^2 - V_i^2}{2} + g(z_e - z_i) \tag{3.41}$$

On the other hand, Euler's equation states

$$0 = \int_i^e v \, dp + \frac{V_e^2 - V_i^2}{2} + g(z_e - z_i) \tag{3.42}$$

Equation (3.41) expresses energy relationships in terms of properties of the fluid at the boundaries of the control volume, without regard for properties within the control volume. Euler's equation, on the other hand, expresses the relationship of the fluid flow during the entire course of the process.

Comparison of Eq. (3.41) with Eq. (3.42) indicates that Euler's equation applies only if no work is done. Furthermore, if reversible work is done, then unsteady-state conditions must exist within the control volume although steady-state conditions may prevail at the inlet and exit of the control volume.

In a control volume undergoing a steady-state process, the reversible work is $\int v \, dp$. This assumes that the kinetic changes in the potential energies at the inlet and exit of the control volume are negligible. In a closed system, the reversible work is equal to $-\int p\,dv$. These are represented on a p-v diagram in Figure 3.19, where points 1 and 2 are the end states of a reversible process. Area 1-2-a-b-1 represents the work $-\int p\,dv$ and area 1-2-c-d-1 represents the work $\int v\,dp$. The difference between these two is the flow work resulting from the force of the fluid exerted against existing pressures at the entrance and exit. This relation can be seen from the following:

$$\int_1^2 d(pv) = \int_1^2 p\,dv + \int_1^2 v\,dp$$

Therefore,

$$p_2 v_2 - p_1 v_1 = \int_1^2 v\,dp - \left(-\int_1^2 p\,dv\right) \tag{3.43}$$

The terms in the preceding equation are represented, respectively, by the areas c-2-a-o-c, d-1-b-o-d, c-2-1-d-c, and a-2-1-b-a of Figure 3.19. It is also evident from the figure that $(c$-2-a-o-$c) + (a$-2-1-b-$a) = (d$-1-b-o-$d) + (c$-2-1-d-$c)$ in accordance with the previous equation. In the case of an irreversible compression, the path followed is described in Figure 3.19 by process 1-2′. In such cases, the work required is more than in the reversible process because of such factors as friction and turbulence.

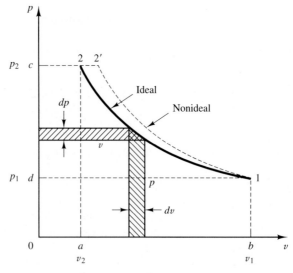

Figure 3.19 $\int v\,dp$ and $-\int p\,dv$.

Example 3.10

Air is compressed reversibly and adiabatically from 100 kPa and 290 K to 600 kPa according to the relation $pv^{1.4} = C$. Neglecting changes in kinetic and potential energies, compute the work done in compression for the following cases: (a) nonflow process; (b) steady-state flow process. Show that the flow work accounts for the difference between (a) and (b). Also compute the work done if the compression were accomplished isothermally at 290 K.

SOLUTION

(a) Referring to Figure 3.19, the specific volume at state 1 is

$$v_1 = \frac{RT_1}{p_1} = \frac{(0.287 \text{ kJ/kg K})(290 \text{ K})}{100 \text{ kPa}} = 0.832 \text{ m}^3\text{/kg}$$

From the process relation,

$$\frac{v_2}{v_1} = \left(\frac{p_1}{p_2}\right)^{1/1.4}$$

the specific volume at state 2 is

$$v_2 = (0.832 \text{ m}^3\text{/kg})\left(\frac{100}{600}\right)^{1/1.4} = 0.231 \text{ m}^3\text{/kg}$$

Also

$$p_1 v_1^{1.4} = (100 \text{ kPa})(0.832 \text{ m}^3\text{/kg})^{1.4} = 77.30$$

The work interaction for the nonflow process is

$$-\int_1^2 p\,dv = -\int_1^2 \frac{77.30}{v^{1.4}}\,dv = 193.25\left[\frac{1}{v^{0.4}}\right]_{v_1}^{v_2}$$

$$= 193.25\left[\left(\frac{1}{0.231}\right)^{0.4} - \left(\frac{1}{0.832}\right)^{0.4}\right] = 139.3 \text{ kJ/kg}$$

(b) The work interaction for the steady-flow process is

$$\int_1^2 v\,dp = \int_1^2 \left(\frac{77.30}{p}\right)^{1/1.4}\,dp = 78.12\left[p^{0.2857}\right]_{p_1}^{p_2}$$

$$= 78.12\left[600^{0.2857} - 100^{0.2857}\right] = 194.7 \text{ kJ/kg}$$

The difference between $\int_1^2 v\,dp$ and $\left(-\int_1^2 p\,dv\right)$ is

$$\int_1^2 v\,dp - \left(-\int_1^2 p\,dv\right) = 194.7 - 139.3 = 55.4 \text{ kJ/kg}$$

and the difference between p_2v_2 and p_1v_1 is

$$p_2v_2 - p_1v_1 = (600 \text{ kPa})(0.231 \text{ m}^3/\text{kg}) - (100 \text{ kPa})(0.832 \text{ m}^3/\text{kg})$$
$$= 138.6 - 83.2 = 55.4 \text{ kJ/kg}$$

which means that the flow work accounts for the difference between the nonflow work and the work in the steady-flow process. If the compression were isothermal, the work transfer per unit mass would be

$$w_{1\text{-}2} = -\int_1^2 p\,dv = \int_1^2 v\,dp = RT \ln\frac{p_2}{p_1}$$
$$= (0.287 \text{ kJ/kg K})(290 \text{ K}) \ln\frac{600}{100} = 149.13 \text{ kJ/kg}$$

3.10 Applications of the Steady-State Flow Energy Equation

In applying the steady-state energy equation to a control volume, it is important to know which forms of energy play dominant roles. Those energies that have only negligible effects can then be ignored relative to other energy transfers, as is illustrated in the following applications.

3.10.1 Heat Exchangers

In heat exchangers such as boilers, condensers, radiators, and evaporators, heat is transferred from a high-temperature stream to a low-temperature stream. In the steam generator shown in Figure 3.20, the chemical energy of the fuel is converted into enthalpy energy by the combustion process. Most of this energy is transferred through the boiler heating surface to water, converting the water to steam. Finally, the gaseous combustion products leave through the stack. The water at the entrance and the steam at the exit differ markedly in enthalpy, but their kinetic and potential energies are small compared to enthalpy changes and are usually neglected. The piping system is ordinarily designed, in this type of

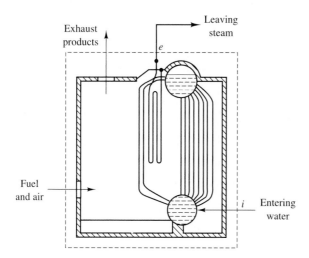

Figure 3.20   Flow through a steam generator.

equipment, for a low velocity flow in order to minimize frictional losses. There-fore, these two types of energy can usually be neglected. Since no work is done on the steam generator, application of the first law leads to

$$q_{i\text{-}e} + \sum h = 0 \qquad \text{or} \qquad q_{i\text{-}e} = h_e - h_i$$

where $q_{i\text{-}e}$ is the net heat transferred to the control volume per unit mass between the inlet and exit.

3.10.2 Turbines and Compressors

A turbine converts the enthalpy of a fluid into work. In the control volume shown in Figure 3.21, some of the energy of the fluid exists as kinetic energy and potential energy. Only minor differences in these forms of energy occur as the fluid passes from section i to section e. Since the heat interaction between the turbine and surroundings is relatively small compared to the enthalpy change and the work interaction, the expansion may be assumed adiabatic so that the first law reduces to

$$w_{i\text{-}e} + \sum h = 0 \qquad \text{or} \qquad w_{i\text{-}e} = (h_e - h_i)$$

Compressors are used to raise the pressure of a fluid while consuming work.

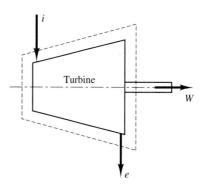

Figure 3.21   Flow through a turbine.

Example 3.11

Steam enters a turbine operating at steady state with a mass rate of flow of 600 kg/hr. The enthalpies at the inlet and exit are $h_i = 3120$ kJ/kg and $h_e = 2330$ kJ/kg, respectively. The velocity of the steam at entrance is $V_i = 100$ m/s and at exit is $V_e = 130$ m/s. If the inlet to the turbines at an elevation 3 m higher than the exit, and the heat transfer from the turbine casing to the surroundings is 630 kJ/hr, find the power developed by the turbine. Assume steady flow conditions.

SOLUTION

The steam turbine is shown in Figure 3.22. Substituting the values given in the problem into the first law, Eq. (3.38), gives

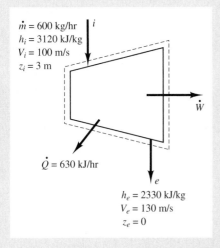

$\dot{m} = 600$ kg/hr
$h_i = 3120$ kJ/kg
$V_i = 100$ m/s
$z_i = 3$ m

$\dot{W}$

$\dot{Q} = 630$ kJ/hr

$h_e = 2330$ kJ/kg
$V_e = 130$ m/s
$z_e = 0$

Figure 3.22

$$\frac{\dot{Q}}{\dot{m}} + \frac{\dot{W}}{\dot{m}} + \left(h + \frac{V^2}{2} + g\,z \right)_i - \left(h + \frac{V^2}{2} + g\,z \right)_e = 0$$

$$-\frac{(630\ \text{kJ/hr})}{(600\ \text{kg/hr})} + \frac{\dot{W}}{(600\ \text{kg/hr})} + \left(3120\ \text{kJ/kg} + \frac{(100\ \text{m/s})^2}{2(1000\ \text{J/kJ})} + \frac{(9.81\ \text{m/s}^2)(3\ \text{m})}{(1000\ \text{J/kJ})} \right)$$

$$- \left(2330\ \text{kJ/kg} + \frac{(130\ \text{m/s})^2}{2(1000\ \text{J/kJ})} + 0 \right) = 0$$

Note that the division by 1000 is to convert the kinetic and potential energy terms to kJ/kg. This equation reduces to

$$-1.05\ \text{kJ/kg} + \frac{\dot{W}}{(600\ \text{kg/hr})} + (3120 + 5 + 0.029)\ \text{kJ/kg} - (2330 + 845 + 0)\ \text{kJ/kg} = 0$$

from which

$$\dot{W} = -471,317 \text{ kJ/hr} = \frac{-471,317 \text{ kJ/hr}}{3600 \text{ s/hr}} = -130.93 \text{ kW}$$

If the potential, kinetic, and heat energies were neglected in these calculations, the power calculated in this example would amount to -131.67 kW. Note the magnitude of these terms is small relative to the enthalpy change or the work output.

Example 3.12

A pump raises the pressure of water in a line by 280 kPa. The pump exit is 6 m above the inlet. Neglecting the changes in internal and kinetic energies between inlet and exit, compute the shaft work needed to drive the pump. (The density of water is 1000 kg/m^3.)

SOLUTION

A sketch of the pump is shown in Figure 3.23. The first law applied to this pump assuming adiabatic steady-state operation reduces to

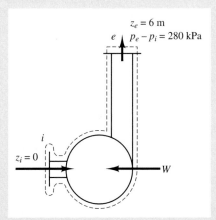

Figure 3.23

$$w_{i\text{-}e} + \sum (pv + gz) = 0$$

$$w_{i\text{-}e} - \frac{(280 \times 10^3 \text{ Pa})}{(1000 \text{ kg/m}^3)} - (6 \text{ m})(9.8067 \text{ m/s}^2) = 0$$

from which

$$w_{i\text{-}e} = 338.84 \text{ J/kg}$$

This is the amount of shaft work needed if the pump were 100 percent efficient. Note that heat interaction with the surroundings has been neglected.

3.10.3 Nozzles and Diffusers

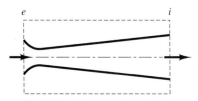

Figure 3.24   Flow through a convergent-divergent nozzle.

A nozzle is a passageway of varying cross-sectional area through which the velocity of a fluid increases as it expands to a lower pressure. A diffuser is a passageway through which the pressure of a fluid increases as its velocity decreases in the direction of flow. In an adiabatic nozzle in which the potential energy can be neglected and the only work interaction is the flow work, enthalpy energy is converted into kinetic energy. For the control volume shown in Figure 3.24, the steady-state energy equation indicates that the enthalpy change is equal to the increase in kinetic energy:

$$\sum \left(h + \frac{V^2}{2} \right) = 0 \qquad \text{or} \qquad \left(h + \frac{V^2}{2} \right)_i - \left(h + \frac{V^2}{2} \right)_e = 0$$

When V_i is small, then

$$V_e = \sqrt{2(h_i - h_e)}$$

Example 3.13

Steam enters a nozzle operating at steady state at a pressure of 2.5 MPa and a temperature of 300°C ($h_i = 3008.8$ kJ/kg) and leaves at a pressure of 1.7 MPa with a velocity of 470 m/s. The rate of flow of steam through the nozzle is 1360 kg/hr. Neglecting the inlet velocity of the steam and considering the flow in the nozzle to be adiabatic, find (a) the enthalpy h_e, (b) the nozzle exit area if $v_e = 0.132$ m³/kg.

SOLUTION

The steam nozzle is shown in Figure 3.25. Assuming one-dimensional flow, properties are uniform across the sections at inlet and outlet of the nozzle. Since the change in potential energy between entrance and exit is small compared with the changes in enthalpy and kinetic energy,

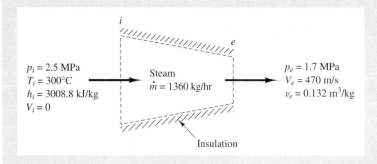

$p_i = 2.5$ MPa
$T_i = 300$°C
$h_i = 3008.8$ kJ/kg
$V_i = 0$

Steam
$\dot{m} = 1360$ kg/hr

$p_e = 1.7$ MPa
$V_e = 470$ m/s
$v_e = 0.132$ m³/kg

Insulation

Figure 3.25

(a) $h_i = h_e + \dfrac{V_e^2}{2}$

Substituting values gives

$$h_e = 3008.8 \text{ kJ/kg} - \frac{(470 \text{ m/s})^2}{2(1000 \text{ J/kJ})} = 2898.4 \text{ kJ/kg}$$

(b) From the continuity equation,

$$A_e = \frac{\dot{m} v_e}{V_e} = \frac{[(1360/3600) \text{ kg/s}](0.132 \text{ m}^3/\text{kg})}{470 \text{ m/s}} = 1.06 \times 10^{-4} \text{m}^2$$

3.10.4 Throttling Devices

Throttling is an irreversible process in which a fluid, flowing across a restriction, undergoes a drop in pressure. Such a process occurs in the flow through a porous plug, a partially closed valve, or a small orifice. Throttling is irreversible, but flow through a frictionless nozzle is internally reversible. Joule and Thomson performed the basic throttling experiments in the period 1852–1862, and their experiments clarified the process and led to the use of throttling as a method for determining certain properties of gaseous substances.

The apparatus used in the Joule–Thomson experiment is shown in Figure 3.26. A steady stream of gas flows through a porous plug contained in a horizontal tube. The control volume is thermally insulated ($q_{i\text{-}e} = 0$), and there is no energy transfer in the form of work other than flow work with the surroundings ($w_{i\text{-}e} = 0$). At sections i and e, both the temperature and the pressure are measured. If the kinetic energy does not change significantly as the fluid passes through the porous plug, the steady-state energy equation reduces to

$$h_i = h_e \tag{3.44}$$

Hence, in an adiabatic throttling process, the enthalpy remains constant. When a series of Joule-Thomson experiments is performed with the same initial temper-

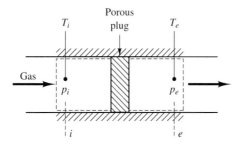

Figure 3.26   Throttling process, the Joule-Thomson porous-plug experiment.

ature T_i and pressure p_i but with a different downstream pressure, it is found that the temperature T_e changes. Results from these experiments can be plotted as a constant-enthalpy curve on a T-p plane. If the rate of flow of the gas is varied at each condition of T_i and p_i, a series of isenthalpic (constant enthalpy) curves is obtained. These are shown in Figure 3.27. The maximum point on each curve is called the *inversion point,* and the locus of the inversion points is called the *inversion curve.* The slope of an isenthalpic curve is called the *Joule–Thomson coefficient* μ_h and is expressed by

$$\mu_h = \left(\frac{\partial T}{\partial p}\right)_h \tag{3.45}$$

At the left side of an inversion point, μ_h is positive; at the right side of an inversion point, μ_h is negative; at the inversion point, μ_h is zero. Since there is always a pressure drop in a throttling process, Δp is always negative. Therefore when μ_h is positive, the temperature change is negative, and throttling produces cooling. Similarly, when μ_h is negative, the temperature change is positive, and throttling produces a rise in temperature even though the process itself is adiabatic. At any particular pressure, μ_h is positive only within a certain temperature range. The limiting temperatures of this range are called the *upper* and *lower* inversion

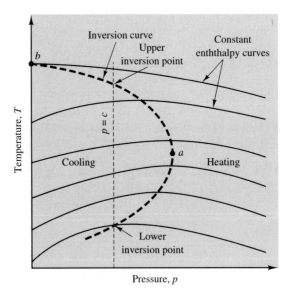

Figure 3.27   Inversion and constant-enthalpy curves of a real gas.

points, and are indicated in Figure 3.27. Between these two temperatures, throttling causes a drop in the temperature of the gas; outside this temperature range, throttling results in a temperature rise. Above a certain pressure, throttling can cause only a heating effect. This is point a in Figure 3.27. Similarly, above a certain temperature throttling can cause only a heating effect. This is point b in Figure 3.27. In the case of an ideal gas, enthalpy is a function of temperature alone, and therefore the temperature of the gas remains constant in a throttling process so that $\mu_h = 0$. In the case of a real gas, the temperature generally does change with pressure, and μ_h is a function of both temperature and pressure.

The Joule–Thomson experiments can be used to determine the specific heat of a gas. Let h be expressed in terms of p and T so that

$$dh = \left(\frac{\partial h}{\partial p}\right)_T dp + \left(\frac{\partial h}{\partial T}\right)_p dT$$

But for an adiabatic throttling process, $dh = 0$, so that

$$0 = \left(\frac{\partial h}{\partial p}\right)_T \left(\frac{\partial p}{\partial T}\right)_h + \left(\frac{\partial h}{\partial T}\right)_p$$

and

$$
\begin{aligned}
c_p &= \left(\frac{\partial h}{\partial T}\right)_p = -\left(\frac{\partial h}{\partial p}\right)_T \left(\frac{\partial p}{\partial T}\right)_h \\
&= -\frac{1}{\mu_h}\left(\frac{\partial h}{\partial p}\right)_T
\end{aligned}
\tag{3.46}
$$

The property $(\partial h / \partial p)_T$ is called the *isothermal Joule–Thomson coefficient*. It can be easily determined by a Joule–Thomson experiment in which the fluid in the test section is maintained at a constant temperature but is subjected to different pressures. The test section is immersed in a constant-temperature bath, and as the pressure is changed, the energy (such as electrical energy) required to maintain the bath at a constant temperature is measured. In this case, the enthalpy does not remain constant, and Δh is given by

$$\Delta h = h_e - h_i = q_{i\text{-}e}$$

where $q_{i\text{-}e}$ is the heat interaction per unit mass between the control volume and the surroundings. As will be shown in Chapter 8, $(\partial h / \partial p)_T$ can be evaluated from measurable quantities. Equation (3.46) can then be used to calculate c_p and consequently other properties, such as internal energy, enthalpy, and specific volume.

3.11 The Carnot Cycle

In this section, the first law is applied to a classical thermodynamic cycle called the Carnot cycle. The Carnot cycle is a sequence of thermodynamically reversible processes carried out between two temperatures. The p-V and the T-V diagrams of a power Carnot cycle using an ideal gas as a working substance are shown in Figure 3.28. The gas, initially at a temperature T_H, and a volume V_a, expands isothermally ($T = $ C) and reversibly to a volume V_b. It then expands reversibly and adiabatically ($pV^\gamma = $ C) from V_b to V_c, as its temperature decreases from T_H to T_L. The gas is then compressed from V_c to V_d reversibly and isothermally at a temperature T_L, and finally the gas is brought back to its initial state by a reversible adiabatic compression from V_d to V_a.

The heat transfer at T_H is

$$Q_{a\text{-}b} = -W_{a\text{-}b} = mRT_H \ln \frac{V_b}{V_a}$$

Where m is the mass of the gas, the heat transfer at T_L is

$$Q_{c\text{-}d} = -W_{c\text{-}d} = mRT_L \ln \frac{V_d}{V_c}$$

From Eq. (3.26)

$$\frac{T_H}{T_L} = \left(\frac{V_d}{V_a}\right)^{R/c_v} = \left(\frac{V_c}{V_b}\right)^{R/c_v} \qquad \text{so that}$$

$$\frac{V_d}{V_a} = \frac{V_c}{V_b} \qquad \text{or} \qquad \frac{V_d}{V_c} = \frac{V_a}{V_b}$$

Hence

$$W_{c\text{-}d} = -mRT_L \ln \frac{V_a}{V_b}$$

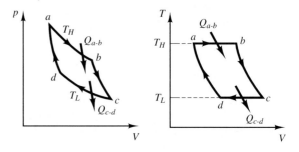

Figure 3.28 *p-V* and *T-V* diagrams for the Carnot cycle.

The work interaction for the cycle is the sum of work done during its four processes:

$$W = W_{a\text{-}b} + W_{b\text{-}c} + W_{c\text{-}d} + W_{d\text{-}a}$$

$$= -mRT_H \ln \frac{V_b}{V_a} + \frac{mR(T_L - T_H)}{\gamma - 1} - mRT_L \ln \frac{V_d}{V_c} + \frac{mR(T_H - T_L)}{\gamma - 1}$$

$$= -mRT_H \ln \frac{V_b}{V_a} - mRT_L \ln \frac{V_d}{V_c} = -(Q_{a\text{-}b} + Q_{c\text{-}d})$$

Hence $\Delta U = 0$ for the cycle, which *fulfills* the conservation of energy principle.

The efficiency of cycle is the ratio of the work done to the heat transfer to the cycle:

$$\eta = \frac{\left| mRT_H \ln \dfrac{V_b}{V_a} + mRT_L \ln \dfrac{V_a}{V_b} \right|}{mRT_H \ln \dfrac{V_b}{V_a}} = \frac{T_H - T_L}{T_H}$$

In the following chapter, the Carnot cycle is discussed in the context of the second law.

3.12 First Law Applied to Steady-State Chemical Systems

Another application of the first law of thermodynamics lies in the analysis of chemical reactions, particularly where steady-state flow at constant pressure occurs. Consider reactants a and b undergoing a chemical process in which they are transformed into products c and d. According to the first law,

$$Q + W + m_a h_a + m_b h_b = m_c h_c + m_d h_d \tag{3.47}$$

Note that changes in kinetic energy and potential energy are assumed to be negligible, and the main energy changes occur in chemical energy, which is included in the enthalpy terms. Because the chemical energies of the various substances must be expressed on a relative basis, all energies are based on an arbitrary reference standard state (25°C and 1 atm). Identifying the reference state by superscript $^\circ$, Eq. (3.47) becomes

$$Q + W + m_a(h - h^\circ)_a + m_b(h - h^\circ)_b \tag{3.48}$$
$$= m_c(h - h^\circ)_c + m_d(h - h^\circ)_d + [(m_c h_c^\circ + m_d h_d^\circ) - (m_a h_a^\circ + m_b h_b^\circ)]$$

where $(h - h^\circ)_i$ represents sensible energy of species i that is due to temperature, and h_i° represents its chemical energy at the reference state. The latter term in square brackets of Eq. (3.48) represents the enthalpy of reaction at the reference state. It can be determined by calorimetric measurements and is given the symbol ΔH° so that

$$\Delta H^\circ = [(m_c h_c^\circ + m_d h_d^\circ) - (m_a h_a^\circ + m_b h_b^\circ)]$$

Equation (3.48) then becomes

$$Q + W + m_a(h - h^\circ)_a + m_b(h - h^\circ)_b + (-\Delta H^\circ)$$
$$= m_c(h - h^\circ)_c + m_d(h - h^\circ)_d \tag{3.49}$$

where ΔH° is the *enthalpy of reaction* or, in the case of combustion of a fuel, *enthalpy of combustion* at the standard state. The enthalpy of combustion represents the energy evolved in the complete oxidation of a fuel, provided that both reactants and products are at the standard state. The constant-pressure heating value of a fuel (defined in Chapter 10) and the enthalpy of combustion have the same magnitude but are of opposite sign.

When chemical reactions occur at constant volume, rather than at constant pressure, a similar analysis can be applied. From the first law of thermodynamics,

$$Q + W + m_a(u - u^\circ)_a + m_b(u - u^\circ)_b + (-\Delta U^\circ)$$
$$= m_c(u - u^\circ)_c + m_d(u - u^\circ)_d \tag{3.50}$$

where ΔU° is called the *internal energy of reaction*, or the *internal energy of combustion* at the standard state. The constant-volume heating value of a fuel (defined in Chapter 10) and the internal energy of combustion are of the same magnitude but of opposite sign.

The internal energy of reaction is related to the enthalpy of reaction as follows:

$$\Delta H = \Delta U + \Delta(pV)$$

and, at the reference state,

$$(\Delta H^\circ) = (\Delta U^\circ) + \Delta(pV)^\circ$$

Using subscripts R and P to refer to reactants and products, this may be written as

$$(\Delta H^\circ)_{RP} = (\Delta U^\circ)_{RP} + [(pV)_P^\circ - (pV)_R^\circ] \tag{3.51}$$

Since the volume occupied by solids and liquids is negligible compared with gases, the last bracket in Eq. (3.51) would represent only the volumes of gaseous products and reactants when a combustion process is involved. Furthermore, if the gaseous reactants and products follow the ideal-gas law then,

$$(pV)_R^\circ = n_R \bar{R} T_0$$
$$(pV)_P^\circ = n_P \bar{R} T_0$$

Equation (3.51), expressed on the basis of 1 mole of fuel, then becomes

$$(\Delta \bar{h}^\circ)_{RP} = (\Delta \bar{u}^\circ)_{RP} + (n_P - n_R) \bar{R} T_0 \qquad (3.52)$$

where

$$(\Delta \bar{h}^\circ)_{RP} = \text{enthalpy of combustion per mole of fuel at the reference state}$$
$$(\Delta \bar{u}^\circ)_{RP} = \text{internal energy of combustion per mole of fuel at the reference state}$$
$$n_P = \text{number of moles of products per mole of fuel}$$
$$n_R = \text{number of moles of reactants per mole of fuel}$$

Equation (3.52) can also be expressed per unit mass of fuel:

$$(\Delta h^\circ)_{RP} = (\Delta u^\circ)_{RP} + (n_P - n_R) \frac{\bar{R} T_0}{M_f} \qquad (3.53)$$

where M_f is the molar mass of the fuel. Chemical processes are discussed in more detail in Chapter 10.

Example 3.14

A fuel has an enthalpy of combustion ΔH° of $-44{,}200$ kJ/kg at a reference temperature of 25°C. The fuel is supplied to a burner at 25°C with 14 kg of air at 40°C per kg of fuel. If the products of combustion leave at a temperature of 300°C, what is the amount of heat transfer per kg of fuel? Assume steady flow. Enthalpy values are as follows:

Temperature (°C)	Enthalpy of Air (kJ/kg)	Enthalpy of Products (kJ/kg)
25	289.19	303.5
40	313.26	322.6
300	578.71	612.9

SOLUTION

A sketch of the combustion chamber is shown in Figure 3.29. For one kilogram of fuel, substituting in Eq. (3.49) yields

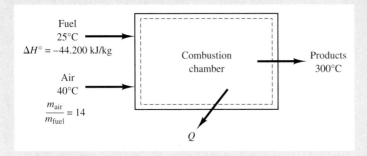

Fuel
25°C
$\Delta H° = -44.200$ kJ/kg

Air
40°C
$\dfrac{m_{air}}{m_{fuel}} = 14$

Combustion
chamber

Products
300°C

Q

Figure 3.29

$$Q + 0 + (14 \text{ kg})[(313.26 - 298.19) \text{ kJ/kg}] + (44{,}200 \text{ kJ}) = [(14 + 1) \text{ kg}][(612.9 - 303.5) \text{ kJ/kg}]$$

from which

$$Q = -39.770 \text{ kJ (per kg of fuel)}$$

Example 3.15

The energy of combustion of n-heptane (C_7H_{16}) at constant volume and 25°C is −47947 kJ/kg. Find the enthalpy of combustion at the standard state.

SOLUTION

The chemical equation may be written as

$$C_7H_{16}(l) + 11O_2(g) \longrightarrow 7CO_2(g) + 8H_2O(l)$$

Since $\Delta u° = -47947$ kJ/kg, Eq. (3.53) gives

$$(\Delta h°)_{RP} = (\Delta u°)_{RP} + (n_P - n_R)\frac{\overline{R}T_0}{M_f}$$

$$= -47947 \text{ kJ/kg} + (7 - 11)\frac{(8.3144 \text{ kJ/kg-mol K})(298.15 \text{ K})}{100 \text{ kg/kg-mol}}$$

$$= -48071 \text{ kJ/kg}$$

3.13 Unsteady or Transient Flow Processes

Transient flow phenomena are encountered during the startup or shutdown of equipment such as turbines and compressors. Other examples occur when a fluid fills or is withdrawn from a vessel. In a transient flow process, both the mass and the state of the fluid within a control volume change with time. Consider the flow of gas through a pipeline and into a vessel as shown in Figure 3.30. Let subscripts 1 and 2 refer to the initial and final conditions in the vessel, respectively. While the gas is filling the vessel, conditions in the pipeline, denoted by subscript p, remain constant.

We present two methods of solution: (a) a system analysis and (b) a control-volume analysis. First consider a system so that no mass crosses the boundaries of the system. The system then includes not only the vessel but also that portion of the fluid in the pipeline that eventually will be introduced into the vessel. This system has variable boundaries, and in the final state, the boundaries are the same as those of the vessel. According to the first law of thermodynamics,

$$Q_{1\text{-}2} + W_{1\text{-}2} = (E_2 - E_1)$$

The initial energy of the system, E_1, consists of the energy of the mass initially in the vessel plus the energy of fluid that will eventually flow from the pipe into the vessel. Since changes in potential energies can be neglected, the change in internal energy is

$$\Delta E = E_2 - E_1 = m_2 u_2 - \left[m_1 u_1 + (m_2 - m_1)\left(u_p + \frac{V_p^2}{2}\right)\right]$$

where subscript p refers to pipeline.

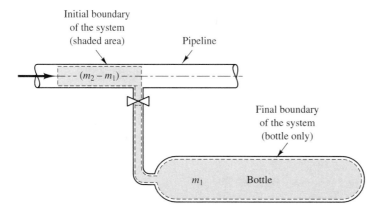

Figure 3.30   Unsteady flow process (fixed-mass system).

Work must be done on the system to introduce the mass $(m_2 - m_1)$ into the vessel, thus reducing the volume of the system to its final state, that is, the volume of the vessel. The mass in the pipeline, $(m_2 - m_1)$, is subjected to constant pressure, p_p, and therefore the work of compression is

$$W_{1\text{-}2} = -p_p \, \Delta V_p = p_p(m_2 - m_1)v_p$$

where ΔV_p is the change in the volume of the system, and v_p is the specific volume of the fluid in the pipeline. Substituting these expressions for ΔE and $W_{1\text{-}2}$ in the first law, yields

$$Q_{1\text{-}2} + p_p(m_2 - m_1)v_p = m_2u_2 - m_1u_1 - (m_2 - m_1)\left(u_p + \frac{V_p^2}{2}\right)$$

Since $h_p = u_p + p_p v_p$,

$$Q_{1\text{-}2} = m_2u_2 - m_1u_1 - (m_2 - m_1)\left(h_p + \frac{V_p^2}{2}\right) \tag{3.54}$$

The problem may also be treated as a control volume, as shown in Figure 3.31. The first law, according to Eq. (3.33b), takes the form,

$$Q + W + \sum m\left(h + \frac{V^2}{2} + gz\right) = \Delta E_{cv}$$

or expressed verbally:

transitional energy (W and Q)	+	net energy entering the vessel and associated with the entering matter	=	change of energy in the control volume

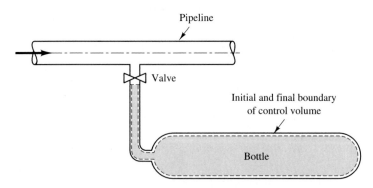

Figure 3.31　Unsteady flow process (control volume).

In this case, there is no work interaction, and the first law reduces to

$$\underbrace{Q_{1\text{-}2} + (m_2 - m_1)\left(h_p + \frac{V_p^2}{2}\right)}_{\text{energy transfer}} = \underbrace{m_2 u_2 - m_1 u_1}_{\text{energy change}}$$

Here, too, the potential energies have been considered negligible. Rearranging the preceding equation gives

$$Q_{1\text{-}2} = m_2 u_2 - m_1 u_1 - (m_2 - m_1)\left(h_p + \frac{V_p^2}{2}\right)$$

which is the same result obtained previously.

Example 3.16

An evacuated chamber is connected through a valve with a large pipeline containing steam at 350 kPa and 200°C ($h = 2863.1$ kJ/kg). The valve is opened, and steam flows into the chamber until the pressure within the chamber is 350 kPa. At the same time heat is transferred from the chamber at the rate of 700 kJ/kg of steam introduced. Neglecting the kinetic and potential energies of the steam, determine the final specific internal energy of the steam in the chamber.

SOLUTION

A sketch of the chamber is shown in Figure 3.32. Treating this as a system or as a control volume, Eq. (3.54) gives

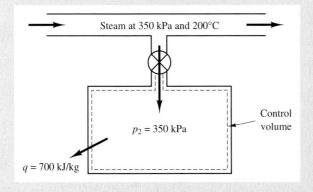

Steam at 350 kPa and 200°C

$p_2 = 350$ kPa

Control volume

$q = 700$ kJ/kg

Figure 3.32

$$m q_{1\text{-}2} = m u_2 - 0 - m(h_p + 0)$$

Dividing by m,

$$q_{1\text{-}2} = u_2 - h_p$$

Substituting values in the preceding equation gives

$$-700 \text{ kJ/kg} = u_2 - 2863.1 \text{ kJ/kg} \qquad \text{or} \qquad u_2 = 2163.1 \text{ kJ/kg}$$

Example 3.17

An insulated tank of volume V_1 (see Figure 3.33) contains an ideal gas at temperature T_1 and pressure p_1. The tank is connected by a valve to a surroundings at a pressure p_2. If the valve is opened and the pressure in the tank drops to that of the surroundings, find the ratio of the final to the initial temperature in the tank. Neglect changes in kinetic and potential energy, and solve the problem by considering the system as a closed system, then as a control volume.

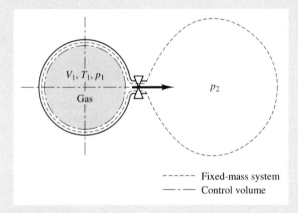

Figure 3.33

SOLUTION

(a) System: Since the tank is insulated, assume no heat interaction so that the first law can be written as

$$\delta W = dU$$

Noting that the surroundings pressure p_2 remains constant, the work interaction when the gaseous system expands from p_1 and V_1 to p_2 and V_2 is

$$-p_2(V_2 - V_1) = mc_v(T_2 - T_1)$$

But

$$m = \frac{p_1 V_1}{R T_1}$$

and

$$c_v = \frac{R}{\gamma - 1}$$

where

$$\gamma = \frac{c_p}{c_v}$$

therefore,

$$-p_2 V_1 \left(\frac{V_2}{V_1} - 1 \right) = \frac{p_1 V_1}{R T_1} \frac{R}{\gamma - 1} (T_2 - T_1)$$

$$-\frac{p_2}{p_1} \left(\frac{p_1 T_2}{T_1 p_2} - 1 \right) = \frac{1}{\gamma - 1} \left(\frac{T_2}{T_1} - 1 \right)$$

$$-\frac{T_2}{T_1} + \frac{p_2}{p_1} = \frac{1}{\gamma - 1} \left(\frac{T_2}{T_1} - 1 \right)$$

$$\frac{T_2}{T_1} \left(\frac{1}{\gamma - 1} + 1 \right) = \frac{p_2}{p_1} + \frac{1}{\gamma - 1}$$

$$\frac{T_2}{T_1} = \frac{\gamma - 1}{\gamma} \left[\frac{p_2}{p_1} + \frac{1}{\gamma - 1} \right]$$

(b) Control volume: Energy balance indicates that the energy leaving the control volume must be equal to the decrease in energy of the control volume. The first law then becomes

$$-h(-dm) = dU$$

Noting that $U = um$, then

$$h \, dm = m \, du + u \, dm$$

But

$$h = u + pv = u + RT$$

Therefore,

$$(u + RT)dm = m\, du + u\, dm$$

or

$$RT\, dm = m\, du \qquad \text{and} \qquad \frac{dm}{m} = \frac{c_v\, dT}{RT}$$

Integrating,

$$\ln \frac{m_2}{m_1} = \frac{c_v}{R} \ln \frac{T_2}{T_1}$$

But

$$m_1 = \frac{p_1 V}{RT_1} \qquad \text{and} \qquad m_2 = \frac{p_2 V}{RT_2}$$

Therefore,

$$\ln \frac{p_2 T_1}{T_2 p_1} = \frac{c_v}{R} \ln \frac{T_2}{T_1}$$

or

$$\frac{p_2 T_1}{p_1 T_2} = \left(\frac{T_2}{T_1}\right)^{c_v/R} = \left(\frac{T_2}{T_1}\right)^{(1/\gamma - 1)}$$

so that

$$\frac{T_2}{T_1} = \left(\frac{p_2}{p_1}\right)^{(\gamma - 1)/\gamma} \qquad \text{(reversible adiabatic relation)}$$

Evidently, the results obtained in treating the problem as a closed system differ from those obtained by the control-volume approach. The closed system method is based on the assumption that equilibrium is maintained between the gas in the tank and the gas outside meaning that T_2 is the same for the gas inside and outside the tank. This is nonrealistic because the process occurs very rapidly and because heat interaction takes place between the surroundings and the gaseous system. Also the heat interaction between the gas leaving the tank and the gas in the tank takes place through a pipe of small diameter so that the temperature of the gas in the tank is different from the temperature of the gas leaving the control volume. The two results are, however, not too dissimilar. For, if the reversible adiabatic relation, obtained by the control volume procedure, is expanded by the binomial series, we obtain

$$\frac{T_2}{T_1} = \left(\frac{p_2}{p_1}\right)^{(\gamma-1)/\gamma} = \left(1 - \frac{p_1 - p_2}{p_1}\right)^{(\gamma-1)/\gamma}$$

$$= \left[1 - \frac{\gamma - 1}{\gamma}\left(\frac{p_1 - p_2}{p_1}\right) - \frac{\gamma - 1}{2!\,\gamma^2}\left(\frac{p_1 - p_2}{p_1}\right)^2 - \cdots\right]$$

$$= \frac{\gamma - 1}{\gamma}\left[\frac{1}{\gamma - 1} + \frac{p_2}{p_1} - \frac{1}{2!\,\gamma}\left(\frac{p_1 - p_2}{p_1}\right)^2 - \cdots\right]$$

The first two terms in this expansion are the same as the result obtained by the system approach.

3.14 Comparison of the Steady-State Energy Equation for a Control Volume with the Equations of Motion of Fluid Mechanics

The steady-state energy equation derived in this chapter is a general equation. But Euler's equation and Bernoulli's equation, which were derived in Chapter 2, apply only under restricted conditions. Euler's equation was derived by applying the momentum principle to a fluid particle along a streamline and is

$$v\,dp + V\,dV + g\,dz = 0 \qquad (3.55)$$

This equation applies to flow under steady-state conditions, in which there is no shaft work, no friction, and no electrical, magnetic, or capillary forces.
  Bernoulli's equation is

$$p_i v + \frac{V_i^2}{2} + gz_i = p_e v + \frac{V_e^2}{2} + gz_e \qquad (3.56)$$

The same restrictions that apply to Euler's equation also apply to Bernoulli's equation; in addition, Bernoulli's equation applies only to the flow of incompressible fluid along a streamline.
  The steady-state energy equation for a control volume is

$$q_{i\text{-}e} + w_{i\text{-}e} + \left(p_i v_i + u_i + \frac{V_i^2}{2} + gz_i\right) = \left(p_e v_e + u_e + \frac{V_e^2}{2} + gz_e\right) \qquad (3.57)$$

Although all the terms in Bernoulli's equation (3.56) appear in the steady-state equation (3.57), Bernoulli's equation is not a special case of the steady-state energy equation. Bernoulli's equation, which applies to reversible incompressible flow, is based on Newton's second law of motion and is independent of the first law of thermodynamics.

The differential form of the steady-state energy equation is

$$\delta q + \delta w = p \, dv + v \, dp + du + V \, dV + g \, dz \tag{3.58}$$

The steady-state equation may be applied to a control volume in which the restrictions specified by Euler's equation exist. Then, when Eq. (3.55) is subtracted from Eq. (3.58) and when δw is set equal to zero, since Euler's equation requires that no shaft work be done, the following is obtained:

$$\delta q = p \, dv + du$$

This is the first law of thermodynamics applied to a system consisting of a one-component substance in which electrical, magnetic, and capillary effects are absent. The only work is the reversible $-p \, dv$ work.

The difference in total head between two sections in the flow due to frictional effects is

$$\left(p_i v + \frac{V_i^2}{2} + gz_i \right) - \left(p_e v + \frac{V_e^2}{2} + gz_e \right) = g \, H_{\ell_{i\text{-}e}} \tag{3.59}$$

where $H_{\ell_{i\text{-}e}}$ represents the total head loss due to energy dispersion between sections i and e. If the foregoing equation is subtracted from the steady-state equation (3.57), then, in the absence of q and w, the following is obtained:

$$g \, H_{\ell_{i\text{-}e}} = u_e - u_i \tag{3.60}$$

In the steady adiabatic flow of an incompressible fluid, energy dispersion due to friction, turbulence, and so on results in an increase in the internal energy of the fluid.

3.15 Summary

The first law for a system is

$$Q + W = \Delta E$$

where

$$\Delta E = E_2 - E_1 = (U_2 - U_1) + \frac{1}{2} m (V_2^2 - V_1^2) + mg(z_2 - z_1)$$

The first law for a control volume is

$$Q + W + \sum_{net=in-out} m\left(h + \frac{V^2}{2} + gz\right) = \Delta E$$

The left-hand side of both equations is "energy transfer," and the right-hand side is "energy change." The enthalpy term includes the work associated with the flow across the boundary of the control volume. For steady flow, $\Delta E = 0$. In addition, $\sum m_i = \sum m_e$ for a control volume.

The work interaction in a reversible steady-state flow process is

$$w_{rev} = \int_i^e v \, dp + \frac{V_e^2 - V_e^2}{2} + g(z_e - z_i)$$

For a closed system, the reversible work is $-\int p \, dv$. The difference between $\int v \, dp$ and $-\int p \, dv$ is the flow work.

For unsteady flow process, the first law in the absence of work and potential energy is

$$\underbrace{Q + (m_2 - m_1)\left(h_p + \frac{V_p^2}{2}\right)}_{\text{energy transfer}} = \underbrace{m_2 u_2 - m_1 u_1}_{\text{energy change}}$$

For an ideal gas, the internal energy, enthalpy, and the specific heats at constant volume and constant pressure are all a function of temperature only so that

$$du = c_v \, dT \quad \text{and} \quad dh = c_p \, dT$$

also,

$$c_v = \frac{R}{\gamma - 1} \qquad c_p = \frac{\gamma R}{\gamma - 1} \quad \text{and} \quad c_p - c_v = R$$

Problems

3.1 A paddle wheel supplies work at the rate of 0.75 kW to a system, as shown in Figure 3.34. During a period of 1 min, the system expands in volume from 0.03 m³ to 0.09 m³ while the pressure remains constant at 500 kPa. Find the net work interaction during this 1-min period.

3.2 Five kg of an ideal gas are heated by supplying 180 kJ. During the process, the volume is held constant at 4 m³ and the pressure increases from 100 kPa to 120 kPa. Compute (a) the work done; (b) internal energy change of the gas; (c) density of the gas before and after the process.

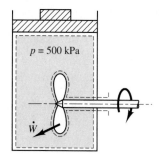

$p = 500$ kPa

$\dot{W}$

Figure 3.34

3.3   The pressure of a gas in a piston-cylinder assembly is given by the expression $p = 9V^2$; $p =$ pressure in kPa, and $V =$ total volume in m³. (a) Find the work interaction if the volume increases from 0.15 m³ to 0.3 m³ in a reversible manner. (b) If 7 kJ of heat are transferred to the gas in the cylinder, what is the change of internal energy?

3.4   In a reversible process, a gas expands from an initial volume of 0.2 m³ to a final volume of 0.4 m³. If the internal energy of the system decreases 30 kJ during the process, find the heat transferred. The relation between pressure and volume during the process is $p = 800 - 1500V$ (p is in kPa, V in m³).

3.5   The latent enthalpy of vaporization of water at 275 kPa is 2172.4 kJ/kg, and the volume changes from 0.001 m³/kg to 0.657 m³/kg. What is the change in internal energy due to evaporation?

3.6   The specific heat at constant volume, c_v, is given by the expression $c_v = A + BT$, where A and B are constants. Suppose that in general c_v is a function of both temperature and pressure. Determine whether the foregoing expression is a path or a point function by means of the properties of an exact differential.

3.7   A cylinder with a vertical axis has a frictionless piston at its upper end. The piston weight is such as to exert a pressure of 350 kPa on the 0.5 kg of gas enclosed. When the gas is cooled, its volume decreases from 0.9 to 0.3m³ and its internal energy decreases from 120 to 50 kJ/kg. Compute (a) work interaction; (b) heat interaction.

3.8   Calculate the heat interaction, the work interaction, and the change in internal energy when one m³ of air at 150 kPa and 20°C undergoes a constant-pressure process until its temperature reaches 320°C.

3.9   Helium contained in a cylinder fitted with a piston expands reversibly according to the relation $pV^{1.5} =$ constant. The initial volume of the helium is 0.1 m³, the initial pressure is 500 kPa, and the initial temperature is 200 K. After expansion, the pressure is 200 kPa. Calculate (a) the work done during expansion and (b) the heat transfer during expansion. For helium, the following values apply:

$$\text{molar mass} = 4.00$$
$$c_p = 5.1926 \text{ kJ/kg K}$$
$$\gamma = 1.66$$

3.10   Air in a piston-cylinder assembly expands from an initial volume, pressure, and temperature of 0.06 m³, 3 MPa, and 700 K to a final volume of 0.2 m³ and a final temperature of 400 K according to the law $pv^n =$ C. Find (a) the exponent n; (b) the final pressure; (c) the work done; (d) the heat transfer.

3.11   One kg of air is compressed in a piston-cylinder assembly from 80 kPa and 300 K to 120 kPa according to the relation $p(v + 0.2) =$ constant, where p is in kPa and v in m³/kg. Calculate the heat transfer. Assume air to be an ideal gas with constant specific heats.

3.12   One kilogram of air at 20°C and 100 kPa is compressed to one-tenth of its original volume. Assuming quasi-equilibrium process, determine the work and heat interaction if the compression is according to the relation $pv^{1.2} =$ C. Assume air to be an ideal gas with constant specific heats.

3.13   0.1 kg of helium ($M = 4$) initially at $p_1 = 100$ kPa and a specific volume $v_1 = 5.8$ m³/kg is compressed quasi-statically to 200 kPa according to the relation $pv^{1.3} =$ constant. The internal energy of helium in kJ/kg is given by $u = 1.5 pv$, where p is in kPa, and v in m³/kg. Determine the work interaction and the heat transfer during the compression process.

3.14 A gas at an initial pressure of 80 kPa and an initial volume of 2 m³ is compressed in a quasi-equilibrium process to a final pressure of 200 kPa. During the compression process, the product pV = constant.
(a) Determine the work done on the gas.
(b) If the gas were an ideal gas, what would be the change in internal energy and the heat transfer?

3.15 One kg of an ideal gas with c_p = 1.042 kJ/kg K and c_v = 0.743 kJ/kg K, initially at a pressure of 350 kPa and occupying a volume of 0.4 m³, undergoes a change of state during which the pressure and volume are both doubled. Find the change in internal energy.

3.16 The internal energy per kg for a certain gas is given by $u = 0.17T + C$, where u is in kJ/kg, T is K, and C is a constant. If the gas is heated in a rigid container from a temperature of 40°C to 315°C, compute the work and the heat interactions per kg.

3.17 Air initially at 40°C and 100 kPa is compressed reversibly and isothermally in a nonflow process to a final pressure of 700 kPa. Find (a) final temperature; (b) work required per kg of air; (c) heat transferred per kg of air.

3.18 The p-v-u relation for a certain gas is given by the equation

$$u = 831 + 0.617pv$$

where u is in kJ/kg, p is in kPa, and v is in m³/kg. Find the work done and the heat transferred to 1 kg of this gas in the following reversible, nonflow processes:

	From		To	
Process	p (kPa)	v (m³/kg)	p (kPa)	v (m³/kg)
(a) Constant pressure	850	0.25	850	0.5
(b) Constant internal energy	1380	0.125	690	—

Find the work done in any nonflow process between the same end states as in process (b) if the heat transfer is 25 kJ/kg.

3.19 One kilogram of air expands adiabatically in a piston-cylinder assembly. If the temperature of the air changes during the expansion from 25°C to 10°C, compute the work done on a frictionless piston during the process (c_v = 0.7165 kJ/kg K).

3.20 In a reversible nonflow process in which the pressure is maintained constant, the heat interaction is equal to (check the correct answer): (a) Δu; (b) Δh; (c) W; (d) $p\Delta V$.

3.21 The gas of a system is subjected to a process in which the pressure and volume change according to the relative $pV^{1.4}$ = C. The initial pressure and volume are, respectively, 700 kPa and 0.09 m³, and the final pressure is 140 kPa. (a) Find the $-p\,dV$ work for this process. (b) Would this be the work for a real process? Why?

3.22 Air expands against a piston in such a way that the p-v relation is described by a circular arc (one-quarter of a circle), as shown in Figure 3.35. If the heat interaction during the expansion process is 0.5 kJ, find the work done by the air and the change in internal energy.

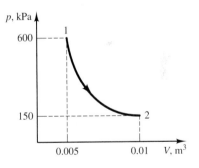

Figure 3.35

3.23 An uninsulated frictionless piston is initially constrained in a rigid insulated cylinder. As shown in Figure 3.36, air (assumed to be ideal gas) is trapped on both sides of the piston. The piston is freed, and equilibrium is reestablished. (a) What is the final pressure? (b) If the system boundary is imagined to pass through the piston, will heat be transferred? Explain.

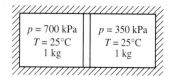

Figure 3.36

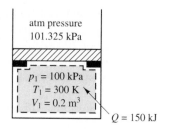

Figure 3.38

3.24 A well-insulated, frictionless piston-cylinder assembly contains 0.5 kg of air initially at 75°C and 300 kPa. An electric-resistance heating element inside the cylinder is energized and causes the air temperature to reach 150°C. The pressure of the air is maintained constant throughout the process. Determine the expansion work and the amount of electrical energy input. Assume air to be an ideal gas with constant specific heats.

3.25 Air at a pressure of 400 kPa and a temperature of 500 K occupies a volume of 0.01 m³ in a piston-cylinder assembly as shown in Figure 3.37. The surrounding pressure is 1 atm (101.325 kPa). The air suddenly expands to double its volume. If the expansion is adiabatic, calculate the final pressure and temperature of the air.

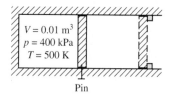

Pin

Figure 3.37

3.26 Nitrogen initially at 100 kPa, 250 K, and 0.5 m³ is compressed reversibly in a piston-cylinder assembly to 500 kPa and 450 K. If the compression process is polytropic ($pV^n = C$), determine the heat interaction. Assume nitrogen is an ideal gas with constant specific heats.

3.27 Air initially at a pressure of 100 kPa and a temperature of 300 K is contained in a piston-cyclinder assembly, as shown in Figure 3.38. The cylinder is open to the atmosphere at the top, and the piston rests on stops such that the air volume is 0.2 m³. The piston is frictionless, has an area of 0.1 m² and weighs 12 kN. Determine the final temperature and the work interaction when 150

kJ of heat is transferred to the system. Assume air to be an ideal gas with constant specific heats.

3.28 A vertical piston-cylinder arrangement contains air at 150 kPa and 300 K. Initially, the piston is resting on a pair of stops, and the enclosed volume is 0.4 m³. The mass of the piston is such that a 350 kPa pressure is required to move it. If the air is heated until its volume doubles, determine the final temperature, and the work and heat interactions.

3.29 Air undergoes a reversible thermodynamic cycle 1-2-3-1, as shown in Figure 3.39. Process 1-2 is isothermal expansion, 2-3 is constant pressure, and process 3-1 is constant volume. Determine the work and the heat interactions and the change of internal energy for each process and for the entire cycle.

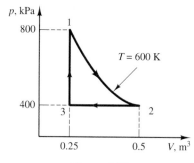

Figure 3.39

3.30 One kilogram of helium undergoes a three-process reversible cycle, as shown in Figure 3.40. Process 1-2 is constant volume, process 2-3 is constant pressure, and process 3-1 is constant temperature. Assuming helium to be an ideal gas, determine (a) the pressure, temperature,

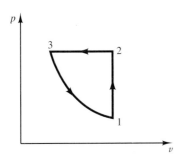

Figure 3.40

and volume at states 1, 2, and 3; (b) the work done during the cycle; (c) the heat transfer during process 3-1.

Data:

$$p_1 = 100 \text{ kPa} \qquad T_1 = 300 \text{ K} \qquad \frac{v_1}{v_3} = 5$$

$$c_v = 3.1156 \text{ kJ/kg K} \quad R = 2.077 \text{kJ/kg K}$$

3.31 One kilogram of air initially at 25°C and 100 kPa undergoes a cycle made up of three processes. The air is first compressed reversibly and adiabatically from its initial state to 0.5 MPa according to the relation $pv^{1.4} = C$. It then receives heat in a constant-pressure process. The cycle is completed by a constant-volume process. Determine Q, W, and ΔU for each process of the cycle. What is the net work output and the thermal efficiency of the cycle? Assume air to be an ideal gas with constant specific heats.

3.32 Figure 3.41 shows a quasi-static cycle executed by 0.5 kg of air. Determine the work, the heat transfer,

and the change of internal energy for processes 1-2, 2-3, and 3-1 and for the entire cycle. Assume air to be an ideal gas with $c_v = 0.7165$ kJ/kg K and $c_p = 1.0035$ kJ/kg K.

3.33 A thermodynamic cycle is composed of the following three quasi-equilibrium processes: isothermal compression (1-2), constant-volume heating (2-3), and a polytropic process with $n = 1.45$. The system is one kilogram of helium initially at 100 kPa and 20°C. During the compression process 300 kJ of energy is transferred from the system. Represent the cycle on a p-V diagram and determine (a) p, V, and T at states 2 and 3, (b) W and Q for each process.

3.34 A thermodynamic cycle uses 1 kg of air and is composed of the following three reversible processes: a polytropic compression ($pv^{1.2} = C$) from state 1, where $p_1 = 120$ kPa and $T_1 = 350$ K, to state 2, where $p_2 = 600$ kPa; a constant-pressure cooling from state 2 to state 3; and a constant temperature heating from state 3 to state 1. (a) Represent the cycle on a p-v diagram. (b) Determine the temperature, pressure, and volume at the end of each process of the cycle. (c) Determine the work and heat for each process. Tabulate your results.

Assume air is an ideal gas with constant specific heats ($c_p = 1.0035$ kJ/kg K and $c_v = 0.7165$ kJ/kg K).

3.35 One kg of air is used as a working fluid in a thermodynamic cycle that consists of three irreversible processes, as shown in Figure 3.42. Process 1-2 is a constant-pressure process, process 2-3 is a polytropic process

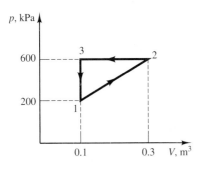

Figure 3.41

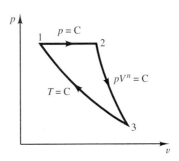

Figure 3.42

(pV^n = constant), and process 3-1 is at constant temperature. The following information is given:

work: $w_{1-2} = -120$ kJ/kg $w_{2-3} = -80$ kJ/kg
         and $w_{3-1} = 50$ kJ/kg

heat: $q_{1-2} = 300$ kJ/kg

Assuming the air to be treated as an ideal gas, determine the heat interaction in process 2-3 and in process 3-1.

3.36   A power cycle using an ideal gas is made up of three processes. Process 1-2 is a constant-volume process where heat is supplied and the temperature rises from T_1 to T_2. Process 2-3 is a reversible adiabatic expansion, and process 3-1 is a reversible isothermal process. Prove that the efficiency of this cycle is

$$\eta = 1 - \frac{T_1}{T_2 - T_1} \ln \frac{T_2}{T_1}$$

3.37   A rigid tank with a volume of 0.1 m³ is connected to a cylinder fitted with a frictionless piston by a pipe-and-valve arrangement, as shown in Figure 3.43. Initially, the valve is shut and the air pressure and temperature in the tank are 350 kPa and 300°C. Initially, the air volume in the cylinder is 0.05 m³, the pressure is 100 kPa, and the temperature is 200°C. An air pressure of 100 kPa is required to support the weight of the piston and the external ambient pressure. The valve is then opened, and air flows from the tank to the cylinder until the pressure and temperature are uniform throughout at 100 kPa and 50°C, respectively. If the temperature of the surroundings is 20°C, determine the work and heat interactions with the surroundings.

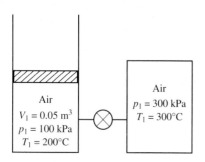

Figure 3.43

3.38   Two chambers A and B each of volume 0.01 m³ containing air are separated by a fixed-heat conducting wall, as shown. A piston performs work on chamber B resulting in heat transfer between the two chambers. The initial conditions are shown in Figure 3.44.

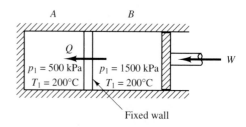

Figure 3.44

If the whole system is adiabatic and the final pressure in chamber A is 1600 kPa, determine (a) the final temperature, (b) the heat interaction between the two chambers, (c) the work done.

3.39   An insulated frictionless piston separates two masses of air (1 kg each) in a rigid tank. The initial pressure and temperature in each side are equal, as shown in Figure 3.45. The tank is insulated except for the left end. When an amount of heat of 100 kJ is transferred through the tank wall, the temperature in side A becomes 400 K. Determine the final pressure and temperature of the air in side B ($c_v = 0.7165$ kJ/kg K, $R = 0.287$ kJ/kg K).

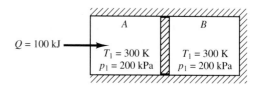

Figure 3.45

3.40   A vertical cylinder fitted with a frictionless piston contains 1 kg of air at a pressure of 400 kPa and 300 K. Stops restrict the position of the piston to a maximum cylinder volume of 0.4 m³. Determine the work interaction and the change in internal energy if the heat input is (a) 200 kJ, (b) 300 kJ.

3.41 For a reversible polytropic process of an ideal gas, prove that the heat interaction is given by

$$Q = mc_v \left(\frac{\gamma - n}{1 - n} \right)(T_2 - T_1)$$

where $\gamma = \dfrac{c_p}{c_v}$ and n is the polytropic exponent.

3.42 A thermodynamic cycle uses 0.05 kg of air and is composed of the following three reversible processes: an adiabatic compression from state 1, where $p_1 = 100$ kPa and $V_1 = 0.03$ m^3, to state 2, where $V_2 = 0.003$ m^3; an isothermal expansion from state 2 to state 3; and a constant-volume process from state 3 to state 1. Represent the cycle on a p-V diagram, and determine the change in internal energy, the work, and heat interaction in each process.

3.43 Which combination of the following energies may be encountered in a nonflow process: (a) kinetic energy, shaft work, and heat; (b) flow energy, heat, and internal energy; (c) potential energy, flow energy, and internal energy; (d) internal energy, heat, and shaft work.

3.44 Show that the increase in kinetic energy per unit decrease in pressure in a reversible adiabatic steady flow of a fluid is equal to the specific volume. Neglect changes in elevation.

3.45 In a reversible steady flow compressor (neglecting kinetic, potential, and heat energies), the work supplied is equal to (check the correct answer):
(a) the decrease in enthalpy.
(b) the increase in internal energy.
(c) the increase in flow work.
(d) the increase in enthalpy.

3.46 Water flows steadily and adiabatically in a constant-diameter pipe past a throttling valve that is partially open. Subscripts i and e refer to conditions before and after the valve.
(a) is $(h_e - h_i)$ positive, negative, or zero?
(b) is $(p_e - p_i)$ positive, negative, or zero?
(c) is $(u_e - u_i)$ positive, negative, or zero?
(d) is $(T_e - T_i)$ positive, negative, or zero?

3.47 The following values for steam are given for specific volume (m^3/kg) and enthalpy (kJ/kg):

	v	h
Water at 5 MPa and 200°C	0.001153	853.9
Dry steam at 5 MPa and 263.99°C	0.03944	2794.3

Compute the amount of energy supplied to a boiler per kg of dry steam delivered at 5 MPa if feedwater entered at 200°C. Separate this energy into the amount of increase in internal energy per kg of the fluid, and the flow energy required to deliver the steam and that required for the entry of the water. Assume steady state.

3.48 Air at 450 kPa and 1000 K enters a turbine operating at steady state at the rate of 0.12 kg/s. The pressure and temperature at the exit are 100 kPa and 500 K. The inlet velocity and changes in potential energy are negligible, but the exit velocity is 120 m/s. If the heat transfer from the turbine is 35 kJ/kg, determine the power developed by the turbine.

3.49 Air flows through an air compressor in steady-state operation. The temperature and pressure at the inlet are 300 K and 100 kPa, whereas the exit temperature and pressure are 525 K and 900 kPa. The compression process may be modeled by a reversible polytropic process ($pv^n =$ constant). Determine the heat interaction per unit mass of air.

3.50 An ideal gas with $c_p = 1.0035$ kJ/kg K and a molar mass $= 28.97$ kg/kg-mole undergoes a reversible steady-state flow compression from 100 kPa and 25°C to 500 kPa. The relationship $pV^{1.3} = C$ describes the compression process. Determine (a) the work input per kg of gas; (b) the power necessary to drive the compressor for a flow rate of 1.5 kg/min; (c) the cooling rate for the compressor in kJ/min; (d) the work input if the compression is done for a system rather than a control volume.

3.51 Compressed nitrogen expands in a turbine operating at steady state. The conditions at the inlet are 600 kPa and 30°C, and at the exit are 100 kPa and −40°C. If the heat transfer to the turbine is 8 kJ/kg, calculate the mass rate of flow per kW of power produced. Assume nitrogen to be an ideal gas with constant specific heats (c_p = 1.0416 kJ/kg K and c_v = 0.7448 kJ/kg K).

3.52   An air preheater raises the temperature of the air supplied to a furnace operating at steady state from 15°C to 200°C, the pressure remaining constant. Considering c_p = 1.0035 kJ/kg K and c_v = 0.7165 kJ/kg K for air to be constant, find (a) the change in internal energy of each kg of air; (b) the heat interaction per kg of air.

3.53   Test-stand measurements indicate that the hot gases flowing through a rocket motor increase in velocity from effectively zero to 1860 m/s. What is the accompanying enthalpy change if the process is adiabatic, and what additional information is required to compute the accompanying temperature change?

3.54   In a test of a water-jacketed air compressor operating at steady state, it was found that the shaft work required to drive the compressor was 180 kJ/kg of air delivered, that the enthalpy of the air leaving was 70 kJ/kg greater than that entering, and that the energy removed by the circulating water was 95 kJ/kg of air. From these data, compute the amount of energy that must have been transferred as heat to the atmosphere from the bearings, cylinder walls, and so forth.

3.55   Steam expands reversibly through a turbine operating at steady state. The pressure of the steam at the inlet is 1.7 MPa and is 100 kPa at the exit. The relation between pressure and volume during the process is p = 1850 − 1000 v (p is in kPa, v is in m³/kg). Neglecting kinetic and potential energies, find the work interaction. If the turbine has a heat loss of 120 kJ/kg during the process, what is the change of internal energy?

3.56   Air is compressed at the rate of 30 kg/min in a steady flow process. The inlet air temperature is 300 K, and the delivery temperature is 370 K. What is the shaft work required per kg of air, and what is the total power required? Heat effects may be neglected.

T (K)	h (kJ/kg)	u (kJ/kg)
300	300.19	214.09
370	370.67	264.47

3.57   A certain gas flows steadily through a centrifugal compressor operating at a steady-state. The following data apply to this process:

	p (kPa)	v (m³/kg)	h (kJ/kg)
Intake	70	0.95	300
Discharge	350	0.31	510

(a) Considering the compression to be reversible and $-\int p\, dv$ = 290 kJ/kg, find the work done. Neglect kinetic and potential energies.
(b) Find the heat interaction during the process.

3.58   Air enters a compressor operating at steady state. The inlet conditions are p_1 = 100 kPa, v_1 = 0.85 m³/kg, V_1 = 15 m/s, and u_1 = 120 kJ/kg. During the compression, an amount of heat of 140 kJ/kg of air is transferred from the air to a cooler, and the air was delivered at p_2 = 500 kPa, v_2 = 0.17 m³/kg, V_2 = 15 m/s, and u_2 = 210 kJ/kg. Determine the minimum work required to accomplish the compression. What is the ratio of the inlet to delivery areas?

3.59   Steam expands in a nozzle from an initial pressure of 2.75 MPa to a final pressure of 10 kPa in a steady-state operation. If the initial and final enthalpies are 2900 kJ/kg and 2500 kJ/kg, respectively, calculate the final velocity in m/s if the initial velocity is negligible. Neglect potential energy changes and heat losses.

3.60   An adiabatic nozzle is supplied with a fluid whose enthalpy is 3300 kJ/kg at an inlet velocity of 250 m/min. At the exit, the velocity is 700 m/s. Neglecting potential energy changes and assuming steady-state operation, find the final enthalpy.

3.61   Find the heat of reaction at 25°C and 1 atm when gaseous propane C_3H_8 is burned with oxygen in a constant-volume process. Assume that the water vapor produced is in the vapor phase.

$$(\Delta H°)_{RP} = -46{,}355 \text{ kJ/kg}$$

3.62   In a steady-state flow process, 45 kg/min of water at 80°C enters a mixing tank where it mixes with 23 kg/min of water at 15°C. If the enthalpy of water is given by $h = 4.18(T - 273.15)$ kJ/kg, where T is in degrees Kelvin. What is the temperature of the exit mixture?

3.63 Two air streams are mixed in a large chamber before passing through an air turbine, as shown in Figure 3.46. The exhaust of the turbine passes through a diffuser before it is discharged to the atmosphere. Assuming steady-state adiabatic flow and neglecting changes in kinetic and potential energies, determine (a) the temperature of the air at the turbine inlet; (b) the temperature at the discharge of the turbine if the rotor of the turbine is stalled, that is, not rotating; (c) the power developed by the turbine if the temperature at discharge of the diffuser is 410 K. What is the function of the diffuser?

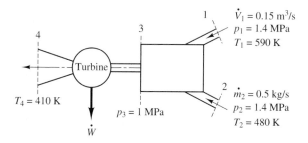

Figure 3.46

3.64 Air is compressed in a steady-state flow process. Upon leaving the compressor, it is cooled in a heat exchanger to a temperature of 300 K. Using the data shown in Figure 3.47, and assuming air to be ideal gas with constant specific heats, determine (a) the compressor power input; (b) the mass rate of flow of water. The specific

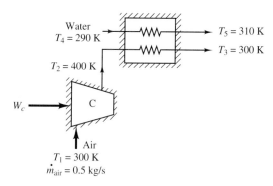

Figure 3.47

heat for water $c = 4.18$ kJ/kg K. The specific heats for air are $c_v = 0.7165$ kJ/kg K and $c_p = 1.0035$ kJ/kg K.

3.65 An air jet strikes a blade, as shown in Figure 3.48. Determine the magnitude and sense of R_x and R_y to hold the blade in place. Solve the problem for the following expansions in the nozzle: (a) adiabatic, (b) isothermal. (Assume air to be an ideal gas, and consider the system to be ideal.)

3.66 A gas is compressed in a reversible steady-state flow process from an initial temperature of 25°C and an initial pressure of 90 kPa to a final pressure of 140 kPa according to the relations $p(v + 0.2) = $ constant, where p is in kPa and v in m³/kg. The inlet velocity is negligible, and the discharge velocity is 100 m/s. If the flow rate is 0.1 kg/s, what is the power input to the compressor and the heat transfer?

3.67 Air is compressed in a steady-state flow process in a water-cooled air compressor. From the data given in

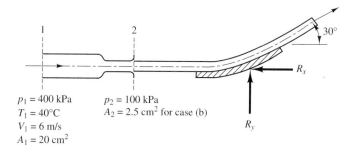

Figure 3.48

Figure 3.49, determine the discharge temperature of the air and the rate at which heat is transferred to the cooling water.

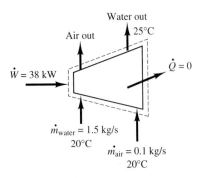

Figure 3.49

3.68   An air compressor is designed to compress atmospheric air (assumed to be at 100 kPa, 20°C) to a pressure of 1 MPa. Heat losses to the surroundings are anticipated to be 15 percent of the power input to the compressor. The air enters at a velocity of 40 m/s, where the inlet area is 100 cm², and leaves at a velocity of 100 m/s through an area of 6 cm². Determine the temperature of the air at the exit and the power input to the compressor. Assume air to be an ideal gas with constant specific heats ($c_p = 1.0035$ k J/kg K, $c_v = 0.7165$ k J/kg K).

3.69   A centrifugal air compressor operating at steady state has an air intake of 1.2 kg/min. The pressure and temperature conditions at inlet are 100 kPa, 0°C, and at outlet, 200 kPa, 50°C. If the heat losses are negligibly small, what is the power input to the compressor?

From tables of air properties, the internal energy and specific volume at inlet and exit are

$$u_i = 195.12 \text{ kJ/kg} \qquad v_i = 0.784 \text{ m}^3/\text{kg}$$
$$u_e = 230.99 \text{ kJ/kg} \qquad v_e = 0.464 \text{ m}^3/\text{kg}$$

What is the answer if air were considered an ideal gas with constant specific heats?

3.70   Air is compressed in a steady-state flow process in a water-cooled compressor. The pressure, specific volume, and internal energy at the inlet of the compressor are 100 kPa, 0.86 m³/kg, and 214 kJ/kg. The corresponding values at the discharge are 725 kPa, 0.17 m³/kg, and 308 kJ/kg. If the cooling water removes 60 k J/kg and the air flow rate is 0.2 kg/s, what is the power of the compressor?

3.71   An ideal gas with $c_p = 1.0035$ k J/kg K and molar mass $= 28.97$ kg/kg-mol undergoes a reversible steady flow compression from 100 kPa and 25°C to 500 kPa. The relationship $pV^{1.3} = C$ describes the compression process. Determine
(a) the work input per kg of gas.
(b) the power necessary to drive the compressor for a flow rate of 1.5 kg/min.
(c) the cooling rate for the compressor in kJ/min.

3.72   The specific heat at constant pressure of N_2 (in k J/kg K) varies with temperature according to the relation

$$c_p = 1.415 - 287.9 \, T^{-1} + 5.35 \times 10^4 T^{-2}$$

where T is in K.

If one kg of N_2 is heated at constant volume from 1000 K to 2000 K, what will be the change in enthalpy and the change in internal energy? Assume N_2 ($M = 28$ k J/kg-mol) to behave like an ideal gas.

Will the answers be different if the process were at constant pressure between the same end temperatures? Explain.

3.73   An uninsulated piston-cylinder system contains helium gas ($c_p = 5.1926$ kJ/kg K, $c_v = 3.1156$ kJ/kg K). The initial state is 800 kPa, 190°C, and 0.06 m³. The gas undergoes a quasi-static process governed by the equation $pv^{1.8} = C$. If the final pressure of the gas is 250 kPa, determine (a) the work done; (b) the change in internal energy; (c) the work done if the expansion were treated as a steady-state reversible process.

3.74   It is found for effectively dry air at atmospheric conditions that the change of enthalpy depends solely on change of temperature according to the relation

$$\frac{\Delta h}{\Delta T} = 1.0035 \text{ kJ/kg K}$$

Using this information, find the final temperature of air when it rises adiabatically from sea level to 5000 m altitude. The temperature of the air at sea level is 25°C. Change in velocity of the air is negligible. (Hint: Steady flow.)

3.75 Water is flowing in a 0.25 m diameter, 150 m long pipe at 3 m/s. The pressure drop per unit pipe length is given by

$$p_1 - p_2 = 0.01 \frac{V^2}{v}$$

where

p_1 = inlet pressure (kPa)
p_2 = outlet pressure (kPa)
V = water velocity (m/sec)
v = specific volume (0.001 m³/kg)

(a) If the pipe is perfectly insulated, determine the change in internal energy of the fluid during its flow through the pipe (assume that water is incompressible).
(b) If the flow had been isothermal, determine the heat interaction.

3.76 It is required to compress one kilogram of air initially at 300 K and 100 kPa to a pressure of 500 kPa in a steady flow process. Which of the following reversible processes requires the least work: (a) isothermal, (b) adiabatic ($pv^{1.4}$ = C). Show the two processes on the same *p-v* diagram indicating values (T, p, and v) at the inlet and exit states. What is the work and heat interactions and the change of internal energy of fluid during each compression process? Assume air is an ideal gas with constant specific heats:

$$(c_p = 1.0035 \text{ kJ/kg K}, \quad c_v = 0.7165 \text{ kJ/kg K})$$

3.77 A vessel with a volume of 0.05 m³ contains air at 350 kPa and 250°C. The vessel is connected to a cylinder and a movable piston on which the external pressure is maintained constant at 110 kPa.

The initial volume of the cylinder is zero, and the final pressure is 110 kPa. If the final temperature of the air occupying the vessel and the cylinder is 100°C, determine the heat and work interactions.

3.78 A 0.3 m³ rigid vessel contains air initially at a temperature of 50°C and a pressure of 0.6 MPa. Heat is transferred to the vessel at the rate of 30 W.
(a) What is the rate of temperature rise in the vessel?
(b) If the air is allowed to leak out of the tank so that the temperature is maintained constant at 50°C, what is the rate of air flow out of the tank?

3.79 Air (c_p = 1.0035 kJ/kg K) at a pressure of 550 kPa and 50°C flows through a pipeline. A tank that is connected to this line through a valve initially contains air at a pressure of 100 kPa and 25°C and has a volume of 0.2 m³. Determine the mass of air that flows into the tank when the valve is opened if the final pressure in the tank is 550 kPa. (Assume an adiabatic process.)

3.80 A 0.5 m³ evacuated tank develops a small crack through which air leaks from a surrounding atmosphere at a pressure of 0.1 MPa. The air leakage is so slow that the temperature in the tank remains essentially constant, equal to that of the surroundings of 20°C. Determine the amount of heat interaction when the pressure in the tank reaches 0.05 MPa.

3.81 A 0.5 m³ tank contains air at 100 kPa and 300 K. The tank is connected to a constant air pressure line through a valve. The pressure and the temperature in the supply line are 2 MPa and 320 K. The valve is opened till the pressure in the tank reached 2 MPa. If the process is considered adiabatic, calculate: (a) the final temperature in the tank; (b) the final mass of air in the tank.

3.82 A rigid tank with a volume of 0.08 m³ contains 5 kg of refrigerant-12 at 30°C. A valve on the top of the tank is opened, and saturated vapor is vented to the atmosphere. The valve is closed when only vapor remains in the tank. If during the venting process the temperature remains constant at 30°C, calculate the heat transfer. (See Appendix for the properties of R-12.)

3.83 An insulated tank of volume 5 m³, containing air at 600 kPa and 25°C, is *rapidly* bled until its pressure drops to 105 kPa. Determine the final temperature on the air in the tank. Assume that air is an ideal gas with constant specific heats and the bled air is at 25°C.

3.84   A tank is connected to a pipe by a valve. The tank has a volume of 1 m³, and it initially contains air at 150 kPa and 300 K. The air in the pipe has a pressure of 750 kPa and a temperature of 300 K. The tank is uninsulated. The valve is opened and left open until the tank reaches pressure equilibrium with the air in the pipe and the air in the tank reaches 300 K. Determine the mass of air that has entered the tank and the heat transfer during the process.

3.85   A tank with a volume of 1.2 m³ contains air at a pressure of 250 kPa and a temperature of 300 K. The air is allowed to escape to the atmosphere until the pressure drops to 120 kPa. Determine the heat interaction if the temperature is maintained constant during this process.

C3.1   Write a computer program to find the final temperature, pressure, work and heat interactions, changes in internal energy and enthalpy when air is compressed reversibly and polytropically from an initial state defined by V_1, T_1, and p_1 to a final volume V_2 for different values of the index of compression. Assume air to be an ideal gas with constant values of specific heats.

C3.2   Modify Problem C2.3 to include the calculation of internal energies of the gas as it expands. Assume that the specific heat does not change with temperature.

C3.3   A rigid tank with volume of 0.2 m² contains nitrogen gas at 200 kPa and 340 K. Heat is transferred to the tank at 200 kJ increments, and the resulting temperature and pressure are measured until the total heat transferred to the tank equals 2000 kJ. It is desired to check the accuracy of the measured data using the fundamental equations. Write a computer program so that the temperature and pressure at each interval can be calculated. Assume constant specific heat.

C3.4   Modify Problem C3.3 to include variation of the specific heat with temperature. The equation for c_v is given in the Appendix. Note that

$$\Delta u = \int_{T_1}^{T_2} c_v \, dT$$

C3.5   Combustion gases at 1500 K pass through a gas turbine for production of work. Write a computer program to investigate the effect of the inlet gas pressure on the amount of work produced. Let the inlet pressure vary from 4 to 40 atm while the exit pressure remains constant at 1 atm. Assume that no losses, such as friction and heat transfer, take place and that the process is reversible. Use air in place of the combustion gases. Plot the work transfer versus the inlet pressure.

C3.6   Modify Problem C3.5 to investigate the effect of inlet temperature on the amount of work produced. Let the inlet pressure be 4 atm while the inlet temperature varies from 1500 to 3000 K. Plot the work transfer versus the inlet temperature.

Chapter 4

Properties of Pure Substances

4.1 Introduction

The thermodynamic analysis of a system requires a knowledge of a set of properties to describe its state. Experience, however, indicates that properties are interrelated. This means that not all the properties of the system are necessary, but rather a limited number is sufficient to define the state. The number of the required properties depends on the complexity of the system and is determined by the *state principle*. This principle involves both the concept of energy and stable equilibrium and implies the existence of interrelations among the properties of the system. It states that the number of independent intensive properties required to define the stable state of a simple compressible system exceeds by one the reversible modes of work that can change the state of the system. For example, reversible work interaction, which is equal to $-\int p \, dV$, can change the state of a simple compressible system such as an ideal gas. In this case two independent intensive properties are sufficient to uniquely define the state.

A system is called a *simple compressible system* in the absence of motion and external force fields such as gravitational, electric, and magnetic fields, as well as in the absence of surface tension and shear forces.

Once the system is defined by a set of independent properties, functional relationships can be used to determine other properties. The result is given in terms of numerical tables, graphically, or in the form of algebraic equations known as equations of state. In the ideal-gas equation, for example, the properties pressure, volume, and temperature are functionally related, and any one of these properties can be determined in terms of the other two independent properties.

This chapter is concerned with the study of the thermodynamic properties of pure substances and the phases in which they may exist. It is introduced early in this book to afford the reader the use of physical properties in the application of the first law.

4.2 One-Component System

The one-component system has a homogeneous and invariable chemical composition irrespective of the phase or phases in which it exists. The *phase* of a substance is the homogeneous, chemical and physical state of aggregation of its molecules. A mixture of ice, water, and steam, for example, is a one-component system since the chemical composition of these three different phases is the same. A mixture of different gases may be treated as a one-component system, provided that its composition is uniform. If one of the gases condenses, the mixture can no longer be considered a one-component system since the condensed phase has a different composition than the original mixture.

Every substance can theoretically exist in at least three phases: solid, liquid, and gas. In addition, many substances, such as water and sulfur, have several crystalline structures. Ice, for example, exists in some seven different forms under high pressures.

All one-component systems exhibit similar qualitative behavior as far as coexisting phases and phase changes. Water is used in this discussion as a typical example. Consider the transfer of heat at constant pressure to a system of unit mass of water. Let the initial state of the water be at point *a* of Figure 4.1. The constant pressure line *a-b-c-d* represents the relationship between the temperature and the volume during the course of heat interaction. From *a* to *b*, heat transfer produces an increase in temperature and a small increase in volume. The amount of heat transfer is equal to the increase in enthalpy of the water. When point *b* is reached, part of the water begins to vaporize and a rapid increase in volume takes place. Although both the temperature and the pressure remain unaltered, the change of state is noted by the increase of the mass of the vapor and the decrease of the mass of the liquid. The intensive properties of both the liquid and the vapor states remain unchanged, and the liquid is in equilibrium with its vapor. The liquid state corresponds to point *b*, whereas the vapor state corresponds to point *c*. When point *c* is reached, all the liquid has been converted into vapor. The vapor and liquid coexisting in equilibrium are called *saturated vapor* and *saturated liquid,* respectively. Points *b* and *c* are called *saturated liquid state* and *saturated vapor state,* respectively. The temperature at which vaporization takes place at a given pressure is called the *saturation temperature.* The pressure corresponding to

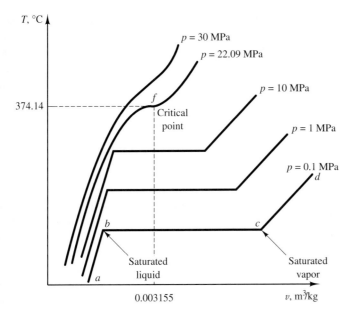

Figure 4.1   *T-v* diagram for water showing constant-pressure lines.

the saturation temperature is called the *saturation pressure*. At point *c*, the change of phase from liquid to vapor is complete, and further transfer of heat to the system increases both the temperature and the volume along portion *c-d* of the constant pressure curve. When the temperature of the vapor is higher than the saturation temperature at a given pressure, it is called *superheated vapor*. The term *superheat* denotes that the temperature of the vapor is in excess of the saturated value corresponding to the given pressure. In the superheated vapor region, the temperature and the pressure are independent and are, therefore, sufficient to determine the intensive state of the system. In the case of a liquid, if the pressure is greater than the saturation pressure corresponding to its temperature, it is called *compressed liquid*. It is also called *subcooled liquid* because the liquid temperature is lower than the saturation temperature for the given pressure.

If the foregoing heating process is repeated at different pressures, curves similar to *a-b-c-d* are obtained. It is noted that there are two discontinuities in the slope of each of the constant-pressure lines (points *b* and *c*), with a constant temperature portion between them as shown in Figure 4.1. If the pressure is increased beyond a certain value called the critical pressure, the discontinuities as well as the constant-temperature portion disappear. In this case, there is no definite transition between the liquid and vapor phases, and the liquid changes to vapor without any discontinuity. The limiting state at which no discontinuities are noticeable is called the *critical state* and is represented by point *f* in Figure

4.1. Properties, such as temperature, pressure, and volume at the critical point, are identified as critical properties. For water, the critical temperature, pressure, and specific volume are 374.14°C, 22.09 MPa, and 0.003155 m³/kg. Critical constants for several substances are given in Appendix A7.

It is conventional to call the substance in the superheated region *vapor* if its temperature is below that of the critical point, and *gas* if above the critical point. In Figure 4.2, line *e-b-f* connecting the saturated-liquid states is called the *saturation liquid line,* and line *f-c-g* connecting the saturated-vapor states is called the *saturation vapor line*. Both lines meet at the *critical point f.* Compressed-liquid states are located to the left of the saturated-liquid line, and superheated-vapor states are located to the right of the saturated-vapor line. Liquid and vapor phases in equilibrium are located under the dome *e-f-g.*

Another diagram similar to the *T-v* diagram is the *p-v* diagram. A typical *p-v* diagram of a one-component system is shown in Figure 4.3. The saturated liquid and saturated-vapor curves meet at the critical point. The region to the left of the saturated-liquid line is compressed liquid, and the region to the right of the saturated-vapor line is superheated vapor. In between the two lines is the two-phase liquid–vapor region. Several isotherms are shown in the figure. The critical isotherm has an inflection point and a zero slope at the critical point. Isotherms below the critical point experience two discontinuities in slope as they cross the saturated-liquid and saturated-vapor lines. In between these two lines, the slope of the isotherms is zero, and a state in this region can be entirely saturated liquid, entirely saturated vapor, or a combination of the two.

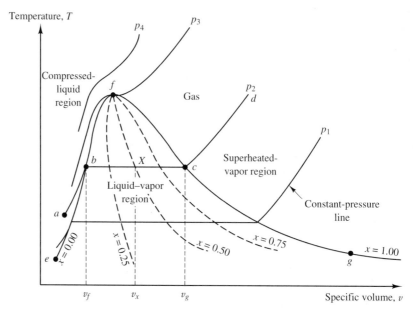

Figure 4.2   Temperature–volume diagram of a one-component substance.

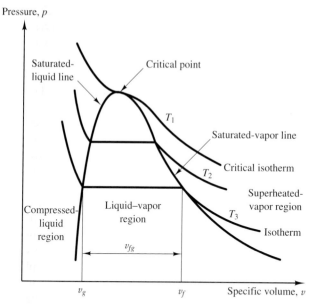

Figure 4.3 Pressure–volume diagram of a one-constant substance.

4.3 The Liquid–Vapor Mixture

A state represented by a point under the dome *e-b-f-c-g* (Figure 4.2), such as *X*, corresponds to an equilibrium state of a homogenous mixture of liquid and vapor that is called *wet vapor*. The intensive states of the vapor and liquid phases in the wet vapor region are represented by points *c* and *b* irrespective of the relative masses of each phase, that is, irrespective of the location of point *X* on line *b-c*. Let *x* be the ratio of the mass of the vapor to the total mass of the mixture, then each kilogram of mixture contains *x* kg of vapor and $(1 - x)$ kg of liquid. The property *x* is called the *quality* or the *dryness fraction* of the mixture. It is an intensive property defined by the equation:

$$x = \frac{m_g}{m_{\text{mixture}}} = \frac{m_g}{m_f + m_g} \qquad (4.1)$$

where m_g and m_f are the masses of the vapor and the liquid in the mixture. Subscripts *g* and *f* refer to saturated vapor and saturated liquid, respectively. If $x = 0$, the mixture is completely liquid, and its state is represented by point *b*, and if $x = 1$, the mixture is completely vapor, and its state is represented by point *c*. Let v_f be the specific volume of the saturated liquid, v_g be the specific volume of the coexisting saturated vapor, and v_{fg} be the increase in the specific volume when the saturated liquid changes to saturated vapor. The specific volume v_x of a mixture is equal to the total volume of the mixture divided by the total mass. Its value depends on the relative masses of the liquid and vapor present and on the specific volume of each phase. It is equal to the sum of the relative specific volumes of the liquid and the vapor.

$$v_x = (1 - x)v_f + xv_g \tag{4.2}$$

But, since

$$v_g = v_f + v_{fg} \tag{4.3}$$

then

$$v_x = v_f + xv_{fg} \tag{4.4}$$

or

$$v_x = v_g - (1 - x)v_{fg} \tag{4.5}$$

Note that the specific volume of a two-phase mixture depends on the intensive properties of its phases as well as the mass proportions of each phase. Furthermore, Eq. (4.4) indicates that the specific volume v_x of the two-phase mixture increases linearly from saturated liquid ($x = 0$) to saturated vapor ($x = 1$).

By means of Eq. (4.2) or its equivalents Eq. (4.4) or (4.5), constant quality lines may be plotted, as shown in Figure 4.1. These lines meet at the critical point. Note that x can be used as one of the parameters to define the state of a wet vapor.

Expressions analogous to Eqs. (4.2), (4.4), and (4.5) can be derived for enthalpy, internal energy, and entropy.[1] These expressions per unit mass are

$$h_x = (1 - x)h_f + xh_g = h_f + xh_{fg} = h_g - (1 - x)h_{fg} \tag{4.6}$$
$$u_x = (1 - x)u_f + xu_g = u_f + xu_{fg} = u_g - (1 - x)u_{fg} \tag{4.7}$$
$$s_x = (1 - x)s_f + xs_g = s_f + xs_{fg} = s_g - (1 - x)s_{fg} \tag{4.8}$$

Note also that,

$$h_g = h_f + h_{fg} \tag{4.9}$$
$$u_g = u_f + u_{fg} \tag{4.10}$$
$$s_g = s_f + s_{fg} \tag{4.11}$$

---

[1] The property entropy is discussed in Chapter 5.

4.4 Extension to the Solid Phase

The foregoing liquid–vapor transition can be represented on the p-T plane. In this section, phase transition is extended to include the solid phase. Figure 4.4 shows how the different phases may exist in equilibrium. Consider, for example, a solid in state a at a pressure p_a and temperature T_a. The temperature T_a is below the fusion temperature T_b. Point b is located on the solid–liquid line (fusion line) and determines the state at which the solid is in equilibrium with its liquid.

Heat transfer to the solid at constant pressure results in a temperature increase to T_b. With further transfer of heat, the solid begins to liquify at a constant temperature T_b until all the substance is transformed to the liquid phase. During the melting process, the temperature is a function of pressure and remains constant at T_b. The change of the state of the system is accomplished by the increase of the amount of liquid and the decrease of the amount of solid during the heating process. The intensive state of each phase is, however, constant during the melting process. The transfer of heat to the liquid will increase its temperature to T_c, at which evaporation begins, and the temperature remains constant at T_c until all the liquid is evaporated. Point c is located on the liquid–vapor line (vaporization line) and determines the state at which the liquid is in equilibrium with its vapor. When all the liquid has evaporated, the temperature will rise again upon further transfer of heat along c-d.

If the preceding process is repeated at different pressures, the corresponding points b and c can be determined and the solid–liquid and liquid–vapor curves

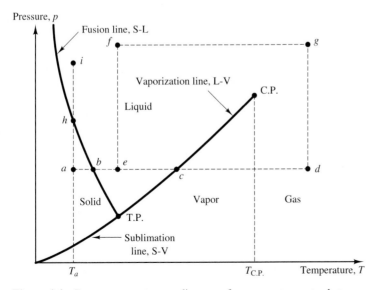

Figure 4.4   Pressure–temperature diagram of a one-component substance.

can be traced. These two curves meet at a point on the *p-T* plane called the *triple point*. Similarly, the solid–vapor line can be determined for pressure values below the pressure of the triple point. Note that it is not possible to obtain a liquid state at a pressure lower than that of the triple point. In this case, heat transfer to a solid will result first in raising its temperature to the solid–vapor line and then changing it directly into vapor. The solid–vapor line (sublimation line) represents the states in which the vapor is in equilibrium with its solid. The triple point defines a state in which the solid, liquid, and vapor phases of a substance coexist in equilibrium. It may be noted that, when two phases of a one-component system exist in equilibrium, a single intensive property is sufficient to define the intensive state of each phase. If this property is chosen as the saturation temperature, then the saturation pressure is a function of this temperature or

$$p = f(T) \qquad (4.12)$$

The foregoing relation is represented by the solid–liquid, liquid–vapor, and solid–vapor curves in Figure 4.4 for the corresponding phases in equilibrium. For water vaporization, this relation is shown in Figure 4.5. Note that the liquid–vapor curve extends from the triple point to the critical point and that three curves meet at the triple point. The intensive state of the phases of a one-component system at the triple point is fixed. This means that in addition to temperature and pressure being fixed at the triple point, the specific volumes of the solid, liquid, and vapor phases are also fixed. To determine the intensive state of the system at the triple point, the relative proportions of two of the solid, liquid, and vapor phases are required. For pure water, the temperature and pressure at the triple point are 0.01°C and 0.6113 kPa. The *triple point* of water should not be

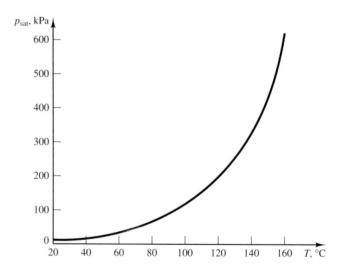

Figure 4.5   Saturated pressure vs. saturated temperature for H$_2$O.

confused with the ice point, the latter being defined as the state at which ice melts at a pressure of 101.325 kPa and a temperature of 0°C.

An alternative path to change a substance from the liquid to the gaseous state with a gradual change of phase is by following path *e-f-g-d*, as shown in Figure 4.4. Along path *e-f*, the temperature is maintained constant while the pressure is raised to a value above the critical pressure. Path *f-g* is a constant pressure line in which the liquid changes to vapor without discontinuity. The final state is attained by lowering the pressure at constant temperature along path *g-d*. Note also that a change from the solid to the liquid state may be achieved by following path *a-h-i* in which an increase of pressure at constant temperature results, first in melting the solid at point *h*, and then proceeding to the final liquid state at point *i*.

The amount of energy necessary to change a unit mass from saturated liquid to saturated vapor at the same temperature and pressure is called the *latent enthalpy of vaporization*. It is equal to the change in enthalpy between the liquid and vapor phases at the same temperature and pressure. The latent enthalpy depends on the value of temperature (or pressure) at which the change of phase takes place. If the change is from solid to liquid, it is called the *latent enthalpy of fusion;* and if the change is from solid directly to vapor, it is called the *latent enthalpy of sublimation*. It must be emphasized that both the temperature and the pressure remain constant during all phase transformations.

An expression of the latent enthalpy of vaporization is

$$\begin{aligned} \text{latent enthalpy} \\ \text{of vaporization} = h_g - h_f &= (u_g + pv_g) - (u_f + pv_f) \\ &= (u_g - u_f) + p(v_g - v_f) \end{aligned} \qquad (4.13)$$

Therefore, the latent enthalpy can be considered the sum of the change in internal energy and the work of expansion of the vapor against the existing equilibrium pressure. Similar expressions for the latent enthalpy of fusion and sublimation may be derived.

4.5 Thermodynamic Surfaces

In addition to providing values of thermodynamic properties, graphical display of property relations permits one to visualize the progress of the change in the state of a system. Figure 4.6 shows a three-dimensional *p-V-T* surface for a substance that expands on freezing. On the same figure, the projections on the *p-T* and *p-V* planes are also shown. Figure 4.7 is the *p-V-T* surface for a substance that contracts on freezing. Also shown are the projections on the *p-T* and *p-V* planes. Any point on the *p-V-T* surface represents an equilibrium state of the substance and is determined by any two independent properties. In certain regions of the figures, the substance exists in a single phase. These are labeled *solid, liquid,* and *gas* (or

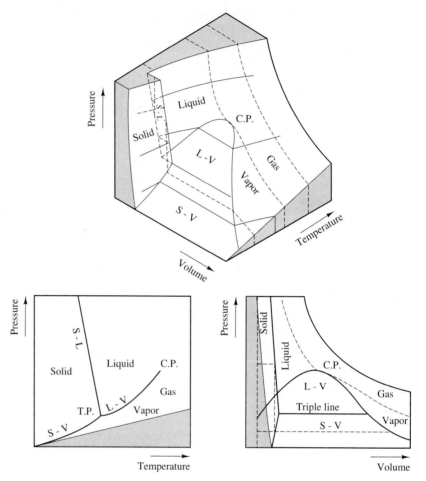

Figure 4.6   *p-V-T* surface and projections for a substance that expands on freezing.

vapor). When the substance exists in two phases, the region is labeled solid–liquid, solid–vapor, and liquid–vapor. These regions are ruled surfaces made up of straight lines parallel to the *V*-coordinate. Therefore, when these surfaces are projected on the *p-T* plane, they appear as lines. The triple line shown on the *p-V-T* surface appears as a point on the *p-T* plane. Note that although both the pressure and the temperature are constant at the triple point, the specific volume can vary depending on the proportions of the three phases existing at that point.

Several isotherms and constant-pressure lines are shown on the *p-V-T* surfaces of Figures 4.6 and 4.7. The different states of the substance carry the same label on the corresponding *p-T* and *p-V* diagrams.

Other thermodynamic properties, besides pressure, volume, and temperature, are often used to describe the state of a system. These properties are usually given in the form of tables and charts.

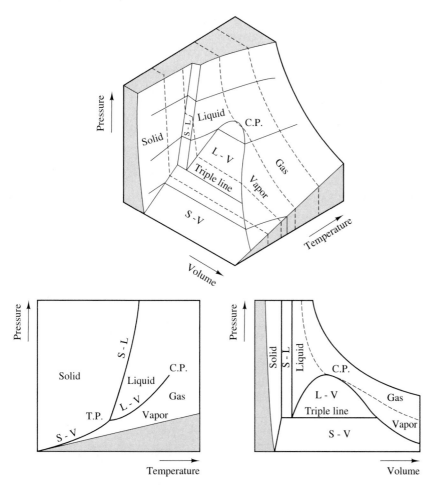

Figure 4.7   *p-V-T* surface and projections for a substance that contracts on freezing.

4.6 Tables of Properties

Properties of many substances are conveniently tabulated in thermodynamic tables. The most common examples are steam[2] and refrigerant tables,[3] in which the properties of the substance are given as functions of temperature and pressure. Tables A3.1 and A3.2 of the steam tables give the saturated properties of the liquid and vapor phases as a function of temperature or pressure. These include specific volume, internal energy, enthalpy, and entropy of the saturated liquid

[2] J. H. Keenan, F. G. Keyes, P. G. Hill, and J. G. Moore, *Steam Tables* (New York: John Wiley & Sons, Inc., 1978). An abstract of steam tables is given in the Appendix.

[3] Tables of properties of ammonia, refrigerant-12, and refrigerant-134a are given in Tables A4, A5, and A6.

and saturated vapor states. Note that both the internal energy and entropy values are calculated relative to an arbitrary temperature of 0.01°C, where both properties for saturated water are taken to be zero at this temperature. Different reference temperatures are used for refrigerants. For the saturated states, the temperature and pressure are dependent and either one, together with a third property, such as specific volume or quality, is sufficient to determine the state. Table A3.3 of the steam tables gives the specific volume, internal energy, enthalpy, and entropy of superheated steam at different pressures and temperatures. Here, the temperature and pressure are independent properties and can therefore be used to determine the state. Table A3.4 gives the specific volume, internal energy, enthalpy, and entropy of compressed liquid as a function of pressure and temperature. Finally, Table A3.5 gives properties of saturated solid and saturated vapor existing in equilibrium. Thermodynamic property data can be determined using the computer software provided with this text. The following examples illustrate the use of the steam tables.

Example 4.1

One kilogram of a water–vapor mixture initially at a pressure of 500 kPa and a quality 20 percent is heated at constant pressure. If the final state is superheated steam at 300°C, determine:

(a) v_f and v_g at 500 kPa
(b) the initial specific volume
(c) the final specific volume of the superheated steam

SOLUTION

A T-v diagram of the process is shown in Figure 4.8.

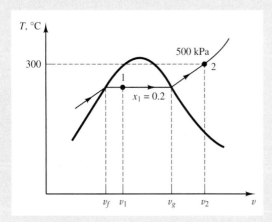

Figure 4.8

(a) From Table A3.2 at 500 kPa,

$$v_f = 0.001093 \text{ m}^3\text{/kg} \qquad \text{and} \qquad v_g = 0.3749 \text{ m}^3\text{/kg}$$

(b) The initial state is in the saturated liquid–vapor region and the specific volume is

$$v_x = v_f + x_1 v_{fg}$$
$$= 0.001093 + (0.2)(0.3749 - 0.001093)$$
$$= 0.0759 \text{ m}^3/\text{kg}$$

(c) From Table A3.3 at 500 kPa and 300°C,

$$v = 0.5226 \text{ m}^3/\text{kg}$$

Table 4.1 gives the solution to this problem using the software provided with this text.

Table 4.1   Solution of Example 4.1 Using the Computer Software

Several calculated properties are displayed on the screen and can be stored in the file called Results.tmp. This file will be overwritten each time the software is executed.

```
              Thermodynamic Properties
          ****************************

              A- Ideal Gases
              B- Steam (Solid/Compressed/Saturated/Vapor)
              C- Ammonia (Saturated/Vapor)
              D- Freon 12 (Saturated/Vapor)
              F- Solids
              H- Help
              E- Exit to main menu
              Q- QUIT
```

The substance in this example is water, so we enter choice (B).

```
        Properties of Steam                        Properties of Steam
     ****************************                ****************************

  A- Given (P & x)     K- Given (T & u)       A- Given (P & x)     K- Given (T & u)
  B- Given (P & T)     L- Given (T & h)       B- Given (P & T)     L- Given (T & h)
  C- Given (P & v)     M- Given (T & s)       C- Given (P & v)     M- Given (T & s)
  D- Given (P & u)     N- Given (s & v)       D- Given (P & u)     N- Given (s & v)
  F- Given (P & h)     O- Given (s & u)       F- Given (P & h)     O- Given (s & u)
  G- Given (P & s)     Q- Given (s & h)       G- Given (P & s)     Q- Given (s & h)
  I- Given (T & x)     R- Given (v & h)       I- Given (T & x)     R- Given (v & h)
  J- Given (T & v)     S- Given (v & u)       J- Given (T & v)     S- Given (v & u)

        E- Exit to Previous menu                  E- Exit to Previous menu
        H- Help                                   H- Help
        P- Print Results                          P- Print Results
        Q- Quit                                   Q- Quit
        U- Change System of Units                 U- Change System of Units

  ---- Please Enter Your Selection ----> A      ---- Please Enter Your Selection ----> A
```

Enter Pressure (in kPa), Quality (0=<x%<=100) ----> 500,0 Enter Pressure (in kPa), Quality (0=<x%<=100) ----> 500,100

(continued)

Table 4.1   Solution of Example 4.1 Using the Computer Software *(continued)*

(a)   Since the two independent properties given are the pressure and quality, we enter choice (A) and after entering the values, the results will be displayed on the screen. Remember that we have to run this part twice, once for the saturated liquid and then again for the saturated vapor.

```
        Calculated Steam Properties                    Calculated Steam Properties
    **********************************              **********************************


P =   .50000E+03 kPa      x =    .00 %        P =   .50000E+03 kPa      x = 100.00 %
T =   .15186E+03 Degree C v =   .10930E-02 m3/kg  T =   .15186E+03 Degree C v =   .37490E+00 m3/kg
u =   .63968E+03 kJ/kg    h =   .64023+03 kJ/kg   u =   .25612E+04 kJ/kg    h =   .27487+04 kJ/kg
s =   .18607E+01 kJ/kg-K  Phase = Saturated liquid wat  s =   .68213E+01 kJ/kg-K  Phase = Saturated Steam
```

(b)   The pressure and the initial quality are given, so again we enter choice (A) and type in the values of the given properties.

```
                        Properties of Steam
                ******************************


            A- Given (P & x)        K- Given (T & u)
            B- Given (P & T)        L- Given (T & h)
            C- Given (P & v)        M- Given (T & s)
            D- Given (P & u)        N- Given (s & v)
            F- Given (P & h)        O- Given (s & u)
            G- Given (P & s)        Q- Given (s & h)
            I- Given (T & x)        R- Given (v & h)
            J- Given (T & v)        S- Given (v & u)


                    E- Exit to Previous menu
                    H- Help
                    P- Print Results
                    Q- Quit
                    U- Change System of Units

            ---- Please Enter Your Selection ----> A

      Enter Pressure (in kPa), Quality (0=<x%<=100) ----> 500,20


                    Calculated Steam Properties
                ***********************************
            P =   .50000E+03 kPa      x = 20.00 %
            T =   .15186E+03 Degree C v =   .75854E-01 m3/kg
            u =   .10240E+04 kJ/kg    h =   .10619E+04 kJ/kg
            s =   .28528E+01 kJ/kg-K  Phase = Liquid-Vapor Mix
```

(continued)

Table 4.1   *(continued)*

(c)   The two independent properties given are the pressure and the temperature, so enter choice (B).

```
                        Properties of Steam
                ******************************

            A- Given (P & x)        K- Given (T & u)
            B- Given (P & T)        L- Given (T & h)
            C- Given (P & v)        M- Given (T & s)
            D- Given (P & u)        N- Given (s & v)
            F- Given (P & h)        O- Given (s & u)
            G- Given (P & s)        Q- Given (s & h)
            I- Given (T & x)        R- Given (v & h)
            J- Given (T & v)        S- Given (v & u)


                E- Exit to Previous menu
                H- Help
                P- Print Results
                Q- Quit
                U- Change System of Units


        ---- Please Enter Your Selection ----> B

    Enter Pressure (in kPa), Temperature (in C) ----> 500,300


                    Calculated Steam Properties
            **********************************
    P =   .50000E+03 kPa        x = 100.00 %
    T =   .30000E+03 Degree C   v =   .52260E+00 m3/kg
    u =   .28029E+04 kJ/kg      h =   .30642E+04 kJ/kg
    s =   .74599E+01 kJ/kg-K    Phase = Superheated Steam
```

Example 4.2

A rigid vessel of volume 3 m^3 contains saturated water vapor at 350 kPa. If the pressure drops to 250 kPa owing to heat transfer from the system, calculate:

(a) the quality of the mixture
(b) the mass of the vapor and liquid in the final state
(c) the amount of heat transferred from the vessel

SOLUTION

A T-v diagram of the process is shown in Figure 4.9.

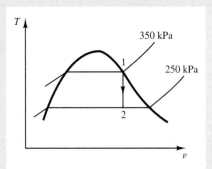

Figure 4.9

The following properties at 250 and 350 kPa are extracted from Table A3.2 of the Appendix:

p (kPa)	T (°C)	Sat. liquid v_f (m³/kg)	Sat. vapor v_g (m³/kg)
250	127.44	0.001067	0.7187
350	138.88	0.001079	0.5243

(a) From the preceding table, the specific volume of saturated water vapor at 350 kPa is 0.5243 m³/kg. Therefore,

$$m = \frac{V}{v_g} = \frac{3 \text{ m}^3}{0.5243 \text{ m}^3/\text{kg}} = 5.722 \text{ kg}$$

and

$$v_2 = v_1 = v_f + x_2 v_{fg}$$
$$0.5243 \text{ m}^3/\text{kg} = 0.001067 \text{ m}^3/\text{kg} + x(0.7187 - 0.001067) \text{ m}^3/\text{kg}$$

from which

$$x_2 = \frac{0.5232}{0.7167} = 72.9 \text{ percent}$$

(b)

$$m_{v_2} = 0.729(5.722 \text{ kg}) = 4.171 \text{ kg}$$

and

$$m_{f_1} = (0.271)(5.722 \text{ kg}) = 1.551 \text{ kg}$$

(c) Since the heat transfer takes place at constant volume, $W = 0$, so that

$$
\begin{aligned}
Q = U_2 - U_1 &= m[(u_f + xu_{fg})_2 - (u_g)_1] \\
&= (5.722 \text{ kg})[(535.10 + 0.729 \times 2002.1) - 2548.9] \text{ kJ/kg} \\
&= -3171.5 \text{ kJ}
\end{aligned}
$$

Example 4.3

A rigid vessel of volume 0.5 m³ initially contains a water–vapor mixture at 400 kPa.

(a) If the quality of the mixture is 40 percent, calculate the mass of the mixture.
(b) If the pressure in the vessel is raised to 700 kPa by the transfer of heat, what will be the mass of the vapor and the mass of the liquid?

SOLUTION

A sketch of the system and a T-v diagram of the process are shown in Figure 4.10.

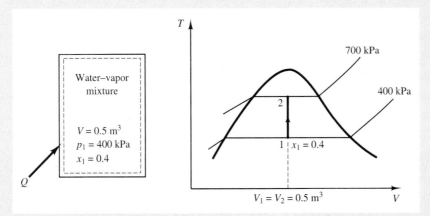

Figure 4.10

The following properties at 400 and 700 kPa are obtained from Table A3.2 of the Appendix:

p (kPa)	T (°C)	Sat. liquid v_f (m³/kg)	Sat. vapor v_g (m³/kg)
400	143.63	0.001084	0.4625
700	164.97	0.001108	0.2729

(a) The specific volume of the mixture at 400 kPa is given by

$$v_x = v_f + xv_{fg}$$
$$= 0.001084 \text{ m}^3/\text{kg} + 0.4(0.4625 - 0.001084) \text{ m}^3/\text{kg}$$
$$= 0.185650 \text{ m}^3/\text{kg}$$

The mass of the mixture is

$$m = \frac{V}{v_x} = \frac{0.5 \text{ m}^3}{0.185650 \text{ m}^3/\text{kg}} = 2.693 \text{ kg}$$

(b) Since the volume and the mass of the water in the vessel remain unchanged, the specific volume must also remain constant. Therefore, at 700 kPa,

$$v_x = 0.185650 \text{ m}^3/\text{kg}$$
$$= 0.001108 \text{ m}^3/\text{kg} + x(0.2729 - 0.001108) \text{ m}^3/\text{kg}$$

from which $x = 67.9$ percent. The respective masses of the vapor and liquid are

$$m_v = (0.679)(2.693 \text{ kg}) = 1.829 \text{ kg}$$

and

$$m_f = (0.321)(2.693 \text{ kg}) = 0.864 \text{ kg}$$

Example 4.4

Wet steam at a pressure of 3.0 MPa is throttled adiabatically in a steady-state operation to a pressure of 100 kPa. If the temperature after throttling is 150°C, find the quality of the steam.

SOLUTION

A sketch of the system and a T-v diagram of the process are shown in Figure 4.11. The enthalpies of the steam before and after throttling are equal. At 150°C and 100 kPa, Table A3.3 gives

$$h = 2776.4 \text{ kJ/kg}$$

which is equal to the enthalpy of the wet steam at 3.0 MPa. The latter is given by

$$h_x = h_f + xh_{fg}$$

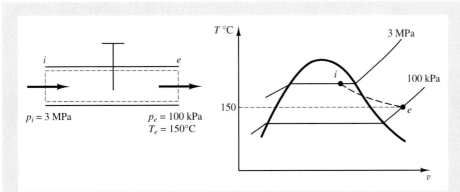

Figure 4.11

Therefore, using Table A3.2 of the Appendix,

$$h_x = 2776.4 \text{ kJ/kg} = 1008.42 \text{ kJ/kg} + x(1795.7) \text{ kJ/kg}$$

from which

$$x = \frac{2776.4 - 1008.42}{1795.7} = 98.456 \text{ percent}$$

4.7 The Ideal-Gas Model

The ideal gas discussed previously in Section 1.16 is a hypothetical substance that obeys the laws of Boyle and Gay-Lussac (or Charles). Real gases at low pressures and high temperatures follow these laws very closely. At room temperatures and pressures, the representation of real gases by the ideal-gas equation is satisfactory for monatomic and diatomic gases of low molar masses.

Robert Boyle in 1662 observed that, when the temperature of a certain mass of a gas was maintained constant, the product of the pressure and the volume of the gas was a constant.

Mathematically, Boyle's law is expressed as

$$pV = \varphi(T) \tag{4.14}$$

where p is the absolute pressure, V is the volume, T is the absolute temperature, and φ is a function.

Boyle's law is represented graphically in Figure 4.12. The isotherms are rectangular hyperbolas on the p-V diagram. For any two points on an isotherm,

$$\frac{p_1}{p_2} = \frac{V_2}{V_1} \tag{4.15}$$

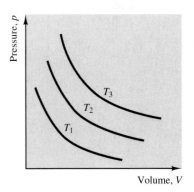

Figure 4.12 p-V relation of an ideal gas at constant temperature.

Equation (4.15) states that, at constant temperature, the volume occupied by a certain mass of an ideal gas is inversely proportional to its pressure.

Charles in 1787, about a century after Boyle, observed that, at constant pressure, the volume of a given mass of a gas was proportional to its absolute temperature, or

$$\frac{V}{T} = \psi(p) \tag{4.16}$$

where ψ is a function.

In order to determine the nature of both the functions φ and ψ of Eqs. (4.14) and (4.16), divide Eq. (4.14) by Eq. (4.16):

$$pT = \frac{\varphi(T)}{\psi(p)}$$

Separating the variables in the preceding equation gives

$$\frac{\varphi(T)}{T} = p\psi(p)$$

But since T and p are independent variables and each appears on one side of the foregoing equation, each side of the equation must be equal to the constant C. Therefore,

$$\varphi(T) = CT \qquad \text{and} \qquad \psi(p) = \frac{C}{p}$$

The foregoing result indicates that the product pV in Boyle's law varies linearly with temperature, and, similarly, the quotient T/p in Charles' law varies linearly with volume.

Gay-Lussac's law (1802) states that, at constant pressure, the volumetric coefficient of expansion is constant for all gases. The variation of the volume of a gas with temperature at constant pressure is given by

$$V_T = V_0(1 + \beta T)\big|_p \tag{4.17}$$

where

V_T = volume occupied by the gas at temperature T
V_0 = volume occupied at some reference temperature ($T = 0°C$)
β = coefficient of volumetric expansion; it is a measure of the change in volume with temperature at constant pressure

Equation (4.17), as shown in Figure 4.13, is a straight-line relationship. The value of β varies slightly for different gases, but has the same value for all gases at very

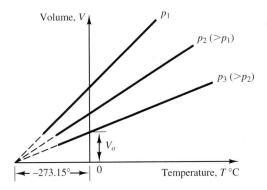

Figure 4.13   *V-T* relation of an ideal gas at constant pressure.

low pressure, a condition that approximates to the behavior of an ideal gas. In an ideal gas, it is assumed that there is no interaction between the molecules. Only in this case is β considered to be constant. The value of β is found experimentally by plotting the volume versus temperature at low pressures and extrapolating to zero pressure. In this case, $1 + \beta T = 0$ and if T is in °C, then

$$T = -\frac{1}{\beta} = -\frac{1}{0.0036609} = -273.15°C$$

This means that $-273.15°C$ is the lowest temperature that can be attained; otherwise, a negative pressure will result. Actually, the volume of the gas will never be zero at this temperature because the gas will liquify before such low temperatures are reached and Eq. (4.17) will no longer be valid.

The absolute temperature is equal to

$$TK = 273.15°C + T°C$$

which is in agreement with what was mentioned in Chapter 1. The volume of a gas at a temperature $T°C$ is given by

$$V_T = V_0\left(1 + \frac{T°C}{273.15}\right)$$

Therefore, an alternative statement of Gay-Lussac's law is that at constant pressure, the volume of a fixed mass of a gas increases by $1/273.15$ of its volume at 0°C for every degree of Celsius temperature rise. The preceding relation can be written in terms of absolute temperature as

$$\frac{V_T}{V_0} = 1 + \frac{T°C}{273.15} = \frac{273.15 + T°C}{273.15} = \frac{T}{T_0}$$

and for two states at the same pressure,

$$\frac{V_1}{V_2} = \frac{T_1}{T_2}$$

(4.18)

which is Charles' law. Therefore Gay-Lussac's and Charles' laws are equivalent.

So far, variations of only two properties of an ideal gas have been considered. In order to explore the relation among temperature, pressure, and volume, consider a gas to change its state in two steps: the first at constant pressure and the second at constant temperature. For the first step, Charles' law gives

$$\frac{V_1}{V_2'} = \frac{T_1}{T_2}$$

where V_2' is the volume of the gas if the pressure is maintained constant while its temperature is increased from T_1 to T_2. The second step is to change the state of the gas at constant temperature, for which Boyle's law gives

$$\frac{V_2'}{V_2} = \frac{p_2}{p_1} \qquad \text{or} \qquad V_2' = \frac{V_2 p_2}{p_1}$$

Substitution in the previous equation yields

$$\frac{V_1}{V_2} = \frac{T_1}{T_2} \frac{p_2}{p_1}$$

or

$$\frac{pV}{T} = \text{constant}$$

(4.19)

Equation (4.19) can also be obtained mathematically as follows: If the volume of a gas is expressed in terms of temperature and pressure, then

$$V = V(p, T)$$

and

$$dV = \left(\frac{\partial V}{\partial p}\right)_T dp + \left(\frac{\partial V}{\partial T}\right)_p dT$$

(4.20)

Substituting

$$\left(\frac{\partial V}{\partial p}\right)_T = -\frac{V}{p} \qquad \text{(from Boyle's law)}$$

and

$$\left(\frac{\partial V}{\partial T}\right)_p = \frac{V}{T} \qquad \text{(from Charles' law)}$$

into Eq. (4.20) and integrating gives,

$$\ln V + \ln p = \ln T + \ln \text{(constant)}$$

which is equivalent to Eq. (4.19). The preceding equation can be written as

$$\frac{p_1 V_1}{T_1} = \frac{p_2 V_2}{T_2} = mR$$

or

$$pV = mRT \qquad\qquad\qquad (4.21)$$

where R is the *specific gas constant*. Its value depends on the particular gas and the units used for pressure, specific volume, and absolute temperature. Equation (4.21) is called the *equation of state of an ideal gas,* and is represented on the p-V-T coordinates, as shown in Figure 4.14. If both sides of Eq. (4.21) are divided

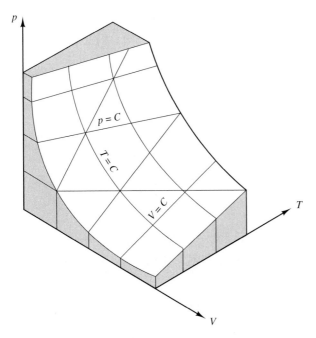

Figure 4.14   *p-V-T* surface for an ideal gas.

by the mass, and since the volume V occupied by a mass m is equal to the product mv, then

$$pv = RT \qquad (4.22)$$

It is often convenient to express the equation of state on a mole basis. The mass m is equal to the product of the number of moles n and the molar mass M. Therefore, substituting for m in Eq. (4.21) gives

$$pV = n\,MRT$$

If $n = 1$, V is the volume occupied by 1 mole or the molar volume. Note that at equal temperature and pressure, ideal gases occupy the same molar volume (Avogadro's law). Considering two ideal gases a and b of equal volumes at the same temperature and pressure then,

$$p_a V_a = n_a M_a R_a T_a$$

and

$$p_b V_b = n_b M_b R_b T_b$$

or

$$\frac{p_a V_a}{n_a T_a} = \frac{p_b V_b}{n_b T_b} = M_a R_a$$

$$= M_b R_b = \text{constant}$$

The product MR is a constant for all ideal gases. It is called the *universal gas constant* and is denoted by $\bar{R}$. Since pV has the dimensions of energy, $\bar{R}$ has the dimensions of energy per mole divided by the absolute temperature, and its value depends on the units chosen. Therefore, the general form of the ideal-gas law on a mole basis is

$$pV = n\,\bar{R}T \qquad (4.23)$$

Table A1 gives values of $\bar{R}$ in different units.

4.8 Internal Energy, Enthalpy, and Specific Heats for an Ideal Gas

For an ideal gas, the internal energy and enthalpy are functions of temperature only so that

$$du = c_v\,dT$$

and

$$\Delta u = u_2 - u_1 = \int_{T_1}^{T_2} c_v(T)dT \qquad (4.24)$$

Similarly,

$$dh = c_p \, dT$$

and

$$\Delta h = h_2 - h_1 = \int_{T_1}^{T_2} c_p(T)dT \qquad (4.25)$$

At low pressure and at temperatures above the critical temperature, the behavior of real gases approaches that of ideal gases, and specific heats become mainly a function of temperature. For example, an expression for $c_p(T)$ that correlates data with high accuracy over a wide range of temperature is given by

$$c_p = a + bT + cT^2 + dT^3 + eT^4 \qquad (4.26)$$

where a, b, c, d, and e are constants. Table A10 of the Appendix gives values of these constants for a number of real gases, and Figure 4.15 shows the variation of c_p with temperature for several monatomic, diatomic, and triatomic gases.

Using specific heat data, Eqs. (4.24) and (4.25) can be integrated to give the values of internal energy and enthalpy as a function of temperature. The results are tabulated in the Appendix. Table A11 gives properties for air, and Tables A12 to A17 are similar tables for various gases. The tabulated properties take into consideration the variation of the specific heat with temperature, and the values are referenced to an arbitrary reference temperature. Table A11, for example, is based on $h = 0$ at $T = 0$ K; other tables are based on $h = 0$ at $T = 298$ K. Note that the reference state is not relevant since applications require only differences rather than absolute values of these properties.

The enthalpy of an ideal gas is

$$h(T) = u(T) + pv = u(T) + RT$$

Noting that both h and u are functions of temperature only, differentiation with respect to T gives

$$\frac{dh}{dT} = \frac{du}{dT} + R$$

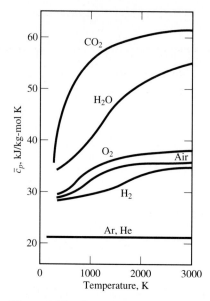

Figure 4.15   Constant-pressure specific heats for some gases.

Substituting the definitions of specific heats for an ideal gas gives

$$c_p - c_v = R \qquad (4.27)$$

Note that for an ideal gas both c_p and c_v are functions of temperature, but their difference is independent of temperature; it is a constant equal to R. Dividing Eq. (4.27) by c_p and noting that the specific heat ratio $c_p/c_v = \gamma$, then

$$c_p = \frac{\gamma R}{\gamma - 1} \qquad (4.28)$$

Similarly, when Eq. (4.27) is divided by c_v,

$$c_v = \frac{R}{\gamma - 1} \qquad (4.29)$$

Average values of specific heats in a certain range of temperature are given by

$$(c_v)_{\text{avg}} = \frac{\displaystyle\int_{T_1}^{T_2} c_v(T)\,dT}{T_2 - T_1} \qquad \text{and} \qquad (c_p)_{\text{avg}} = \frac{\displaystyle\int_{T_1}^{T_2} c_p(T)\,dT}{T_2 - T_1}$$

Example 4.5

An ideal gas occupies a volume of 36 m³ at a pressure of 90 kPa and a temperature of 25°C. Calculate the volume of the gas when the pressure is 101.3 kPa and the temperature is 4°C.

SOLUTION

Using the relation

$$\frac{p_1 V_1}{T_1} = \frac{p_2 V_2}{T_2}$$

the volume V_2 of the gas is

$$V_2 = V_1 \left(\frac{p_1}{p_2}\right)\left(\frac{T_2}{T_1}\right)$$

$$= (36 \text{ m}^3)\left(\frac{90}{101.3}\right)\left(\frac{4 + 273.15}{25 + 273.15}\right) = 29.73 \text{ m}^3$$

Example 4.6

Find the volume in liter/gm-mol and ft³/lbm-mol occupied by an ideal gas at a pressure of 1 atm and a temperature of 273.15 K (491.67 R).

SOLUTION

$$\bar{v} = \frac{\bar{R}T}{p} = \frac{(0.082056 \text{ lit. atm/gm-mol K})(273.15 \text{ K})}{1 \text{ atm}}$$

$$= 22.413 \text{ liters/gm-mol}$$

$$= \frac{(1545.33 \text{ ft-lbf/lbm-mol R})(491.67 \text{ R})}{(14.7 \text{ lbf/in}^2)(144 \text{ in}^2/\text{ft}^2)}$$

$$= 359 \text{ ft}^3/\text{lbm-mol}$$

Example 4.7

The constant-pressure specific heat for water vapor as a function of temperature is given by

$$\frac{\bar{c}_p}{\bar{R}} = 4.07 - 1.108 \times 10^{-3}T + 4.152 \times 10^{-6}T^2 - 2.964 \times 10^{-9}T^3 + 0.807 \times 10^{-12}T^4$$

where the units of $\bar{c}_p$ and $\bar{R}$ are kJ/kg-mol K and T in K. Using this expression, determine the change in enthalpy when the water vapor is heated from an initial state of $T_1 = 500$ K and $p_1 = 100$ kPa to a final state of $T_2 = 1000$ K and $p_2 = 300$ kPa. Compare the result with the value found from steam tables. What is the average value c_p in that range of temperature?

SOLUTION

Assuming ideal gas, the enthalpy change is

$$h_2 - h_1 = \int_{T_1}^{T_2} c_p\, dT = \left(\frac{\bar{R}}{M}\right)\left[4.07(T_2 - T_1) - \frac{1.108 \times 10^{-3}}{2}(T_2^2 - T_1^2) + \frac{4.152 \times 10^{-6}}{3}(T_2^3 - T_1^3) \right.$$

$$\left. - \frac{2.964 \times 10^{-9}}{4}(T_2^4 - T_1^4) + \frac{0.807 \times 10^{-12}}{5}(T_2^5 - T_1^5) \right]$$

Substituting the values of T_1 and T_2 gives

$$h_2 - h_1 = \left(\frac{8.3144 \text{ kJ/kg-mol K}}{18.02 \text{ kg/kg-mol}}\right)\left[(2035 - 415.5 + 1211 - 631.4 + 156.36) \text{ K}\right]$$

$$= 1086.81 \text{ kJ/kg}$$

From superheated steam tables,

$$h_2 - h_1 = (4639.7 - 3488.1) \text{ kJ/kg} = 1151.6 \text{ kJ/kg}$$

which is 5.6 percent greater than the value obtained using the ideal-gas specific heat. The average c_p is

$$(c_p)_{\text{avg}} = \frac{\int_{T_1}^{T_2} c_p\, dT}{T_2 - T_1} = \frac{1086.81 \text{ kJ/kg}}{(1000 - 500) \text{ K}} = 2.7136 \text{ kJ/kg K}$$

4.9 Molecular Interpretation of the Ideal-Gas Model

The equation of state of an ideal gas can be derived by making use of a mechanical model dictated by the kinetic theory of gases. As shown in Chapter 1, the product pV, according to Eq. (1.18), is

$$pV = \frac{1}{3} Nmv_{\text{rms}}^2$$

where N is the number of molecules, m is the mass of each molecule, and v_{rms} is the root mean square speed.

Comparing the foregoing equation with Eq. (4.23) indicates that the absolute temperature and, consequently, the internal energy of a gas, is proportional to, and can be interpreted as, a measure of the translational kinetic energy of the molecules. The justification of the statement made in Chapter 1 regarding temperature not being a primary concept is now established. It should be remarked that reference to properties, such as temperature or pressure, apply to a large number of molecules, and it is meaningless to refer to such properties for a single molecule.

Avogadro's number N_a was defined in Chapter 1 as the number of molecules per kg-mol. The value of Avogadro's number is $N_a = 6.0221 \times 10^{26}$ molecules/kg-mol. Noting that the number of moles n is equal to N/N_a, Eq. (4.23) can be written as

$$pV = \frac{N}{N_a}\overline{R}T = N\frac{\overline{R}}{N_a}T = N\,kT \qquad (4.30)$$

where k is the Boltzmann's constant.

The total kinetic energy of translation is the sum of the kinetic energies of the molecules or

$$U_{trans} = \frac{1}{2}\sum_i m_i v_i^2$$

but since,

$$n\overline{R}T = \frac{1}{3}\sum_i m_i v_i^2$$

and

$$nN_a = N$$

then

$$\frac{1}{2}\sum_i m_i v_i^2 = \frac{3}{2}n\overline{R}T = \frac{3}{2}N\,kT$$

Therefore, the average kinetic energy of translation per molecule is

$$\varepsilon_{trans} = \frac{\frac{1}{2}\sum_i m_i v_i^2}{N} = \frac{3}{2}kT = \frac{3}{2}\frac{\overline{R}T}{N_a} \qquad (4.31)$$

Equation (4.31) states that the average translational kinetic energy of a molecule of an ideal gas is a sole function of its absolute temperature. This means that in a mixture of ideal gases in thermal equilibrium, the value of ε_{trans} is the same for all the molecules of the mixture. Furthermore, at the same temperature, molecules of larger masses have a lower value of root mean square speed than those of lower masses. Note from Eq. (4.31) that the translational motion of the molecules ceases at zero absolute temperature.

The ideal-gas law states that at constant temperature the product pV is a constant. Under this condition, the preceding equations indicate that the mean translational kinetic energy of the molecules is independent of the pressure; it depends only on the temperature of the gas. Hence, the internal energy of an ideal gas, which is equal to the sum of the kinetic energies of its molecules, is a sole function of temperature. This is in accordance with Joule's law.

If the molecules of the gas are considered as point masses, then the translational energy per molecule, as given by Eq. (4.31), is equal to the internal energy of the molecules. Therefore, the molar specific heat at constant volume $\bar{c}_v$ can be determined by differentiating Eq. (4.31) with respect to the temperature to give

$$\bar{c}_v = \frac{3}{2}\bar{R} \qquad \text{(for monoatomic gases)} \qquad (4.32)$$

The ratio of specific heats is

$$\gamma = \frac{\bar{c}_p}{\bar{c}_v} = \frac{(3/2)\bar{R} + \bar{R}}{(3/2)\bar{R}} = \frac{5}{3} = 1.66 \qquad (4.33)$$

A detailed discussion of the specific heats of ideal gases is postponed until Chapter 13; it suffices here to indicate that the value cited is in agreement with those observed for monatomic gases.

4.10 Compressibility Factor

From the ideal-gas law, it can be seen that the specific volume of a gas becomes very large when the pressure is low or the temperature is high. Hence, specific volume cannot be conveniently used to represent the behavior of real gases at low pressure or high temperature. To surmount this difficulty, the properties of real gases are indicated by means of a compressibility factor, which expresses the extent of deviation of the gas from an ideal gas. The compressibility factor Z is defined by

$$Z \equiv \frac{pv}{RT} \qquad (4.34)$$

The compressibility factor of an ideal gas has a value of unity under all conditions. For a real gas, the value of Z may be less or more than unity, depending on the temperature and pressure of the gas, but must be finite. As the pressure is reduced, the compressibility factor of any gas approaches unity since a gas acts more like an ideal gas as the pressure is lowered:

$$\lim_{p \to 0} Z = 1 \qquad (4.35)$$

Values of the compressibility factor of any gas may be determined experimentally. But it is also possible to determine values of Z for many substances, based on a limited amount of data. This is accomplished by describing gases in terms of reduced properties, rather than in terms of properties alone. For example, by dividing pressure by the critical pressure of the gas under consideration, the reduced pressure is obtained:

$$p_R \equiv \frac{p}{p_c} \qquad (4.36)$$

Similarly,

$$T_R \equiv \frac{T}{T_c} \qquad (4.37)$$

and

$$v_R \equiv \frac{v}{v_c} = \frac{v}{RT_c/p_c} \qquad (4.38)$$

The compressibility factor of any one-component substance is a function of only two properties, usually temperature and pressure so that

$$Z = f(T_R, p_R) \qquad (4.39)$$

Equation (4.39) is called the *law of corresponding states* and is the basis for the generalized compressibility chart shown in Figure 4.16.

Both temperature and pressure introduce deviations from ideal behavior, and on this chart, the compressibility factor Z is plotted as a function of reduced temperature and reduced pressure. According to this chart, the value of Z at the critical point is approximately 0.29, and one-component substances that are at the same reduced pressure and temperature have the same compressibility factor. Note that real gases or vapors behave like ideal gases at low pressures and high temperatures.

Generalized compressibility charts for different ranges of p_R and T_R are given in the Appendix (Figures A6 to A8). As seen from these charts, gases behave as an ideal gas at very low values of p_R regardless of temperature. Similarly, ideal-gas behavior can be assumed at high values of T_R with moderate values of p_R.

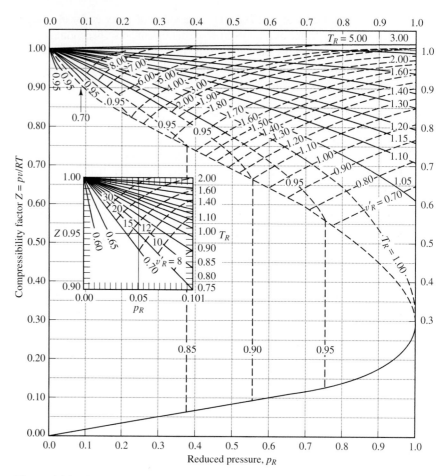

Figure 4.16   Generalized compressibility chart. (Modified by Peter E. Liley, *Chemical Engineering*, 94 (1987). *Original Source:* E. F. Obert, *Concepts of Thermodynamics.* McGraw-Hill, Inc., 1960.)

Note also that the deviation from ideal gas is greatest near the critical point. To further illustrate these observations, consider the *T-v* diagram for water, as shown in Figure 4.17. The percentage error in volume in treating water vapor as an ideal gas is indicated at different temperatures and pressures. In the shaded area, the percentage error is less than 0.5 percent but at high pressures the error increases and is worse in the vicinity of the critical point.

When the specific volume and either the pressure or temperature are given, the remaining property can be determined using the generalized compressibility chart. The procedure, however, is a trial-and-error solution, in which $v_R = v/v_c$ in combination with either p_R or T_R is used to determine the third variable. A more convenient method is to use a pseudo-reduced specific volume defined as

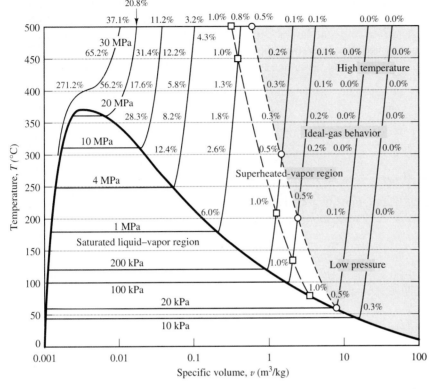

Figure 4.17   Steam behaves like an ideal gas at (a) a very low pressure regardless of temperature and (b) a high temperature and moderate pressure. Maximum deviation near the critical point.

$$v_R' = \frac{v}{\dfrac{RT_c}{p_c}} \tag{4.40}$$

Lines of constant v_R' are indicated in Figure 4.16, and any pair of the variables p_R, T_R, and v_R' can be used as independent parameters in entering the compressibility chart.

The generalized compressibility chart provides one of the best means of expressing deviation from ideal behavior and gives results with an accuracy of within 5 percent. A gas that has a compressibility factor less than unity is more compressible than an ideal gas.

The compressibility factor can be expressed in terms of the compressibility factor Z at the critical point. This relationship is derived by substituting Eqs. (4.36), (4.37), and (4.38) into Eq. (4.34):

$$Z = \frac{pv}{RT} = \frac{p_R v_R}{T_R} \frac{p_c v_c}{RT_c} = \frac{p_R v_R}{T_R} Z_c \tag{4.41}$$

The compressibility factor at the critical point has a value between 0.22 and 0.31; the exact value depends on the substance under consideration. Because Z is not affected by the law of corresponding states, it can be treated as an independent variable. Therefore, the compressibility factor is a function of the critical compressibility factor and the reduced pressure and temperature so that

$$Z = f(p_R, T_R, Z_c) \qquad (4.42)$$

Example 4.8

Determine the specific volume of steam at a pressure of 100 kPa and a temperature of 500°C by use of (a) steam tables; (b) the ideal-gas law; (c) van der Waals equation; $(p + a/v^2)(v - b) = RT$; and (d) compressibility chart.

SOLUTION

A sketch of the system is shown in Figure 4.18.

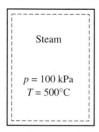

Steam

$p = 100$ kPa
$T = 500°C$

Figure 4.18

(a) From superheated steam tables:

$$v = 3.565 \text{ m}^3/\text{kg}$$

(b) Ideal-gas law:

$$v = \frac{RT}{p} = \frac{(8.3144 \text{ kJ/kg-mol K})/(18.015 \text{ kg/kg-mol})(773.15 \text{ K})}{100 \text{ kPa}}$$

$$= 3.567 \text{ m}^3/\text{kg}$$

(c) van der Waals equation: For water, the values of a and b are

$$a = 550.7(\text{kPa})(\text{m}^3/\text{kg-mol})^2$$
$$b = 0.0304(\text{m}^3/\text{kg-mol})$$

$$\left(p + \frac{a}{v^2}\right)(v - b) = RT$$

$$\left[\left(100 + \frac{550.7/(18.015)^2}{v^2}\right) \text{kPa}\right]\left[\left(v - \frac{0.0304}{18.015}\right) \text{m}^3\text{/kg}\right] = \left(\frac{8.3144 \text{ kJ/kg-mol K}}{18.015 \text{ kg/kg-mol}}\right)(773.15 \text{ K})$$

From which $v = 3.567$ m^3/kg.

(d) Compressibility chart:

$$T_R = \frac{T}{T_c} = \frac{773.15}{647.29} = 1.1944$$

$$p_R = \frac{p}{p_c} = \frac{100}{22090} = 0.00453$$

From the chart,

$$Z = 1.0 \qquad \text{and} \qquad v = ZRT/p = 3.567 \text{ m}^3\text{/kg}$$

Note that results obtained by these different methods are practically identical. The ideal-gas law proved valid because of the low density of the steam in this case.

Example 4.9

Determine the specific volume of N$_2$ at 8 MPa and 200 K using the generalized compressibility chart, and compare the result with the value calculated from the ideal-gas equation of state.

SOLUTION

$$pv = ZRT$$

For N$_2$,

$$p_c = 3.39 \text{ MPa}, \qquad T_c = 126.2 \text{ K}$$

and

$$R = \frac{8.3144 \text{ kJ/kg-mol K}}{28.013 \text{ kg/kg-mol}} = 0.2968 \text{ kJ/kgK}$$

$$p_R = \frac{p}{p_c} = \frac{8}{3.39} = 2.36$$

$$T_R = \frac{T}{T_c} = \frac{200}{126.2} = 1.585$$

At these values of p_R and T_R, Figure A7 gives $Z = 0.85$, so that

$$v = \frac{ZRT}{p} = \frac{(0.85)(0.2968 \text{ kJ/kg K})(200 \text{ K})}{8000 \text{ kPa}} = 0.0063 \text{ m}^3/\text{kg}$$

The ideal-gas equation of state gives

$$v = \frac{RT}{p} = \frac{(0.2968 \text{ kJ/kg K})(200 \text{ K})}{8000 \text{ kPa}} = 0.0074 \text{ m}^3/\text{kg}$$

Example 4.10

Determine the pressure of superheated steam at $T = 300°C$ and $v = 0.04532$ m³/kg using (a) steam tables, (b) ideal-gas equation of state, (c) the generalized compressibility chart.

SOLUTION

(a) From superheated steam tables at $T = 300°C$ and $v = 0.04532$ m³/kg, the pressure is 5 MPa.

(b) $p = \dfrac{RT}{v} = \left(\dfrac{8.3144 \text{ kJ/kg-mol K}}{18.015 \text{ kg/kg-mol}}\right)\dfrac{(300 + 273.15) \text{ K}}{0.04532 \text{ m}^3/\text{kg}}$

$= 5836.8 \text{ kPa}$

$= 5.8368 \text{ MPa}$

(c) For H_2O,

$$p_c = 22.09 \text{ MPa}, \qquad T_c = 647.3 \text{ K}$$

and

$$R = \frac{8.3144 \text{ kJ/kg-mol K}}{18.015 \text{ kJ/kg-mol}} = 0.4615 \text{ kJ/kg K}$$

The pseudo-reduced volume is

$$v_R' = \frac{v}{\dfrac{RT_c}{p_c}} = \frac{(0.04532 \text{ m}^3/\text{kg})(22090 \text{ kPa})}{(0.4615 \text{ kJ/kg K})(647.3 \text{ K})} = 3.3513$$

and

$$T_R = \frac{T}{T_c} = \frac{573.15}{647.3} = 0.8854$$

At these values of v'_R and T_R, Figure 4.16 gives $p_R = 0.24$. Therefore,

$$p = p_R p_c = (0.24)(22.09 \text{ MPa}) = 5.3016 \text{ MPa}$$

4.11 Summary

Properties of a single-component system are presented by means of two- or three-dimensional diagrams. Areas on a three-dimensional p-v-T diagram represent a single- or a two-phase system and are separated by the saturation lines. At a given pressure, a pure substance boils at a fixed temperature called the saturation temperature. Likewise, at a given temperature, it boils at a fixed pressure called the saturation pressure. The saturated-liquid line and the saturated-vapor line meet at the critical point. Beyond this point on a p-v-T diagram, liquid–vapor transformations are not possible. The sublimation line separates the solid phase from the vapor phase. The triple-state line represents the existence of the three phases in equilibrium. For most substances, the freezing temperature increases as the pressure increases. Water expands upon freezing, and increasing the pressure lowers the freezing temperature.

Properties of saturated liquid–vapor, superheated vapor, compressed liquid, and saturated solid–vapor states are tabulated in the Appendix as a function temperature and pressure. Two independent properties define the state of a pure substance. Properties such as u, h, and s are indicated, and since changes in these properties are needed, the datum of the numerical values is arbitrary.

During change of phase, the temperature and pressure of a pure substance are dependent. The quality x of a liquid–vapor mixture is the ratio of the mass of the vapor to the total mass of the mixture. The value of x varies from 0 (saturated liquid) to 1 (saturated vapor); it has no meaning in the compressed-liquid or superheated-vapor regions. Intensive properties such as specific volume, internal energy, and enthalpy of a two-phase mixture (say liquid–vapor) are given by

$$v_x = v_f + x v_{fg}$$
$$u_x = u_f + x u_{fg}$$
$$h_x = h_f + x h_{fg}$$

where subscript f indicates the liquid phase, and subscript fg indicates the difference in the property value between the vapor and liquid phases at the same temperature.

The equation of state of an ideal gas is

$$pv = RT$$

Deviation from the ideal-gas behavior can be indicated by the equation

$$pv = ZRT$$

where the compressibility factor Z is a function of the reduced temperature (T/T_c) and reduced pressure (p/p_c), where T_c and p_c are the critical temperature and pressure, respectively.

For an ideal gas, the internal energy, enthalpy, and specific heats are functions of temperature only. Changes of internal energy and enthalpy are given by

$$\Delta u = \int_{T_1}^{T_2} c_v(T)\, dT$$

and

$$\Delta h = \int_{T_1}^{T_2} c_p(T)\, dT$$

When either p or T is unknown, Z can be evaluated from a pseudo-reduced specific volume defined by

$$v_R' = \frac{v_{actual}}{RT_c/p_c}$$

Interpretation of the behavior of an ideal gas from a molecular point of view indicates that the molecular translational energy of monatomic gases is a function of temperature only and

$$\varepsilon_{tran} = \frac{3}{2}kT$$

where k is Boltzmann's constant.

Problems

4.1 Sketch the pressure–temperature diagram for a pure substance identifying the solid, liquid, and vapor regions. Use the sketch to describe in detail the transition from the solid state to the vapor state as the temperature is raised along a constant-pressure line. Consider the two cases when the pressure is below and above the triple point.

4.2 Sketch a T-v diagram identifying the solid, liquid, and vapor regions, and show several constant-pressure lines. Assume the substance (a) contracts upon freezing, (b) expands upon freezing.

4.3 Sketch a T-v diagram for H_2O at atmospheric pressure for the range of temperature from $T = -10°C$ to $100°C$, indicating the values of T and v at the initial and final points and at each change of phase.

4.4 Using steam tables, plot the density of saturated water and saturated steam as a function of pressure up to the critical pressure.

4.5 Using steam tables, plot the enthalpy–temperature relation for several constant-pressure lines. Identify the saturated-liquid and saturated-steam lines.

4.6 A rigid vessel of volume 0.04 m^3 contains saturated water at 2.5 MPa. If the pressure is reduced to 100 kPa, what are the volumetric proportions of the saturated liquid and saturated vapor in the final state?

4.7 Determine whether the state of the following water systems is compressed liquid, a mixture of saturated liquid and saturated vapor, or superheated vapor:
(a) $T = 200°C$, $p = 2.0$ MPa
(b) $T = 20°C$, $p = 100$ kPa
(c) $T = 100°C$, $v = 1.2$ m^3/kg
(d) $p = 50$ kPa, $v = 3.6$ m^3/kg
(e) $p = 10$ kPa, $v = 12$ m^3/kg

4.8 Determine the volume occupied by 1 kg of H_2O at 5 MPa if the temperature is (a) 40°C, (b) 400°C.

4.9 A closed rigid vessel has a volume of 0.005 m^3 and contains a liquid–vapor mixture of refrigerant-12 at 20°C with a quality of 90 percent. Determine the masses of the saturated liquid and saturated vapor in the tank.

4.10 A closed rigid tank contains 2 kg of saturated water vapor at 4 MPa. The tank is cooled until the quality reaches 80 percent. Determine the volume of the tank and the final pressure. What fraction of the total volume is occupied by each phase?

4.11 Complete the following table of properties of H_2O:

T, °C	p, kPa	v, m^3/kg	h, kJ/kg	u, kJ/kg	s, kJ/kg K	x (if applicable)
100	5000					
100	10					
100		1.0				
100		1.6729				
100				2000		
100					2511.6	
100					5.0	
100						0.8

4.12 Complete the following table of properties of H_2O:

T, °C	p, kPa	v, m^3/kg	h, kJ/kg	u, kJ/kg	s, kJ/kg K	x (if applicable)
	500		2000			
	50			1800		
	750					0.6
160				800		
1000					9.4690	
	400				7.8985	
0.01				−333.4		
0.01				2375.2		

4.13 Calculate the specific volume and specific enthalpy of 35 percent quality steam if the pressure is 20 kPa.

4.14 Steam at 550 kPa and a quality of 92 percent occupies a rigid vessel of volume 0.4 m^3. Calculate the mass, internal energy, and enthalpy of the steam.

4.15 A rigid vessel of volume 0.2 m^3 contains 1 kg of wet steam. What are the pressure and internal energy of the steam if the temperature is 250°C?

4.16 A rigid vessel with a volume of 0.03 m^3 contains 1.5 kg of refrigerant-12 (R-12) at 800 kPa. Determine the temperature of the refrigerant.

4.17 A tank of 0.4 m^3 capacity is filled with steam at 2 MPa and 250°C. The tank and contents are then cooled with negligible change in volume to 120°C.

(a) What will be the final amounts of saturated vapor and saturated water (in kg), and what is the corresponding quality of the mixture?

(b) How much energy as heat was transferred during this process?

4.18   Determine the amount of heat necessary to raise the temperature of 1 kg of saturated steam to 250°C at constant pressure of 1 MPa.

4.19   A 0.05 m³ rigid vessel contains a water–vapor mixture at 100 kPa. Determine the amount of heat transfer if the final state of the system is at the critical point. What is the quality of the mixture at the initial state?

4.20   A water–vapor mixture at 1.2 MPa with a quality of 0.8 is heated to a temperature of 250°C. Determine the heat interaction per kg of mixture if the process takes place at (a) constant volume, (b) constant pressure.

4.21   Saturated water at a pressure of 400 kPa expands reversibly along a constant-pressure path. If the final temperature is 300°C, determine the work and the heat interactions per unit mass of fluid.

4.22   Superheated refrigerant-12 (R-12) at 500 kPa and 80°C is condensed along a constant-pressure path to the saturated-liquid state. Determine the work and heat interaction per unit mass of fluid.

4.23   A rigid closed vessel contains dry saturated steam at 150 kPa. Determine the amount of heat interaction necessary to raise the temperature to 200°C.

4.24   A rigid closed tank of volume 0.1 m³ contains steam at 100 kPa and 200°C. Calculate the amount of heat transfer in order to lower the pressure to 50 kPa.

4.25   A 0.5 kg of air at 200 kPa and 25°C is contained in a vertical cylinder fitted with a frictionless piston, as shown in Figure 4.19. The piston rests on a set of stops, and a pressure of 400 kPa is required to move it. Heat is transferred to the system till the volume is doubled. Determine the final temperature and the work and heat interactions.

Air
0.5 kg
200 kPa
25°C

Figure 4.19

4.26   A steam main closed at both ends has an internal volume of 1.6 m³, which is filled with dry saturated steam at 140°C.

(a) How much heat must be transferred from the steam to reduce the quality to 0.5?

(b) What will be the pressure in the main when the quality is 0.5?

(c) Show the process and end points on a p-V diagram.

4.27   Water vapor initially at 200°C and 300 kPa expands reversibly in a piston–cylinder assembly to 100 kPa. If the expansion takes place according to the relation $pv^{1.2} = C$, determine the work and heat interactions per kg.

4.28   Determine the work required to compress one kilogram of refrigerant-12 in a quasi-equilibrium process according to the equation

$$pv = \text{constant}$$

The initial state is 400 kPa and 20°C, and the final pressure is 800 kPa.

4.29   A piston–cylinder assembly contains H_2O at a temperature of 200°C and a specific volume of 0.025 m³/kg. The H_2O expands isothermally until its internal energy reaches 1500 kJ/kg. Determine the final specific volume and the work and heat interactions.

4.30   One kilogram of steam undergoes a cycle consisting of three processes represented by straight lines, as shown in Figure 4.20. Sketch the cycle on a T-v diagram, and determine the change in internal energy, work, and heat interactions for each process.

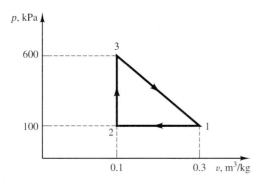

Figure 4.20

4.31 One kilogram of steam is confined in a piston–cylinder assembly at 1 MPa and 98 percent quality. The piston is permitted to move in such a way that when the contained volume is six times as great as the initial volume, the pressure has dropped to 100 kPa. At the same time, the heat transfer amounts to −4.5 kJ. Calculate the work interaction during the process.

4.32 A rigid tank of total volume of 0.8 m³ contains a water–vapor mixture at 150 kPa. The volumes occupied by the saturated liquid and saturated vapor are 0.004 and 0.796 m³. Determine the amount of heat required to vaporize the liquid and the pressure at the completion of the heat transfer process.

4.33 One kilogram of steam at a quality of 40 percent and a pressure of 1.3 MPa is heated at a constant pressure to a temperature of 250°C. Neglecting changes in kinetic and potential energies, calculate the heat interaction and the work interaction.

4.34 Superheated steam at a pressure of 800 kPa and a temperature of 300°C is cooled in a piston–cylinder assembly at constant pressure to the saturated vapor state. It then undergoes constant volume cooling to a pressure of 400 kPa. Neglecting changes in kinetic and potential energy, determine the work and heat interactions in kJ/kg for the overall process.

4.35 Two kilograms of ammonia at 1.2 MPa and 100°C are cooled at constant volume to 0.275 MPa. Find the

change in internal energy and the amount of heat and work interactions.

4.36 A system contains 1 kg of steam at 1.2 MPa, and a quality of 0.2 is heated until the temperature is 250°C. Calculate the heat transfer if the process is carried out at (a) constant pressure, (b) constant volume.

4.37 One kilogram of water at 20°C in a vertical piston–cylinder assembly is initially subjected to a pressure of 500 kPa by the combined weight of a frictionless piston and the atmospheric pressure. If the expansion of the system is limited to a volume of 0.227 m³/kg by stops in the cylinder, determine the amount of heat transfer to the system if the final state is saturated vapor. What is the quality of the mixture when the piston first reaches the stops in the cylinder? Represent the process on a *T-v* diagram.

4.38 Find the heat transfer, change in internal energy, and work transfer when 1 kg of a water–vapor mixture at 1.0 MPa and a quality of 0.2 undergoes one of the following processes: (a) the mixture is heated in a constant-volume process to 4.0 MPa, (b) the mixture is heated in a constant-pressure process to the saturated-vapor line.

4.39 A mass of 1.8 kg of saturated liquid water at 200°C is mixed with 7.2 kg of saturated water vapor at the same pressure. What is the quality of the resulting mixture? What is the volume and the specific enthalpy of the mixture?

4.40 A rigid, thermally insulated vessel contains 0.2 kg of a water–vapor mixture at 75 kPa and a quality of 30 percent. An electric heater supplies energy to the vessel till the pressure in the vessel reaches 200 kPa. Determine (a) the volume of the vessel, (b) the quality of the mixture in the final state, (c) the electric energy supplied in kJ.

4.41 A rigid vessel of volume of 0.5 m³ initially contains a water–vapor mixture at 0.6 MPa.
(a) If the quality of mixture is 40 percent, calculate the mass of the mixture.
(b) If the pressure in the vessel is raised to 0.9 MPa by the transfer of heat, what will be the mass of the vapor and the mass of the liquid?

4.42 A piston–cylinder assembly contains 1 kg of substance at 100 kPa. The initial volume is 0.5 m³. Heat is transferred to the substance in an amount necessary to cause a quasi-equilibrium expansion until the final volume is twice the initial volume. If the temperature is maintained constant, determine the magnitude of the heat transfer required if the substance is (a) nitrogen, (b) H_2O.

4.43 A rigid container has a volume of 0.35 m³ and contains water vapor at 250°C and 100 kPa. The water vapor is allowed to cool to 50°C. Determine the energy lost by the water and the final state of the system.
    What would be the heat transfer if the cooling took place at constant pressure from the initial state to the saturated vapor state?

4.44 A 0.5 kg water–vapor mixture is enclosed in a rigid insulated tank. Initially, the temperature is 60°C, and the quality is 0.53. If the system is heated electrically causing the pressure to rise to 0.1 MPa, determine the amount of electrical work.

4.45 A closed rigid vessel contains 2 percent liquid and 98 percent water vapor by volume in equilibrium at 150°C. Determine the quality of the mixture. If the mixture is cooled to 100°C, what will be the quality and the liquid and vapor percentages by volume at the new state?

4.46 A vertical piston–cylinder assembly contains 1 kg of steam. The piston is frictionless and rests on two stops. The steam is initially at 1 MPa and 75% quality. Heat is transferred to the system until the temperature is 500°C. However, the piston does not move until the system pressure reaches 2 MPa, and then it moves at constant pressure. Determine (a) the work, (b) the total heat transferred.

4.47 A closed rigid vessel contains 10 percent liquid and 90 percent vapor of refrigerant-12 by volume. If the temperature of the mixture is 0°C, determine the quality of the mixture. If the mixture is cooled to −30°C, what will then be the quality?

4.48 Steam at 100°C and 20 percent quality and an initial volume of 0.5 m³ undergoes a constant-pressure process till its quality becomes 90 percent. Calculate (a) the work transfer, (b) the heat transfer.

4.49 A water–vapor mixture at a pressure of 0.1 MPa undergoes three processes. First it is heated at constant volume to a pressure of 0.5 MPa and a quality of 20 percent. This is followed by a constant-pressure process until the mixture becomes saturated vapor. Finally, the system is cooled at constant volume to the original pressure of 0.1 MPa. Determine (a) the quality at the initial and final states, (b) the work transfer, (c) the heat transfer from the initial to the final state.

4.50 A piston–cylinder assembly contains 1.0 kg of steam of quality 0.2 and an initial temperature of 100°C. At this state, a spring is just touching the piston. Heat is transferred causing the piston to move against the spring force. If the force exerted by the spring is proportional to the distance moved, and the spring constant is 10 kN/m, determine (a) the volume of the cylinder when the pressure reaches 300 kPa, (b) the work transfer, (c) the heat transfer.

4.51 One kilogram of water initially at 5 MPa and 100°C is heated at constant pressure. If the final state is superheated steam at 300°C. Determine:
(a) the initial specific volume of the water
(b) v_f and v_g at 5 MPa
(c) the final specific volume of the superheated steam
(d) if the final state were wet stream at 50 percent quality, what is v_x?
(e) the work done for case (d)

4.52 Saturated water undergoes process 1-2-3-4, as shown in Figure 4.21. The initial pressure is 0.2 MPa and final pressure is 0.5 MPa. If the quality at point 2 is 30 percent, find the work and the heat interactions for process 1-4.

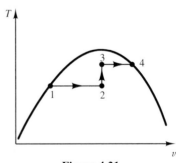

Figure 4.21

4.53 One kilogram of superheated water vapor at 500 kPa and 300°C is cooled first at constant pressure to the saturated vapor state and then at constant volume till the pressure reaches 100 kPa. Represent the processes on a T-v and a p-v diagram, and determine the work and heat interactions for each process.

4.54 Compare the work required to compress reversibly 1 kg of air from an initial state of 100 kPa and 300 K to a final pressure of 1000 kPa for the following processes:
(a) polytropic with $n = 1.2$
(b) adiabatic with $n = 1.4$
(c) isothermal

4.55 The steam flow through a turbine operating at steady state is 0.2 kg/s. At the inlet, the pressure is 800 kPa, and the temperature is 250°C. At the exit, the pressure is 10 kPa, and the quality is 0.94. If the power output of the turbine is 15 kW, determine the rate of heat transfer.

4.56 Steam at 2 MPa and 350°C enters an adiabatic turbine operating at steady state. At the exit, the pressure is 10 kPa, and the quality is 0.90. Determine the mass rate of flow for a power output of 1 MW.

4.57 Refrigerant-12 at 100 kPa enters an adiabatic compressor as saturated vapor and leaves at 400 kPa and 60°C. The mass rate of flow is 0.2 kg/s. Assuming steady-state operation, determine the power input to the compressor and the volumetric flow rate at the exit.

4.58 Ammonia at 98 kPa and a quality of 95 percent, enters an adiabatic compressor operating at steady state. If the exit conditions are 1.0 MPa and 90°C, determine the flow rate of ammonia if the power consumption is 12 kW.

4.59 Saturated liquid ammonia at a pressure 1.2 MPa is throttled in a steady-flow process to a pressure of 80 kPa. What is the quality of the resulting mixture if the process is adiabatic?

4.60 An adiabatic steam turbine, operating at steady state, develops 1500 kW of power. At the inlet, the pres-sure is 400 kPa, and the temperature is 300°C, and at the exit, the pressure is 10 kPa. The flow rate of the steam is 2.5 kg/s. Neglecting kinetic and potential energy changes, determine the quality (if wet) or temperature (if super-heated) of the exhaust steam.

4.61 Steam at 600 kPa and a quality of 0.8 enters a chamber where it is mixed with a water stream entering at 600 kPa and 40°C. The resulting mixture leaves as sat-urated liquid at 600 kPa. For steady-state operation, de-termine the ratio of the mass flow rates of the steam and water entering the chamber.

4.62 A throttling valve operating at steady state re-duces the pressure of steam as it flows through a pipeline from 1 MPa to 0.3 MPa. The superheat is 5°C upstream of the valve, and the velocity is low. Some distance after the valve, the steam fills the pipe and proceeds at negli-gible velocity. Find the temperature and the degree of superheat of the steam downstream of the valve. (As-sume negligible heat loss from the pipe.)

4.63 A sampling tube and a throttling calorimeter are installed in order to find the enthalpy of steam flowing in a steam main. The steam in the main is 210°C and after passing through the calorimeter is at 10 cm Hg gauge pressure and 140°C. The barometric reading is 76 cm Hg abs. Find (a) the enthalpy of steam in the main, (b) the specific volume of steam in the main.

4.64 Saturated water vapor at 0.65 MPa and saturated water at 0.65 MPa are delivered from separate lines (through throttling valves) to a 0.18 m³ tank, which is initially evacuated. It is desired to obtain, as a final condition in the tank, saturated vapor at a pressure of 0.3 MPa. Calculate the mass of the water and the mass of steam to be admitted from the separate lines. (As-sume the tank is insulated and the velocity terms are negligible.)

4.65 An evacuated chamber is connected through a valve with a very large tank containing steam at 400 kPa and 200°C. The valve is opened, and steam flows into the chamber until the pressure within the chamber is 340 kPa. At the same time, heat is transferred from the chamber at the rate of 300 kJ/kg of steam introduced.

Determine the temperature and the quality of the steam in the chamber when the flow stops.

4.66   A 0.5 m³ rigid tank contains steam at 1.8 MPa and 300°C. A valve on the tank is opened, and steam is allowed to escape until the pressure drops to 200 kPa. If during the venting process, the temperature is maintained constant, determine the amount of heat transfer.

4.67   A 0.7 m³ rigid vessel contains a water–vapor mixture at 300°C. The volume of the vapor is 0.4 m³, and the volume of the liquid is 0.3 m³. If the temperature remains constant as liquid is withdrawn from the bottom of the tank, calculate the amount of heat transfer when half the initial mass in the tank has been withdrawn.

4.68   What is the density of air at atmospheric pressure and 25°C?

4.69   Determine the specific volume of superheated steam at 8 MPa and 300°C using (a) the ideal-gas equation, (b) the compressibility chart. Compare these values with the value indicated in the steam tables.

4.70   Determine the pressure of steam at 450°C and a specific volume of 0.0335 m³/kg using (a) the ideal-gas equation, (b) the generalized compressibility chart, (c) the steam tables.

4.71   Determine the specific volume of refrigerant-12 at 1.4 MPa and 100°C using (a) the ideal-gas equation, (b) the compressibility chart. Compare the results with the value indicated in refrigerant-12 tables.

4.72   Calculate the specific volume of oxygen at a pressure of 20 atm and a temperature of 140 K. Compare the result with that obtained by the ideal-gas law. The critical temperature and pressure of oxygen are $T_c = 154.78$ K and $p_c = 50.1$ atm.

4.73   A spherical tank 30 cm in diameter contains nitrogen initially at a pressure of 20 MPa and a temperature of −75°C. If the system receives heat at the rate of 5 kJ/s, find the time elapsed till the temperature reaches −15 °C. What is the final pressure?

4.74   Using the compressibility chart, calculate:
(a) the density of nitrogen at 27 MPa and 20°C
(b) the temperature of 3 kg of CO_2 gas in a container of volume 0.5 m³ at a pressure of 400 kPa

4.75   A rigid vessel contains superheated steam initially at a pressure of 16 MPa and a temperature of 500°C. Heat interaction with the surroundings reduces the temperature of the steam to 360°C. Using the compressibility chart, determine the specific volume at the initial state and the final pressure in the vessel. Compare your results with values obtained from steam tables.

C4.1   Write an interactive computer program to calculate the specific volume for the following gases: (a) oxygen, (b) carbon dioxide, (c) helium, (d) hydrogen. Each gas is contained in a separate rigid tank and behaves like an ideal gas. The user inputs temperature and pressure for each gas.

C4.2   Using the software provided with the book, determine the following properties at 500°C and 500 kPa for the gases listed in problem C4.1: (a) specific heat at constant pressure, (b) specific heat at constant volume, (c) enthalpy, (d) entropy.

C4.3   Using the software provided with the book, determine the following properties for H_2O at a temperature of 500°C and pressures of 110, 210, 310, and 510 kPa: (a) specific volume, (b) enthalpy, (c) internal energy, (d) entropy.

C4.4   Using the software provided with the book, determine the following properties for a water–vapor mixture having a quality of 35 percent at a temperature of 175°C: (a) specific volume, (b) enthalpy, (c) internal energy, (d) entropy.

C4.5   Using the software provided with the book, determine the following properties for H_2O having an enthalpy of 770 kJ/kg and temperatures of 50, 110, 400, and 1000°C: (a) specific volume, (b) internal energy, (c) entropy.

Chapter 5

The Second Law
of Thermodynamics

5.1 Introduction

Processes that occur spontaneously in nature proceed toward equilibrium states. Water flows from high to low levels; heat flows from hot to cold bodies; and gases expand from high pressures to low pressures. Spontaneous processes can be reversed, but they will not reverse themselves spontaneously even though energy balance is satisfied. Energy must be supplied to the system for a non-spontaneous process to occur. Energy from an external source is required to pump water from a low to a high level, or to compress a gas from a low to a high pressure, or to transfer heat from a cold body to a hot body. This means that a permanent change in the surroundings would occur.

A spontaneous process can proceed only in a definite direction. The first law of thermodynamics gives no information about direction; it merely states that when one form of energy is converted into another, identical quantities of energy are involved regardless of the feasibility of the process. In that sense, events could be envisioned that would not violate the first law such as the transfer of a certain quantity of heat from a low-temperature body to a high-temperature body, without the expenditure of work. But experience indicates that this process is not possible and therefore the first law alone is inadequate in depicting completely energy transfer. Similarly, experiments demonstrated that when energy in the form of heat is transferred to a system, only a portion of the heat can be converted into work. On the other hand, Joule's experiments showed that when energy is supplied to a system in the form of work, it can be converted completely into heat. Evidently, heat and work are not completely interchangeable methods of energy transfer. Furthermore, when energy is transformed from one form into another, there often is also a degradation of the supplied energy into a less "useful" form. In fact, experience indicates that natural processes are accompanied by the dispersal of energy.

Consider the system shown in Figure 5.1a. The engine receives a quantity of heat, Q_H,[1] from a high-temperature thermal energy reservoir at T_H and performs an equal amount of work W_e, in accordance with the first law of thermodynamics. Let a heat pump, operating between a low-temperature reservoir at T_L and the high-temperature reservoir at T_H, extract heat Q_L, deliver heat Q'_H, and require work W_p. If the pump could be adjusted so that the heat supplied by the pump and the heat transferred to the engine are identical, that is, $Q'_H = Q_H$, the high-temperature reservoir becomes superfluous and can be eliminated without affecting the operation of either the engine or the pump. According to the first law, the energy required to perform work W_p would be less than the heat interaction Q_H. Therefore, as shown in Figure 5.1b, the engine driving the pump would also be providing net external work, such that $W_{net} = W_e - W_p$. In the system enclosed by the dotted boundaries, a quantity of heat, Q_L, would be transferred from the low-temperature reservoir, and an equal amount of work, W_{net}, would be delivered to the surroundings. If this were feasible, then it would be possible to convert the internal energy of low-temperature bodies in the uni-

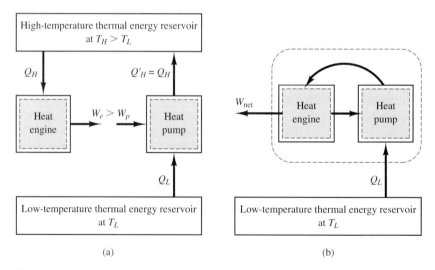

Figure 5.1   Impossibility of a system executing a cycle to convert heat from a single thermal reservoir completely into work.

---

[1] The subscript does not pertain to a state of a system but rather serves to identify a quantity of heat.

verse, such as the earth or the ocean, into useful work. This has never been achieved and is not consistent with human experience. We must therefore conclude that our original assumption was invalid and that an engine extracting heat from a single thermal reservoir cannot convert all the heat to work. On the other hand, there is no reason from the standpoint of thermodynamics why work cannot be completely converted into heat, and, in fact, this conversion is readily achieved.

The second law of thermodynamics establishes the difference in quality between different forms of energy and explains why some processes can spontaneously occur, whereas others cannot. It indicates a trend of change and is usually expressed as an inequality. Like other physical laws of nature, the second law of thermodynamics has been confirmed by experimental evidence.

Following are some definitions that will be used in subsequent discussion of the second law.

Thermal reservoir: A thermal (energy) reservoir is a sufficiently large system in stable equilibrium to which and from which finite amounts of heat can be transferred without any change in its temperature. In dealing with heat-engine cycles, for example, a high-temperature reservoir, from which heat is transferred (heat source), and a low-temperature reservoir, to which heat is transferred (heat sink), will be considered.

Heat engine: A heat engine is a thermodynamic system operating in a cycle to which net heat is transferred and from which net work is delivered. The system undergoes a series of processes that constitute the heat-engine cycle.

An index of performance of a work-producing machine or a heat engine is expressed by its *thermal efficiency,* defined as the ratio of the net work output to the heat input. Only a fraction of the heat input is converted into work, and the rest is rejected. Figure 5.2 shows a system that executes a cycle while interacting thermally with two thermal reservoirs, and developing net work. Let

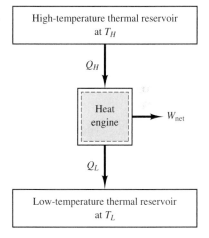

Figure 5.2   Principle of a heat engine.

Q_H = heat transferred to the system from the high-temperature thermal reservoir

Q_L = heat transferred from the system to the low-temperature thermal reservoir

W_{net} = net work developed by the system

Then the thermal efficiency of the cycle is given by

$$\eta_{th} = \frac{W_{net}}{Q_H}$$

but the first law of thermodynamics applied to a cycle requires that

$$\oint \delta Q + \oint \delta W = 0 \qquad \text{or} \qquad Q_H - Q_L = W_{net}$$

Therefore,

$$\eta_{th} = \frac{W_{net}}{Q_H} = \frac{Q_H - Q_L}{Q_H} = 1 - \frac{Q_L}{Q_H} \qquad (5.1)$$

Note that the efficiency of any power cycle operating between two thermal reservoirs is always less than unity. Only a portion of Q_H can be transformed into work, and the remainder Q_L is discarded to the low-temperature reservoir.

A thermoelectric engine converts heat into electric energy. As shown in Figure 5.3, it consists of two dissimilar electric conductors connected at two junctions. When the two junctions are maintained at different temperatures, an electric current flows in the conductors. An electric motor placed in the circuit may be used to supply mechanical work.

With the same notation used before, the efficiency of the thermoelectric device is

$$\eta = \frac{W_{net}}{Q_H} = \frac{\dot{W}_{net}}{\dot{Q}_H} = \frac{\mathcal{E}I}{\dot{Q}_H}$$

where $\mathcal{E}$ and I are the electric potential and current, respectively.

Heat pump: A heat pump is a thermodynamic system operating in a cycle that removes heat from a low-temperature body and delivers heat to a high-temperature body. In accomplishing this, the heat pump receives external energy in the form of work. The system or working substance undergoes a series of processes that constitute the heat-pump cycle.

The heat pump may be used as a refrigerator where the primary function is the transfer of heat from a low-temperature system. In a heating system utilizing a heat pump, the primary function is the transfer of heat to a high-temperature system.

An index of performance of a refrigeration or a heat-pump cycle is the *coefficient of performance, β*. The definition of β depends on whether the primary

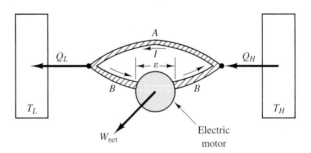

Figure 5.3   Thermoelectric heat engine.

function is to remove energy Q_L from the low-temperature reservoir or to deliver energy Q_H to the high-temperature reservoir. Referring to Figure 5.4, let

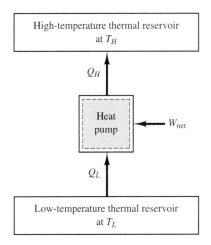

Q_H = heat transferred from the system to the high-temperature thermal reservoir

Q_L = heat transferred to the system from the low-temperature thermal reservoir

W_{net} = net work required by the system

Applying the first law of thermodynamics to the cycle gives

$$\oint \delta Q + \oint \delta W = 0 \qquad \text{or} \qquad Q_H - Q_L = W_{net}$$

Therefore, the coefficient of performance of a refrigeration cycle is

Figure 5.4   Principle of a heat pump.

$$\beta_{ref} = \frac{Q_L}{W_{net}} = \frac{Q_L}{Q_H - Q_L} \tag{5.2}$$

and for a heat pump cycle,

$$\beta_{heat\ pump} = \frac{Q_H}{W_{net}} = \frac{Q_H}{Q_H - Q_L} = 1 + \beta_{ref} \tag{5.3}$$

It is clear from Eqs. (5.1), (5.2), and (5.3) that, in expressing the effectiveness of a power cycle in terms of efficiency or refrigeration and a heat-pump cycle in terms of coefficient of performance, the energy of primary interest is divided by the energy expended. Note also that since W_{net} is finite, the coefficient of performance can never be infinite.

A thermoelectric heat pump transfers heat from a low- to a high-temperature thermal reservoir by supplying electric energy. As shown in Figure 5.5, an amount of heat Q_L is transferred from the low-temperature reservoir, and an amount of heat Q_H is transferred to the high-temperature reservoir. The coefficient of performance of the thermoelectric heat pump is

$$\beta_{heat\ pump} = \frac{Q_H}{W} = \frac{\dot{Q}_H}{\mathcal{E}I}$$

and if the thermoelectric device is operated as a refrigerator then

$$\beta_{ref} = \frac{Q_L}{W} = \frac{\dot{Q}_L}{\mathcal{E}I}$$

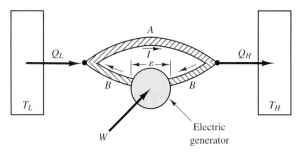

Figure 5.5   Thermoelectric heat pump.

It was mentioned in this section that the efficiency of a power cycle is less than 100% and the coefficient of performance of a refrigerator or a heat pump is less than infinity. Experimental evidence supports these statements, and the second law formalizes these experimental findings.

5.2 The Second Law of Thermodynamics

There are several statements of the second law of thermodynamics. This section deals with the two that are most frequently used in classical thermodynamics: the *Kelvin–Planck* statement and the *Clausius* statement. Each statement involves the use of a cyclic process. The first considers the transformation of heat into work; the second considers the transfer of heat between two thermal reservoirs.

Kelvin–Planck statement: It is impossible to construct a device that, operating continuously (in a cycle), will produce no effect other than the transfer of heat from a single thermal reservoir and the performance of an equal amount of work. This statement means that only part of the heat transferred to the cycle from a high-temperature reservoir can be converted to work; the rest must be rejected to a low-temperature reservoir. Therefore, at least two thermal reservoirs of different temperature are necessary for a heat engine to operate. The statement further means that no power cycle can have a 100% thermal efficiency. The word *continuous* in the Kelvin–Planck statement has an important implication. Consider the isothermal expansion of an ideal gas in a piston–cylinder arrangement as a result of heat interaction. Since the internal energy remains constant, the heat transfer, according to the first law, is completely converted into work. The motion of the piston results in an increase in the volume and a decrease in the pressure of the gas. The process continues till the pressure of the gas equals that of the surroundings, at which point no further work can be produced. This example indicates that continuous conversion of heat into work requires a *cyclic* process.

Clausius statement: It is impossible to construct a device that, operating continuously, will produce no effect other than the transfer of heat from a low-temperature body to a high-temperature body. This statement means that energy (in the form of work) must be supplied to the device in order to transfer heat from a cold body to a hot body. Therefore, the coefficient of performance can never be infinity.

Although the Kelvin–Planck and Clausius statements appear to be unrelated, they turn out to be equivalent in the sense that a violation of either statement entails the violation of the other. Both statements, however, identify a dissymmetry in nature. In the Kelvin–Planck statement, this dissymmetry is between work and heat; in the Clausius statement, the dissymmetry is in the direction of natural change.

Assume that the heat engine shown in Figure 5.6 is violating the Kelvin–Planck statement by absorbing heat from a single reservoir and producing an equal amount of work. The work output of the engine is used to drive a heat pump that transfers an amount of heat Q_L from a low-temperature reservoir and an amount of heat $(Q_H + Q_L)$ to a high-temperature reservoir. The combined system of the heat engine and heat pump, shown by a dotted line on Figure 5.6, then acts like a heat pump transferring an amount of heat Q_L from the low-temperature to the high-temperature reservoir without any external work. This, of course, is a violation of the Clausius statement.

Similarly, to prove that a violation of the Clausius statement implies the violation of the Kelvin–Planck statement, consider a heat pump that is violating the Clausius statement. Figure 5.7 shows a heat pump that requires no work and transfers an amount of heat Q_L from a low-temperature to a high-temperature reservoir. Let an amount of heat Q_H greater than Q_L be transferred from the high-temperature reservoir to a heat engine that develops a net work $W = Q_H - Q_L$.

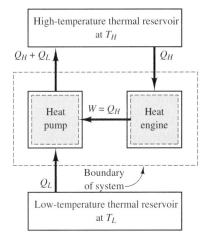

Figure 5.6   Equivalence of the Kelvin–Planck statement to the Clausius statement.

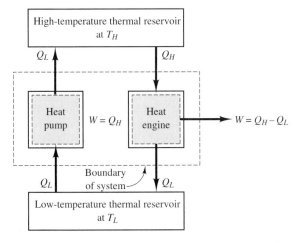

Figure 5.7   Equivalence of the Clausius statement to the Kelvin–Planck statement.

and rejects Q_L to the low-temperature reservoir. Since there is no net heat interaction with the low-temperature reservoir, it can be eliminated. The combined system of the heat engine and heat pump acts then like a heat engine exchanging heat with a single reservoir, which is a violation of the Kelvin–Planck statement.

The foregoing analyses establish the equivalence of the Kelvin–Planck and Clausius statements of the second law of thermodynamics.

5.3 Perpetual Motion Machine of the Second Kind

As shown in Section 3.2, a machine that violates the first law of thermodynamics was called a perpetual machine of the first kind. *A perpetual motion machine of the second kind* is one that operates in a cycle and delivers an amount of work equal to heat extracted from a single reservoir at a uniform temperature. Such a machine (efficiency 100%) obviously violates the second law of thermodynamics and does not exist.

5.4 The Carnot Cycle

French military engineer Nicolas Sadi Carnot (1796–1832), in his paper, "Réflections sur la puissance motrice du feu et les moyens propres à la développer,"[2] was among the first to study the principles of the second law of thermodynamics. Carnot was the first to introduce the concept of cyclic operation in which the working substance, after passing through a sequence of events, is brought back to its initial state. In his treatise, published in 1824, he devised a classical ideal cycle named after him. The Carnot cycle was based on the principles of the first law formulated later by Joule, and was the heuristic step in the evolution of the second law of thermodynamics 25 years later by Rudolf Clausius and William Thomson.

The Carnot cycle was introduced briefly in Section 3.11. As shown in Figure 5.8, it consists of an alternate series of two reversible isothermal processes and two reversible adiabatic processes. All processes are individually reversible, and therefore the Carnot cycle as a whole is a reversible cycle. Let the system be an arbitrary but homogeneous working substance. Starting at an initial state 1 and referring to Figures 5.8(a) and 5.9, the system undergoes a Carnot cycle in the following manner:

(a) During process 1-2, heat is transferred reversibly and isothermally to the working substance from a high-temperature reservoir at T_H. This process is

---

[2] "Reflections on the motive power of heat and the proper ways to develop it."

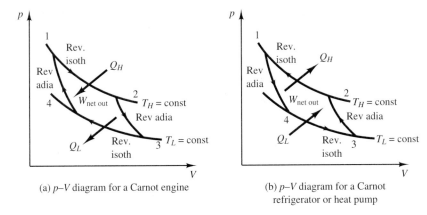

(a) *p–V* diagram for a Carnot engine

(b) *p–V* diagram for a Carnot refrigerator or heat pump

Figure 5.8   The Carnot cycle.

accomplished reversibly by bringing the system in contact with a thermal reservoir whose temperature is equal to or infinitesimally higher than the working substance. During this process, the system performs an amount of work equal to the area underneath path 1-2 of the *p-V* diagram.

(b) Process 2-3 is a reversible adiabatic expansion during which the system is thermally insulated and the temperature of the working substance decreases from the high temperature T_H to the low temperature T_L. The work done in this process is represented by the area underneath path 2-3 of the *p-V* diagram.

(c) During process 3-4, the system is brought in contact with a low-temperature reservoir, and heat is transferred reversibly and isothermally from the working substance to the low-temperature reservoir. This process requires that the temperature of the working substance be equal to or infinitesimally higher than the low-temperature reservoir. The work done on the system during this isothermal process is equal to the area underneath path 3-4 of the *p-V* diagram.

(d) The final process 4-1, which completes the cycle, is a reversible adiabatic process, and the working substance is returned to its original state at 1. During this process, the temperature of the working substance is raised from the low temperature T_L to the initial temperature T_H. The work done in this process is represented by the area underneath path 4-1 of the *p-V* diagram.

Since an amount of heat Q_H is transferred to the working substance at the higher temperature, while Q_L is transferred from the working substance at the lower temperature, the net heat transfer $Q_H - Q_L$ is equal to the net work output of the cycle. This net work output is represented by the closed area 1-2-3-4-1 on the *p-V* diagram of Figure 5.8 and is equal to $\oint p \, dV$.

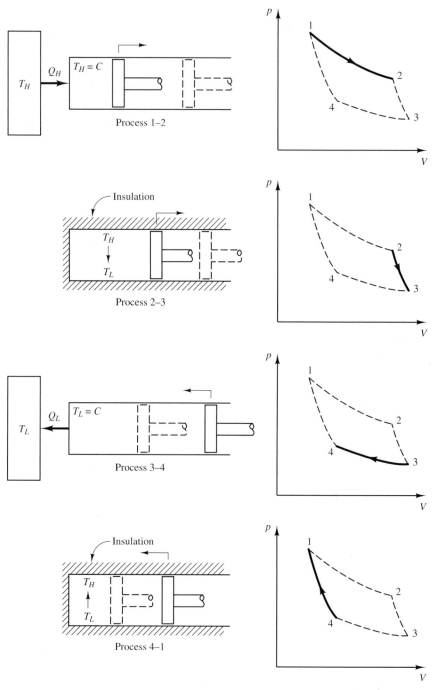

Figure 5.9   Processes of a Carnot power cycle.

The thermal efficiency of the cycle is given by

$$\eta_{th} = \frac{W_{net}}{Q_H} = \frac{\oint p\,dV}{Q_H} = \frac{Q_H - Q_L}{Q_H} \qquad (5.4)$$

If the Carnot cycle for a heat engine is carried out in the reverse direction, the magnitudes of all energy transfers remain the same but their sign changes. The result will be a Carnot cycle for a refrigerator or a heat pump. Such a cycle is shown in Figure 5.8(b). The individual reversible processes are performed in the following manner:

(a) Process 4-3 is a reversible isothermal expansion during which heat is transferred from the low-temperature reservoir to the working substance.
(b) Process 3-2 is a reversible adiabatic compression during which the temperature of the working substance is raised from the low-temperature T_L to the high-temperature T_H.
(c) During process 2-1, the working substance is compressed as heat is transferred reversibly and isothermally from the working substance to the high-temperature reservoir.
(d) To complete the cycle, the working substance is returned to its initial state 4 by the reversible adiabatic expansion 1-4. During this process, the temperature of the working substance decreases from T_H to T_L.

Similar to the heat engine cycle, the net work done on the system is represented by the area 1-2-3-4-1 of Figure 5.8 and is equal to $-\oint p\,dV$. When the system is a refrigerator, the coefficient of performance of the cycle is

$$\beta_{ref} = \frac{Q_L}{W_{net}} = \frac{Q_L}{-\oint p\,dV} = \frac{Q_L}{Q_H - Q_L} \qquad (5.5)$$

Figures 5.10 and 5.11 show two idealized examples, a simple vapor power plant and a heat pump operating on the Carnot cycle. During the heat transfer processes, the working fluid changes phase, resulting in isothermal heat interaction.

A final comment regarding the cycle is that any reversible cycle that operates between two constant but different temperature reservoirs and comprises two reversible isothermals and two reversible adiabatics is a Carnot cycle. Reversible processes do not occur in nature, and therefore the Carnot cycle is an ideal cycle. It serves as a limit of optimum performance of real cycles.

So far, in analyzing the Carnot cycle, no specific characteristics were imposed regarding the working substance. As will be shown in the next section, the Carnot efficiency (and the coefficient of performance) is independent of the amount or the nature of the working substance.

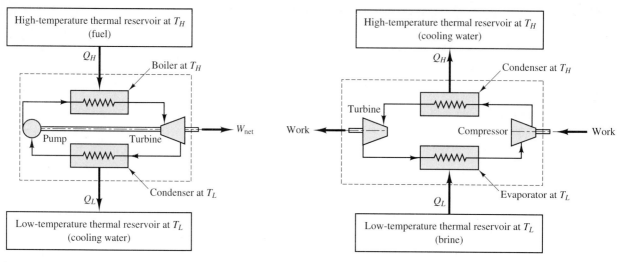

Figure 5.10   Carnot cycle for a heat engine.   **Figure 5.11**   Carnot cycle for a heat pump.

5.5 The Carnot Principles

The Carnot principles include three propositions or corollaries that are of great use in comparing the performance of cycles. The method of proving Carnot propositions is to assume first that the opposite of the proposition is true, and then prove that the result leads to a violation of the second law of thermodynamics. The Carnot propositions are as follows:

(a) No heat engine operating between two constant-temperature reservoirs can be more efficient than a reversible engine operating between the same two reservoirs. As shown in Figure 5.12, let an irreversible engine *I* and a reversible engine *R* operate between high- and low-temperature reservoirs. Engine *R*, which operates on the Carnot cycle, can be reversed to operate as a heat pump, as shown by the dotted arrows in Figure 5.12. For simplicity, assign numerical values to Q_H, Q_L, and *W* such as 100, 70, and 30. Operating as a heat engine, engine *R* will have an efficiency of 30%; as a heat pump, it will have a coefficient of performance of $100/30 = 3.33$.

Assume that engine *I* is more efficient than engine *R* and has an efficiency of more than 30 percent, say, 40 percent. Let engine *I* operate to develop 30 units of work so that it can be used to drive the Carnot heat pump. The heat input to engine *I* in this case should be $30/0.4 = 75$ units, and the heat output is $75 - 30 = 45$ units.

If the foregoing is possible, then the combined system of the Carnot heat pump and engine *I* has the net effect of transferring 25 units of heat from the low- to the

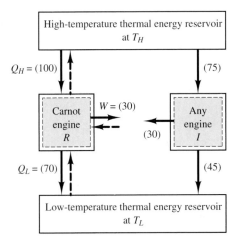

Figure 5.12   The Carnot engine is the most efficient engine that could operate between two thermal reservoirs.

high-temperature reservoir without any external work. This is obviously a violation of the Clausius statement of the second law of thermodynamics. Therefore, the assumption that engine I has a higher efficiency than a reversible engine is invalid, and the reversible engine, operating on the Carnot cycle, is the most efficient engine that could operate between the two thermal reservoirs. Note that it was not proved explicitly that the efficiency of engine I is lower than that of a reversible engine, but rather no engine can be more efficient than a reversible one.

A similar argument that leads to a violation of the Kelvin–Planck form of the second law, and therefore leads to the same conclusion cited before, is

$$\eta_R = \frac{W_R}{Q_{HR}} = \frac{W_R}{W_R + Q_{LR}} = \frac{1}{1 + (Q_{LR}/W_R)}$$

and

$$\eta_I = \frac{W_I}{Q_{HI}} = \frac{W_I}{W_I + Q_{LI}} = \frac{1}{1 + (Q_{LI}/W_I)}$$

Assume $\eta_I > \eta_R$, then

$$\frac{Q_{LR}}{W_R} > \frac{Q_{LI}}{W_I}$$

and for comparison, if

$$Q_{LR} = Q_{LI} \qquad \text{then} \qquad W_I > W_R$$

Now reversing the Carnot engine to operate as a heat pump, W_I can be used to supply W_R to the pump, and a net work output $(W_I - W_R)$ is developed. Since $Q_{LR} = Q_{LI}$, the low-temperature reservoir can be eliminated, and the net heat leaving the high-temperature reservoir is

$$Q_{HI} - Q_{HR} = (W_I + Q_{LI}) - (W_R + Q_{LR}) = W_I - W_R$$

This means that heat is exchanged with a single reservoir and an equal amount of work is performed, which is contradictory to the Kelvin–Planck statement.

  (b) The second corollary of Carnot states that all reversible engines operating between two constant temperature reservoirs have the same efficiency irrespective of the working substance.

  Let two reversible engines R and R' operate between the same constant-high and constant-low temperature reservoirs at temperature T_H and T_L, respectively. Assume that engine R is more efficient than R'. Similar to the procedure outlined in proposition (a), if R is reversed to run as a heat pump using the output work of R', we come to the conclusion that the combined system of engine R and heat pump R' is removing heat continuously from the low- to the high-temperature reservoirs without any external work. This is contrary to the second law of thermodynamics. Therefore, R cannot be more efficient than R'. Similarly, if engine R' is assumed to be more efficient than R, it can be concluded that there is a violation of the second law. Therefore, neither engine R nor R' can be more efficient than the other, which means that they must have equal efficiencies, irrespective of the type of engine, so that

$$\eta_R = \eta_{R'}$$

Hence

$$1 - \frac{Q_L}{Q_H} = 1 - \frac{Q_L'}{Q_H'}$$

where subscripts H and L refer to the higher and lower temperatures. The previous equation gives

$$\frac{Q_H}{Q_L} = \frac{Q_H'}{Q_L'}$$

The ratio Q_H/Q_L is therefore a constant for all reversible cycles operating between the two reservoirs, and since no restrictions were imposed on the working substance, the efficiency of a reversible engine is independent of the nature of the working substance. This ratio is a function of only the two temperatures of the reservoirs, that is,

$$\frac{Q_H}{Q_L} = \varphi(T_H, T_L) \tag{5.6}$$

and

$$\eta_R = 1 - \frac{Q_L}{Q_H} = 1 - \frac{1}{\varphi(T_H, T_L)} \tag{5.7}$$

The third corollary of Carnot, presented in Section 5.6, defines the nature of the function in Eq. (5.6).

5.6 The Thermodynamic Temperature Scale

With the aid of the second Carnot principle, Lord Kelvin (W. Thomson) in 1848 used energy as a thermometric property to define temperature. He devised a temperature scale that is independent of the nature of the thermometric substance. It was shown in the previous section that the efficiency of a Carnot cycle operating between two thermal reservoirs is a function of the temperatures of the reservoirs. To determine the nature of the function appearing in Eqs. (5.6) and (5.7), consider three reversible engines operating between the pairs of temperature (T_1, T_2), (T_2, T_3), and (T_1, T_3), as shown in Figure 5.13. Figure 5.14 shows the cycles on a p-V diagram where the three isotherms T_1, T_2, and T_3 $(T_1 > T_2 > T_3)$ intersect the two adiabatics a-f and b-e. If Q_1, Q_2, and Q_3 represent the amounts of heat interaction at temperatures T_1, T_2, and T_3, then, for cycle a-b-c-d-a,

$$\frac{Q_1}{Q_2} = \varphi(T_1, T_2)$$

for cycle d-c-e-f-d,

$$\frac{Q_2}{Q_3} = \varphi(T_2, T_3)$$

and for cycle a-b-e-f-a,

$$\frac{Q_1}{Q_3} = \varphi(T_1, T_3)$$

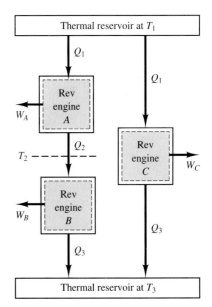

Figure 5.13 Definition of absolute temperature by means of Carnot engines.

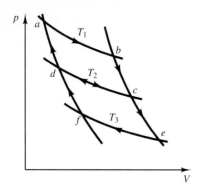

Figure 5.14 *p-V* diagram of Carnot engines.

From the previous three equations, since

$$\frac{Q_1}{Q_2} = \frac{Q_1/Q_3}{Q_2/Q_3}$$

then

$$\varphi(T_1, T_2) = \frac{\varphi(T_1, T_3)}{\varphi(T_2, T_3)} \tag{5.8}$$

but since T_1, T_2, and T_3 are independent, and the efficiency of the engine operating between temperature T_1 and T_2 is unaffected by T_3, Eq. (5.8) can be satisfied only when the function $\varphi(T_1, T_2)$ is of the form $\varphi'(T_1)/\varphi'(T_2)$ so that[3]

$$\frac{Q_1}{Q_2} = \varphi(T_1, T_2) = \frac{\varphi'(T_1)}{\varphi'(T_2)}$$

---

[3] Consider Eq. (5.8)

$$\varphi(T_1, T_2)\varphi(T_2, T_3) = \varphi(T_1, T_3) \tag{5.8}$$

Differentiate with respect to T_3,

$$\varphi(T_1, T_2)\frac{\partial}{\partial T_3}\varphi(T_2, T_3) = \frac{\partial}{\partial T_3}\varphi(T_1, T_3)$$

Substituting for $\varphi(T_1, T_2)$ from Eq. (5.8) gives,

$$\frac{1}{\varphi(T_2, T_3)}\frac{\partial}{\partial T_3}\varphi(T_2, T_3) = \frac{1}{\varphi(T_1, T_3)}\frac{\partial}{\partial T_3}\varphi(T_1, T_3)$$

or

$$\frac{\partial \ln \varphi(T_2, T_3)}{\partial T_3} = \frac{\partial \ln \varphi(T_1, T_3)}{\partial T_3}$$

Each of the two sides of the preceding equation is a function of T_3 because only T_3 appears on both sides. Therefore,

$$\frac{\partial \ln \varphi(T_2, T_3)}{\partial T_3} = \frac{\partial \ln \varphi(T_1, T_3)}{\partial T_3} = \psi(T_3)$$

Integrating

$$\ln \varphi(T_2, T_3) = \psi_1(T_3) + \ln \varphi'(T_2) \qquad \text{or} \qquad \varphi(T_2, T_3) = \varphi'(T_2)e^{\psi_1(T_3)}$$

and

$$\ln \varphi(T_1, T_3) = \psi_1(T_3) + \ln \varphi'(T_1) \qquad \text{or} \qquad \varphi(T_1, T_3) = \varphi'(T_1)e^{\psi_1(T_3)}$$

where $\ln \varphi'(T_2)$ and $\ln \varphi'(T_1)$ are constants of integration. Dividing the foregoing two equations gives

$$\frac{\varphi(T_1, T_3)}{\varphi(T_2, T_3)} = \frac{\varphi'(T_1)}{\varphi'(T_2)} = \varphi(T_1, T_2)$$

If T_1 is higher than T_2, Q_1 is larger than Q_2, and therefore, $\varphi'(T_1)$ is larger than $\varphi'(T_2)$. This means that the function $\varphi'(T)$ increases with T and can be used for temperature measurement. If the arbitrary monotonic function $\varphi'(T)$ is denoted simply by T, an absolute temperature scale may be defined to satisfy the relation

$$\frac{Q_1}{Q_2} = \frac{T_1}{T_2} \qquad (5.9)$$

Equation (5.9) states that the ratio of any two temperatures on the absolute scale is equal to the ratio of the amounts of heat transferred to and from a system undergoing a reversible cycle while interacting thermally with reservoirs at these temperatures. Note in this case that the heat transfer Q is used as a thermometric property to define temperature. Therefore, Eq. (5.9) may be used to measure the temperature of a system by imagining a Carnot engine to operate between the system and a thermal reservoir at a standard fixed point of temperature. By measuring the heat interaction between the engine and the system and between the engine and the reservoir, the temperature of the system can be determined in accordance with Eq. (5.9). The absolute temperature scale is *independent of the thermometric substance* since it is defined in terms of heat quantities that do not depend on the working substance. It is called absolute since, as Q approaches zero, the lowest possible temperature is approached. This may be explained by referring to Figure 5.13. It is clear that Q_2 is smaller than Q_1, and Q_3 is smaller than Q_2, and so on. Therefore, every successive Carnot engine serves to define an ever-decreasing temperature at which an amount of heat is rejected. Thus it is conceivable that a lower limit of temperature is approached such that the heat rejected becomes zero. This temperature is called the *absolute zero*.

Consider a Carnot engine operating between two thermal reservoirs at temperatures T_H and T_L. From the first law, the work output is

$$W = Q_H - Q_L$$

but from Eq. (5.9),

$$Q_L = Q_H \frac{T_L}{T_H}$$

Hence

$$W = Q_H - Q_H \frac{T_L}{T_H}$$

or

$$\frac{T_L}{T_H} = \left(1 - \frac{W}{Q_H}\right) \tag{5.10}$$

The second law of thermodynamics requires that the work W must be less than the heat input Q_H. Therefore, T_L/T_H is always > 0, or the lowest value of T_L is greater than zero. Thus the absolute zero is the lowest conceivable temperature, and absolute temperatures are positive numbers.

The next step is to determine the temperature interval on the absolute scale. Imagine a set of Carnot engines so arranged that each engine receives the heat rejected by the preceding one and each engine develops the same amount of work. The temperature intervals, in this case, are equal. For the first engine,

$$W = Q_1 - Q_2 = Q_1\left(1 - \frac{Q_2}{Q_1}\right) = Q_1\left(1 - \frac{T_2}{T_1}\right) = \frac{Q_1}{T_1}(T_1 - T_2)$$

Similarly, for the second engine,

$$W = \frac{Q_2}{T_2}(T_2 - T_3)$$

and so on. Since each engine develops the same amount of work,

$$\frac{Q_1}{T_1}(T_1 - T_2) = \frac{Q_2}{T_2}(T_2 - T_3) = \frac{Q_3}{T_3}(T_3 - T_4) = \cdots, \text{etc.}$$

But from Eq. (5.9),

$$\frac{Q_1}{T_1} = \frac{Q_2}{T_2} = \frac{Q_3}{T_3} = \cdots, \text{etc.}$$

Therefore,

$$T_1 - T_2 = T_2 - T_3 = T_3 - T_4 = \cdots, \text{etc.}$$

which means that equal temperature intervals on the absolute scale are the temperature intervals between which a series of Carnot engines operate in such a way that each engine absorbs the heat rejected by the preceding one and all engines develop the same amount of work.

The remaining item in the absolute scale is to assign a numerical value to one standard fixed point of temperature. In 1954, the triple point of water, 273.16 K, was selected as the standard fixed point of temperature. Kelvin had suggested that the absolute scale be based on this standard fixed point, instead of the previous procedure of fixing the size of the degree by means of two fixed points (the ice point = 273.15 K, and the steam point = 373.15 K).

Assigning a numerical value of 273.16 to the triple point of water results in the absolute Celsius scale or Kelvin scale. Similarly, assigning a numerical value of 491.69 to the triple point of water results in the absolute Fahrenheit scale or the Rankine scale.

Equation (5.9) may be used to express the efficiency of a Carnot engine in terms of the absolute temperatures of the high-temperature and low-temperature thermal reservoirs as

$$\eta_{th} = 1 - \frac{Q_L}{Q_H} = 1 - \frac{T_L}{T_H} \qquad (5.11)$$

Similarly, for a Carnot refrigerator,

$$\beta_{ref} = \frac{1}{T_H/T_L - 1} = \frac{T_L}{T_H - T_L} \qquad (5.12)$$

and for a Carnot heat pump,

$$\beta_{heat\ pump} = \frac{1}{1 - T_L/T_H} = \frac{T_H}{T_H - T_L} \qquad (5.13)$$

These expressions give the maximum values for any cycle operating between two thermal reservoirs and can be used as standards for comparison for actual cycles.

5.7 Equivalence of the Absolute and the Ideal-Gas Scales of Temperature

The ideal-gas scale was defined in Chapter 1. Figure 5.15 shows a p-V diagram of a Carnot cycle in which the working fluid is an ideal gas. Neglecting changes of potential and kinetic energy, the first law can be written as

$$\delta Q + \delta W = dU$$

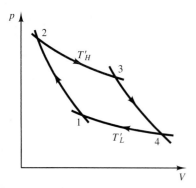

Figure 5.15 Equivalence of the absolute and the ideal-gas temperature scales.

Since the working substance is an ideal gas and all the processes in the Carnot cycle are reversible,

$$\delta Q - p \, dV = mc_v \, dT'$$

where c_v is the specific heat at constant volume, m is the mass of the gas, and T' is the temperature of the gas based on the ideal-gas scale. For the isothermal process 2-3, the heat transfer to the system is

$$Q_H = \int_{V_2}^{V_3} p \, dV = mRT'_H \ln \frac{V_3}{V_2}$$

Similarly, for the isothermal process 4-1, the heat transfer from the system is

$$Q_L = mRT'_L \ln \frac{V_4}{V_1}$$

Therefore,

$$\frac{Q_H}{Q_L} = \frac{T'_H \ln (V_3/V_2)}{T'_L \ln (V_4/V_1)}$$

Since processes 1-2 and 3-4 are reversible adiabatics, the temperature and volumes of the end states of each process are given by the following relations:

$$\left(\frac{V_1}{V_2}\right)^{\gamma - 1} = \frac{T'_H}{T'_L} = \left(\frac{V_4}{V_3}\right)^{\gamma - 1}$$

from which

$$\frac{V_1}{V_2} = \frac{V_4}{V_3}$$

The ratio of Q_H to Q_L then becomes

$$\frac{Q_H}{Q_L} = \frac{T'_H}{T'_L}$$

But the absolute temperature scale is defined by

$$\frac{Q_H}{Q_L} = \frac{T_H}{T_L}$$

Therefore,

$$\frac{T'_H}{T'_L} = \frac{T_H}{T_L}$$

Since the numerical value assigned to the fixed point (triple point) may be chosen to be the same on both scales, $T' = T$, or the absolute thermodynamic temperature is *numerically equal* to the absolute temperature as measured by the ideal-gas thermometer.

The absolute thermodynamic temperature scale is independent of the thermometric substance and is therefore the most fundamental temperature scale.

Example 5.1

A Carnot power cycle using air as a working fluid has a thermal efficiency of 40 percent. At the beginning of the isothermal expansion, the pressure is 620 kPa, and the specific volume of the air is 0.1 m³/kg. If the heat input to the cycle is 50 kJ/kg, determine:
  (a) the highest and lowest temperature for the cycle
  (b) the work and heat interactions per unit mass for each process of the cycle
  Assume air to be an ideal gas with constant specific heats.

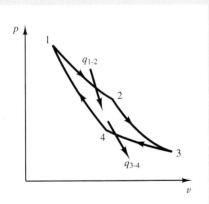

Figure 5.16

SOLUTION

  (a) Referring to Figure 5.16, the highest temperature in the cycle T_H can be determined from the ideal-gas law:

$$T_1 = \frac{p_1 v_1}{R} = \frac{(620 \text{ kPa})(0.1 \text{ m}^3/\text{kg})}{(0.287 \text{ kJ/kg K})} = 216.028 \text{ K}$$

The thermal efficiency of the Carnot cycle is

$$\eta_{th} = \frac{|w_{net}|}{q_{3\text{-}4}} = 1 - \frac{T_L}{T_H}$$

Therefore, $0.4 = (w_{net}/50)$, from which

$$|w_{net}| = 20 \text{ kJ/kg} \qquad \text{and} \qquad q_{3\text{-}4} = -30 \text{ kJ/kg}$$

Also,

$$0.4 = 1 - \frac{T_L}{T_H} = 1 - \frac{T_L}{216.028 \text{ K}}$$

from which $T_L = 129.617$ K.

(b) The first law applied to the different processes gives Process 1-2:

$$q_{1\text{-}2} + w_{1\text{-}2} = 0 \qquad \text{or} \qquad q_{1\text{-}2} = -w_{1\text{-}2} = +50 \text{ kJ/kg}.$$

Process 2-3:

$$q_{2\text{-}3} = 0 \qquad \text{and} \qquad w_{2\text{-}3} = u_3 - u_2$$
$$= c_v(T_3 - T_2)$$
$$= (0.7165 \text{ kJ/kg K})(129.617 - 216.028) \text{ K}$$
$$= -61.914 \text{ kJ/kg}$$

Process 3-4:

$$q_{3\text{-}4} + w_{3\text{-}4} = 0 \qquad \text{or} \qquad q_{3\text{-}4} = -w_{3\text{-}4} = -30 \text{ kJ/kg}$$

Process 4-1:

$$q_{4\text{-}1} = 0 \qquad \text{and} \qquad w_{4\text{-}1} = u_1 - u_3$$
$$= c_v(T_1 - T_4)$$
$$= (0.7165 \text{ kJ/kg K})(86.411) \text{ K} = 61.914 \text{ kJ/kg}$$

Example 5.2

Determine the coefficient of performance for the refrigeration cycle obtained by reversing the power cycle described in Example 5.1.

SOLUTION

$$\beta_{\text{ref}} = \frac{q_L}{w_{\text{net}}} = \frac{q_L}{q_H - q_L} = \frac{T_L}{T_H - T_L} = \frac{129.617 \text{ K}}{(216.028 - 129.617) \text{ K}} = 1.5$$

5.8 Clausius Theorem

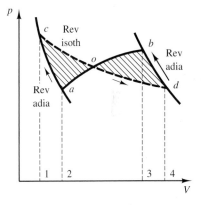

Figure 5.17 Replacement of a reversible process by two reversible adiabatics and a reversible isothermal.

Consider a system undergoing a reversible process a-b, as shown on the p-V diagram of Figure 5.17. The same change of states of both system and surroundings may be achieved if process a-b is replaced by an adiabatic process a-c, an isothermal process c-d, and an adiabatic process d-b, provided that the areas a-c-o-a, and o-d-b-o are equal. The validity of this statement is established if both the heat and the work interactions during processes a-b and a-c-d-b are the same. In the cycle a-c-d-b-a, since a-c and b-d are adiabatics, heat interaction takes place during processes a-b and c-d only. Also the net work interaction in this cycle is zero since the areas a-c-o-a and o-b-d-o are equal. Therefore,

$$Q_{a\text{-}b} = Q_{c\text{-}d}$$

But since processes a-c and b-d are adiabatics,

$$Q_{a\text{-}b} = Q_{a\text{-}c\text{-}d\text{-}b}$$

Substituting for $Q_{a\text{-}b}$ and $Q_{c\text{-}d}$ in the first law gives

$$E_b - E_a - W_{a\text{-}b} = E_b - E_a - W_{a\text{-}c\text{-}d\text{-}b}, \quad \text{or} \quad W_{a\text{-}b} = W_{a\text{-}c\text{-}d\text{-}b}$$

The reversible process a-b can thus be replaced by a sequence of reversible processes consisting of an adiabatic, an isothermal, and an adiabatic such that the heat interaction during the isothermal process is equal to the heat interaction during the original process.

Consider a system specified by two independent variables undergoing a reversible cycle as represented on the p-V diagram of Figure 5.18. The cycle is subdivided by a family of reversible adiabatics, and every two adjacent adiabatics are connected by two reversible isothermals so that the heat transferred during all isothermal processes is equal to the heat transferred during the original cycle. Since no two adiabatics cross each other (Example 5.11), the cycle can be subdivided into a large number of Carnot cycles that provide the same amount of

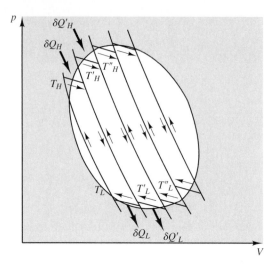

Figure 5.18   A reversible cycle can be subdivided into a large number of Carnot cycles.

work as the original cycle. If magnitudes and directions of heat interactions are considered, Eq. (5.9) for the first Carnot cycle becomes

$$\frac{\delta Q_H}{T_H} = -\frac{\delta Q_L}{T_L} \qquad \text{or} \qquad \frac{\delta Q_H}{T_H} + \frac{\delta Q_L}{T_L} = 0$$

Similarly, for the second cycle,

$$\frac{\delta Q'_H}{T'_H} + \frac{\delta Q'_L}{T'_L} = 0, \text{ etc.}$$

Adding the preceding equations gives

$$\frac{\delta Q_H}{T_H} + \frac{\delta Q_L}{T_L} + \frac{\delta Q'_H}{T'_H} + \frac{\delta Q'_L}{T'_L} + \cdots = 0$$

or

$$\oint_R \frac{\delta Q}{T} = 0$$

where the subscript R refers to a reversible cycle. The foregoing equation states that the algebraic sum of the quantity $\delta Q/T$ vanishes around a reversible cycle, which means that $(\delta Q/T)_{\text{rev}}$ is a *property* of the system. This property is called *entropy* (see Section 5.9).

In the case of an irreversible engine I, it was shown in Section 5.5 that $\eta_I < \eta_R$. Substituting for the efficiency in terms of heat quantities and absolute temperatures gives

$$1 - \frac{\delta Q_{LI}}{\delta Q_{HI}} < 1 - \frac{\delta Q_{LR}}{\delta Q_{HR}}$$

$$1 - \frac{\delta Q_{LI}}{\delta Q_{HI}} < 1 - \frac{T_L}{T_H}$$

or

$$\frac{\delta Q_{HI}}{T_H} - \frac{\delta Q_{LI}}{T_L} < 0$$

Considering directions of heat interactions and noting that δQ_{LI} is heat rejection,

$$\frac{\delta Q_{HI}}{T_H} + \frac{\delta Q_{LI}}{T_L} < 0$$

and summing up all the formulas analogous to the preceding procedure yields

$$\oint_I \frac{\delta Q}{T} < 0$$

where the subscript I refers to an irreversible cycle, and T is the temperature of the thermal reservoir.

From the foregoing analysis, we can write for any cycle,

$$\oint \frac{\delta Q}{T_{\text{reservoir}}} \leq 0 \qquad (5.14)$$

Equation (5.14) is called the *Clausius inequality,* which states that when a system undergoes a complete cycle, the integral of $\delta Q / T_{\text{res}}$ around the cycle is less than or equal to zero. The equality and inequality signs apply to internally reversible and to internally irreversible cycles, respectively. Note that the temperatures T in Clausius inequality pertain to the temperature at the boundary where δQ is transferred.

Example 5.3

Prove that the steam cycle shown in Figure 5.19 is consistent with the Clausius theorem, and state the condition for the cycle to be reversible. Data:

heat transfer to the boiler (at $T_{\text{av}} = 200°C$) = 2600 kJ/kg
heat rejected from the condenser (at $T_2 = 50°C$) = 2263 kJ/kg
adiabatic turbine and pump

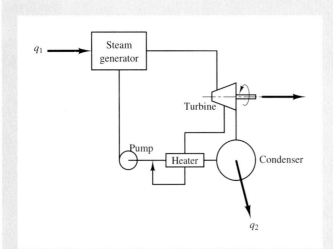

Figure 5.19

SOLUTION

Since heat interaction takes place only in the boiler and condenser and the temperature remains constant in both

$$\oint \frac{\delta q}{T} = \frac{2600 \text{ kJ/kg}}{473.15 \text{ K}} - \frac{2260 \text{ kJ/kg}}{323.15 \text{ K}}$$
$$= (5.495 - 6.994) \text{ kJ/kg K} = -1.499 \text{ kJ/kg K}$$

Since the result is negative, the cycle is irreversible according to the Clausius inequality (Eq. 5.14). For this cycle to be reversible, $\oint \delta q / T$ must be equal to zero.

5.9 Entropy

The law of conservation of energy was developed first for a cycle and by introducing a property, internal energy, it was subsequently cast in a form applicable to a process. The development of the second law follows the same pattern; it is first considered for a cycle and then developed for a process by introducing a property, entropy. In 1854, Clausius attempted to make possible the application of the second law in a quantitative way. To do this, he introduced a mathematical function, which he named *entropy*.[4] Entropy provides a method, based on calculations, of identifying the direction of spontaneous processes and explaining why certain energy transformations are impossible. Kelvin proposed the concept of the entropy of the universe, which states that entropy continuously in-

[4] The word *entropy* comes from Greek and means "transformability."

creases when spontaneous processes occur in nature. The concept of entropy will now be developed.

In Figure 5.20, two equilibrium states, 1 and 2, are shown. They are connected by two internally reversible processes a and b. If the direction of process b is reversed, then the reversible cycle 1-a-2-b-1 is formed. Applying the Clausius theorem to this cycle we have,

$$\oint_R \frac{\delta Q}{T} = 0$$

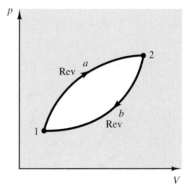

Figure 5.20   An internally reversible cycle.

The cyclic integral can be expressed as the sum of two integrals, one along process a and the second along process b:

$$\oint_R \frac{\delta Q}{T} = \int_{1(a)}^{2} \frac{\delta Q}{T} + \int_{2(b)}^{1} \frac{\delta Q}{T} = 0$$

From this it follows that

$$\int_{1(a)}^{2} \frac{\delta Q}{T} = -\int_{2(b)}^{1} \frac{\delta Q}{T}$$

But since process b is reversible, the limits of the integral can be reversed and therefore

$$\int_{1(a)}^{2} \frac{\delta Q}{T} = \int_{1(b)}^{2} \frac{\delta Q}{T}$$

No restriction was imposed on processes a and b, except that they must both be internally reversible. Consequently, the expression $\int (\delta Q/T)_{rev}$ between two states is independent of the path, and hence it represents a property. This property is called entropy, S, and is defined by the equation

$$dS \equiv \left(\frac{\delta Q}{T}\right)_{int\ rev} \tag{5.15}$$

The subscript in this equation indicates internally reversible process. This method of defining entropy is analogous to the method of defining total energy E in Eq. (3.3). Both equations express differences in a property between two states of a system, but they do not provide a measure of absolute energy or absolute entropy. The amount of work interaction or heat interaction in a process connecting two fixed states depends on how the process is carried. On the other hand, the change in entropy and the change in total energy are invariant.

It should be noted that the change of entropy can be calculated from Eq. (5.15) provided the process is internally reversible. The entropy change for an irreversible process can be calculated by devising a reversible process (or a series of reversible processes) between the initial and the final states of the system. Since entropy is a property and the end states of both the reversible and irreversible processes are identical, the entropy change will be the same for the two processes. Thus Eq. (5.15) allows the determination of the change in entropy, and this entropy change is the same for all processes between the end states.

Consider next a reversible process 1-2 along path a, as shown in Figure 5.21. Let the cycle be completed by an irreversible process 2-1 along path b, so that processes 1-a-2 and 2-b-1 together form an irreversible cycle. The net change of entropy when the system changes from state 1 to state 2 and back to state 1 is zero or

$$\int_{1(a)}^{2} dS + \int_{2(b)}^{1} dS = 0$$

Note that, if heat interaction during the reversible process results in an increase of entropy, heat must be rejected during the irreversible process to decrease the entropy such that the total change of entropy is zero according to the preceding equation. The Clausius inequality gives

$$\int_{1(a)}^{2} \frac{\delta Q}{T} + \int_{2(b)}^{1} \frac{\delta Q}{T_{surr}} < 0$$

but since path a is reversible,

$$\int_{1(a)}^{2} dS = \int_{1(a)}^{2} \frac{\delta Q}{T} = -\int_{2(b)}^{1} dS$$

Therefore,

$$-\int_{2(b)}^{1} dS + \int_{2(b)}^{1} \frac{\delta Q}{T_{surr}} < 0$$

or

$$\int_{2(b)}^{1} dS > \int_{2(b)}^{1} \frac{\delta Q}{T_{surr}}$$

If process b were reversible, an equality sign replaces the inequality sign in the foregoing equation. Therefore, a general form of that equation, which can be considered a statement of the second law for a system, is

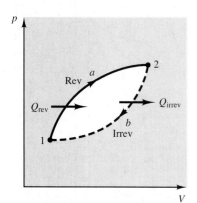

Figure 5.21   The entropy change between two end states is the same whether the process is reversible or irreversible.

$$\underbrace{\int_1^2 dS}_{\substack{\text{entropy change}\\\text{(property)}}} \quad \geq \quad \underbrace{\int_1^2 \frac{\delta Q}{T}}_{\substack{\text{entropy transfer}\\\text{(nonproperty)}}} \qquad (5.16)$$

where the equality and inequality signs apply to internally reversible and internally irreversible processes, respectively. Note that the temperature T in Eq. (5.16) is the temperature of the system for an internally reversible process. However, it is the temperature of the surroundings (at the boundary of the system) when internal irreversibilities are present.

For an irreversible process,

$$\delta Q_{\text{irr}} < T_{\text{surr}}\, dS_{\text{sys}}$$

where δQ_{irr} is the heat interaction with the system, and T_{surr} is the temperature of the surroundings. Although the entropy change between two states is the same for any process, reversible or not, the heat transfer to the system is less (more, if heat is transferred from the system) for the irreversible path than for the reversible one. This means that these two processes will not cause the same change of state in the surroundings.

For an isolated system that experiences no heat or work interaction with the surroundings, the total energy of all possible states remains constant. The second law, however, dictates that only those states for which the entropy increases or remains unchanged are possible. Thus, according to Eq. (5.16) and since $\delta Q = 0$,

$$dS_{\text{isolated}} \geq 0 \qquad (5.17)$$

Therefore, for any irreversible process, there is a creation of entropy, and only those states resulting in entropy increase may be attained from the initial state of the system.

Equation (5.17) is called the *principle of increase of entropy,* which may be considered another version of the second law. It states: The entropy of an isolated system either increases or, in the limit, remains constant. That is, the inequality in Eq. (5.17) rules out as physically impossible any real process that violates this inequality.

Following the principle of increase of entropy, the entropy of the universe (assumed an isolated system) increases, owing to natural processes such that

$$\Delta S_{\text{univ}} = \Delta S_{\text{sys}} + \Delta S_{\text{surr}} > 0 \qquad (5.18)$$

Note that for a reversible process,

$$\Delta S_{\text{sys}} = \left(\frac{Q}{T}\right)_{\text{rev}}$$

$$\Delta S_{\text{surr}} = -\left(\frac{Q}{T}\right)_{\text{rev}}$$

Hence, the entropy generated S_{gen} or the entropy change of the universe is

$$S_{\text{gen}} = \Delta S_{\text{univ}} = 0 \qquad \text{(reversible process)}$$

But for an irreversible process, $\Delta S_{\text{sys}} > Q/T_{\text{surr}}$ owing to entropy generation within the system as a result of internal irreversibilities. Hence, although the change in entropy of the system and surroundings may individually increase, decrease, or remain constant, the total entropy change or the entropy generation cannot be negative for any process:

$$S_{\text{gen}} = \Delta S_{\text{sys}} + \Delta S_{\text{surr}} = \Delta S_{\text{sys}} - \frac{Q}{T_{\text{surr}}} \geq 0 \qquad (5.19)$$

Note that S_{gen} is zero for reversible processes and is positive for irreversible processes in accordance with the principle of increase of entropy.

To illustrate the previous relations, consider the free expansion of an ideal gas. For the system shown in Figure 5.22, when the diaphragm is removed, the gas expands and occupies the total volume. During this process, $W = 0$, $Q = 0$, and according to the first law, $\Delta U = 0$. In order to calculate the entropy change, this irreversible process is replaced by an internally reversible process connecting the initial and final states of the system. Since the initial and final temperatures are the same, choose a reversible isothermal process during which the system is expanded in a controlled fashion. This may be accomplished by letting the system do work on a piston till the final state is reached. To maintain isothermal conditions, heat must be transferred to the system. Since the temperature remains constant,

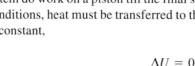

$$\Delta U = 0$$

Figure 5.22   Free expansion of an ideal gas.

and

$$Q_{\text{rev}} = -W_{\text{rev}} = \int_{V_1}^{V_2} p\, dV = mRT \ln \frac{V_2}{V_1}$$

where subscripts 1 and 2 refer to the initial and final conditions. The entropy change for the reversible process is given by

$$\Delta S = \left(\frac{Q}{T} \right)_{\text{rev}} = mR \ln \frac{V_2}{V_1}$$

which is the same for the irreversible process. Note that the entropy change has been evaluated independent of the details of the process and only by the end states. Also the change of entropy is positive in accordance with the principle of increase of entropy.

Entropy is an extensive property, as can be easily seen from Eq. (5.15). If the heat transfer per unit mass is constant, the amount of heat transfer to a system is proportional to its mass and so is the entropy. In the case of irreversible processes, the particles of the system follow different paths between the initial and the final states. But since the mass of each particle in the system experiences the same change of entropy per unit mass between the end states, the total change of entropy is equal to the specific entropy change multiplied by the mass of the system or

$$\Delta S = \sum (dm \, \Delta s) = m \, \Delta s \tag{5.20}$$

where m is the mass of the system.

The unit of entropy in the SI system is J/K, and for specific entropy is J/kg K.

5.10 Temperature–Entropy Diagram

The choice of temperature and entropy as two parameters to describe a thermodynamic system has a number of advantages. On a T–S diagram, isotherms and reversible adiabatics are horizontal and vertical lines. For an internally reversible process, the heat interaction is given by

$$\delta Q_{\text{rev}} = T \, dS$$

The amount of heat δQ_{rev} is represented by the shaded area of height T and width dS under the process path, as shown in Figure 5.23. The total amount of heat interaction during the reversible process 1-2 is equal to the integral of $T \, dS$ between states 1 and 2, or

$$Q_{\text{rev}} = \int_1^2 T \, dS$$

and is represented by area a-1-2-b-a on the T-S diagram. This is true only if the process 1-2 is internally reversible, and in this case the process is represented by a continuous curve on the T-S diagram. If the process is irreversible, it is usually

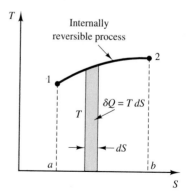

Figure 5.23   Temperature–entropy diagram.

represented by a dotted line, and the area under an irreversible process path is greater than the amount of heat interaction. The dotted representation of irreversible processes serves also to indicate that, owing to the nonequilibrium conditions encountered in irreversible processes, a single continuous path is insufficient to identify the state of the substance at any point during the process.

Constant entropy processes are called *isentropic* processes and are represented by vertical lines on the *T-S* diagram. For a reversible adiabatic process, $\delta Q_{rev} = 0$, and since $dS = (\delta Q/T)_{rev}$, $dS = 0$. Thus, a reversible adiabatic process is an isentropic process. The opposite is not always true in the sense that an isentropic process does not necessarily mean a reversible adiabatic process. For example, fluid flow accompanied by friction results in an increase in entropy. If, at the same time, heat is transferred from the system at such a rate to offset the entropy increase due to irreversibilities, then the process is isentropic. Such a process is obviously neither reversible nor adiabatic and is accompanied by an increase in entropy of the surroundings. The same conclusion may be reached by noting that since $dS = 0$ then $\delta Q/T_{surr} \leq 0$; that is, $\delta Q \leq 0$. Thus an irreversible isentropic process requires heat transfer from the system.

The Carnot power cycle is represented graphically on the *T-S* diagram of Figure 5.24. Process 1-2 is a reversible isothermal process during which heat is transferred to the system at constant temperature, and the entropy change is given by

$$S_2 - S_1 = \int_1^2 \left(\frac{\delta Q}{T}\right)_{rev} = \frac{Q_H}{T_H}$$

The heat transfer to the system, $Q_{1\text{-}2}$, is represented by the area 1-2-*b*-*a*-1. Process 2-3 is a reversible adiabatic process (isentropic) during which the entropy remains constant. Process 3-4 is a reversible isothermal compression during which heat is transferred from the system at constant temperature, and the entropy change is given by

$$S_4 - S_3 = \int_3^4 \left(\frac{\delta Q}{T}\right)_{rev} = \frac{Q_L}{T_L}$$

The heat rejection $Q_{3\text{-}4}$ is represented by the area 3-4-*a*-*b*-3. Note that the increase of entropy of the system during process 1-2 is equal to the decrease in entropy during process 3-4. The final process 4-1 is an isentropic process that restores the system to its original state and thus completes the cycle.

The first law applied to the cycle gives

$$\oint \delta Q + \oint \delta W = 0$$

or the net heat interaction during the cycle is equal to the net work output, and since each process is reversible,

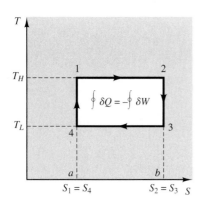

Figure 5.24   Carnot cycle on a *T-S* diagram.

$$\oint T \, dS = \oint p \, dV$$

Therefore, the net heat or work interactions during the cycle may be represented by either the area 1-2-3-4-1 of Figure 5.23 or the corresponding area on a p-V diagram. The thermal efficiency of the cycle expressed in terms of areas is

$$\eta = \frac{|W_{\text{cycle}}|}{Q_{1\text{-}2}} = \frac{\text{area } (1\text{-}2\text{-}3\text{-}4\text{-}1)}{\text{area } (1\text{-}2\text{-}b\text{-}a\text{-}1)}$$

$$= \frac{(T_H - T_L)(S_2 - S_1)}{T_H(S_2 - S_1)} = 1 - \frac{T_L}{T_H}$$

5.11 Relation to Other Thermodynamic Properties

Consider a system undergoing an internally reversible process during which there are heat and work interactions. The first law is

$$\delta Q_{\text{rev}} + \delta W_{\text{rev}} = dU \qquad (5.21)$$

Substituting $T \, dS$ and $-p \, dV$ for δQ_{rev} and δW_{rev} in the preceding equation and rearranging gives

$$T \, dS = dU + p \, dV \qquad (5.22)$$

But the differential of enthalpy is

$$dH = dU + p \, dV + V \, dp$$

which when substituted in Eq. (5.22) gives

$$T \, dS = dH - V \, dp \qquad (5.23)$$

Equations (5.22) and (5.23) are called the *TdS* equations. They are fundamental thermodynamic relations combining the first and second laws into a single equation.

Although the foregoing equations have been derived for an internally reversible process, they are valid for any process reversible or not because they express relationships between properties. Note that $\delta Q = T\,dS$ and $\delta W = -p\,dV$ are true only for a reversible process; hence, for any irreversible process the difference between $T\,dS$ and δQ is equal to the difference between $-p\,dV$ and δW.

Equations (5.22) and (5.23) can be written for a unit mass as

$$T\,ds = du + p\,dv$$

and

$$T\,ds = dh - v\,dp$$

For an ideal gas, $pv = RT$, $du = c_v\,dT$, and $dh = c_p\,dT$, which, when combined with the preceding two equations, give the following expressions for the differential of entropy per unit mass:

$$ds = c_v\frac{dT}{T} + \frac{p}{T}dv = c_v\frac{dT}{T} + R\frac{dv}{v}$$

and

$$ds = c_p\frac{dT}{T} - \frac{v}{T}dp = c_p\frac{dT}{T} - R\frac{dp}{p}$$

Assuming c_v and c_p are constants, integrating the foregoing equations between states 1 and 2 gives

$$s_2 - s_1 = c_v\ln\frac{T_2}{T_1} + R\ln\frac{v_2}{v_1} \qquad (5.24)$$

and

$$s_2 - s_1 = c_p\ln\frac{T_2}{T_1} - R\ln\frac{p_2}{p_1} \qquad (5.25)$$

If c_v and c_p are functions of temperature, as in the case of an ideal gas, it is necessary to express c_v and c_p in terms of temperature before the integration of the first term in Eqs. (5.24) and (5.25).

Entropy, being a thermodynamic property, can be used as one of the parameters to define the state of a system. It can be expressed as a function of any two of the independent variables, such as pressure, volume, temperature, and internal energy. Therefore

$$s = s(u, v)$$
$$= s(T, p) \qquad\qquad (5.26)$$
$$= s(p, v), \text{ etc.}$$

If, for example, the first equation is considered, ds can be expressed as

$$ds = \left(\frac{\partial s}{\partial u}\right)_v du + \left(\frac{\partial s}{\partial v}\right)_u dv$$

but

$$ds = \left(\frac{1}{T}\right) du + \left(\frac{p}{T}\right) dv$$

Since u and v are independent variables, the coefficients of du and dv in the two preceding equations may be equated to give

$$\left(\frac{\partial s}{\partial u}\right)_v = \frac{1}{T} \qquad\qquad (5.27)$$

and

$$\left(\frac{\partial s}{\partial v}\right)_u = \frac{p}{T} \qquad\qquad (5.28)$$

Since $(\partial s/\partial u)_v$ and $(\partial s/\partial v)_u$ are expressed in terms of properties, they themselves are also properties. Note that the slope of a constant volume line on an *s-u* diagram is equal to $1/T$, and the slope of a constant internal energy line on an *s-v* diagram is equal to p/T. Similarly, the second and third equations (5.26) provide useful property relations.

For incompressible substances, the specific volume is constant, and the specific heat depends only on temperature. Accordingly, the differential change entropy is given by

$$ds = \frac{du}{T} = \frac{c\, dT}{T}$$

and if the specific heat is assumed constant, integration gives

$$s_2 - s_1 = c \ln \frac{T_2}{T_1} \qquad\qquad (5.29)$$

Note that for an isentropic process of an incompressible substance, the temperature also remains constant.

Example 5.4

Two kilograms of water at 90°C are mixed with 3 kg of water at 10°C in an isolated system. Calculate the change of entropy due to the mixing process. The specific heat for water is 4.18 kJ/kg K.

SOLUTION

Referring to the isolated system shown in Figure 5.25, there is no heat or work interaction with the surroundings. Therefore, from the first law of thermodynamics, the change of the internal energy of the system is zero, or

$$m_1 c(T_f - T_1) + m_2 c(T_f - T_2) = 0$$

where c is the specific heat of water, and T_f is the final temperature. Substituting values gives

$$2c(T_f - 90) + 3c(T_f - 10) = 0$$

or

$$T_f = \frac{180 + 30}{5} = \frac{210}{5} = 42°C$$

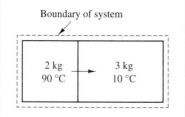

Figure 5.25

The total change of entropy of the system is the sum of the entropy changes of its components, or

$$\Delta S = \Delta S_1 + \Delta S_2$$
$$= m_1 c \ln \left(\frac{T_f}{T_i}\right)_1 + m_2 c \ln \left(\frac{T_f}{T_i}\right)_2 \qquad \text{for constant } c$$

Substituting values gives

$$\Delta S = (2 \text{ kg})(4.18 \text{kJ/kg K}) \ln \frac{(42 + 273.15)}{(90 + 273.15)} + (3 \text{ kg})(4.18 \text{ kJ/kg K}) \ln \frac{(42 + 273.15)}{(10 + 273.15)}$$
$$= -1.1852 + 1.3427 = 0.1575 \text{ kJ/K}$$

Note that the entropy of the cold body has increased, whereas the entropy of the hot body has decreased. There is, however, a net increase in entropy of the total system without heat transfer, a criterion of internally irreversible processes. Therefore, the mixing process is irreversible.

Example 5.5

Two kilograms of water at 90°C are mixed with 5 kg of ice at 0°C in an isolated system. Calculate (a) the change of entropy; (b) the change of entropy if 1 instead of 5 kg of ice were present. The specific heat for water is 4.18 kJ/kg K, and the latent energy of melting the ice is 333.5 kJ/kg.

SOLUTION

  (a) The first step is to find the final temperature after the mixing process. Two cases arise: If the decrease in the internal energy of the water is less than the latent energy required to melt the ice, the final temperature will be 0°C, and some ice will remain unmelted. If, on the other hand, the decrease in the internal energy of the water is more than the latent energy required to melt the ice, the final temperature of the mixture lies between 0°C and 90°C. To find the final temperature, the energy balance gives

$$(2 \text{ kg})(4.18 \text{ kJ/kg K})[(90 - T_f) \text{ K}] = (5 \text{ kg})(333.5 \text{ kJ/kg}) + (5 \text{ kg})(4.18 \text{ kJ/kg K})[(T_f - 0) \text{ K}]$$

or

$$T_f = \frac{2 \times 4.18 \times 90 - 5 \times 333.5}{(4.18)(7)} = -31.28°C$$

This value of T_f is unrealistic since the minimum temperature of the system can only be 0°C. This indicates that the ice did not melt completely, and the temperature of the mixture is therefore 0°C. To find the amount of ice that melted,

$$(2 \text{ kg})(4.18 \text{ kJ/kg K})[(90 - 0) \text{ K}] = x(333.5 \text{ kJ/kg})$$

or

$$x = \frac{(2 \text{ kg})(4.18 \text{ kJ/kg K})(90 \text{ K})}{333.5 \text{ kJ/kg}} = 2.256 \text{ kg of ice}$$

$$\Delta S_{\text{water}} = (2 \text{ kg})(4.18 \text{ kJ/kg K}) \ln \frac{273.15}{363.15} = -2.381 \text{ kJ/K}$$

$$\Delta S_{\text{ice}} = \frac{(2.256 \text{ kg})(333.5 \text{ kJ/kg})}{273.15 \text{ K}} = 2.754 \text{ kJ/K}$$

$$\Delta S_{\text{total}} = -2.381 \text{ kJ/K} + 2.754 \text{ kJ/K} = 0.373 \text{ kJ/K}$$

Since the process is adiabatic but involves an entropy change, the process must be irreversible.
  (b) If only 1 kg of ice were present, then

$$(2 \text{ kg})(4.18 \text{ kJ/kg K})[(90 - T_f) \text{ K}] = (1 \text{ kg})(333.5 \text{ kJ/kg}) + (1 \text{ kg})(4.18 \text{ kJ/kg K})[(T_f - 0) \text{ K}]$$

and

$$T_f = \frac{2 \times 4.18 \times 90 - 1 \times 333.5}{4.18(2 + 1)} = 33.4°C$$

$$\Delta S_{\text{water}} = (2 \text{ kg})(4.18 \text{ kJ/kg K}) \ln \frac{33.4 + 273.15}{90 + 273.15} = -1.415 \text{ kJ/K}$$

$$\Delta S_{\text{ice}} = \frac{(1 \text{ kg})(333.5 \text{ kJ/kg})}{273.15 \text{ K}} + (1 \text{ kg})(4.18 \text{ kJ/kg}) \ln \frac{33.4 + 273.15}{273.15}$$

$$= 1.704 \text{ kJ/K}$$

$$\Delta S = -1.415 \text{ kJ/K} + 1.704 \text{ kJ/K} = 0.289 \text{ kJ/K}$$

The entropy change in this case is smaller because the final temperature is higher, and the high final temperature means more available work can be obtained from the system.

5.12 Property Tables and Diagrams Involving Entropy

In previous chapters, properties such as temperature, pressure, specific volume, quality, and enthalpy have been used to define the state of a system. Relations among these properties were depicted in the form of diagrams such as the p-v and T-v diagrams or in the form of tables such as the steam tables. Now that the second law has introduced the property entropy, the list of independent properties that can be used to define the state of a system can be extended to include entropy. Representing thermodynamic processes on diagrams for which one of the coordinates is entropy is convenient especially when the system is analyzed from the second-law perspective. Representation of the change of state on diagrams in which entropy is one coordinate is not limited to single-phase substances. For a one-component substance in one or more phases, several diagrams such as the T-s and h-s diagrams have been developed. On these diagrams, lines of constant volume, pressure, quality, and enthalpy serve to indicate states and plot processes. Actual compression and expansion processes can be compared with the ideal isentropic processes, which are represented by vertical lines on these diagrams. Figure 5.26 shows a T-s diagram for the liquid and vapor regions of a single-component substance. For reversible processes, the heat interaction can be represented by an area on this diagram $(q_{\text{rev}} = \int T \, ds)$. Another diagram commonly used in analyzing flow devices such as turbines and compressors is the enthalpy–entropy diagram, also called the Mollier diagram.[5] Figure 5.27 is a Mollier diagram for the liquid–vapor regions of a single-component substance. As shown in this

[5] A Mollier chart for steam is given in Figure A1 of the Appendix.

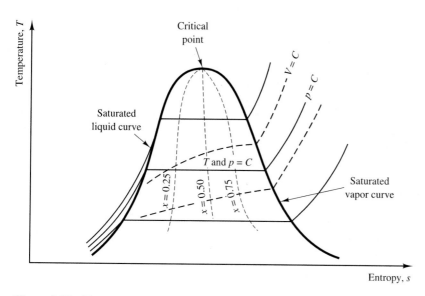

Figure 5.26   Temperature–entropy diagram.

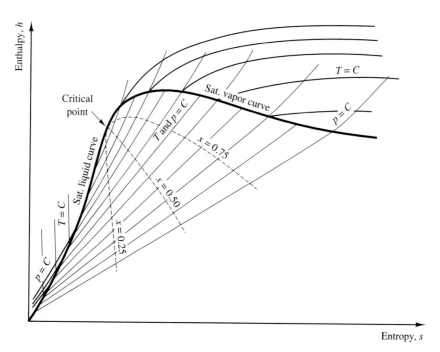

Figure 5.27   Enthalpy–entropy diagram.

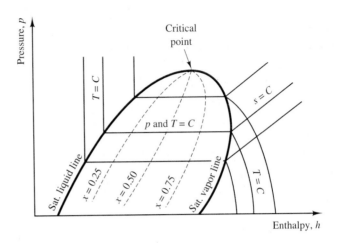

Figure 5.28 Pressure–enthalpy diagram.

figure, the constant-pressure and constant-temperature lines are identical in the liquid–vapor region. In the liquid and vapor regions, the temperature and pressure are independent; hence, constant-temperature curves and constant-pressure curves are shown as independent curves. The slope of the constant-pressure curves at any point is a measure of the temperature. This is easily seen by noting that $(\partial h/\partial s)_p = T$. In the wet region, constant-pressure lines are straight lines since the temperature remains the same at constant pressure. In the superheated-vapor region, constant-temperature lines are almost horizontal as pressure is reduced. Note that isentropic processes are represented by vertical lines, whereas constant-enthalpy lines are horizontal lines on this diagram. Figure 5.28 is a *p-h* diagram commonly used in representing refrigeration cycles, and Figures A2, A3, and A4 of the Appendix are detailed *p-h* diagrams for ammonia, refrigerant-12, and refrigerant-134a.

Specific entropy values are indicated in the property tables of the Appendix. They are tabulated in the same manner as *v*, *u*, and *h*. For compressed liquid and superheated vapor, the entropy can be read directly from the tables. For a mixture of liquid and vapor, it can be calculated from the relation

$$s = s_f + x\,s_{fg} = s_f + x(s_g - s_f) \tag{5.30}$$

where s_f, s_{fg}, and s_g are listed in the saturation tables, and x is the quality of the mixture. In the compressed liquid and superheated tables, the value of the specific entropy is given in terms of temperature and pressure. Note that the tabulated values of entropy indicated in the tables are referenced to s_f at 0.01°C for steam and to s_f at -40°C for ammonia, and refrigerant-12. These reference states are arbitrary and do not affect calculations of entropy differences.

Example 5.6

Compressed water initially at 25°C and 200 kPa is heated at constant pressure. If the amount of heat interaction is 1.8 MJ/kg, determine the entropy change per kg of fluid. Show the process on a *T-s* diagram indicating the final state.

SOLUTION

The entropy and enthalpy of the compressed liquid at the initial state is approximated as s_f and h_f at 25°C so that

$$s_1 = 0.3674 \text{ kJ/kg K}$$
$$h_1 = 104.89 \text{ kJ/kg}$$

To determine the final state of the system, another independent property, in addition to pressure, is required. At constant pressure, the heat interaction for a quasi-equilibrium process is equal to the change in enthalpy so that

$$q = h_2 - h_1$$

Therefore,

$$1800 = h_2 - 104.89 \qquad \text{or} \qquad h_2 = 1904.89 \text{ kJ/kg}$$

The properties needed to identify the final state are

$$p_2 = 200 \text{ kPa} \qquad \text{and} \qquad h_2 = 1904.89 \text{ kJ/kg}$$

But from steam tables, $h_f < h_2 < h_g$, indicating that the final state is in the liquid–vapor region

$$h_2 = h_f + x_2 h_{fg}$$
$$1904.89 \text{ kJ/kg} = 504.7 \text{ kJ/kg} + x_2(2201.9 \text{ kJ/kg})$$

from which

$$x_2 = 0.6359$$

The entropy at state 2 is

$$s_2 = s_f + x_2 s_{fg} = 1.5301 \text{ kJ/kg K} + 0.6359(5.5970 \text{ kJ/kg K})$$
$$= 5.4458 \text{ kJ/kg K}$$

and the entropy change is

$$s_2 - s_1 = (5.4458 - 0.3674) \text{ kJ/kg K} = 5.0784 \text{ kJ/kg K}$$

Figure 5.29 shows the process on a *T-s* diagram.

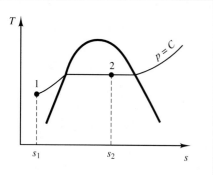

Figure 5.29

Example 5.7

One kilogram of water at $T_1 = 20°C$ and $p_1 = 500$ kPa is heated at a constant pressure to $T_2 = 400°C$. Calculate the change of entropy assuming constant specific heat for both the liquid and the superheated steam. Compare your results with the steam tables.

SOLUTION

Referring to Figure 5.30, at 500 kPa, $T_{sat} = 151.86°C$. The water is heated from 20°C to the saturation temperature; it vaporizes at this temperature, and then it is superheated to 400°C. Assuming the specific heat for water $c = 4.184$ kJ/kg K,

$$\Delta s_{water} = c \ln \frac{T_{sat}}{T_1} = (4.184 \text{ kJ/kg K}) \ln \left(\frac{151.86 + 273.15}{20 + 273.15} \right)$$

$$= 1.55406 \text{ kJ/kg K}$$

and

$$s_{fg} = \frac{h_{fg}}{T} = \frac{2108.5 \text{ kJ/kg}}{(151.86 + 273.15) \text{ K}} = 4.96106 \text{ kJ/kg K}$$

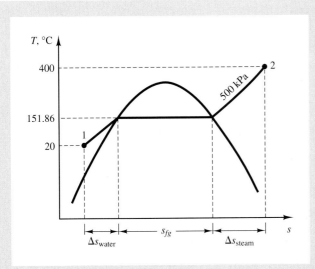

Figure 5.30

Assuming the specific heat for superheated steam is $c_p = 1.8723$ kJ/kg K,

$$\Delta s_{steam} = c_p \ln \frac{T_2}{T_{sat}} = (1.8723 \text{ kJ/kg K}) \ln \left(\frac{400 + 273.15}{151.86 + 273.15} \right)$$
$$= 0.8610 \text{ kJ/kg K}$$

The corresponding values from steam tables are

$$\Delta s_{water} = (1.8607 - 0.2956) \text{ kJ/kg K} = 1.5652 \text{ kJ/kg K}$$
$$s_{fg} = 4.9606 \text{ kJ/kg K}$$
$$\Delta s_{steam} = (7.7983 - 6.8213) \text{ kJ/kg K} = 0.9725 \text{ kJ/kg K}$$

The values of Δs_{water} and s_{fg} are practically the same by the two methods. The value of $(c_p)_{steam} = 1.8723$ kJ/kg K was assumed constant in the range of 151.86°C to 400°C, and this accounts for the difference in Δs_{steam}.

Example 5.8

Steam at a pressure of 2 MPa and a quality of 0.98 is throttled adiabatically in a steady-state operation to a pressure of 100 kPa in a steady flow process. Neglecting kinetic and potential energy changes, determine the entropy change.

SOLUTION

A sketch of the throttling process and a T-s diagram is shown in Figure 5.31. In an adiabatic throttling process, the enthalpy remains constant so that $h_i = h_e$.

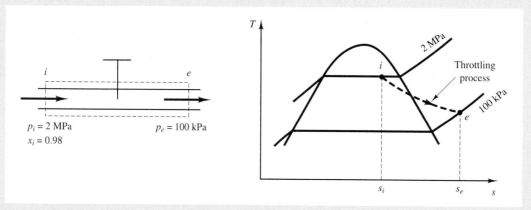

Figure 5.31

At the initial state,

$$h_i = h_f + x_i\, h_{fg} = 908.79\ \text{kJ/kg} + 0.98(1890.7\ \text{kJ/kg})$$
$$= 2761.676\ \text{kJ/kg}$$

and

$$s_i = s_f + x_i\, s_{fg} = (2.4474\ \text{kJ/kg K}) + 0.98(3.8935\ \text{kJ/kg K})$$
$$= 6.2630\ \text{kJ/kg K}$$

$$h_e = h_i = 2761.676\ \text{kJ/kg}$$

At $h_e = 2761.676$ kg/kg and $p_e = 100$ kPa, the steam is superheated and

$$s_e = 7.5764\ \text{kJ/kg K}$$

The specific change of entropy is

$$s_e - s_i = (7.5764 - 6.2630)\ \text{kJ/kg K} = 1.3134\ \text{kJ/kg K}$$

5.13 Gas Tables

Gas tables[6] may conveniently be used to calculate property relations between different states of a system. Properties of a number of gases, including air at low

---

[6] For example, see JANAF Thermochemical Tables, third edition, Thermal Group. Dow Chemical U.S.A., Midland, MI, 1985. An abstract of these tables is given in the Appendix, Tables A11 to A17.

pressure (based on ideal-gas relations), are tabulated in gas tables for a wide range of temperature. The tabulated values take into consideration the variation of specific heats with temperature.

Gas tables may be used also to relate properties of systems undergoing isentropic processes. The change of entropy of a unit mass of an ideal gas in terms of temperature and pressure is

$$\Delta s = \int_{T_1}^{T_2} c_p \frac{dT}{T} - R \ln \frac{p_2}{p_1} \tag{5.31}$$

In gas tables, a function s_T° is defined as

$$s_T^\circ = \int_{T_0}^{T} c_p \frac{dT}{T}$$

Since c_p for an ideal gas is a function of temperature, s_T° is also a function of temperature only. s_T° is called the *entropy function*. Its value is tabulated as a function of temperature and is chosen zero at 0 K. The change of entropy can then be written

$$\Delta s = \left(s_{T_2}^\circ - s_{T_1}^\circ \right) - R \ln \frac{p_2}{p_1} \tag{5.32}$$

If a compression or expansion process is isentropic, $\Delta s = 0$ so that

$$\ln \left(\frac{p_2}{p_1} \right)_s = \frac{s_{T_2}^\circ - s_{T_1}^\circ}{R} \tag{5.33}$$

A *relative pressure* p_r is defined as the ratio of the pressure p to a reference pressure p_0 such that

$$p_r \equiv \frac{p}{p_0} \tag{5.34}$$

Substituting in Eq. (5.33) gives

$$\ln \left(\frac{p_2}{p_1} \right)_s = \ln \left(\frac{p_2/p_0}{p_1/p_0} \right)_s = \ln \left(\frac{p_{r_2}}{p_{r_1}} \right)$$
$$= \frac{s_{T_2}^\circ - s_{T_1}^\circ}{R} = f(T)$$

Also

$$\left(\frac{p_2}{p_1} \right)_s = \frac{p_{r_2}}{p_{r_1}} \tag{5.35}$$

Equation (5.35) states that the isentropic pressure ratio is equal to the relative pressure ratio. Values of p_r are tabulated as a function of temperature for air in Table A11.

A similar relation may be obtained for specific volumes. Starting with the equation

$$ds = c_v \frac{dT}{T} + R \frac{dv}{v}$$

and following the same procedure just outlined, leads to the following relation of relative specific volumes:

$$\left(\frac{v_2}{v_1}\right)_s = \frac{v_{r_2}}{v_{r_1}} \tag{5.36}$$

where the *relative volume* v_r is defined as

$$v_r \equiv \frac{RT}{p_r} \tag{5.37}$$

Another method of arriving at Eq. (5.36) is substituting Eq. (5.37) into Eq. (5.35).

Example 5.9

One kilogram of air expands reversibly and adiabatically in a steady-state flow process from an initial state of $p_i = 650$ kPa and $T_i = 530$ K to a final state of $p_e = 100$ kPa. If the kinetic and potential energies are negligible, calculate the final temperature, the volume, and the work developed.

SOLUTION

Figure 5.32 shows the process on a T-s diagram. From Table A11, at $T_i = 530$ K,

$$p_{r_i} = 8.3578 \qquad v_{r_i} = 42.4293 \qquad \text{and} \qquad h_i = 534.33 \text{ kJ/kg}$$

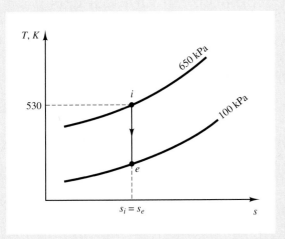

Figure 5.32

From Eq. (5.35),

$$p_{r_e} = p_{r_i}\left(\frac{p_e}{p_i}\right) = 8.3578\left(\frac{100}{650}\right) = 1.2858$$

From gas tables, at $p_{r_e} = 1.2858$,

$$T_e = 312.12 \text{ K} \qquad v_{r_e} = 163.275 \qquad \text{and} \qquad h_e = 312.65 \text{ kJ/kg}$$

From Eq. (5.36),

$$v_e = v_i\left(\frac{v_{r_e}}{v_{r_i}}\right) = \frac{RT_i}{p_i}\left(\frac{v_{r_e}}{v_{r_i}}\right) = \frac{(0.287 \text{ kJ/kg K})(530 \text{ K})}{650 \text{ kPa}}\left(\frac{163.275}{42.4293}\right)$$
$$= 0.9 \text{ m}^3/\text{kg}$$

The work interaction per unit mass can be determined from the first law:

$$q + w + \left(h_2 + \frac{V_2^2}{2} + gz_i\right) - \left(h_e + \frac{V_e^2}{2} + gz_e\right) = 0$$

Neglecting kinetic and potential energies, the first law for adiabatic flow reduces to

$$w = h_e - h_i = (312.12 - 534.33) \text{ kJ/kg} = -222.21 \text{ kJ/kg}$$

5.14 Isentropic Relations for an Ideal Gas

Relations describing an isentropic process of an ideal gas may be obtained by setting $s_2 - s_1 = 0$ in either Eq. (5.24) or Eq. (5.25). Using Eq. (5.24), and assuming constant values of specific heats,

$$c_v \ln \frac{T_2}{T_1} = -R \ln \frac{v_2}{v_1}$$

from which

$$\frac{T_2}{T_1} = \left(\frac{v_1}{v_2}\right)^{R/c_v}$$

but, for an ideal gas,

$$c_p = c_v + R \qquad \text{or} \qquad \frac{R}{c_v} = \gamma - 1$$

Therefore,

$$\frac{T_2}{T_1} = \left(\frac{v_1}{v_2}\right)^{\gamma - 1}$$

The preceding equation relates the temperatures and specific volumes of two states connected by an isentropic process. The relation between pressures and volumes may be obtained by substituting $p_2 v_2 / p_1 v_1$ for T_2/T_1 in the preceding equation. Hence,

$$\frac{p_2 v_2}{p_1 v_1} = \left(\frac{v_1}{v_2}\right)^{\gamma - 1}$$

or

$$\left(\frac{p_2}{p_1}\right) = \left(\frac{v_1}{v_2}\right)^{\gamma} \tag{5.38}$$

Any state along the isentropic path between states 1 and 2 satisfies Eqs. (5.38) and (5.39). Combining these two equations gives

$$\frac{T_2}{T_1} = \left(\frac{p_2}{p_1}\right)^{(\gamma - 1)/\gamma} = \left(\frac{v_1}{v_2}\right)^{\gamma - 1} \tag{5.39}$$

Example 5.10

During a certain process, the volume of 1 kg of air is doubled and its pressure is tripled. Find the change of entropy per kg. Assume air to be an ideal gas, with constant specific heats ($c_p = 1.0035$ and $c_v = 0.7165$ kJ/kg K).

SOLUTION

Using the equation of state of an ideal gas to eliminate the temperatures in Eq. (5.24) gives

$$
\begin{aligned}
s_2 - s_1 &= c_v \ln \frac{p_2 v_2}{p_1 v_1} + R \ln \frac{v_2}{v_1} \\
&= (0.7165 \text{ kJ/kg K}) \ln (3 \times 2) + (0.287 \text{ kJ/kg K}) \ln 2 \\
&= (1.2838 + 0.19889) \text{ kJ/kg K} \\
&= 1.4827 \text{ kJ/kg K}
\end{aligned}
$$

Example 5.11

For a system of an ideal gas, prove that reversible adiabatic lines, plotted on any thermodynamic coordinates, do not intersect. Prove also that reversible adiabatic lines are steeper than reversible isothermal lines on a p-V plane.

SOLUTION

To prove that reversible adiabatic lines do not intersect, assume that the opposite is true, as shown on the p-V diagram of Figure 5.33. Let an isotherm intersect both adiabatic lines, and let the three lines form a cycle in which an amount of work equal to the enclosed area on the p-V diagram is produced. Since heat transfer to the system can take place during only one process of the cycle, namely, the isothermal process, no other process is available for heat rejection. Therefore, the system is exchanging heat with a single reservoir and yet performing work. This, of course, is impossible since the cycle forms a perpetual motion machine of the second kind. Therefore, two reversible adiabatic lines may not intersect. To prove that an adiabatic line is steeper than an isothermal line, consider a reversible adiabatic process of an ideal gas ($pV^\gamma = C$). The slope of an adiabatic line on the p-V diagram is

$$
\frac{dp}{dV} = -\gamma \frac{p}{V}
$$

and for an isothermal process ($pV = C$), the slope is

$$
\frac{dp}{dV} = -\frac{p}{V}
$$

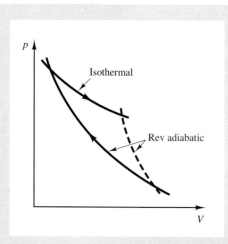

p

Isothermal

Rev adiabatic

V

Figure 5.33

It is clear from the two preceding equations that the adiabatic line is steeper than the isothermal line on a p-V diagram since $\gamma > 1$. This further means that isothermal and adiabatic lines cannot touch without intersection.

Example 5.12

Prove that the internal energy of an ideal gas is a function of temperature only.

SOLUTION

Let the internal energy of an ideal gas be expressed in terms of two properties, such as T and v. If the dependence of u on v can be eliminated, then u becomes a function of T only:

$$u = u(T, v)$$

The differential of u is

$$du = \left(\frac{\partial u}{\partial T}\right)_v dT + \left(\frac{\partial u}{\partial v}\right)_T dv$$

but

$$ds = \left(\frac{1}{T}\right) du + \frac{p}{T} dv$$

$$= \frac{1}{T}\left[\left(\frac{\partial u}{\partial T}\right)_v dT + \left(\frac{\partial u}{\partial v}\right)_T dv\right] + \frac{p}{T} dv$$

From the ideal-gas relation $p/T = R/v$,

$$ds = \frac{1}{T}\left(\frac{\partial u}{\partial T}\right)_v dT + \left[\frac{1}{T}\left(\frac{\partial u}{\partial v}\right)_T + \frac{R}{v}\right] dv \qquad\qquad \text{(a)}$$

Entropy may be expressed in terms of T and v as

$$s = s(T, v)$$

The differential of entropy is

$$ds = \left(\frac{\partial s}{\partial T}\right)_v dT + \left(\frac{\partial s}{\partial v}\right)_T dv \qquad\qquad \text{(b)}$$

Comparing Eqs. (a) and (b) and noting that T and v are independent, the coefficients of dT and dv, in both equations, must be equal or

$$\left(\frac{\partial s}{\partial T}\right)_v = \frac{1}{T}\left(\frac{\partial u}{\partial T}\right)_v \qquad\qquad \text{(c)}$$

and

$$\left(\frac{\partial s}{\partial v}\right)_T = \frac{1}{T}\left(\frac{\partial u}{\partial v}\right)_T + \frac{R}{v} \qquad\qquad \text{(d)}$$

Since entropy is a property,

$$\frac{\partial^2 s}{\partial v \partial T} = \frac{\partial^2 s}{\partial T \partial v}$$

Differentiating Eqs. (c) and (d) with respect to v and T, respectively, and equating gives

$$\frac{1}{T}\frac{\partial^2 u}{\partial v \partial T} = \frac{1}{T}\frac{\partial^2 u}{\partial T \partial v} + \left(\frac{\partial u}{\partial v}\right)_T\left(-\frac{1}{T^2}\right)$$

from which

$$\left(\frac{\partial u}{\partial v}\right)_T = 0$$

Therefore, u is independent of v; it depends only on T.

Example 5.13

Air initially at 920 K and 500 kPa expands adiabatically and reversibly to 100 kPa. Determine the final temperature of the air using gas tables, and compare the result assuming an average value of the specific heat ratio.

SOLUTION

Figure 5.34 shows the process on a T-s diagram. From Table A11, at 920 K, $p_{r_1} = 65.95261$.

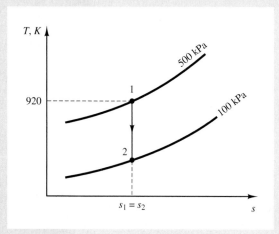

Figure 5.34

Since the process is adiabatic reversible (isentropic),

$$\frac{p_{r_2}}{p_{r_1}} = \frac{p_2}{p_1}$$

Therefore,

$$p_{r_2} = \left(\frac{100}{500}\right)(65.95261) = 13.1905$$

At this value of p_{r_2}, interpolation gives $T_2 = 601.17$ K. The final temperature can also be determined using the relation

$$\frac{T_2}{T_1} = \left(\frac{p_2}{p_1}\right)^{(\gamma-1)/\gamma}$$

Since the value of γ varies with temperature, the previous result can be used to estimate an average temperature of $(920 + 601.17)/2 = 760.6$ K, yielding $\gamma = 1.358$.

Therefore,

$$T_2 = (920 \text{ K})\left(\frac{100}{500}\right)^{0.258/1.358} = 601.92 \text{ K}$$

which is close to the value obtained using the air tables. For a wider range of temperature, the two values can differ considerably.

5.15 Entropy Change for a Control Volume

In the previous sections, it was shown that the entropy of an isolated system can only increase or remain constant. For a system, the change of entropy is due to internal irreversibilities and to energy interactions in the form of heat transfer with the surroundings. For a control volume, there is an additional change of entropy owing to the mass crossing the boundaries of the control volume. The change of entropy of a control volume due to mass transport is equal to the difference between the product of the mass and its specific entropy at the inlet and at the exit from the control volume. Therefore, the change of entropy of a control volume during a small time interval is given by

$$\underbrace{dS_{cv}}_{\text{entropy change}} \geq \underbrace{\frac{\delta Q}{T_{\text{surr}}} + \sum s_i \, dm_i - \sum s_e \, dm_e,}_{\text{entropy transfer}} \qquad (5.40)$$

The total entropy change or the entropy generation S_{gen} is given by

$$S_{\text{gen}} = \Delta S_{cv} + \Delta S_{\text{surr}}$$
$$= \Delta S_{cv} - \frac{Q}{T_{\text{surr}}} + \sum m_e s_e - \sum m_i s_i \geq 0 \qquad (5.41)$$

where T_{surr} is the temperature of the surroundings, and subscripts i and e refer to inlet and exit conditions. In Eq. (5.40), entropy flow into the control volume is considered positive, and entropy outflow is considered negative. The equality sign in the preceding two equations applies to reversible processes in which the heat interaction, as well as the mass transport to and from the control volume, is accomplished reversibly. The inequality sign applies to irreversible processes. Equations (5.40) and (5.41) can be expressed as rate equations in the form

$$\dot{S}_{cv} \geq \frac{\dot{Q}}{T_{surr}} + \sum \dot{m}_i s_i - \sum \dot{m}_e s_e \tag{5.42}$$

and

$$\dot{S}_{gen} = \Delta \dot{S}_{cv} - \frac{\dot{Q}}{T_{surr}} + \sum \dot{m}_e s_e - \sum \dot{m}_i s_i \geq 0 \tag{5.43}$$

Note that Eqs. (5.40) to (5.43) become applicable to a system when the summation terms in these equations are eliminated.

Irreversibilities within a system or a control volume result in a dispersion of energy accompanied by an increase in internal entropy. The total entropy change or the entropy generation may be expressed as

$$S_{gen} = \Delta S_{total} = \Delta S_{ext} + \Delta S_{int} \tag{5.44}$$

where subscripts ext and int refer to external and internal effects, respectively. Changes in entropy arise from external sources when there is mass transfer or heat transfer due to interaction with the surroundings. Changes in entropy arise internally when there is dispersion of energy as a result of internal nonequilibrium processes. Whereas ΔS_{ext} can be positive or negative depending on the direction of entropy transfer, ΔS_{int} must be either positive or zero, but can never be negative. The change of internal entropy, or the rate of internal entropy generation, indicates the magnitude of irreversibility in comparing irreversible processes.

In the case of a steady-state flow process, the time rate of change of entropy of the control volume is zero, and the time rate of the mass entering is equal to that leaving the control volume. Under these conditions, Eqs. (5.42) and (5.43) become

$$\frac{\dot{Q}}{T_{surr}} + \sum \dot{m}_i s_i - \sum \dot{m}_e s_e \leq 0$$

and

$$\dot{S}_{gen} = -\frac{\dot{Q}}{T_{surr}} + \sum \dot{m}_e s_e - \sum \dot{m}_i s_i \geq 0$$

and for an adiabatic steady flow process,

$$\sum \dot{m}s \leq 0$$

If, in addition, the process were reversible, then

$$\sum \dot{m}s = 0$$

The preceding equations can also be obtained using the procedure outlined in Section 2.20. Note that for a control volume undergoing a steady-state process, both mass and energy are conserved, but entropy increases unless the process is reversible.

Example 5.14

In an adiabatic desuperheater, water is sprayed into superheated steam in the proper amount to cause it to become saturated. Consider the process to be an adiabatic steady-state process and the following data to apply:

    steam flow 1000 kg/hr
    steam entering at 2.7 MPa and 300°C
      ($h = 3002.7$ kJ/kg, $s = 6.6019$ kJ/kg K)
    water entering at 2.7 MPa and 40°C
      ($h = 174$ kJ/kg, $s = 0.571$ kJ/kg K)
    steam leaving at 2.5 MPa, 223.99°C dry and saturated
      ($h = 2803.1$ kJ/kg, $s = 6.2575$ kJ/kg K)

Neglecting changes in kinetic and potential energy, calculate the mass rate of flow of water necessary for desuperheating. What is the change of entropy of the control volume? Show that the process is irreversible.

SOLUTION

A schematic diagram of the desuperheater is shown in Figure 5.35. The first law is

$$\dot{Q} + \dot{W} + \sum \dot{m}_i \left(h + \frac{V^2}{2} + gz \right)_i - \sum \dot{m}_e \left(h + \frac{V^2}{2} + gz \right)_e = \Delta \dot{E}$$

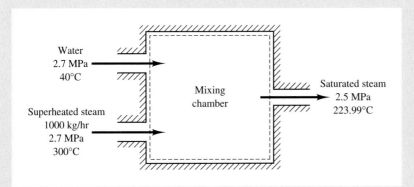

Figure 5.35

For the assumptions of this problem, this equation reduces to

$$\dot{m}_w h_w + \dot{m}_{sh} h_{sh} - \dot{m}_s h_s = 0$$

where subscripts w, sh, and s indicate water, superheated steam, and saturated steam. Substituting values and noting that $\dot{m}_s = \dot{m}_w + \dot{m}_{sh}$ gives

$$\dot{m}_w(174 \text{ kJ/kg}) + (1000 \text{ kg/hr})(3002.7 \text{ kJ/kg}) - (\dot{m}_w + 1000 \text{ kg/hr})(2803.1 \text{ kJ/kg}) = 0$$

from which $\dot{m}_w = 75.92$ kg/hr.

The change of entropy of the control volume is zero since the process takes place under steady-state conditions. From Eq. (5.42),

$$\frac{\dot{Q}}{T} \leq \dot{m}_e s_e - \sum \dot{m}_i s_i$$

so that

$$\frac{\dot{Q}}{T} \leq \left[(1000 + 75.92) \text{ kg/hr}\right](6.2575 \text{ kJ/kg K}) - \left[(1000 \text{ kg/hr})(6.6019 \text{ kJ/kg K}) + (75.92 \text{ kg/hr})(0.571 \text{ kJ/kg K})\right]$$

$$\leq 87.32 \text{ kJ/hr K}$$

But since $\dot{Q} = 0$, the inequality sign applies, and therefore, the mixing process is irreversible. This further means that the rate of entropy transfer from the control volume exceeds the rate of entropy at which it enters, the difference being the rate of entropy generation within the control volume.

Example 5.15

Air at a pressure of 400 kPa and a temperature of 700°C expands in a turbine operating at steady state to a pressure of 100 kPa. The work output is 260 kJ/kg, and the heat transfer from the turbine is 30 kJ/kg. Determine the entropy generation within the turbine per unit mass of air. Assume air to be an ideal gas with constant specific heats, and neglect changes in kinetic and potential energy.

SOLUTION

A schematic diagram of the turbine and a *T-s* diagram of the process is shown in Figure 5.36. For a steady-state irreversible expansion, Eqs. (5.40) and (5.41) per unit mass are

$$\frac{q}{T_{\text{surr}}} + (s_i - s_e) < 0$$

and

$$s_{\text{gen}} = -\frac{q}{T_{\text{surr}}} + (s_e - s_i) > 0$$

where s_{gen} is the internal entropy generated per unit mass so that

$$s_{\text{gen}} = -\frac{q}{T_{\text{surr}}} + (s_e - s_i) = -\frac{q}{T_{\text{surr}}} + \left(c_p \ln \frac{T_e}{T_i} - R \ln \frac{p_e}{p_i} \right)$$

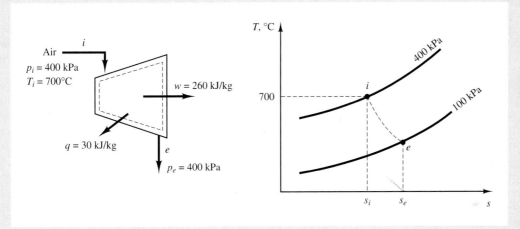

Figure 5.36

Applying the first law gives

$$q + w + h_i = h_e$$

so that

$$-30 \text{ kJ/kg} - 260 \text{ kJ/kg} + (1.0035 \text{ kJ/kg K})[(973.15 - T_e) \text{ K}] = 0$$

from which $T_e = 684.16$ K.

The internal entropy generation is

$$s_{\text{gen}} = -\frac{-30 \text{ kJ/kg}}{298.15 \text{ K}} + \left[(1.0035 \text{ kJ/kg K}) \ln \frac{684.16}{973.15} - (0.287 \text{ kJ/kg K}) \ln \frac{100}{400} \right]$$

$$= 0.1006 + (-0.3536 + 0.3979) = 0.1449 \text{ kJ/kg K}$$

5.16 Isentropic Efficiency

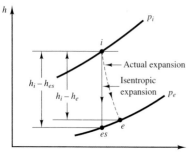

Figure 5.37 Comparison of actual and isentropic expansions.

Maximum work interaction can be developed if the expansion in an adiabatic turbine is isentropic. Similarly, minimum work is required when compression is accomplished isentropically in an adiabatic compressor. The concept of isentropic efficiency provides a comparison between the actual and the ideal performance of devices such as compressors and turbines. Figure 5.37 shows an actual and an isentropic expansion of a fluid in a turbine on an *h-s* diagram. The isentropic expansion is along process *i-es*, and the actual expansion is between states *i* and *e*. The state of the fluid at the turbine inlet and the exit pressure are the same for both cases. The entropy increase at state *e* is in accordance with the second law of thermodynamics, which dictates that the change of entropy in actual adiabatic processes cannot be negative, and only those states that correspond to an increase in entropy are accessible. Note also that the enthalpy at state *e* is more than that of state *es,* and the increase in both entropy and enthalpy is attributed to internal irreversibilities. Since the enthalpy drop in an adiabatic turbine operating at steady state, neglecting changes in kinetic and potential energy, is equal to the work developed, it is obvious that the actual work is less than the isentropic work. The isentropic turbine efficiency indicates such a difference and is defined by

$$\eta_T = \frac{|w_{\text{act}}|}{|w_{\text{isent}}|} = \frac{h_i - h_e}{h_i - h_{es}} \qquad (5.45)$$

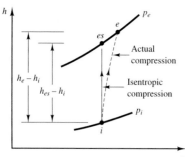

Figure 5.38 Comparison of actual and isentropic compressions.

Figure 5.38 shows an adiabatic steady-state compression process on an *h-s* diagram. The actual compression takes place between states *i* and *e* and results in an increase in entropy. Process *i-es* is an isentropic process. The enthalpy at state *e* is higher than at state *es* owing to the increase in entropy in the actual compression process. Hence the actual work input, neglecting changes in kinetic and potential energy, is larger than the isentropic work. The isentropic work corresponds to the minimum work input. The isentropic compressor efficiency is defined by

$$\eta_C = \frac{w_{\text{isent}}}{w_{\text{act}}} = \frac{h_{es} - h_i}{h_e - h_i} \qquad (5.46)$$

If the fluid is considered an ideal gas with constant specific heats, the isentropic turbine efficiency can be written as

$$\eta_T = \frac{c_p(T_i - T_e)}{c_p(T_i - T_{es})} = \frac{T_i\left(1 - \dfrac{T_e}{T_i}\right)}{T_i\left(1 - \dfrac{T_{es}}{T_i}\right)}$$

from which

$$\eta_T\left(1 - \frac{T_{es}}{T_i}\right) = 1 - \frac{T_e}{T_i} \quad \text{or} \quad \frac{T_i}{T_e} = \frac{1}{1 - \eta_T\left(1 - \frac{T_{es}}{T_i}\right)}$$

But $T_{es}/T_i = (p_e/p_i)^{(\gamma-1)/\gamma}$ so that the actual temperature ratio can be expressed in terms of the pressure ratio as

$$\frac{T_i}{T_e} = \frac{1}{1 - \eta_T\left[1 - \dfrac{1}{(p_i/p_e)^{(\gamma-1)/\gamma}}\right]} \tag{5.47}$$

In the case of a compressor, similar procedure leads to

$$\frac{T_e}{T_i} = 1 + \frac{\left[(p_e/p_i)^{(\gamma-1)/\gamma} - 1\right]}{\eta_C} \tag{5.48}$$

Figures 5.39 and 5.40 represent Eqs. (5.47) and (5.48) in graphical form. They show why, when gas turbines were being developed, it was necessary to have designs with high values of isentropic efficiency for turbine and compressor in order to obtain a net work output.

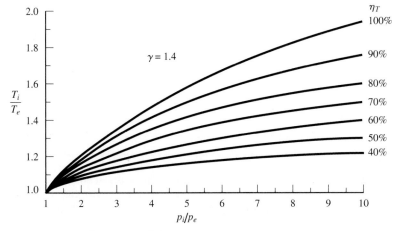

Figure 5.39 T_i/T_e versus p_i/p_e for various expansion efficiencies (Eq. 5.47).

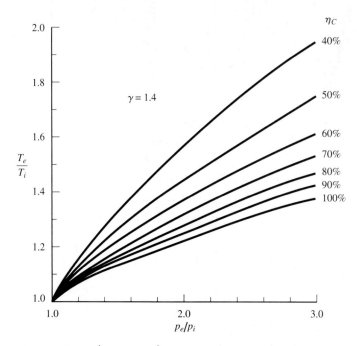

Figure 5.40 T_e/T_i versus p_e/p_i for various compression efficiencies (Eq. 5.48).

In a similar manner, the actual performance of a nozzle in accelerating the flow can be compared with ideal performance. For steady-state operation, the isentropic efficiency of a nozzle is defined as the actual kinetic energy of the fluid leaving the nozzle to the kinetic energy at the exit if the expansion were isentropic:

$$\eta_{\text{nozzle}} = \frac{\left(V^2/2\right)_{\text{act}}}{\left(V^2/2\right)_{\text{isent}}} \qquad (5.49)$$

Example 5.16

Air at a pressure of 100 kPa and a temperature of 300 K enters an adiabatic compressor operating at steady state. The pressure ratio across the compressor is 3.5, and changes in kinetic and potential energy are negligible. If the isentropic efficiency of the compressor is 80 percent, determine the temperature of the air at the exit. Assume a constant value of γ for air equal to be 1.4.

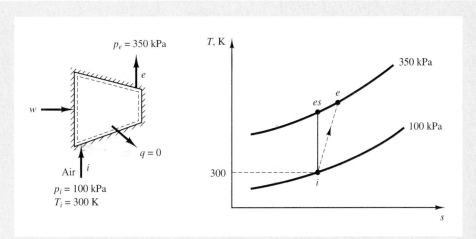

Figure 5.41

SOLUTION

A schematic of the compressor and a *T-s* diagram of the process is shown in Figure 5.41. Referring to Figure 5.38, the isentropic compressor efficiency is

$$\eta_C = \frac{h_{es} - h_i}{h_e - h_i} = \frac{T_{es} - T_i}{T_e - T_i}$$

but process *i-es* is isentropic so that $T_{es}/T_i = (p_e/p_i)^{(\gamma-1)/\gamma}$ from which

$$T_{es} = (300 \text{ K})(3.5)^{(1.4-1)/1.4} = (300 \text{ K})(3.5)^{0.2857} = 429.1 \text{ K}$$

Hence, $0.8 = (429.1 - 300)/(T_e - 300)$, from which $T_e = 461.38$ K. The same result can also be obtained using Figure 5.40.

Example 5.17

Air enters an adiabatic nozzle operating at steady state at a pressure of 350 kPa, a temperature of 1200 K, and with negligible velocity. At the exit, the pressure is 100 kPa, and the temperature is 850 K. For steady-state operation, determine:
   (a) the velocity at the exit
   (b) the isentropic efficiency of the nozzle
   (c) the entropy change per unit mass of flow.
   Assume air to be an ideal gas with constant specific heats.

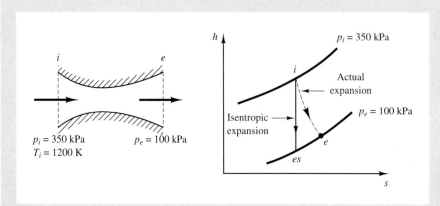

Figure 5.42

SOLUTION

(a) Referring to Figure 5.42, the first law is

$$q + w + \left(h + \frac{V^2}{2} + gz\right)_i - \left(h + \frac{V^2}{2} + gz\right)_e = \Delta e$$

which, for the conditions and assumptions of this problem, reduces to

$$h_i = h_e + \frac{V_e^2}{2}$$

or

$$V_e = \sqrt{2c_p(T_i - T_e)} = \sqrt{2(1003.5 \text{ J/kg K})[(1200 - 850) \text{ K}]}$$
$$= 838.12 \text{ m/s}$$

(b) From the isentropic relation,

$$\frac{T_{es}}{T_i} = \left(\frac{p_e}{p_i}\right)^{(\gamma-1)/\gamma}$$

$$T_{es} = (1200 \text{ K})\left(\frac{100}{350}\right)^{(1.4-1)/1.4} = 839 \text{ K}$$

The isentropic efficiency is

$$\eta_{\text{nozzle}} = \frac{V_e^2/2}{V_{es}^2/2} = \frac{h_i - h_e}{h_i - h_{es}} = \frac{T_i - T_e}{T_i - T_{es}} = \frac{1200 - 850}{1200 - 839} = \frac{350}{361} = 96.95\%$$

(c) The entropy change is

$$\Delta s = c_p \ln \frac{T_e}{T_i} - R \ln \frac{p_e}{p_i}$$

$$= (1.0035 \text{ kJ/kg K}) \ln \frac{850}{1200} - (0.287 \text{ kJ/kg K}) \ln \frac{100}{350}$$

$$= -0.34605 + 0.35954 = 0.01349 \text{ kJ/kg K}$$

5.17 Mathematical Formulation of the Second Law

In Section 5.9, the concept of entropy and the second law of thermodynamics were introduced by the Carnot–Clausius use of cycles, which is actually an application of the first law. This presentation, although simple and instructive, does not illustrate the broad scope of the second law. In an attempt to provide a basis for the concept of entropy and to formulate the second law by a formal mathematical procedure, Carathéodory, in 1909, used an integrating factor $(1/T)$ that renders δQ an exact differential. He postulated that, for reversible (physically impossible) processes between states in the vicinity of an arbitrary initial state of a system, there exists a function, S, and another, T, that satisfy the equation $\delta Q_{rev} = T\, dS$.

Before introducing Carathéodory's principle, an investigation of the condition for a linear differential equation to be integrable is in order.[7] Consider a linear differential equation of the form

$$df = X\, dx + Y\, dy + Z\, dz \qquad (5.50)$$

where X, Y, and Z are continuous functions of the coordinates x, y, and z. The condition $df = 0$ defines a family of surfaces passing through a given point. If the foregoing equation has an integrating factor $\lambda(x, y, z)$ such that λdf is a differential of a function and therefore is integrable, then the solutions of the differential equation $df = 0$ are functions of the form $\varphi(x, y, z) = $ constant, called *solution curves*.

The condition for integrability, that is, the existence of unique solution curves, follows.

---

[7] This procedure is an extension of the condition for an exact differential for two variables outlined in Section 1.4.

Consider the exact differential $d\varphi$ as given by

$$d\varphi = \lambda df = \lambda X \, dx + \lambda Y \, dy + \lambda Z \, dz$$

then,

$$\frac{\partial \varphi}{\partial x} = \lambda X \qquad \frac{\partial \varphi}{\partial y} = \lambda Y \qquad \frac{\partial \varphi}{\partial z} = \lambda Z$$

Therefore,

$$\frac{\partial(\lambda X)}{\partial y} = \frac{\partial^2 \varphi}{\partial y \partial x} = \frac{\partial(\lambda Y)}{\partial x}$$

Performing the differentiation and arranging gives

$$\lambda\left(\frac{\partial X}{\partial y} - \frac{\partial Y}{\partial x}\right) = Y\frac{\partial \lambda}{\partial x} - X\frac{\partial \lambda}{\partial y} \qquad (5.51a)$$

Similarly,

$$\lambda\left(\frac{\partial Y}{\partial z} - \frac{\partial Z}{\partial y}\right) = Z\frac{\partial \lambda}{\partial y} - Y\frac{\partial \lambda}{\partial z} \qquad (5.51b)$$

and

$$\lambda\left(\frac{\partial Z}{\partial x} - \frac{\partial X}{\partial z}\right) = X\frac{\partial \lambda}{\partial z} - Z\frac{\partial \lambda}{\partial x} \qquad (5.51c)$$

Multiplying Eq. (5.51a) by Z, Eq. (5.51b) by X, and Eq. (5.51c) by Y and adding gives the condition for an integrating factor to exist as

$$X\left(\frac{\partial Z}{\partial y} - \frac{\partial Y}{\partial z}\right) + Y\left(\frac{\partial X}{\partial z} - \frac{\partial Z}{\partial x}\right) + Z\left(\frac{\partial Y}{\partial x} - \frac{\partial X}{\partial y}\right) = 0 \qquad (5.52)$$

This means that, if an integrating factor exists, two neighboring points can be joined by one of the solution curves of the linear differential equation.

Consider the equation

$$X(x, y, z) \, dx + Y(x, y, z) \, dy + Z(x, y, z) \, dz = 0 \qquad (5.53)$$

where x, y, and z are functions of two variables u and v so that

$$x = x(u, v) \qquad y = y(u, v) \qquad z = z(u, v)$$

Then

$$dx = \frac{\partial x}{\partial u} du + \frac{\partial x}{\partial v} dv$$

$$dy = \frac{\partial y}{\partial u} du + \frac{\partial y}{\partial v} dv$$

$$dz = \frac{\partial z}{\partial u} du + \frac{\partial z}{\partial v} dv$$

Substituting in Eq. (5.53) and rearranging gives

$$\left(X\frac{\partial x}{\partial u} + Y\frac{\partial y}{\partial u} + Z\frac{\partial z}{\partial u} \right) du + \left(X\frac{\partial x}{\partial v} + Y\frac{\partial y}{\partial v} + Z\frac{\partial z}{\partial v} \right) dv = 0$$

which is an equation in two variables u and v. It may be integrated to give a unique curve in the space (u, v), which satisfies Eq. (5.53) and passes through the initial point (u_o, v_o). Therefore, points not lying on this curve cannot be reached from the initial point.

Carathéodory's principle states that in the vicinity of any arbitrary initial state of a system neighboring states exist that are not accessible from the initial state by means of a reversible adiabatic process. In other words, they are not accessible along solution curves of Eq. (5.53), unless Eq. (5.53) is integrable, which requires the existence of a function $\lambda(x, y, z)$ and another $F(x, y, z)$ such that

$$X\,dx + Y\,dy + Z\,dz = \lambda dF \qquad (5.54)$$

where $dF = df/\lambda$. As an example, states of a system that lie to the left of an isentropic line on a T-S diagram are adiabatically inaccessible from any initial state lying on the isentropic line. The irreversible adiabatic stirring of a fluid is another example of the inaccessibility of the original state. In this case, the reversal of the process cannot be achieved along an adiabatic path.

Therefore, the second law of thermodynamics stated mathematically implies that the accessibility of neighboring states from a given initial state along a path is possible only if

$$\delta Q = X\,dx + Y\,dy + Z\,dz = T\,dS \qquad (5.55)$$

where $T(= \lambda)$ and $S(= F)$ are functions of the state.

Statistical mechanics, as indicated in Chapter 12, show that entropy is a measure of the distribution of the energy of the system. This approach is also considered of broader scope in comparison to the definition deduced from the classical Carnot–Clausius treatment.

5.18 Physical Interpretation of Entropy

It is not sufficient in discussing entropy to treat it only as a mathematical parameter. It is necessary to clarify the significance of entropy and of its continuous creation in irreversible processes. For this reason, the physical interpretations of entropy will be discussed.

When a differential amount of heat δQ is transferred to a system operating in a cycle, only a portion of the heat is available for work. This portion is called the available energy. Consider a reversible power cycle operating between two thermal reservoirs, and let T_0 and T be the absolute temperatures of the low- and high-temperature reservoirs, respectively. The efficiency of the cycle is $(1 - T_0/T)$; the work done then is the efficiency multiplied by the heat input. Therefore, the available energy is given by

$$\text{available energy} \leq \delta Q \left(1 - \frac{T_0}{T} \right) \tag{5.56}$$

In a reversible process, the equality applies; in an irreversible process, the inequality holds. When heat δQ is transferred reversibly to an engine, some of this energy is unavailable for work. This amounts to $\delta Q(T_0/T)$. But $\delta Q/T$ represents the increase of entropy in a reversible process. Therefore,

$$\text{unavailable energy} = T_0 \, dS \tag{5.57}$$

When irreversibilities occur in a process, the entropy of the system increases, and less energy is then available for work. This means that changes in entropy may be used to describe quantitatively any changes in the amount of available energy. Entropy is therefore associated with the dispersion or degradation of energy into forms that are more difficult to utilize in the production of work.[8]

Consider the transfer of a quantity of heat energy δQ from a thermal reservoir at T_1 to another at a lower temperature T_2. The maximum available energy before the heat transfer process is $\delta Q(1 - T_0/T_1)$, where T_0 is the temperature of the environment to which heat energy may be rejected. After the heat transfer process to the lower-temperature reservoir, the maximum available energy is only $\delta Q(1 - T_0/T_2)$. The difference constitutes a dispersion of energy into less useful forms for producing work.

There is a tendency for all physical systems to proceed to states in which they become less ordered. This suggests that entropy can be regarded as a measure of the *disorder* of a system, or of the randomness of motion of its microscopic particles. Consider the heat transfer to a homogeneous system. As the temperature rises, the molecules of the system move in a larger variety of ways in each of the

---

[8] This statement, however, cannot be generalized since, in the case of isothermal expansion of an ideal gas, there is no dispersion of energy, yet there is an increase of entropy.

various vibrational, rotational, translational, electronic, and other modes. Conversely, heat transfer from a system decreases the disorder of the molecules, although an equivalent amount of disorder, or greater disorder, results in the surroundings. The molecular disorder of the universe is therefore continuously increasing owing to the irreversible processes of nature so that the second law of thermodynamics is sometimes called the *law of increase of disorder.*

From the standpoint of probability, a system is less likely to be in an orderly state than in a disordered state. Spontaneous changes in nature proceed from the less probable to the more probable states. At thermodynamic equilibrium, the degree of molecular disorder is maximum, and the state of the system is the state of highest probability.[9] Hence, entropy, being a function of molecular disorder, may be used to measure the statistical probability of a macrostate of a system. In Section 5.9, it was shown that the entropy of an isolated system may increase or remain the same, but can never decrease:

$$dS_{\text{isolated}} \geq 0 \qquad (5.58)$$

An isolated system tends to approach complete randomness. When equilibrium is reached, no changes in the properties of the system are observed, and the system is in a state corresponding to the macrostate of maximum probability. Through statistical thermodynamics, a relationship is developed between entropy and the number of microstates[10] in the dominating macrostate of a system. A preliminary account is given in this section, whereas a more detailed discussion will be presented in Chapter 12.

The meaning of thermodynamic probability can be clarified by means of an illustration. Consider the free expansion of a gas of volume V_a into an evacuated chamber. Let the total volume, after expansion, be V_b. If probability is represented by W, the probability that a certain molecule of the gas will be found in volume V_b is unity, or $W_b = 1$. The probability that it will be found in V_a is proportional to the ratio of the two volumes so that $W_a = V_a/V_b$. When several independent events are considered, the probability that they will all occur simultaneously is the product of their individual probabilities. Hence, for N molecules, the probability that they all are located in V_a at any particular time is $(V_a/V_b)^N$. Therefore,

$$\frac{W_a}{W_b} = \left(\frac{V_a}{V_b}\right)^N \qquad (5.59)$$

---

[9] The thermodynamic probability of a macrostate is equal to the number of microstates in a certain macrostate. It indicates the relative frequency with which a macrostate occurs. This is different from mathematical probability, which may be defined as the ratio of the number of favorable events to the total number of events. These two probabilities are proportional to each other.

[10] The ways that the particles of a system are distributed over the available energy states defines a microstate of the system.

Boltzmann proposed that a relationship exists between probability and entropy. From a probability point of view, the state of equilibrium is the most probable state; from the classical thermodynamic point of view, the state of equilibrium represents the state of maximum entropy. During an irreversible process, both the thermodynamic probability and the entropy of a system increase. Entropy is, however, an additive (that is, extensive) property, whereas thermodynamic probability is a multiplicative property. Boltzmann, therefore, suggested that the relationship takes the form

$$S = k \ln W \qquad (5.60)$$

where $k = \overline{R}/N_a$ is Boltzmann's constant and represents the gas constant based on one molecule, and N_a is Avogadro's number. Consider a system made up of several parts, each of fixed energy, volume, and composition. The entropy of such a system is equal to the sum of the entropies of its components

$$S = S_1 + S_2 + \ldots$$

The thermodynamic probability of a state of a system is equal to the product of the individual probabilities of the components of the system

$$W = W_1 W_2 \ldots$$

The entropy of the system, according to Eq. (5.60), therefore, is

$$S = S_1 + S_2 + \ldots = k \ln W_1 + k \ln W_2 + \ldots = k \ln W$$

Let us now return to the example of an ideal gas undergoing free expansion. The differential change of entropy is

$$dS = mc_v \frac{dT}{T} + mR \frac{dV}{V}$$

Since the temperature does not change, this equation, when integrated, becomes

$$S_b - S_a = mR \ln \frac{V_b}{V_a} = N \frac{\overline{R}}{N_a} \ln \frac{V_b}{V_a} = k \ln \frac{W_b}{W_a} \qquad (5.61)$$

According to this equation, the change of entropy is proportional to the logarithm of the ratio of the statistical probabilities of the end states of the system. Note that the final state of the process is of greater probability than is its initial state.

Mention should be made of the philosophical implications of entropy. According to the first and second laws of thermodynamics, the energy of the universe is constant, but the entropy of the universe is increasing continuously. Can the entropy of the universe increase indefinitely? Or is there a maximum possible entropy for the universe? If such a limit exists, when will it be reached? Since all real processes are accompanied by an increase in entropy, the quality of energy available for mankind continuously declines. At some time in the future, no more energy will be available for use. At that time, the entropy of the universe will be at its maximum, and complete disorder and randomness will prevail. At that state of the universe, all matter will be at the same temperature level, and the universe will reach, according to Clausius, the state of "thermal death." No energy will be available for use in raising local temperatures above that of the environment, and performance of work will be impossible. Nevertheless, the total energy of the universe will be no different from its present value. The challenge then is to reduce the production of entropy and to realize that the need to conserve the quality of energy creates order in the universe.

Another paradox that arises is centered about the theory of probability. According to this theory, all possible microstates of a macrostate are equally probable. That is to say, a system can exist in a particular microstate at any time so that no particular time for its existence is more probable than any other time. All possible events and states of the universe occur randomly. On the other hand, the second law of thermodynamics states that only those processes that result in an increase in the entropy of the universe can occur. Therefore, the second law may be interpreted as one of very high statistical probability but not one of absolute certainty. Perhaps the idea to keep in mind in this discussion is that "Nature cannot be ordered about, except by obeying her" (Francis Bacon, English philosopher, 1561–1626). Nature acts in a random way, but it also tends to act rationally.

5.19 Summary

Processes in nature proceed spontaneously in one direction but never in the opposite direction. Heat transfer takes place along a negative temperature gradient, and cyclic devices such as engines are limited in converting heat transfer into work transfer (no engine is 100% efficient). The second law, by means of entropy, assesses the possibility of an adiabatic process to take place.

The Carnot cycle is the most efficient cycle. It consists of an alternate series of two reversible isothermal processes and two reversible adiabatic processes. The efficiency of the Carnot power cycle is a function of the temperatures of the thermal reservoirs across which the cycle operates:

$$\eta_{\text{Carnot}} = 1 - \frac{T_L}{T_H}$$

For a Carnot refrigeration or a heat pump,

$$\beta_{\text{ref}} = \frac{T_L}{T_H - T_L} \qquad \beta_{\text{ht pump}} = \frac{T_H}{T_H - T_L}$$

The thermodynamic temperature scale can be defined in terms of quantities of heat energy, independent of the properties of the thermometric substance. This scale is equivalent to the ideal-gas temperature scale.

The Clausius inequality states that the cycle integral of $\delta Q/T_{\text{reservoir}}$ is less or equal to zero. For internally reversible processes between two states, the $\int_{\text{rev}} \delta Q/T$ is the same regardless of the path. This defines the entropy change of the system undergoing an internally reversible process by

$$dS = \left(\frac{\delta Q}{T}\right)_{\text{int. rev.}}$$

A general statement of the second law for a system is $dS \geq \delta Q/T$.

The entropy of a system can increase or decrease depending on the direction of heat transfer; however, it can only increase owing to internal irreversibilities. When an isolated system changes from one equilibrium state to another, only those states for which the entropy increases or remains constant are possible.

Combining the first and second laws of thermodynamics leads to the following $T\,ds$ equations:

$$T\,ds = du + p\,dv \qquad \text{and} \qquad T\,ds = dh - v\,dp$$

These equations are derived for a simple substance and may be used to evaluate the entropy change. For an ideal gas of a unit mass,

$$\Delta s = \int_{T_1}^{T_2} c_v \frac{dT}{T} + R \ln \frac{v_2}{v_1} \qquad \text{and} \qquad \Delta s = \int_{T_1}^{T_2} c_p \frac{dT}{T} - R \ln \frac{p_2}{p_1}$$

For an incompressible substance,

$$\Delta s = \int_{T_1}^{T_2} c \frac{dT}{T}$$

Heat interaction with a thermal reservoir causes infinitesimal changes in its properties and can therefore be considered reversible. It follows that the entropy change of a thermal reservoir is

$$\Delta S = \frac{Q}{T}$$

Specific entropy values for several one-component substances such as water, ammonia, refrigerant-12, and refrigerant-134a are indicated in property tables of the Appendix.

Processes can be represented on *T-s* and *h-s* diagrams. Isentropic processes are vertical lines on these diagrams, and the area underneath an internally reversible process on a *T-s* diagram is equal to the heat interaction. Isentropic relations for ideal gases are obtained by setting $ds = 0$ in the $T \, ds$ equations. For an isentropic path between states 1 and 2,

$$\frac{T_2}{T_1} = \left(\frac{p_2}{p_1}\right)^{(\gamma-1)/\gamma} = \left(\frac{v_1}{v_2}\right)^{\gamma-1}$$

where $\gamma = c_p/c_v$ is assumed constant.

Gas tables relate properties of gases taking into consideration the variation of specific heats with temperature.

For isentropic processes of ideal gases with variable specific heats,

$$\frac{p_2}{p_1} = \frac{p_{r_2}}{p_{r_1}} \quad \text{and} \quad \frac{v_2}{v_1} = \frac{v_{r_2}}{v_{r_1}}$$

The relative pressure p_r and the relative volume v_r are tabulated as a function of temperature.

The second law for a control volume is

$$dS \geq \frac{\delta Q}{T} + \sum s_i \, dm_i - \sum s_e \, dm_e$$

Actual processes are compared to isentropic processes by means of the isentropic efficiency.

$$\text{For a turbine:} \quad \eta_T = \frac{w_{\text{act}}}{w_{\text{isent}}}$$

$$\text{For a compressor:} \quad \eta_C = \frac{w_{\text{isent}}}{w_{\text{act}}}$$

$$\text{For a nozzle:} \quad \eta_{\text{nozzle}} = \frac{(V^2/2)_{\text{act}}}{(V^2/2)_{\text{isent}}}$$

Processes in nature have to satisfy the energy and entropy relations introduced by the first and second laws, respectively.

Entropy is a measure of irreversibility. It is a property that relates to the degree of disorder of a system. In contrast to energy, it is not conserved except in reversible processes.

Table 5.1 is a summary of the first and second laws of thermodynamics for a system and a control volume. Note that the first law treats all forms of energy as equivalent without differentiating between their quality. The second law assesses the quality of energy and provides information about internal losses.

Table 5.1   Equations of the First and Second Laws for a System and a Control Volume

	Energy Transfer	*Energy Change*
First Law		
(a) System:	$Q + W$	$= \Delta E_{sys}$
(b) Control Volume:	$Q + W + \sum m_i\left(h + \dfrac{V^2}{2} + gz\right)_i$	
	$- \sum m_e\left(h + \dfrac{V^2}{2} + gz\right)_e$	$= \Delta E_{cv}$

	Entropy Transfer	*Entropy Change*
Second Law		
(a) System:	$\displaystyle\int \dfrac{\delta Q}{T_{surr}}$	$\leq \Delta S_{sys}$
	or	
	$\displaystyle\int \dfrac{\delta Q}{T_{surr}} + S_{gen}$	$= \Delta S_{sys}$
(b) Control Volume:	$\displaystyle\int \dfrac{\delta Q}{T_{surr}} + \sum m_i s_i - \sum m_e s_i$	$\leq \Delta S_{cv}$
	or	
	$\displaystyle\int \dfrac{\delta Q}{T_{surr}} + \sum m_i s_i - \sum m_e s_e + S_{gen}$	$= \Delta S_{cv}$

Problems

5.1   A refrigerator driven by a 0.75 kW motor transfers 200 kJ/min from a cold body. What is the coefficient of performance of this refrigerator? At what rate is heat rejected to the hot body?

5.2   By the use of the second law of thermodynamics show that (a) mechanical friction is a source of irreversibility, (b) throttling is an irreversible process, (c) the process of dropping a body into a well containing water is irreversible.

5.3 A Carnot engine operating between temperatures of 200°C and 20°C receives 172 kJ of heat.

(a) Compute the change in entropy during the process in which the engine rejects heat to its receiver.

(b) Draw the Carnot cycle on a temperature–entropy diagram and determine the magnitude of the area representing the heat transfer process.

5.4 A Carnot engine operates between two thermal reservoirs at 700 K and 300 K. If the engine output is 20 kW, determine the rate of heat interactions with the thermal reservoirs and the engine efficiency. What would be the coefficient of performance if the engine is reversed to operate as a refrigerator?

5.5 A refrigeration system exhibits a coefficient of performance one-half that of a Carnot cycle operating between the same temperature limits. It receives 600 kJ/min from a thermal reservoir at −30°C while the upper temperature is maintained at 120°C.

(a) What is the rate of energy transfer to the high-temperature reservoir?

(b) If the refrigerator were reversed to operate as a power cycle (all data remaining the same), what power would be developed?

5.6 A reversible power cycle receives heat at the rate of 120 kJ/min and produces 1.0 kW of power. If the heat rejection is at 20°C, at what temperature does the heat input occur?

5.7 The thermal efficiency of a reversible engine is 50%. The engine operates between a high-temperature reservoir and an ambient atmosphere at 10°C.

(a) What is the temperature of the high-temperature source?

(b) If the engine consisted of a closed system and a 25°C temperature difference is necessary to cause heat flow between the thermal reservoirs and the engine, what would be the temperature of the source and receiver?

(c) What would be the maximum thermal efficiency of the irreversible engine mentioned in (b)?

5.8 Which is more effective in increasing the efficiency of a reversible engine operating between two

thermal reservoirs: (a) increasing the temperature of the high-temperature reservoir by an amount ΔT, keeping the temperature of the low-temperature reservoir constant, or (b) decreasing the temperature of the low-temperature reservoir by the same amount ΔT, keeping the temperature of the high-temperature reservoir constant?

5.9 Repeat Problem 5.8 for a reversible heat pump, and compare the coefficients of performance.

5.10 A reversible heat engine operates between three constant temperature reservoirs, as shown in Figure 5.43. If the thermal efficiency is 40 percent and $W = 2000$ kJ, calculate the temperature T_1 and the heat quantity Q_2.

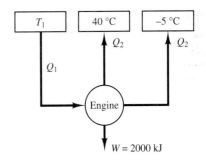

Figure 5.43

5.11 An inventor claims that his new heat engine develops 2 kW when heat at the rate of 300 kJ/min is transferred to the engine. The cycle operates between a maximum temperature of 1100°C and a minimum temperature of 130°C. Evaluate this claim.

5.12 A Carnot engine operates between two thermal reservoirs 600°C and 25°C. The heat input to the engine is 90 kJ. The work output of the engine is used to drive a Carnot refrigerator operating between −20°C and 25°C. Determine the refrigeration load.

5.13 Two Carnot heat engines operate in series between a source at $T_1 = 780$ K and a sink at $T_3 = 290$ K. The first engine rejects 400 kJ to the second engine through an intermediate thermal reservoir at T_2. If both engines have the same efficiency, calculate:

(a) the temperature T_2 of the intermediate reservoir
(b) the heat input to the first engine from the 780 K source
(c) the work done by each engine
(d) the efficiency of each engine
(e) represent the cycles on the same T-s diagram

5.14 A system undergoes a reversible cycle while exchanging heat with three thermal reservoirs, as shown in Figure 5.44. Compute the direction and magnitude of Q_3 and W.

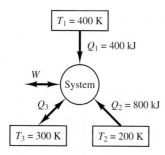

Figure 5.44

5.15 Compare the thermal efficiencies of the cycles a-b-d-e-a and a-b-c-e-a of Figure 5.45 by means of areas. Account for the difference in efficiencies. (All processes are reversible except for b-c.)

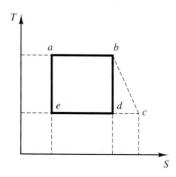

Figure 5.45

5.16 A quantity of water may be heated to increase its temperature from 0°C to 100°C in either of two ways: It may be placed in contact with a large thermal reservoir

at 100°C, or it may be placed first in contact with a thermal reservoir at 50°C until its temperature rises to this value and then in contact with a second reservoir at 100°C. What is the total change in entropy of the water, the thermal reservoirs, and the local universe by each method? The specific heat of water is 4.18 kJ/kg K.

5.17 Two reversible engines A and B reject heat to a common reservoir at a temperature T. Engine A receives heat from a reservoir at T_1, whereas B receives heat from a reservoir at $T_2 (T_2 < T_1)$. If both engines receive the same amount of heat, for engine A compared with engine B:

	Greater	Same	Less
The work is	_____	_____	_____
The work rejected is	_____	_____	_____
$\oint dS$ is	_____	_____	_____
$\oint \delta Q/T$ is	_____	_____	_____
$\oint T\, dS$ is	_____	_____	_____

5.18 In Figure 5.46, the circle represents a reversible power cycle. During one cycle of operation, 1200 kJ of heat is withdrawn from the 200 K reservoir and 200 kJ of work is developed. Determine:
(a) the amount and directions of heat interactions with the other two reservoirs
(b) the entropy change due to each of the heat interactions with the engine
(c) the entropy production during the cycle

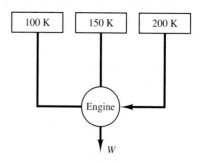

Figure 5.46

5.19 Figure 5.47 shows a Carnot engine operating between two thermal reservoirs. The work output of the engine is used to drive a "reversible device" that operates in a cycle. From the data given in the figure, determine the heat input to the Carnot engine and the amount and direction of Q_5.

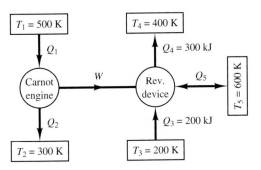

Figure 5.47

5.20 A Carnot engine cycle receives 1000 kJ of energy at 280°C and rejects energy at a lower cycle temperature of 40°C.
(a) What is the entropy change of the engine due to the heat rejection process?
(b) What is the total energy rejected and the work output of the cycle for each 1000 kJ input?

5.21 Two large bodies at temperatures 10°C and 200°C are exchanging heat at the rate of 4 kJ/s. Assuming that the two bodies are large enough so that the temperature change in either one is negligible, compute the entropy change per minute of (a) the warmer body, (b) the cooler body, (c) the entire system.

5.22 Ten kilograms of water at a temperature of 30°C are heated at a pressure of 450 kPa to a temperature of 147.93°C and are then evaporated at this temperature and pressure by heat interaction equal to 2120.7 kJ/kg. The vapor is then further heated at 450 kPa to a final temperature of 250°C. Compute the change in entropy of the water.

$$c = 4.184 \text{ kJ/kg K (liquid)}$$
$$c_p = 1.87 \text{ kJ/kg K (vapor)}$$

Compare your results with those found by using steam tables.

5.23 It is required to determine the total change of entropy (system and surroundings) as the temperature of 1 kg of water is raised from 300 K to 500 K according to the following process:
(a) The water is brought in contact with a thermal reservoir at 500 K.
(b) The water is first brought in contact with a thermal reservoir at 400 K and then with a thermal reservoir at 500 K.
Comment on the results. Assume the specific heat of water is 4.18 kJ/kg K.

5.24 Two equal masses of the same liquid initially at temperatures T_1 and T_2 are mixed adiabatically. Prove that the change of entropy is positive and is given by

$$\Delta S = 2mc \ln \frac{(T_1 + T_2)}{2\sqrt{T_1 T_2}}$$

where c is the specific heat of the liquid.

5.25 For a system consisting of a one-component substance in the absence of gravity, electricity, capillarity, magnetism, and motion, indicate by placing a check mark in the proper column under what condition the following equations apply:

	Reversible Process or Cycle	Irreversible Process or Cycle	Any Process or Cycle
$\delta Q + \delta W = dE$	_____	_____	_____
$\oint \delta Q/T = 0$	_____	_____	_____
$dh = T\,ds + v\,dp$	_____	_____	_____
$(s_2 - s_1)_p = c_p \ln T_2/T_1$ (ideal gas)	_____	_____	_____
$\oint T\,dS = \oint p\,dV$	_____	_____	_____
$\oint T\,dS = -\oint \delta W$	_____	_____	_____
$\oint dS = 0$	_____	_____	_____
First law of thermodynamics	_____	_____	_____
Second law of thermodynamics	_____	_____	_____

5.26   One kilogram of ice at 0°C is contained initially in a beaker holding 2 kg of water at 40°C. The beaker in turn is in a large air reservoir that remains substantially at 10°C. There is free exchange of heat between all the systems. The latent enthalpy of fusion of ice is 333.5 kJ/kg. Determine the net change of entropy for all the systems involved in the foregoing process in going from the initial to a final state.

5.27   One kilogram of ice at 0°C is melted by placing it into an insulated container with 2 kg of water at 50°C.
(a) Determine the change in entropy of the mass that was originally ice.
(b) Determine the change in entropy of the mass that was originally water.
(c) Is the process reversible? Why? (Note: Enthalpy of fusion of ice is 333.5 kJ/kg.)

5.28   In a certain reversible process, the specific heat of a system is constant and is given by

$$c = \frac{\delta q}{dT} = 2.0 \text{ kJ/kg K}$$

(a) Find the increase in entropy of the system if its temperature rises from 250 K to 300 K.
(b) In a second process between the same end states, the temperature rise is accomplished by stirring accompanied by a heat transfer half as great as in (a). What is the increase in entropy in this case?

5.29   A car of mass 1500 kg is brought to rest from a speed of 50 km/hr by application of its brakes. The initial temperature of the brakes is 20°C. Calculate the entropy change if the temperature of the brakes rises at the rate of 2°C/kJ of energy absorbed. Assume the braking process to be adiabatic.

5.30   A Carnot engine utilizes steam as the working fluid and has an efficiency of 20 percent. Heat is transferred to the working fluid at 200°C, and during this process, the working fluid changes from saturated liquid to saturated vapor.

(a) Show this cycle on a *T-s* diagram that includes the saturated-liquid and saturated-vapor lines.
(b) Calculate the quality at the beginning and end of the heat-rejection process.
(c) Calculate the work per kg of steam.

5.31   A noncondensable gas is used in the cycle shown in Figure 5.48. Is this possible? Sketch the *T-S* diagram.

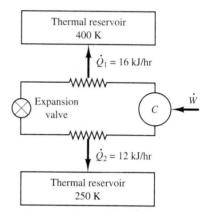

Figure 5.48

5.32   A Carnot cycle uses 1 kg of steam as a working fluid. During the isothermal expansion, the working fluid changes from 1.2 MPa and 20 percent quality to dry saturated steam. If the lowest pressure in the cycle is 50 kPa, calculate (a) the thermal efficiency of the cycle. (b) the work done per cycle.

5.33   A rigid container has two chambers separated by a partition. Each chamber contains 1 kg of air. The temperature and pressure in one of the chambers are 650 K and 30 kPa, and in the second chamber are 300 K and 10 kPa. The partition is removed, and during the mixing process, the system rejects 200 kJ of heat to the surroundings, which is at 290 K. Determine (a) the final volume, temperature, and pressure; (b) the change of entropy of the universe.

5.34   One kilogram of air at 1.4 MPa occupies a volume of 0.15 m³ and undergoes a constant pressure process until the volume is doubled. The volume is then held constant, and the pressure is reduced to 0.7 MPa. This is followed by isothermal compression to the original volume of the air. Compute the change of entropy for each of the three processes.

5.35 An insulated rigid vessel is divided into two chambers of equal volumes. One chamber contains air at 500 K and 2 MPa. The other chamber is evacuated. If the two chambers are connected, what would be the entropy change?

5.36 Each of two rigid vessels contains 1 kg of air. Initially, the state of the air in vessel A is identical with the state of the air in vessel B. A heat pump transfers heat from vessel A and rejects heat to vessel B. During a certain integral number of cycles of the heat pump, the work input to the pump is 1.0 kJ. At the completion of the cycles, it is found that the air in vessel A has decreased in temperature by 27°C.

Assume air to be an ideal gas ($c_p = 1.0035$ kJ/kg K, $c_v = 0.7165$ kJ/kg K), and neglect heat interaction between the vessels and the surroundings.
(a) What is the increase in temperature of the air in vessel B?
(b) What is the entropy change for the entire system comprising the contents of vessels A and B and the heat pump during the process described?
(c) What is the initial temperature of the air in each vessel?

5.37 An ideal gas undergoes isothermal expansion from state 1 to state 2, as shown in Figure 5.49. It then undergoes an increase in pressure at constant volume from state 2 to state 3. Find the change of entropy between states 1 and 3. Data:

$$\frac{p_2}{p_1} = \frac{1}{2}, \quad \frac{p_3}{p_2} = 2$$

$$T_3 = 1000 \text{ K}$$

$$T_2 = 500 \text{ K}$$

$$R = 0.287 \text{ kJ/kg K}$$

$$\gamma = \frac{c_p}{c_v} = 1.4$$

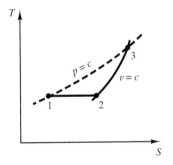

Figure 5.49

5.38 A vessel contains 2.2 kg of wet steam at a pressure of 250 kPa. Heat is transferred at constant volume until the steam becomes saturated at a pressure of 450 kPa. The steam is then compressed isentropically to a pressure of 750 kPa. Sketch the process on a p-v and a T-s diagram, and determine the heat and work interactions.

5.39 Air initially at a temperature of 520 K and a pressure of 400 kPa undergoes a process in which the entropy decreases by 0.09 kJ/kg K. If the final pressure is 100 kPa, determine the final temperature assuming air to be an ideal gas with constant specific heats.

5.40 A rigid insulated container holds 5 kg of an ideal gas. The gas is stirred so that its state changes from 5 kPa and 300 K to 15 kPa. Assuming $c_p = 1.0$ kJ/kg K and $\gamma = 1.4$, determine the change of entropy of the system.

5.41 A mass of 1 kg of an ideal gas at 21 kPa and 550 K is mixed with 1 kg of the same gas at 7 kPa and 320 K. The mixing takes place at constant volume, and during the process, the system rejects 300 kJ of heat to the surroundings, which is at 300 K. If the molar mass of the gas is 32 kg/kg-mol and $\gamma = 1.33$, determine (a) final volume, temperature, and pressure; (b) change in the entropy of the universe.

5.42 For the Clausius equation $p(v - b) = RT$, prove that the change of entropy is given by

$$\Delta s = c_v \ln \frac{T_2}{T_1} + R \ln \frac{v_2 - b}{v_1 - b}$$

5.43 An ideal gas initially at a pressure p_1 and a volume V_1 expands adiabatically and reversibly according to the law $pV^{1.5} = C$ until it doubles its initial volume. The gas is then compressed to its initial state by a sequence of constant-pressure and constant-volume processes. Compare the efficiency of the cycle to the maximum efficiency obtainable. Explain.

5.44 One kilogram of air at $T_1 = 400$ K, $v_1 = 0.2$ m³/kg expands in a process to $T_2 = 320$ K, and $v_2 = 0.4$ m³/kg. Find the change of entropy and the work done per kg of air for the following cases:
(a) if the expansion were adiabatic irreversible
(b) if the expansion were reversible polytropic
Assume air to be an ideal gas with constant specific heats.

5.45 A Carnot engine uses air as a working fluid and operates between two thermal reservoirs at 580 K and 400 K. If the pressure at the beginning of the isothermal compression process is 100 kPa and the pressure at the beginning of the isothermal expansion process is 1000 kPa, calculate (a) the net work produced, (b) the thermal efficiency of the cycle.

5.46 The highest and lowest pressures in a Carnot power cycle utilizing air as a working fluid are 600 kPa and 100 kPa. If the lowest temperature in the cycle is 300 K and the pressures at the end of the heat transfer processes are identical, calculate (a) the work developed during the cycle per kg of air, (b) the efficiency of the cycle.

5.47 A Carnot engine utilizes 1 kg of air as a working substance and operates between temperature limits of 1500 K and 500 K. If during the heat rejection process, the pressure is tripled, determine:
(a) the heat input per cycle
(b) the work interaction
(c) the ratio of the volumes at the beginning of the heat interaction processes

5.48 A Carnot power cycle utilizes air as a working fluid. During the heat transfer to the engine, the pressure decreases from 500 kPa to 200 kPa. If the minimum pressure in the cycle is 100 kPa, calculate (a) the thermal efficiency of the cycle, (b) the increase in entropy during the heat transfer to the engine.

5.49 A Carnot refrigerator uses 1.0 kg of air as a working substance and operates between temperature limits of 400 K and 500 K. If during the process of heat rejection from the refrigerator, the pressure is tripled, determine:
(a) the coefficient of performance of the refrigerator
(b) the heat input to the cycle (refrigeration effect)
(c) the ratio of the volume before to the volume after the heat input process

5.50 A Carnot engine operates between two thermal reservoirs at 700 K and 300 K and utilizes air as a working fluid. The pressure at the beginning of the heat transfer process to the engine is 350 kPa, and the volume doubles during this process. Determine (a) the mass rate of flow for a power output of 1 kW, (b) the lowest pressure in the cycle.

5.51 A Carnot engine utilizes air as a working fluid. During the heat transfer to the engine, the volume increases from 0.01 m³ to 0.015 m³. If the maximum volume in the cycle is 0.16 m³, calculate the thermal efficiency of the cycle.

5.52 For a Carnot cycle utilizing an ideal gas as working fluid, compare the magnitude of the work interactions in the adiabatic processes. Prove that $V_1 V_3 = V_2 V_4$, where the subscripts refer to consecutive points at the beginning of each process in the cycle.

5.53 A Carnot engine operates between two thermal reservoirs at 830 K and 280 K. During the heat rejection process, the volume is reduced to one-third its value. If the working substance is 1 kg of air, calculate (a) the heat input to the engine, (b) the net work developed by the engine.

5.54 A Carnot engine operates between temperature limits of 1200 K and 400 K, using 0.4 kg of air and running at 500 rev/min. The pressure at the beginning of heat input is 1500 kPa, and at the end of heat rejection is

750 kPa. Determine (a) the heat input per cycle, (b) the heat rejected, (c) the power, (d) the thermal efficiency.

5.55 A Carnot heat pump supplies 30,000 kJ/hr to a room to maintain its temperature at 22°C. If the surrounding atmosphere is at 10°C, determine the power input to the heat pump.

5.56 A Carnot engine receives energy in the form of heat at 500 K and rejects 100 kJ of heat at 400 K. The Carnot engine provides the work input to a heat pump. The heat pump receives 100 kJ of heat at a temperature of 250 K and rejects heat at 300 K. Determine if all the stated performance is possible.

5.57 Three Carnot engines operate between the three thermal reservoirs, as shown in Figure 5.50. Engines 1 and 2 have the same efficiency. Compare the work output of engine 3 with the combined work developed by engines 1 and 2.

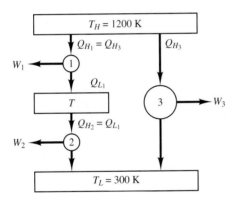

Figure 5.50

5.58 A Carnot power cycle using air has a thermal efficiency of 55 percent. The maximum pressure in the cycle is 650 kPa, and the minimum volume is 0.2 m³/kg. If the heat transfer to the cycle is 65 kJ/kg, determine:
(a) the maximum and minimum temperatures of the cycle
(b) the minimum pressure in the cycle
(c) the maximum volume in the cycle in m³/kg
(d) the work developed in kJ/kg

5.59 A Carnot engine uses H_2O as a working substance. During the heat transfer to the cycle, the state of the fluid changes from 250°C and 20 percent quality to 2.5 MPa. If the heat rejection from the cycle takes place at 80°C, represent the cycle on p-v and T-s diagrams, indicating the corner points of the cycle relative to the saturated-liquid and saturated-vapor lines, and determine (a) the thermal efficiency, (b) the heat input per unit mass, (c) the work developed per unit mass.

5.60 Figure 5.51 shows a Carnot cycle utilizing a water–vapor mixture as a working fluid. During the isothermal expansion, the fluid changes from saturated liquid to saturated vapor. If the temperature and volume at the end of the adiabatic expansion is 100°C and 1.2 m³/kg, determine (a) the thermal efficiency, (b) the work interaction.

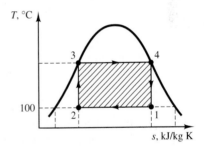

Figure 5.51

5.61 A Carnot engine using air operates between a high-temperature thermal reservoir at 800 K and a low-temperature reservoir at T_L. If the pressure ratio for the adiabatic compression is 12 to 1 and the volume during the heat transfer process is tripled, determine:
(a) the thermal efficiency of the cycle
(b) the work produced
(c) the ratio of the highest to the lowest volume in the cycle
(d) the ratio of the highest to the lowest pressure in the cycle

5.62 A Carnot engine utilizing air as the working fluid produces a net work of 84 kJ/kg of air. During the heat input process, the pressure decreases from 6 to 2.26 MPa. If the pressure at the start of the heat rejection process is

200 kPa, determine the maximum and minimum temperatures of the cycle.

5.63   One kilogram of air expands in a reversible polytropic process according to the law $pV^{1.25} = C$. The initial pressure and volume are 600 kPa and 0.15 m^3. The final volume is 1.0 m^3. Find (a) the final temperature, (b) the work done, (c) the heat interaction, (d) the entropy change of the system.

5.64   One kilogram of air at a pressure of 500 kPa and a temperature of 600 K expands isentropically to 100 kPa. Heat in the amount of 65 kJ is then transferred to the air at constant pressure. Determine the work and heat interactions in each process.

If the two processes were replaced by a single process between the initial and final states according to the law $pV^n = C$, what will then be the work and heat interactions?

5.65   Hydrogen at $T_1 = 300$ K, $V_1 = 100$ cm^3 expands adiabatically in a piston–cylinder assembly to $T_2 = 270$ K and $V_2 = 200$ cm^3. Find the change of entropy and the work done per kg of hydrogen.

If the expansion were adiabatic reversible (isentropic) between the same volumes, what would be the final temperature and work done?

Assume hydrogen to be an ideal gas with constant specific heats ($c_v = 10.21$ kJ/kg K, $R = 4.124$ kJ/kg K).

5.66   A 1.5 kg block of copper initially at 250°C is immersed into an insulated tank containing 2 kg of water at 25°C. Determine the entropy change of the system ($c_{cu} = 0.39$ kJ/kg K, $c_w = 4.18$ kJ/kg K). If the tank were not insulated, what would be the total entropy change? Assume the temperature of the surroundings to be 25°C.

5.67   An initially evacuated tank of volume 0.2 m^3 is filled with air from a main supply line. The temperature and the pressure in the line are 50°C and 500 kPa. If the properties in the supply line are maintained constant and the filling process is adiabatic, determine the final temperature in the tank when the pressure reaches 300 kPa. What is the entropy change?

5.68   Air, having a 0.1 kg mass, is contained in a rigid insulated tank at 500 K and 1 MPa. A valve is opened allowing the air to escape until the pressure decreases to 0.3 MPa. Assuming the air that remains in the tank has undergone a reversible adiabatic expansion, determine the final mass of air in the tank.

5.69   One kilogram of air at 800 kPa and 500 K expands in a reversible process according to the relation $pV^{1.2} = C$ until the final temperature is 320 K. If the surrounding temperature is 450 K, calculate the net entropy change of the system plus surroundings for this process. Assume air to be an ideal gas with constant specific heats.

5.70   One kilogram of air expands from a pressure 600 kPa and a temperature of 700 K to a pressure of 50 kPa. If the expansion were isentropic, determine the final temperature of the air (a) assuming constant specific heats, (b) using air tables.

If the expansion to the same final pressure were irreversible and the final temperature were 400 K, what would be the change of entropy?

5.71   Air expands irreversibly but with positive heat transfer from 300 kPa and 180°C to 150 kPa and 100°C. What is the change in entropy of the system?

If the process were carried out reversibly between the same end states, would the entropy be more or less? Would the heat be more or less?

5.72   In a reversible polytropic process ($pv^{1.3} = C$), 1 kg of air expands from an initial state of 300 kPa and 400 K to a final state of 100 kPa. Calculate the change of entropy of the air.

In another process, the air at the same initial state is throttled to 100 kPa. What then is the change of entropy? Assume air to be an ideal gas.

5.73   Air initially at a pressure of 100 kPa and a temperature of 20°C is compressed reversibly and adiabatically to 1.0 MPa in a steady flow process. If the power input is 1.5 kW and the changes in kinetic and potential energies are negligible, determine (a) the mass rate of flow of air, (b) the enthalpy change during the compression process.

5.74 The components of a gas turbine plant using air as a working fluid and operating at steady state are shown in Figure 5.52. The air enters the compressor at 300 K and 100 kPa and is compressed to 800 kPa. The isentropic compressor efficiency is 80 percent. The air is heated to 1200 K, and expands in the turbine to 100 kPa. If there is no net power output, calculate the isentropic efficiency of the turbine.

Assume air to be an ideal gas with constant specific heats.

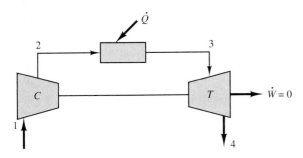

Figure 5.52

5.75 Deduce Eq. (5.41) using the procedure outlined in Section 2.21.

5.76 Air expands adiabatically through a turbine operating at steady state from 800 K and 1.4 MPa to 480 K. If the process had been reversible, the final temperature would have been 450 K at the same exit pressure. Calculate the change in entropy and the work produced per unit mass of air.

5.77 It is proposed to compress a liquid–vapor mixture, initially at 125 kPa and 90 percent quality, to a final pressure of 2 MPa using a reversible adiabatic process.
(a) Show this process schematically on a *T*-*s* diagram.
(b) If this process is carried out in a piston–cylinder device having an initial total volume of 0.01 m³, determine the work required.
(c) What will be the work interaction if the process is carried out under steady flow conditions?

5.78 Air enters a nozzle at 400 kPa and 350 K with a velocity of 30 m/s and exits at 120 kPa and 300 m/s. If the heat loss from the nozzle to the surroundings is 4.0 kJ/kg, determine (a) the exit temperature, (b) the total entropy change. Assume steady-state operation and T_{surr} = 25°C.

5.79 Superheated steam at a pressure of 100 kPa and a temperature of 150°C is compressed reversibly and isothermally until its specific volume is 0.2 m³/kg. Determine the heat and work interactions per kg of steam. What would be your answers if the fluid were nitrogen?

5.80 Steam expands in an adiabatic turbine operating at steady state from 500 kPa and 250°C to 10 kPa. If the mass rate of flow is 2 kg/s and the power output is 1 MW, determine: (a) the isentropic efficiency of the turbine, (b) the rate of change of entropy per second.

5.81 Steam enters a turbine operating at steady state at 0.8 MPa, 400°C, and exhausts at 0.2 MPa, 200°C. Any heat transfer is with the surroundings at 25°C, and changes in kinetic and potential energies are negligible. It is claimed that this turbine produces 100 kW with a mass flow of 0.3 kg/s. Does this process violate the second law of thermodynamics?

5.82 Air enters an adiabatic turbine operating at steady state at a pressure of 600 kPa and a temperature of 900 K. The pressure and temperature at the exit are 110 kPa and 650 K. Is this process possible? If so, determine the work developed per unit mass of air flowing through the turbine and the isentropic efficiency.

5.83 Air expands in a turbine operating at steady state if the inlet pressure and temperature are 650 kPa and 1000 K and the exit pressure is 100 kPa. Determine the work developed per unit mass of air flowing through the turbine if the isentropic efficiency is 80 percent. Use gas tables.

5.84 Air is compressed isothermally in a steady-state flow process from a pressure of 100 kPa and a temperature of 20°C to a pressure of 620 kPa. If the mass flow rate is 0.3 kg/s, what is the minimum power required? What is the entropy change?

If the process took place isentropically, what would then be the power required?

5.85 Air is compressed in a steady flow process from 100 kPa and 300 K to 400 kPa and 350 K. Is this process adiabatic? Explain.

5.86 Air is compressed in a reversible steady-state flow process from 101.3 kPa and 30°C to 0.6 MPa. Calculate the change of entropy and heat transfer per kg of air if compression takes place according to the following processes: (a) isothermal (b) polytropic ($n = 1.25$) (c) adiabatic.

Show all these processes on p-v and T-s diagrams.

5.87 Air in a cylinder ($V_1 = 0.03$ m³, $p_1 = 0.1$ MPa, $T_1 = 10°C$) is compressed reversibly in a constant-temperature process to a pressure of 0.42 MPa. Determine the heat transfer, the work done, and the entropy change if the process is (a) steady flow (b) nonflow.

5.88 Air is compressed in a steady-state flow polytropic process from 0.1 MPa and 25°C to 0.8 MPa. Calculate the work done and the change of entropy if the compression takes place such that $pv^{1.25} = C$. Assume air to be an ideal gas with constant specific heats.

5.89 Air is compressed reversibly according to the relation $pv^{1.3} = C$ in a compressor operating at steady state. The air enters the compressor at 100 kPa and 300 K and leaves at 620 kPa. Determine the work and heat interaction per unit mass of flow. Assume air to be an ideal gas, and neglect changes in kinetic and potential energy between the inlet and the exit.

What is the entropy generation if the heat interaction takes place with a surroundings at a temperature of 300 K?

5.90 An adiabatic air compressor operates at steady state but with an isentropic efficiency of 78 percent. The temperature and pressure at the inlet are 300 K and 100 kPa, and the pressure at the discharge is 0.5 MPa. Making use of gas tables, find:
(a) the temperature of the leaving air
(b) the power required if the flow rate at the inlet is 40 m³/min
(c) the rate of the entropy generation
(d) the volumetric flow rate at the discharge

Assume kinetic and potential energy changes to be negligible.

5.91 A throttling valve operating at steady state reduces the pressure of steam as it flows through a pipeline from 1 MPa to 0.3 MPa. The superheat is 5°C upstream of the throttling valve, and the velocity is low. Some distance after the valve, the steam fills the pipe and proceeds at negligible velocity. Find the temperature and the degree of superheat of the steam downstream of the valve. (Assume negligible heat loss from the pipe.) Show the process on h-s and T-s diagrams.

5.92 Steam enters a turbine at 3 MPa and 400°C and expands adiabatically to 10 kPa. The isentropic efficiency of the turbine is 80 percent. If the turbine produces 500 kW of power, determine the mass rate of flow of the steam.

5.93 Helium expands in an insulated nozzle operating at steady state from 300 kPa and 450 K to 180 kPa. The velocity of the helium at the nozzle inlet is small compared with exit velocity.
(a) Determine the maximum exit velocity that can be attained by the helium.
(b) If the temperature of the helium is measured to be 373 K at the nozzle exit, determine the actual exit velocity.
(c) Determine the total increase in entropy per unit mass of helium for the actual process.

Assume helium to be an ideal gas with constant specific heats ($c_p = 5.1926$ kJ/kg K, $c_v = 3.1156$ kJ/kg K).

5.94 A refrigeration system operating at steady state maintains a freezer compartment at $-5°C$. If the refrigeration load is 1200 kJ/hr and the system rejects heat to the surroundings at 25°C, determine the minimum power input.

5.95 A steady flow of saturated liquid ammonia enters an expansion valve at 22°C. If the ammonia is throttled adiabatically to 160 kPa, determine the increase in entropy. Assume changes in kinetic and potential energies are negligible.

5.96 Steam at 600 kPa and a quality of 0.8 enters an insulated chamber where it is mixed with a water stream entering at 600 kPa and 40°C. The resulting mixture leaves the mixing chamber as saturated liquid at 600 kPa. For steady-state operation, determine the ratio of the mass flow rates of the steam and water entering the chamber. What is the entropy generation per unit mass of the entering steam?

5.97 Air expands isentropically in a steady flow operation in nozzle. The inlet conditions are 800 kPa, 900 K, and negligible velocity. If the exit pressure is 350 kPa, calculate the exit velocity.

What would be the change of entropy if the expansion were adiabatic irreversible resulting in an exit velocity equal to 90 percent of its maximum value?

C5.1 Water temperature in the Pacific Ocean varies substantially from the surface to the bottom of the ocean. It is proposed to operate a Carnot engine to produce power using the temperature differential of the ocean. Assume that the water temperature as a function of depth of the ocean is given by $T(z) = 3 + 20e^{-z}$ °C, where $z = 0$ m at the surface of the ocean. Write a computer program to determine the engine efficiencies at various depths. Plot the efficiencies and the water temperature versus depth of water. A quick library search will provide you with an approximate maximum depth of the ocean. How would you evaluate this proposal for its economic and technical merits?

C5.2 The power plant of Problem C5.1 is located on the surface of the ocean and is to produce 1.0 MW of power. Water is drawn in by a pump from some depth and its energy extracted. Assume that 30 percent of the power produced is used to operate the pump and that the extractable energy from the water at any depth is given by $Q = 100[T(z) - 23]$ kW. Determine the depth from which the water should be pumped into the plant in order to produce the required power. Discuss the feasibility of this project and any technical problems that you foresee.

C5.3 Write an interactive computer program to determine the work output or the exit temperature of the working fluid for a gas turbine. All the necessary data should be obtained interactively. The output should include the input data as well as all the information and data relating to the problem. Note that the turbine is operating adiabatically but not reversibly, thus the user should also input the desired efficiency for the turbine.

C5.4 Write an interactive computer program to determine the actual and isentropic work for a given pressure rise across a compressor. The user supplies all the required data interactively, and the program outputs all the properties at the inlet as well as the outlet. The program should be capable of including heat loss from the compressor if the user desires. The heat loss should be entered as a percentage of the output work.

C5.5 Consider 0.5 kg of air contained in an insulated rigid tank initially at 4 kPa and 300 K. The air is stirred, causing a change in its state due to the added energy. Write a computer program to determine the entropy change of the system for pressure rise of 2 kPa increments up to 20 kPa of final pressure. Program output should include the change in temperatures as well as entropies.

Chapter 6

Irreversibility, Availability, and Criteria of Equilibrium

6.1 Introduction

By withdrawing energy in the form of heat transfer from a high temperature source, work can be performed through the transformation of heat energy into mechanical energy. But as a consequence of the second law of thermodynamics, a continuous process for accomplishing this transformation will not produce a quantity of work that is exactly equivalent to the heat provided. Losses occur in such transformations so that the problem lies in determining how to attain maximum efficiencies in converting internal energy into work and how to evaluate the quantity of energy available for work. As shown in Chapter 5, the most efficient cycle to produce work is a reversible power cycle (Carnot cycle). But even with the use of the reversible Carnot cycle, the efficiency of conversion is always less than unity. To establish a scale for comparison of the work-energy that can be developed in actual processes, it is first important to determine the maximum theoretical work obtainable with respect to some datum. In most practical cases, the environmental condition (pressure 1 atm and temperature 25°C) is chosen as the reference datum.

The concept of availability provides the information needed to determine the maximum useful work that a system in a given state can perform when it comes

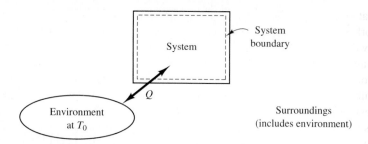

Figure 6.1   The surroundings includes everything outside the system boundary.

to equilibrium with an idealized system called the *environment*. The term *surroundings* refers to everything outside the system boundary and includes the environment, as shown in Figure 6.1. The intensive properties of the surroundings may vary, but those of the environment remain constant.

The internal energy of a system may be divided into two parts: available energy, which under ideal conditions may be completely converted to work, and unavailable energy, which is usually discarded as heat. In the case of the Carnot engine, the quantity $[(T - T_0)/T]Q_{supplied}$ is the available energy, and the quantity $(T_0/T)\,Q_{supplied}$ is the unavailable energy. As previously shown, these two quantities are represented by rectangular areas when the Carnot cycle is drawn on the *T-S* diagram, as shown in Figure 6.2. In actual engines, which are less efficient than the Carnot engine, only part of the available energy is converted into work so that the degree of conversion depends also on the efficiency of the engine.

The concept of availability addresses the problem of better energy utilization and is an important tool in the analysis, design, and optimization of thermal systems. J. W. Gibbs is credited as the originator of the availability concept. He

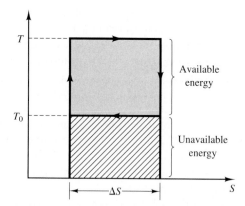

Figure 6.2   Available and unavailable energies for a Carnot cycle operating between T and T_0.

indicated that the environment plays an important part in evaluating the amount of work that can be performed by a system in any given state. According to Gibbs, the availability analysis determines the work potential when a system interacts with an environment at ambient conditions of temperature and pressure. This environment may be thought of as a large stable, homogeneous system whose temperature, pressure, and composition are uniform and virtually unaffected by any process experienced by the system. The availability or the potential to do work exists if the state of the system differs from that of the environment. The availability, however, diminishes as the state of the system approaches that of the environment. At equilibrium, there can be no interaction whatsoever to develop work, and the value of availability is zero. For this reason, the state of the environment (usually at $T_0 = 25°C$ and $p_0 = 1$ atm) is referred to as the *dead state*.

6.2 Reversible Versus Irreversible Work

As a preliminary step in the process of developing an expression that shows the relationship between availability and the state of a system, it is first necessary to prove that a system produces maximum work during a process if it pursues a reversible path. Or, conversely, it would be necessary to prove that, when work is done on a system, reversible work represents the minimum work required to attain a predetermined end state of the system. In other words, the reversible work is the maximum amount of work output (or the minimum amount of work input) that could be produced for a given change of state.

Consider a system in thermal contact with an environment at T_0 and p_0. When the system experiences a differential change from an initial state 1 to a final state 2, the change in energy dE is independent of the process. Thus for two processes, one reversible and the other irreversible, between states 1 and 2, the change in internal energy is a function of only states 1 and 2:

$$dE_{rev} = dE_{irrev}$$

But according to the first law, the change in internal energy dE is a result of interactions involving both heat and work, with the environment or,

$$dE = \delta Q + \delta W$$

The sum would be identical for reversible and irreversible processes:

$$\delta Q_{rev} + \delta W_{rev} = \delta Q_{irrev} + \delta W_{irrev} \qquad (6.1)$$

In a reversible process, the differential change of entropy is given by

$$dS = \frac{\delta Q_{rev}}{T_0} \qquad \text{or} \qquad \delta Q_{rev} = T_0 \, dS$$

But in an irreversible process, the same entropy change is obtained with a smaller amount of heat supplied:

$$dS > \frac{\delta Q_{\text{irrev}}}{T_0} \qquad \text{or} \qquad \delta Q_{\text{irrev}} < T_0 \, dS$$

Hence, since the quantities of heat are different,

$$\delta Q_{\text{rev}} > \delta Q_{\text{irrev}}$$

therefore, from Eq. (6.1), the work required is different:

$$\delta W_{\text{rev}} < \delta W_{\text{irrev}} \tag{6.2}$$

The work done on a system during a reversible process is less than the work done during an irreversible process even though the same end states are involved in the two processes. In an irreversible process, some work must be done in over-coming the dissipative effects caused by factors such as friction. If work is done by the system, the sense of the inequality is reversed so that

$$-\delta W_{\text{rev}} > -\delta W_{\text{irrev}} \tag{6.3}$$

The work done by a system during a reversible process then is greater than that done in an irreversible process connecting the same end states.

It may also be shown that all reversible processes operating between the same two states will produce identical amounts of work, provided that the system interacts thermally with the environment only. This can be proved by considering two reversible processes connecting states 1 and 2. If one of the processes is reversed, a complete cycle results. If the two reversible processes do not produce equal amounts of work, then reversal of one of the processes could result in a net amount of work done by the system. But since the environment is the only thermal reservoir with which the system interacts thermally, such a cycle would constitute a perpetual-motion machine of the second kind. Therefore, the two reversible processes must produce equal amounts of work.

6.3 Reversible Work in a Nonflow Process

Consider a system undergoing a change of state by means of a process along a reversible path. This process is equivalent to a reversible change of state that takes place along two separate processes, one a reversible adiabatic process and the other a reversible isothermal process in which heat is exchanged with the environment at a temperature T_0. Two cases are of particular interest, the non-flow process and the steady-state flow process.

Consider a system at an initial state defined by pressure p_1 and temperature T_1 surrounded by an environment at a pressure p_o and a temperature T_o. Let the system undergo a change from state 1 to state 2 along either of the two paths shown in Figures 6.3(a) and 6.3(b). In both cases, the process involves a reversible adiabatic expansion to a pressure p' and then reversible isothermal energy transfer that brings the system to point 2 (at the pressure p_0). The intermediate pressure p' may be either less than the final pressure, as in Figure 6.3(a), or greater than the final pressure, as in Figure 6.3(b). The total work involved is the algebraic sum of the work interactions in the two processes and represents the maximum work done by the system between states 1 and 2. For the reversible adiabatic process 1-a, the work done on the system according to the first law is

$$W_{1\text{-}a} = E_a - E_1$$

For the reversible isothermal process a-2, the work done on the system is

$$W_{a\text{-}2} = (E_2 - E_a) - T_o(S_2 - S_a)$$

The total work is the sum of the work done during the two processes, or

$$(W_{1\text{-}2})_{\text{rev}} = W_{1\text{-}a} + W_{a\text{-}2}$$

But since process 1-a is a reversible adiabatic (isentropic) process, $S_1 = S_a$. Hence the preceding equation becomes

$$(W_{1\text{-}2})_{\text{rev}} = (E_2 - E_1) - T_o(S_2 - S_1) \qquad (6.4)[1]$$

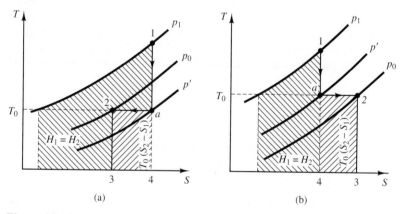

(a)   (b)

Figure 6.3 Maximum work in a nonflow process.

[1] The subscript max or min may be used instead of the subscript rev, depending upon whether work is done by or on the system.

Another method of arriving at Eq. (6.4) is to apply the first law directly to the overall process 1-2, assuming reversible heat interaction with the environment at a temperature T_0. In the case of compression, Eq. (6.4) indicates the minimum work necessary for the entire process; and in the case of expansion, Eq. (6.4) shows the maximum work realized during the process. Note that both the change in the energy of the system and the heat interaction contribute to the magnitude of W_{rev}.

If a system in a gravitational field of strength g has a velocity V, then the total internal energy neglecting electrical, magnetic, and surface effects is given by

$$E = U + \frac{mV^2}{2} + m\,gz \tag{6.5}$$

When Eq. (6.5) is substituted into Eq. (6.4), the expression for the total work interaction for a reversible process becomes

$$(W_{1\text{-}2})_{rev} = (U_2 - U_1) - T_0(S_2 - S_1) + m\frac{V_2^2 - V_1^2}{2} + m\,g(z_2 - z_1) \tag{6.6}$$

and per unit mass

$$(w_{1\text{-}2})_{rev} = (u_2 - u_1) - T_0(s_2 - s_1) + \frac{V_2^2 - V_1^2}{2} + g(z_2 - z_1) \tag{6.7}$$

6.4 Reversible Work in a Flow Process

To develop an expression for reversible work in a flow process, consider first a control volume undergoing an actual process, as shown in Figure 6.4. The control volume interacts with the environment as mass, heat, and work cross the

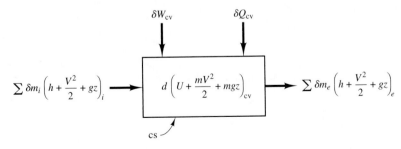

Figure 6.4   Energy flow through a control volume.

control surface. Let δQ_{cv} and δW_{cv} be the heat and work interactions with the environment. The total work done on the system may consist of electrical, mechanical, and magnetic work, in addition to any flow work done on the fluid as it flows through the control surface. From the first law, the work done on the control volume other than flow work is

$$\delta W_{cv} = -\delta Q_{cv} + \sum dm_e \left(h + \frac{V^2}{2} + gz \right)_e$$
$$- \sum dm_i \left(h + \frac{V^2}{2} + gz \right)_i + d \left(U + m\frac{V^2}{2} + m\,gz \right)_{cv} \tag{6.8}$$

Now let the control volume exchange energy with the environment producing reversible work. This can be conceived only when the change of state of the control volume and all interactions with the environment take place reversibly. A comparison between this idealized situation and the actual case can be made meaningful if the control volume undergoes the same change of state for both cases and the mass and state of the fluid entering and leaving the control volume are also the same for the actual and ideal cases.

Reversible heat interaction between the control volume and the environment can be accomplished in the following manner. Although the temperature may vary at different locations within the control volume, heat is transferred only at those locations where the system and the environment are at the same temperature, thereby reversible heat interaction is achieved. When this is not feasible and there is a difference in temperature between the control volume and the environment, heat interaction can take place through a reversible heat engine or a reversible heat pump. The energy transfer as work from or to the Carnot heat engine or pump is credited to or deducted from the reversible work of the control volume, and the total reversible work is labeled W_{rev}. The reversible heat interaction is given by

$$\delta Q_0 = T_0\, dS$$

where T_0 is the temperature of the environment, and dS is the entropy change of the environment. In a control volume where both heat and mass transfer occur across the control surface, the total entropy change for reversible conditions is

$$dS_{cv} = \frac{\delta Q_0}{T_0} + \sum dm_i\, s_i - \sum dm_e\, s_e$$

Hence the reversible heat interaction is

$$\delta Q_0 = T_0 \left(dS_{cv} - \sum dm_i\, s_i + \sum dm_e\, s_e \right) \tag{6.9}$$

Substituting for δQ_0 from Eq. (6.9) into the first law gives the reversible work as

$$\delta W_{rev} = -T_0\left(dS_{cv} - \sum dm_i\, s_i + \sum dm_e\, s_e\right) + \sum dm_e\left(h + \frac{V^2}{2} + gz\right)_e$$
$$- \sum dm_i\left(h + \frac{V^2}{2} + gz\right)_i + d\left(U + m\frac{V^2}{2} + m\,gz\right)_{cv}$$

Rearranging terms, this equation can be written as

$$\delta W_{rev} = \sum dm_e\left(h_e - T_0\, s_e + \frac{V_e^2}{2} + gz_e\right)$$
$$- \sum dm_i\left(h_i - T_0\, s_i + \frac{V_i^2}{2} + gz_i\right) + d\left(U - T_0 S + m\frac{V^2}{2} + m\,gz\right)_{cv}$$

If the properties are uniform at the inlet, exit, and within the control volume at any instant of time, integration of the preceding equation gives

$$W_{rev} = \sum m_e\left(h_e - T_0\, s_e + \frac{V_e^2}{2} + gz_e\right)$$
$$- \sum m_i\left(h_i - T_0\, s_i + \frac{V_i^2}{2} + gz_i\right)$$
$$+ \left[m_2\left(u_2 - T_0\, s_2 + \frac{V_2^2}{2} + gz_2\right)\right.$$
$$\left.- m_1\left(u_1 - T_0\, s_1 + \frac{V_1^2}{2} + gz_1\right)\right]_{cv} \tag{6.10}$$

Equation (6.10) gives the reversible work done on the control volume as a function of the inflow and outflow fluid properties, the initial and final states of the control volume, and the temperature of the environment.

Equation (6.10) may be readily adapted to a closed system by noting that there is no mass flow across the boundary of the system so that

$$\sum m_i\left(h_i - T_0\, s_i + \frac{V_i^2}{2} + gz_i\right) = 0$$

and

$$\sum m_e \left(h_e - T_0 s_e + \frac{V_e^2}{2} + gz_e \right) = 0$$

and for $m_1 = m_2$, Eq. (6.10) reduces to

$$W_{\text{rev}} = m \left(u_2 - T_0 s_2 + \frac{V_2^2}{2} + gz_2 \right) - m \left(u_1 - T_0 s_1 + \frac{V_1^2}{2} + gz_1 \right)$$

which is the same as Eq. (6.6).

6.5 Reversible Work in a Steady-State Flow Process

Application of Eq. (6.10) to steady-state flow processes imposes two requirements. These are

$$\sum m_i = \sum m_e \qquad (\Delta m_{\text{cv}} = 0)$$

and

$$d \left(U - T_0 S + m \frac{V^2}{2} + m \, gz \right)_{\text{cv}} = 0$$

From Eq. (6.10), the reversible work in a steady-state flow process is then

$$W_{\text{rev}} = \sum m_e \left(h_e - T_0 s_e + \frac{V_e^2}{2} + gz_e \right)$$
$$- \sum m_i \left(h_i - T_0 s_i + \frac{V_i^2}{2} + gz_i \right) \qquad (6.11)$$

When there is a single flow of fluid entering and leaving the control volume, this equation can be written for a unit mass as

$$(w_{i\text{-}e})_{\text{rev}} = (h_e - h_i) - T_0(s_e - s_i) + \frac{V_e^2 - V_i^2}{2} + g(z_e - z_i) \qquad (6.12)$$

As in the nonflow process, the reversible work in a steady-state flow process is defined by the inlet and outlet states of the fluid entering and leaving the control volume and by the temperature of the environment.

Example 6.1

Find the maximum work developed when air expands in a piston–cylinder assembly from an initial state of 530 kPa and $T_1 = 400$ K to a final state of 150 kPa and 300 K. Neglect changes in potential and kinetic energies, and assume $T_0 = 25°C$.

SOLUTION

The expansion process is shown in Figure 6.5. Using Eq. (6.7) for a nonflow process and assuming ideal gas yields

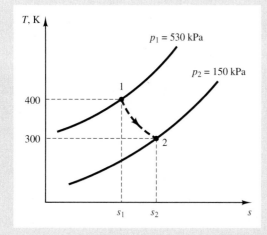

Figure 6.5

$$(w_{1\text{-}2})_{\text{rev}} = u_2 - u_1 - T_0(s_2 - s_1) = c_v(T_2 - T_1) - T_0\left(c_p \ln \frac{T_2}{T_1} - R \ln \frac{p_2}{p_1}\right)$$

$$= (0.7165 \text{ kJ/kg K})(300 - 400) \text{ K}$$

$$- (298.15 \text{ K})\left[(1.0035 \text{ kJ/kg K}) \ln \frac{300}{400} - (0.287 \text{ kJ/kg K}) \ln \frac{150}{530}\right]$$

$$= -93.583 \text{ kJ/kg}$$

Example 6.2

The flow rate through an air compressor operating at steady state is 25 kg/min. The tabulated data give the temperature, pressure, internal energy, and enthalpy of the air at the inlet and at the delivery. What is the actual work and the minimum work required per kg of air? Find the actual and reversible power. (Neglect heat interaction and changes in potential and kinetic energies between inlet and delivery. Assume the environment to be at a temperature of 20°C.) Data:

	p (kPa)	T (K)	u (kJ/kg)	h (kJ/Kg)
Inlet	100	310	221.27	310.24
Delivery	200	380	271.72	380.77

SOLUTION

Figure 6.6 shows a schematic of the air compressor. For steady-state flow process, the work done per unit mass according to the first law is

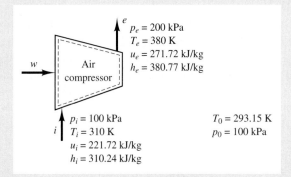

Figure 6.6

$$w_{\text{actual}} = h_e - h_i = (380.77 - 310.24)\ \text{kJ/kg} = 70.53\ \text{kJ/kg}$$

The power input to the compressor is

$$\dot{W}_{\text{actual}} = (70.53\ \text{kJ/kg})(25\ \text{kg/min})\left(\frac{1}{60\ \text{s/min}}\right)$$

$$= 29.39\ \text{kW}$$

The minimum work of compression is equal to the reversible work. It is given by Eq. (6.12) as

$$w_{\min} = w_{\text{rev}} = (h_e - T_0\,s_e) - (h_i - T_0\,s_i)$$

$$= (h_e - h_i) - T_0\left(c_p \ln\frac{T_e}{T_i} - R\ln\frac{p_e}{p_i}\right)$$

$$= (380.77 - 310.24)\ \text{kJ/kg} - (293.15\ \text{K})\left[(1.0035\ \text{kJ/kg K})\ln\frac{380}{310} - (0.287\ \text{kJ/kg K})\ln\frac{200}{100}\right]$$

$$= 68.95\ \text{kJ/kg}$$

The minimum power is

$$\dot{W}_{rev} = 68.95 \times 25 \times \frac{1}{60} = 28.73 \text{ kW}$$

The difference between the actual power and the reversible power is attributed to internal irreversibilities within the compressor.

6.6 Irreversibility

Irreversible processes result in lost opportunities for developing work because maximum work can be produced only if the system interacts reversibly with the environment. The actual work done on a system when it experiences the same change of state is always more than the idealized reversible work. *Irreversibility I* is the difference between the reversible work and the actual work for the same change of state so that

$$I \equiv W_{act} - W_{rev} \tag{6.13}$$

An expression of the irreversibility for a nonflow process is obtained as follows: Let a system receive an amount of heat Q from the environment and perform an amount of work W. The first law is

$$W = (E_2 - E_1) - Q$$

Substituting for W_{rev} from Eq. (6.4) and for W from the preceding equation into Eq. (6.13) gives the following expression for irreversibility:

$$I = T_0(S_2 - S_1) - Q \tag{6.14}$$

If the system receives heat from its environment at a temperature T_0, the entropy change of the environment is

$$\Delta S_{envir} = -\frac{Q}{T_0}$$

Substituting in the previous equation gives

$$I = T_0(S_2 - S_1) + T_0 \, \Delta S_{\text{envir}} = T_0(\Delta S_{\text{sys}} + \Delta S_{\text{envir}}) \quad (6.15)$$

According to this equation, the irreversibility of a process is the product of the temperature of the environment and the increase in entropy of both the system and the environment. Since neither the absolute temperature T_0 nor the total change of entropy is negative,

$$I \geq 0 \quad (6.16)$$

where the equality and inequality signs apply to reversible and irreversible processes, respectively. Irreversibility cannot be a negative value.

In a similar way, an expression for irreversibility in a flow process can be derived. Consider the control volume of Figure 6.2. The work done on the control volume is

$$W_{\text{cv}} = \sum m_e \left(h_e + \frac{V_e^2}{2} + gz_e \right) - \sum m_i \left(h_i + \frac{V_i^2}{2} + gz_i \right)$$
$$+ \left[m_2 \left(u_2 + \frac{V_2^2}{2} + gz_2 \right) - m_1 \left(u_1 + \frac{V_1^2}{2} + gz_1 \right) \right]_{\text{cv}} - Q_{\text{cv}} \quad (6.17)$$

By substituting for W_{rev} from Eq. (6.10) and for W_{cv} from Eq. (6.17), Eq. (6.13) becomes

$$I = \sum m_e \, s_e \, T_0 - \sum m_i \, s_i \, T_0 + m_2 \, s_2 \, T_0 - m_1 \, s_1 \, T_0 - Q_{\text{cv}}$$

Since T_0 is constant, the preceding equation can be written as

$$I = T_0 \left[\sum m_e \, s_e - \sum m_i \, s_i + m_2 \, s_2 - m_1 \, s_1 - \frac{Q_{\text{cv}}}{T_0} \right] \quad (6.18)$$

Equation (6.18) is the general expression for irreversibility. In the case of a system ($m_i = m_e = 0$ and $m_1 = m_2 = m$), this equation becomes

$$I_{1\text{-}2} = mT_0(s_2 - s_1) - Q_{1\text{-}2} \quad (6.19)$$

which is the same as Eq. (6.14).

For a control volume undergoing a steady-state flow process, Eq. (6.18) reduces to

$$I = \sum m_e \, T_0 \, s_e - \sum m_i \, T_0 \, s_i - Q_{\text{cv}} \quad (6.20)$$

The general expression for irreversibility can also be related to the principle of increase of entropy. The entropy generated within the control volume is

$$\Delta S_{cv} = m_2 \, s_2 - m_1 \, s_1$$

and the entropy change of the environment is

$$\Delta S_{envir} = \sum m_e \, s_e - \sum m_i \, s_i - \frac{Q_{cv}}{T_0}$$

When these terms are substituted into Eq. (6.18), the expression for irreversibility becomes

$$I = T_0(\Delta S_{cv} + \Delta S_{envir}) \tag{6.21}$$

Evidently, the same expression for irreversibility applies to both flow and nonflow processes. Note that the total change of entropy ($\Delta S_{cv} + \Delta S_{envir}$) is the entropy generation, and hence the effect of irreversibilities is embodied in the entropy generation. A reduction in entropy generation decreases the irreversibility and increases the work output (or decreases the work input). The quantity $T_0(\Delta S_{cv} + \Delta S_{envir})$ also represents a decrease in available energy. Note that the increase in the unavailable energy in a process is equal in magnitude to the decrease in the available energy, and both are a result of net entropy generation.

Energy transfer and entropy generation can be represented graphically on a temperature–energy diagram. Figure 6.7 shows a comparison between an actual and a reversible power cycle on the basis of equal heat input. Both cycles operate between two thermal reservoirs at T_H and T_L. For the reversible cycle, the first and second laws are

$$W_{rev} + Q_{H,rev} - Q_{L,rev} = 0$$

$$\Delta S = S_{gen} = -\frac{Q_{H,rev}}{T_H} + \frac{Q_{L,rev}}{T_L} = 0$$

For the actual cycle,

$$W_{act} + Q_H - Q_L = 0$$

$$S_{gen} = -\frac{Q_H}{T_H} + \frac{Q_L}{T_L} > 0$$

But since $Q_{H,rev} = Q_H$, the lost work is

$$W_{lost} = W_{act} - W_{rev} = Q_L - Q_{L,rev}$$

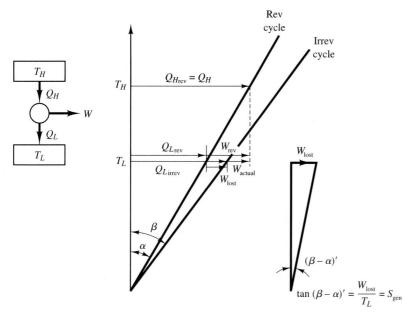

Figure 6.7   Energy transfer and entropy generation in a power cycle.

Substituting for $Q_{L,\,\mathrm{rev}}$ from the relation,

$$Q_{L,\,\mathrm{rev}} = \frac{Q_{H,\,\mathrm{rev}}}{T_H} T_L$$

gives

$$W_{\mathrm{lost}} = T_L\!\left(\frac{Q_L}{T_L} - \frac{Q_{H,\,\mathrm{rev}}}{T_H}\right) = T_L\, S_{\mathrm{gen}}$$

The lost work is proportional to entropy generation. Note that the W_{lost} is equal to the irreversibility I when the low-temperature reservoir is at T_0. The energy transfers in the form of heat and work for the reversible and the irreversible cycles and the lost work are indicated in Figure 6.7.

The entropy transfer to the reversible cycle is equal to $\tan a = Q_{H,\,\mathrm{rev}}/T_H$ and $\tan \beta = Q_{L,\,\mathrm{irrev}}/T_L$. The difference $(\tan \beta - \tan a) = (Q_{L,\,\mathrm{irrev}}/T_L - Q_{L,\,\mathrm{rev}}/T_L)$ is a graphical representation of the entropy generation. Figure 6.8 shows a similar representation of energy transfer and entropy generation when reversible and actual refrigeration cycles are compared based on equal refrigeration load $(Q_{L,\,\mathrm{rev}} = Q_L)$.

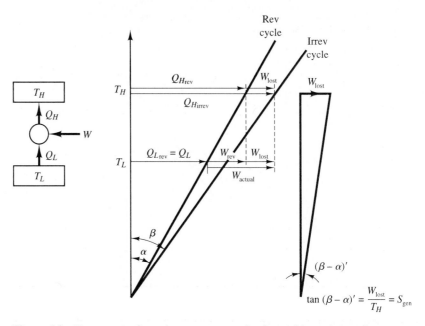

Figure 6.8   Energy transfer and entropy generation in a refrigeration cycle.

Example 6.3

A metal block of mass 10 kg at a temperature of 600 K is allowed to cool to the ambient temperature of 25°C. If the specific heat of the metal is 0.9 kJ/kg K, determine the irreversibility of this process.

SOLUTION

A sketch of the system is shown in Figure 6.9. The irreversibility, according to Eq. (6.14), is

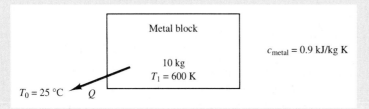

Figure 6.9

$$I_{1\text{-}2} = T_0 \, \Delta S - Q_{1\text{-}2}$$

The heat transfer, according to the first law, is

$$Q_{1\text{-}2} = (U_2 - U_1) = mc(T_2 - T_1)$$
$$= (10 \text{ kg})(0.9 \text{ kJ/kg K})(298.15 - 600) \text{ K} = -2716.65 \text{ kJ}$$

The entropy change of the system is

$$\Delta S = \int_1^2 \frac{\delta Q}{T} = \int_1^2 \frac{dU}{T} = \int_1^2 \frac{mc\, dT}{T} = mc \ln \frac{T_2}{T_1}$$

$$= (10 \text{ kg})(0.9 \text{ kJ/kg K}) \ln \frac{298.15}{600} = -6.294 \text{ kJ/K}$$

Hence, the irreversibility of the heat transfer process is

$$I_{1\text{-}2} = (298.15 \text{ K})(-6.294 \text{ kJ/K}) - (-2716.65 \text{ kJ}) = 840.094 \text{ kJ}$$

This is the work potential wasted during this process.

Example 6.4

Air enters a turbine at 400 kPa, 150°C and leaves at 100 kPa, 60°C. The rate of flow is 0.6 kg/s, and the power developed is 45 kW. If the environment temperature is 10°C, calculate the rate of the heat transfer and the rate of irreversibility of the process. Assume steady-state operation.

SOLUTION

A schematic of the turbine is shown in Figure 6.10. Assuming air to be an ideal gas with constant specific heats, the heat interaction according to the first law is

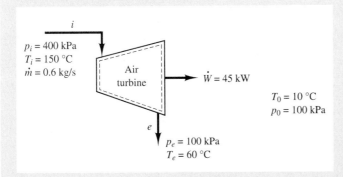

$p_i = 400$ kPa
$T_i = 150$ °C
$\dot{m} = 0.6$ kg/s

Air turbine $\dot{W} = 45$ kW

$T_0 = 10$ °C
$p_0 = 100$ kPa

$p_e = 100$ kPa
$T_e = 60$ °C

Figure 6.10

$$\dot{Q} = \dot{m}(h_e - h_i) - \dot{W} = \dot{m}\, c_p(T_e - T_i) - \dot{W}$$

$$= (0.6 \text{ kg/s})(1.0035 \text{ kJ/kg K})[(60 - 150) \text{ K}] + 45 \text{ kW}$$

$$= (-54.189 + 45) \text{ kW} = -9.189 \text{ kW}$$

The rate of irreversibility, according to Eq. (5.20), is

$$\dot{I} = \dot{m}T_0(s_e - s_i) - \dot{Q}_{cv}$$

$$= (0.6 \text{ kg/s})(283.15 \text{ K})\left[(1.0035 \text{ kJ/kg K}) \ln \frac{333.15}{423.15} - (0.287 \text{ kJ/kg K}) \ln \frac{100}{400} \right] - (-9.189 \text{ kW})$$

$$= 0.6(283.15)(-0.23997 + 0.39787) + 9.189 = 36.015 \text{ kW}$$

Example 6.5

Calculate the rate of irreversibility involved in the heat transfer between a hot and a cold fluid with the following characteristics:

Fluid	$\dot{m}$ (kg/s)	T_i (°C)	T_e (°C)	c_p (kJ/kg K)
Hot	0.06	135	38	4.31
Cold	—	28	105	4.65

The temperature of the environment $T_0 = 25°C$. What is the rate of entropy production? (Assume steady-state operation and constant values of specific heats.)

SOLUTION

Figure 6.11 is a schematic of the heat exchanger. Assuming no pressure or heat losses, the mass flow rate of the cold fluid may be determined by the simple energy balance

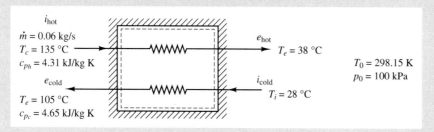

Figure 6.11

$$\dot{m}_c\, c_{p_c}(T_e - T_i)_c = \dot{m}_h\, c_{p_h}(T_i - T_e)_h$$

so that

$$\dot{m}_c = \frac{0.06 \times 4.31(135 - 38)}{4.65(105 - 28)} = 0.0701 \text{ kg/s}$$

The rate of increase of irreversibility, according to Eq. (6.21), is

$$\dot{I} = T_0(\Delta \dot{S}_c + \Delta \dot{S}_h) = T_0\left(\dot{m}_c \, c_{p_c} \ln \frac{T_{e_c}}{T_{i_c}} + \dot{m}_h \, c_{p_h} \ln \frac{T_{e_h}}{T_{i_h}}\right)$$

$$= (298.15 \text{ K})\left[(0.0701 \text{ kg/s})(4.65 \text{ kJ/kg K}) \ln \frac{378.15}{301.15} + (0.06 \text{ kg/s})(4.31 \text{ kJ/kg K}) \ln \frac{311.15}{408.15}\right] = 1.205 \text{ kW}$$

The rate of entropy generation is

$$(\Delta \dot{S}_c + \Delta \dot{S}_h) = \frac{\dot{I}}{T_0} = \frac{1.205 \text{ kW}}{298.15 \text{ K}} = 4.04 \times 10^{-3} \text{ kW/K}$$

Note that entropy generation is a quantitative measure of irreversibility.

Example 6.6

An insulated rigid tank contains 1.5 kg of air at 25°C and 100 kPa. The temperature of the air is raised to 80°C by work interaction, as shown in Figure 6.12(a). Determine the reversible work and the irreversibility for this process.

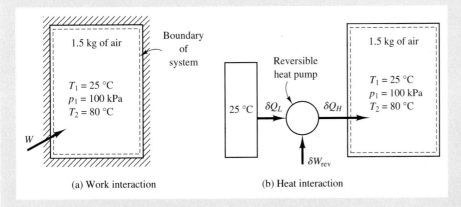

(a) Work interaction   (b) Heat interaction

Figure 6.12

Compare this process if the same change of state is achieved by means of heat interaction supplied by a reversible heat pump, as shown in Figure 6.12(b).

SOLUTION

The reversible work is

$$(W_{1\text{-}2})_{\text{rev}} = (U_2 - U_1) - T_0(S_2 - S_1)$$

Assuming air to be an ideal gas with constant specific heats and noting that the volume is constant,

$$(W_{1-2})_{rev} = m\,c_v(T_2 - T_1) - T_0\left(m\,c_v \ln\frac{T_2}{T_1}\right)$$

$$= (1.5\text{ kg})(0.7165\text{ kJ/kg K})[(80 - 25)\text{ K}] - (298.15\text{ K})(1.5\text{ kg})(0.7165\text{ kJ/kg K})\ln\left(\frac{80 + 273.15}{25 + 273.15}\right)$$

$$= (59.1113\text{ kJ}) - (54.2487\text{ kJ}) = 4.8626\text{ kJ}$$

The actual work is determined by the first law,

$$W_{1-2} = (U_2 - U_1) = (1.5\text{ kg})(0.7165\text{ kJ/kg K})[(80 - 25)\text{ K}] = 59.1113\text{ kJ}$$

The irreversibility is

$$I = W_{act} - W_{rev} = (59.1113\text{ kJ}) - (4.8626\text{ kJ}) = 54.2487\text{ kJ}$$

or

$$I = T_0(\Delta S_{sys} + \Delta S_{envir}) = T_0(\Delta S_{sys}) = 54.2487\text{ kJ}$$

In order to accomplish the same change of state, the energy input must still be the same so that the heat input $Q_2 = 59.1113$ kJ.

The coefficient of performance of a Carnot heat pump operating between the environment at $T_0 = 25°C$ and the system temperature T is

$$\beta_{ht\ pump} = \frac{T}{T - T_0} = \frac{\delta Q_H}{\delta W_{rev}}$$

Therefore,

$$\delta W_{rev} = \left(\frac{T - T_0}{T}\right)\delta Q_H = \left(1 - \frac{T_0}{T}\right)(m\,c_v\,dT)$$

Integration gives

$$W_{rev} = \int_{T_1}^{T_2}\left(1 - \frac{T_0}{T}\right)m\,c_v\,dT = m\,c_v(T_2 - T_1) - m\,c_v\,T_0 \ln\frac{T_2}{T_1}$$

$$= 4.8626\text{ kJ}\qquad \text{as found previously}$$

This means that through the use of a reversible heat pump, only 4.8626 kJ of work is needed to supply the energy required. This compares with 59.1113 kJ of work needed. The difference, which is equal to the irreversibility, could have been saved.

6.7 Availability (Exergy)

When heat interaction takes place between a system and the environment only, the maximum useful work that can be performed by the system indicates the energy availability of the system. Availability, also referred to as *exergy,* can be viewed as the useful work potential of an amount of energy at a specified state. It is an extensive composite property that depends on the state of both the system and the environment. In performing maximum work, the system undergoes only reversible processes before it finally reaches mutual equilibrium with the environment. At this state, the system is incapable of producing any further work, and the value of availability is zero. Note that work can always be produced when there is a difference in temperature between the system and the environment. Figure 6.13 depicts how maximum work can be obtained through a reversible engine when the temperature of the system is higher or lower than that of the environment. Note that total availability, unlike total energy, is not conserved except in reversible processes. In real processes, availability is reduced as a result of the irreversibilities of these processes.

The expressions derived for reversible work in nonflow and in steady-state flow processes are used in defining availability of energy for a system and for a control volume. Availability, however, is associated with a state, in contrast with reversible work or irreversibility, which relate to a process.

Consider a nonflow process in which the system exchanges heat with the environment only, which is at a temperature T_0 and a pressure p_0. Let the state of the

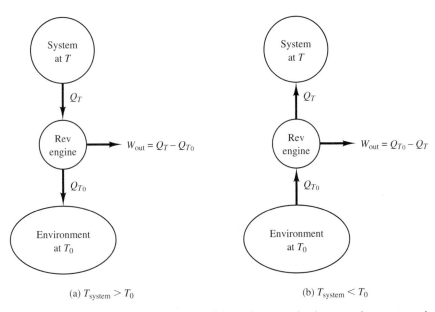

(a) $T_{\text{system}} > T_0$ (b) $T_{\text{system}} < T_0$

Figure 6.13   Work developed by a reversible engine operating between the system and the environment.

system change from an initial state 1 to a final state 0, at which it is in thermal and mechanical equilibrium with the environment. In the absence of surface, electrical, and magnetic effects, the maximum reversible work according to Eq. (6.4) is

$$W_{rev} = (W_{1\text{-}0})_{rev} = (E_0 - T_0 S_0) - (E_1 - T_0 S_1) \tag{6.22}$$

The availability is equal to the reversible work minus the work interaction with the environment as the system expands or contracts:

$$W_{rev,\,useful} = W_{rev} - W_{envir} \tag{6.23}$$

But the work interaction with the environment[2] at a pressure p_0 is

$$W_{envir} = -p_0(V_0 - V_1) \tag{6.24}$$

Substituting into Eq. (6.23) and rearranging gives

$$\begin{aligned} W_{rev,useful} &= (E_0 + p_0V_0 - T_0S_0) - (E_1 + p_0V_1 - T_0S_1) \\ &= \Phi_0 - \Phi_1 \end{aligned} \tag{6.25}$$

where Φ is called the *availability function for a closed system.* It is defined as

$$\Phi \equiv E + p_0V - T_0S \equiv U + p_0V - T_0S + KE + PE \tag{6.26}$$

and per unit mass,

$$\varphi = u + p_0v - T_0s + \frac{V^2}{2} + gz \tag{6.27}$$

The reversible useful work is equal to the change in the availability function. In differential form,

$$d\Phi = dE + p_0\,dV - T_0\,dS \tag{6.28}$$

The availability of the energy of a system at state 1 to do useful work is equal to the maximum decrease in the availability function. Thus

$$\text{availability} = \Phi_1 - \Phi_0 \tag{6.29}$$

---

[2] Note that the work interaction with the environment has significance only when systems expand or contract to reach equilibrium with the environment. It is zero in steady-state flow processes in which we are dealing with a control volume whose boundaries remain fixed.

Substituting from Eq. (6.25) and noting that at the dead state the velocity is zero, the reversible useful work per unit mass becomes

$$(w_{1\text{-}0})_{\substack{\text{rev} \\ \text{useful}}} = (u_0 - u_1) + p_0(v_0 - v_1)$$

$$- T_0(s_0 - s_1) - \frac{V_1^2}{2} + g(z_0 - z_1) \tag{6.30}$$

Note that availability is a property that depends only on the state of the system, provided that the state of the environment remains unchanged and that the system reaches equilibrium with the environment. It is a measure of the departure of the state of the system from that of the environment, and unlike energy it is not generally conserved. It decreases owing to irreversibilities incurred by any spontaneous changes of the system.

When a system undergoes a change from state 1, representing an availability of $(\Phi_1 - \Phi_0)$, to state 2, representing an availability of $(\Phi_2 - \Phi_0)$, the reversible work for the given change of state of the system is the difference between the two availability functions:

$$(W_{1\text{-}2})_{\substack{\text{rev} \\ \text{useful}}} = (\Phi_0 - \Phi_1) - (\Phi_0 - \Phi_2) = \Phi_2 - \Phi_1 \tag{6.31}$$

$$(W_{1\text{-}2})_{\substack{\text{rev} \\ \text{useful}}} = (u_2 - u_1) + p_0(v_2 - v_1)$$

$$- T_0(s_2 - s_1) + \frac{V_2^2 - V_1^2}{2} + g(z_2 - z_1) \tag{6.32}$$

which is the same as Eq. (6.7) except for the work interaction with the environment. In general,

$$W_{1\text{-}2} \geq \Phi_2 - \Phi_1 \tag{6.33}$$

and in differential form,

$$\delta W \geq d\Phi \tag{6.34}$$

where the inequality sign applies to irreversible processes.

The second case to be considered is a control volume undergoing a steady-state flow process. According to Eq. (6.12), the reversible work output is the maximum (or minimum in case of compression) work interaction. In the absence of surface, electrical, and magnetic effects, the reversible work per unit mass, for a single fluid stream, is

$$w_{\text{rev}} = (h_0 - T_0 s_0 + g z_0) - \left(h - T_0 s + \frac{V^2}{2} + gz\right) \tag{6.35}$$

The *availability function for steady flow* is defined as

$$\Psi \equiv H - T_0 S + m\frac{V^2}{2} + mgz \tag{6.36}$$

and per unit mass,

$$\psi = h - T_0 s + \frac{V^2}{2} + gz \tag{6.37}$$

In differential form,

$$d\Psi = dH - T_0\, dS + d(KE) + d(PE) \tag{6.38}$$

The reversible work in a flow process then is equal to the maximum decrease in the availability function:

$$W_{\text{rev}} = (W_{i\text{-}0})_{\text{rev}} = \text{availability change} = \Psi_0 - \Psi_i \tag{6.39}$$

As in the nonflow process, the reversible work in a steady-state flow process between state i and state e, while exchanging heat with the environment, is

$$(W_{i\text{-}e})_{\text{rev}} = (\Psi_0 - \Psi_i) - (\Psi_0 - \Psi_e) = \Psi_e - \Psi_i \tag{6.40}$$

Substituting for Ψ gives

$$
\begin{aligned}
(W_{i\text{-}e})_{\text{rev}} = {} & \left(H_e - T_0 S_e + m\frac{V_e^2}{2} + mgz_e\right) \\
& - \left(H_i - T_0 S_i + m\frac{V_i^2}{2} + mgz_i\right) \\
= {} & (H_e - H_i) - T_0(S_e - S_i) + m\frac{V_e^2 - V_i^2}{2} + mg(z_e - z_i)
\end{aligned}
\tag{6.41}
$$

and per unit mass,

$$(W_{i\text{-}e})_{\text{rev}} = (h_e - h_i) - T_0(s_e - s_i) + \frac{V_e^2 - V_i^2}{2} + g(z_e - z_i) \tag{6.42}$$

In general,

$$W_{i\text{-}e} \geq \Psi_e - \Psi_i \tag{6.43}$$

or, in a differential form,

$$\delta W \geq d\Psi \tag{6.44}$$

where the inequality sign applies to irreversible processes. When spontaneous changes in a system cannot occur, the system is at its minimum value of availability. Spontaneous changes can occur only if a system can go to a state of lower availability, and work can be obtained only if changes of state occur that result in a decrease in availability. As shown earlier, the maximum reversible work interaction with a system or a control volume exchanging heat with the environment only is equal to the change in the availability function:

$$W_{\substack{rev \\ useful}} = \Phi_{min} - \Phi \quad \text{for a nonflow process} \tag{6.45}$$

and

$$W_{rev} = \Psi_{min} - \Psi \quad \text{for a steady-state flow process} \tag{6.46}$$

Here, Φ and Ψ are the availability functions at some particular state, whereas Φ_{min} and Ψ_{min} are the availability functions at the most stable state of equilibrium (dead state). Stable equilibrium will be discussed in Section 6.14.

6.8 Availability Transfer

This section considers the availability transfer as a system undergoes a process during which it interacts with its environment in the form of heat and work. The heat transfer from the system has a potential use in performing work, and the availability transfer of the heat interaction is equal to the work that could have been developed if a reversible engine is inserted to operate between the boundary temperature T_b and the environment temperature T_0 so that

$$\begin{array}{c}\text{availability transfer} \\ \text{associated with heat interaction}\end{array} = Q\left(1 - \frac{T_0}{T_b}\right) \tag{6.47}$$

Note that when energy is transferred by heat some energy must be permanently degraded. When work interaction takes place between a system and its environment, the availability transfer is equal to the work done by the system minus any expansion or contraction work. In other words, availability transfer is equal to the net work transfer so that

$$\begin{array}{c}\text{availability transfer} \\ \text{associated with work interaction}\end{array} = (-W) - p_0(V_2 - V_1) \tag{6.48}$$

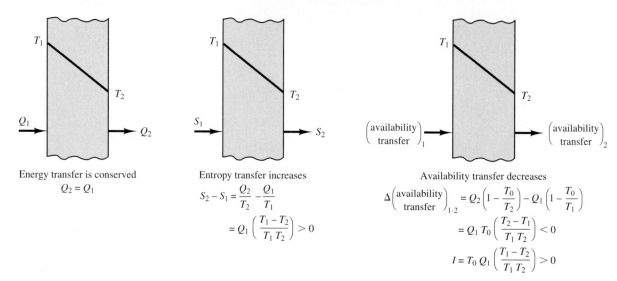

Figure 6.14   Energy, entropy, and availability transfer associated with heat transfer across a temperature difference.

Since energy transfer by work can be stored as kinetic or potential energies, which are completely recoverable as work, these two types of energy equate to availability transfer of the same magnitude. Figure 6.14 compares energy, entropy, and availability in a heat transfer process.

Availability reduction takes place owing to internal and external irreversibilities when a system undergoes a real process so that

$$\text{the decrease in availability} = I = T_0(\Delta S_{int} + \Delta S_{ext}) \tag{6.49}$$

Note that the decrease in availability, which is equal to the irreversibility, indicates the lost potential to perform work.

Example 6.7

A hydraulic torque converter operates at steady state with an input and output power of 100 and 95 kW. The converter is maintained at 55°C. If the environment temperature is 25°C, determine the availability transfer due to heat interaction. What is the internal rate of change of availability?

SOLUTION

Referring to Figure 6.15, the first law is

$$\dot{Q} + \dot{W} = \Delta \dot{U}$$

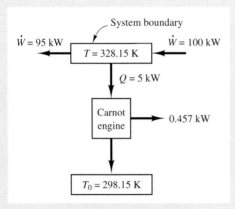

Figure 6.15

But at steady state, $\Delta \dot{U} = 0$; therefore,

$$\dot{Q} + (100 - 95) \text{ kW} = 0 \quad \text{or} \quad \dot{Q} = -5 \text{ kW}$$

The rate of availability transfer associated with heat interaction is equal to the reversible power developed by a Carnot engine inserted between the system and the environment. It is equal to

$$\dot{Q}\left(1 - \frac{T_0}{T_b}\right) = (-5 \text{ kW})\left(1 - \frac{298.15}{328.15}\right) = -0.457 \text{ kW}$$

Since there is no displacement work, the rate of change of availability transfer associated with the work interaction is

$$(-100 \text{ kW}) - (-95 \text{ kW}) = -5 \text{ kW}$$

and the internal rate of decrease of availability, or $\dot{I}$, is

$$\dot{I} = \dot{Q}\left(1 - \frac{T_0}{T_b}\right) - \dot{W} = (-0.457 \text{ kW}) - (-5 \text{ kW}) = 4.543 \text{ kW}$$

Example 6.8

Calculate the availability in the initial and final states of Example 6.1. Check that the difference is equal to the reversible work for that process. Assume the temperature and pressure of the environment are 25°C and 101.325 kPa.

SOLUTION

The availability per unit mass at the initial state according to Eq. (6.30) is

$$\varphi_1 - \varphi_0 = (u_1 - u_0) + p_0(v_1 - v_0) - T_0(s_1 - s_0)$$

$$= c_v(T_1 - T_0) + p_0(v_1 - v_0) - T_0\left(c_p \ln \frac{T_1}{T_0} - R \ln \frac{p_1}{p_0}\right)$$

where

$$v_1 = \frac{RT_1}{p_1} = \frac{(0.287 \text{ kJ/kg K})(400 \text{ K})}{530 \text{ kPa}} = 0.217 \text{ m}^3/\text{kg}$$

and

$$v_0 = \frac{RT_0}{p_0} = \frac{(0.287 \text{ kJ/kg K})(298.15 \text{ K})}{101.325 \text{ kPa}} = 0.845 \text{ m}^3/\text{kg}$$

Hence,

$$\varphi_1 - \varphi_0 = (0.7165 \text{ kJ/kg K})[(400 - 298.15) \text{ K}] + (101.325 \text{ kPa})(0.217 \text{ m}^3/\text{kg} - 0.845 \text{ m}^3/\text{kg})$$

$$- (298.15 \text{ K})\left[(1.0035 \text{ kJ/kg K}) \ln \frac{400}{298.15} - (0.287 \text{ kJ/kg K}) \ln \frac{530}{101.325}\right] = 63.001 \text{ kJ/kg}$$

The availability at the final state is

$$\varphi_2 - \varphi_0 = (u_2 - u_0) + p_0(v_2 - v_0) - T_0(s_2 - s_1)$$

$$= c_v(T_2 - T_0) + p_0(v_2 - v_0) - T_0\left(c_p \ln \frac{T_2}{T_0} - R \ln \frac{p_2}{p_0}\right)$$

where

$$v_2 = \frac{RT_2}{p_2} = \frac{(0.287 \text{ kJ/kg K})(300 \text{ K})}{150 \text{ kPa}} = 0.574 \text{ m}^3/\text{kg}$$

Hence,

$$\varphi_2 - \varphi_0 = (0.7165 \text{ kJ/kg K})[(300 - 298.15) \text{ K}] + (101.325 \text{ kPa})(0.574 \text{ m}^3/\text{kg} - 0.845 \text{ m}^3/\text{kg})$$

$$- (298.15 \text{ K})\left[(1.0035 \text{ kJ/kg K}) \ln \frac{300}{298.15} - (0.287 \text{ kJ/kg K}) \ln \frac{150}{101.325}\right] = 5.587 \text{ kJ/kg}$$

The reversible useful work is equal to the change in the availability or

$$(w_{1\text{-}2})_{\substack{\text{rev} \\ \text{useful}}} = \varphi_2 - \varphi_1 = 5.587 \text{ kJ/kg} - 63.001 \text{ kJ/kg} = -57.414 \text{ kJ/kg}$$

The reversible work is

$$(w_{1\text{-}2})_{\text{rev}} = (w_{1\text{-}2})_{\substack{\text{rev} \\ \text{useful}}} + [-p_0(v_2 - v_1)]$$

$$= -57.414 \text{ kJ/kg} - (101.325 \text{ kPa})(0.574 \text{ m}^3/\text{kg} - 0.217 \text{ m}^3/\text{kg})$$

$$= -93.587 \text{ kJ/kg}$$

which agrees with the result of Example 6.1.

Example 6.9

Find the unavailable energy resulting from the transfer of a quantity of heat Q between a thermal reservoir at temperature T and the environment at T_0.

SOLUTION

The increase of entropy due to the heat transfer process is

$$\Delta S = \frac{Q}{T_0} - \frac{Q}{T} = Q\frac{T - T_0}{TT_0}$$

and the increase in unavailable energy (irreversibility) is $I = T_0\Delta S = Q[(T - T_0)/T]$. This energy represents the maximum work obtainable from a reversible engine operating between temperature T and the temperature of the environment, T_0. This work is referred to as *lost work* because it becomes unavailable as a result of the heat transfer process. Thus the *availability of a thermal reservoir* can be defined as the reversible work that can be obtained when a quantity of energy is withdrawn from the reservoir. It is equal to the work that could be developed by a reversible power cycle receiving heat at a temperature T and discharging energy in the form of heat to an environment at T_0. Figure 6.16 shows the increase in the unavailable energy as a result of heat interaction. Note that all the energy originally available is transformed in the process into unavailable energy.

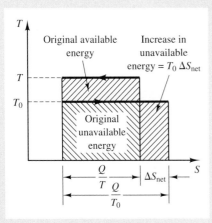

Figure 6.16

Example 6.10

A power cycle comprises two reversible isothermal processes and two irreversible adiabatic processes. The cycle receives 1000 kJ of heat at the upper temperature of 1000 K. The low-temperature receiver is at 500 K. Owing to fluid friction during the adiabatic compression and expansion processes, there is an increase in entropy in each of these processes equal to 5 percent of the entropy increase during the heat transfer to the engine.

(a) Sketch a *T-S* diagram for the cycle, and label the areas that represent the available and the unavailable energies.
(b) Calculate the thermal efficiency of the cycle.

SOLUTION

(a) The increase in entropy is due to heat interaction and fluid friction. The entropy increase per cycle due to heat transfer to the engine is

$$\Delta S = (1000 \text{ kJ})/(1000 \text{ K}) = 1 \text{ kJ/K}$$

As shown in Figure 6.17, the available energy is 500 kJ. The unavailable energy due to heat transfer at 500 K is also 500 kJ/kg. The unavailable energy due to heat transfer and friction is $500 \times 1.1 = 550$ kJ/kg. The work done is -450 kJ, and the irreversibility is

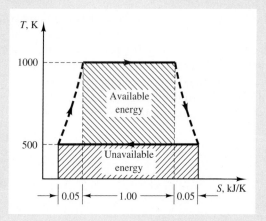

Figure 6.17

$$I = T_0 \Delta S = (500 \text{ K})(2)(0.05 \text{ kJ/K}) = 50 \text{ kJ}$$

(b) The thermal efficiency of the cycle is

$$\eta = \frac{W_{\text{net out}}}{Q_{\text{in}}} = \frac{Q_{\text{in}} - Q_{\text{out}}}{Q_{\text{in}}} = \frac{(1000 \text{ kJ}) - 1.1(500 \text{ kJ})}{1000 \text{ kJ}} = 45 \text{ percent}$$

Example 6.11

For the power cycle shown in Figure 6.18, indicate the magnitude and the corresponding area of the available energy, the unavailable energy of the heat input, and the irreversibility of the cycle. The temperature of the environment is 300 K.

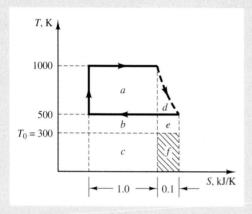

Figure 6.18

SOLUTION

The available energy of the heat input is 700 kJ (area $a + b$). The unavailable energy due to heat transfer at 300 K is

$$(300 \text{ K})(1.0 \text{ kJ/K}) = 300 \text{ kJ} \qquad (\text{area } c)$$

The irreversibility is

$$I = T_0 \Delta S_{\text{net}} = (300 \text{ K})(0.1 \text{ kJ/K}) = 30 \text{ kJ} \qquad (\text{area } f)$$

The unavailable energy due to both heat transfer and friction is 330 kJ/kg. Note that the heat input is 1000 kJ (area $a + b + c$), the heat output is 550 kJ (area $b + c + e + f$), and the work output is 450 kJ (area $a - e - f$). The efficiency of this cycle is 0.45, and its effectiveness, as measured by the actual work to the maximum possible work, is $450/700 = 0.643$.

Example 6.12

The steady-state heat transfer by conduction through a wall is 4 kW. If the temperature of the inner and outer surfaces of the wall are 1200 and 600 K, determine the availability of the heat rate at both surfaces. What is the rate of irreversibility resulting from the heat transfer process? Assume the temperature of the environment $T_0 = 300$ K.

SOLUTION

Figure 6.19 is a schematic of the heat transfer through the wall. The availability is equal to the work that could be obtained when an amount of heat Q is supplied to a reversible power cycle. The rate of availability at the inner surface is

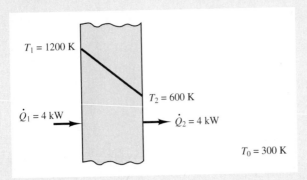

$T_1 = 1200$ K

$T_2 = 600$ K

$\dot{Q}_1 = 4$ kW

$\dot{Q}_2 = 4$ kW

$T_0 = 300$ K

Figure 6.19

$$\dot{Q}\left(1 - \frac{T_0}{T_1}\right) = 4\left(1 - \frac{300}{1200}\right) = 3 \text{ kW}$$

Similarly, the rate of availability at the outer surface is

$$\dot{Q}\left(1 - \frac{T_0}{T_2}\right) = 4\left(1 - \frac{300}{600}\right) = 2 \text{ kW}$$

The irreversibility rate due to the heat transfer process is equal to the decrease in the availability rates:

$$\dot{I} = \dot{Q}\left(1 - \frac{T_0}{T_1}\right) - \dot{Q}\left(1 - \frac{T_0}{T_2}\right) = (3 \text{ kW}) - (2 \text{ kW}) = 1 \text{ kW}$$

Example 6.13

Determine the availability for the heat transfer process of Example 6.3.

SOLUTION

Figure 6.20 shows the T-S history of the metal block as it cools from state 1 to state 0. It is possible to use the prevailing thermal potential to perform work by operating a series of Carnot engines between the block at temperature T and the environment at temperature T_0. Note that in each Carnot cycle the entropy of the block decreases while the entropy of the environment increases by exactly the same amount. When a sufficiently large number of Carnot cycles are executed, the total entropy change of the block is equal and opposite to the change in entropy of the environment.

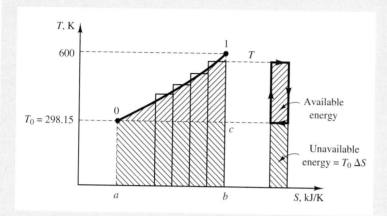

Figure 6.20

The amount of heat that can be transferred from the block as indicated in Example 6.3 is

$$Q_{1\text{-}0} = mc(T_0 - T_1)$$

This is represented by area *a-0-1-b-a* in Figure 6.20. But not all this energy is available for conversion to work. Area *0-1-c-0* is available work, whereas area *a-0-c-b-a* represents energy discarded to the environment and is unavailable to perform work.

Since the change of entropy of the environment is equal and opposite to the change of entropy of the metal block,

$$(\Delta S)_{\text{envir}} = -(\Delta S)_{\text{block}} = \int_{T_1}^{T_0} \frac{\delta Q}{T} = -m \int_{T_1}^{T_0} \frac{c\,dT}{T} = -mc \ln \frac{T_0}{T_1}$$

$$= -(10 \text{ kg})(0.9 \text{ kJ/kg K}) \ln \frac{298.15}{600}$$

$$= 6.294 \text{ kJ/K}$$

The availability or work potential for the process is

$$-W_{\text{rev}} = Q - T_0 \Delta S = (2716.65 \text{ kJ}) - (298.15 \text{ K})(6.294 \text{ kJ/K})$$
$$= 840.094 \text{ kJ}$$

Note that this is equal to the irreversibility of the process, and all the work potential is wasted.

Example 6.14

Two kg/s of water at 40°C are mixed with 3 kg/s of water at 96°C in a steady-state flow adiabatic process.

(a) What is the temperature of the resulting mixture?
(b) Is the mixing isentropic? If not, what is the rate of entropy change?
(c) What is the rate of change of the unavailable energy if the environment temperature is 25°C?

SOLUTION

Figure 6.21 is a schematic of the mixing chamber. (a) From the first law,

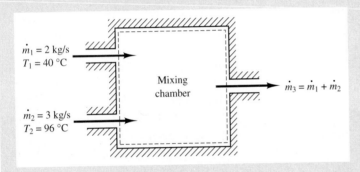

Figure 6.21

$$2h_1 + 3h_2 = 5h_3$$

where subscripts 1 and 2 refer to the two water streams, and subscript 3 refers to the mixture. Assuming the specific heat of water is constant,

$$(2 \text{ kg/s})c(40°C) + (3 \text{ kg/s})c(96°C) = (5 \text{ kg/s})c \, T_3$$

from which the temperature of the mixture is

$$T_3 = 73.6°C = 346.75 \text{ K}$$

(b) The rate of increase of entropy due to the mixing process is

$$\Delta \dot{S} = \dot{m}_1 c \ln \frac{T_3}{T_1} + \dot{m}_2 c \ln \frac{T_3}{T_2}$$

$$= (2 \text{ kg/s})(4.18 \text{ kJ/kg K}) \ln \frac{346.75}{313.15} + (3 \text{ kg/s})(4.18 \text{ kJ/kg K}) \ln \frac{346.75}{369.15} = 0.0671 \text{ kJ/K s}$$

(c) The rate of change of the unavailable energy is

$$T_0 \Delta \dot{S} = (25 + 273.15) \text{ K}(0.0671 \text{ kJ/K s}) = 20 \text{ kW}$$

Example 6.15

Water is evaporated from saturated liquid to saturated vapor as a result of heat interaction with combustion gases in a steady-state flow process. The gases are cooled from 1600 K to 600 K at constant pressure, and the vaporization temperature of the water is 498.15 K (h_{fg} = 1836.5 kJ/kg). If the environment temperature is 298.15 K and c_p of the gas is 1.0035 kJ/kg K, determine:

(a) the irreversibility of the process per kg of gas
(b) the availability of the energy per kg of gas before and after the heat interaction

SOLUTION

(a) The heat transfer per kg of gas is

$$q = c_p(1600 \text{ K} - 600 \text{ K}) = (1.0035 \text{ kJ/kg K})(1000 \text{ K})$$
$$= 1003.5 \text{ kJ/kg of gas}$$

Therefore, $(1003.5/1836.5)$ = 0.54642 kg of water is evaporated per kg of gas. The decrease in available energy, or irreversibility, per kg of gas is

$$\frac{I}{m_g} = T_0\left(\frac{m_w}{m_g}\Delta s_w + \Delta s_g\right) = (298.15 \text{ K})\left[(0.54642 \text{ kg/kg})\left(\frac{1836.5 \text{ kJ/kg K}}{498.15 \text{ K}}\right) - (1.0035 \text{ kJ/kg K})\ln\frac{1600}{600}\right]$$
$$= (298.15 \text{ K})(2.0145 - 0.9843) \text{ kJ/kg K} = 307.15 \text{ kJ/kg of gas}$$

Note that the irreversibility can be determined either from an availability balance or from $T_0\Delta S_{net}$, that is, from an entropy balance. The unavailable energy before the heat transfer process is represented by area *a-b-c-d-a* of Figure 6.22 and is equal to

$$T_0\Delta S_{gas} = (298.15 \text{ K})(0.9843 \text{ kJ/kg K}) = 293.46 \text{ kJ/kg of gas}$$

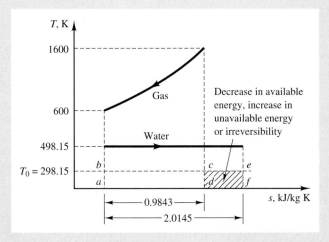

Figure 6.22

After the heat transfer process, the unavailable energy (area a-b-e-f-a) is

$$T_0 \Delta S_{\text{water}} = (298.15 \text{ K})(2.0145 \text{ kJ/kg K}) = 600.62 \text{ kJ/kg of gas}$$

Hence, the increase in the unavailable energy due to the heat interaction (area d-c-e-f-d) is 307.16 kJ/kg by difference. Note that the increase in the unavailable energy is the same as the decrease in the available energy obtained before.

(b) The availability of energy per kg of gas before the heat transfer is

$$\psi_i - \psi_0 = c_p(T_i - T_0) - T_0\left(c_p \ln \frac{T_i}{T_0}\right)$$

$$= (1.0035 \text{ kJ/kg K})[(1600 - 298.15) \text{ K}] - (298.15 \text{ K})(1.0035 \text{ kJ/kg K}) \ln \frac{1600}{298.15}$$

$$= 1306.41 - 502.69 = 803.72 \text{ kJ/kg}$$

The availability after the heat transfer is

$$\psi_e - \psi_0 = c_p(T_e - T_0) - T_0\left(c_p \ln \frac{T_e}{T_0}\right)$$

$$= (1.0035 \text{ kJ/kg K})[(600 - 298.15) \text{ K}] - (298.15 \text{ K})(1.0035 \text{ kJ/kg K}) \ln \frac{600}{298.15}$$

$$= 93.67 \text{ kJ/kg}$$

6.9 Heat Interaction with Thermal Reservoirs

In the previous sections, we have considered heat interaction between the system and environment only. Now consider a system (Figure 6.23) that exchanges heat with one or more thermal reservoirs at different temperatures T_R in addition to thermal interaction with the environment. To determine an expression for the total reversible work, we add all the reversible works due to interactions between the system and the thermal reservoirs, including the environment. This implies that heat interactions must occur with infinitesimal temperature difference at the boundary of the system, which can be accomplished by inserting a reversible heat engine (or heat pump) between the thermal reservoir and the system, as shown in Figure 6.24. The work developed by the engine is $Q_R(1 - T/T_R)$, and the heat rejected is $Q_R(T/T_R)$, where Q_R is the heat input to the engine. The heat rejected can in turn be transferred reversibly to the environment by a second

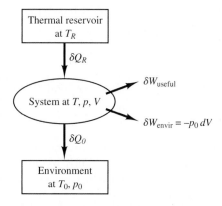

Figure 6.23   Heat interaction with thermal reservoirs and the environment.

reversible engine that develops work in the amount of $Q_R(T/T_R)(1 - T_0/T)$. The total amount of work developed by the two engines represents the increase in the reversible work due to the heat interaction with the thermal reservoir and is equal to

$$Q_R\left(1 - \frac{T}{T_R}\right) + Q_R\left(\frac{T}{T_R}\right)\left(1 - \frac{T_0}{T}\right) = Q_R\left(1 - \frac{T_0}{T_R}\right)$$

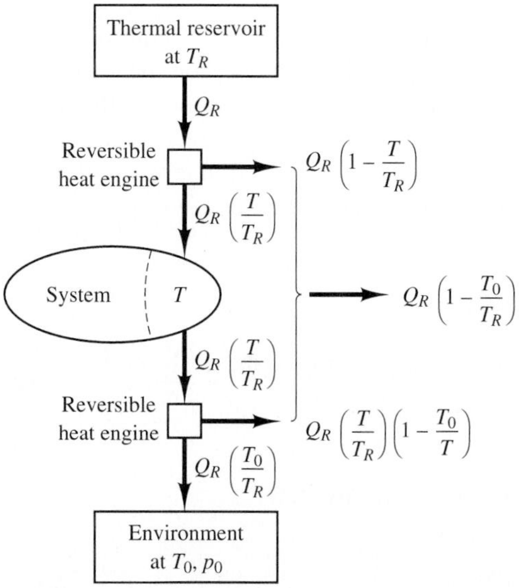

Figure 6.24   Availability of energy when a system interacts with thermal reservoirs and the environment.

which is the same as if the thermal reservoir is interacting directly with the environment. Equation (6.4) for a system can thus be modified as

$$(W_{1\text{-}2})_{\text{rev}} = (E_2 - E_1) - T_0(S_2 - S_1) + \sum_R Q_R\left(1 - \frac{T_0}{T_R}\right) \qquad (6.50)$$

where the summation sign has been included to account for interaction with several thermal reservoirs. The sign of Q_R is *relative to the thermal reservoir*.

Following the same analysis, the reversible work interaction for a control volume at steady state is modified as

$$(W_{i\text{-}e})_{\text{rev}} = \left[\sum m_e\left(h_e + \frac{V_e^2}{2} + gz_e\right)\right.$$
$$\left. - \sum m_i\left(h_i + \frac{V_i^2}{2} + gz_i\right)\right] \qquad (6.51)$$
$$- T_0(S_e - S_i) + \sum_R Q_R\left(1 - \frac{T_0}{T_R}\right)$$

The expressions of availability for a system or a control volume remain the same as before because availability depends only on the state of the system and the environment and is not affected by the interaction with other thermal reservoirs. The reversible work and irreversibility, however, depend on the interaction with thermal reservoirs. They can be expressed in terms of the availability when the effect of thermal reservoirs is included. For a system, the reversible useful work is given by

$$(W_{1\text{-}2})_{\text{rev useful}} = \Phi_2 - \Phi_1 + \sum_R Q_R\left(1 - \frac{T_0}{T_R}\right) \qquad (6.52)$$

and for a control volume at steady state,

$$(W_{i\text{-}e})_{\text{rev}} = \Psi_e - \Psi_i + \sum_R Q_R\left(1 - \frac{T_0}{T_R}\right) \qquad (6.53)$$

The irreversibility of a process is equal to the product of the temperature of the environment and the entropy generation that results in all systems participating in the process:

$$I = T_0 \left(\Delta S_{\substack{\text{sys} \\ \text{or cv}}} + \Delta S_{\text{envir}} + \sum_R \frac{Q_R}{T_R} \right) \tag{6.54}$$

where the last two terms in the parentheses account for the thermal entropy flux of the environment and the thermal reservoirs.

Example 6.16

Steam at a pressure of 2.5 MPa and a temperature of 400°C enters a turbine operating at steady state. The turbine develops 650 kJ/kg, and the steam leaves at 50 kPa and a quality of 0.95. Heat transfer between the turbine and the environment takes place at an average temperature of 200°C. Neglecting changes in kinetic and potential energies and assuming $T_0 = 25°C$ and $p_0 = 1$ atm, determine the following per unit of mass of steam:

(a) the heat transfer
(b) the decrease in the flow availability
(c) the reversible work
(d) the irreversibility

SOLUTION

Figure 6.25(a) is a schematic of the turbine and a *T-s* diagram of the expansion process.

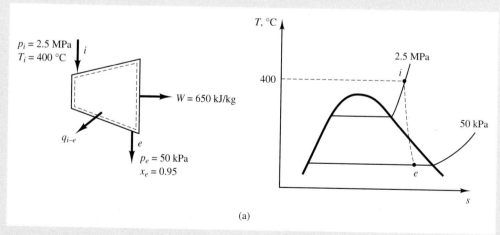

(a)

Figure 6.25(a)

(a) Neglecting kinetic and potential energies, the first law of thermodynamics is

$$q_{i\text{-}e} + w_{i\text{-}e} = h_e - h_i$$

At $p_i = 2.5$ MPa and $T_i = 400°C$,

$$h_i = 3239.3 \text{ kJ/kg} \qquad \text{and} \qquad s_i = 7.0148 \text{ kJ/kg K}$$

At $p_e = 50$ kPa and $x_e = 0.95$,

$$h_e = h_f + x_e h_{fg} = 340.49 + (0.95)(2305.4) = 2530.62 \text{ kJ/kg}$$

and

$$s_e = s_f + x_e s_{fg} = 1.0910 + (0.95)(6.5029) = 7.2688 \text{ kJ/kg K}$$
$$q_{i\text{-}e} - 650 \text{ kJ/kg} = (2530.62 - 3239.3) \text{ kJ/kg}$$

so that $q_{i\text{-}e} = -58.68$ kJ/kg

(b)

$$\begin{aligned}\psi_i - \psi_e &= (h_i - h_e) - T_0(s_i - s_e)\\ &= [(3239.3 - 2530.62) \text{ kJ/kg}] - (298.15 \text{ K})[(7.0148 - 7.2688) \text{ kJ/kg K}]\\ &= 708.68 + 75.73 = 784.41 \text{ kJ/kg}\end{aligned}$$

(c) The reversible work according to Eq. (6.53) is

$$\begin{aligned}(w_{i\text{-}e})_{\text{rev}} &= \psi_e - \psi_i + q_0\left(1 - \frac{T_0}{T}\right)\\ &= (-784.41 \text{ kJ/kg}) + (58.68 \text{ kJ/kg})\left(1 - \frac{298.15}{473.15}\right)\\ &= -762.7065 \text{ kJ/kg}\end{aligned}$$

(d) The irreversibility according to Eq. (6.54) is

$$\frac{I}{m} = T_0\left(\Delta s_{\text{cv}} + \Delta s_{\text{envir}}\right)$$
$$\therefore \frac{I}{m} = (298.15 \text{ K})\left[(7.2688 - 7.0148) \text{ kJ/kg K} + \frac{58.68 \text{ kJ/kg}}{473.15 \text{ K}}\right]$$
$$= 112.7066 \text{ kJ/kg}$$

Check: Availability loss due to heat transfer is

$$q_0\left(1 - \frac{T_0}{T}\right) = (58.68 \text{ kJ/kg})\left(1 - \frac{298.15}{473.15}\right) = 21.7035 \text{ kJ/kg}$$

$$
\begin{aligned}
\text{actual work transfer} &= 650 \text{ kJ/kg} \\
\text{irreversibility} &= 112.7066 \text{ kJ/kg} \\
\text{availability loss due to heat transfer} &= \underline{\ \ 21.7035 \text{ kJ/kg}} \\
&\ \ \ 784.4101 \text{ kJ/kg}
\end{aligned}
$$

which is equal to the decrease in the flow availability as obtained in part (b). Figure 6.25(b) shows the availability balance for the problem.

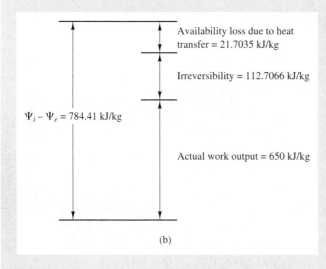

Availability loss due to heat transfer = 21.7035 kJ/kg

Irreversibility = 112.7066 kJ/kg

$\Psi_i - \Psi_e = 784.41$ kJ/kg

Actual work output = 650 kJ/kg

(b)

Figure 6.25(b)

6.10 Second-Law Efficiency

The efficiency of a thermodynamic system in converting energy from one form to another assesses the effectiveness of converting an input into an end use. The first-law efficiency indicates such conversion based on energy input whether or not it is possible to utilize this energy. The first-law efficiency (for a heat engine) is defined as the ratio of the energy utilized to the input energy:

$$\eta_1 = \frac{\text{energy used}}{\text{input energy}} = \frac{W}{Q_H} \tag{6.55}$$

where Q_H is the heat transfer from the high-temperature reservoir. The short-coming of the first-law efficiency is that only a portion of the energy input is actually available that can be effectively utilized to perform work. The first law merely asserts that energy cannot be created or destroyed. But thermodynamic inefficiencies cannot be detected by energy balance alone. Processes such as heat transfer or throttling lead to a decrease in energy quality without any energy loss. A better assessment of the energy-use effectiveness is to compare the actual performance of a device with the maximum *possible* performance.

The concept of availability is used to determine the *effectiveness* of a system to perform work. The effectiveness of a system is the ratio of the actual useful work produced to the reversible useful or *available* work. Note that in this definition, available input energy, rather than input energy, is used because available energy input represents the maximum work potential for the system to perform work. Effectiveness, also called *second-law efficiency,* can thus be considered the proper measure of the potential to do work. It is defined as

$$\eta_{II} = \frac{\text{useful energy produced}}{\text{decrease in available input energy}} = \frac{W}{W_{rev}} \quad (6.56)$$

The first-law efficiency η_I is related to the second-law efficiency η_{II} by the relation

$$\eta_I = \frac{W}{Q_H} = \left(\frac{W}{W_{rev}}\right)\left(\frac{W_{rev}}{Q_H}\right) = \eta_{II}\left(1 - \frac{T_c}{T_H}\right) = \eta_{II}\eta_{Carnot} \quad (6.57)$$

This means that the second-law efficiency is the ratio of the thermal efficiency of an actual engine to the thermal efficiency of a reversible engine operating between the same temperature limits so that

$$\eta_{II} = \eta_{I\,act}/\eta_{I\,Carnot} \quad (6.58)$$

Similarly, the second-law effectiveness for work-consuming devices such as compressors or pumps is defined as the ratio of the reversible work input to the actual work input:

$$\eta_{II} = \frac{W_{rev}}{W} \quad (6.59)$$

For refrigerators and heat pumps, the effectiveness is defined in terms of the coefficients of performance as

$$\varepsilon_{\substack{\text{refrigerator} \\ \text{or heat pump}}} = \frac{|\text{ change in availability of a thermal reservoir }|}{\text{work input}}$$

$$= \frac{\beta}{\beta_{\text{rev}}} \tag{6.60}$$

Figure 6.26 shows energy and available energy distribution for an engine and a heat pump.

In the case of energy transfer in the form of heat, the second-law efficiency is the ratio of the amount of heat used (Q_u) to the amount of heat input (Q_H). But the availability of the heat input is only $Q_H(1 - T_0/T_H)$, whereas the availability of the heat that can be utilized is $Q_u(1 - T_0/T_u)$, where T_u is the use temperature. Based on these availabilities, the effectiveness of heat interaction is

$$\eta_{\text{II}} = \frac{Q_u\left(1 - \dfrac{T_0}{T_u}\right)}{Q_H\left(1 - \dfrac{T_0}{T_H}\right)} = \eta_{\text{I}}\frac{\left(1 - \dfrac{T_0}{T_u}\right)}{\left(1 - \dfrac{T_0}{T_H}\right)} \tag{6.61}$$

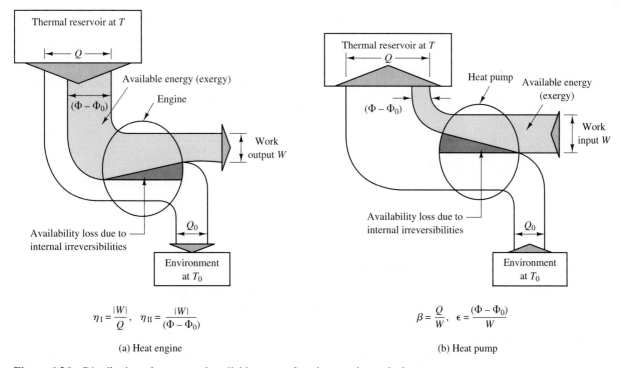

Figure 6.26   Distribution of energy and available energy for a heat engine and a heat pump.

The second-law efficiency is always less than unity; it attains the value of unity for an ideal reversible process. It is more than the first-law efficiency because of the utilization factor stipulated by the availability of energy. Note that a decrease in the thermal potential $(T_H - T_u)$ between the source and end use results in a decrease in irreversibility and a corresponding increase in η_{II}. This means that inefficient use of a fuel occurs when the temperature of the products of combustion is significantly higher than the use temperature of the system to which heat is transferred. A frequently cited example is the burning of fossil fuel to heat water. There is an excessive temperature gap between the high temperature of combustion and the low temperature of the end use. This represents an unexploited energy supply, and the fuel should be first used in high-temperature applications. The heat rejected from these applications can then be cascaded to lower-temperature applications.

In the case of work-producing or work-demanding machines such as adiabatic turbines and compressors operating at steady state, the second-law efficiency takes the form

$$\eta_{II} = \frac{W_{cv}}{\Psi_e - \Psi_i} \qquad \text{(turbine)} \qquad (6.62)$$

$$\eta_{II} = \frac{\Psi_e - \Psi_i}{W_{cv}} \qquad \text{(compressor)} \qquad (6.63)$$

where $(\Psi_e - \Psi_i)$ is the change in the availability between the inlet and exit of the control volume.

In a heat exchanger, heat is transferred from the hot fluid to the cold fluid, resulting in an increase in the availability of the cold fluid and a decrease in the availability of the hot fluid. Referring to Figure 6.27, the effectiveness of such a transaction under steady-state conditions is

$$\eta_{II} = \frac{\dot{m}_c(\psi_e - \psi_i)_c}{\dot{m}_h(\psi_i - \psi_e)_h} \qquad (6.64)$$

where subscripts c and h apply to cold and hot fluids, and subscripts i and e indicate inlet and exit conditions.

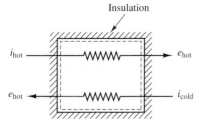

Insulation

i_{hot} ——— e_{hot}

e_{hot} ——— i_{cold}

Figure 6.27 In an adiabatic heat exchanger, the availability of the hot fluid decreases, and the availability of the cold fluid increases.

Example 6.17

Air at a pressure of 550 kPa and a temperature of 500 K enters an adiabatic turbine operating at steady state where it expands to 100 kPa and 350 K. Assuming the change in kinetic and potential energies is negligible, determine:

(a) the actual work developed
(b) the irreversibility of the process
(c) the second-law efficiency of the turbine

Assume air to be an ideal gas with constant specific heats.

SOLUTION

(a) A schematic of the turbine is shown in Figure 6.28. The actual work per unit mass is

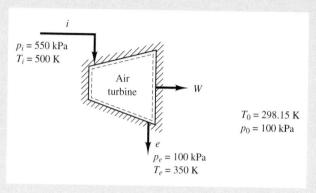

Figure 6.28

$$w_{\text{act}} = h_e - h_i = c_p(T_e - T_i)$$
$$= (1.0035 \text{ kJ/kg K})[(350 - 500) \text{ K}] = -150.525 \text{ kJ/kg}$$

(b) The irreversibility per unit mass is

$$\frac{I}{m} = T_0 \Delta s = T_0 \left[c_p \ln \frac{T_e}{T_i} - R \ln \frac{p_e}{p_i} \right]$$
$$= (298.15 \text{ K}) \left[(1.0035 \text{ kJ/kg K}) \ln \frac{350}{500} - (0.287 \text{ kJ/kg K}) \ln \frac{100}{550} \right]$$
$$= (298.15 \text{ K})[(-0.3579 + 0.4893) \text{ kJ/kg K}] = 39.17 \text{ kJ/kg}$$

(c) The second-law efficiency is

$$\eta_{\text{II}} = \frac{w_{\text{act}}}{w_{\text{rev}}} = \frac{150.525}{150.525 + 39.17} = 0.7943$$

Example 6.18

Steam at a pressure of 2 MPa and a temperature of 400°C expands in a turbine operating at steady state to 10 kPa and a quality of 93 percent. The heat transfer from the turbine to the environment is 85 kJ/kg. Neglecting the kinetic and potential energies at the inlet and exit of the turbine, determine:

(a) the actual and reversible work
(b) the irreversibility of the process

(c) the availability before and after expansion
(d) the second-law efficiency

Assume the environment is at a temperature of 25°C and a pressure of 100 kPa.

SOLUTION

A schematic of the turbine is shown in Figure 6.29.

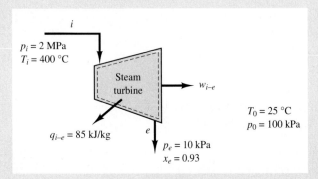

Figure 6.29

(a) The first law is

$$q_{i\text{-}e} + w_{i\text{-}e} = h_e - h_i$$

At 2 MPa and 400°C,

$$h_i = 3247.6 \text{ kJ/kg} \quad \text{and} \quad s_i = 7.1271 \text{ kJ/kg K}$$
$$h_e = h_f + x_e h_{fg} = (191.83 \text{ kJ/kg}) + (0.93)(2392.8 \text{ kJ/kg})$$
$$= 2417.13 \text{ kJ/kg}$$

Therefore,

$$(-85 \text{ kJ/kg}) + w_{i\text{-}e} = (2417.13 \text{ kJ/kg}) - (3247.6 \text{ kJ/kg})$$

from which the actual work per unit mass is $w_{i\text{-}e} = -745.47 \text{ kJ/kg}$.
   The reversible work per unit mass is

$$(w_{i\text{-}e})_{\text{rev}} = (h_e - h_i) - T_0(s_e - s_i) = (2417.13 - 3247.6) \text{ kJ/kg} - (298.15 \text{ K})[(7.6251 - 7.1271) \text{ kJ/kg K}]$$
$$= -830.47 - 148.49 = -978.96 \text{ kJ/kg}$$

(b) The irreversibility per unit mass is

$$\frac{I}{m} = (w_{i\text{-}e})_{\text{act}} - (w_{i\text{-}e})_{\text{rev}} = -745.47 \text{ kJ/kg} - (-978.96 \text{ kJ/kg}) = 233.49 \text{ kJ/kg}$$

(c)

$$\psi_i - \psi_0 = (h_i - h_0) - T_0(s_i - s_0)$$
$$h_0 = h_{f \text{ at } 25°C} = 104.89 \text{ kJ/kg}$$
$$s_0 = s_{f \text{ at } 25°C} = 0.3674 \text{ kJ/kg K}$$

Therefore,

$$\psi_i - \psi_0 = (3247.6 - 104.89) \text{ kJ/kg}$$
$$- (298.15 \text{ K})[(7.1271 - 0.3674) \text{ kJ/kg K}]$$
$$= 3142.71 - 2015.40 = 1127.31 \text{ kJ/kg}$$
$$\psi_e - \psi_0 = (h_e - h_0) - T_0(s_e - s_0)$$
$$= (2417.13 - 104.89) \text{ kJ/kg}$$
$$- (298.15 \text{ K})[(7.6251 - 0.3674) \text{ kJ/kg K}]$$
$$= 2312.24 - 2163.88 = 148.36 \text{ kJ/kg}$$

Note that

$$(w_{i\text{-}e})_{\text{rev}} = \psi_e - \psi_i = 148.36 \text{ kJ/kg} - 1127.31 \text{ kJ/kg} = -978.9 \text{ kJ/kg}$$

which is the same as before.

(d) The second-law efficiency is the ratio of the work produced to the reversible work

$$\eta_{\text{II}} = \frac{|w_{i\text{-}e}|_{\text{act}}}{|w_{i\text{-}e}|_{\text{rev}}} = \frac{745.47}{978.96} = 76.15\%$$

Example 6.19

In a heat exchanger, operating at steady flow, 0.3 kg/s of oil is cooled from 150 to 60°C using a water stream entering the heat exchanger at 30°C. It is required to compare the effectiveness of the heat exchanger when it operates in the parallel-flow mode and the counter-flow mode.

For parallel flow, the exit temperature of the water is 50°C, and for counterflow, it is 70°C. There is no heat interaction with the environment, and the pressure drop for both the oil and water streams is negligible. Assuming the specific heats of the oil and water are 2.1 and 4.18 kJ/kg K, respectively, calculate:

(a) the mass rate of flow of the water
(b) the second-law efficiency
(c) the rate of irreversibility

SOLUTION

(a) Referring to Figure 6.30, energy balance gives,

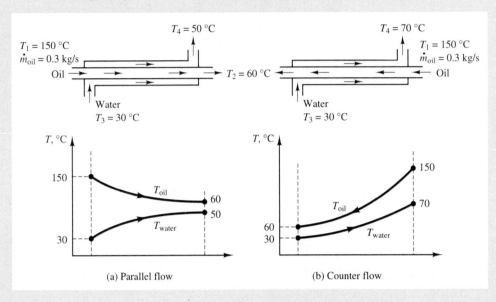

(a) Parallel flow   (b) Counter flow

Figure 6.30

$$\dot{m}_{water}(h_4 - h_3)_{water} = \dot{m}_{oil}(h_1 - h_2)$$
$$\dot{m}_w c_w (T_4 - T_3)_w = \dot{m}_o c_o (T_1 - T_2)$$

For parallel flow:

$$\dot{m}_w = \frac{(0.3 \text{ kg/s})(2.1 \text{ kJ/kg K})[(150 - 60)°\text{C}]}{(4.18 \text{ kJ/kg K})[(50 - 30)°\text{C}]} = 0.678 \text{ kg/s}$$

For counter flow:

$$\dot{m}_w = \frac{(0.3 \text{ kg/s})(2.1 \text{ kJ/kg K})[(150 - 60)°\text{C}]}{(4.18 \text{ kJ/kg K})[(70 - 30)°\text{C}]} = 0.339 \text{ kg/s}$$

(b) The decrease in the availability of the oil stream is

$$\dot{\Psi}_1 - \dot{\Psi}_2 = \dot{m}_o[h_1 - h_2 - T_0(s_1 - s_2)]$$

The increase in the availability of the water stream is

$$\dot{\Psi}_4 - \dot{\Psi}_3 = \dot{m}_w[h_4 - h_3 - T_0(s_4 - s_3)]$$

The second-law efficiency is the ratio of the increase in the availability of the cold fluid to the decrease of the availability of the hot fluid:

$$\eta_{II} = \frac{\dot{m}_w[h_4 - h_3 - T_0(s_4 - s_3)]}{\dot{m}_o[h_1 - h_2 - T_0(s_1 - s_2)]}$$

For parallel flow:

$$\eta_{II} = \frac{\left(0.678 \, \frac{kg}{s}\right)\left[\left(4.18 \, \frac{kJ}{kg \, K}\right)[(50 - 30) \, K] - (298.15 \, K)\left(4.18 \, \frac{kJ}{kg \, K}\right)\ln \frac{323.15}{303.15}\right]}{\left(0.3 \, \frac{kg}{s}\right)\left[\left(2.1 \, \frac{kJ}{kg \, K}\right)[(150 - 60) \, K] - (298.15 \, K)\left(2.1 \, \frac{kJ}{kg \, K}\right)\ln \frac{432.15}{333.15}\right]} = 0.2289 = 22.89\%$$

For counter flow:

$$\eta_{II} = \frac{\left(0.339 \, \frac{kg}{s}\right)\left[\left(4.18 \, \frac{kJ}{kg \, K}\right)[(70 - 30) \, K] - (298.15 \, K)\left(4.18 \, \frac{kJ}{kg \, K}\right)\ln \frac{343.15}{303.15}\right]}{\left(0.3 \, \frac{kg}{s}\right)\left[\left(2.1 \, \frac{kJ}{kg \, K}\right)[(150 - 60) \, K] - (298.15 \, K)\left(2.1 \, \frac{kJ}{kg \, K}\right)\ln \frac{423.15}{333.15}\right]} = 0.3665 = 36.65\%$$

(c) The rate of irreversibility is

$$\dot{I} = T_0(\dot{m}_{water}\Delta s_{water} + \dot{m}_{oil}\Delta s_{oil})$$

For parallel flow:

$$\dot{I} = (298.15 \, K)\left[(0.678 \, kg/s)(4.18 \, kJ/kg \, K)\ln \frac{323.15}{303.15} + (0.3 \, kg/s)(2.1 \, kJ/kg \, K)\ln \frac{333.15}{423.15}\right]$$

$$= (298.15 \, K)[(0.678 \, kg/s)(0.2671 \, kJ/kg \, K) + (0.3 \, kg/s)(-0.5022 \, kJ/kg \, K)]$$

$$= 9.0757 \, kW$$

For counter flow:

$$\dot{I} = (298.15 \, K)\left[(0.339 \, kg/s)(4.18 \, kJ/kg \, K)\ln \frac{343.15}{303.15} + (0.3 \, kg/s)(2.1 \, kJ/kg \, K)\ln \frac{333.15}{423.15}\right]$$

$$= (298.15 \, K)[(0.1756 - 0.1507) \, kW/K]$$

$$= 7.4239 \, kW$$

6.11 Helmholtz and Gibbs Free Energies

Systems undergoing physical and chemical reactions often exist in an environment of constant temperature. These systems exchange energy in the form of heat with the environment during the course of the reaction, and their initial and final states eventually attain thermal equilibrium with the environment. But when a system is in thermal and mechanical equilibrium with the environment, it does not necessarily mean that the system is in chemical equilibrium. Difference in composition due to chemical nonequilibrium can be exploited to obtain additional work.

Consider the heat interaction between a system and its environment at a pressure p_0 and temperature T_0. The work done on this system, in the absence of kinetic, potential, electrical, magnetic, and surface energies, is given by the first law:

$$\delta W = dU - \delta Q \tag{6.65}$$

where δW includes any p-V work interaction with the environment. If the system exchanges heat with the environment only, the change of entropy of the system, according to the second law, is

$$dS \geq \frac{\delta Q}{T_0}$$

Substituting for δQ in the first law gives

$$\delta W \geq dU - T_0\, dS \tag{6.66}$$

Integration of Eq. (6.66) for a finite process between two equilibrium states 1 and 2 gives

$$W_{1\text{-}2} \geq (U_2 - U_1) - T_0(S_2 - S_1) \tag{6.67}$$

If the initial and final temperatures of the system are equal to the temperature of the environment, Eq. (6.67) becomes

$$\begin{aligned} W_T &\geq [(U_2 - T_2 S_2) - (U_1 - T_1 S_1)]_T \\ &\geq (F_2 - F_1)_T \end{aligned} \tag{6.68}$$

where F is called the *Helmholtz function* or *Helmholtz free energy*[3] and is defined as

---

[3] The term *free energy* refers to maximum energy that can be *freed* to do work.

$$F \equiv U - TS \qquad (6.69)$$

The equality and inequality signs in Eq. (6.68) apply to a reversible process and an irreversible process, respectively. According to this equation, the work done on a system whose initial and final temperatures are equal to that of the environment and that exchanges heat with the environment only is either equal to or greater than the change in the Helmholtz function. Hence, when a system undergoes a change of state while in temperature equilibrium with the environment, the reversible work is given by

$$(W_{1\text{-}2})_{\text{rev}} = F_2 - F_1 \qquad (6.70)$$

Note that if work is done by the system, Eq. (6.68) is written as

$$-W_T \leq (F_1 - F_2)_T$$

and the work done by the system will be equal to or less than $-\Delta F_T$.

If, besides the foregoing restrictions, the initial and final volumes of the system are the same, the p-V work interaction with the environment is zero. Then Eq. (6.68) becomes

$$W_{T,V} \geq (F_2 - F_1)_{T,V} \qquad (6.71)$$

Equation (6.70), which is more restrictive than Eq. (6.68), applies to a constant-volume process in which the temperature at the beginning of the process, the temperature at the end of the process, and the temperature of the surrounding environment are all identical.

Consider next a control volume undergoing a steady-state flow process. In this case, too, heat is exchanged only with the environment, and the control volume exists in thermal equilibrium with the environment, both initially and ultimately. The *Gibbs free energy function* is defined as

$$G \equiv H - TS = U + pV - TS \qquad (6.72)$$

An equation analogous to Eq. (6.66), which applies to the nonflow process, applies to the steady-state flow process. Since the net flow work ($p_i V_i - p_e V_e$) at entry to and exit from the control volume affects the amount of work done, the following expression indicates work done in the steady-state flow process:

$$\begin{aligned} W_{\text{cv}} &\geq (U_e - U_i) - T_0(S_e - S_i) - (p_i V_i - p_e V_e) \\ &\geq (U_e + p_e V_e - T_0 S_e) - (U_i + p_i V_i - T_0 S_i) \end{aligned} \qquad (6.73)$$

where changes in kinetic and potential energies are neglected. Since the initial and final temperatures of the control volume were specified to be equal to the environment temperature,

$$
\begin{aligned}
(W_T)_{cv} &\geq (U_e + p_e V_e - T_e S_e)_T - (U_i + p_i V_i - T_i S_i)_T \\
&\geq (G_e - G_i)_T
\end{aligned}
\tag{6.74}
$$

In a steady-state flow process, when the inlet and exit temperatures are equal to that of the environment and when the control volume exchanges heat with the environment only, the work done is equal to or greater than the change in the Gibbs function. The equality and inequality signs refer to reversible and irreversible processes.

If, in addition to the preceding restrictions, the control volume is also in pressure equilibrium with the environment at the initial and final states, the flow work $(p_i V_i - p_e V_e)$ is equal to the work done by the environment on the control volume, $p_0(V_i - V_e)$. Thus the work done in this case is

$$
(W_{T,p})_{cv} \geq (G_e - G_i)_{T,p}
\tag{6.75}
$$

The expression for work done described by Eq. (6.75) also applies to a nonflow, constant-temperature, constant-pressure process. If $-p_0 \Delta V$ is subtracted from the expression given by Eq. (6.68), then

$$
\begin{aligned}
W_{T,p} &\geq [(F_e - F_i) + p_0(V_e - V_i)]_{T,p} \\
&\geq [(F_e + p_e V_e) - (F_i + p_i V_i)]_{T,p} \\
&\geq (G_e - G_i)_{T,p}
\end{aligned}
$$

Both the Gibbs and the Helmholtz free energy functions are properties of the system. They are used to establish criteria for thermodynamic equilibrium. At equilibrium, these functions are at their minimum values.

Example 6.20

Steam expands isothermally from an initial state of 350°C and 3.5 MPa to a final state of 350°C and 100 kPa. The properties of steam at the initial and final states are given in the following table:

T (°C)	p (kPa)	u (kJ/kg)	h (kJ/kg)	s (kJ/kg K)
350	3500	2835.3	3104.0	6.6579
350	100	2889.2	3176.3	8.3797

What is the reversible work developed per unit mass of steam if the expansion process is (a) nonflow, (b) steady flow? Neglect changes in kinetic and potential energies.

SOLUTION

(a) The reversible work developed in a nonflow process is the change in the Helmholtz free energy:

$$
\begin{aligned}
w_{rev} &= (f_2 - f_1) \\
&= [(u_2 - u_1) - T(s_2 - s_1)]_T \\
&= [(2889.2 - 2835.3) \text{ kJ/kg}] - (623.15 \text{ K})[8.3797 - 6.6579) \text{ kJ/kg K}] \\
&= -1019.0 \text{ kJ/kg}
\end{aligned}
$$

The reversible work in the absence of changes in kinetic and potential energies is also given by

$$
w_{rev} = -\int p \, dv
$$

No information, however, is provided on the variation of p with respect to v.

(b) The maximum work (other than p-V work) developed in a steady-state flow process is the change in the Gibbs free energy:

$$
\begin{aligned}
(w_{rev})_T &= (g_e - g_i)_T \\
&= [(h_e - h_i) - T(s_e - s_i)] \\
&= [(3176.3 - 3104.0) \text{ kJ/kg}] - (623.15 \text{ K})[8.3797 - 6.6579) \text{ kJ/kg K}] \\
&= -1000.6 \text{ kJ/kg}
\end{aligned}
$$

The foregoing result might also be obtained form the formula $\int v \, dp$ if p-v data were available.

6.12 Chemical Potential

The intensive state of a simple (one-component) system is defined by two independent intensive properties. In the case of systems where changes in composition occur, or where phases are open for mass transfer, information is needed about concentrations of the components of the system. The composition of a homogeneous phase of a multicomponent system may be defined either by indicating the number of moles of each component or by indicating the relative molar concentration of each constituent and the total number of moles. In either case, if a system has k components, the composition is not adequately defined unless k independent variables are known. This means that, in order to define the thermodynamic state of a k-component system adequately, it is necessary to specify a total of $(k + 2)$ parameters. For example, the internal energy of a homogeneous phase of a system with k components is defined by

$$U = U(S, V, n_1, n_2, ..., n_k) \tag{6.76}$$

The differential of U is

$$dU = \left(\frac{\partial U}{\partial S}\right)_{V,n_i} dS + \left(\frac{\partial U}{\partial V}\right)_{S,n_i} dV + \sum_{i=1}^{i=k} \left(\frac{\partial U}{\partial n_i}\right)_{S,V,n_j} dn_i \tag{6.77}$$

where $n_1, n_2, ..., n_k$ are the number of moles of the components of the system, and where the use of the subscript j implies that only one of the components is varied at a time. If the composition of the system does not change, the last term of the equation drops out so that the differential of internal energy becomes

$$dU = \left(\frac{\partial U}{\partial S}\right)_{V,n_i} dS + \left(\frac{\partial U}{\partial V}\right)_{S,n_i} dV$$

The combined first and second laws, however, give dU as

$$dU = T\,dS - p\,dV$$

The coefficients of the independent variables S and V can be equated, yielding

$$\left(\frac{\partial U}{\partial S}\right)_{V,n_i} = T \tag{6.78}$$

and

$$\left(\frac{\partial U}{\partial V}\right)_{S,n_i} = -p \tag{6.79}$$

The chemical potential μ of component i is defined as

$$\mu_i \equiv \left(\frac{\partial U}{\partial n_i}\right)_{S,V,n_j} \tag{6.80}$$

When changes in composition are considered in the statements of the first and second laws, the change in internal energy, according to Eq. (6.77), becomes

$$dU = T\,dS - p\,dV + \sum_{i=1}^{i=k} \mu_i\,dn_i \tag{6.81}$$

Equation (6.81) shows the effect of the reversible heat interaction ($T\,dS$), the reversible work interaction ($-p\,dV$), and the reversible mass transfer or change in composition ($\sum_i \mu_i dn_i$) on the internal energy of the system.

Note that T, p, and μ are driving potentials since a difference in T gives rise to heat transfer, a difference in p tends to produce a change in volume, and a difference in μ gives rise to mass transfer.

In a similar manner, the Gibbs function ($G = H - TS$) is given by

$$G = G(T, p, n_1, n_2,\ldots, n_k) \qquad (6.82)$$

Changes in the free energy function result from changes in pressure, temperature, and composition. Hence, the differential of the Gibbs function is

$$dG = \left(\frac{\partial G}{\partial p}\right)_{T,n_i} dp + \left(\frac{\partial G}{\partial T}\right)_{p,n_i} dT + \sum_{i=1}^{i=k} \left(\frac{\partial G}{\partial n_i}\right)_{p,T,n_j} dn_i \qquad (6.83)$$

But the Gibbs free energy can also be expressed as

$$dG = V\,dp - S\,dT + \sum_{i=1}^{i=k} \mu_i\,dn_i \qquad (6.84)$$

Equating the coefficients of dp, dT, and dn_i yields

$$\left(\frac{\partial G}{\partial p}\right)_{T,n_i} = V \qquad (6.85)$$

and

$$\left(\frac{\partial G}{\partial T}\right)_{p,n_i} = -S \qquad (6.86)$$

$$\left(\frac{\partial G}{\partial n_i}\right)_{p,T,n_j} = \mu_i \qquad (6.87)$$

The chemical potential μ_i, as indicated by Eqs. (6.80) and (6.87), can be expressed in terms of either the internal energy or the Gibbs function. The functions $(\partial U/\partial n_i)_{S,V,n_j}$ and $(\partial G/\partial n_i)_{p,T,n_j}$ can be shown to be equivalent by substituting $(U + pV - TS)$ for G in Eq. (6.84), then performing the following operations:

$$d(U + pV - TS) = V\,dp - S\,dT + \sum_{i=1}^{i=k} \mu_i\,dn_i$$

Therefore,

$$dU + p \, dV + V \, dp - T \, dS - S \, dT = V \, dp - S \, dT + \sum_{i=1}^{i=k} \mu_i \, dn_i$$

And finally,

$$dU = T \, dS - p \, dV + \sum_{i=1}^{i=k} \mu_i \, dn_i$$

which is Eq. (6.81).

The chemical potential μ measures the tendency of a substance to change composition or to give rise to diffusion. It is used as an index of phase or chemical equilibrium in the same manner as temperature and pressure are used as indices of thermal and mechanical equilibria. It is an intensive property and may be expressed in terms of independent properties, such as

$$\mu_i = \mu_i(U, V, n_1, n_2, \ldots, n_k) \tag{6.88}$$

Analogous to U and G, the differentials of enthalpy and Helmholtz functions can be written in the forms

$$dH = T \, dS + V \, dp + \sum_i \mu_i \, dn_i \tag{6.89}$$

$$dF = -S \, dT - p \, dV + \sum_i \mu_i \, dn_i \tag{6.90}$$

The chemical potential μ can be expressed in terms of the preceding functions as

$$\mu_i = \left(\frac{\partial H}{\partial n_i}\right)_{S,p,n_j} = \left(\frac{\partial F}{\partial n_i}\right)_{T,V,n_j} = \left(\frac{\partial U}{\partial n_i}\right)_{V,S,n_j} = \left(\frac{\partial G}{\partial n_i}\right)_{T,p,n_j} \tag{6.91}$$

The chemical potential of a species is the partial differential of the internal energy, enthalpy, Helmholtz function, or Gibbs function with respect to the number of moles of that species, subject to the conditions indicated by subscripts in Eq. (6.91). It accounts for the effects of the variation in the amount of a chemical species on the energy functions. The term *potential* arises from the analogous situation in which a force is exerted as a result of a force potential. Chemical potential is an intensive property of a phase that depends on its composition. It can be expressed on the basis of energy per mole or per unit mass.

6.13 Criteria of Equilibrium

When unrestrained thermodynamic potentials exist between a system and its surroundings, a flow of energy will occur. Energy may be expressed, as shown in Chapter 2, as the product of an intensive property and the differential of an extensive property as a capacity term. Conversely, if a system is in equilibrium, the intensive properties and thermodynamic potentials of both the system and the surroundings must be in complete balance. Thermodynamic equilibrium also depends on constraints that prevent variations in the state of a system. A high-pressure gas confined in a rigid container, for example, is in a pressure equilibrium with the surroundings, for the constant-volume container represents a constraint imposed on the variation of the system. It is obvious, of course, that if the container is removed, the system will not remain in equilibrium with the surroundings. A state of stable equilibrium is attained when spontaneous changes cannot occur, within the limits of constraint imposed on the system.

The equilibrium state of a system may be identified by the means of one or more of several criteria. The criteria may involve extensive properties, such as entropy, internal energy, enthalpy, Gibbs function, and Helmholtz function, or intensive properties, such as pressure, temperature, and chemical potential. Although the Gibbs function is the most important criterion in chemical thermodynamics, other potentials also help describe aspects of the equilibrium state of a system, subject to certain conditions to be discussed. Chemical equilibrium is treated in more detail in Chapter 10.

According to the second law of thermodynamics, the entropy of an isolated system ($dE = 0$ and $dV = 0$) always increases or remains constant in the absence of irreversibilities, but never decreases, so that

$$dS_{\text{isolated}} \geq 0 \qquad (6.92)$$

This means that when an isolated system undergoes a real process, only those states that have a higher entropy than the initial state are possible. The second law further indicates that the isolated system never returns to the same state, but continuously proceeds in the same direction until it attains a state at which its entropy is a maximum. At this state, no further change is possible, and the system is then in a state of stable equilibrium. Thus the state of stable equilibrium of an isolated system is attained by irreversible processes and is characterized by maximum entropy, consistent with the initial internal energy and the initial volume of the system. An additional criterion of the equilibrium state of an isolated system is that, under conditions of constant entropy and volume, the internal energy is at a minimum.

Suppose an isolated system is divided into two parts, a and b. If energy is transferred from one part to the other as they approach equilibrium, the total entropy change is the sum of the entropy changes of the two parts, or

$$dS = dS_a + dS_b$$

Expressing S in terms of U and V, as given by the combined first and second laws, then

$$dS = \frac{1}{T_a}(dU_a + p_a\, dV_a) + \frac{1}{T_b}(dU_b + p_b\, dV_b)$$

Since the system is isolated

$$dU_a = -dU_b \qquad \text{and} \qquad dV_a = -dV_b$$

therefore

$$dS = \left(\frac{1}{T_a} - \frac{1}{T_b}\right)dU_a + \left(\frac{p_a}{T_a} - \frac{p_b}{T_b}\right)dV_a$$

But since $dS = 0$ when the system is in equilibrium, and since U and V are independent of each other,

$$T_a = T_b \qquad \text{and} \qquad p_a = p_b$$

The foregoing result indicates that both the temperature and pressure must be uniform throughout the system, which, in effect, are the conditions for thermal and mechanical equilibrium.

Example 6.21

An isolated system is divided by a stationary heat-conducting wall into two equal compartments, as shown in Figure 6.31. Each compartment contains 1 kg of air. Initially, the temperature of the air in compartment A is 700 K and in compartment B is 300 K. Plot the change of entropy as a function of the temperature difference between the two compartments as the system proceeds to the equilibrium state. Assume air to be an ideal gas with constant specific heats.

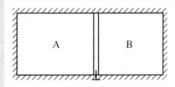

Figure 6.31

SOLUTION

For the isolated system, the first law indicates that $\Delta E = 0$ so that

$$mc_v(T_f - T_A) + mc_v(T_f - T_B) = 0$$

from which

$$T_f = \frac{T_A + T_B}{2} = \frac{(300 \text{ K}) + (700 \text{ K})}{2} = 500 \text{ K}$$

The entropy change for the system when it reaches the equilibrium state is

$$\Delta S = mc_v \ln \frac{T_f}{T_A} + mc_v \ln \frac{T_f}{T_B} = mc_v \frac{T_f^2}{T_A T_B}$$

$$= (1 \text{ kg})(0.7165 \text{ kJ/kg K}) \ln \frac{(500)^2}{(700)(300)} = 0.1249 \text{ kJ/K}$$

Denoting the temperatures of compartments A and B during the intermediate steps toward equilibrium to be T_A' and T_B', then

$$T = \frac{T_A' + T_B'}{2}$$

The change of entropy relative to the final equilibrium state is

$$\Delta S = mc_v \ln \frac{T^2}{T_A' T_B'} = (1 \text{ kg})(0.7165 \text{ kJ/kg K}) \ln \frac{(500)^2}{T_A' T_B'}$$

Values of ΔS, based on the preceding equation, are shown in the following table, and Figure 6.32 is a plot of ΔS versus ΔT:

T_A' (K)	T_B' (K)	ΔT (K)	ΔS (kJ/K)
700	300	400	0.1249
675	325	350	0.0936
650	350	300	0.0676
625	400	225	0.0462
600	425	175	0.0292
575	450	125	0.0163
550	450	100	0.0072
525	475	50	0.0018
500	500	0	0

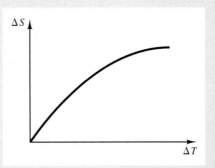

Figure 6.32

The state of maximum entropy is a convenient criterion for equilibrium in systems that are isolated from their surroundings. When the system is not isolated, entropy changes of both the system and its surrounding must be considered. Under these conditions, other criteria for equilibrium are found to be much more useful, as discussed next.

Consider a multicomponent system defined by internal energy U, volume V, and molar concentrations. A change in the entropy of the system, according to Eq. (6.81), is

$$dS = \frac{1}{T}\left(dU + p\,dV - \sum_i \mu_i\,dn_i\right)$$

The overall change in entropy of the universe, that is, system plus surroundings, is

$$dS_{sys} + dS_{surr} \geq 0$$

If the temperature of the surroundings remains constant, the entropy change of the surroundings is given by

$$dS_{surr} = -\frac{\delta Q}{T_{surr}} = -\frac{dU + p\,dV}{T_{surr}}$$

Hence,

$$dS_{sys} - \frac{dU + p\,dV}{T_{surr}} \geq 0$$

In an isothermal, isobaric system, this relationship, when combined with Eq. (6.81), leads to

$$\sum_i \mu_i\,dn_i \leq 0 \tag{6.93}$$

A multicomponent system, then, may proceed toward the equilibrium state through changes in composition that cause a decrease in the chemical potential of the system.

The relationship between the Helmholtz function and work interaction, as expressed in Eq. (6.71), can be indicated in differential form:

$$dF_{T,V} \leq \delta W_{T,V} \tag{6.94}$$

The change in the Helmholtz function is less than, or (under reversible conditions) equal to, the work done on the system. If there is no work interaction and the system undergoes an isothermal process, then

$$dF_{T,V} \leq 0 \tag{6.95}$$

The value of the Helmholtz function then decreases as the state of equilibrium is approached. At equilibrium, subject to constant-volume, constant-temperature conditions, the infinitesimal change in the Helmholtz function is zero:

$$dF_{T,V} = 0 \qquad (6.96)$$

At this state, the system is in equilibrium with its surroundings.

Conditions for equilibrium in an isothermal, isobaric system can be determined by following a similar procedure. The differential of the Gibbs function, based on Eq. (6.75), is

$$dG_{T,p} \leq \delta W_{T,p} \qquad (6.97)$$

and if $\delta W_{T,p} = 0$, then

$$dG_{T,p} \leq 0 \qquad (6.98)$$

The Gibbs function of a system at constant temperature and pressure then decreases as the state of equilibrium is approached. The minimum value of G that is possible under the imposed temperature and pressure conditions of the system corresponds to the state of equilibrium. This means that a system not in equilibrium will tend to change irreversibly, thereby lowering its free energy until no further change is possible. Thus infinitesimal changes in a system that is in chemical equilibrium at constant temperature and pressure produce no change in the Gibbs free energy:

$$dG_{T,p} = 0 \qquad (6.99)$$

This condition applies equally well to multicomponent, multiphase systems.

Note that the decrease in the free energy of a system as it changes spontaneously (irreversibly) is consistent with physical intuition. For example, the change in the free energy in an isothermal process is

$$\Delta G = \Delta H - T \Delta S$$

Common experience indicates that systems tend to proceed spontaneously from a high to a low energy state (ΔH is negative). At the same time, spontaneous processes result in an increase in entropy (ΔS is positive). Note that the more negative the ΔH and the more positive the ΔS, the more negative ΔG is. Hence, a system, in seeking equilibrium, tends to move toward minimum enthalpy and maximum entropy, and the Gibbs function provides a convenient criterion for equilibrium because it assesses these two tendencies simultaneously.

Equation (6.99) implies certain restrictions in addition to the specified constant-temperature and constant-pressure conditions. Consider, for example, a system consisting of a pure liquid and its vapor in equilibrium. The temperature and pressure of both phases are identical since they must be in thermal and pressure equilibrium. Since the Gibbs function of the mixture depends on the proportions of liquid and vapor, the change in the Gibbs free energy of the mixture is a function of the change in quantity of each constituent:

$$dG_{T,p} = \left(\frac{\partial G}{\partial n_v}\right)_{n_l,T,p} dn_v + \left(\frac{\partial G}{\partial n_l}\right)_{n_v,T,p} dn_l$$

The changes in quantity, however, are identical:

$$dn_v + dn_l = 0$$

Solving the two preceding equations simultaneously and noting that $dG_{T,p} = 0$ at equilibrium, then

$$\left(\frac{\partial G}{\partial n_v}\right)_{n_l,T,p} = \left(\frac{\partial G}{\partial n_l}\right)_{n_v,T,p}$$

or

$$\mu_v = \mu_l \qquad\qquad (6.100)$$

As will be shown in Chapter 9, when two or more phases are in equilibrium, the chemical potential μ of each component has the same value in every phase.

Similar considerations can be applied in analyzing the role of other criteria of equilibrium, such as internal energy and enthalpy. The following table summarizes the thermodynamic potentials used as criteria for equilibrium, together with the corresponding constraints imposed on the variation of the system. The equilibrium state corresponds to the minimum value of the potential.

Properties Held Constant	*Thermodynamic Potential*
S, V	*U*
S, p	*H*
T, V	*F*
T, p	*G*

6.14 Types of Equilibrium

The thermodynamic potential that controls equilibrium in a system depends on the particular constraints that exist. This section discusses several types of equilibrium using the Gibbs function as the typical criterion for equilibrium. When a finite change of state occurs in a system at constant temperature and pressure, the Gibbs free energy may increase, remain unchanged, or decrease. This trend of change of the Gibbs function establishes four types of equilibrium: (a) stable, (b) neutral, (c) unstable, and (d) metastable equilibrium which are defined as follows:

(a) A system is in a state of *stable equilibrium* if for any finite variation of the system at constant temperature and pressure the Gibbs free energy increases. This means that the stable equilibrium state of a system at constant temperature and pressure corresponds to the minimum value of Gibbs function so that

$$\delta G_{T,p} > 0 \tag{6.101}$$

The symbol δ in the foregoing equation represents a small but finite change. Analogous criteria of stable equilibrium, corresponding to thermodynamic conditions indicated by subscripts, are

$$\delta F_{T,V} > 0 \tag{6.102}$$

$$\delta U_{S,V} > 0 \tag{6.103}$$

$$\delta H_{S,p} > 0 \tag{6.104}$$

(b) A system is in a state of *neutral equilibrium* when the thermodynamic criterion of equilibrium remains at a constant value for all possible variations of finite magnitudes. For a system at constant temperature and pressure, the criterion of neutral equilibrium is

$$\delta G_{p,T} = 0 \tag{6.105}$$

This means that if the system can exist in two or more different states without any change in its free energy, the system is in neutral equilibrium. The change in state of the system may take place at an infinitesimally slow rate, and it involves no change in the value of the free energy. Analogous criteria for neutral equilibrium of other systems are

$$\delta F_{T,V} = 0 \tag{6.106}$$

$$\delta U_{S,V} = 0 \tag{6.107}$$

$$\delta H_{S,p} = 0 \tag{6.108}$$

(c) A system is in a state of *unstable equilibrium* when the appropriate thermodynamic criterion of equilibrium is neither a minimum nor a constant value for all possible variations of the system. In a system at constant temperature and pressure, the Gibbs free energy, for at least one variation, can be reduced:

$$\delta G_{p,T} < 0 \tag{6.109}$$

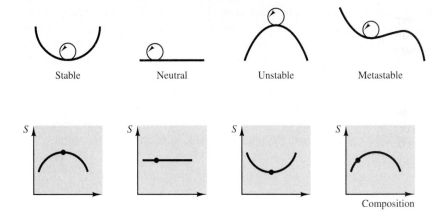

Figure 6.33　Types of equilibrium.

Criteria for unstable equilibrium in other systems are

$$\delta F_{T,V} < 0 \tag{6.110}$$

$$\delta U_{S,V} < 0 \tag{6.111}$$

$$\delta H_{S,p} < 0 \tag{6.112}$$

(d) A system is in a state of *metastable equilibrium* if it is stable to small but not to large disturbances. An example of metastable equilibrium is a mixture of oxygen and hydrogen. Since a spark may start a chemical reaction, such a mixture is not in its most stable thermodynamic equilibrium even though, in the absence of a spark, it appears to be stable. Figure 6.33 shows the different types of equilibrium, together with their mechanical analogies. As shown, entropy has been used as the criterion for equilibrium.

6.15　Summary

The expression for work per unit mass when a system undergoes a reversible process between two equilibrium states while exchanging heat with the environment only is

$$(w_{1\text{-}2})_{\text{rev}} = (u_2 - u_1) - T_0(s_2 - s_1) + \frac{V_2^2 - V_1^2}{2} + g(z_2 - z_1)$$

and for a steady-state flow process

$$(w_{i\text{-}e})_{\text{rev}} = (h_e - h_i) - T_0(s_e - s_i) + \frac{V_e^2 - V_i^2}{2} + g(z_e - z_i)$$

Irreversible processes result in lost opportunities for producing work (or in increasing the required work input), and the difference between reversible and actual work is the irreversibility:

$$I = W_{\text{act}} - W_{\text{rev}} = T_0(\Delta S)_{\text{total}}$$

The maximum useful work output (or minimum useful work input) when a system or a fluid undergoes a noncyclic process from a specified state 1 to a second state 0 in equilibrium with an environment at T_0 and p_0 is called the availability. For a system, the availability at state 1 is

$$\varphi_1 - \varphi_0 = -(w_{1\text{-}0})_{\substack{\text{rev} \\ \text{useful}}}$$

$$= (u_1 - u_0) + p_0(v_1 - v_0) - T_0(s_1 - s_0) + \frac{V_1^2}{2} + g(z_1 - z_0)$$

and for the steady flow of a fluid at a state i,

$$\psi_i - \psi_0 = -(w_{i\text{-}0})_{\text{rev}} = (h_i - h_0) - T_0(s_i - s_0) + \frac{V_i^2}{2} + g(z_i - z_0)$$

Availability of thermal energy from a thermal reservoir at a temperature T_R is

$$Q\left(1 - \frac{T_0}{T_R}\right)$$

This is equivalent to the work output of a Carnot engine operating between the thermal reservoir and the environment.

For a system which exchanges heat between the environment and thermal reservoirs, the maximum useful work is

$$(w)_{\substack{\text{rev} \\ \text{useful}}} = \Delta e + p_0 \Delta v - T_0 \Delta s + \sum_R q_R\left(1 - \frac{T_0}{T_R}\right)$$

$$= (\varphi_2 - \varphi_1) + \sum_R q_R\left(1 - \frac{T_0}{T_R}\right)$$

where the sign of q_R is determined relative to the thermal reservoir and for a control volume at steady state,

$$(w)_{\text{rev}} = \Delta h + \Delta KE + \Delta PE - T_0 \Delta s + \sum q_R\left(1 - \frac{T_0}{T_R}\right)$$

$$= \psi_e - \psi_i + \sum q_R\left(1 - \frac{T_0}{T_R}\right)$$

The second-law efficiency for a work-producing device is the ratio of the work output to the work potential input or

$$\eta_{II} = \frac{\text{useful work output}}{\text{decrease in available input energy}} = \frac{W}{W_{rev}}$$

for a work-consuming device (such as compressors),

$$\eta_{II} = \frac{W_{rev}}{W}$$

The Helmholtz and Gibbs functions are defined as

$$F = U - TS$$

and

$$G = H - TS$$

Changes in these two functions can be related to work interaction under certain conditions

$$W_{T,V} \geq (F_2 - F_1)_{T,V}$$

and

$$(W_{T,p})_{cv} \geq (G_e - G_i)_{T,p}$$

The chemical potential is an index of chemical equilibrium. It is the molar Gibbs function and is defined as

$$\mu_i = \left(\frac{\partial G}{\partial n_i}\right)_{T,\, p,\, n_j}$$

Complete equilibrium is attained when the thermal, pressure, and chemical potentials are zero.

Problems

In the following problems, assume the temperature and pressure of the environment to be 25°C and 1 standard atmosphere, unless otherwise stated.

6.1   (a) What is the reversible work that can be obtained when an amount of heat equal to 100 kJ is transferred from a thermal reservoir at 500 K to an environment at

100 K? (b) Suppose that the same quantity of heat flows by conduction from the reservoir at 500 K to a second reservoir at 250 K. What is the reversible work that can now be obtained when the same amount of heat is withdrawn from the second reservoir? The temperature of the environment remains the same.

6.2 What is the reversible work that can be obtained from 1 kg of air at 1000 K and atmospheric pressure?

6.3 A body of mass m has a constant-pressure specific heat c_p and is in equilibrium with the atmosphere at temperature T_a. What is the minimum possible work required to cool this body to a lower temperature T_b?

6.4 A vessel with a volume of 0.1 m^3 contains air initially at 700 K and 300 kPa. The temperature of the air decreased to 500 K owing to heat interaction with the environment, which is at 25°C and 100 kPa. Determine the reversible work and the irreversibility of this process.

6.5 An aluminum block of mass 2 kg and temperature 200°C is brought into thermal contact with a copper block of mass 6 kg and temperature −10°C. The total system of the two blocks is isolated. Determine the final temperature and the irreversibility when the two blocks come to thermal equilibrium. The specific heat of aluminum and copper are 0.95 and 0.39 kJ/kg K, respectively.

6.6 An 8-kg aluminum block initially at 200°C is dropped into an insulated tank containing 1.5 kg of water initially at 30°C. Determine (a) the final equilibrium temperature, (b) the irreversibility of the process.

6.7 An ideal gas that has a molar mass of 28.013 kg/kg-mol and a constant-pressure specific heat of 1.0416 kJ/kg K flows steadily through a control volume. The inlet conditions are 660 kPa and 370°C, and the exit conditions are 130 kPa and 94°C. If heat is exchanged only with the environment, what is the maximum work that can be obtained from this process?

6.8 An ideal gas having a constant-pressure specific heat of 1.6 kJ/kg K and a molar mass of 30 kg/kg-mol, initially at a pressure of 6.7 MPa and a temperature of 425°C, undergoes a steady flow process until it reaches a pressure of 1.3 MPa and a temperature of 150°C. If the environment is at 100 kPa and 25°C, find the reversible work that could be obtained in this process.

6.9 One kilogram of nitrogen inside a rigid vessel is heated from 25 to 250°C by means of a heat pump. Heat is transferred to the pump from an environment at 25°C. Determine the minimum work required for the heating process, and compare it with the energy absorbed by the nitrogen.

6.10 Air at atmospheric pressure is heated in a heat exchanger from 30°C to 70°C by condensing steam at 150 kPa in a steady-state flow process. The volumetric flow rate of the air is 0.8 m^3/s, and the steam enters the heat exchanger at 90 percent quality and leaves as saturated water. Determine the rate of irreversibility of the process. Assume $T_0 = 25$°C.

6.11 Steam at a pressure of 2 MPa and a temperature of 500°C expands adiabatically in a steam turbine operating at steady state to a pressure of 10 kPa. If the isentropic efficiency is 95 percent, determine:
(a) the actual work and the reversible adiabatic work
(b) the reversible work and the irreversibility per kg
when the steam expands between the end states
Assume the temperature of the environment is 25°C.

6.12 Steam enters an adiabatic turbine operating at steady state at 5 MPa and 550°C with a negligible velocity. It exhausts at a pressure of 100 kPa with a quality of 99 percent and a velocity 140 m/s. Calculate (a) the turbine efficiency, (b) the irreversibility, and (c) the reversible work per unit mass of steam flowing through the turbine. Assume $T_0 = 25$°C.

6.13 Air enters an adiabatic nozzle operating at steady state with negligible velocity. At the inlet, the pressure is 180 kPa, and the temperature is 65°C. The mass flow rate is 0.15 kg/s, and the exit pressure is 100 kPa. If the exit velocity is 300 m/s, determine (a) the isentropic efficiency of the nozzle, (b) the exit temperature, (c) the irreversibility of the process, (d) the exit area of the nozzle. Assume $T_0 = 25$°C and $p_0 = 100$ kPa.

6.14   Calculate the irreversibility when air is throttled adiabatically from 2 MPa to 100 kPa in a steady-state flow process. Neglect changes in kinetic energy.

6.15   Air expands adiabatically in a turbine operating at steady state from an initial state $T_1 = 810$ K, $p_1 = 1.38$ MPa, to a final state $T_2 = 480$ K and a pressure p_2. If the expansion had been adiabatic reversible to the same final pressure, the final temperature would have been 450 K. Calculate (a) the final pressure, (b) the change in entropy, (c) the actual work, (d) the reversible work, (e) the irreversibility. Assume the air to be an ideal gas with constant specific heats.

6.16   Air at a pressure of 750 kPa enters a throttling valve operating at a steady state. If the pressure downstream of the valve is 140 kPa, determine the specific irreversibility of the process. Assume air to be an ideal gas with constant specific heats, and neglect changes in kinetic and potential energy.

6.17   The flow rate of an air compressor operating at steady state is 0.3 kg/s. The following data give the temperature and pressure of the air at the inlet and delivery. Assuming adiabatic compression, determine (a) the actual power, (b) the reversible power, (c) the isentropic power, (d) the isentropic efficiency, (e) the irreversibility. Assume air to be an ideal gas with constant specific heats and the temperature of the environment is 300 K. Data:

	p (kPa)	T (K)
Inlet	100	290
Delivery	200	370

6.18   Nitrogen gas expands adiabatically in a steady flow process from an initial state of 750 kPa and 500 K to a final state of 120 kPa and 350 K. Calculate (a) the actual work, (b) the isentropic work, (c) the reversible work, (d) the irreversibility of the expansion. Assume nitrogen to be an ideal gas ($c_p = 1.0416$ kJ/kg K and $c_v = 0.7448$ kJ/kg K) and $T_0 = 300$ K.

6.19   Air at ambient conditions of 100 kPa and 10°C enters the compressor of a gas turbine and exits at 1.2

MPa. The compressor is insulated and has an isentropic efficiency of 85 percent. Determine (a) the work of compression, (b) the reversible work for the same change of state, and (c) the irreversibility of this process per kilogram of air. Assume steady-state operation.

6.20   Air at a pressure of 1 MPa and 600 K expands adiabatically through a turbine operating at steady state to 214 kPa and 500 K. Neglecting changes in the kinetic and potential energies of the air at the inlet and exit of the turbine, determine (a) the actual work, (b) the reversible work, (c) the isentropic work, (d) the irreversibility. Consider air to be an ideal gas with constant specific heats, and assume the environment temperature $T_0 = 298$ K.

6.21   Air is compressed isothermally in a steady-state flow process from a pressure of 100 kPa and at a temperature of 20°C to a pressure of 620 kPa. If the mass flow rate is 0.3 kg/s, what is the minimum power required? What is the entropy change? If the process took place isentropically, what would then be the power required?

6.22   Air at a temperature of 30°C and a pressure of 100 kPa enters an adiabatic compressor operating at steady state and exits at 180 kPa. The air then enters a heat exchanger, where its temperature drops 40°C. If the isentropic efficiency of the compressor is 80 percent, what is the irreversibility of the compressor alone and the compressor–heat exchanger combination?

6.23   Air enters an insulated nozzle operating at steady state at a temperature of 1000 K, a pressure of 450 kPa, and a negligible velocity. The air expands in the nozzle to a pressure of 160 kPa and a temperature of 830 K. Determine the exit velocity and the irreversibility of the process.

6.24   Two streams of H_2O are mixed in an insulated control volume in a steady-state flow process. Using the data shown in Figure 6.34, determine the rate of irreversibility of the process.

6.25   A tank with a volume of 1 m³ contains air at a pressure of 600 kPa and a temperature of 400 K. What is

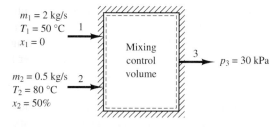

$m_1 = 2$ kg/s
$T_1 = 50\,°C$ 1
$x_1 = 0$

Mixing control volume   3 $\longrightarrow p_3 = 30$ kPa

$m_2 = 0.5$ kg/s   2
$T_2 = 80\,°C$
$x_2 = 50\%$

Figure 6.34

the availability of air? Assume $T_0 = 298$ K, $p_0 = 100$ kPa, and air to be an ideal gas with $c_p = 1.0035$ kJ/kg K and $c_v = 0.7165$ kJ/kg K. Neglect the effects of motion and gravity.

6.26 A rigid vessel contains 1 kg of air at a pressure of 800 kPa and a temperature of 600 K. The air is cooled by heat interaction with the environment until it reaches 25°C. Determine:
(a) the final pressure in the vessel
(b) the availability of the air at the initial state
(c) the reversible work interaction in the cooling process
(d) the irreversibility of the process

6.27 A rigid vessel of volume of 0.5 m^3 initially contains a water–vapor mixture of 0.6 MPa.
(a) If the quality of mixture is 40 percent, calculate the mass of mixture.
(b) If the pressure in the vessel is raised to 0.9 MPa by the transfer of heat, what will be the mass of the vapor and the mass of the liquid?
(c) What is the availability at the initial state? Assume the dead state to be at 25°C and 100 kPa.

6.28 One kg of oxygen and 1 kg of hydrogen at the same temperature and pressure are allowed to mix adiabatically without change in total volume. Find the decrease in availability and the irreversibility of the process.

6.29 A Carnot cycle operates between two thermal reservoirs at 550 K and 390 K. If the increase in entropy during the heat interaction process is 3.6 kJ/K, calculate the available and the unavailable energies. If the temperature of the cold reservoir were reduced to 340 K, other factors being the same, what would be the decrease in

the unavailable energy? (Assume the environment temperature to be 298 K.)

6.30 Air is confined in a piston–cylinder device, which is completely immersed in a fluid bath. The fluid bath is considered to represent an environment at a constant temperature of 27°C and a constant pressure of 150 kPa. The gas system can interact with the environment only.
What is the maximum possible work that could be obtained in a process when the gas goes from an initial state of 260°C and 3.5 MPa to a final condition of 150°C and 0.7 MPa? What is the availability in the initial state?

6.31 Calculate the availability per kg of air at 1100 K and 200 kPa with respect to an environment at 25°C and 101.325 kPa. What will be the maximum efficiency of an engine using the energy of this gas and rejecting part of it to the environment?

6.32 Find the change in the availability of 1 kg of oxygen at atmospheric pressure and 10°C when it expands isothermally to double its volume.

6.33 A tank with a volume of 0.5 m^3 contains air at 800 kPa and 500 K. Determine the availability of the air in the tank. What would be the change in the availability if the air expands freely and adiabatically to double its volume? Assume air to be an ideal gas with constant specific heats and $T_0 = 25°C$ and $p_0 = 100$ kPa.

6.34 A power plant utilizes geothermal energy as an energy source. If the temperature and pressure of the hot water are 180°C and 2 MPa, determine the availability of energy per unit mass if used in a steady-state flow process.

6.35 A 0.6 m^3 rigid vessel contains air at a pressure of 600 kPa and a temperature of 180°C. There is an exchange of energy in the form of heat between the vessel and the environment till the temperature of the air in the vessel drops to 25°C. Determine the availability of the air at the initial and final states and the irreversibility of the process.

6.36 Six hundred kJ of heat are exchanged between two constant-temperature reservoirs at temperatures of 1000 K and 400 K. Calculate:
(a) the total change of entropy due to the heat transfer process
(b) the available energy before and after the heat transfer
(c) the increase in the unavailable energy
Show parts (a) and (b) on a *T-s* diagram.

6.37 The increase of entropy due to stirring water at constant pressure is 0.48 kJ/kg K. If the water is initially at 28°C and the process is adiabatic, calculate (a) the irreversibility of the process, (b) the final temperature of the water, (c) the change in the available energy. Assume the specific heat of water $c = 4.184$ kJ/kg K and the temperature of the environment $T_0 = 10°C$.

6.38 One kilogram of water is heated at constant pressure from 20°C to 170°C. ($c = 4.184$ kJ/kg K). Assuming steady flow, determine:
(a) the heat interaction
(b) the unavailable energy based upon an environment temperature of $-10°C$
Show the process on a *T-s* diagram, and indicate the various energies designated.

6.39 Air is heated at constant pressure from 4°C to 280°C. If the temperature of the environment is 4°C, what percentage of the heat transfer to the air represents an increase in the available energy of the air?

6.40 Air at 20°C and 101.3 kPa is compressed isentropically to a temperature of 150°C and then heated at constant pressure to 700°C. The air then expands isentropically to the initial pressure of 101.3 kPa in a turbine, after which it is permitted to cool in the atmosphere to complete the cycle. Assuming steady state, determine:
(a) the energy supplied during the heating process from 150°C to 700°C
(b) how much of this energy is available
(c) how much of this energy is unavailable
(d) the thermal efficiency of this ideal cycle
(e) the power developed if the mass flow rate is 0.1 kg/s

6.41 A cylinder of an internal combustion engine contains gaseous combustion products at a pressure of 550 kPa and a temperature of 800°C. The volume of the cylinder is 0.8 liter. Calculate the availability of the products just before the exhaust valve opens. Assume $T_0 = 25°C$, $p_0 = 100$ kPa, and the properties of the products to be the same as those of air.

6.42 A metal bar of mass 1.2 kg initially at a temperature of 950°C is quenched by immersing it in a tank containing 15 kg of water at 20°C. If the heat transfer from the tank contents is negligible, determine the decrease in availability. Assume $c_{metal} = 0.4$ kJ/kg K, $c_w = 4.18$ kJ/kg K, and $T_0 = 25°C$.

6.43 Energy in the form of heat is transferred across a wall in a steady-state operation. The temperature of wall surfaces are 1000 K and 400 K. If 10 kJ of heat are transferred by conduction through the wall, calculate (a) the availability of heat energy at each surface of the wall, (b) the irreversibility of the process.

6.44 A rigid insulated vessel contains 0.6 kg of air initially at a pressure of 1 atm and a temperature of 300 K. The air is stirred by a paddle wheel until its temperature reaches 500 K. Determine (a) the work, (b) the change in availability, and (c) the irreversibility of the process.

6.45 An ideal gas at 680 kPa and 540°C expands freely to atmospheric pressure. Calculate the decrease in availability ($R = 0.287$ kJ/kg K).

6.46 One kilogram of a fluid is warmed and vaporized at constant pressure. If the initial temperature of the fluid is 27°C, find (a) the temperature of vaporization, (b) the entropy change during warming, (c) the total entropy change, (d) the total heat transfer, (e) the increase in the available energy.
Specific heat of fluid = 2.4 kJ/kg K
Enthalpy for vaporization = 1540 kJ/kg
Entropy change during vaporization = 1.9 kJ/kg K

6.47 Water is heated at constant pressure from 25°C to 100°C, vaporized by supplying energy in the amount of 2442.3 kJ/kg of water at 100°C and then superheated from 100°C to 150°C. The specific heat of the liquid is 4.18 kJ/kg K, and the specific heat of the vapor is 1.87 kJ/kg K. The water receives the energy from furnace gas that is cooled from 815°C to 480°C (c_p of the gas 1.0035 kJ/kg K).

(a) Determine the entropy change of the water per kg.
(b) Determine the entropy change of the furnace gas per kg of water.
(c) What is the net increase in unavailable energy per kg of water if $T_0 = 0°C$?

6.48 A vessel contains air at a pressure of 800 kPa and a temperature of 600 K. The air is cooled by heat interaction with the environment until it reaches 25°C. Determine:

(a) the availability of the air at the initial state
(b) the reversible work interaction in the cooling process
(c) the irreversibility of the process

6.49 Show that the irreversibility of a power cycle is given by

$$I_{cycle} = |Q_h|(\eta_{Carnot} - \eta_{actual})$$

where $|Q_h|$ is the heat interaction with the high-temperature thermal reservoir.

6.50 Calculate the availability per kg of ice at 0°C if the environment is at a temperature of 25°C and a pressure of 1 atm.

6.51 A rigid tank of volume 1 m³ contains air initially at 25°C and 1 atm. The temperature of the air is raised to 120°C by two methods:

(a) stirring with a paddle wheel
(b) heat transfer from a 200°C thermal reservoir

Determine the irreversibility in each method, and indicate which one is better.

6.52 A piston–cylinder device contains 0.05 kg of steam initially at 1 MPa and 300°C. The steam expands to a final state of 200 kPa and 150°C. If the heat losses from the system to the environment are 2 kJ and assuming the environment to be at $T_0 = 25°C$ and $p_0 = 100$ kPa, determine:

(a) the availability of the steam at the initial and the final states
(b) the reversible work
(c) the irreversibility of the process

6.53 In a combustion process, 50 kW are generated when a fuel burns at a temperature of 1500 K. Determine the maximum power that can be developed from this energy.

6.54 An air-to-air heat exchanger operates at steady state. The mass flow rate of the cold fluid is 1 kg/s. It enters at 230°C and 300 kPa and leaves at 270°C and 250 kPa. The mass flow rate of the hot fluid is 0.8 kg/s. It enters at 500°C and 150 kPa and leaves at 100 kPa. Assuming no heat interaction with the environment, calculate:

(a) the rate of the irreversibility in kW
(b) the availability of the hot fluid when it leaves the heat exchanger in kW

6.55 Air at a temperature of 1700 K and 800 kPa expands in a piston–cylinder assembly according to the relation $pv^{1.5} = C$. The air is cooled by heat transfer to the environment ($T_0 = 25°C$) to a final temperature of 1000 K. Calculate the reversible work developed per unit mass of air. What is the heat interaction and entropy generation for the process?

6.56 Determine the minimum power required to compress air at the rate of 1.2 m³/s from 100 kPa and 25°C to 320 kPa and 100°C. Heat interaction takes place with the environment at 25°C. Neglect changes in kinetic and potential energies and assume steady flow process.

6.57 Gas from the combustion chamber of a gas turbine plant enters the turbine at 1000 K and 500 kPa and expands adiabatically to 150 kPa. Assuming steady-state operation, calculate:

(a) the availability of the gas after isentropic expansion in the turbine
(b) the availability of the gas after expanding with an isentropic efficiency of 80 percent
(c) the irreversibility of the expansion in (b)

6.58 Gas from the combustion chamber of a gas turbine plant operating at steady state enters the turbine at 800°C and 500 kPa and expands to 100 kPa. Assuming the gas can be treated as air, and the environment is at 25°C and 0.1 MPa, calculate:

(a) the availability of the gas entering the turbine
(b) the availability of the gas after isentropic expansion

(c) the availability of the gas after expanding with an expansion efficiency of 80 percent

(d) the irreversibility of the expansion in (b)

(e) the irreversibility of the expansion in (c)

6.59　One kg/s of air is heated in a steady-state operation by hot gases in the heat exchanger shown in Figure 6.35. If the temperature of the environment $T_0 = 300$ K, determine:

(a) the rate of the irreversibility of the process per kg of air

(b) the rate of availability of the heated air

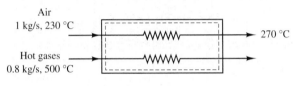

Figure 6.35

Assume the hot gases to have the same properties as air and all pressures are atmospheric.

6.60　In the heat exchanger shown in Figure 6.36, products of combustion are used to heat air. From the data given, calculate:

(a) the initial and final availability of the products

(b) the rate of the irreversibility of the process

(c) if the process is made reversible through the use of a reversible heat engine, what would be the final temperature of the air?

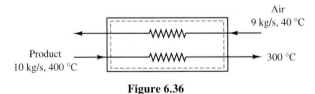

Figure 6.36

$$c_{P(air)} = 1.004 \text{ kJ/kg K}$$

$$c_{P(prod)} = 1.1 \text{ kJ/kg K}$$

All pressures are atmospheric (0.1 MPa).

6.61　Air at an inlet temperature of 1100 K and an inlet pressure of 450 kPa flows adiabatically through a nozzle

operating at a steady state. The inlet velocity is negligible, and the exit pressure is 230 kPa. If the exit velocity is 500 m/s, determine (a) the availability of energy at the exit, (b) the irreversibility of the process. Assume the temperature and pressure of the environment are at 298.15 K and 100 kPa.

6.62　Air enters an adiabatic nozzle operating at steady state with negligible velocity. At the inlet, the pressure is 180 kPa, and the temperature is 65°C. The mass flow rate is 0.15 kg/s, and the exit pressure is 100 kPa. If the exit velocity is 300 m/s, determine (a) the exit temperature, (b) the isentropic efficiency of the nozzle, (c) the rate of irreversibility of the process. Assume $T_0 = 10$°C and $p_0 = 100$ kPa.

6.63　Air enters a compressor at 100 kPa and 300 K and exits at 700 kPa and 550 K. The heat transfer to the environment is 35 kJ/kg. Assuming steady-state operation and $T_0 = 300$ K and $p_0 = 100$ kPa, determine: (a) the actual work, (b) the reversible work, (c) the irreversibility, (d) the availability at the exit.

6.64　Air flows through an adiabatic compressor operating at steady state at the rate of 2 kg/s. The inlet conditions are 100 kPa and 310 K, and the exit conditions are 700 kPa and 600 K. Neglecting changes in kinetic and potential energy, determine:

(a) the minimum power required

(b) the reversible power required

(c) the rate of irreversibility

(d) the availability of energy at the exit state

Assume air to be an ideal gas with constant specific heats.

6.65　Air at a pressure of 1 MPa and 1300 K expands through a turbine operating at steady state to 100 kPa and 780 K. Neglecting changes in kinetic and potential energy of the air at the inlet and exit of the turbine, determine:

(a) the availability before and after expansion

(b) the actual and the maximum work developed

(c) the isentropic work

Consider air to be an ideal gas with constant specific heats, and the environment temperature and pressure are 298.15 K and 0.1 MPa.

6.66 In a heat exchanger operating at steady state, products of combustion are used to heat air. The flow does not experience any significant drop in pressure. The following data are given:

	$\dot{m}$ (kg/s)	T_{in} (K)	T_{out} (K)	c_p (kJ/kg K)
Products	5	670	500	1.100
Air	4	320	—	1.004

(a) What is the availability of the products at the initial state?
(b) What is the rate of the irreversibility?
(c) Indicate the available energy and irreversibility on a *T-s* diagram.

6.67 Air at a pressure of 1 MPa and a temperature of 400 K expands adiabatically through a turbine operating at steady state to 0.15 MPa and 320 K. Neglecting the kinetic and potential energies of the air at inlet and exit of the turbine, determine:
(a) the availability before and after expansion
(b) the actual and the reversible work
(c) the isentropic work
(d) the irreversibility of the process
 Consider air to be an ideal gas with constant specific heats, and the environment temperature and pressure are 298 K and 0.1 MPa.

6.68 A rigid vessel contains 1.0 kg of air initially at a pressure of 300 kPa and a temperature of 40°C. Heat is transferred to the vessel from a thermal reservoir at a temperature of 800°C. If the temperature of the air is increased to 400°C, determine: (a) the change in the availability of the air, (b) the reversible work, (c) the irreversibility of the process. Assume $T_0 = 300$ K and $p_0 = 1$ atm.

6.69 A thermal reservoir at a temperature of 300°C is used to supply energy in the form of heat to vaporize water at 1 MPa from the saturated liquid state to the saturated vapor state. If the environment temperature and pressure are $T_0 = 300$ K and $p_0 = 0.1$ MPa, determine: (a) the heat transfer, (b) the actual work, (c) the reversible work, (d) the irreversibility of the process.

6.70 Steam at 2.5 MPa and 400°C enters a turbine operating at steady state and exits as saturated vapor at 0.2 MPa. The heat transfer to the surroundings is 25 kJ/kg. Neglecting kinetic and potential energy changes determine:
(a) the irreversibility
(b) the availability of the steam at the inlet and exit conditions
(c) the second-law efficiency

6.71 For the turbine of Problem 6.20, determine the second-law efficiency.

6.72 For the compressor of Problem 6.63, determine the second-law efficiency.

6.73 Air expands in a turbine operating at steady state from 400 kPa and 500°C to 100 kPa and 340°C. The heat transfer to the environment is 12 kJ/kg of air. If the net change in kinetic and potential energy is negligible, determine:
(a) the decrease in the available energy
(b) the second-law efficiency of the turbine

6.74 Air is compressed adiabatically from 100 kPa, 5°C to 300 kPa at the rate of 0.2 kg/s. The power input to the compressor is 23 kW. Assuming steady-state operation and neglecting changes in kinetic and potential energies, determine:
(a) the rate of increase of irreversibility
(b) the second-law efficiency of the compressor

6.75 Prove that the chemical potential of an ideal gas is given by

$$\mu = \bar{R}T \ln p + f(T)$$

where $f(T)$ is a function of temperature only. What is the effect of pressure on μ when the temperature is maintained constant?

6.76 Prove that the chemical potential of a component i in a mixture of ideal gases is equal to the Gibbs function per mole of i evaluated at the temperature of the mixture and the partial pressure of the component i in the mixture so that

$$\mu_i = g_i(T, p_i)$$

6.77   Show that

$$U = F - T\left(\frac{\partial F}{\partial T}\right)_v = G - T\left(\frac{\partial G}{\partial T}\right)_p - p\left(\frac{\partial G}{\partial p}\right)_T$$

6.78   Two equal blocks of aluminium at temperatures 300 and 500 K are brought together in an isolated system. If the specific heat of aluminium is 0.9 kJ/kg K, plot the entropy change as a function of the temperature difference between the two blocks as the isolated system proceeds to equilibrium.

C6.1   Modify the program of the Problem C5.3 to include calculation of the reversible work if the environment is at 25°C and 101.325 kPa.

C6.2   Modify the program of the Problem C5.4 to include calculation of the irreversibility of the process. Consider the environment to be at 25°C and 101.325 kPa.

C6.3   Two kilograms of oxygen gas contained in a rigid compressed gas bottle are to be heated from an initial condition of 400 K and 105 kPa to a final condition of 2100 K. Write an interactive computer program to calculate the irreversibility of the process and the amount of heat required. The environment temperature is 25°C, and the pressure is 101.325 kPa.

C6.4   Determine the change in irreversibility in Problem C6.3 if the specific heat of the oxygen is to vary with temperature. Note that the specific heat as a function of temperature is given in the Appendix. Perhaps the user could be given a choice of a constant or variable specific heat.

C6.5   Design a problem and write an interactive program that you think best describes or illustrates the subject of irreversibility. A graphical animation of some sort is preferred.

Chapter 7

Analysis of Thermodynamic Cycles

7.1 Introduction

In Chapter 5, it was shown that the Carnot cycle yields the best possible performance of all cycles operating between a high-temperature heat source and a low-temperature heat sink. In addition to being reversible, the Carnot cycle imposes no limitations on the behavior of the working substance.

This chapter is concerned with the analysis of actual cycles. The Carnot cycle serves as the starting point of an analysis, but then models are devised to represent more closely what realistically happens in actual cycles. Actual cycles may approach standard or idealized cycles as an upper limit of performance; however, the performance of the idealized cycle is still inferior to that of the Carnot cycle.

Five basic types of idealized cycles will be considered: (a) the reciprocating compressor cycle; (b) the Rankine vapor-power cycle; (c) the reciprocating internal combustion engine cycle; (d) the Brayton gas turbine cycle; (e) the Rankine vapor-compression cycle. These cycles will be discussed from the standpoint of both the thermodynamic analysis and the principles of operation. The treatment will not extend to a discussion of physical equipment. Advanced energy systems and direct-energy conversion systems are presented in Chapter 11.

7.2 Ideal Processes

Idealized reversible cycles employ ideal processes, such as the reversible adiabatic process, the reversible isothermal process, and the constant-pressure process. The performance obtained in many actual processes is almost the same as that of ideal processes. An actual process becomes essentially an adiabatic process when the system is thermally insulated or when the actual process is so rapid that heat interaction between the system and its surroundings is negligible. Similarly, reversible isothermal processes may almost be attained when the system is in contact with a thermal reservoir that allows heat interaction to take place with very small temperature differences.

An important process encountered in ideal model cycles is the reversible adiabatic (isentropic) process. The compression of water in a pump or the expansion of steam in a turbine is idealized by an isentropic process. Actual processes deviate from isentropic processes because internal irreversibilities, such as fluid or mechanical friction, inevitably occur during compression or expansion. A real adiabatic process, therefore, is accompanied by an increase in entropy. In a preliminary analysis, the pressure drop due to fluid flow is usually neglected, and emphasis is placed on the deviation of a compression process or an expansion process from the ideal process.

In a compression process, the actual work done on a system or control volume is more than the ideal work. The difference is the work necessary to overcome the internal irreversibilities in the actual process. For similar reasons, the actual work done by a system in an expansion process is less than the ideal work. The deviation of an actual process from the isentropic process is indicated by *isentropic efficiency* defined previously in Chapter 5. For a pump or a compressor, the isentropic compression efficiency is

$$\eta_{\text{isentropic}} = \frac{W_{\text{isentropic}}}{W_{\text{actual}}} \qquad (\text{ pump or compressor})$$

Similarly, the isentropic expansion efficiency of a turbine or an engine is expressed as

$$\eta_{\text{isentropic}} = \frac{W_{\text{actual}}}{W_{\text{isentropic}}} \qquad (\text{turbine})$$

Note that although ideal cycles are internally reversible, they are not totally reversible because of heat interaction with the surroundings across a finite temperature difference. These external irreversibilities reduce the performance of ideal cycles below comparable Carnot cycles operating between the same temperature limits.

7.3 The Reciprocating Compressor Cycle

The reciprocating compressor operates on a two-stroke open cycle. Gas, admitted into the cylinder during the suction stroke, is compressed during the compression stroke and then delivered to a receiver. An inlet valve and a discharge valve admit and discharge the gas from the cylinder. To allow for the operation of these valves, a clearance volume between the piston and cylinder head is necessary. Figure 7.1 shows a *p-V* diagram of an air-standard cycle. During the suction stroke 0-1, air is drawn through an inlet valve into the cylinder at a pressure slightly lower than atmospheric. At the end of the suction stroke, the inlet valve is closed, and the air is compressed to a pressure slightly higher than the discharge pressure. The discharge valve is then opened, and the compressed air is delivered to a receiver. At the end of the delivery stroke (state 3), the discharge valve is closed, leaving some residual high-pressure gas in the cylinder. This gas expands, as the piston moves on the suction stroke of the next cycle, until the pressure within the cylinder equals the suction pressure (state 4). The inlet valve then opens, and a fresh charge of gas enters the cylinder during process 4-1. This type of cycle is called an *open* cycle in contrast to a closed cycle, in which the same fluid is constantly recirculated. The clearance volume results in induction

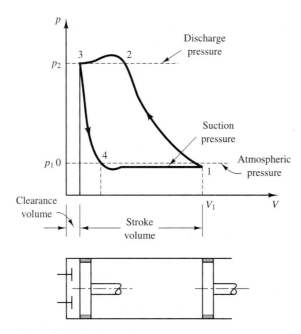

Figure 7.1   Reciprocating compressor.

of the gas only during part of the suction stroke. The compression process 1-2 and the expansion process 3-4 are assumed polytropic, whereas processes 4-1 and 2-3 are constant-pressure processes.

Several indexes are used to evaluate the performance of an air compressor. For example, the aspirating capacity of an air compressor is indicated by the *volumetric efficiency* defined as the mass of gas actually compressed per cycle divided by the ideal mass that can be compressed:

$$\eta_{vol} = \frac{(V_1 - V_4)/v_1}{(V_1 - V_3)/v_s} \tag{7.1}$$

where v_1 is the specific volume of air at the end of the suction stroke, and v_s is the specific volume in the suction line. The volumetric efficiency can be expressed in terms of the *clearance ratio* c, defined as the clearance volume divided by the stroke volume or piston displacement:

$$\eta_{vol} = \left[\frac{(V_1 - V_3) + (V_3 - V_4)}{V_1 - V_3} \right] \left(\frac{v_s}{v_1} \right)$$
$$= \left[1 + c - \frac{V_3}{V_1 - V_3} \frac{V_4}{V_3} \right] \left(\frac{v_s}{v_1} \right) \tag{7.2}$$

But if the expansion from 3 to 4 is assumed polytropic, so that

$$\left(\frac{V_4}{V_3} \right) = \left(\frac{p_3}{p_4} \right)^{1/n}$$

the volumetric efficiency can then be written as

$$\eta_{vol} = \left[1 + c - c \left(\frac{p_3}{p_4} \right)^{1/n} \right] \left(\frac{v_s}{v_1} \right) \tag{7.3}$$

Note that when $c = 0$ and $v_1 = v_s$, $\eta_{vol} = 1.0$. Typically, the value of c is between 3 and 10 percent. An increase in the clearance volume results in a decrease in the mass that can be drawn into the compressor and a corresponding decrease in η_{vol}. Also an increase in the pressure ratio decreases η_{vol}.

An expression for the work input to a compressor can be determined by considering an internally reversible steady-state flow process. The work done in the absence of kinetic and potential energy changes is

$$W = \int V \, dp$$

The work is represented by the area to the left of process 1-2 of Figure 7.1. Since $pV^n = C$,

$$W = \frac{n}{n-1}(p_2 V_2 - p_1 V_1) = \frac{n p_1 V_1}{n-1}\left[\left(\frac{p_2}{p_1}\right)^{(n-1)/n} - 1\right] \qquad (7.4)$$

Note that the clearance volume has no effect on the work input to the compressor per unit mass of flow. The clearance volume, however, affects the work input per cycle because the trapped gas at state 3 expands doing work on the piston until state 4 is reached.

The expansion work is small compared to the compression work, and the error involved in assuming the same exponent for the two polytropic processes is very small. The cycle work for a gas compressor with clearance is

$$W_{cycle} = \frac{n}{n-1}p_1 V_1\left[\left(\frac{p_2}{p_1}\right)^{(n-1)/n} - 1\right] - \frac{n}{n-1}p_4 V_4\left[\left(\frac{p_3}{p_4}\right)^{(n-1)/n} - 1\right]$$

Noting that $p_3 = p_2$ and $p_4 = p_1$, then

$$W_{cycle} = \frac{n}{n-1}p_1(V_1 - V_4)\left[\left(\frac{p_2}{p_1}\right)^{(n-1)/n} - 1\right] \qquad (7.5)$$

Another index of evaluating the performance of a compressor is by comparing the work input to the compressor to the work input if the compression took place isothermally. Figure 7.2 shows three processes of compression: isentropic, polytropic, and isothermal. Since the work input per cycle is represented by the area

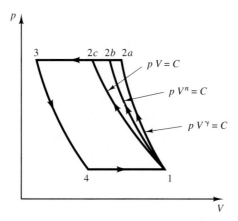

Figure 7.2 Three possible compression processes.

1-2-3-4-1, isothermal compression indicates minimum work requirement. The *isothermal efficiency* is defined as

$$\eta_{\text{isoth}} = \frac{\text{isothermal work of compression}}{\text{actual work of compression}}$$

Since isothermal work of compression without clearance is

$$W_{1\text{-}2} = \int_{1}^{2} V \, dp = p_1 V_1 \ln \frac{p_2}{p_1}$$

the isothermal efficiency is

$$\eta_{\text{isoth}} = \frac{p_1 V_1 \ln \dfrac{p_2}{p_1}}{W_{\text{actual}}} \tag{7.6}$$

In order to reduce the work input to the compressor, and thereby increase the efficiency of the compressor, isothermal compression is recommended. This is, however, impractical because a substantial amount of energy transfer as heat must be rejected during the course of compression in order to maintain constant temperature. This difficulty is partially surmounted when a gas is to be compressed through a large pressure ratio, which is usually carried out in two or more stages. This allows intercooling between stages, as indicated by process *a-b* of Figure 7.3. The compression process is along 1-*a*-*b*-*c* rather than 1-*a*-2, which indicates a reduction in the work input equal to area *a-b-c-2-a*. An opti-

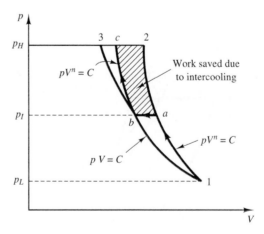

Figure 7.3   Effect of intercooling in a two-stage compressor.

mum value of the intermediate pressure can be obtained as follows. The work interaction during process l-*a*-*b*-*c* is

$$W = \frac{n}{n-1}(p_I V_a - p_L V_1) + \frac{n}{n-1}(p_H V_c - p_I V_b)$$

$$= \frac{n p_L V_1}{n-1}\left[\left(\frac{p_I}{p_L}\right)^{(n-1)/n} - 1\right] + \frac{n p_I V_b}{n-1}\left[\left(\frac{p_H}{p_I}\right)^{(n-1)/n} - 1\right]$$

but $p_L V_1 = p_I V_b$ (isothermal), hence

$$W = \frac{n p_I V_1}{n-1}\left[\left(\frac{p_I}{p_L}\right)^{(n-1)/n} + \left(\frac{p_H}{p_I}\right)^{(n-1)/n} - 2\right] \qquad (7.7)$$

Minimum work can be obtained by differentiating this expression with respect to p_I and equating the result to 0, noting that p_L, p_H, V_1, and n are constants. Differentiation gives

$$\frac{dW}{dp_I} = \frac{n p_1 V_1}{n-1}\left[\frac{n-1}{n p_L}\left(\frac{p_I}{p_L}\right)^{-1/n} - \frac{n-1}{n p_I}\left(\frac{p_H}{p_I}\right)^{1-1/n}\right] = 0$$

hence

$$\left(\frac{p_I}{p_L}\right)^{-1/n} = \frac{p_L p_H}{p_I^2}\left(\frac{p_H}{p_I}\right)^{-1/n}$$

or

$$\left(\frac{p_I^2}{p_L p_H}\right)^{-1/n} = \frac{p_L p_H}{p_I^2}$$

so that

$$p_I^{2-2/n} = p_L p_H (p_L p_H)^{-1/n}$$

from which

$$\frac{p_I}{p_L} = \frac{p_H}{p_I}$$

or

$$p_I = \sqrt{p_H p_L} \tag{7.8}$$

If the gas were compressed in three stages, it can be shown that the intermediate pressures for minimum work are

$$p_{I,1} = \sqrt[3]{p_H^2 p_L} \qquad \text{and} \qquad p_{I,2} = \sqrt[3]{p_H p_L^2}$$

Example 7.1

Atmospheric air at a temperature of 20°C and a pressure of 100 kPa is compressed in a single-acting reciprocating compressor to a pressure of 800 kPa. The piston displacement is 522 cm³, the clearance volume is 20 cm³, and the compressor operates at 550 rpm. Assuming steady-state operation and the compression is polytropic with $n = 1.25$, determine:

(a) the volumetric efficiency
(b) the mass of air compressed per hour
(c) the work input to the compressor per unit mass of flow
(d) the cycle power required if the clearance volume is not neglected
(e) the isothermal efficiency neglecting clearance volume

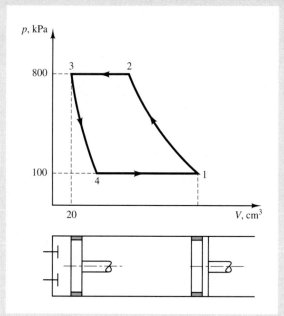

Figure 7.4

SOLUTION

(a) A p-V diagram of the cycle is shown in Figure 7.4. Assuming no drop in pressure during suction,

$$\eta_{\text{vol}} = \left[1 + c - c\left(\frac{p_3}{p_4}\right)^{1/n}\right]$$

$$= \left[1 + \frac{20}{522} - \frac{20}{522}\left(\frac{800}{100}\right)^{1/1.25}\right] = 83.61\%$$

(b) The mass of air compressed per unit time is

$$\dot{m} = (\text{piston displacement})(\text{number of cycles per unit time}) \times \frac{\eta_{\text{vol}}}{v_1}$$

where

$$v_1 = \frac{RT_1}{p_1} = \frac{(0.287 \text{ kJ/kg K})(293.15 \text{ K})}{100 \text{ kPa}} = 0.8413 \text{ m}^3\text{/kg}$$

Noting that the number of cycles is equal to the number of revolutions,

$$\dot{m} = \frac{(522 \times 10^{-6} \text{ m}^3)(550 \text{ rpm})(0.8361)}{0.8413 \text{ m}^3\text{/kg}} = 0.2853 \text{ kg/min}$$

$$= 17.12 \text{ kg/hr}$$

(c) The work input to the compressor per unit mass of flow is

$$\frac{n p_1 v_1}{n - 1}\left[\left(\frac{p_2}{p_1}\right)^{(n-1)/n} - 1\right] = \frac{1.25[(100 \text{ kPa})(0.8413 \text{ m}^3\text{/kg})}{1.25 - 1}\left[\left(\frac{800}{100}\right)^{(1.25-1)/1.25} - 1\right]$$

$$= 216.936 \text{ kJ/kg}$$

(d) If the clearance volume is not neglected, the work input per cycle is given by Eq. (7.5). Noting that $(V_1 - V_4) = \eta_{\text{vol}}(V_1 - V_3)$, the cyclic power is

$$\dot{W}_{\text{cyclic}} = (N)(W_{\text{cycle}})$$

$$= \left(\frac{550 \text{ rpm}}{60 \text{ s/min}}\right)\left(\frac{1.25}{0.25}\right)(100 \text{ kPa})(0.8361)[(542 - 20) \times 10^{-6} \text{ m}^3] \times \left[\left(\frac{800}{100}\right)^{0.25/1.25} - 1\right] = 1.0318 \text{ kW}$$

(e) Neglecting the clearance volume, the isothermal efficiency is

$$\eta_{\text{isoth}} = \frac{p_1 v_1 \ln \dfrac{p_2}{p_1}}{w} = \frac{(100 \text{ kPa})(0.8413 \text{ m}^3\text{/kg}) \ln \frac{800}{100}}{216.9362 \text{ kJ/kg}} = 0.8064$$

7.4 The Rankine Vapor-Power Cycle

In vapor power plants, the working fluid is alternately vaporized and condensed as it recirculates in a closed cycle. A steam power plant is a typical example in which water is used as a working fluid. Besides being plentiful and low in cost, water has a relatively large value of enthalpy of vaporization at ordinary steam generator pressures. This results in a reduced mass rate of flow for a given power plant output.

A standard vapor-power cycle that excludes internal irreversibilities is called the *ideal Rankine cycle.* The components of the Rankine cycle are shown in Figure 7.5. Although the present analysis deals with water as a working substance, it is equally applicable to any fluid. In the vapor-power cycle, water at low pressure and low temperature is compressed reversibly and adiabatically to the boiler pressure by the feed pump. Heat is then transferred to the water in the boiler at constant pressure. The heat interaction raises the temperature of the water to the saturation

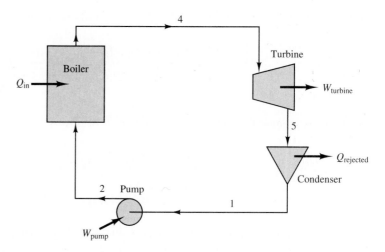

Figure 7.5   Components of the Rankine cycle.

temperature. After evaporation, further transfer of heat superheats the vapor to a higher temperature. The superheated vapor is then allowed to expand reversibly and adiabatically in a steam engine or turbine to the condenser pressure. Condensation occurs at constant pressure in the condenser as heat is rejected from the hot vapors to cold water. The condensed steam then enters the boiler feed pump, and the cycle is repeated. The *T-s* and *h-s* diagrams of the Rankine cycle are shown in Figure 7.6. In cycle 1-2-3-4-5-1, there is no superheating of the vapor, whereas in cycle 1-2-3-4-6-7-1, superheating does occur. Note that heat is transferred at constant pressure in the steam generator and the condenser, and the flow is isentropic in the pump and the turbine. Since constant-pressure and constant-temperature lines are identical in the liquid–vapor region, a Carnot power cycle can be repre-

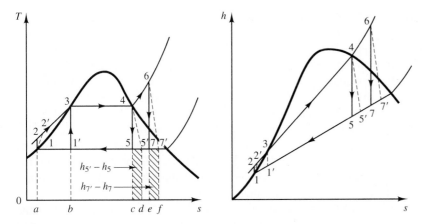

Figure 7.6   *T-s* and *h-s* diagrams of the Rankine cycle.

sented on this diagram if all four processes of the Rankine cycle are located in that region. Thus cycle 1′-3-4-5-1′ is a Carnot cycle.

The first law of thermodynamics may be applied in the analysis of the Rankine cycle 1-2-3-4-5-1. Neglecting changes in kinetic and potential energies, the following information is obtained, based on a unit mass of fluid flowing under steady-state conditions:

Heat interaction in the boiler $q_{2\text{-}4} = h_4 - h_2$
Work done by the turbine $w_{4\text{-}5} = h_4 - h_5$
Heat rejected in the condenser $q_{5\text{-}1} = h_5 - h_1$
Work done on the pump $w_{1\text{-}2} = h_2 - h_1$

The thermal efficiency of the Rankine cycle is expressed as

$$\eta_{th} = \frac{w_{net,\,out}}{q_{in}} = \frac{w_{4\text{-}5} - w_{1\text{-}2}}{q_{2\text{-}4}} = \frac{(h_4 - h_5) - (h_2 - h_1)}{h_4 - h_2}$$

$$= 1 - \frac{q_{out}}{q_{in}} = 1 - \frac{h_5 - h_1}{h_4 - h_2}$$

(7.9)

An expression for thermal efficiency may be obtained in terms of the average temperatures during the heat interaction processes. Noting that all processes in the Rankine cycle are internally reversible, the areas underneath these processes on the *T-s* diagram represent heat energy. The efficiency can then be written as

$$\eta_{th} = \frac{q_{in} - q_{out}}{q_{in}} = 1 - \frac{q_{5\text{-}1}}{q_{2\text{-}4}} = 1 - \frac{-\int_5^1 T \, ds}{\int_2^4 T \, ds}$$

$$= 1 - \frac{T_{avg,\,5\text{-}1}(s_5 - s_1)}{T_{avg,\,2\text{-}4}(s_4 - s_2)} = 1 - \frac{T_{avg,\,5\text{-}1}}{T_{avg,\,2\text{-}4}}$$

(7.10)

Equation (7.10) leads to the general conclusion that for reversible (and irreversible) cycles, the average temperatures of heat interactions, rather than the extreme temperatures, govern the cycle efficiency. Equation (7.10) also provides a basis for comparing the thermal efficiencies of the Rankine and Carnot cycles. In the Carnot cycle 1'-3-4-5-1' shown on the T-s diagram of Figure 7.6, the ratio of the heat rejected to the heat supplied is equal to the area 1'-5-c-b-1' to area 3-4-c-b-3. The corresponding ratio, for the Rankine cycle 1-2-3-4-5-1, is the ratio of area a-1-5-c-a to area a-2-3-4-c-a. Noting that a small value of this ratio means a higher value of efficiency, it becomes evident that the efficiency of the Rankine cycle is less than that of the Carnot cycle operating between the same thermal reservoirs. Note also the average temperature at which heat is supplied in the Rankine cycle is less than in the Carnot cycle.

A high Carnot efficiency requires a greater range for temperature between the low- and high-temperature thermal reservoirs. The lowest temperature is fixed by the temperature of the surroundings, and for a Rankine cycle, the highest temperature is limited to the critical temperature. The latter is equal to 374.14°C in the case of water. For a fixed $T_{avg, 5-1}$, Eq. (7.10) indicates that an increase in the thermal efficiency can be attained by raising the average temperature at which heat is supplied. This can be accomplished by superheating the steam to a high temperature (up to metallurgical limitations). The average temperature, on the other hand, does not increase appreciably because most of the heat interaction takes place during the constant-temperature vaporization process. Thus, in general, it may be stated that higher boiler temperatures (and pressures) and lower condenser pressures improve the thermal efficiency. Superheating has also the added advantage of limiting the moisture content at the exhaust of the turbine to its maximum allowable value of 10 percent ($x = 0.90$. Moisture contents in excess of this value lead to excessive turbine blading erosion. Note that increasing the boiler pressure or decreasing the condenser pressure may result in a reduction of the steam quality at the turbine exit.

In Figure 7.6, the actual expansion in the turbine is represented by the irreversible line 4-5' instead of by the isentropic expansion 4-5. Area c-5-5'-d-c is the resulting increase in the unavailable energy. The same conclusion can be reached by noting that the change of availability per unit mass during the expansion process is

$$-\Delta \psi = -w_{rev} = (h_4 - h_{5'}) - T_5(s_4 - s_{5'}) \qquad \text{(for } T_0 = T_5)$$

and the irreversibility per unit mass is

$$\frac{I}{m} = w_{act} - w_{rev} = T_5(s_{5'} - s_4) = T_5(s_{5'} - s_5)$$

which is equal to area c-5-5'-d-c.

Another way of expressing the performance of a vapor-power cycle is in terms of the *heat rate,* defined as the heat input divided by the net work output; it is the reciprocal of the thermal efficiency. Choosing 1 kW-hr as a unit output, then 1 kW-hr = 3600 kJ, so that

$$\text{heat rate} = \frac{3600}{\eta_{th}} \text{ kJ/kW-hr} \qquad (7.11)$$

The *steam rate* is defined as the amount of steam that is required to produce 1 kW-hr,

$$\text{steam rate} = \frac{3600}{w_{net,\,out}} \text{ kg/kW-hr} \qquad (7.12)$$

where $w_{net,\,out}$ is in kJ/kg.

A major impracticality of the Carnot cycle when applied to steam power plants is the state of the working substance at the pump inlet. That state is in the mixed phase region, which has a high value of specific volume and therefore requires a large pump to handle the wet compression process. This difficulty is surmounted in the Rankine cycle, since condensation continues to the saturated liquid line, thus reducing the volume appreciably.

It is clear from the foregoing discussion that the Rankine cycle represents a realistic model for the actual vapor-power cycle.

Example 7.2

Superheated steam at 2.5 MPa and 300°C enters the turbine of a steam power plant operating at steady state and expands to a condenser pressure of 10 kPa. Assuming the isentropic efficiencies of the turbine and pump are 85 and 80 percent, respectively, calculate

(a) the thermal efficiency of the cycle
(b) the heat rate in kJ/kW-hr
(c) the steam supply in kg/hr to deliver 1000 kW
(d) the corresponding Rankine cycle efficiency

SOLUTION

(a) Referring to Figure 7.7, the work input to an ideal pump is

$$\text{ideal } w_p = h_{2s} - h_1 = \int_{p_1}^{p_2} v\, dp \simeq v_{f_1}(p_2 - p_1)$$
$$= (0.00101 \text{ m}^3/\text{kg})[(2500 - 10) \text{ kPa}] = 2.515 \text{ kJ/kg}$$

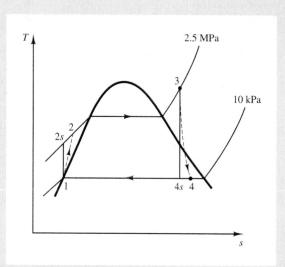

Figure 7.7

The actual work input is

$$\text{actual } w_p = h_2 - h_1 \simeq \frac{v_{f_1}(p_2 - p_1)}{\eta_p} = \frac{2.515 \text{ kJ/kg}}{0.8} = 3.144 \text{ kJ/kg}$$

From steam tables, $h_1 = 191.83$ kJ/kg. The enthalpy at point $2s$ is

$$h_{2s} = 191.83 + 2.515 = 194.34 \text{ kJ/kg}$$

and

$$h_2 = 191.83 + 3.144 = 194.97 \text{ kJ/kg}$$

To calculate h_{4s}, it is first necessary to determine the quality at state $4s$. At 2.5 MPa and 300°C, $s_3 = s_{4s} = 6.6438$ kJ/kg K. But

$$s_{4s} = s_f + x_{4s} s_{fg}$$

then

$$6.6438 \text{ kJ/kg K} = 0.6493 \text{ kJ/kg K} + x_{4s}(7.5009 \text{ kJ/kg K})$$

or

$$x_{4s} = \frac{6.6438 - 0.6493}{7.5009} = 79.92 \text{ percent}$$

Hence h_{4s} is

$$h_{4s} = h_f + x_{4s}h_{fg} = 191.83 + 0.7992(2392.8) = 2104 \text{ kJ/kg}$$

The preceding value of h_{4s} can also be determined from the Mollier diagram. For the turbine,

$$\text{ideal } w_{\text{turbine}} = h_{4s} - h_3 = (2104 - 3008.8) \text{ kJ/kg} = -904.8 \text{ kJ/kg}$$

$$\text{actual } w_{\text{turbine}} = (h_{4s} - h_3)\eta_{\text{tur}} = h_4 - h_3$$

$$= (-904.8 \text{ kJ/kg}) \times 0.85 = -769.1 \text{ kJ/kg}$$

The thermal efficiency of the cycle 1-2-3-4-1 is

$$\eta_{\text{th}} = \frac{(h_3 - h_4) - (h_2 - h_1)}{h_3 - h_2}$$

$$= \frac{769.1 - 3.144}{3008.8 - 194.97} = \frac{765.96 \text{ kJ/kg}}{2813.83 \text{ kJ/kg}} = 27.22 \text{ percent}$$

(b)

$$\text{heat rate} = \frac{3600}{0.2722} = 13.23 \text{ MJ/kW-hr}$$

(c)

$$\text{steam flow} = 1000\left(\frac{3600}{w_{\text{net, out}}}\right) = 1000\left(\frac{3600}{765.96}\right) = 4700 \text{ kg/hr}$$

(The steam rate = 4.7 kg/kW-hr.)

(d) The Rankine efficiency is

$$\eta_{\text{Rankine}} = \frac{(h_3 - h_{4s}) - (h_{2s} - h_1)}{h_3 - h_{2s}}$$

$$= \frac{904.8 - 2.515}{3008.8 - 194.34} = \frac{902.3 \text{ kJ/kg}}{2814.5 \text{ kJ/kg}} = 32.06\%$$

Note that approximately two-thirds of the energy transferred to the water in the boiler is rejected to the surroundings during the condensation process. The availability of this energy is low because of the low condensation temperature. Note also that the work input to the pump is small (less than 1 percent of the work output of the turbine).

Example 7.3

Using the data of Example 7.2, compare the Carnot and Rankine efficiencies. Assume saturated vapor to enter the turbine.

SOLUTION

Referring to Figure 7.8, the Carnot and Rankine efficiencies are

$$\eta_{\text{Carnot}} = \frac{T_{2\text{-}3} - T_{4\text{-}1}}{T_{2\text{-}3}} = \frac{223.99 - 45.81}{223.99 + 273.15} = \frac{178.18}{497.14}$$

$$= 35.84 \text{ percent.}$$

$$\eta_{\text{Rankine}} = \frac{(h_3 - h_4) - (h_6 - h_5)}{h_3 - h_6} = \frac{(2803.1 - 1980.9) - 2.515}{2803.1 - 94.34}$$

$$= \frac{819.68 \text{ kJ/kg}}{2608.76 \text{ kJ/kg}} = 31.42\%$$

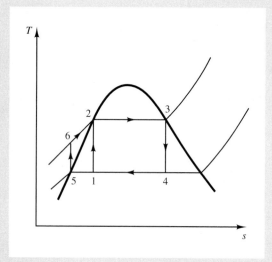

Figure 7.8

7.5 Reheat Cycle

Two modifications of the Rankine cycle that improve the performance of steam power plants are the *reheat* and *regenerative* cycles.

The components of the reheat cycle are shown in Figure 7.9. Figure 7.10 shows this cycle on a *T-s* and an *h-s* diagram. The vapor, after expanding in the high-pressure turbine, is returned to the boiler for further reheating before expanding in the low-pressure turbine.

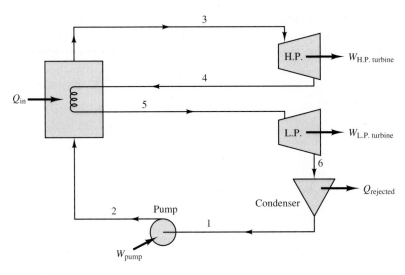

Figure 7.9   Components of the reheat cycle.

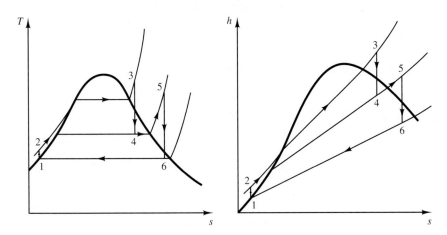

Figure 7.10   *T-s* and *h-s* diagrams of the reheat cycle.

Reheating occurs essentially at constant pressure and raises the temperature of the steam to T_5 (usually equal to T_3). Reheating has a twofold advantage. First, it reduces the moisture content in the low-pressure end of the turbine. With low-quality steam, the impact of the liquid droplets erodes the turbine blades, causing a decrease in turbine efficiency. Second, although more energy is expended in the reheat cycle than in the Rankine cycle, more work is usually developed. The net result is a small improvement in the thermal performance because heat input to the cycle takes place at a higher average temperature than the nonreheat cycle. An increase in efficiency may thus be attained without an increase in the maximum pressure or temperature of the cycle. Note that the availability of energy is improved by reheating, and a higher percentage of heat is supplied at a high temperature. In calculating the thermal efficiency of the reheat cycle, note that the work output of the turbines as well as the heat input in the boiler take place in two stages. Referring to Figure 7.10, the efficiency of the reheat cycle is

$$\eta_{th} = \frac{(h_3 - h_4) + (h_5 - h_6) - (h_2 - h_1)}{(h_3 - h_2) + (h_5 - h_4)} \tag{7.13}$$

Example 7.4

In a reheat cycle, steam at 8 MPa and 350°C is supplied to a high-pressure turbine. After expansion to 1.4 MPa, it is reheated at constant pressure to 350°C before expanding in the low-pressure turbine to 10 kPa. Determine:
  (a) the work output for the high- and low-pressure turbines
  (b) the thermal efficiency and the steam rate of the cycle
Assume isentropic expansion and compression for the turbines and pump.

SOLUTION

  (a) Refering to Figure 7.10, the pump work is

$$w_p \simeq v_1(p_2 - p_1)$$
$$= (0.00101 \text{ m}^3/\text{kg})[(8 \times 10^3 - 10) \text{ kPa}] = 8.070 \text{ kJ/kg}$$

At 10 kPa,

$$h_f = 191.83 \text{ kJ/kg}$$

so that the enthalpy at state 2 is

$$h_2 = 191.83 + 8.07 = 199.90 \text{ kJ/kg}$$

Using the steam tables or Mollier chart and the foregoing value of h_2, the values of the enthalpy and entropy at the successive states of the cycle may be determined. These are shown in Figure 7.11. The work interactions in the high- and low-pressure turbines are

$$w_{hp} = h_4 - h_3 = 2631.2 - 2987.3 = -356.1 \text{ kJ/kg}$$
$$w_{lp} = h_6 - h_5 = 2261.1 - 3149.5 = -888.4 \text{ kJ/kg}$$

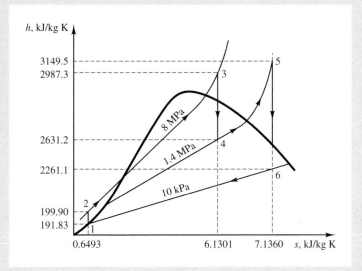

Figure 7.11

(b) The efficiency of the cycle, according to Eq. (7.13), is

$$\eta_{th} = \frac{356.1 + 888.4 - 8.07}{(2987.5 - 199.90) + (3149.5 - 2631.2)}$$
$$= \frac{1236.43}{3305.90} = 37.401\%$$

The steam rate $= 3600/1236.43 = 2.912$ kg/kW-hr.

Example 7.5

Calculate the irreversibility for each process and for the cycle of Example 7.4. Assume that the heat transfer to the working fluid in the boiler takes place from a thermal reservoir at 700 K and that the fluid in the condenser rejects heat to the surroundings at 300 K.

SOLUTION

The irreversibility per unit mass for each process is given by

$$i = T_0\left[(s_e - s_i) + \frac{q_{\text{surr}}}{T_{\text{surr}}}\right]$$

so that

$$i_{2\text{-}3} = (300 \text{ K})\left(5.4808 \text{ kJ/kg K} - \frac{2787.40 \text{ kJ/kg}}{700 \text{ K}}\right) = 449.64 \text{ kJ/kg}$$

$$i_{4\text{-}5} = (300 \text{ K})\left(1.0059 \text{ kJ/kg K} - \frac{518.30 \text{ kJ/kg}}{700 \text{ K}}\right) = 79.64 \text{ kJ/kg}$$

$$i_{6\text{-}1} = (300 \text{ K})\left(-6.4867 \text{ kJ/kg K} + \frac{2069.27 \text{ kJ/kg}}{300 \text{ K}}\right) = 123.26 \text{ kJ/kg}$$

The following table summarizes the results of this and the previous example:

Process	q (kJ/kg)	w (kJ/kg)	$(s_e - s_i)$ (kJ/kg K)	i (kJ/kg)
1-2	0.0	8.07	0.0	0.0
2-3	2787.40	0.0	5.4808	449.64
3-4	0.0	−356.1	0.0	0.0
4-5	518.30	0.0	1.0059	79.64
5-6	0.0	−388.4	0.0	0.0
6-1	−2069.27	0.0	−6.4867	123.26
Cycle	1236.43	1236.43	0.0	652.54

Note that the largest irreversibility occurs during the heating process 2-3 because of the large temperature difference between the water in the boiler and the high-temperature thermal reservoir.

7.6 Regenerative Cycle

The regenerative cycle is another modification of the Rankine cycle that results in an improvement in thermal efficiency. In actual cycles, the feed water enters the boiler at a temperature well below the saturation temperature corresponding to the boiler pressure (subcooled state). Upon heat interaction, the temperature rises first to the saturated liquid state before vaporization begins. This initial heating process constitutes a major irreversibility in the cycle because of the wide range of thermal potential between the combustion products and the water in the boiler. In the regenerative cycle, the degree of external irreversibility is reduced by decreasing the temperature difference during this heat interaction process.

Figure 7.12 shows an ideal regenerative cycle. Part of the steam is extracted or bled from the turbine at different stages and utilized to heat the feed water before it enters the boiler. Several heaters may be used. In each one, water is heated to within a few degrees from the saturation temperature corresponding to the extraction pressure. Ideally, if an infinite number of heaters are used, each heating the feed water an infinitesimal amount, the temperature of the feed water reaches the saturation temperature corresponding to the boiler pressure. The heating process thus approaches a reversible isothermal heat interaction process. The boiler in this case supplies only the energy necessary for vaporization and superheat. In actual cycles with a finite number of heaters, heat interaction cannot be reversible because of the temperature difference between the extracted steam and the water entering the heater. Note, however, that as the number of heaters is increased, the heat transfer process takes place with a smaller temperature difference, and consequently, the external irreversibility incurred in the heating process is decreased.

Analysis of the regenerative cycle is accomplished by performing a mass balance and an energy balance on each component of the system. A common practice is to perform calculations based on a unit mass of fluid entering the turbine.

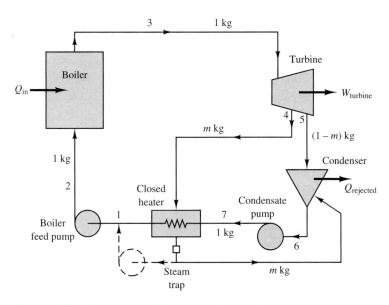

Figure 7.12   Components of the regenerative cycle.

Referring to the T-s and h-s diagrams of Figure 7.13, consider 1 kg of steam entering the turbine at state 3. Let m kg be the mass (or mass fraction) extracted from the turbine at state 4 for feed water heating. The remaining mass (1-m) continues to expand in the turbine till the pressure drops to the condenser pressure. As the extracted steam of mass m condenses in the heater, it transfers its latent enthalpy to the feed water flowing through the heater. The condensed steam then leaves the heater and flows to the condenser, where it mixes with the (1-m) kg of steam leaving the turbine. The total mass (1 kg) then enters the heater as feed water. Under ideal and adiabatic conditions, the temperature of the condensed mass m and the feed water leaving the heater are the same and are represented by state 1. The feed water is then pumped to the boiler pressure and enters the boiler where it is heated to state 3, and the cycle is repeated.

Compared to the Rankine cycle operating between the same two extreme pressures, the regenerative cycle produces less work per unit mass of steam entering the turbine because only a portion of the flow, (1-m), expands between states 4 and 5. On the other hand, the energy used for evaporating the feed water is reduced so that the amount of heat transfer from state 2 to state 3 is less than in the case of the Rankine cycle. When these two differences are taken into consideration and the two cycles analyzed, the regenerative cycle shows an improvement in the thermal efficiency over the Rankine cycle; that is, the reduction in heat input more than offsets the work developed. This increase in efficiency, as previously explained, is attributed to the decrease in irreversibility in the process of feed water heating. Referring to Figure 7.13, the thermal efficiency of the regenerative cycle is

$$\eta_{th} = \frac{(1)(h_3 - h_4) + (1 - m)(h_4 - h_5) - w_{pump}}{(h_3 - h_2)} \qquad (7.14)$$

Feed water heating takes place in either an *open* or a *closed* heater. In the open heater shown in Figure 7.14(a), the extracted steam is allowed to mix with the

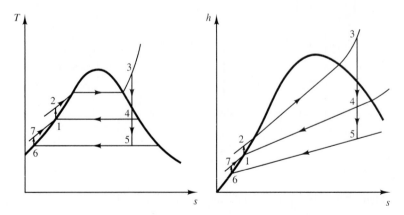

Figure 7.13 T-s and h-s diagrams of the regenerative cycle.

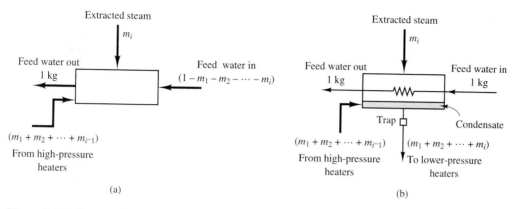

Figure 7.14 Open and closed heaters.

incoming feed water. The mixture then proceeds to the boiler feed pump. In the closed heater shown in Figure 7.14(b), the incoming feed water does not mix with the extracted steam, and both streams flow separately through the heater. This allows the two streams to be at different pressures. A steam trap, which allows only liquid to pass, is inserted in the condensed steam line, leaving the heater to ensure that all the extracted steam condenses before it leaves the heater. Most of the energy of the steam is thus delivered to the feed water. After passing through the steam trap, the condensate either is pumped into the feed water line leaving the heater or is connected to a lower-pressure region, such as the condenser or a low-pressure heater. The choice of either scheme is a matter of economics in which the pumping cost is weighed against the cost of energy loss if the condensate goes to the condenser.

Example 7.6

An ideal regenerative feed water heating cycle operates with steam supplied at 2.5 MPa and 300°C, and condenser pressure of 10 kPa. Extraction points for two heaters (one closed and one open) are at 400 kPa and 50 kPa, respectively. Calculate the thermal efficiency of the plant, neglecting pump work.

SOLUTION

Figure 7.15 shows the components of the cycle together with an *h-s* diagram. Assuming 1 kg of steam enters the turbine, an energy balance on the closed heater gives

$$m_1 = \frac{h_8 - h_6}{h_2 - h_7} = \frac{604.74 - 340.49}{2633.5 - 604.74} = \frac{264.25}{2028.8} = 0.13025 \text{ kg/kg}$$

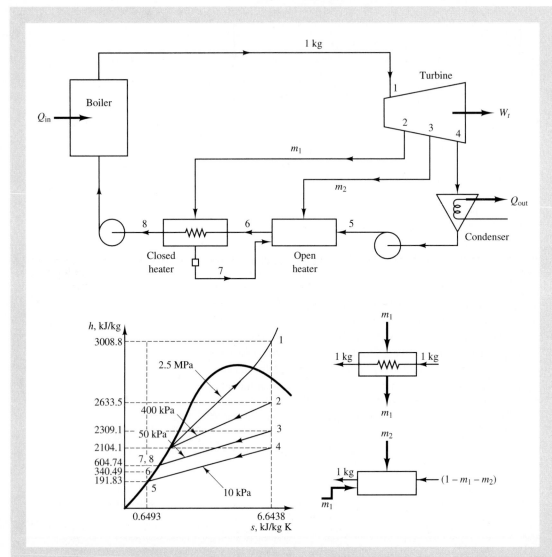

Figure 7.15

An energy balance on the open heater gives

$$m_1 h_7 + (1 - m_1 - m_2)h_5 + m_2 h_3 = (1)h_6$$
$$(0.13025)(604.74 \text{ kJ/kg}) + (1 - 0.13025 - m_2)(191.83 \text{ kJ/kg}) + m_2(2309.1 \text{ kJ/kg}) = 340.49$$

from which $m_2 = 0.044812$ kg/kg.

The work done per unit mass of steam entering the turbine is

$$w = (1)(h_2 - h_1) + (1 - m_1)(h_3 - h_2) + (1 - m_1 - m_2)(h_4 - h_3)$$
$$= (1)(2633.5 - 3008.8) \text{ kJ/kg} + (1 - 0.13025)(2309.1 - 2633.5) \text{ kJ/kg}$$
$$+ (1 - 0.13025 - 0.044812)(2104.1 - 2309.1) \text{ kJ/kg}$$
$$= -826.56 \text{ kJ/kg}$$

The thermal efficiency is

$$\eta_{th} = \frac{w_{net, out}}{h_1 - h_8} = \frac{826.56 \text{ kJ/kg}}{(3008.8 - 604.74) \text{ kJ/kg}} = \frac{826.56}{2404.1} = 34.38\%$$

Example 7.7

Calculate the irreversibility for each process and for the cycle of Example 7.6. Assume that the heat transfer to the working fluid in the boiler takes place from a thermal reservoir at 1000 K and that the fluid in the condenser rejects heat to the surroundings at 290 K.

SOLUTION

The irreversibility for each process is evaluated based on a unit mass entering the turbine:

$$I = T_0\left[\sum_e ms - \sum_i ms + \frac{Q_{surr}}{T_{surr}}\right]$$

where m is the mass involved in the process per unit mass entering the turbine (mass fraction). Referring to Figure 7.15, and since the flow in the turbine is isentropic, then,

$$I_{1\text{-}2} = I_{2\text{-}3} = I_{3\text{-}4} = 0$$

$$I_{4\text{-}5} = (1 - m_1 - m_2)T_0\left[(s_5 - s_4) - \frac{q_{4\text{-}5}}{T_L}\right]$$

$$q_{4\text{-}5} = h_5 - h_4 = 191.83 - 2104.1 = -1912.27 \text{ kJ/kg}$$

so that

$$I_{4\text{-}5} = (1 - 0.13025 - 0.044812)(290 \text{ K})\left[(0.6493 - 6.6438) \text{ kJ/kg K} + \frac{1912.27 \text{ kJ/kg}}{290 \text{ K}}\right]$$

$$= (0.8249)(290 \text{ K})[(-5.9945 + 6.5940) \text{ kJ/kg K}] = 143.4212 \text{ kJ/kg}$$

$$q_{4\text{-}5} = (-1912.27)(0.8249) = -1577.43 \text{ kJ}$$

$$q_{8\text{-}1} = h_1 - h_8 = 3008.8 - 604.74 = 2404.06 \text{ kJ/kg}$$

$$I_{8-1} = mT_0 \left[(s_1 - s_8) - \frac{q_{8-1}}{T_H} \right]$$

$$I_{8-1} = (1)(290 \text{ K}) \left[(6.6438 - 1.7766) \text{ kJ/kg K} - \frac{2404.06 \text{ kJ/kg}}{1000 \text{ K}} \right]$$

$$= 714.31 \text{ kJ/kg}$$

Closed heater: $I = (T_0)[\{m_1 s_7 + (1)s_8\} - \{m_1 s_2 + (1)s_6\}]$

$$I = (290 \text{ K})[\{(0.13025)(1.7766 \text{ kJ/kg K}) + (1)(1.7766 \text{ kJ/kg K})\}$$
$$- \{(0.13025)(6.6438 \text{ kJ/kg K}) + (1)(1.0910 \text{ kJ/kg K})\}]$$
$$= (290 \text{ K})(2.0080 - 1.9564 \text{ kJ/kg K}) = 14.965 \text{ kJ/kg}$$

Open heater: $I = T_0[(1)s_6 - \{m_1 s_7 + m_2 s_3 + (1 - m_1 - m_2)s_5\}]$

$$I = (290 \text{ K})[(1)(1.0910 \text{ kJ/kg K}) - \{(0.13025)(1.7766 \text{ kJ/kg K})$$
$$+ (0.044812)(6.6438 \text{ kJ/kg K}) + 0.8249(0.6493 \text{ kJ/kg K})\}]$$
$$= 290[1.0910 - (0.2314 + 0.2977 + 1.5356)]$$
$$= 290(1.0910 - 1.0647) = 7.625 \text{ kJ/kg}$$

The following table summarizes the characteristics of the cycle based on a unit mass of steam entering the turbine:

Process	q (kJ/kg)	w (kJ/kg)	$(\Sigma_e ms - \Sigma_i ms)$ (kJ/K)	i (kJ/kg)
1-2	0.0	−375.30	0.0	0.0
2-3	0.0	−282.15	0.0	0.0
3-4	0.0	−169.10	0.0	0.0
4-5	−1577.43	0.0	−4.9449	143.42
Closed heater	0.0	0.0	0.0516	14.965
Open heater	0.0	0.0	0.0263	7.625
8-1	2404.06	0.0	4.8672	714.3
Cycle	826.63	−826.56	~0.0	880.32

Example 7.8

From the data of Example 7.2, compare the efficiencies of the reheat and regenerative cycles. The reheat pressure in the reheat cycle is 300 kPa, and the extraction pressure in the regenerative cycle is also 300 kPa. Assume isentropic efficiencies for the turbines and boiler feed water pumps to be 100 percent. (Neglect the power required for the condensate pump.)

SOLUTION

(a) Regenerative cycle: Referring to Figure 7.16(a), an energy balance on the feed water heater gives

$$m(h_4 - h_1) = (1 - m)(h_1 - h_6)$$
$$m(2583.7 - 561.47) \text{ kJ/kg} = (1 - m)(561.47 - 191.83) \text{ kJ/kg}$$

from which $m = 0.15454$ kg/kg. The work done on the boiler feed pump is

$$w_{1\text{-}2} = h_2 - h_1 \simeq v_1 \Delta p$$
$$= (0.001073 \text{ m}^3/\text{kg})[(2500 - 300) \text{ kPa}] = 2.361 \text{ kJ/kg}$$

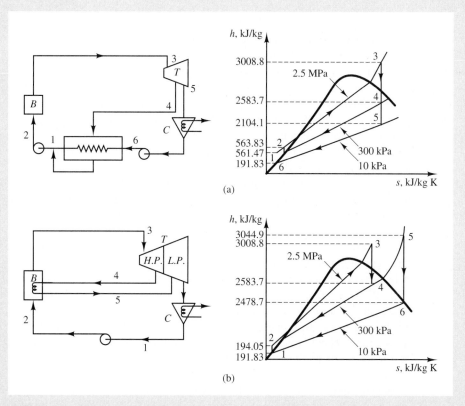

(a)

(b)

Figure 7.16

The thermal efficiency, according to Eq. (7.14), is

$$\eta_{\text{th}} = \frac{(3008.8 - 2583.7) \text{ kJ/kg} + (1 - 0.15454)(2583.7 - 2104.1) \text{ kJ/kg} - 2.361 \text{ kJ/kg}}{3008.8 \text{ kJ/kg} - (561.47 + 2.361) \text{ kJ/kg}}$$

$$= 33.8 \text{ percent}$$

(b) Reheat cycle: Referring to Figure 7.16(b), the pump work is

$$w_{1\text{-}2} \simeq v_1 \Delta p = (0.00101 \text{ m}^3/\text{kg})[(2500 - 300) \text{ kPa}] = 2.222 \text{ kJ/kg}$$

The thermal efficiency, according to Eq. (7.12), is

$$\eta_{\text{th}} = \frac{(3008.8 - 2583.7) \text{ kJ/kg} + (3044.9 - 2478.7) \text{ kJ/kg} - 2.222 \text{ kJ/kg}}{(3008.8 - 194.05) \text{ kJ/kg} + (3044.9 - 2583.7) \text{ kJ/kg}}$$

$$= 30.19\%$$

7.7 Reciprocating Internal Combustion Engines

In internal combustion engines, chemical energy is converted into thermal energy as a result of the combustion of an air–fuel mixture. Reciprocating internal combustion engines operate on either a *four-stroke cycle* (suction, compression, expansion, and exhaust) or a *two-stroke cycle* (compression and expansion). The fuel is ignited by a spark (spark-ignition engines), as in most automotive engines; others are ignited by compression (compression-ignition engines), as in diesel engines. Intermittent combustion in the cylinders makes it possible to cope with high temperatures that are now unapproachable in continuous combustion systems such as the gas turbine.

Although internal combustion engines undergo mechanical cycles, the working fluid does not go through a thermodynamic cycle. When the air–fuel mixture is introduced into the cylinders of a spark-ignition engine, the vapors are compressed and ignited. The resulting combustion products are allowed to expand before they are discharged to the surrounding atmosphere. This open cycle is then repeated with a fresh charge of an air–fuel mixture.

In order to analyze internal combustion engines, we devise an ideal closed cycle that approximates the performance of the open cycle. A closed air-standard cycle is based on the following conditions: (a) the working fluid is an ideal gas having the same properties as those of air; (b) all processes are internally reversible; (c) heat interaction occurs between the system and a high-temperature source (instead of combustion) and between the system and a low-temperature reservoir (in place of exhaust).

Even under these conditions, the conversion of the chemical energy of the fuel into mechanical energy is incomplete. This is shown by the second law of thermodynamics. In actual engines, the degree of conversion is less than in the ideal case because irreversible losses occur. These are associated with mechanical and fluid frictions. In addition, there is incomplete combustion as well as energy loss when the exhaust gases are discharged to the surroundings. An overall index of performance of internal combustion engines is *thermal efficiency,* which is defined as the ratio of the net work output to the energy supplied to the engine. Other methods indicating certain aspects of engine performance include *mechanical efficiency* and *volumetric efficiency*.

The power developed within the cylinders of an engine as a result of the combustion process is measured by an engine indicator and is called *indicated power*. The power delivered by the engine shaft is called *brake power* and is less than the indicated power. Some of the power is lost in mechanical friction and fluid friction during the transmission process; some of the power is consumed driving engine auxiliaries, such as fuel pump, generator, and water pump. The ratio of the brake power to the indicated power therefore indicates the mechanical efficiency of the engine:

$$\eta_{mech} = \frac{\text{brake power}}{\text{indicated power}} \qquad (7.15)$$

The *mean effective pressure (mep)* is defined as that constant pressure, if acting on the piston during one stroke, would produce the same net amount of work as in the actual cycle. It is equal to the net work output divided by the displacement volume. The mean effective pressure is an index that relates work output of an engine to its size (displacement volume).

$$mep = \frac{-\oint p \, dV}{V_{stroke}} \qquad (7.16)$$

The performance of an engine also depends upon its aspirating capacity. The mass of air actually induced into the cylinders is less than ideal because of the short time for induction and the air's resistance to the motion. During the suction stroke, the piston creates a partial vacuum in the cylinder, and air is therefore drawn into the cylinder. However, the inertia of the air and the frictional resistance at the inlet ports tend to impede the flow of air and reduce the mass of air that enters the cylinder. Insufficient air in the cylinders results in incomplete combustion, which means less engine power. A measure of this aspirating effect is given by the *volumetric efficiency;*

$$\eta_{vol} = \frac{m_{actual}}{m_{ideal}} = \frac{\dot{m}_{actual}}{\dot{m}_{ideal}} \qquad (7.17)$$

where

m_{actual} = actual mass introduced in the cylinders
m_{ideal} = ideal mass introduced in the cylinders

In high-speed engines, volumetric inefficiencies can represent large power losses. Improvement in volumetric efficiency can be achieved by a *supercharger* (or a turbocharger). The supercharger, which is essentially a low-pressure compressor, forces the air through the inlet ports into the cylinders at a pressure slightly higher than atmospheric. To justify the use of a supercharger, its power consumption must be less than the gain in power due to the increase in volumetric efficiency.

Although the Carnot cycle is the most efficient cycle, the area bounded by the cycle on a *p-v* diagram and the corresponding *mep* are extremely small. When this is added to the difficulty of transferring heat during the isothermal compression and expansion processes, a Carnot engine becomes an abstract, impractical design. For these reasons, other ideal cycles are devised to describe more realistically the operation of viable internal combustion engines.

The Otto and Diesel air-standard cycles, which are idealized models for the spark-ignition and compression-ignition engines, will be discussed in the following sections.

7.8 The Air-Standard Otto Cycle

Figure 7.17 shows the standard Otto cycle on *p-v* and *T-s* diagrams. All the processes of the cycle are internally reversible. Considering air (an ideal gas), with constant specific heats in a piston–cylinder assembly, the Otto cycle is performed according to the following succession of nonflow processes:

(a) Process 1-2 is a reversible adiabatic (isentropic) compression in which the air is compressed by the inward stroke of the piston. State 2 is at a higher temperature and pressure than state 1.

(b) During process 2-3, heat is transferred to the system at constant volume, resulting in an increase in temperature, pressure, and entropy. The heat interac-

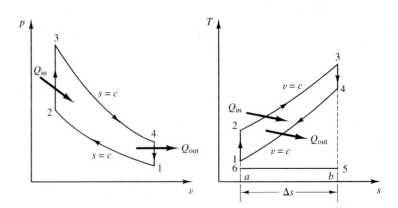

Figure 7.17   Standard Otto cycle.

tion is accomplished by allowing the system to come in contact with a high-temperature reservoir. The amount of heat transfer to the system is

$$Q_{2\text{-}3} = mc_v(T_3 - T_2)$$

(c) Process 3-4 is a reversible adiabatic (isentropic) expansion that takes place as the piston moves on its outward stroke. Both the temperature and pressure decrease.

(d) During process 4-1, heat is transferred from the system at a constant volume. The heat interaction between the system and a low-temperature reservoir (or surroundings) results in a decrease in temperature, pressure, and entropy. The amount of heat transferred from the system is

$$Q_{4\text{-}1} = mc_v(T_4 - T_1)$$

The thermal efficiency of the Otto cycle is

$$\eta_{\text{th}} = \frac{Q_{2\text{-}3} - |Q_{4\text{-}1}|}{Q_{2\text{-}3}} = 1 - \frac{|Q_{4\text{-}1}|}{Q_{2\text{-}3}} = 1 - \frac{mc_v(T_4 - T_1)}{mc_v(T_3 - T_2)}$$

$$= 1 - \frac{T_1}{T_2}\left[\frac{(T_4/T_1) - 1}{(T_3/T_2) - 1}\right]$$

The volumes and temperatures of the end states of the compression and expansion processes are given by the following isentropic relations:

$$\frac{T_2}{T_1} = \left(\frac{V_1}{V_2}\right)^{\gamma-1} = \left(\frac{V_4}{V_3}\right)^{\gamma-1} = \frac{T_3}{T_4}$$

Thus

$$\frac{T_4}{T_1} = \frac{T_3}{T_2}$$

The expression for efficiency then becomes

$$\eta_{\text{th}} = 1 - \frac{T_1}{T_2} = 1 - \frac{1}{r_v^{\gamma-1}} \qquad \text{(constant } \gamma) \qquad (7.18)$$

where r_v is called the *compression ratio* $= V_1/V_2 = V_4/V_3$. Note that the thermal efficiency is a function of the compression ratio r_v and specific heat ratio γ. For constant γ, an increase in the compression ratio results in an increase in

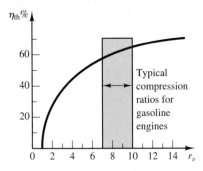

Figure 7.18 Thermal efficiency of the Otto cycle as a function of compression ratio ($\gamma = 1.4$).

thermal efficiency. A plot of η_{th} versus r_v of the air-standard cycle given by Eq. (7.18) is shown in Figure 7.18. In spark-ignition engines, an upper limit of the compression ratio is set by the ignition temperature of the fuel. The temperature of the air–fuel mixture at the end of the compression stroke must be below the fuel ignition temperature. If this limit is exceeded, a high-speed, high-pressure combustion wave (detonation wave) will propagate prematurely in the engine cylinder causing "engine knock."

Referring to the *T-s* diagram of Figure 7.17, area *a-2-3-b-a* is equal to the amount of heat transferred to the system during the constant-volume process 2-3. Since T_6 is the lowest temperature at which heat may be rejected, process 5-6 corresponds to the minimum possible heat rejection. The area above process 5-6 (area 6-2-3-5-6) is available energy; the area below it (area *a-6-5-b-a*) is unavailable energy. The rejection of heat along process 5-6 is, however, impractical since it requires expansion to state 5 corresponding to a low pressure and to a temperature equal to that of the surroundings, followed by a constant-temperature heat rejection along process 5-6. In the Otto cycle, rejection of heat is along process 4-1. This process divides the amount of heat input into work energy (area 1-2-3-4-1) and heat energy (area *a-1-4-b-a*). Note that area 6-1-4-5-6 represents the increase in unavailable energy due to heat rejection along 4-1 instead of along the constant-temperature process 5-6. The enclosed area 1-2-3-4-1 of the *T-s* or *p-v* diagram is equal to the net work output (or net heat input).

Example 7.9

An air-standard Otto cycle has a compression ratio of 8. At the start of the compression process, the temperature is 300 K, and the pressure is 100 kPa. If the maximum temperature of the cycle is 1200 K determine:
   (a) the heat supplied per kg of air
   (b) the net work done per kg of air
   (c) the thermal efficiency of the cycle

SOLUTION

Referring to Figure 7.17, the temperatures at states 2 and 4 are determined from isentropic process relations with $\gamma = 1.4$.

$$\frac{T_2}{T_1} = \left(\frac{V_1}{V_2}\right)^{\gamma-1} \qquad \text{or} \qquad T_2 = (300 \text{ K})(8)^{0.4} = 689.2 \text{ K}$$

and

$$\frac{T_4}{T_3} = \left(\frac{V_3}{V_4}\right)^{\gamma-1} \qquad \text{or} \qquad T_4 = (1200 \text{ K})\left(1/8\right)^{0.4} = 522.3 \text{ K}$$

(a) For an average temperature of $1200 + 689.2/2 = 944.6$ K, the specific heat value at constant volume $c_v = 0.754$ kJ/kg K. The heat supplied per kg of air (along process 2-3) is equal to the change in internal energy so that

$$q_{2\text{-}3} = c_v(T_3 - T_2) = (0.754 \text{ kJ/kg K})(1200 - 689.2) \text{ K} = 385.1 \text{ kJ/kg}$$

(b) The work done per kg of air is given by

$$w_{\text{net}} = -q_{4\text{-}1} - q_{2\text{-}3}$$

During the heat rejection process, $c_v = 0.7165$ kJ/kg K so that

$$q_{4\text{-}1} = c_v(T_1 - T_4) = (0.7165 \text{ kJ/kg K})[(300 - 522.3) \text{ K}] = -159.3 \text{ kJ/kg}$$

Therefore,

$$w_{\text{net}} = 159.3 - 385.1 = -225.8 \text{ kJ/kg}$$

(c) The thermal efficiency is

$$\eta_{\text{th}} = \frac{w_{\text{net, out}}}{q_{2\text{-}3}} = \frac{225.8 \text{ kJ/kg}}{385.1 \text{ kJ/kg}} = 0.586 \qquad \text{or} \qquad 58.6 \text{ percent}$$

The preceding solution assumes constant specific heats throughout the wide temperature range of the cycle. More accurate answers may be obtained by considering taking the variation of the specific heats with temperature into account. Tables of isentropic relations given in Appendix A11 may be used to give the following more accurate solution:

At $T_1 = 300$ K

$$v_{r_1} = 179.4906 \qquad u_1 = 214.36 \text{ kJ/kg}$$

For the isentropic process 1-2,

$$v_{r_2} = v_{r_1}\left(\frac{V_2}{V_1}\right) = 179.4906\left(\frac{1}{8}\right) = 22.4363$$

at $v_{r_2} = 22.4363$,

$$T_2 = 673.3 \text{ K} \qquad u_2 = 491.71 \text{ kJ/kg}$$

At 1200 K,

$$v_{r_3} = 4.18586 \qquad u_3 = 933.37 \text{ kJ/kg}$$

For the isentropic process 3-4,

$$v_{r_4} = v_{r_3}\left(\frac{V_4}{V_3}\right) = 4.18586(8) = 33.4869$$

at $v_{r_4} = 33.4869$,

$$u_4 = 419.87 \text{ kJ/kg} \qquad T_4 = 580 \text{ K}$$

(a) The heat supplied per kg of air is

$$q_{2\text{-}3} = u_3 - u_2 = (933.37 - 491.71) \text{ kJ/kg} = 441.66 \text{ kJ/kg}$$

(b) The net work done per kg of air is

$$w_{\text{net}} = -q_{2\text{-}3} - q_{4\text{-}1} = -441.66 \text{ kJ/kg} + (419.87 - 214.36) \text{ kJ/kg}$$
$$= -236.15 \text{ kJ/kg}$$

(c) The thermal efficiency is

$$\eta_{\text{th}} = \frac{w_{\text{net, out}}}{q_{\text{in}}} = \frac{236.15 \text{ kJ/kg}}{441.66 \text{ kJ/kg}} = 53.5 \text{ percent}$$

Example 7.10

Calculate the irreversibility for each process and for the cycle of the previous example.

SOLUTION

Referring to Figure 7.17, processes 1-2 and 3-4 are each internally reversible and adiabatic (isentropic) and hence the irreversibility of each of these processes is zero. Processes 2-3 and 4-1 are also internally reversible, but they are externally irreversible because they involve heat interaction with a finite temperature difference. The irreversibility per unit mass for process 2-3 is given by

$$i_{2\text{-}3} = T_0(\Delta s_{\text{sys}} + \Delta s_{\text{surr}})$$

Noting that the volume is constant during this process, then

$$i_{2\text{-}3} = T_0\left[\left(c_v \ln \frac{T_3}{T_2}\right) - \frac{q_{2\text{-}3}}{T_3}\right]$$

$$= (300 \text{ K})\left[(0.754 \text{ kJ/kgK}) \ln \frac{1200 \text{ K}}{689.2 \text{ K}} - \frac{385.1 \text{ kJ/kg}}{1200 \text{ K}}\right]$$

$$= (300 \text{ K})(0.4181 \text{ kJ/kgK} - 0.3209 \text{ kJ/kgK}) = 29.16 \text{ kJ/kg}$$

Similarly, the irreversibility for process 4-1 is

$$i_{4\text{-}1} = T_0\left(c_v \ln \frac{T_1}{T_4} - \frac{q_{4\text{-}1}}{T_1}\right)$$

$$= (300 \text{ K})\left[(0.7165 \text{ kJ/kgK}) \ln \frac{300 \text{ K}}{522.3 \text{ K}} - \frac{-159.3 \text{ kJ/kg}}{300 \text{ K}}\right]$$

$$= (300 \text{ K})(-0.3973 + 0.5310)\text{kJ/kgK} = 40.11 \text{ kJ/kg}$$

The irreversibility for the cycle is the sum of the irreversibilities of the individual processes so that

$$i_{\text{cycle}} = i_{1\text{-}2} + i_{2\text{-}3} + i_{3\text{-}4} + i_{4\text{-}1}$$

$$= 0 + 29.16 + 0 + 40.11 = 69.27 \text{ kJ/kg}$$

Note that the net entropy change for this or any cycle (reversible or not) is zero because entropy is a property. The irreversibility in the present cycle is not zero because the two nonadiabatic processes 2-3 and 4-1 involve external irreversibilities. Note also that work interaction takes place only during the isentropic processes 1-2 and 3-4, and therefore, the work of the actual cycle is equal to the reversible work. The irreversibility during these two processes is zero.

7.9 The Air-Standard Diesel Cycle

Compression-ignition reciprocating engines operate on the Diesel cycle. Figure 7.19 shows the standard Diesel cycle on *p-v* and *T-s* diagrams. During the compression stroke 1-2, an ideal gas (air) with constant specific heats is compressed reversibly and adiabatically to a high temperature and pressure. Process 2-3 is a constant pressure heat interaction. The amount of heat transferred to the system is

$$Q_{2-3} = mc_p(T_3 - T_2)$$

In actual cycles, the energy supplied to the engine is accomplished by fuel injection during process 2-3. The temperature after compression exceeds the

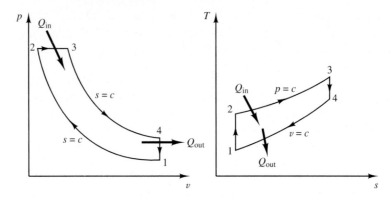

Figure 7.19   Standard Diesel cycle.

fuel ignition temperature so that fuel ignites spontaneously upon injection into the combustion chamber.

The ratio V_3/V_2 is called the *cutoff ratio*. It indicates the percentage of the stroke during which heat is supplied. Process 3-4 is a reversible adiabatic expansion in which the pressure and temperature decrease. Process 4-1 is a constant-volume heat rejection accompanied by a further decrease in temperature and pressure. The amount of heat rejected from the system is

$$Q_{4\text{-}1} = mc_v(T_4 - T_1)$$

The thermal efficiency of the Diesel cycle is

$$\eta_{th} = \frac{Q_{2\text{-}3} - |\,Q_{4\text{-}1}\,|}{Q_{2\text{-}3}} = 1 - \frac{|\,Q_{4\text{-}1}\,|}{Q_{2\text{-}3}} = 1 - \frac{mc_v(T_4 - T_1)}{mc_p(T_3 - T_2)}$$

$$= 1 - \frac{T_1[(T_4/T_1) - 1]}{\gamma T_2[(T_3/T_2) - 1]}$$

Using isentropic relations between temperatures and volumes for processes 1-2 and 3-4, the foregoing expression becomes

$$\eta_{th} = 1 - \frac{1}{r_v^{\gamma-1}}\left[\frac{r_c^{\gamma} - 1}{\gamma(r_c - 1)}\right] \tag{7.19}$$

where r_c is the cutoff ratio $= V_3/V_2$.

Comparison of the expressions of the thermal efficiencies of the Otto and Diesel cycles, Eqs. (7.18) and (7.19), shows that, for the same compression ratio, the Otto cycle is more efficient than the Diesel cycle. This conclusion may be verified by noting that when $r_c = 1$, Eq. (7.19) reduces to Eq. (7.18), and, when $r_c > 1$, the bracketed term of Eq. (7.19) is more than unity. Diesel cycles have, however, a higher compression ratio than Otto cycles, resulting in a higher efficiency. The same result can be obtained by referring to the p-v and T-s diagrams of Figure 7.20. The Otto cycle 1-2-3′-4-1 and the Diesel cycle 1-2-3-4-1 have the same compression ratio and the same heat rejection from 4 to 1, but the work developed by the Otto cycle is more than that of the Diesel cycle. The Otto cycle 1-2′-3-4-1 and the Diesel cycle 1-2-3-4-1 have the same heat rejection and the same maximum pressure and temperature (state 3). Under these conditions, the efficiency of the Diesel cycle exceeds that of the Otto cycle. It may also be remarked that increasing the maximum temperature in the Diesel cycle to state 3″ results in a decrease in the efficiency. This may be explained by the T-s diagram where the increase in work is area 3-3″-4″-4-3, and the increase in heat input is represented by area 3-3″-b-a-3. The ratio of these two quantities decreases because the slope of constant-pressure curve 2-3-3″ is less than the constant-volume curve 1-4-4″ on the T-s diagram.

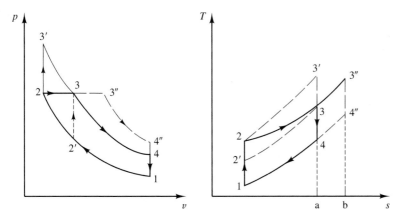

Figure 7.20   Comparison of the Otto and Diesel cycles.

Example 7.11

The engine of an air-standard Diesel cycle has six cylinders of 10 cm bore (diameter) and 12 cm stroke. The engine speed is 2000 rpm. At the beginning of compression, the air is at 100 kPa and 300 K. If the clearance volume is 6.667 percent of the stroke volume, compute:

   (a) compression ratio
   (b) pressure and temperature of the air after compression
   (c) thermal efficiency and power output if the air is heated to 1750 K

SOLUTION

(a) Referring to Figure 7.21, V_2 is the clearance volume, thus

$$0.0667 = \frac{V_2}{V_1 - V_2} = \frac{1}{(V_1/V_2) - 1} = \frac{1}{r_v - 1}$$

from which $r_v = 16$.

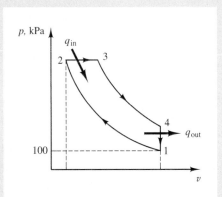

Figure 7.21

(b) The specific volume at state 1 according to the ideal-gas law is

$$v_1 = \frac{RT_1}{p_1} = \frac{(0.287 \text{ kJ/kg K})(300 \text{ K})}{100 \text{ kPa}} = 0.861 \text{ m}^3/\text{kg}$$

Therefore,

$$v_2 = \frac{v_1}{r_v} = \frac{0.861 \text{ m}^3/\text{kg}}{16} = 0.0538 \text{ m}^3/\text{kg}$$

From gas tables at $T_1 = 300$ K,

$$v_{r_1} = 179.4906 \qquad p_{r_1} = 1.11458 \qquad u_1 = 214.36 \text{ kJ/kg}$$

Therefore,

$$v_{r_2} = v_{r_1}\left(\frac{v_2}{v_1}\right) = 179.4906 \left(\frac{1}{16}\right) = 11.2182$$

and from gas tables at $v_{r_2} = 11.2182$,

$$T_2 = 862.5 \text{ K} \qquad p_{r_2} = 51.3042 \qquad h_2 = 891.29 \text{ kJ/kg}$$

The pressure at state 2 is

$$p_2 = p_1\left(\frac{p_{r_2}}{p_{r_1}}\right) = (100 \text{ kPa})\left(\frac{51.3042}{1.11458}\right) = 4603 \text{ kPa}$$

or from the ideal-gas law

$$p_2 = \frac{RT_2}{v_2} = \frac{(0.287 \text{ kJ/kg K})(862.5 \text{ K})}{0.0538 \text{ m}^3/\text{kg}} = 4601 \text{ kPa}$$

(c) At $T_3 = 1750$ K,

$$h_3 = 1941.28 \text{ kJ/kg} \qquad v_{r_3} = 1.2533$$

Therefore,

$$v_3 = \frac{RT_3}{p_3} = \frac{(0.287 \text{ kJ/kg K})(1750 \text{ K})}{4601 \text{ kPa}} = 0.109 \text{ m}^3/\text{kg}$$

$$v_{r_4} = v_{r_3}\left(\frac{v_4}{v_3}\right) = 1.2533\left(\frac{0.861}{0.109}\right) = 9.8999$$

and from gas tables at $v_{r_4} = 9.8999$,

$$T_4 = 900.6 \text{ K} \qquad u_4 = 675.29 \text{ kJ/kg}$$

The thermal efficiency of the cycle is

$$\eta_{\text{th}} = 1 - \frac{q_{\text{out}}}{q_{\text{in}}} = 1 - \frac{u_4 - u_1}{h_3 - h_2} = 1 - \frac{(675.29 - 214.36) \text{ kJ/kg}}{(1941.28 - 891.29) \text{ kJ/kg}}$$

$$= 1 - \frac{460.93}{1049.99} = 56.1\%$$

The volume V_1 of each cylinder is

$$V_1 = V_{\text{clearance}} + V_{\text{displacement}}$$

where

$$V_{\text{disp}} = \left(\frac{\pi}{4}\right)(0.10 \text{ m})^2(0.12 \text{ m}) = 942.48 \times 10^{-6} \text{ m}^3$$

hence

$$V_1 = 0.0667(942.28 \times 10^{-6} \text{ m}^3) + 942.48 \times 10^{-6} \text{ m}^3 = 1.005 \times 10^{-3} \text{ m}^3$$

The mass of the air for the six cylinders is

$$m = \frac{6V_1}{v_1} = \frac{6(1.005 \times 10^{-3}\,\text{m}^3)}{0.861\,\text{m}^3/\text{kg}} = 0.007\,\text{kg}$$

The work output is

$$W = mw = (0.007\,\text{kg})[(1049.99 - 460.93)\,\text{kJ/kg}] = 4.123\,\text{kJ}$$

and the power developed is

$$P = W \times \frac{rpm}{2} = \frac{(4.123\,\text{kJ})(1000\,\text{cycles/min})}{60\,\text{s/min}} = 68.72\,\text{kW}$$

Note that air tables are used in this example to account for the variation of specific heats with temperature.

7.10 The Dual Cycle

Combustion in the Otto cycle is based on a constant-volume process; in the Diesel cycle, it is based on a constant-pressure process. But combustion in actual spark-ignition engines requires a finite amount of time if the process is to go to completion. For this reason, combustion in Otto cycle engines does not actually occur under constant-volume conditions. Similarly, in compression-ignition engines, combustion does not occur under constant-pressure conditions because of the rapid, uncontrolled combustion process.

The operation of the reciprocating internal combustion engines represents a compromise between the Otto cycle and the Diesel cycle, and can be described as a dual-combustion or a limited-pressure standard cycle. Figure 7.22 shows the cycle on p-v and T-s diagrams. Heat transfer to the system may be considered to occur first at constant volume and then at constant pressure. In the actual system, injection of fuel is started during the compression stroke and continues during part of the return stroke. The efficiency of the limited-pressure standard cycle may be shown to be

$$\eta_{\text{th}} = 1 - \frac{T_5 - T_1}{T_3 - T_2 + \gamma(T_4 - T_3)} \tag{7.20}$$

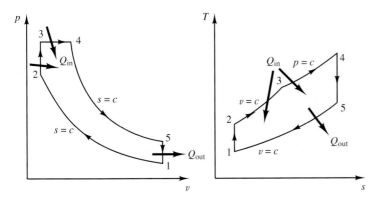

Figure 7.22   Standard dual cycle.

7.11 The Stirling Cycle

The *Stirling* cycle consists of two totally reversible isotherms and two totally reversible constant-volume processes. Figure 7.23 shows the processes of the cycle on a *p-v* and a *T-s* diagram. During the expansion process 3-4, heat is supplied to maintain a constant temperature T_H, and during the compression process 1-2, heat is rejected to maintain a constant temperature T_L. There are also heat interactions along the constant-volume processes 2-3 and 4-1. The quantities of heat in these two processes are essentially equal but opposite in direction, and the exchange process takes place by means of a *regenerator*. The function of the regenerator is to act as a temporary reservoir, being able to absorb heat during process 4-1 and ideally delivering the same quantity of heat during process 2-3. These two quantities of heat are represented by the shaded areas of the *T-s* diagram in Figure 7.23.

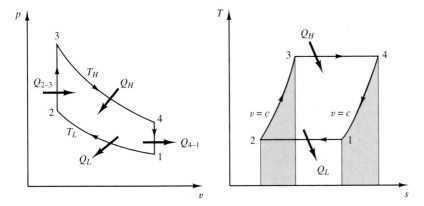

Figure 7.23   *p-v* and *T-s* diagrams of the air-standard Stirling cycle.

Figure 7.24 shows a system that can execute the different processes of the Stirling cycle. By means of linkages, both pistons can be synchronized to maintain constant volume during processes 2-3 and 4-1. During these two processes,

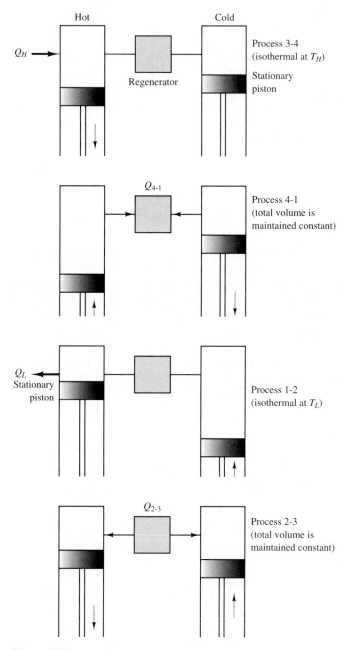

Figure 7.24   The different processes of the Stirling cycle.

the regenerator absorbs and rejects equal amounts of heat to the working fluid. The heat input Q_H and the heat rejection Q_L take place along 3-4 and 1-2, and during these two processes, the temperature is maintained constant. In that respect, the Stirling cycle and the Carnot cycle are similar, and the thermal efficiencies of both cycles are the same so that

$$\eta_{Stirling} = 1 - \frac{T_L}{T_H} \tag{7.21}$$

A difference, however, exists between the two cycles; a larger area is enclosed by the Stirling cycle than the Carnot cycle on the *p-v* diagram, which means more work can be delivered per cycle. Practical attempts to follow the Stirling cycle present difficulties primarily due to the difficulty of achieving isothermal compression and isothermal expansion in a machine operating at a reasonable speed.

Example 7.12

A closed Stirling cycle operates with 1 kg of air as a working fluid between two thermal reservoirs at $T_H = 1000$ K and $T_L = 300$ K. The lowest pressure in the cycle is 0.4 MPa, and the highest pressure is 4 MPa. Calculate:
  (a) the heat and work interactions for each process
  (b) the thermal efficiency of the cycle

SOLUTION

  (a) Referring to Figure 7.23, the ideal-gas law gives

$$v_2 = v_3 = \frac{RT_3}{p_3} \quad \text{and} \quad v_4 = v_1 = \frac{RT_1}{p_1}$$

Therefore,

$$\frac{v_1}{v_2} = \frac{p_3 T_1}{p_1 T_3} = \frac{v_4}{v_3} = \frac{(4 \text{ MPa})(300 \text{ K})}{(0.4 \text{ MPa})(1000 \text{ K})} = 3$$

Energy balance for each process gives the following results:

Process 1-2

$$q_{1\text{-}2} + w_{1\text{-}2} = u_2 - u_1 = 0$$

or

$$q_{1\text{-}2} = -w_{1\text{-}2} = -\left(-\int p \, dv\right) = RT_L \ln \frac{v_2}{v_1}$$

$$= (0.287 \text{ kJ/kg K})(300 \text{ K}) \ln \left(\frac{1}{3}\right) = -94.59 \text{ kJ/kg}$$

Process 2-3

$$q_{2\text{-}3} + w_{2\text{-}3} = u_3 - u_2$$

but $w_{2\text{-}3} = 0$ so that

$$q_{2\text{-}3} = u_3 - u_2 = 759.19 - 214.36 = 544.83 \text{ kJ/kg}$$

Process 3-4

$$q_{3\text{-}4} + w_{3\text{-}4} = 0$$

so that

$$q_{3\text{-}4} = q_H = -w_{3\text{-}4} = -\left(-\int p \, dv\right) = RT_H \ln \frac{v_4}{v_3}$$

$$= (0.287 \text{ kJ/kg K})(1000 \text{ K}) \ln 3 = 315.3 \text{ kJ/kg}$$

Process 4-1

$$q_{4\text{-}1} + w_{4\text{-}1} = u_1 - u_4 \qquad \text{but} \qquad w_{4\text{-}1} = 0$$

Therefore,

$$q_{4\text{-}1} = u_1 - u_4 = 214.09 - 859.02 = -544.83 \text{ kJ/kg}$$

These results are indicated in the following table:

Process	*q* (kJ/kg)	*w* (kJ/kg)	Δu (kJ/kg)
1-2	−94.59	94.59	0.0
2-3	544.83	0.0	544.83
3-4	315.3	−315.3	0.0
4-1	−544.83	0.0	−544.83
Cycle	220.71	−220.71	0.0

(b) The thermal efficiency of the cycle is

$$\eta_{th} = \frac{|w_{net}|}{q_{input}} = \frac{|w_{net}|}{q_{3\text{-}4}} = \frac{220.71 \text{ kJ/kg}}{315.3 \text{ kJ/kg}} = 70\%$$

The same result can be obtained using Eq. (7.21):

$$\eta_t = 1 - \frac{T_L}{T_H} = 1 - \frac{300}{1000} = 70\%.$$

7.12 The Ericsson Cycle

The *Ericsson* cycle differs from the Stirling cycle in that the constant-volume processes are replaced by constant-pressure processes. Similar to the Stirling cycle, all processes are totally reversible. The *p-v* and the *T-s* diagrams of the Ericsson cycle are shown in Figure 7.25. Similar to the Stirling cycle, the heat interaction during the constant-pressure processes is internal and can be accomplished by a regenerator. For an ideal regenerator, $Q_{1\text{-}2} = -Q_{3\text{-}4}$, and the shaded areas underneath processes 1-2 and 3-4 on the *T-s* diagram are identical. The heat interactions with the surroundings Q_H and Q_L during processes 4-1 and 3-2 take place at constant temperature T_H and T_L, respectively. The thermal efficiency of the cycle is therefore the same as that of a Carnot cycle operating between the temperature limits T_H and T_L.

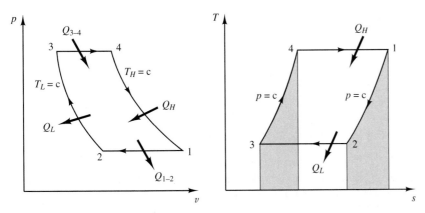

Figure 7.25   *p-v* and *T-s* diagrams of the air-standard Ericsson cycle.

Figure 7.26 shows a system that can execute an Ericsson cycle. This, in essence, is a gas turbine plant in which the internal heat exchange during processes 1-2 and 3-4 is made equal by an ideal regenerator. But similar to the Stirling engine, difficulties still remain in accomplishing isothermal compression and isothermal expansion in an engine operating at a reasonable speed. In Section 7.15, the gas turbine cycle, with multistage compression with intercooling and multistage expansion with reheating, approaches the Ericsson cycle.

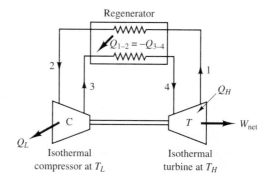

Figure 7.26   A system operating on the Ericsson cycle.

7.13 The Gas-Turbine Cycle

The gas turbine is a rotary type of internal combustion engine that usually operates on an open cycle. As shown in Figure 7.27(a), the gas turbine plant consists of a compressor, a combustion chamber, and a turbine. After compression, air enters the combustion chamber into which fuel is injected. The mixture of fuel and air burns, and the combustion occurs at a fairly constant pressure. The resulting products of combustion then expand and drive the turbine wheel, before they are discharged into the atmosphere.

The ideal cycle for the gas turbine is the *Brayton* or *Joule* cycle, shown in Figure 7.27(b). This cycle is of the closed type utilizing an ideal gas with constant specific heats as a working fluid. The combustion and heat rejection processes of actual cycles are idealized by heat interactions between thermal reservoirs and the working fluid. The Brayton cycle is shown in Figure 7.28 on p-v and T-s diagrams. It consists of the following processes:

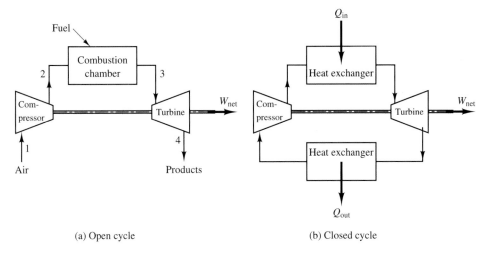

Figure 7.27   Simple gas turbine plant.

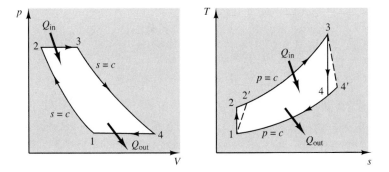

Figure 7.28   Brayton cycle.

(a) Process 1-2 is a reversible adiabatic (isentropic) compression in which both the temperature and the pressure of the system increase.

(b) During process 2-3, the system is brought in contact with a high-temperature thermal reservoir, and heat interaction takes place at constant pressure. The heat interaction process increases both the temperature and the entropy of the system. The amount of heat transferred to the system is

$$Q_{2\text{-}3} = mc_p(T_3 - T_2)$$

(c) Process 3-4 is a reversible adiabatic (isentropic) expansion in which temperature and pressure decrease.

(d) Process 4-1 is a constant-pressure heat rejection. The amount of heat rejected from the system is

$$Q_{4-1} = mc_p(T_4 - T_1)$$

The thermal efficiency of the Brayton cycle is

$$\eta_{th} = \frac{Q_{2-3} - |Q_{4-1}|}{Q_{2-3}} = 1 - \frac{|Q_{4-1}|}{Q_{2-3}} = 1 - \frac{mc_p(T_4 - T_1)}{mc_p(T_3 - T_2)}$$

$$= 1 - \frac{T_1(T_4/T_1 - 1)}{T_2(T_3/T_2 - 1)}$$

The pressures and temperatures of the end states of the compression and expansion processes are given by the following isentropic relations:

$$\frac{T_2}{T_1} = \left(\frac{p_2}{p_1}\right)^{(\gamma-1)/\gamma} = \left(\frac{p_3}{p_4}\right)^{(\gamma-1)/\gamma} = \frac{T_3}{T_4}$$

Hence

$$\frac{T_4}{T_1} = \frac{T_3}{T_2}$$

The expression for the thermal efficiency of the Brayton cycle then becomes

$$\eta_{th} = 1 - \frac{T_1}{T_2} = 1 - \frac{1}{(p_2/p_1)^{(\gamma-1)/\gamma}}$$

$$= 1 - \frac{1}{r_p^{(\gamma-1)/\gamma}} \qquad (\gamma = \text{constant})$$

(7.22)

where r_p is called the *pressure ratio* $(r_p = p_2/p_1 = p_3/p_4)$. Figure 7.29 is a plot of η_{th} versus the pressure ratio. Note that the efficiency of the Brayton cycle increases as the pressure ratio is increased. Equation (7.22) also indicates that, for the same pressure ratio and specific heat ratio, all Brayton cycles have the same thermal efficiency. An expression for the optimum pressure ratio corresponding to maximum work output (Problem 7.40) is given by

$$r_p = \left(\frac{T_3}{T_1}\right)^{\gamma/[2(\gamma-1)]}$$

(7.23)

and the maximum work done per unit mass is

$$w = c_p T_1 \left[r_p^{(\gamma-1)/\gamma} - 1 \right] + c_p T_3 \left[\frac{1}{r_p^{(\gamma-1)/\gamma}} - 1 \right] \qquad (7.24)$$

Gas turbines may operate either on a closed or on an open cycle. Closed cycles are typical in power plants using nuclear reactors as an energy source and in high-temperature solar energy systems. A suitable fluid transfers the energy from the reactor or a solar collector to the gas turbine either directly or by means of a secondary fluid through a heat exchanger. The majority of gas turbines currently in use operate on the open cycle in which the working fluid, after completing the cycle, is exhausted to the atmosphere. The air–fuel ratio used in these gas turbines is approximately 60:1.

The compressor and turbine in gas-turbine plants are usually mounted on the same shaft, and a large percentage of the power developed by the turbine is utilized to drive the compressor. The compressor power requirement varies from 40 to 80 percent of the power output of the turbine. The remainder is net power output. This high power requirement of the compressor is typical when a gas is compressed because of the large specific volume of gases in comparison to that of liquids. In the case of steam power plants, for example, the pump requirement is only a very small percentage of the turbine output. This is not true in gas-turbine power plants. Hence, careful design of the compressor is mandatory in order to minimize the power required for compression. Even with this drawback, the gas-turbine power plant has many advantages: It can operate with different fuels; it is simple in construction and easier to maintain; it has a small weight-to-power ratio; and it can handle large volumes of gas compared to reciprocating engines. Improvement in efficiency, on the other hand, is limited by the gas-turbine inlet temperature, currently in the range from 1130°C to 1260°C. Higher temperatures result in difficulties in cooling the turbine blades, and additional protection of the hot-section superalloy components by ceramic coatings becomes an absolute necessity.

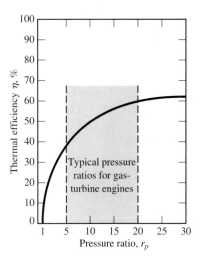

Figure 7.29   Thermal efficiency of the Brayton cycle as a function of pressure ratio ($\gamma = 1.4$).

Example 7.13

A gas turbine power plant operating at steady state receives air at 100 kPa and 300 K. The air is compressed to 500 kPa and reaches a maximum temperature of 920 K in the combustion chamber. The products of combustion expand in the turbine back to 100 kPa. Assuming an air-standard cycle, compute the thermal efficiency of the plant if the compressor and turbine are each 83 percent efficient. What is the ratio of the work required to drive the compressor to the work developed by the turbine?

SOLUTION

Figure 7.30 shows the different processes of the cycle. At 300 K, the gas tables give

$$p_{r_1} = 1.11458 \qquad h_1 = 300.47 \text{ kJ/kg}$$

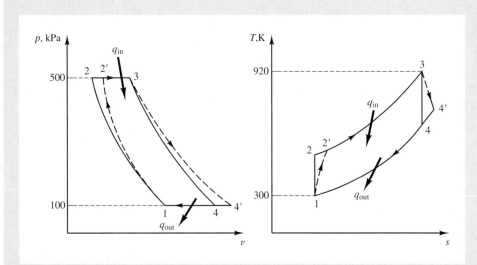

Figure 7.30

Therefore,

$$p_{r_2} = p_{r_1}\left(\frac{p_2}{p_1}\right) = 1.11458\left(\frac{500}{100}\right) = 5.5729$$

From gas tables at $p_{r_2} = 5.5729$,

$$h_2 = 475.84 \text{ kJ/kg}$$

$$h_{2'} = h_1 + \frac{h_2 - h_1}{\eta_{\text{comp}}} = 300.47 \text{ kJ/kg} + \frac{475.84 - 300.47}{0.83} \text{ kJ/kg}$$

$$= 511.76 \text{ kJ/kg}$$

At $T_3 = 920$ K,

$$p_{r_3} = 65.95261 \qquad h_3 = 955.60 \text{ kJ/kg}$$

$$p_{r_4} = p_{r_3}\left(\frac{p_4}{p_3}\right) = 65.95261\left(\frac{100}{500}\right) = 13.1905$$

From gas tables at $p_{r_4} = 13.1905$,

$$h_4 = 608.55 \text{ kJ/kg} \qquad T_4 = 601.17 \text{ K}$$

$$h_{4'} = h_3 - \eta_{\text{turb}}(h_3 - h_4) = 955.60 \text{ kJ/kg} - 0.83(955.60 - 608.55) \text{ kJ/kg}$$

$$= 667.37 \text{ kJ/kg}$$

The thermal efficiency of the cycle is

$$\eta_{th} = \frac{(h_3 - h_{4'}) - (h_{2'} - h_1)}{h_3 - h_{2'}}$$

$$= \frac{(955.60 - 667.55) \text{ kJ/kg} - (511.76 - 300.47) \text{ kJ/kg}}{(955.60 - 511.76) \text{ kJ/kg}}$$

$$= \frac{288.05 - 211.29}{443.84} = 17.29\%$$

The compressor–turbine work ratio is

$$\frac{h_{2'} - h_1}{h_3 - h_{4'}} = \frac{(511.76 - 300.47) \text{ kJ/kg}}{(955.60 - 667.55) \text{ kJ/kg}} = 0.7335$$

Example 7.14

Determine the irreversibility for each process and for the cycle of Example 7.13. Assume the high-temperature and low-temperature reservoirs are at 1000 and 280 K, respectively.

SOLUTION

Referring to Figure 7.30, the increase in entropy during the adiabatic processes 1-2' and 3-4' is due to internal irreversibilities. Processes 2'-3 and 4'-1 are internally reversible, but the change in entropy is due to heat interaction across a finite temperature difference.

The irreversibility per unit mass for each process is given by

$$i = T_0\left(s_e - s_i - \frac{q}{T_{\text{surr}}}\right)$$

Thus

$$i_{1\text{-}2'} = T_0(s_{2'} - s_1) = T_0\left[(s_{2'}^\circ - s_1^\circ) - R \ln \frac{p_2}{p_1}\right]$$

$$i_{2'\text{-}3} = T_0\left(s_3 - s_{2'} - \frac{q_{2'\text{-}3}}{T_H}\right) = T_0\left[(s_3^\circ - s_{2'}^\circ) - \frac{q_{2'\text{-}3}}{T_H}\right]$$

$$i_{3\text{-}4'} = T_0(s_{4'} - s_3) = T_0\left[(s_{4'}^\circ - s_3^\circ) - R \ln \frac{p_4}{p_3}\right]$$

$$i_{4'\text{-}1} = T_0\left(s_1 - s_{4'} - \frac{q_{4'\text{-}1}}{T_L}\right) = T_0\left[(s_1^\circ - s_{4'}^\circ) - \frac{q_{4'\text{-}1}}{T_L}\right]$$

Adding these four equations gives the irreversibility per unit mass for the cycle as

$$i = T_0\left(-\frac{q_{2'\text{-}3}}{T_H} - \frac{q_{4'\text{-}1}}{T_L}\right)$$

Referring to the solution of the previous example and using air tables, the entropy values at the corner points of the cycle can be evaluated:

at $T_1 = 300$ K $h_1 = 300.47$ kJ/kg   and $s_1^\circ = 6.86926$ kJ/kg K

at $h_{2'} = 511.76$ kJ/kg $T_{2'} = 508.15$ K   and $s_{2'}^\circ = 7.4034$ kJ/kg K

at $T_3 = 920$ K $s_3 = s_3^\circ = 8.04048$ kJ/kg K

at $h_{4'} = 667.55$ kJ/kg $T_{4'} = 656.96$ K   and $s_{4'} = s_{4'}^\circ = 7.6722$ kJ/kg K

The heat interactions during the constant-pressure processes are

$$q_{2'\text{-}3} = h_3 - h_{2'} = 955.60 - 511.76 = 443.84 \text{ kJ/kg}$$

and

$$q_{4'\text{-}1} = h_1 - h_{4'} = 300.47 - 667.55 = -367.08 \text{ kJ/kg}$$

Substituting values gives

$$i_{1\text{-}2'} = (300 \text{ K})[(7.4034 - 6.86926) \text{ kJ/kg K} - (0.287 \text{ kJ/kg K}) \ln 5]$$
$$= 300(0.5341 - 0.4619) = 300(0.0722) = 21.66 \text{ kJ/kg}$$

$$i_{2'\text{-}3} = (300 \text{ K})\left[(8.04048 - 7.4034) \text{ kJ/kg K} - \frac{443.84 \text{ kJ/kg}}{1000 \text{ K}}\right]$$
$$= 300(0.6371 - 0.44384) = 300(0.1933) = 57.978 \text{ kJ/kg}$$

$$i_{3\text{-}4'} = (300 \text{ K})\left[(7.6722 - 8.04048) \text{ kJ/kg K} - (0.287 \text{ kJ/kg K}) \ln\left(\frac{1}{5}\right)\right]$$
$$= 300(-0.3683 + 0.4619) = 300(0.0936) = 28.083 \text{ kJ/kg}$$

$$i_{4'\text{-}1} = (300 \text{ K})\left[(6.86926 - 7.6722) \text{ kJ/kg K} - \left(\frac{-367.08 \text{ kJ/kg}}{280 \text{ K}}\right)\right]$$
$$= 300(-0.8029 + 1.311) = 300(0.5081) = 152.418 \text{ kJ/kg}$$

A summary of these results is indicated in the following table:

Process	q (kJ/kg)	w (kJ/kg)	$(s_e - s_i)$(kJ/kg K)	i (kJ/kg)
1-2′	0.0	211.29	0.0722	21.660
2′-3	443.84	0.0	0.6371	57.978
3-4′	0.0	−288.05	0.0936	28.083
4′-1	−367.08	0.0	−0.8029	152.418
Cycle	76.76	−76.76	0.0	260.139

Note that the net heat input to the cycle is equal to the net work output, that the net entropy change is zero, and that the processes involving heat transfer contribute the most to the irreversibility of the cycle. Since the entropy change around the cycle is zero, the irreversibility is equal to $T_0 \Delta s_{\text{surr}}$, or

$$i = (300 \text{ K})\left(-\frac{443.84 \text{ kJ/kg}}{1000 \text{ K}} - \frac{-367.08 \text{ kJ/kg}}{280 \text{ K}}\right)$$

$$= -133.152 + 393.3 = 260.148 \text{ kJ/kg}$$

which agrees with the previously determined value.

7.14 Regenerative Gas-Turbine Cycle

In the simple gas-turbine cycle, the temperature of the exhaust gases leaving the turbine may be higher than the temperature of air leaving the compressor. Utilization of any of the energy of the exhaust gases, which is otherwise discarded to the environment, results in an improvement in the thermal efficiency of the turbine. The exhaust gases can therefore heat the air entering the combustion chamber, thereby reducing the energy requirement of the fuel. Figure 7.31(a) shows a gas turbine plant in which a regenerator is incorporated between the compressor and the combustion chamber. The hot gases, upon leaving the turbine, enter the regenerator where they transfer heat to the compressed air. Figure 7.31(b) shows the *T-s* diagram of a regenerative gas-turbine cycle. In a counterflow regenerator, the temperature difference between the exhaust gases and the air can be made very small, thus approaching reversible conditions. Under ideal

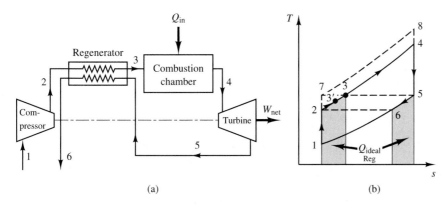

Figure 7.31   Regenerative Brayton cycle.

conditions, the temperature of the air leaving the regenerator is then equal to the temperature of the exhaust gases entering the regenerator ($T_5 = T_3$). This condition forms the basis for calculating *regenerator effectiveness,* which is defined as

$$e_{reg} = \frac{h_{3'} - h_2}{h_5 - h_6}$$

where $h_{3'}$ is the enthalpy of the air leaving the regenerator. Note that an ideal situation exists if the air leaves at state 3; the effectiveness then would be unity.

Since $h_5 = h_3$ and $h_6 = h_2$,

$$e_{reg} = \frac{h_{3'} - h_2}{h_3 - h_2} \tag{7.25}$$

For an ideal gas with constant specific heats, the preceding equation becomes

$$e_{reg} = \frac{T_{3'} - T_2}{T_3 - T_2} \tag{7.26}$$

Typical values of regenerator effectiveness are in the range from 60 to 80 percent. Note that for a high pressure ratio resulting in cycle 1-7-8-5-1, the temperature of the gas T_5 leaving the turbine is equal to the temperature of the air T_7 leaving the compressor. It is obvious in this case that a regenerator becomes useless. An expression of the thermal efficiency for the *ideal* regenerative gas-turbine cycle is

$$\eta_{th} = 1 - \frac{Q_{out}}{Q_{in}} = 1 - \frac{h_6 - h_1}{h_4 - h_3}$$

For an ideal gas with constant specific heats, this expression becomes

$$\eta_{th} = 1 - \frac{T_6 - T_1}{T_4 - T_3}$$

but $T_3 = T_5$ and $T_6 = T_2$; therefore,

$$\eta_{th} = 1 - \frac{T_2 - T_1}{T_4 - T_5} = 1 - \frac{T_1[(T_2/T_1) - 1]}{T_4[1 - (T_5/T_4)]}$$

Substituting for the temperature ratios from the isentropic relations,

$$\frac{T_2}{T_1} = \left(\frac{p_2}{p_1}\right)^{(\gamma-1)/\gamma} \quad \text{and} \quad \frac{T_5}{T_4} = \left(\frac{p_1}{p_2}\right)^{(\gamma-1)/\gamma}$$

gives

$$\eta_{th} = 1 - \left(\frac{T_1}{T_4}\right)\left(\frac{p_2}{p_1}\right)^{(\gamma-1)/\gamma} \tag{7.27}$$

According to Eq. (7.27), the thermal efficiency of a regenerative cycle depends on the ratio of the two extreme temperatures T_1/T_4 as well as on the pressure ratio. Comparison of Eqs. (7.22) and (7.27) shows an improvement in thermal efficiency when a regenerator is used. Figure 7.32 shows a plot of efficiency versus pressure ratio for different values of T_1/T_4. Points a, b, and c indicate that the temperature of the exhaust gases leaving the turbine equals the temperature of air leaving the compressor where utilization of a regenerator becomes useless.

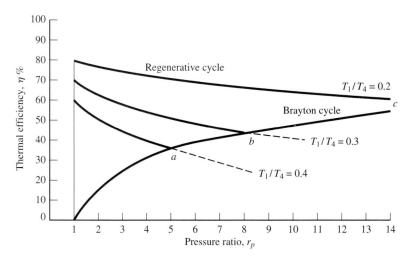

Figure 7.32   Thermal efficiency of the Brayton and regenerative cycles as a function of pressure ratio ($\gamma = 1.4$).

Example 7.15

A regenerator is incorporated in the cycle of Example 7.13. If the air is heated to 600 K before it enters the combustor, determine the thermal efficiency of the cycle and the effectiveness of the regenerator. What is the irreversibility of the cycle? Assume $T_0 = 300$ K.

SOLUTION

Referring to Figure 7.33, at $T_{3'} = 600$ K, $h_{3'} = 607.02$ kJ/kg, where prime indicates actual states. From Example 7.13, the net work output is $= 76.3$ kJ/kg. The heat transferred to the system is

$$q_H = 955.38 - 607.02 = 348.36 \text{ kJ/kg}$$

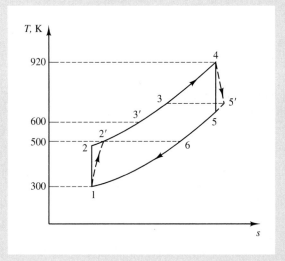

Figure 7.33

The thermal efficiency is

$$\eta_{th} = \frac{76.3 \text{ kJ/kg}}{348.36 \text{ kJ/kg}} = 21.9\%$$

and

$$e_{reg} = \frac{h_3' - h_2'}{h_3 - h_2'} = \frac{(607.02 - 511.9) \text{ kJ/kg}}{(667.37 - 511.9) \text{ kJ/kg}} = 61.2\%$$

The heat interaction in the exhaust process, assuming $h_6 = h_{2'}$, is

$$q_L = h_1 - h_{2'} = 300.19 - 511.9 = -211.71 \text{ kJ/kg}$$

The irreversibility of the cycle per unit mass is

$$i = T_0\left(-\frac{q_H}{T_H} - \frac{q_L}{T_L}\right) = (300\ \text{K})\left(-\frac{348.36\ \text{kJ/kg}}{1000\ \text{K}} - \frac{-211.71\ \text{kJ/kg}}{280\ \text{K}}\right)$$

$$= 122.324\ \text{kJ/kg}$$

Note that the reduction in irreversibility is due to the addition of the regenerator (see Example 7.14). Note also that the compressor–turbine work ratio is unchanged by the introduction of the regenerator.

7.15 Stage Compression and Expansion

In Section 7.14, it was shown that the regenerative gas-turbine cycle reduces the amount of fuel consumed in the combustion chamber and, consequently, results in an improvement in thermal efficiency. Another improvement in thermal efficiency may be obtained by adhering to isothermal rather than to isentropic compression and expansion processes. As shown in Figure 7.34, isothermal compression and expansion result, respectively, in minimum work input and maximum work output during a flow process. Hence, by adhering to isothermal processes, the work input to the compressor is reduced while the work output from the turbine is increased, resulting in an increase in the net work output. But the isothermal compression or expansion requires an infinite number of stages of intercoolers and reheaters, a situation similar to that encountered in regenerative feed water

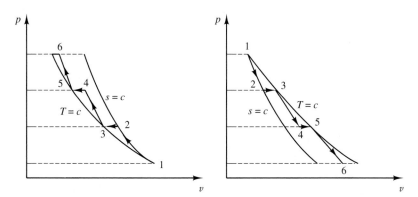

Figure 7.34 Stage compression and stage expansion.

heating in the Rankine vapor power cycles. In practice, a finite number of stage compression with intercooling and stage expansion with heating is used as a means to achieve isothermal processes. Figure 7.34 shows the sequence of processes (represented by 1-2-3-4-5-6) with three stages of compression or expansion. Cooling or heating takes place at constant pressure. A regenerative gas-turbine cycle with a double-stage compression with intercooling and a double-stage expansion with reheating is shown in Figure 7.35. The thermal efficiency of this cycle is

$$\eta_{th} = 1 - \frac{(h_{10} - h_1) + (h_2 - h_3)}{(h_6 - h_5) + (h_8 - h_7)} \tag{7.28}$$

The effect of regeneration and reheating is to increase the average temperature at which heat is transferred to the cycle. Similarly, the effect of regeneration and intercooling is to decrease the average temperature at which heat is transferred from the cycle. Both schemes result in a decrease in the external irreversibility. If, in a gas-turbine cycle, a large number of stages of intercooling, regeneration,

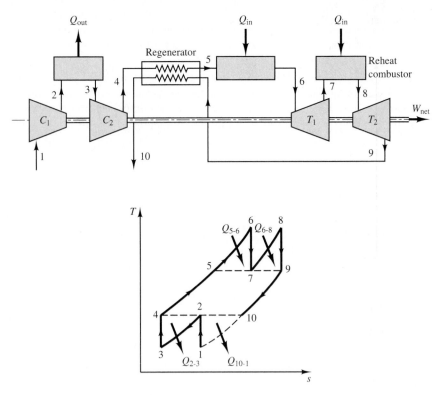

Figure 7.35   Regenerative Brayton cycle with stage compression and stage expansion.

and reheating are used, the Ericsson cycle is approached as shown Figure 7.36. In this cycle, heat interactions occur isothermally, resulting in a thermal efficiency equal to that of a Carnot cycle operating between the same temperature limits.

It is important to note that reheating alone or intercooling alone, when added to the simple Brayton cycle will decrease the cycle efficiency and will decrease the work output per unit mass flow. An increase in efficiency can be attained only when reheating and intercooling are incorporated together and when either or both is added to the regenerative gas-turbine cycle.

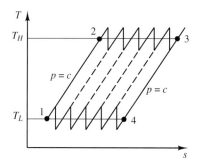

Figure 7.36 Gas turbine cycle with multistage intercooling and reheating approaches the Ericsson cycle.

Example 7.16

Referring to the modified air-standard Brayton cycle shown in Figure 7.37, the temperatures at the inlet of the compressors and turbines are $T_1 = T_3 = 300$ K and $T_6 = T_8 = 1200$ K, and the pressure ratio $p_2/p_1 = p_4/p_3 = p_6/p_7 = p_8/p_9 = 4$. Assuming isentropic flow in the compressors and turbines, and regenerator effectiveness of 80 percent, determine the thermal efficiency of the cycle.

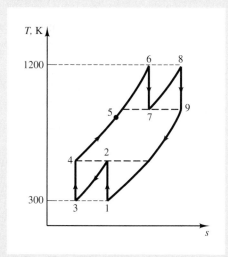

Figure 7.37

SOLUTION

From Table A11 at $T_1 = 300$ K,

$$h_1 = h_3 = 300.47 \text{ kJ/kg} \quad \text{and} \quad p_{r_1} = p_{r_3} = 1.11458$$
$$h_6 = h_8 = 1277.81 \text{ kJ/kg}$$

Since the compression processes 1-2 and 3-4 are isentropic,

$$p_{r_2} = p_{r_4} = p_{r_1}\frac{p_2}{p_1} = (1.11458)(4) = 4.4583$$

Interpolating in Table A11 gives $h_2 = h_4 = 446.64$ kJ/kg. Hence, the work input per unit mass for each compressor is

$$w_{1\text{-}2} = w_{3\text{-}4} = h_2 - h_1 = 446.64 - 300.47 = 147.17 \text{ kJ/kg}$$

Similarly, at $T_6 = T_8 = 1200$ K,

$$h_6 = h_8 = 1277.81 \text{ kJ/kg} \qquad \text{and} \qquad p_{r_6} = p_{r_8} = 191.1735$$

Since the expansion processes 6-7 and 8-9 are isentropic,

$$p_{r_7} = p_{r_9} = p_{r_6}\frac{p_7}{p_6} = (191.1735)\left(\frac{1}{4}\right) = 47.9338$$

Interpolating in Table A11 gives $h_7 = h_9 = 874.52$ kJ/kg, and the work developed per unit mass by each turbine is

$$w_{6\text{-}7} = w_{8\text{-}9} = h_7 - h_6 = 874.521 - 1277.81 = -403.29 \text{ kJ/kg}$$

The net work developed per unit mass is

$$w_{\text{net}} = 2(-403.29 + 147.17) = -512.23 \text{ kJ/kg}$$

The specific enthalpy at state 5 can be determined from the expression of the regenerator effectiveness

$$e_{\text{reg}} = \frac{h_5 - h_4}{h_9 - h_4}$$

from which

$$h_5 = h_4 + e_{\text{reg}}(h_9 - h_4)$$
$$= 446.64 + 0.8(874.52 - 446.64) = 788.94 \text{ kJ/kg}$$

The heat input per unit mass is

$$q = (h_6 - h_5) + (h_8 - h_7)$$
$$= (1277.81 - 788.94) \text{ kJ/kg} + (1277.81 - 874.52) \text{ kJ/kg} = 892.16 \text{ kJ/kg}$$

The thermal efficiency is

$$\eta_t = \frac{w_{\text{net out}}}{q} = \frac{512.23 \text{ kJ/kg}}{892.16 \text{ kJ/kg}} = 0.5741$$

Figure 7.38 presents p-v and T-s diagrams of internal combustion engine cycles discussed in previous sections.

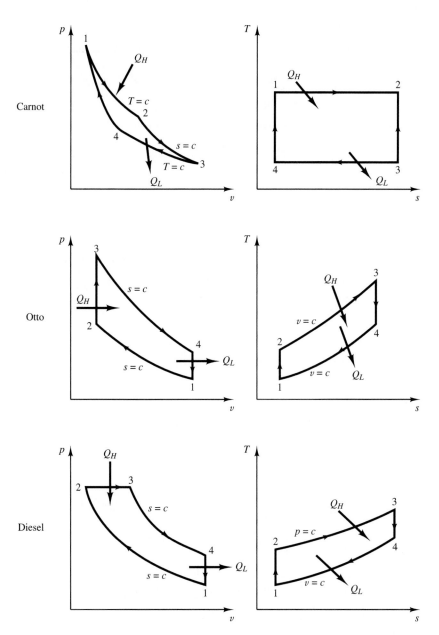

Figure 7.38 p-v and T-s diagrams of internal combustion engine cycles. *(continued)*

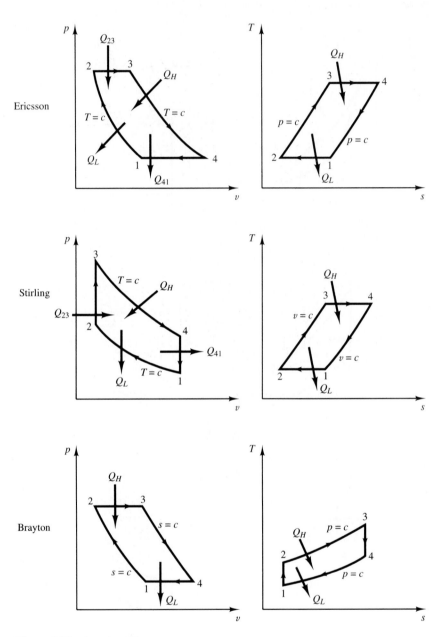

Figure 7.38   (*continued*)

7.16 Turbo-Jet

A turbo-jet engine operates on an open gas-turbine cycle and generates a high-velocity gas stream. The turbo-jet engine consists of the same types of components as in the gas-turbine plant, but in addition, it contains an expansion nozzle. As shown in Figure 7.39, the components are an air intake, compressor, combustor, turbine, and exit nozzle. The cycle of an ideal turbo-jet engine appears on

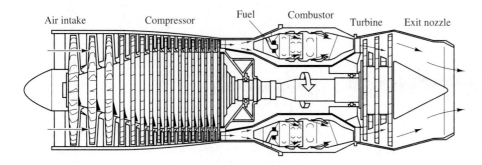

Figure 7.39   Turbo-jet engine.

a *T-s* diagram in Figure 7.40. Air entering the diffuser is decelerated and slightly compressed, changing from state 1 to state 2. The air entering the compressor is further compressed proceeding from state 2 to state 3. The compressed air then enters the combustor, where it is heated at constant pressure from state 3 to state 4. In the turbine, the air expands from state 4 to state 5, and further expansion takes place in the exit nozzle to state 6. As a result of the momentum of the exhaust gases flowing at high velocity from the nozzle, the engine exerts a forward thrust. In actual turbo-jets, a slight pressure drop occurs in the combustor. Further, the compression and expansion processes are not quite isentropic. These nonideal processes are shown as dashed lines in Figure 7.40.

Since the entire power output of the turbine is used to drive the compressor, turbo-jets are usually rated in terms of *specific thrust,* which is defined as the thrust force divided by the air flow rate.[1]

$$F_{\text{specific}} = \frac{F}{\dot{m}_{\text{air}}} \tag{7.29}$$

[1] The *specific thrust* may alternatively be defined as the thrust force divided by the flow rate of the fuel, rather than that of air.

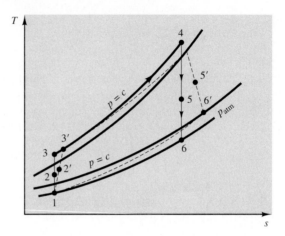

Figure 7.40   *T-s* diagram for a turbo-jet engine.

Assuming that atmospheric pressure acts on the turbo-jet and neglecting the momentum of the fuel at the entrance, the thrust of the turbo-jet, according to the momentum equation, is

$$F = (\dot{m}_a + \dot{m}_f)V_6 - \dot{m}_aV_1 = \dot{m}_a\left[\left(1 + \frac{m_f}{m_a}\right)V_6 - V_1\right] \qquad (7.30)$$

where V_1 and V_6 are the velocities of the flow entering and leaving the system, and subscripts a and f refer to air and fuel, respectively. The *specific thrust* is

$$F_{sp} = \left(1 + \frac{m_f}{m_a}\right)V_6 - V_1 \qquad (7.31)$$

Several modifications of the turbo-jet engine have been made in order to improve its operation within a certain range of speed. In the turboprop engine shown in Figure 7.41(a), the net power developed by the turbine drives a propeller that provides most of the thrust to the aircraft. Turboprop engines are most efficient at speeds up to 600 km/hr. In the turbofan engine shown in Figure 7.41(b), the turbine drives a large-diameter fan that induces flow of air around the core of the engine. The exhaust from the turbine and the fan leaves at a high velocity to provide the necessary thrust. Turbofan engines are commonly used for commercial aircraft with flight speeds up to 100 km/hr. The ramjet, shown in Figure 7.41(c), is a simple engine used for high-speed propulsion. It has no compressor or turbine, and its flow passage decelerates the high-speed, incoming air (ram effect), causing an increase in pressure within the diffuser. Therefore, a ramjet must already be in flight at a high speed before it can operate. The exhaust leaves at high speed and provides the necessary thrust.

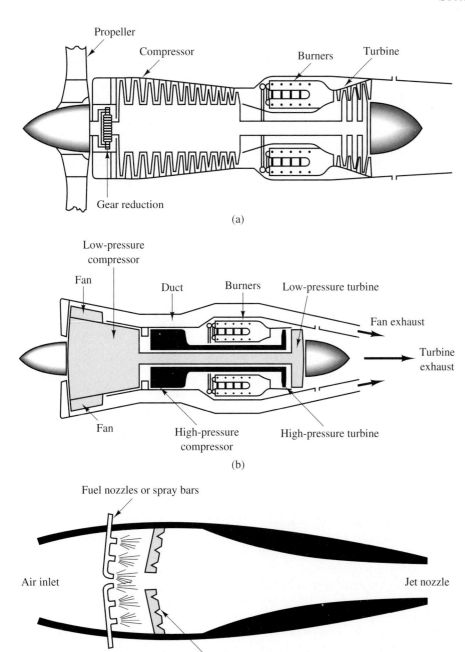

Figure 7.41 Aircraft engines: (a) turboprop, (b) turbofan, (c) ramjet. [*The Aircraft Gas Turbine Engine and Its Operation.* © United Aircraft Corp. (Now United Technologies Corp.), 1951, 1974].

Example 7.17

A jet-propelled plane consuming air at the rate of 18 kg/s is to fly at a Mach number[2] 0.6 at an altitude of 4600 m (p = 55 kPa, T = $-20°C$). The diffuser, which has a pressure coefficient of 0.9, decelerates the flow to a negligible velocity. The compressor pressure ratio is 5, and the maximum temperature in the combustion chamber is 1000°C. After expanding in the turbine, the gases continue to expand in the nozzle to a pressure of 60 kPa. The isentropic efficiencies of the compressor, turbine, and nozzle are, respectively, 0.81, 0.85, and 0.915. The enthalpy of combustion of the fuel $\Delta h° = -46520$ kJ/kg. Assuming that the products of combustion have the same properties as air, find:

(a) the power input to the compressor
(b) the power output of the turbine
(c) the fuel–air ratio on a mass basis
(d) the exit Mach number
(e) the thrust provided by the engine
(f) the thrust power developed.

SOLUTION

(a), (b) Referring to Figures 7.39 and 7.42, the velocity at the diffuser inlet is

$$V_1 = M_1 c_1 = 0.6(20.1\sqrt{253.15}) = 192 \text{ m/s}$$

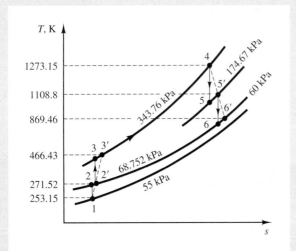

Figure 7.42

From gas tables at T_1 = 253.15 K, h_1 = 253.45 kJ/kg. Since the velocity at point 2 is negligible,

---

[2] The *Mach number* is the ratio of local flow velocity to the local velocity of sound ($M = V/c$), for an ideal gas, $M = V/\sqrt{\gamma RT}$, and $c = 20.1\sqrt{T}$ for air.

$$h_2 = h_{2'} = h_1 + \frac{V_1^2}{2} = 253.45 \text{ kJ/kg} + \frac{(192 \text{ m/s})^2}{2(1000 \text{ J/kJ})}$$
$$= 253.45 + 18.432 = 271.88 \text{ kJ/kg}$$

At $h_2 = 271.88$ kJ/kg, $T_2 = 271.52$ K.
  For isentropic flow in the diffuser,

$$\frac{p_2}{p_1} = \left(\frac{T_2}{T_1}\right)^{\gamma/(\gamma-1)}$$

from which

$$p_2 = (55 \text{ kPa})\left(\frac{271.52}{253.15}\right)^{1.4/0.4} = 70.28 \text{ kPa}$$

$$p_{2'} = p_1 + \eta_{\text{diff}}(p_2 - p_1) = (55 \text{ kPa}) + 0.9(70.28 - 55) \text{ kPa}$$
$$= 68.752 \text{ kPa}$$

At state $2'$, $p_{r_{2'}} = 0.7908$. Therefore,

$$p_{r_3} = \left(\frac{p_3}{p_{2'}}\right)p_{r_{2'}} = 5(0.7908) = 3.9542$$

giving

$$h_3 = 431.48 \text{ kJ/kg of air}$$
$$h_{3'} = h_{2'} + \frac{h_3 - h_{2'}}{\eta_{\text{comp}}} = 271.88 \text{ kJ/kg} + \frac{(431.48 - 271.88) \text{ kJ/kg}}{0.81}$$
$$= 468.92 \text{ kJ/kg of air}$$

At $h_{3'} = 468.92$ kJ/kg,

$$T_{3'} = 466.43 \text{ K} \qquad \text{and} \qquad p_3 = 5(68.752 \text{ kPa}) = 343.76 \text{ kPa}$$

The work done on the compressor is

$$w = h_{3'} - h_2 = (468.92 - 271.88) \text{ kJ/kg} = 197.04 \text{ kJ/kg of air}$$

The power input to the compressor is

$$\dot{m}_{\text{air}}(h_{3'} - h_2) = (18 \text{ kg/s})(197.04 \text{ kJ/kg}) = 3546.72 \text{ kW}$$

which is equal to the power output from the turbine.

(c) At $T_4 = 1273.15$ K, $h_4 = 1364.1$ kJ/kg, $p_{r_4} = 244.3981$. Neglecting changes in kinetic energies, the first law applied to the combustor gives

$$\dot{m}_a(h_{3'} - h°)_a + \dot{m}_f(h_3 - h°)_f + \dot{m}_f(-\Delta h°)_f = (\dot{m}_a + \dot{m}_f)(h_4 - h_4°)_p$$

where superscript ° refers to the standard state at 25°C, and subscripts a, f, and p refer to air, fuel, and products, respectively. Assuming that the fuel is introduced at 25°C, the preceding equation may be written

$$(h_{3'} - h°)_a + \frac{\dot{m}_f}{\dot{m}_a}(-\Delta h°)_f = \left(1 + \frac{\dot{m}_f}{\dot{m}_a}\right)(h_4 - h_4°)_p$$

Assuming that the properties of the products are the same as those of air, then

$$(468.92 - 298.34) \text{ kJ/kg air} + \frac{\dot{m}_f}{\dot{m}_a}(46520 \text{ kJ/kg fuel}) = \left(1 + \frac{\dot{m}_f}{\dot{m}_a}\right)[(1364.1 - 298.34) \text{ kJ/kg prod}]$$

from which

$$\frac{\dot{m}_f}{\dot{m}_a} = 0.0197 \qquad \text{and} \qquad \frac{\dot{m}_a}{\dot{m}_p} = 0.9807$$

(d) The work done on the compressor = $197.04 \times 0.9807 = 193.23$ kJ/kg of products. The enthalpy at the exit of the turbine is

$$h_{5'} = 1364.1 - 193.23 = 1170.87 \text{ kJ/kg of products}$$

corresponding to a temperature $T_{5'} = 1108.35$ K and $p_{r_{5'}} = 138.4833$. For isentropic expansion, the enthalpy at the exit of the turbine is

$$h_5 = h_4 - \frac{h_4 - h_{5'}}{\eta_{\text{turbine}}} = 1364.1 - \left(\frac{193.23}{0.85}\right)$$
$$= 1364.1 - 227.33 = 1136.77 \text{ kJ/kg of products}$$

At $h_5 = 1136.77$ kJ/kg,

$$p_{r_5} = 124.1799$$

$$p_5 = p_{5'} = p_4\left(\frac{p_{r_5}}{p_{r_4}}\right) = (343.76 \text{ kPa})\left(\frac{124.1799}{244.3981}\right) = 174.67 \text{ kPa}$$

$$p_{r_6} = \left(\frac{p_6}{p_{5'}}\right)p_{r_5'} = \frac{60}{174.67}(138.4833) = 47.5697$$

corresponding to $h_6 = 872.69$ kJ/kg of products. The velocity at the exit of the nozzle is

$$
\begin{aligned}
V_{6'} &= \sqrt{\eta_{\text{nozzle}}}\sqrt{2(h_{5'} - h_6)} \\
&= \sqrt{(0.915)(2)(1000 \text{ J/kJ})(1170.87 - 872.69) \text{ kJ/kg}} \\
&= 738.69 \text{ m/s}
\end{aligned}
$$

$$
\begin{aligned}
h_{6'} &= h_{5'} - \frac{V_{6'}^2}{2} = 1170.87 \text{ kJ/kg} - \frac{(738.69 \text{ m/s})^2}{2(1000 \text{ J/kJ})} \\
&= 1170.87 - 272.83 = 899.04 \text{ kJ/kg}
\end{aligned}
$$

corresponding to a temperature $T_{6'} = 869.46$ K. The Mach number at the exit is

$$
M_{6'} = \frac{V_{6'}}{20.1\sqrt{T_{6'}}} = \frac{738.69 \text{ m/s}}{(20.1\sqrt{869.46}) \text{ m/s}} = 1.2464
$$

(e) Neglecting the difference in pressure forces on the turbo-jet and assuming that the momentum of the fuel at entrance is small, the thrust force is

$$
\begin{aligned}
F &= \dot{m}_a \left[\left(1 + \frac{\dot{m}_f}{\dot{m}_a} \right) V_{6'} - V_1 \right] \\
&= (18 \text{ kg/s})[(1 + 0.0197)(738.69 \text{ m/s}) - 192 \text{ m/s}] = 10{,}034.28 \text{ N}
\end{aligned}
$$

(f) The thrust power is

$$
P = FV_1 = (10034.28 \text{ N})(192 \text{ m/s}) = 1926.58 \text{ kW}
$$

7.17 Refrigeration Cycles

A refrigeration system removes thermal energy from a low-temperature region and transfers heat to a high-temperature region. Refrigeration can be accomplished by noncyclic methods, such as by melting ice or by subliming carbon dioxide (dry ice), or by cyclic methods. Analogous to power cycles, refrigeration cycles may be classified as gas-compression cycles or vapor-compression cycles. Of these two, the vapor-compression cycles are more commonly used and will be discussed in the following sections.

The Carnot cycle can serve as an initial model of the ideal refrigeration cycle. This cycle gives the maximum coefficient of performance of any refrigeration cycle operating between the same two thermal reservoirs. Figure 7.43 shows the

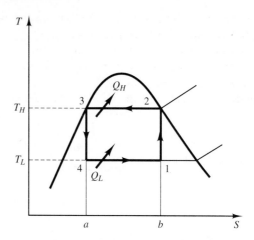

Figure 7.43   Carnot refrigeration cycle.

Carnot refrigeration cycle on a T-S diagram. The working substance is a condensable vapor. Starting at state 1 and following the arrows, the four processes of the cycle are reversible adiabatic compression, reversible isothermal heat rejection, reversible adiabatic expansion, and reversible isothermal heat absorption. Since all processes are reversible, areas on the T-S diagram represent heat transfers. During the heat rejection (process 2-3), condensation takes place at a constant temperature T_H, and the area a-3-2-b-a on the T-S diagram represents the heat transfer Q_H. Similarly, during process 4-1, evaporation takes place as heat is transferred from the surrounding space at temperature T_L, and the area a-4-1-b-a on the T-S diagram represents the refrigeration load Q_L. The enclosed area 1-2-3-4-1 is the net heat transfer from the refrigerant and is equal to the net work done on the refrigerant. The coefficient of performance of the Carnot refrigeration cycle is given by

$$\beta_{\text{ref}} = \frac{Q_L}{Q_H - Q_L} = \frac{T_L}{T_H - T_L} \tag{7.32}$$

It is not practical, however, to operate a refrigeration unit according to the Carnot cycle requirements. The main difficulty lies in the isothermal heat interaction processes. These processes proceed at a very slow rate so that very large heat transfer surfaces are needed. Furthermore, the presence of the liquid phase of the working fluid in the compressor creates problems, for it causes severe erosion of the compressor. This difficulty is avoided in the Rankine refrigeration cycle. Refrigerant entering the compressor in the Rankine cycle consists of saturated vapor, and the subsequent compression process produces only superheated vapor. Another feature that makes the Carnot cycle impractical arises from the reversible expansion (process 3-4). Only a work-producing machine, such as a turbine or an engine, can produce reversible expansion of a fluid. The work output of such a machine is too small to justify its expense. For this reason, an expansion or a throttling valve is used in the Rankine cycle so that a lower cost is achieved, but the expansion is not reversible.

A unit used in refrigeration systems to indicate the rate of cooling is called the *ton*. One ton of refrigeration[3] is equivalent to 200 Btu/min or 12,000 Btu/hr (3.517 kW). Another index of performance of refrigerators is the power input per ton of refrigeration. This may be expressed in terms of the coefficient of performance β as

$$\text{horse power per ton} = \frac{W_{net}/2544.4}{Q_L/12,000} = \frac{4.716}{\beta} \qquad (7.33)$$

or

$$\text{kW per ton} = \frac{W_{net}/3412}{Q_L/12,000} = \frac{3.517}{\beta} \qquad (7.34)$$

7.18 The Rankine Vapor-Compression Refrigeration Cycle

Analogous to the Rankine vapor-power cycle, the Rankine vapor-compression refrigeration cycle is considered the ideal model for actual cycles. As shown in Figure 7.44, it is a closed cycle using a refrigerant,[4] such as ammonia, refrigerant-12, refrigerant-134a, methylchloride, and carbon dioxide. The cycle consists of the following sequence of processes taking place in the compressor, condenser, expansion valve, and evaporator, respectively:

(a) Process 1-2 is a reversible adiabatic (isentropic) compression. The saturated vapor at state 1 is superheated to state 2. The amount of work done on the system (working substance) of mass m is

$$W_{1\text{-}2} = m(h_2 - h_1)$$

(b) Process 2-3 is an internally reversible constant-pressure heat rejection in which the working substance is first desuperheated and then condensed to the saturated liquid state 3. During this process, the working substance rejects most of its energy to the condenser cooling water. The amount of heat interaction is

$$Q_{2\text{-}3} = m(h_3 - h_2)$$

---

[3] A ton of refrigeration is defined as the rate of thermal energy extraction corresponding to the production, in a period of 24 hr, of 1 ton of ice (2000 lbm) at 0°C from water at the same temperature.

[4] Tables of properties and *p-h* charts for ammonia, R-12, and R-134a are given in the Appendix.

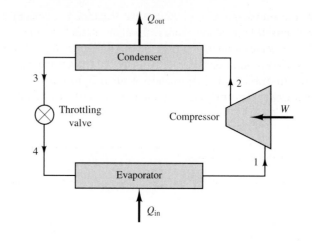

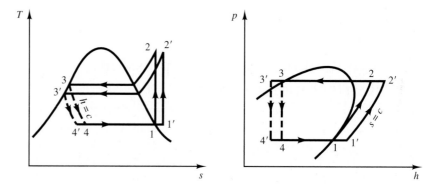

Figure 7.44   Vapor-compression refrigeration cycle.

(c) Process 3-4 is an irreversible throttling process in which the temperature and pressure decrease at constant enthalpy ($h_3 = h_4$).

(d) Process 4-1, which completes the cycle, is an internally reversible constant-pressure heat interaction in which the working substance is evaporated to the saturated vapor state 1. The latent enthalpy necessary for evaporation is supplied by the refrigerated space surrounding the evaporator. The amount of heat transferred to the working substance in the evaporator is called the *refrigeration effect* or *refrigeration load* and is given by

$$Q_{4\text{-}1} = m(h_1 - h_4).$$

The coefficient of performance of the Rankine refrigeration cycle is

$$\beta = \frac{Q_{4\text{-}1}}{W_{1\text{-}2}} = \frac{h_1 - h_4}{h_2 - h_1} \tag{7.35}$$

Actual refrigeration cycles have a lower value of β than that given by Eq. (7.35) because of the inevitable irreversibilities and heat interactions encountered in actual cycles. Also states 3 and 1 do not necessarily lie on the saturated liquid and vapor lines. Note also that frictional effects result in pressure drops as the refrigerant flows through the different components and the interconnection piping. The cycle identified by primes in Figure 7.44 shows these deviations.

Example 7.18

An ideal ammonia refrigerating machine is required to transfer 18 kW from a cold room. The evaporator temperature is $-18°C$, and the condenser temperature $32°C$. Assuming saturated vapor enters the compressor and saturated liquid leaves the condenser, compute:
  (a) evaporator pressure
  (b) condenser pressure
  (c) coefficient of performance
  (d) power requirements
  (e) volumetric flow rate of vapor entering compressor
  What is the coefficient of performance of a Carnot refrigeration cycle operating between the same temperature limits?

SOLUTION

Referring to Figure 7.45, the following pressures may be obtained from ammonia tables:
  (a) Evaporator pressure $p_1 = 207.71$ kPa.
  (b) Condenser pressure $p_3 = 1237.41$ kPa.
  (c) The enthalpies at the key points are

$$h_1 = 1421.7 \text{ kJ/kg} \qquad h_2 = 1684.2 \text{ kJ/kg} \qquad h_3 = h_4 = 332.6 \text{ kJ/kg}$$

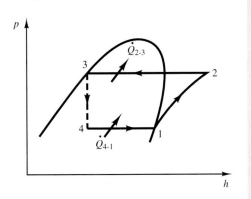

Figure 7.45

The refrigeration load per unit mass is

$$q_{4\text{-}1} = h_1 - h_4 = 1421.7 - 332.6 = 1089.1 \text{ kJ/kg}$$

The work input to the compressor per unit mass is

$$w_{1\text{-}2} = h_2 - h_1 = 1684.2 - 1421.7 = 262.5 \text{ kJ/kg}$$

The coefficient of performance is

$$\beta = \frac{1089.1 \text{ kJ/kg}}{262.5 \text{ kJ/kg}} = 4.15$$

(d) The rate of refrigerant flow is

$$\dot{m} = \frac{18 \text{ kW}}{1089.1 \text{ kJ/kg}} = 0.01653 \text{ kg/s}$$

The power input is

$$\dot{m}w_{1\text{-}2} = (0.01653 \text{ kg/s})(262.5 \text{ kJ/kg}) = 4.339 \text{ kW}$$

(e)

$$\text{volumetric flow rate} = \dot{m}v_1$$
$$= (0.01653 \text{kg/s})(0.5743 \text{m}^3\text{/kg})$$
$$= 0.009493 \text{m}^3\text{/s}$$

The coefficient of performance of a Carnot refrigerator operating between $-18°C$ and $32°C$ is

$$\beta_{\text{Carnot}} = \frac{T_L}{T_H - T_L} = \frac{255.15 \text{ K}}{(305.15 - 255.15) \text{ K}} = 5.10$$

Example 7.19

If the compressor of the vapor compression cycle of Example 7.18 has an efficiency of 80 percent, other data being the same, calculate:

(a) the compressor power
(b) the coefficient of performance
(c) the irreversibility rates of the compressor and expansion valve (assume $T_0 = 300$ K)

SOLUTION

(a) Referring to Figure 7.46 and from the solution of the previous example, the compressor efficiency is

$$\eta_c = \frac{h_{2s} - h_1}{h_2 - h_1}$$

$$0.8 = \frac{(1684.2 - 1421.7) \text{ kJ/kg}}{(h_2 - 1421.7) \text{ kJ/kg}}$$

from which $h_2 = 1749.8$ kJ/kg.

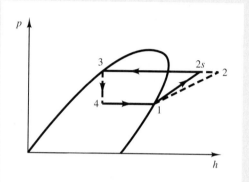

Figure 7.46

The compressor power is

$$\dot{W}_c = \dot{m}(h_2 - h_1) = (0.01653 \text{ kg/s})[(1749.8 - 1421.7) \text{ kJ/kg}]$$
$$= 5.424 \text{ kW}$$

(b) The coefficient of performance is

$$\beta = \frac{h_1 - h_4}{h_2 - h_1} = \frac{(1421.7 - 332.6) \text{ kJ/kg}}{(1749.8 - 1421.7) \text{ kJ/kg}} = 3.32$$

(c) The irreversibility rate for the compressor is

$$\dot{I}_c = \dot{m}T_0(s_2 - s_1)$$

Interpolation in superheated ammonia tables at $h_2 = 1749.8$ kJ/kg and $p_2 = 1237.41$ kPa gives $s_2 \approx 5.75$ kJ/kg K. Substituting values gives

$$\dot{I}_c = (0.01653 \text{ kg/s})(300 \text{ K})\left[(5.75 - 5.59)\frac{\text{kJ}}{\text{kg K}}\right] = 0.7934 \text{ kW}$$

The irreversibility rate for the expansion valve is

$$\dot{I}_{\text{valve}} = \dot{m}T_0(s_4 - s_3)$$

but

$$h_4 = 332.6 = h_{f_4} + x_4 h_{fg4}$$

then

$$x_4 = \frac{h_4 - h_{f_4}}{h_{fg4}} = \frac{322.6 - 98.8}{1322.9} = 0.1767$$

and

$$s_4 = s_{f_4} + x_4 s_{fg4} = 0.4040 + (0.1767)(5.1860) = 1.3204 \text{ kJ/kg K}$$

Hence

$$\dot{I}_{\text{valve}} = (0.01653 \text{ kg/s})(300 \text{ K})[(1.3204 - 1.2343) \text{ kJ/kg K}]$$
$$= 0.427 \text{ kW}$$

Note that the irreversibility rates can also be calculated from the reduction of the availability rates in the compressor and the expansion valve.

7.19 Modifications of the Rankine Refrigeration Cycle

Particular applications dictate modifications of the Rankine refrigeration cycle and evidently result in an improvement of the coefficient of performance. A multiple evaporator refrigeration system using a single compressor and a single condenser is shown in Figure 7.47. This system is used for an application requiring two or more low-temperature regions. The valve at the exit of the high-pressure evaporator regulates the suction pressure of the compressor. Figure 7.48 shows a double-stage compression refrigeration system. Multiple compression systems are justified when the compression ratio is high and stage compression with intercooling reduces the work input to the compressors. The high-pressure receiver (or intercooler) desuperheats the incoming vapor from the low-pressure compressor and also serves to supply refrigerant in the liquid state only to the expansion valve.

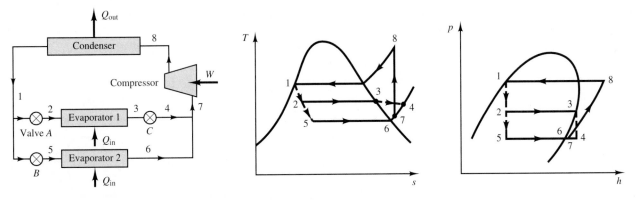

Figure 7.47   Multiple evaporator refrigeration system.

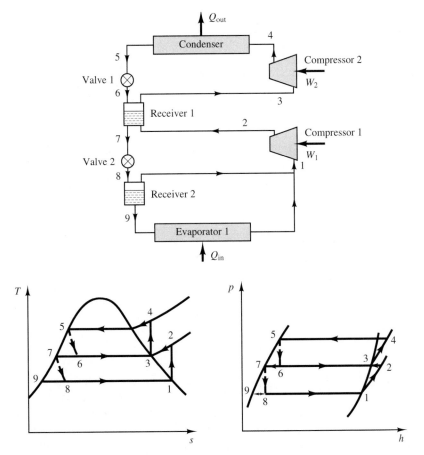

Figure 7.48   Multiple-stage compression and expansion refrigeration system.

The low-pressure receiver serves as a by-pass of the vapor from the expansion valve to the inlet of the low-pressure compressor. Figure 7.49 shows other modifications that improve the performance of vapor-refrigeration cycles.

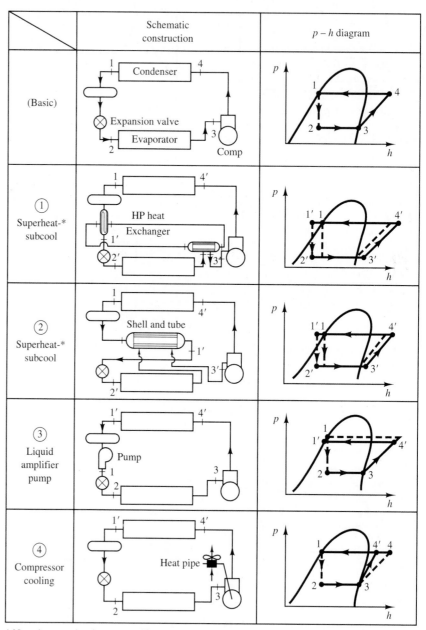

* Note that superheating to state $3'$ results in an increase in the work of the compressor.

Figure 7.49   Schemes to improve the performance of refrigeration cycles.

Example 7.20

A multiple evaporator refrigeration system uses R-12 as a refrigerant. Pressures in the condenser, evaporator 1, and evaporator 2 are 847.7, 308.6, and 219.1 kPa, respectively. Saturated liquid leaves the condenser, and saturated vapor leaves each evaporator.

Valve *C* is an evaporator-pressure regulator that maintains the higher pressure in evaporator 1, whereas the compressor suction pressure is the pressure in evaporator 2. If the loads on evaporators 1 and 2 are, respectively, 10 and 5 tons, and the system is ideal, find:

(a) the mass rate of flow of refrigerant passing through each evaporator
(b) the power required to operate the compressor

SOLUTION

Referring to Figure 7.47, and using R-12 tables, property values of state 1 to 8 are given in the following table:

State	1	2	3	4	5	6	7	8
p (kPa)	847.7	308.6	308.6	219.1	219.1	219.1	219.1	847.7
T (°C)	35	0	0	-3.06	-10	-10	-5.28	47.2
h (kJ/kg)	69.494	69.494	187.397	187.397	69.494	183.058	185.91	210.47
s (kJ/kg K)							0.71385	0.71385

The mass flow rates in evaporators 1 and 2 are

$$\dot{m}_3 = \frac{(10 \text{ ton})(3.517 \text{ kW/ton})}{(187.397 - 69.494) \text{ kJ/kg}} = 0.298 \text{ kg/s}$$

and

$$\dot{m}_6 = \frac{(5 \text{ ton})(3.517 \text{ kW/ton})}{(183.058 - 69.494) \text{ kJ/kg}} = 0.1548 \text{ kg/s}$$

An energy balance gives the enthalpy at the compressor's inlet as

$$\begin{aligned} h_7 &= \frac{\dot{m}_4 h_4 + \dot{m}_6 h_6}{\dot{m}_4 + \dot{m}_6} \\ &= \frac{(0.298 \text{ kg/s})(187.397 \text{ kJ/kg}) + (0.1548 \text{ kg/s})(183.058 \text{ kJ/kg})}{0.4528 \text{ kg/s}} \\ &= 185.93 \text{ kJ/kg} \end{aligned}$$

The enthalpy of the refrigerant at the end of compression $h_8 = 210.47$ kJ/kg. The power input to the compressor is

$$\begin{aligned} \dot{W} &= \dot{m}_7(h_8 - h_7) \\ &= [(0.2980 + 0.1548) \text{ kg/s}](210.47 - 185.91) \text{ kJ/kg} = 11.12 \text{ kW} \end{aligned}$$

7.20 Absorption Refrigeration Systems

Considerable savings in power input can be made if a pump replaces the compressor in a refrigeration system because of the small specific volumes of liquids compared to gases. This may be accomplished by absorbing the refrigerant gas in a liquid such as water. In absorption refrigeration systems, the refrigerant is absorbed and desorbed by varying the temperature, and the degree of absorption increases as the temperature of the solution decreases. The liquid is used as a transport means to carry the refrigerant in solution form through the refrigeration cycle. The most widely used absorption refrigeration system is the ammonia–water system. Other absorption systems are water–lithium bromide and water–lithium chloride systems. In these two systems, water is used as the refrigerant, and thus their applications are limited by the freezing temperature of water.

Figure 7.50 is a diagram of an ammonia–water absorption cycle. In the absorber, ammonia is absorbed by water in an exothermic reaction, and cooling is necessary to maintain the temperature constant. The strong ammonia–water solution is pumped to the generator, where the ammonia separates from the water in an endothermic reaction. Heat is transferred to the generator to enhance the separation process. The weak solution remaining in the generator flows back to the absorber through a valve. A rectifier is used to further separate any water vapor carried by the ammonia vapor. At this stage, the ammonia vapor undergoes the normal vapor-compression refrigeration cycle. It condenses, passes through a throttling valve, and then vaporizes in the evaporator. After accomplishing the cooling effect, it enters the absorber, where it is absorbed by water, and the cycle

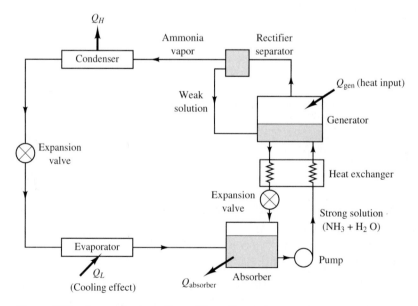

Figure 7.50   Ammonia absorption refrigeration system.

is repeated. To improve the performance of the cycle, a heat exchanger or regenerator is inserted between the absorber and the generator, as shown in the figure.

7.21 Gas Refrigeration Systems

In gas refrigeration systems, the working fluid remains a gas throughout the cycle. Gas refrigeration systems are commonly used to achieve very low temperatures for applications such as the liquifaction of gases. Another common use, with air as a working fluid, is aircraft cabin cooling.

   The ideal gas refrigeration cycle operates on a reversed closed Brayton cycle. A schematic and a *T-s* diagram of an aircycle refrigeration system is shown in Figure 7.51. The refrigerant gas is first compressed isentropically along process 1-2, and after an initial cooling at constant pressure in a heat exchanger along process 2-3, it is allowed to expand isentropically in the turbine along process 3-4, causing a

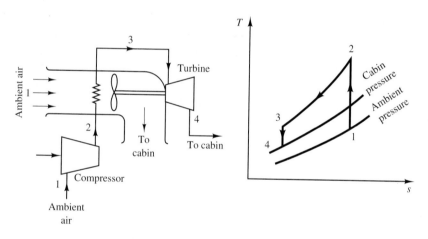

Figure 7.51   Air-refrigeration cycle.

further reduction in temperature. The work developed by the turbine can be used to offset some of the work required by the compressor. Alternatively, the turbine work can be used to circulate the cool air first through the heat exchanger and then into the space to be cooled. Figure 7.52 shows an aircycle unit used in aircraft cabin or cockpit cooling. In this system, air is bled from the jet-engine compressor and serves as a primary source of compressed air. The air is cooled in a heat exchanger and compressed further by a small compressor. It is then cooled before it expands in a small turbine. The unit uses a high-speed compressor and turbine that can handle large volumes of air. The moisture collected in the water separator is used to lower the temperature of the ram air used to cool the engine bleed-air and the compressed air leaving the compressor.

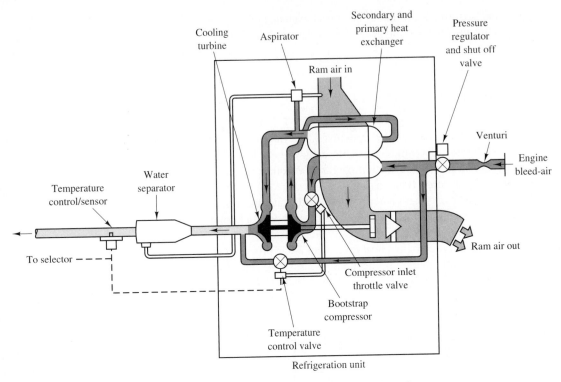

Figure 7.52   Three wheel bootstrap aircycle unit. (*Source:* Airesearch Manufacturing Company of California, a division of Garrett Corporation.)

Example 7.21

Air enters the compressor of a Brayton refrigeration cycle at 1 atm and 26°C. The compressor pressure ratio is 2, and the turbine inlet temperature is 40°C. If the isentropic efficiency of both the compressor and turbine is 80 percent, determine:

  (a) the cooling load
  (b) the mass rate of flow of air per ton of refrigeration
  (c) the power input per ton of refrigeration
  (d) the coefficient of performance

SOLUTION

Referring to Figure 7.53 and assuming a constant specific heat ratio of 1.4,

$$T_{2s} = T_1 \left(\frac{p_2}{p_1}\right)^{(\gamma-1)/\gamma} = (299.15 \text{ K})(2^{0.2857}) = 364.66 \text{ K}$$

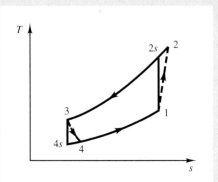

Figure 7.53

The efficiency of the compressor is

$$\eta_c = 0.8 = \frac{T_{2s} - T_1}{T_2 - T_1} = \frac{(364.68 - 299.15) \text{ K}}{T_2 - 299.15 \text{ K}}$$

from which $T_2 = 381.04$ K

$$T_{4s} = T_3 \left(\frac{p_4}{p_3}\right)^{(\gamma-1)/\gamma} = \frac{313.15 \text{ K}}{(2^{0.2857})} = 256.89 \text{ K}$$

The efficiency of the turbine is

$$\eta_T = 0.8 = \frac{T_3 - T_4}{T_3 - T_{4s}} = \frac{313.15 - T_4}{313.15 - 256.89}$$

from which $T_4 = 268.14$ K
  (a) The cooling load per unit mass is

$$q_{4\text{-}1} = c_p(T_1 - T_4) = (1.0035 \text{ kJ/kg K})[(299.15 - 268.14) \text{ K}]$$
$$= 31.12 \text{ kJ/kg}$$

  (b) The mass flow rate is

$$\dot{m} = \frac{3.517 \text{ kJ/(s. ton)}}{31.12 \text{ kJ/kg}} = 0.113 \text{ kg/(s. ton)}$$

(c) The net power input is the difference between the power input to the compressor and the power developed by the turbine

$$\dot{W}_{net} = \dot{m}c_p(T_2 - T_1) - \dot{m}c_p(T_3 - T_4)$$
$$= \dot{m}c_p[(T_2 - T_1) - (T_3 - T_4)]$$
$$= \left(0.113\ \frac{kg}{s.\ ton}\right)\left(1.0035\ \frac{kJ}{kg\ K}\right) \times [(381.04 - 299.15) - (313.15 - 268.14)]\ K$$
$$= 4.182\ kW/ton$$

(d) The coefficient of performance is

$$\beta = \frac{\dot{m}q_{4\text{-}1}}{\dot{W}_{net}} = \frac{\left(0.113\ \dfrac{kg}{s.\ ton}\right)\left(31.12\ \dfrac{kJ}{kg}\right)}{\left(4.182\ \dfrac{kW}{ton}\right)} = 0.841$$

Note the lower value of β inherent to gas-refrigeration cycles compared to vapor-refrigeration cycles. This is mainly due to the relatively significant change in the temperature of the gas in the heat exchangers of the refrigeration system. Further, because of the low density of air, an increase in the refrigeration load can be attained either by increasing the volume flow rate or by increasing the pressure ratio.

7.22 Summary

Although the Carnot cycle is the most efficient cycle, it is not suitable as a model for practical cycles because of its limitations and because of the behaviour of actual working substances. For this reason, more realistic cycles are devised to serve as models for actual cycles.

For steam power plants, the Rankine vapor-power cycle is considered the model cycle. It comprises two isentropic processes, compression and expansion in the pump and turbine, and two constant-pressure processes in the boiler and condenser. Increasing the average temperature of the steam in the boiler and lowering the condensing temperature increases the thermal efficiency. Reheating and regeneration are two modifications to the Rankine cycle. Reheating increases the average temperature of the steam leaving the boiler and also improves the steam quality at turbine exit. In the regenerative cycle, heating of the feed water in open or closed heaters reduces the temperature difference between the feed water entering the boiler and the heat source. This leads to a decrease in this major irreversibility in this cycle.

It is common to indicate the performance of steam power plants by means of the heat rate and steam rate defined as

$$\text{heat rate} = \frac{3600}{\eta_{\text{th}}} \text{ kJ/kW-hr}$$

$$\text{steam rate} = \frac{3600}{w_{\text{net, out}}} \text{ kg/kW-hr}$$

For internal combustion engines, air-standard cycles are devised as model cycles. These are closed cycles using air as a working fluid. All processes are assumed internally reversible, and the combustion and exhaust processes are replaced by heat transfer to and from the cycle, respectively.

The Otto cycle is the ideal cycle for spark-ignition engines; its efficiency is given by

$$\eta_{\text{Otto}} = 1 - \frac{1}{r_v^{\gamma-1}}$$

where $r_v = V_{\text{max}}/V_{\text{min}}$ is the compression ratio.

The Diesel cycle is the ideal cycle for compression-ignition engines; its efficiency is given by

$$\eta_{\text{Diesel}} = 1 - \frac{1}{r_v^{\gamma-1}}\left[\frac{r_c^{\gamma} - 1}{\gamma(r_c - 1)}\right]$$

where r_c is the cutoff ratio.

The Brayton or Joule cycle is the ideal cycle for gas-turbine engines; its efficiency is given by

$$\eta_{\text{Brayton}} = 1 - \frac{1}{r_p^{(\gamma-1)/\gamma}}$$

where $r_p = p_{\text{max}}/p_{\text{min}}$ is the pressure ratio.

Similar to steam regenerative heating, the regenerative gas-turbine cycle reduces the irreversibilities and improves performance. The regenerator is effective provided that the temperature of the turbine exhaust is higher than the temperature of air at the compressor exit.

The thermal efficiency of a Brayton cycle with regeneration is

$$\eta = 1 - \left(\frac{T_{\text{min}}}{T_{\text{max}}}\right)(r_p)^{(\gamma-1)/\gamma}$$

where T_{min} and T_{max} are the minimum and maximum temperatures in the cycle.

Improvement of the performance of gas turbines can also be accomplished by multistage compression with intercooling and multistage expansion with heating. Further improvement is possible by incorporating a regenerator in the cycle.

For refrigeration cycles in which the refrigerant is a condensable fluid, the idealized cycle operates as the reversed Rankine cycle. The performance of refrigerators and heat pumps is expressed in terms of the coefficient of performance, which relates the energy of primary interest in each device to the net work input.

$$\beta_{ref} = \frac{Q_L}{W_{net\ in}} \quad \text{and} \quad \beta_{ht.\ pump} = \frac{Q_H}{W_{net\ in}}$$

In absorption refrigeration, the refrigerant is absorbed by a fluid that is compressed in liquid form, thereby requiring a small amount of work input. The most common example is the ammonia–water system. Another refrigeration system commonly used in aircraft cabins is the gas refrigeration system, which operates on the reversed Brayton cycle.

Problems

7.1 Air at an inlet temperature of 25°C and a pressure of 150 kPa is compressed polytropically in a compressor operating at steady state to an exit pressure 700 kPa. If the polytropic exponent $n = 1.2$, determine the work interaction per unit mass. What will be the decrease in the work if the process takes place isothermally?

7.2 Atmospheric air at a temperature of 25°C and a pressure of 100 kPa is compressed isentropically in an ideal single-acting reciprocating compressor to a pressure of 450 kPa. The clearance volume is 5 percent of the stroke volume. If the compressor is handling 0.1 kg of air per cycle, determine:
(a) the volumetric efficiency
(b) the work interaction per cycle
(c) the isothermal efficiency neglecting the clearance volume.

7.3 A two-stage air compressor with an intercooler operates at steady state. It compresses the air isentropically from 100 kPa and 300 K to 600 kPa. The optimum staging

occurs when $p_I = \sqrt{p_H p_L}$ with intercooling to the initial temperature. Calculate the power input if the mass rate of flow is 0.1 kg/s. What is the heat removed in the intercooler? What is the power required if the air is compressed isentropically in a single stage from the initial state to the final pressure?

7.4 Compare the thermal efficiencies of the Carnot and Rankine cycles operating between two thermal reservoirs at 200°C and 10°C. (Assume water to be the working fluid.)

7.5 Steam at 3.5 MPa and 300°C is delivered to a turbine operating at steady state where it expands irreversibly and adiabatically to an exhaust pressure of 7.5 kPa. Assuming that the kinetic and potential energy changes are negligible and that the work realized is 70 percent of the ideal work, find:
(a) the entropy change in the actual turbine compared with an ideal turbine operating between the same ini-

tial temperature and pressure and the same exhaust pressure

(b) the steam rate in kg/kW-hr

(c) the increase in unavailable energy due to the irreversible expansion

Sketch the process on a *T-s* diagram.

7.6 A Rankine cycle operates with steam at 3.5 MPa and 300°C and a condenser pressure of 7.5 kPa. If the kinetic and potential energy changes are negligible, find:

(a) the ideal work done by the turbine per kg of steam

(b) the amount of steam required per net kW-hr output (steam rate)

(c) the thermal efficiency of the cycle

(d) the amount of heat required per net kW-hr output (heat rate)

(e) the unavailable energy with respect to the lowest temperature in the cycle

7.7 Using Tables A3.1, A3.3, and A3.4 of the Appendix, complete the following tabulation, which pertains to a Rankine cycle:

Entering	p (kPa)	T (°C)	h (kJ/kg)	u (kJ/kg)	v (m³/kg)	S (kJ/kg K)	x
Pump	7.5	40.29					
Boiler	4000	40.29					
Superheater	4000						0.98
Turbine	4000	300					
Condenser	7.5					6.890	

Find:

(a) heat rejected in the condenser

(b) work required by the pump

(c) heat interaction in the boiler

(d) heat interaction in the superheater

(e) work developed by the turbine

(f) work developed by a reversible adiabatic turbine operating on the same cycle.

Sketch the cycle on an *h-s* diagram and a *T-s* diagram.

7.8 The following data are for a steam power plant operating on the Rankine cycle:

Point		p (kPa)	T (°C)	x (%)
1	Leaving boiler	3000	350	
2	Entering turbine	2500	300	
3	Leaving turbine (entering condenser)	15		93
4	Leaving condenser (entering pump)	10	40	
5	Leaving pump (entering boiler)	3500		

Calculate the following per kg of steam flowing through the plant: (a) pump work, (b) turbine work, (c) heat transfer in line between boiler and turbine, (d) heat transfer in boiler, (e) heat transfer in condenser.

7.9 In a steam power plant, operating at steady state, steam is supplied to the high-pressure turbine at a pressure and temperature of 4 MPa and 300°C. After expansion to 600 kPa, the steam is returned to the boiler and reheated at constant pressure to the original temperature. The steam expands in the low-pressure turbine to an exhaust pressure of 10 kPa. For the ideal cycle, find the cycle efficiency. If the steam were not reheated, what would be the efficiency of the cycle?

7.10 A steam power plant operates on the theoretical reheat cycle. Steam is generated at 3.5 MPa and 400°C, and expands through a high-pressure turbine to the saturated vapor condition. It is then reheated at constant pressure to 400°C and expands through a low-pressure turbine to a condenser at 10 kPa. Neglecting pump work, find the thermal efficiency of the cycle. What is the reheat pressure?

If the efficiency of the turbines is 80 percent and reheating takes place at the same reheat pressure, determine the thermal efficiency, heat rate, and steam rate. Neglect pump work.

7.11 In a Rankine cycle with reheat, steam at 8 MPa and 400°C enters a turbine. The steam expands to a pressure of 500 kPa, then is reheated at constant pressure to 400°C. The steam then expands in a low-pressure turbine to 10 kPa. Determine the thermal efficiency of the cycle.

7.12 Superheated steam enters the turbine of a regenerative Rankine cycle at a pressure of 3 MPa and a temperature of 300°C. The condenser pressure is 10 kPa. Steam is extracted at 0.3 MPa to heat the feed water in an open heater. Neglecting pump work, calculate:
(a) the mass of the extracted steam per kg of steam entering the turbine
(b) the thermal efficiency of the cycle

7.13 In a two-heater (closed) regenerative cycle, steam is supplied to the turbine at 2.5 MPa and 300°C. Condenser pressure is 4.0 kPa. If the extraction pressures are 500 and 75 kPa, calculate the thermal efficiency of the cycle. Neglect pump work.

7.14 Compare the thermal efficiency and the steam consumption in kg/kW-hr for the Rankine, regenerative, and reheat cycles under the following conditions:

Initial steam pressure	2.0 MPa
Initial temperature	300°C
Extraction pressure	200 kPa
Exhaust pressure	7.5 kPa

Reheating in the reheat cycle takes place at 200 kPa to the initial superheat before expanding in the low-pressure turbine to 7.5 kPa. The regenerative cycle uses one closed heater. Neglect pump work.

7.15 A steam turbine plant equipped with a single regenerative feed-water heater of the open type operates at steady state under the following conditions:

Initial steam pressure	2.0 MPa
Initial temperature	300°C
Extraction pressure	200 kPa
Exhaust pressure	7.5 kPa

Neglecting work of the pump, compare the regenerative and Rankine cycles with respect to the following:
(a) thermal efficiency
(b) steam consumption in kg/kW-hr
(c) condenser duty (steam condensed per kW-hr)

7.16 Superheated steam at 10 MPa and 350°C enters the high-pressure turbine of a steam power plant and expands to 2 MPa. The steam is reheated at this pressure to 300°C before it expands in a low-pressure turbine to a condenser pressure of 10 kPa. The plant has two closed heaters for feed water heating. Steam for the first heaters is bled at 2 MPa. For the second heater, steam is bled at 200 kPa. If the expansion in the turbines is isentropic, determine the cycle efficiency, and compare it to that of the reheat cycle with the same inlet conditions. Neglect the work required for the condensate pump.

7.17 An air-standard Otto cycle has a compression ratio of 8. At the start of the compression stroke, the temperature is 300 K, and the pressure is 100 kPa. If the maximum pressure in the cycle is 3.0 MPa, determine:
(a) the heat input per kg of air
(b) the thermal efficiency of the cycle
(c) the irreversibility of the heating process assuming the heat transfer takes place from a thermal reservoir at a temperature equal to the highest temperature in the cycle

7.18 Verify Eq. (7.19).

7.19 An air-standard Otto cycle has a compression ratio of 8. At the start of the compression process, the temperature is 300 K, and the pressure is 100 kPa. If the heat transfer to the cycle is 800 kJ/kg of air, determine:
(a) the pressure and temperature at key points of the cycle
(b) the net work
(c) the thermal efficiency
(d) the irreversibility of the cycle

7.20 An air-standard Otto cycle has a compression ratio of 8.5. At the beginning of the compression stroke, the temperature is 300 K, and the pressure is 90 kPa. If the heat input is 800 kJ and the maximum temperature in the cycle is 1100 K, determine:
(a) the mass of air in the cycle
(b) the heat rejection
(c) the thermal efficiency of the cycle
(d) the mean effective pressure. Use air tables.

7.21 An ideal Otto cycle and an ideal Diesel cycle utilize the same quantity of the same working substance. The states of the working substance are identical in both cycles at the beginning of the compression stroke, at the end of heat supply, and at the end of the expansion stroke. Sketch the two cycles, superimposed and identified, on *p-V* and *T-S* diagrams. By referring to the diagrams, show which of the two cycles has the higher thermal efficiency.

7.22 The cyclic operation of a 1.8-liter engine with a compression ratio of 8 is approximated as an Otto cycle. The temperature and pressure at the beginning of the compression stroke are 30°C and 85 kPa. The compression is equivalent to a polytropic process following the relation $pv^{1.3} = C$. If the work done during the expansion process is four times that required for compression, determine the power developed by the engine at 4000 rpm. What is the mean effective pressure?

7.23 Represent the air-standard Otto cycle on *S-V* and *T-S* diagrams, and show that the work ratio of the two isentropic processes is given by

$$r_w = \frac{T_3}{T_1}\left(\frac{1}{r_v^{\gamma-1}}\right)$$

where r_w is the work ratio, r_v is the volumetric compression ratio, and T_1 and T_3 are the minimum and maximum temperatures of the cycle.

7.24 An air-standard Diesel cycle has a compression ratio of 20. The temperature and pressure at the beginning of the compression stroke are 100 kPa and 20°C. If the cutoff ratio is 1.8, determine the thermal efficiency and the mean effective pressure. Use air tables.

7.25 An air-standard 4-stroke Diesel engine has a compression ratio of 18 and a cutoff ratio of 1.5. The pressure and temperature at the beginning of the compression stroke are 100 kPa and 300 K. Determine (a) the maximum pressure and temperature in the cycle, (b) the thermal efficiency, (c) the mean effective pressure.

7.26 An air-standard Diesel cycle that has a compression ratio 18 to 1 receives energy at the rate of 1300 kJ/kg

of air. At the beginning of the compression stroke, the temperature and pressure are 300 K and 100 kPa, respectively. Calculate:
(a) the net work output per kg of air
(b) the thermal efficiency of the cycle
(c) the expansion ratio
(d) the mean effective pressure
(e) the irreversibility of each process in the cycle

7.27 A theoretical air-standard Diesel cycle has an initial pressure and temperature of 100 kPa and 25°C. The compression ratio is 14 to 1, and the temperature at the end of heat interaction is 1700°C. The specific heat ratios $(c_p/c_v = \gamma)$ for the compression, heat supply, and expansion processes are 1.37, 1.34, and 1.31, respectively.
(a) What is the thermal efficiency of the cycle?
(b) What is the specific work for the cycle?
(c) What is the entropy change for the heat supply process?
Sketch *p-V* and *T-S* diagrams.

7.28 An air-standard diesel engine has a compression ratio of 16 and a cutoff ratio of 2. The air enters the engine at 100 kPa and 300 K. Determine the thermal efficiency of the cycle. Assume air to be an ideal gas with constant specific heats ($c_v = 0.7165$ kJ/kg K, $c_p = 1.0035$ kJ/kg K).
   If an Otto cycle matches this cycle at the beginning of the compression stroke, at the end of heat supply, and at the end of the expansion stroke, determine the thermal efficiency of the Otto cycle.

7.29 The heat input to an air-standard Diesel cycle is 1000 kJ/kg. At the beginning of the compression, the temperature is 20°C and the pressure is 150 kPa. If the compression ratio is 20, determine the maximum pressure and temperature in the cycle.

7.30 Plot the thermal efficiency of an air-standard diesel engine versus compression ratio for cutoff ratios of 1, 2, and 3.

7.31 An air-standard Diesel cycle has a compression ratio of 18 and a cutoff ratio of 2.2. The temperature at the beginning of compression is 300 K. Assuming constant values of specific heats, determine (a) the temperatures at the corner points of the cycle, (b) the thermal efficiency of the cycle.

7.32 In an air-standard dual cycle, heat in the amount of 500 kJ/kg is supplied during the constant-volume process, and the 1000 kJ/kg is supplied during the constant-pressure process. The temperature and pressure at the beginning of the compression process are 300 K and 100 kPa, respectively. The compression ratio of the cycle is 9.5. Determine the net work output and the thermal efficiency of the cycle.

7.33 In a dual air-standard cycle, one-half of the heat supply occurs during the constant-volume process, and the remaining half in the constant-pressure process. The total heat supply is 2400 kJ/kg per cycle. The compression ratio is 10. The temperature and pressure at the beginning of the compression process are 300 K and 100 kPa, respectively. Calculate the thermal efficiency of the cycle. What is the irreversibility of each process in the cycle?

7.34 An air-standard dual cycle has a compression ratio of 10. At the beginning of compression, the temperature is 300 K, and the pressure is 95 kPa. The heat input during the constant-volume process causes the pressure to double. If the maximum temperature in the cycle is 1800 K, determine (a) the thermal efficiency, (b) the mean effective pressure.

7.35 Prove that the thermal efficiency of the dual cycle shown in Figure 7.22 is given by

$$\eta = 1 - \frac{1}{r_v^{\gamma-1}} \frac{r_p r_c^{\gamma} - 1}{(r_p - 1) + \gamma r_p (r_c - 1)}$$

where

$$r_p = \frac{p_3}{p_2} \qquad r_c = \frac{v_4}{v_3} \qquad r_v = \frac{v_1}{v_2}$$

and γ is the specific heat ratio.

7.36 A Stirling engine, using an ideal gas as a working substance, operates between a high temperature T_H and a

low temperature T_L. If a regenerator is not used, determine an expression for the cycle efficiency, and compare it with the efficiency of a Carnot cycle operating between T_H and T_L.

7.37 An ideal Stirling cycle with a perfect regenerator uses helium as a working fluid. The cycle operates between 300 and 1100 K. If the maximum and minimum pressures are 400 and 100 kPa, determine (a) the net work output per kg of helium, (b) the cycle thermal efficiency.

7.38 An air-standard Brayton cycle receives air at 0.1 MPa and 300 K. The upper pressure and temperature limits of the cycle are 0.4 MPa and 1000 K, respectively. The compression is isentropic, but the turbine efficiency is 90 percent. Calculate the thermal efficiency of the cycle assuming constant specific heats.

7.39 A gas turbine power plant operates at steady state and produces 10 megawatts. Air enters the compressor at one atmosphere and 10°C. The compressor pressure ratio is 5, and the maximum cycle temperature is 1000°C. The turbine outlet pressure is one atmosphere. Assuming an ideal cycle with air as a working fluid, find:
(a) the total turbine power output
(b) the power required to drive the compressor
(c) the cycle thermal efficiency
(d) the fuel flow if the heating value of the fuel is 50×10^3 kJ/kg of fuel

7.40 A gas turbine operates at steady state between a low and a high temperature limit. The low temperature is dictated by the temperature of the surrounding air; the high temperature is dictated by the maximum temperature the turbine material can withstand
(a) Determine an expression for the work output in terms of T_{min}, r_p, c_p, and γ.
(b) Prove that the maximum work output corresponds to a pressure ratio given by

$$r_p = \left(\frac{T_{max}}{T_{min}}\right)^{\gamma/[2(\gamma-1)]}$$

7.41 A closed regenerative gas-turbine cycle operates with air as a fluid. Data for the proposed cycle are

$p_1 = 130$ kPa	$T_1 = 300$ K
$p_2/p_1 = 5$	$T_{max} = 1300$ K
Effectiveness ratio of regenerator $= 0.6$	
Net output $= 1500$ kW	

Assuming the flow through the compressor and the turbine to be isentropic, calculate (a) the thermal efficiency, (b) the air flow rate. Assume air to behave like an ideal gas with constant specific heats.

7.42 An air-standard regenerative gas turbine unit is to provide a power output of 500 kW. The compressor pressure ratio is 5.0, and the maximum cycle temperature is 1100 K; the system incorporates a regenerator. Assuming all components to be ideal, and the inlet air temperature to be 300 K, determine the mass flow rate of air required and the overall plant thermal efficiency. Assume air to be an ideal gas with $c_p = 1.004$ kJ/kg K and $\gamma = 1.4$.

7.43 Find the efficiency and the work per unit mass of fluid for an air-standard Brayton cycle working between pressures of 100 kPa and 500 kPa. The maximum temperature in the cycle is 1400 K, and the minimum temperature is 300 K.
(a) Solve the problem by (1) gas laws, (2) air tables. State the reasons for differences in answers, if any, by the two methods (Assume a constant value of c_p for air $= 1.0035$ kJ/kg K.)
(b) What would the efficiency and the work developed per kg of air be if regenerative heating is used? The regenerator effectiveness is 70 percent. What is the temperature of the air leaving the regenerator to the atmosphere?

7.44 An air-standard gas-turbine cycle takes in air at 100 kPa and 300 K, and compresses it to 550 kPa in a compressor that is 60 percent efficient. After compression, the air is preheated in a regenerator that has an effectiveness of 80 percent. If the maximum permissible temperature is 1100 K and the turbine unit is ideal, what is the thermal efficiency of this cycle? Assume air to be an ideal gas with constant specific heats.

7.45 A gas-turbine unit is equipped with a regenerator and operates at steady state. The compression process is isentropic with air entering the compressor at 290 K and 100 kPa. The pressure ratio is 8:1. The maximum temperature at the turbine inlet is 1400 K, and the expansion process through the turbine is isentropic. If the regenerator effectiveness is 60 percent, determine (a) the temperature at the entrance of the combustor, (b) the thermal efficiency. Assume air to behave like an ideal gas with constant specific heats.

7.46 A gas-turbine cycle has a double-stage compression with intercooling and a double-stage expansion with reheating. If the air enters both compressors at T_{min} and enters both turbines at T_{max}, prove that maximum work output is obtained when each pressure ratio $r_p = \sqrt{r_{p_{max}}}$. Plot the cycle efficiency versus r_p for different values of T_{max}/T_{min}. Assume air to be an ideal gas with constant specific heats.

7.47 A jet-propelled aircraft is to operate at a speed of 206 m/s relative to the surrounding air at a local pressure and temperature of 55 kPa and 255 K (Mach number 0.644). Steady-state operating features are

Pressure coefficient of diffuser	0.8
Pressure ratio of compressor	5 to 1
Isentropic efficiency of compressor	0.85
Isentropic efficiency of turbine	0.88
Isentropic efficiency of nozzle	0.94
Maximum allowable gas temperature at turbine entry	1060 K

Assuming that the cycle is air standard and that the kinetic energy of the air entering and leaving the compressor and turbine is negligible, determine (a) the states at entry and exit for the compressor and the turbine, (b) the propulsive force per kg/s of air.

7.48 Determine the thermal efficiency for a gas-turbine cycle if the turbine is 83 percent efficient and compression is in two stages each 83 percent efficient (a) if no regeneration is used, (b) if a regenerator heats the compressed air to 300°C.

Assume intercooling between the compression stages reduces the air temperature to 20°C and the intermediate pressure is 220 kPa. Data:

Inlet air at 100 kPa, 20°C.
Pressure at compressor exit is 500 kPa.
Maximum temperature in combustor is 650°C.
Air expands in turbine back to 100 kPa.

7.49 In an air-standard gas turbine cycle, air is compressed from 100 kPa and 300 K to 500 kPa. The turbine inlet temperature is 1080 K. Determine the improvement in efficiency resulting from the installation of the regenerator with an efficiency of 80 percent. Assume $\gamma = 1.4$ and $c_p = 1.03$ kJ/kg K.

7.50 The pressure ratio in an ideal air-standard Brayton cycle is 6:1. The air enters the compressor at 300 K and 95 kPa. If the heat input to the cycle is 700 kJ/kg, determine the thermal efficiency of the cycle. What is the irreversibility of the heat input process?

7.51 The pressure ratio in an ideal air-standard Brayton cycle is 10:1. The air enters the compressor at 300 K and 100 kPa and leaves the turbine at 720 K. Determine:
(a) the net work output per unit mass of air
(b) the thermal efficiency of the cycle
(c) the irreversibility of the cycle per unit mass of air
Assume the environment is at a temperature of 300 K and a pressure of 100 kPa.

7.52 Air enters the diffuser of a turbo-jet engine operating at steady state at 240 K, 80 kPa, with a velocity of 200 m/s. The pressure ratio across the compressor is 8.5. Constant-pressure combustion raises the temperature to 1200 K. The products of combustion expand through the turbine and the nozzle to 80 kPa. If the mass flow rate is 120 kg/s, determine (a) the rate of heat input, (b) the velocity at the exit of the nozzle. Assume combustion products to have the same properties as air and isentropic flow during compression and expansion.

7.53 Air at a temperature of 300 K and a pressure of 100 kPa enters the compressor of an ideal Brayton cycle.

If the temperature and pressure at the turbine inlet are 1100 K and 450 kPa, determine:
(a) the net work done per unit mass of air
(b) the thermal efficiency of the cycle
(c) the thermal efficiency if a regenerator with an effectiveness of 75 percent is added to the cycle
(d) calculate the minimum irreversibility of the heat input process in part (c)

7.54 Using a *T-s* diagram, show that a single stage of intercooling or reheating, when added to the simple Brayton cycle, will decrease the cycle efficiency.

7.55 Show that, for an *n*-stage compression with intercooling in a gas turbine plant, the minimum work input is attained when the pressure ratios of successive stages are equal.

7.56 An ammonia refrigeration cycle operates with an evaporator saturation temperature of $-16°C$ while removing 2.93 kW from a cold room. The saturation temperature in the condenser is 24°C. Assuming that the ammonia is saturated liquid at the entrance of the expansion valve, and saturated vapor at the entrance of the compressor, find:
(a) the coefficient of performance of the cycle and the ammonia circulation rate in kg/s
(b) the coefficient of performance of a Carnot refrigerator operating between the condenser and evaporator temperatures

7.57 An ice-making unit of 15-ton capacity operates at steady state with ammonia as a refrigerant. In order to cool the brine to the proper temperature, the ammonia temperature in the evaporator is maintained at $-16°C$. Cooling water enables the ammonia to condense at 16°C. Assuming dry saturated vapor at the compressor inlet and isentropic compression, find:
(a) the pressure in the evaporator and in the condenser
(b) the actual and ideal coefficient of performance
(c) the actual power required
(d) the circulation rate (kg/min) of refrigerant
Sketch *T-s* and *p-h* diagrams of the cycle.

7.58 The following data were obtained in a test of an ammonia refrigerating unit operating at steady state:

Pressures (kPa):	Evaporator 275, condenser, 1500
Temperatures (°C):	Liquid entering expansion valve, 30
	Vapor leaving evaporator and
	entering compressor, 4
	Vapor leaving compressor and
	entering condenser, 115
Rate of flow:	2.585 kg/min of ammonia
Compressor	
isentropic efficiency:	75 percent

Compute:
(a) the capacity in tons
(b) the coefficient of performance
(c) the power required to drive compressor
(d) the heat transfer per minute from compressor
(e) the heat transfer per minute from condenser

7.59 A vapor-compression refrigerator uses refrigerant-12 as the working fluid. The refrigerator operates at steady state between $-34°C$ and $20°C$. The vapor leaving the evaporator and the liquid leaving the condenser are saturated vapor and saturated liquid. Assuming that the compressor operates reversibly and adiabatically, find:
(a) the coefficient of performance
(b) the power required per ton of refrigeration
(c) the mass flow rate per ton of refrigeration
(d) the power that could have been developed by a reversible process instead of the throttling process per ton of refrigeration

7.60 The following data are obtained in a test of an ammonia refrigeration system operating at steady state:

Pressures:	Condenser = 1.0 MPa
	Evaporator = 250 kPa
Temperatures:	Leaving compressor = 100°C
	Entering condenser = 90°C
	Leaving condenser = 18°C
	Entering expansion valve = 20°C
	Leaving evaporator = $-10°C$
	Entering compressor = $-2°C$
Ammonia circulated per hour = 87 kg	

Calculate:
(a) the capacity of the plant in tons
(b) the heat rejected by ammonia in condenser (kJ/hr)
(c) the power input into the compressor
(d) the coefficient of performance
(e) the heat interaction with pipeline and receiver (kJ/hr)

7.61 An ammonia-compression refrigeration cycle operates so as to maintain a temperature of $-2°C$ in a cold storage room. The condenser for this unit is set in the air, which is at $31°C$. If the efficiency of the compressor is 90 percent, find:
(a) the power necessary for 25 tons of refrigeration
(b) the compressor capacity in m^3/min
(c) the coefficient of performance

7.62 A Carnot-vapor refrigeration cycle uses R-12 as the working fluid. Saturated vapor enters the condenser and leaves as saturated liquid, and the cycle operates between $-10°C$ and $40°C$. If the mass flow rate of refrigerant is 4.5 kg/min, determine:
(a) the cooling effect of the evaporator
(b) the net power input to the cycle
(c) the coefficient of performance

7.63 An ideal-compression refrigeration cycle uses R-12 as the working fluid. Saturated vapor enters the compressor at $-15°C$, and saturated liquid leaves the condenser at $30°C$. For a refrigeration capacity of 5 tons, determine:
(a) the power input to the compressor
(b) the mass flow rate
(c) the coefficient of performance

7.64 A closed-loop refrigeration unit uses nitrogen as the working fluid and operates on a reversed Brayton cycle. The inlet conditions to the compressor are $p = 200$ kPa and $T = 30°C$. There is a fivefold increase in pressure across the compressor, and the inlet temperature to the turbine is $20°C$. Determine the coefficient of performance and the cycle work input for ideal performance of the compressor and expander. Assume $c_p = 1.04$ kJ/kgK and $\gamma = 1.4$.

7.65 Determine the refrigeration effect and the coefficient of performance for an ideal refrigeration cycle operating between an upper and lower temperature of 30°C and −10°C if the refrigerant is (a) ammonia, (b) R-12, (c) R-134a. What is the refrigeration effect per unit volume of refrigerant entering the compressor?

7.66 In an air-standard gas refrigeration cycle (reversed Brayton cycle), the air enters the compressor at 20°C and 100 kPa. The pressure ratio of both the turbine and the compressor is 2.5, and the compression and expansion are isentropic. The air enters the turbine at 35°C, and the mass rate of flow through the cycle is 0.3 kg/s. Determine
(a) the net power required
(b) the refrigeration capacity
(c) the coefficient of performance of the cycle
Solve this problem if the compressor has an isentropic efficiency of 80 percent and the turbine has an isentropic efficiency of 84 percent.

7.67 For a refrigeration cycle, use a *T-s* diagram to show that improvement in performance can be attained if superheating of the vapor leaving the compressor can be avoided.

C7.1 Write a computer program to investigate the effect of maximum cycle temperature on the thermal efficiency of an ideal Rankine cycle using H_2O as the working fluid. Consider a range of maximum temperatures from 400 to 700°C. Let the lowest pressure of the cycle be 20 kPa and the highest pressure be 10 MPa.

C7.2 Write a computer program to investigate the effect of maximum cycle pressure on the thermal efficiency of an ideal Rankine cycle using H_2O as the working fluid. Consider a range of maximum pressures from 2.5 to 8.5 MPa. Let the lowest pressure of the cycle be 20 kPa and the highest temperature be 700°C.

C7.3 Write a computer program for a regenerative steam Rankine power cycle with one closed heater to determine the effect of bleed temperature on the cycle efficiency. Assume saturated steam leaves the boiler at a pressure of 3 MPa and the condenser pressure is 5 kPa. Plot the cycle efficiency versus the bleed temperature, and comment on the results.

C7.4 A power plant is designed to operate on an air-standard Otto cycle. The pressure and temperature of the gas at the beginning of the compression process are 110 kPa and 350 K, respectively. Write a program to determine the cycle efficiency, the maximum cycle temperature, the maximum cycle pressure, and the mep for this cycle under the following operating conditions. The compression ratio ranges from 5 to 18, and the amount of heat input ranges from 1000 to 4000 kJ for each compression ratio. Plot the results against the compression ratio using the heat input as a parameter for the family of curves.

C7.5 Modify the program in Problem C7.4 to study an air-standard Diesel cycle.

C7.6 Write a computer program to determine the cutoff ratio for which the thermal efficiency of a Diesel cycle is 70 percent. Use γ values ranging from 1.125 to 1.4. Plot the cutoff ratios against γ values.

C7.7 Modify the program in Problem C7.4 to study an air-standard Brayton cycle.

C7.8 Write a computer program to determine the maximum thermal efficiency of a regenerative air-standard gas turbine cycle as a function of pressure ratio and the inlet temperature to the turbine. Assume the inlet temperature of the compressor $T_1 = 300K$.

Chapter 8

Thermodynamic Relations and Equations of State

8.1 Introduction

Energy and matter are intimately related and in applying the laws of thermodynamics, property values are needed for numerical calculations. In the preceding chapters, we made use of several tables and equations to determine thermodynamic properties for a given state of a system. Some of these properties such as temperature, pressure, volume, and mass can be directly measured; others such as internal energy, enthalpy, and entropy are not. These, however, can be related to measurable properties through functional relations that afford their evaluation.

This chapter develops useful relations among properties for simple compressible substances, in particular, those relations that allow us to calculate unmeasurable properties in terms of easily measurable properties. The chapter also presents several equations of state and discusses property evaluation of real gases.

8.2 Mathematical Relations

Consider the mathematical relations in which a dependent property is defined in terms of two independent properties. The functional dependence can be expressed as

$$x = x(y, z)$$
$$y = y(x, z)$$

The total differential of the dependent functions x and y can be expressed in terms of their partial derivatives with respect to the independent variables as

$$dx = \left(\frac{\partial x}{\partial y}\right)_z dy + \left(\frac{\partial x}{\partial z}\right)_y dz \qquad (8.1)$$

$$dy = \left(\frac{\partial y}{\partial x}\right)_z dx + \left(\frac{\partial y}{\partial z}\right)_x dz \qquad (8.2)$$

The subscripts in the preceding two equations indicate the independent variable that is held constant in calculating the derivative. Eliminating dy between the foregoing equations gives

$$dx = \left(\frac{\partial x}{\partial y}\right)_z \left[\left(\frac{\partial y}{\partial x}\right)_z dx + \left(\frac{\partial y}{\partial z}\right)_x dz\right] + \left(\frac{\partial x}{\partial z}\right)_y dz$$

or

$$\left[1 - \left(\frac{\partial x}{\partial y}\right)_z \left(\frac{\partial y}{\partial x}\right)_z\right] dx = \left[\left(\frac{\partial x}{\partial y}\right)_z \left(\frac{\partial y}{\partial z}\right)_x + \left(\frac{\partial x}{\partial z}\right)_y\right] dz$$

Since x and z are independent variables, any values may be assigned to them, and the previous equation will be valid. Let $z = $ constant ($dz = 0$) and $dx \neq 0$, then

$$\left(\frac{\partial x}{\partial y}\right)_z \left(\frac{\partial y}{\partial x}\right)_z = 1 \qquad (8.3)$$

Similarly, if x is held constant ($dx = 0$) and $dz \neq 0$, then

$$\left(\frac{\partial x}{\partial y}\right)_z \left(\frac{\partial y}{\partial z}\right)_x + \left(\frac{\partial x}{\partial z}\right)_y = 0$$

or in an alternative form,

$$\left(\frac{\partial x}{\partial y}\right)_z \left(\frac{\partial y}{\partial z}\right)_x \left(\frac{\partial z}{\partial x}\right)_y = -1 \qquad (8.4)$$

Equation (8.3) is called the *reciprocity relation,* and Eq. (8.4) is called the *cycle relation.*

Now let a property P be a single-valued continuous function of two independent variables. First, P may be expressed as a function of x and y; then P may be expressed as a function of z and y:

$$dP = \left(\frac{\partial P}{\partial x}\right)_y dx + \left(\frac{\partial P}{\partial y}\right)_x dy \tag{8.5}$$

and

$$dP = \left(\frac{\partial P}{\partial z}\right)_y dz + \left(\frac{\partial P}{\partial y}\right)_z dy \tag{8.6}$$

The integral of dP between state 1 and state 2 depends only on the value of P at these two states and is independent of the path followed between the two states. After substituting for dx from Eq. (8.1), Eq. (8.5) becomes

$$dP = \left[\left(\frac{\partial P}{\partial x}\right)_y \left(\frac{\partial x}{\partial z}\right)_y\right] dz + \left[\left(\frac{\partial P}{\partial x}\right)_y \left(\frac{\partial x}{\partial y}\right)_z + \left(\frac{\partial P}{\partial y}\right)_x\right] dy$$

When this equation is compared with Eq. (8.6), it becomes evident that the coefficients of dz and dy can be equated, since z and y are independent variables, giving:

$$\left(\frac{\partial P}{\partial z}\right)_y = \left(\frac{\partial P}{\partial x}\right)_y \left(\frac{\partial x}{\partial z}\right)_y$$

or

$$\left(\frac{\partial P}{\partial x}\right)_y \left(\frac{\partial x}{\partial z}\right)_y \left(\frac{\partial z}{\partial P}\right)_y = 1 \tag{8.7}$$

and

$$\left(\frac{\partial P}{\partial y}\right)_z = \left(\frac{\partial P}{\partial x}\right)_y \left(\frac{\partial x}{\partial y}\right)_z + \left(\frac{\partial P}{\partial y}\right)_x \tag{8.8}$$

Equation (8.7) is called the *chain relation;* it applies to systems with three independent variables and is an extension of Eq. (8.3), which applies to systems with only two variables.

Equations (8.3), (8.4), (8.7), and (8.8) are useful in developing property relationships, as will be shown in subsequent sections.

Example 8.1

Show that the equation $p(v - b) = RT$, where b and R are constants, satisfies Eq. (8.4).

SOLUTION

Let $x = p$, $y = v$, and $z = T$. Equation (8.4), in terms of the new variables, is

$$\left(\frac{\partial p}{\partial v}\right)_T \left(\frac{\partial v}{\partial T}\right)_p \left(\frac{\partial T}{\partial p}\right)_v = -1$$

But,

$$\left(\frac{\partial p}{\partial v}\right)_T = -\frac{RT}{(v - b)^2}$$

and

$$\left(\frac{\partial p}{\partial T}\right)_v = \frac{R}{v - b}$$

and since

$$v = \frac{RT}{p} + b$$

then

$$\left(\frac{\partial v}{\partial T}\right)_p = \frac{R}{p}$$

Substitution gives

$$\left(-\frac{RT}{(v - b)^2}\right)\left(\frac{R}{p}\right)\left(\frac{v - b}{R}\right) = -\frac{RT}{p(v - b)} = -1$$

Example 8.2

Determine the partial derivative $(\partial v / \partial T)_p$ for a gas obeying the van der Waals equation of state:

$$\left(p + \frac{a}{v^2}\right)(v - b) = RT$$

What is the differential of p in terms of T and v?

SOLUTION

From Eq. (8.4),

$$\left(\frac{\partial v}{\partial T}\right)_p \left(\frac{\partial T}{\partial p}\right)_v \left(\frac{\partial p}{\partial v}\right)_T = -1$$

From the van der Waals equation, the temperature and pressure can be expressed explicitly as

$$T = \left(\frac{v-b}{R}\right)\left(p + \frac{a}{v^2}\right)$$

and

$$p = \left(\frac{RT}{v-b}\right) - \frac{a}{v^2}$$

Differentiation gives,

$$\left(\frac{\partial T}{\partial p}\right)_v = \frac{v-b}{R}$$

$$\left(\frac{\partial p}{\partial v}\right)_T = -\frac{RT}{(v-b)^2} + \frac{2a}{v^3}$$

Therefore,

$$\left(\frac{\partial v}{\partial T}\right)_p = \frac{-1}{\left(\dfrac{v-b}{R}\right)\left[-\dfrac{RT}{(v-b)^2} + \dfrac{2a}{v^3}\right]}$$

The differential of $p = p(T, v)$ is

$$dp = \left(\frac{\partial p}{\partial T}\right)_v dT + \left(\frac{\partial p}{\partial v}\right)_T dv$$

Substituting for the partial derivatives obtained before gives

$$dp = \left(\frac{R}{v-b}\right)dT + \left[\frac{-RT}{(v-b)^2} + \frac{2a}{v^3}\right]dv$$

8.3 Maxwell Relations

James Clerk Maxwell made many contributions to science; his main contribution to thermodynamics was the "Maxwell relations." To obtain these relations, consider the following differential forms of the expressions of internal energy, enthalpy, Helmholtz function, and Gibbs function for a unit mass of a substance of fixed composition:

$$du = T\,ds - p\,dv \tag{8.9}$$

$$dh = T\,ds + v\,dp \tag{8.10}$$

$$df = -s\,dT - p\,dv \tag{8.11}$$

$$dg = v\,dp - s\,dT \tag{8.12}$$

Noting that u, h, f, and g are properties and recalling that the differential of a function f, $df = M\,dx + N\,dy$ is exact if $(\partial M/\partial y)_x = (\partial N/\partial x)_y$, it follows from the preceding equations that

$$\left(\frac{\partial T}{\partial v}\right)_{s,\,n_i} = -\left(\frac{\partial p}{\partial s}\right)_{v,\,n_i} \tag{8.13}$$

$$\left(\frac{\partial T}{\partial p}\right)_{s,\,n_i} = \left(\frac{\partial v}{\partial s}\right)_{p,\,n_i} \tag{8.14}$$

$$\left(\frac{\partial p}{\partial T}\right)_{v,\,n_i} = \left(\frac{\partial s}{\partial v}\right)_{T,\,n_i} \tag{8.15}$$

$$\left(\frac{\partial v}{\partial T}\right)_{p,\,n_i} = -\left(\frac{\partial s}{\partial p}\right)_{T,\,n_i} \tag{8.16}$$

Equations (8.13) through (8.16) are called the *Maxwell relations*. They interrelate the variables p, v, T, and s. In particular, they express the entropy of a simple system in terms of experimentally measurable properties such as pressure, volume, and temperature.

The properties p, v, T, and s in Eqs. (8.9) through (8.12) can also be expressed in terms of partial derivatives of the dependent variables u, h, f, and g with respect to the independent variables. The differential of internal energy $u = u\,(s, v)$, for example, is

$$du = \left(\frac{\partial u}{\partial s}\right)_{v,\,n_i} ds + \left(\frac{\partial u}{\partial v}\right)_{s,\,n_i} dv \tag{8.17}$$

By equating the coefficients of the independent variables s and v in Eqs. (8.9) and (8.17), then

$$T = \left(\frac{\partial u}{\partial s}\right)_{v,\,n_i} \qquad \text{and} \qquad -p = \left(\frac{\partial u}{\partial v}\right)_{s,\,n_i}$$

In a like manner, the following relations are obtained from Eqs. (8.10) through (8.12):

$$T = \left(\frac{\partial u}{\partial s}\right)_{v,\,n_i} = \left(\frac{\partial h}{\partial s}\right)_{p,\,n_i} \tag{8.18}$$

$$p = -\left(\frac{\partial u}{\partial v}\right)_{s,\,n_i} = -\left(\frac{\partial f}{\partial v}\right)_{T,\,n_i} \tag{8.19}$$

$$v = \left(\frac{\partial h}{\partial p}\right)_{s,\,n_i} = \left(\frac{\partial g}{\partial p}\right)_{T,\,n_i} \tag{8.20}$$

$$s = -\left(\frac{\partial f}{\partial T}\right)_{v,\,n_i} = -\left(\frac{\partial g}{\partial T}\right)_{p,\,n_i} \tag{8.21}$$

From Eqs. (8.18) and (8.7), the specific heat at constant volume of a simple system may be expressed in terms of entropy:

$$c_v = \left(\frac{\partial u}{\partial T}\right)_v = \left(\frac{\partial u}{\partial s}\right)_v \left(\frac{\partial s}{\partial T}\right)_v = T\left(\frac{\partial s}{\partial T}\right)_v \tag{8.22}$$

Similarly,

$$c_p = \left(\frac{\partial h}{\partial T}\right)_p = \left(\frac{\partial h}{\partial s}\right)_p \left(\frac{\partial s}{\partial T}\right)_p = T\left(\frac{\partial s}{\partial T}\right)_p \tag{8.23}$$

8.4 Clapeyron Equation

The Clapeyron equation makes use of the Maxwell relations to determine the enthalpy change during a phase change of a pure substance in terms of the measurable properties T, p, and v.

Consider the change of state of a pure substance from the saturated-liquid to the saturated-vapor state. During the vaporization process, both the temperature and the pressure are independent of the specific volume. Noting that during the change of phase each extensive property is linearly related to the quality x, it follows that extensive properties are also linearly interrelated. The Maxwell relation, Eq. (8.15), is

$$\left(\frac{\partial p}{\partial T}\right)_v = \left(\frac{\partial s}{\partial v}\right)_T$$

or

$$ds = \left(\frac{\partial p}{\partial T}\right)_v dv$$

But during a phase change, the saturation pressure is a function of temperature only so that a total derivative can replace the partial derivative:

$$\left(\frac{\partial p}{\partial T}\right)_v = \left(\frac{dp}{dT}\right)_{sat}$$

Note that $(dp/dT)_{sat}$ is independent of the specific volume and thus can be treated as a constant. Substituting this equation into the preceding relation and integrating gives

$$s_2 - s_1 = \left(\frac{dp}{dT}\right)_{sat} (v_2 - v_1)$$

or

$$\left(\frac{dp}{dT}\right)_{sat} = \frac{s_2 - s_1}{v_2 - v_1} \tag{8.24}$$

For a phase change from the saturated-liquid state to the saturated-vapor state, the change of entropy s_{fg} is equal to the enthalpy of vaporization divided by the absolute temperature,

$$s_{fg} = \frac{h_{fg}}{T_{sat}} \tag{8.25}$$

which, when substituted in the previous equation, gives

$$\left(\frac{dp}{dT}\right)_{sat} = \frac{s_{fg}}{v_g - v_f} = \frac{h_{fg}}{T_{sat}(v_g - v_f)} \tag{8.26}$$

Equation (8.26) is called the *Clapeyron equation* and is conveniently used to determine the enthalpy of vaporization from the data of the rate of change of pressure with temperature, that is, the slope of the p-T curve of the substance and the specific volumes of the saturated liquid and saturated vapor. Equations similar to Eq. (8.26) can be derived for changes from solid to liquid (fusion, $i \rightarrow f$) and from solid to vapor (sublimation, $i \rightarrow g$). A simplification of Eq. (8.26) may be obtained by neglecting v_f in comparison with v_g and substituting for v_g from the ideal-gas equation. Under these conditions, which are valid at low pressures, Eq. (8.26) becomes

$$\left(\frac{dp}{dT}\right)_{sat} = \frac{h_{fg}}{T_{sat}v_g} = \frac{h_{fg}}{RT_{sat}^2/p}$$

or

$$\left(\frac{dp}{p}\right)_{sat} = \frac{h_{fg}}{R}\left(\frac{dT}{T^2}\right)_{sat} \tag{8.27}$$

which can be written in the form

$$\frac{d\ln p_{sat}}{d(1/T_{sat})} = -\frac{h_{fg}}{R} \tag{8.28}$$

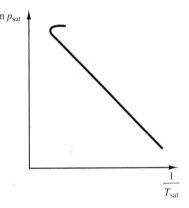

Figure 8.1 $\ln p_{sat}$ versus $1/T_{sat}$.

Figure 8.1 shows a plot of $\ln p_{sat}$ versus $1/T_{sat}$, whose slope can be readily used to determine the enthalpy of vaporization.

Similar approximation can be made for the solid–vapor phase change when $v_{vapor} \gg v_{solid}$. For the solid–liquid phase change, both v_{solid} and v_{liquid} are of the same order of magnitude, and both volumes should be retained in the Clapeyron equation.

Example 8.3

Using the Clapeyron equation, estimate the saturation pressure of water vapor at $-40°C$ using data at $-30°C$ from steam tables.

SOLUTION

The following data are obtained from steam tables:

T (°C)	p (kPa)	v_i (m³/kg)	v_g (m³/kg)	h_{ig} (kJ/kg)
-30	0.0381	1.0858×10^{-3}	2943.0	2839.0

The Clapeyron equation for a solid–vapor phase change when v_i is much smaller than v_g is

$$\frac{dp}{p} = \frac{h_{ig}}{R}\frac{dT}{T^2}$$

where

$$R = \frac{\overline{R}}{M} = \frac{8.3144 \text{ kJ/kg-mol K}}{18.016 \text{ kg/kg-mol}} = 0.4615 \text{ kJ/kg K}$$

Integration between two saturation states gives

$$\ln\left(\frac{p_2}{p_1}\right) = \frac{h_{ig}}{R}\left(\frac{T_2 - T_1}{T_1 T_2}\right)_{\text{sat}}$$

where subscript 1 refers to conditions at $-30°C$, and subscript 2 refers to conditions at $-40°C$. Substituting values gives

$$\ln\frac{p_2}{0.0381} = \frac{2839 \text{ kJ/kg}}{0.4615 \text{ kJ/kg K}}\left[\frac{233.15 \text{ K} - 243.15 \text{ K}}{(233.15 \text{ K})(243.15 \text{ K})}\right]$$
$$= -1.0867$$

so that $p_2 = 0.0129$ kPa which is identical to that given by the steam tables.

Example 8.4

SOLUTION

Using the Clapeyron equation, determine the rate of change of the melting point of ice with pressure. At what temperature will ice melt if the pressure is 1000 atmospheres?

SOLUTION

The Clapeyron equation is

$$\frac{dp}{dT} = \frac{h_{if}}{T(v_f - v_i)}$$

so that

$$\frac{dT}{dp} = \frac{T(v_f - v_i)}{h_{if}}$$

From steam tables, at the triple point,

$$v_f = 0.001 \text{ m}^3/\text{kg}$$
$$v_{ice} = 0.0010908 \text{ m}^3/\text{kg}$$

and the enthalpy of fusion of ice $h_{if} = 333.3$ kJ/kg.

Therefore,

$$\frac{dT}{dp} = \frac{(273.16 \text{ K})(0.001 - 0.0010908) \text{ m}^3/\text{kg}}{333.3 \text{ kJ/kg}}$$

$$= -0.0744 \times 10^{-3} \text{ Km}^3/\text{kJ}$$

$$= -0.0744 \times 10^{-3} \text{ K/kPa} = -0.0074 \text{ K/atm}$$

This means that ice at a pressure of 1000 atm would melt at $-7.4°C$.

Example 8.5

The Gay-Lussac–Joule experiment, outlined in Chapter 3, indicates that the internal energy of an ideal gas is a function of temperature only. Using the appropriate Maxwell relation, give a mathematical proof of this statement.

SOLUTION

The combined first and second laws can be written as

$$T \, dS = dU + p \, dV$$

Dividing by dV and considering an isothermal change gives

$$T\left(\frac{\partial S}{\partial V}\right)_T = \left(\frac{\partial U}{\partial V}\right)_T + p$$

Substituting for $(\partial S/\partial V)_T$ from the Maxwell relation, Eq. (8.15), and noting that for an ideal gas, $T(\partial p/\partial T)_V = p$, then

$$\left(\frac{\partial U}{\partial V}\right)_T = 0$$

which means that the internal energy of an ideal gas is not a function of volume; hence, it is a function of temperature only.

8.5 Internal Energy, Enthalpy, and Entropy of a Simple Substance in a Single Phase

The combined first law and second law for a simple substance is

$$du = T\,ds - p\,dv \tag{8.29}$$

The differential du and ds in this equation can be expressed in terms of temperature and volume as

$$du = \left(\frac{\partial u}{\partial T}\right)_v dT + \left(\frac{\partial u}{\partial v}\right)_T dv = c_v\,dT + \left(\frac{\partial u}{\partial v}\right)_T dv \tag{8.30}$$

and

$$ds = \left(\frac{\partial s}{\partial T}\right)_v dT + \left(\frac{\partial s}{\partial v}\right)_T dv$$

but

$$\left(\frac{\partial s}{\partial v}\right)_T = \left(\frac{\partial p}{\partial T}\right)_v \qquad \text{(Maxwell relation)}$$

Therefore,

$$ds = \left(\frac{\partial s}{\partial T}\right)_v dT + \left(\frac{\partial p}{\partial T}\right)_v dv \tag{8.31}$$

Substituting for du and ds in the combined first and second laws gives

$$c_v\,dT + \left(\frac{\partial u}{\partial v}\right)_T dv = T\left[\left(\frac{\partial s}{\partial T}\right)_v dT + \left(\frac{\partial p}{\partial T}\right)_v dv\right] - p\,dv$$

$$= T\left(\frac{\partial s}{\partial T}\right)_v dT + \left[T\left(\frac{\partial p}{\partial T}\right)_v - p\right]dv$$

Since the specific volume and the temperature can be varied independently, the coefficients of dv and dT in this equation must be equal so that

$$\left(\frac{\partial u}{\partial v}\right)_T = T\left(\frac{\partial p}{\partial T}\right)_v - p \tag{8.32}$$

and

$$\left(\frac{\partial s}{\partial T}\right)_v = \frac{c_v}{T} \tag{8.33}$$

Substituting Eq. (8.32) into Eq. (8.30) gives

$$du = c_v\, dT + \left[T\left(\frac{\partial p}{\partial T}\right)_v - p\right]dv \tag{8.34}$$

and substituting Eq. (8.33) into Eq. (8.31) gives

$$ds = \frac{c_v}{T}\, dT + \left(\frac{\partial p}{\partial T}\right)_v dv \tag{8.35}$$

Equations (8.34) and (8.35) are expressed in terms of the easily measurable properties of T, p, v, and c_v, and can be numerically integrated to give Δu and Δs.

Similar analysis can be applied to determine an expression of enthalpy in terms of two independent properties.

The differential of enthalpy is

$$dh = T\, ds + v\, dp$$

Expressing enthalpy and entropy in terms of temperature and pressure gives

$$dh = \left(\frac{\partial h}{\partial T}\right)_p dT + \left(\frac{\partial h}{\partial p}\right)_T dp = c_p\, dT + \left(\frac{\partial h}{\partial p}\right)_T dp \tag{8.36}$$

and

$$ds = \left(\frac{\partial s}{\partial T}\right)_p dT + \left(\frac{\partial s}{\partial p}\right)_T dp$$

but

$$\left(\frac{\partial s}{\partial p}\right)_T = -\left(\frac{\partial v}{\partial T}\right)_p \qquad \text{(Maxwell relation)}$$

Therefore,

$$ds = \left(\frac{\partial s}{\partial T}\right)_p dT - \left(\frac{\partial v}{\partial T}\right)_p dp \tag{8.37}$$

Substituting for dh and ds in the combined first and second law equation gives

$$c_p\, dT + \left(\frac{\partial h}{\partial p}\right)_T dp = T\left[\left(\frac{\partial s}{\partial T}\right)_p dT - \left(\frac{\partial v}{\partial T}\right)_p dp\right] + v\, dp$$

$$= T\left(\frac{\partial s}{\partial T}\right)_p dT + \left[v - T\left(\frac{\partial v}{\partial T}\right)_p\right] dp$$

Since the pressure and temperature are independent, the coefficients of dp and dT in the preceding equation can be equated so that

$$\left(\frac{\partial h}{\partial p}\right)_T = v - T\left(\frac{\partial v}{\partial T}\right)_p \qquad (8.38)$$

and

$$\left(\frac{\partial s}{\partial T}\right)_p = \frac{c_p}{T} \qquad (8.39)$$

Substituting these differentials into Eqs. (8.36) and (8.37), respectively, gives

$$dh = c_p\, dT + \left[v - T\left(\frac{\partial v}{\partial T}\right)_p\right] dp = c_p\, dT + v(1 - \beta T)dp \qquad (8.40)$$

and

$$ds = \frac{c_p}{T}\, dT - \left(\frac{\partial v}{\partial T}\right)_p dp = \frac{c_p}{T}\, dT - \beta v\, dp \qquad (8.41)$$

Integration of Eqs (8.40) and (8.41) yields Δh and Δs in terms of T, p, v, β, and c_p.

8.6 Specific Heat Relations

The specific heat of a real gas is a function of two independent properties such as the temperature and pressure, whereas the specific heat of an ideal gas is a function of temperature only.

In this section, expressions for the variation of specific heat with pressure or volume are determined. In particular, we seek expressions for $(\partial c_v/\partial v)_T$ and $(\partial c_p/\partial p)_T$ for real gases in terms of directly measurable properties.

The specific heat at constant volume is defined by

$$c_v \equiv \left(\frac{\partial u}{\partial T}\right)_v = T\left(\frac{\partial s}{\partial T}\right)_v$$

The change of c_v with respect to volume at constant temperature is

$$\left(\frac{\partial c_v}{\partial v}\right)_T = \left[\frac{T\partial(\partial s/\partial T)_v}{\partial v}\right]_T$$

Since entropy is a property, the order of differentiation is immaterial, and so this equation can also be written as

$$\left(\frac{\partial c_v}{\partial v}\right)_T = \left[\frac{T\partial(\partial s/\partial v)_v}{\partial T}\right]_v$$

But according to the Maxwell relation, Eq. (8.15),

$$\left(\frac{\partial s}{\partial v}\right)_T = \left(\frac{\partial p}{\partial T}\right)_v$$

Consequently, by substitution

$$\left(\frac{\partial c_v}{\partial v}\right)_T = T\left(\frac{\partial^2 p}{\partial T^2}\right)_v \tag{8.42}$$

In a similar manner, an expression for $(\partial c_p/\partial p)_T$ can be obtained as follows.

The specific heat at constant pressure is

$$c_p \equiv \left(\frac{\partial h}{\partial T}\right)_p = T\left(\frac{\partial s}{\partial T}\right)_p$$

The change of c_p as a function of pressure, at constant temperature, is

$$\left(\frac{\partial c_p}{\partial p}\right)_T = \left[\frac{T\partial(\partial s/\partial T)_p}{\partial p}\right]_T$$

The order of differentiation is immaterial, and this equation can also be written as

$$\left(\frac{\partial c_p}{\partial p}\right)_T = \left[\frac{T\partial(\partial s/\partial p)_T}{\partial T}\right]_p$$

Substituting from the Maxwell relation, Eq. (8.16), gives

$$\left(\frac{\partial c_p}{\partial p}\right)_T = -T\left(\frac{\partial^2 v}{\partial T^2}\right)_p \tag{8.43}$$

Equations (8.42) and (8.43) relate specific heat values to p-v-T information.

An expression of the difference between c_p and c_v can be obtained by equating Eqs. (8.35) and (8.41):

$$c_v\frac{dT}{T} + \left(\frac{\partial p}{\partial T}\right)_v dv = c_p\frac{dT}{T} - \left(\frac{\partial v}{\partial T}\right)_p dp$$

Solving for dT,

$$dT = \frac{T(\partial p/\partial T)_v}{c_p - c_v}dv + \frac{T(\partial v/\partial T)_p}{c_p - c_v}dp$$

But the differential of temperature can be expressed in terms of v and p as

$$dT = \left(\frac{\partial T}{\partial v}\right)_p dv + \left(\frac{\partial T}{\partial p}\right)_v dp$$

Equating the coefficients of dv and dp in the preceding two equations gives

$$\left(\frac{\partial T}{\partial v}\right)_p = \frac{T(\partial p/\partial T)_v}{c_p - c_v}$$

and

$$\left(\frac{\partial T}{\partial p}\right)_v = \frac{T(\partial v/\partial T)_p}{c_p - c_v}$$

These two relations yield the same expression for $(c_p - c_v)$ so that

$$c_p - c_v = T\left(\frac{\partial v}{\partial T}\right)_p\left(\frac{\partial p}{\partial T}\right)_v \tag{8.44}$$

But, according to Eq. (8.4),

$$\left(\frac{\partial p}{\partial T}\right)_v = -\left(\frac{\partial v}{\partial T}\right)_p\left(\frac{\partial p}{\partial v}\right)_T \tag{8.45}$$

Substituting this relation in Eq. (8.44) gives

$$c_p - c_v = -T\left(\frac{\partial v}{\partial T}\right)_p^2 \left(\frac{\partial p}{\partial v}\right)_T \qquad (8.46)$$

The ratio of specific heats is

$$\gamma = \frac{c_p}{c_v} = \frac{(\partial s/\partial T)_p}{(\partial s/\partial T)_v} = \frac{(\partial p/\partial v)_s}{(\partial p/\partial v)_T} = \left(\frac{\partial p}{\partial v}\right)_s \left(\frac{\partial v}{\partial p}\right)_T \qquad (8.47)$$

The coefficient of volumetric expansion β and isothermal compressibility κ have been defined in Chapter 1 as

$$\beta \equiv \frac{1}{v}\left(\frac{\partial v}{\partial T}\right)_p \qquad \text{and} \qquad \kappa \equiv -\frac{1}{v}\left(\frac{\partial v}{\partial p}\right)_T$$

Substituting these definitions into Eq. (8.46) gives

$$c_p - c_v = \frac{\beta^2 v T}{\kappa} \qquad (8.48)$$

Note that the right-hand side of this equation is always positive (β^2 is positive, and the value of κ is positive for all substances) implying that c_p is greater than c_v. When $\beta = 0$, as in the case of water at its maximum density (at 1 atm and about 4°C), $c_p = c_v$. In general, for liquids and solids, β is relatively small, and the specific heat c_p and c_v are approximately equal and are designated by c.

Example 8.6

The density, coefficient of volumetric expansion, coefficient of compressibility, and the specific heat for water at 0°C and 50°C are as follows:

T (°C)	ρ (kg/m^3)	$\beta \times 10^6 (K)^{-1}$	$\kappa \times 10^8 (kPa)^{-1}$	c (kJ/kg K)
0	999.84	−68.14	50.89	4.217
50	988.04	457.8	44.18	4.181

Determine the percent error in using the same tabulated value for c_p and c_v at these two temperatures.

SOLUTION

$$c_p - c_v = \frac{\beta^2 vT}{\kappa} = \frac{\beta^2 T}{\rho \kappa}$$

At 0°C,

$$c_p - c_v = \frac{(-68.14 \times 10^{-6} \text{ K}^{-1})^2 (273.15 \text{ K})}{(999.84 \text{ kg/m}^3)[50.89 \times 10^{-8} \text{ (kPa)}^{-1}]}$$

$$= -2.49 \times 10^{-3} \text{ kJ/kg K}$$

$$\text{error} = \frac{-2.49 \times 10^{-3}}{4.217} = -0.059\%$$

At 50°C,

$$c_p - c_v = \frac{(457.8 \times 10^{-6} \text{ K}^{-1})^2 (323.15 \text{ K})}{(988.04 \text{ kg/m}^3)[44.18 \times 10^{-8} \text{ (kPa)}^{-1}]} = 0.1552 \text{ kJ/kg K}$$

$$\text{error} = \frac{0.1552}{4.181} = 3.71\%$$

8.7 Properties in Terms of the Joule–Thomson Coefficient

The Joule–Thomson experiment described in Section 3.10 provides a precise method of determining properties. During an adiabatic steady-state throttling process, the enthalpy remains constant if only negligible changes in kinetic energy occur. Further, if the gas is ideal, the temperature remains unchanged. The *isenthalpic Joule–Thomson coefficient* is defined as

$$\mu_h \equiv \left(\frac{\partial T}{\partial p} \right)_h \qquad (8.49)$$

whereas the *isothermal Joule–Thomson coefficient* is defined as

$$\mu_T \equiv \left(\frac{\partial h}{\partial p} \right)_T \qquad (8.50)$$

The differential of enthalpy, as a function of temperature and pressure, is

$$dh = \left(\frac{\partial h}{\partial T}\right)_p dT + \left(\frac{\partial h}{\partial p}\right)_T dp$$

In a throttling process, enthalpy does not change, and therefore

$$\left(\frac{\partial h}{\partial T}\right)_p dT = -\left(\frac{\partial h}{\partial p}\right)_T dp$$

or

$$\left(\frac{\partial h}{\partial T}\right)_p = -\left(\frac{\partial h}{\partial p}\right)_T \left(\frac{\partial p}{\partial T}\right)_h$$

This equation is also the cyclic equation derived in Section 8.2. By substituting the coefficients μ_h and μ_T given by Eqs. (8.49) and (8.50), and noting that $c_p = (\partial h / \partial T)_p$, the preceding equation becomes

$$c_p = -\frac{\mu_T}{\mu_h} \tag{8.51}$$

This equation provides a method of determining specific heat values from h-p-T data obtained by means of Joule–Thomson experiments.

An expression of the Joule–Thomson coefficient can be obtained from Eq. (8.40) by setting $dh = 0$ so that

$$\mu_h = \frac{1}{c_p}\left[T\left(\frac{\partial v}{\partial T}\right)_p - v\right] \tag{8.52}$$

Equation (8.52) can be used to determine c_p using p-v-T data and the value of the Joule–Thomson coefficient. For an ideal gas, $\mu_h = 0$, and the gas experiences no change in temperature upon throttling.

For a real gas following an equation of state $pv = ZRT$, Eq. (8.52) becomes

$$\mu_h = \frac{RT^2}{p\,c_p}\left(\frac{\partial Z}{\partial T}\right)_p \tag{8.53}$$

and the sign of μ_h is governed by the sign of the derivative $(\partial Z / \partial T)_p$.

Example 8.7

Determine the Joule–Thompson coefficient for a van der Waals gas given by the equation

$$\left(p + \frac{a}{v^2}\right)(v - b) = RT$$

Prove that for large volume (or low pressure), the inversion temperature is equal to $2a/bR$.

SOLUTION

$$\mu = \left(\frac{\partial T}{\partial p}\right)_h = \frac{1}{c_p}\left[T\left(\frac{\partial v}{\partial T}\right)_p - v\right]$$

Differentiating the van der Waals equation with respect to T at constant p gives

$$\left(p + \frac{a}{v^2}\right)\left(\frac{\partial v}{\partial T}\right)_p + (v - b)\left[-\frac{2a}{v^3}\left(\frac{\partial v}{\partial T}\right)_p\right] = R$$

$$\left(\frac{\partial v}{\partial T}\right)_p\left(p + \frac{a}{v^2} - \frac{2a}{v^2} + \frac{2ab}{v^3}\right) = R$$

$$\left(\frac{\partial v}{\partial T}\right)_p\left(p - \frac{a}{v^2} + \frac{2ab}{v^3}\right) = R$$

Substituting in the expression of μ gives

$$\left(\frac{\partial T}{\partial p}\right)_h = \frac{1}{c_p}\left[\frac{RT}{p - \frac{a}{v^2} + \frac{2ab}{v^3}} - v\right] = \frac{1}{c_p}\left[\frac{\left(p + \frac{a}{v^2}\right)(v - b)}{p - \frac{a}{v^2} + \frac{2ab}{v^3}} - v\right] = \frac{1}{c_p}\left[\frac{-bp + \frac{2a}{v} - \frac{3ab}{v^2}}{p - \frac{a}{v^2} + \frac{2ab}{v^3}}\right]$$

For large values of v, corresponding to a gas at low pressure expanding into a near vacuum,

$$\left(\frac{\partial T}{\partial p}\right)_h = \frac{1}{c_p}\left[\frac{-bp + \frac{2a}{v}}{p}\right] = \frac{1}{c_p}\left[-b + \frac{2a}{pv}\right] = \frac{1}{c_p}\left(-b + \frac{2a}{RT}\right)$$

When

$$\left(\frac{\partial T}{\partial p}\right)_h = 0$$

then

$$b = \frac{2a}{RT_{\text{inversion}}} \quad \text{or} \quad T_{\text{inversion}} = \frac{2a}{bR}$$

But

$$T_c = \frac{8a}{27bR} \quad \text{(see Section 8.8)}$$

so that

$$\frac{T_{\text{inversion}}}{T_c} = \frac{27}{4}$$

Therefore,

$$\left(\frac{\partial T}{\partial p}\right)_h = \frac{b}{c_p}\left(\frac{1}{b}\frac{2a}{RT} - 1\right) = \frac{b}{c_p}\left(\frac{T_{\text{inversion}}}{T} - 1\right)$$

For $T \ll T_{\text{inversion}}$, $(\partial T/\partial p)_h$ is proportional to $1/T$, and the colder the gas before throttling, the larger the temperature drop due to throttling.

8.8 Equations of State

Properties, such as pressure, temperature, and volume, may be used to specify the state of a thermodynamic system in equilibrium. It is convenient to correlate the values of these properties as found experimentally through functional relations called *equations of state*. Properties may be related by means of graphs, tables of properties, or an equation of the form

$$f(p, V, T) = 0 \tag{8.54}$$

It has been shown previously that the thermodynamic state of a one-component system in the absence of electrical, magnetic, gravitational, surface, and motion effects is two-dimensional; that is, two independent properties are sufficient to describe the state of the system. A third property may be related to the two aforementioned independent properties by an equation of state.

In the case of an ideal gas, the equation of state relating pressure, temperature, and specific volume is expressed by the simple relation

$$pv = RT \tag{8.55}$$

The ideal-gas gas equation was first deduced from the experiments of Charles and Boyle and can also be derived from the kinetic theory of gases. The latter case uses a highly idealized model of gas made up of perfectly elastic particles of negligible volume and no interparticle forces. Therefore, the ideal-gas law is an approximation describing the behavior of real gases at nominal pressures and is an expression of the limiting behavior of real gases at high molar volumes. The ideal-gas law becomes inadequate at low molar volumes or high pressure because molecular interaction cannot be neglected at high particle density.

Realizing the inherent inaccuracies in describing real gases by the ideal-gas law, investigators have proposed numerous equations of state for real gases. The derivations are either empirical, according to experimental data, or have a physical interpretation based on the kinetic theory. Investigation of the gaseous state has received more attention than other states and has been an important factor in the development of present knowledge of the magnitude and nature of intermolecular forces. It has also helped in the analytical description of the behavior of high-pressure gases encountered in many industrial processes.

Deviations of a real gas from an ideal gas are readily displayed by plotting the deviation function $Z = pv/RT$ versus temperature, or pressure. Figure 8.2 shows the behavior of two real gases near room temperature. Note that as the pressure approaches zero, Z approaches unity. The ratio pv/RT is called the *compressibility factor* and was discussed in detail in Section 4.10. Although graphical methods are used extensively for calculation, analytical methods seem to be superior where equations of state give accurate representation of the compressibility. Each equation, however, holds in a specified range of density variations. Equations of state permit calculation of the variables involved along with their derivatives. Other thermodynamic properties, as well as conditions of equilibrium, can also be calculated with an accuracy comparable to the accuracy of the equation of state.

The following paragraphs discuss a few of these equations and investigate the principles of their derivations.

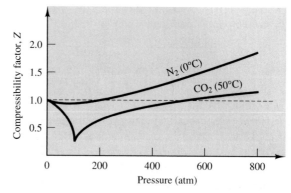

Figure 8.2   Compressibility factor versus pressure for CO_2 and N_2.

8.8.1 Van der Waals Equation of State

At low pressures, the compressibility of real gases has a higher value than predicted by ideal-gas analysis. Similarly, at high pressures, real gases are found to
be less compressible. The first observation is attributed to the attraction forces
between the molecules and is more noticeable at low pressures. The second observation is attributed to the volume occupied by the molecules themselves. If the
molecules of a gas are considered rigid spheres, only the space between the molecules is available for compression or expansion. At low pressures, where the
mean free path of the molecules is large compared to the size of the molecules,
the volume occupied by the molecules is relatively small and can be neglected.
At high pressure, this volume cannot be neglected in comparison to the total volume occupied by the gas.

   In order to find a correction term to compensate for the attraction forces between the molecules, consider the force of attraction on a single molecule hitting
the wall of the confining vessel. When the molecule is not near the wall, it is surrounded by other molecules; consequently, it is equally attracted in all directions, resulting in a zero net force on the molecule. This balance of forces is
offset, however, when the molecule hits the wall since it is then exposed to a net
attraction force away from the wall toward the surrounding molecules. This
attraction force reduces the pressure that otherwise would have been exerted by
the molecule on the wall. The attraction force is proportional, among other factors, to the number of molecules per unit volume of the gas. The pressure is also
proportional to the frequency of collision with the wall, which in turn is proportional to the number of molecules per unit volume. Therefore, the attraction
force is proportional to $(N/V)^2$ and, since $N = nN_a$, it is proportional to $1/\bar{v}^2$,
where $\bar{v}$ is the molar volume of the gas. The pressure exerted by the gas then
equals a kinetic pressure p' as calculated by molecular change of momentum
minus a correction term to compensate for the attraction forces so that

$$p = p' - \frac{a}{\bar{v}^2}$$

or

$$p' = p + \frac{a}{\bar{v}^2} \tag{8.56}$$

where a is a proportionality constant.

   The second correction due to the molecular sizes can be investigated by considering the collision of two identical molecules of the gas. If the molecular
radius is r, the distance between the centers of identical molecules at the instant
of collision is $2r$. Therefore, any molecule cannot come closer than a distance $2r$
to other identical molecules. This sets up a forbidden volume of $\frac{4}{3}\pi(2r)^3$ around
each molecule, which is eight times the volume of the molecule itself. Only half
of this volume, that is, only the hemisphere facing the incoming molecule, is

effective in excluding the molecule from the forbidden volume. The latter is only four times the volume of the molecule. Denoting this volume per mole of the gas by the symbol b, the effective volume of the gas becomes $(\bar{v} - b)$.

Van der Waals, in 1873, realizing the foregoing deviations from the ideal-gas law, proposed the following equation of state for a real gas:

$$\left(p + \frac{a}{\bar{v}^2}\right)(\bar{v} - b) = \bar{R}T \tag{8.57}$$

When n moles of the gas are present, Eq. (8.57) becomes

$$\left(p + \frac{n^2 a}{V^2}\right)(V - nb) = n\bar{R}T \tag{8.58}$$

Values of a and b for a few gases are listed in Table 8.1. Simplicity and ease of the constants' evaluation from limited data are the advantages of van der Waals equation; however, it gives only approximate results under many conditions. In general, the accuracy with which van der Waals equation fits experimental data decreases at high densities.

At low pressures, where the mean free path is large in comparison to the molecule's dimensions, the quantity b in van der Waals equation may be neglected. Hence, Eq. (8.57) reduces to

$$\left(p + \frac{a}{\bar{v}^2}\right)(\bar{v}) = \bar{R}T$$

or

$$p\bar{v} = \bar{R}T - \frac{a}{\bar{v}} \tag{8.59}$$

Table 8.1   Values of a and b in van der Waals Equation of State

	$a(kPa[m^3/kg\text{-}mol]^2)$	$b(m^3/kg\text{-}mol)$
Air	136.8	0.0367
Ammonia	423.3	0.0373
Carbon dioxide	364.3	0.0427
Carbon monoxide	146.3	0.0394
Freon-12	1049	0.0971
Helium	3.41	0.0234
Hydrogen	24.7	0.0265
Nitrogen	136.1	0.0385
Oxygen	136.9	0.0315
Water	550.7	0.0304

Equation (8.59) states that the product $p\bar{v}$ is less than $\bar{R}T$. This accounts for the shape of the curves of Figure 8.2 at low pressures. Similarly, at high pressures, the term $a/\bar{v}^2$ can be neglected compared to p, and van der Waals equation reduces to

$$p(\bar{v} - b) = \bar{R}T$$

or

$$p\bar{v} = \bar{R}T + bp \qquad (8.60)$$

Equation (8.60) is called the *Clausius equation of state*. Here the product $p\bar{v}$ is greater than $\bar{R}T$ and increases linearly with pressure.

Note that the pressure and temperature in the van der Waals equation can be expressed explicity. The equation, however, is cubic in V and, therefore, has three roots at each pressure. Figure 8.3 shows several isotherms plotted on a p-V diagram. Only one value of volume is obtained at temperatures equal to or above the critical isotherm; three values of specific volumes can be obtained at temperatures below the critical temperature. If thermodynamic equilibrium prevails,

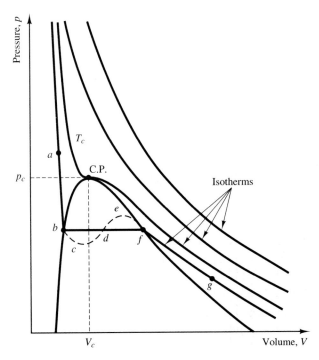

Figure 8.3   Pressure–volume diagram for a one-component substance.

isothermal expansion will proceed from point *a* along path *a-b-f-g*. Along *b-f*, both the liquid and vapor have the same pressure. If thermodynamic equilibrium does not prevail, the liquid state follows the metastable path *b-c*, and vaporization is delayed. Note that the curve *b-c-d-e-f* is an isotherm but has different values of pressure. Along *b-c*, the temperature of the liquid exceeds the saturation temperature corresponding to the prevailing pressure. The liquid along *b-c* is then superheated liquid. In a similar fashion, isothermal compression of a gas under metastable conditions follows path *g-f-e*, and condensation is delayed. Along *e-f*, the temperature of the vapor is less than the saturation temperature corresponding to the prevailing pressure, and the vapor is thus subcooled. Note that the stable conditions to which the metastable states at *c* or *e* eventually tend are along line *b-f*. The part of the curve *c-d-e* is unstable since its slope is positive, indicating that an increase in pressure results in an increase in volume.

The values of the constants *a* and *b* in van der Waals equation can be calculated in terms of the properties at the critical point. To determine the critical values of van der Waals equation, the first and second derivatives of the pressure with respect to volume are equated to zero. This follows from the fact that the critical isotherm has an inflection point at the critical point and the tangent to the isotherm is horizontal. Therefore,

$$\left(\frac{\partial p}{\partial \bar{v}}\right)_c = 0 \qquad \text{and} \qquad \left(\frac{\partial^2 p}{\partial \bar{v}^2}\right)_c = 0$$

where subscript *c* denotes critical point. The van der Waals equation at the critical point is

$$p_c = \frac{\bar{R}T_c}{\bar{v}_c - b} - \frac{a}{\bar{v}_c^2} \tag{8.61}$$

Differentiating with respect to $\bar{v}$ and noting that T_c is a constant gives

$$\left(\frac{\partial p}{\partial \bar{v}}\right)_c = -\frac{\bar{R}T_c}{(\bar{v}_c - b)^2} + \frac{2a}{\bar{v}_c^3} = 0$$

and

$$\left(\frac{\partial^2 p}{\partial \bar{v}^2}\right)_c = \frac{2\bar{R}T_c}{(\bar{v}_c - b)^3} - \frac{6a}{\bar{v}_c^4} = 0$$

Combining the last three equations, the critical pressure, volume, and temperature, in terms of the coefficients *a* and *b* of the van der Waals equation, are

$$p_c = \frac{a}{27b^2} \qquad \bar{v}_c = 3b \qquad \text{and} \qquad T_c = \frac{8a}{27\bar{R}b}$$

Alternatively, the coefficients of van der Waals equation can be determined in terms of the critical properties. The result is as follows:

$$a = 3p_c \bar{v}_c^2 = \frac{9}{8} \bar{R} T_c \bar{v}_c = \frac{27}{64} \frac{\bar{R}^2 T_c^2}{p_c}$$

$$b = \frac{\bar{v}_c}{3} = \frac{\bar{R} T_c}{8 p_c}$$

Checking the foregoing results is left as an exercise to the reader. Table A6 gives the critical properties for several gases.

8.8.2 Redlich–Kwong Equation

This equation is more accurate than the van der Waals equation over a wide range of property values. Like the van der Waals equation, it is cubic in specific volume and involves two-dimensional constants. The constants are expressed in terms of the properties at the critical point. The Redlich–Kwong equation is

$$p = \frac{\bar{R} T}{(\bar{v} - b)} - \frac{a}{\bar{v}(\bar{v} + b) T^{1/2}} \tag{8.62}$$

where a and b are given by

$$a = 0.42748 \frac{\bar{R}^2 T_c^{2.5}}{p_c} \quad \text{and} \quad b = 0.08664 \frac{\bar{R} T_c}{p_c}$$

8.8.3 The Beattie–Bridgeman Equation of State

This equation has more constants than other equations of state; therefore, it is more successful in representing the compressibility of gases. The Beattie–Bridgeman equation is

$$p = \frac{\bar{R} T}{\bar{v}^2} (1 - e)(\bar{v} + B) - \frac{A}{\bar{v}^2} \tag{8.63}$$

where

$$A = A_o \left(1 - \frac{a}{\bar{v}} \right)$$

$$B = B_o \left(1 - \frac{b}{\bar{v}} \right)$$

$$e = \frac{c}{\bar{v} T^3}$$

Table 8.2   Constants in the Beattie–Bridgeman Equation of State*

Gas	A_0	a	B_0	b	$c \times 10^{-4}$
Air	131.8441	0.01931	0.04611	−0.001101	4.34
Argon	130.7802	0.02328	0.03931	0.0	5.99
Carbon dioxide	507.2836	0.07132	0.10476	0.07235	66.00
Helium	2.1886	0.05984	0.01400	0.0	0.0040
Hydrogen	20.0117	−0.00506	0.02096	−0.04359	0.0504
Nitrogen	136.2315	0.02617	0.05046	0.00691	4.20
Oxygen	151.0857	0.02562	0.04624	0.004208	4.80

*Constants for pressure in kPa; specific volume in m³/kg-mol; and temperature in K.

The five constant a, b, c, A_0, and B_0 in Eq. (8.63) are given for several gases in Table 8.2. Data for many gases can be fitted to within 0.5 percent accuracy over a wide range of temperatures and pressures. The Beattie–Bridgeman equation is, however, inaccurate near the critical point.

8.8.4 Berthelot Equation of State

Another classical equation of state is

$$\left(p + \frac{an^2}{TV^2} \right)(V - nb) = n\bar{R}T \tag{8.64}$$

where a and b are constants. Berthelot made the additive pressure term proportional to $1/T$, and his equation reduces to van der Waals equation at high molar volumes. Both equations, however, fail at the critical point.

8.8.5 Dieterici Equation of State

In 1899, Dieterici proposed the following equation of state:

$$\left(p e^{an/\bar{V}\bar{R}T} \right)(V - nb) = n\bar{R}T \tag{8.65}$$

where a and b are constants.

The Berthelot and Dieterici equations do not give accurate results over wide ranges of pressure and temperature. They give some insight, however, into the p-V-T relationships and have generated experimental and theoretical work in the field of intermolecular forces. Note that the Dieterici equation reduces to van der Waals equation of state at high molar volumes. The proof is left to the reader as an exercise.

8.8.6 Virial[1] Equation of State

Kammerlingh Onnes, in 1901, introduced the virial equation of state in the form of an infinite expansion of the product pV. Two forms are available:

In terms of V,

$$pV = n\overline{R}T\left(1 + \frac{nB}{V} + \frac{n^2C}{V^2} + \cdots\right) \qquad (8.66)$$

In terms of p,

$$pV = n\overline{R}T(1 + nB'p + n^2C'p^2 + \cdots) \qquad (8.67)$$

The coefficients B, C... and B', C'... are called the second, third, and so on, *virial coefficients*. These coefficients are functions of temperature, and they relate the deviation of actual gases from ideal gases in terms of intermolecular forces. The virial equation applies to gases at low or medium densities only. Note that in the limit when the gas molecules do not interact in any way, all virial coefficients vanish, and Eq. (8.67) reduces to the ideal-gas equation.

Comparison of accuracy of the various equations of state is difficult since accuracy of any equation depends on the nature of the gas under consideration, as well as on the method of constants' evaluation and the ranges of pressure and temperature. Quite often, the equation of state is suitable for thermodynamic calculations of functions such as p, V, and T but unsuitable for the evaluation of properties that depend on first and second derivatives of functions of p, V, and T.

Example 8.8

Find the second and third virial coefficients of the van der Waals equation of state when expressed in the expansion forms of Eq. (8.66).

SOLUTION

The van der Waals equation is

$$\left(p + \frac{an^2}{V^2}\right)(V - nb) = n\overline{R}T$$

Therefore,

$$pV + \frac{an^2}{V} - pnb - \frac{abn^3}{V^2} = n\overline{R}T$$

---

[1] The word *virial* comes from the Latin word for "force," and it refers to interaction forces between molecules.

or

$$pV = n\overline{R}T\left(1 + \frac{pb}{\overline{R}T} - \frac{an}{V\overline{R}T} + \frac{abn^2}{V^2\overline{R}T}\right)$$

but

$$p = \frac{n\overline{R}T}{V - nb} - \frac{an^2}{V^2}$$

Therefore,

$$pV = n\overline{R}T\left[1 + \frac{b}{\overline{R}T}\left(\frac{n\overline{R}T}{V - nb} - \frac{an^2}{V^2}\right) - \frac{an}{V\overline{R}T} + \frac{abn^2}{V^2\overline{R}T}\right]$$

$$= n\overline{R}T\left[1 + \frac{nb}{V - nb} - \frac{an}{V\overline{R}T}\right]$$

$$= n\overline{R}T\left[1 + \frac{nb}{V}\left(1 - \frac{nb}{V}\right)^{-1} - \frac{an}{V\overline{R}T}\right]$$

$$= n\overline{R}T\left[1 + \frac{nb}{V}\left(1 + \frac{nb}{V} + \frac{n^2b^2}{V^2} + \cdots\right) - \frac{an}{V\overline{R}T}\right]$$

$$= n\overline{R}T\left[1 + \frac{n}{V}\left(b - \frac{a}{\overline{R}T}\right) + \frac{n^2b^2}{V^2} + \frac{n^3b^3}{V^3} + \cdots\right]$$

Comparison with Eq. (8.66) gives the following virial coefficients:

second coefficient $B = b - a/\overline{R}T$
third coefficient $C = b^2$, etc.

A similar procedure may be used to expand any equation of state into a series form.

8.9 Enthalpy of Real Gases

In previous sections, equations of state were developed that provided methods of correlating various properties of single-component gases. In this section, a method for evaluating thermodynamic properties of gases from p-V-T data is discussed.

A real gas behaves like an ideal gas when it is at a pressure much lower than its critical pressure, and at a temperature much higher than its critical temperature.

The further the state point of the gas is away from the critical point, the more nearly does the gas obey the ideal-gas law. The internal energy or enthalpy of a real gas is a function of two properties, such as temperature and pressure. Thus the enthalpy of a real gas may be expressed as

$$h = h(T, p) \tag{8.68}$$

and its differential is

$$dh = \left(\frac{\partial h}{\partial T}\right)_p dT + \left(\frac{\partial h}{\partial p}\right)_T dp$$

or

$$dh = c_p\, dT + \left(\frac{\partial h}{\partial p}\right)_T dp \tag{8.69}$$

The first term, which is equal to $c_p\, dT$, indicates the change of enthalpy of an ideal gas as the temperature changes. The second term is the change of enthalpy due to deviation of a real gas, at finite pressures, from ideal gas behavior.

According to the first and second laws, the change of enthalpy for a simple system can be expressed as

$$dh = T\, ds + v\, dp$$

This can be written in the form

$$\left(\frac{\partial h}{\partial p}\right)_T = T\left(\frac{\partial s}{\partial p}\right)_T + v \tag{8.70}$$

But $(\partial s/\partial p)_T$, according to the Maxwell relation, Eq. (8.16), is

$$\left(\frac{\partial s}{\partial p}\right)_T = -\left(\frac{\partial v}{\partial T}\right)_p$$

Using the two preceding equations, Eq. (8.69) becomes

$$dh = c_p\, dT + \left[v - T\left(\frac{\partial v}{\partial T}\right)_p\right] dp \tag{8.71}$$

For constant pressure, this equation reduces to

$$dh_p = c_p\, dT_p \tag{8.72}$$

and for constant temperature,

$$dh_T = \left[v - T\left(\frac{\partial v}{\partial T}\right)_p \right] dp_T \tag{8.73}$$

Integration of Eq. (8.71) gives

$$h_2 - h_1 = \int_1^2 c_p \, dT + \int_1^2 \left[v - T\left(\frac{\partial v}{\partial T}\right)_p \right] dp \tag{8.74}$$

Since the enthalpy deviation between a real and an ideal gas is zero under vacuum conditions, integration of Eq. (8.73) leads to

$$h - h^* = \int_0^p \left[v - T\left(\frac{\partial v}{\partial T}\right)_p \right] dp \tag{8.75}$$

where the asterisk denotes the ideal-gas property value. The enthalpy deviation $(h - h^*)$ can be readily evaluated from experimental p-V-T data. First, specific volume is plotted against temperature, and from the slopes, values of $(\partial v/\partial T)$ at constant pressure are obtained. Then the quantity $[v - T(\partial v/\partial T)_p]$ is calculated, and values are plotted against pressure. Numerical integration yields values of the enthalpy deviation at any desired pressure.

Consider the enthalpy change of a real gas as it changes its state from state 1 to state 2, as shown in the T-s diagram of Figure 8.4. Since the change of enthalpy $(h_2 - h_1)$ is independent of the path, it is possible to evaluate that change along three processes 1-1*, 1*-2*, and 2*-2 so that

$$h_2 - h_1 = (h_2 - h_2^*) + (h_2^* - h_1^*) + (h_1^* - h_1)$$

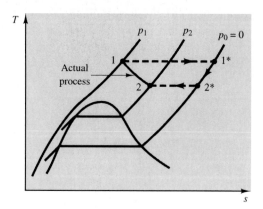

Figure 8.4   Alternative process path to evaluate the enthalpy change of a real gas.

Path 1-1* is a constant-temperature process from the initial state 1 to a low-pressure state 1*. The enthalpy change is due to a pressure change and is given by

$$h_1^* - h_1 = \int_{p_1}^{p_1^*} \left[v - T\left(\frac{\partial v}{\partial T}\right)_p \right] dp$$

Path 1*-2* is a constant-pressure process at such a low pressure that the ideal-gas equation applies so that

$$h_2^* - h_1^* = \int_{T_1}^{T_2} c_p \, dT_p$$

Path 2*-2 is a constant-temperature process to the final state 2 so that

$$h_2 - h_2^* = \int_{p_2^*}^{p_2} \left[v - T\left(\frac{\partial v}{\partial T}\right)_p \right] dp$$

In the absence of *p-V-T* relationship for the gas, it is possible to express the enthalpy deviation in terms of the compressibility factor. Specific volume is given by

$$v = \frac{ZRT}{p} \tag{8.76}$$

Hence, by differentiation,

$$\left(\frac{\partial v}{\partial T}\right)_p = \frac{R}{p}\left[Z + T\left(\frac{\partial Z}{\partial T}\right)_p \right] \tag{8.77}$$

Substitutions can now be made in Eq. (8.73):

$$dh_T = \left\{ \frac{ZRT}{p} - \frac{RT}{p}\left[Z + T\left(\frac{\partial Z}{\partial T}\right)_p \right] \right\} dp_T$$

This reduces to

$$dh_T = -\frac{RT^2}{p}\left(\frac{\partial Z}{\partial T}\right)_p dp_T$$

But it was shown that

$$dh_T = \left(\frac{\partial h}{\partial p}\right)_T dp_T$$

Therefore,

$$\left(\frac{\partial h}{\partial p}\right)_T = -\frac{RT^2}{p}\left(\frac{\partial Z}{\partial T}\right)_p \tag{8.78}$$

This can be expressed as an integral equation:

$$\int_{h^*}^{h} dh_T = h - h^*$$
$$= -RT^2 \int_0^p \left(\frac{\partial Z}{\partial T}\right)_p \frac{dp_T}{p} \tag{8.79}$$

To evaluate $(h - h^*)$, the compressibility factor is first plotted against temperature at various selected pressures. From this graph, values of $(\partial Z/\partial T)_p$ are obtained. Then values of $(RT^2/p)(\partial Z/\partial T)$ are plotted against pressure, and the area under the curve between the origin and a chosen pressure represents the enthalpy deviation at that pressure.

The enthalpy deviation $(h - h^*)$ can be expressed in a more general form as a function of reduced temperature and pressure, rather than in terms of absolute values. Since $T_R = T/T_c$ and $p_R = p/p_c$, substitution in Eq. (8.79) leads to

$$h - h^* = -RT_R^2 T_c^2 \int_0^{PR} \frac{1}{T_c}\left(\frac{\partial Z}{\partial T_R}\right)_{PR} \frac{dp_R}{p_R}$$

From this, it follows that

$$\frac{h^* - h}{RT_c} = T_R^2 \int_0^{PR}\left(\frac{\partial Z}{\partial T_R}\right)_{PR} \frac{dp_R}{p_R} \tag{8.80}$$

Note that the right-hand side of Eq. (8.80) depends only on the reduced temperature and reduced pressure, and so $(h^* - h)/RT_c$ is also a function of T_R and p_R. The integral in the preceding equation is evaluated by using numerical and graphical techniques. The result is shown in Figure 8.5.

The enthalpy deviation can also be expressed in terms of the isothermal coefficient. The enthalpy deviation at constant temperature is

$$dh_T = \left(\frac{\partial h}{\partial p}\right)_T dp_T$$

Substituting the isothermal coefficient, μ_T for $(\partial h/\partial p)_T$, in the preceding equation gives

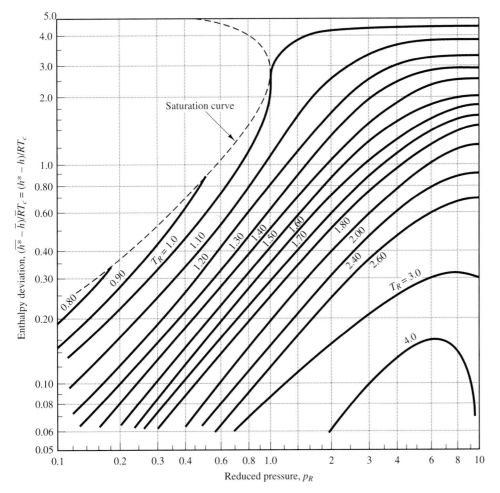

Figure 8.5   Generalized enthalpy deviation chart. (K. Wark, *Thermodynamics,* 5th ed., McGraw-Hill, Inc., New York, 1988. Based on data from A. L. Lydersen, R. A. Greenkorn, and O. A. Hougen. "Engineering Experiment Station Report No. 4." University of Wisconsin, 1955)

$$dh_T = \mu_T dp_T$$

Since the enthalpy deviation is zero at zero pressure, integration leads to

$$h_T = \int_o^p \mu_T dp_T \qquad (8.81)$$

8.10 Entropy of Real Gases

The entropy deviation of a real gas from an ideal gas may be evaluated by a procedure similar to that used for enthalpy deviation. Note, however, that for an ideal gas, entropy, unlike enthalpy, is not a function of temperature only.

The differential entropy, according to Eq. (8.41), can be expressed as

$$ds = \frac{c_p \, dT}{T} - \left(\frac{\partial v}{\partial T}\right)_p dp \tag{8.82}$$

However, for an ideal gas

$$ds^* = \frac{c_p \, dT}{T} - \frac{R}{p} \, dp$$

At constant temperature, the difference between the entropy of an ideal gas and an actual gas is a result of pressure change, and entropy deviation can be expressed in the form

$$(ds - ds^*)_T = -\left[\left(\frac{\partial v}{\partial T}\right)_p - \frac{R}{p}\right] dp_T$$

In integral form, this becomes

$$(s_p - s_0^*)_T - (s_p^* - s_0^*)_T = s_{p,\,T} - s_{p,\,T}^*$$

$$= -\int_{p \to 0}^{p} \left[\left(\frac{\partial v}{\partial T}\right)_p - \frac{R}{p}\right] dp_T \tag{8.83}$$

Equation (8.83) gives the difference in entropy between the real gas and the gas if it were ideal at the same temperature and pressure.

When the equation of state of the gas is $pv = ZRT$, an expression of the entropy difference between an actual gas and an ideal gas is obtained as follows. Substituting the Maxwell relation

$$\left(\frac{\partial s}{\partial p}\right)_T = -\left(\frac{\partial v}{\partial T}\right)_p$$

into Eq. (8.77) gives

$$\left(\frac{\partial s}{\partial p}\right)_T = -\frac{R}{p}\left[Z + T\left(\frac{\partial Z}{\partial T}\right)_p\right] \tag{8.84}$$

When both the temperature and the pressure are expressed in terms of reduced properties with respect to the values at the critical point, the change of entropy of a real gas at constant temperature becomes

$$ds = -R \left[Z + T_R \left(\frac{\partial Z}{\partial T_R} \right)_{P_R} \right]_{T_R} \frac{dp_R}{p_R} \tag{8.85}$$

For an ideal gas, however, the change of entropy at constant temperature is

$$ds^* = -R \left(\frac{dp_R}{p_R} \right)_{T_R} \tag{8.86}$$

Therefore, entropy deviation takes the form

$$ds - ds^* = -R \left[(Z - 1) + T_R \left(\frac{\partial Z}{\partial T_R} \right)_{P_R} \right]_{T_R} \frac{dp_R}{p_R} \tag{8.87}$$

In integral form, this becomes

$$\frac{s_p^* - s_p}{R} = \int_0^{P_R} \left[(Z - 1) + T_R \left(\frac{\partial Z}{\partial T_R} \right)_{P_R} \right]_{T_R} \frac{dp_R}{p_R} \tag{8.88}$$

Note that the right-hand side of this equation depends only on the reduced temperature T_R and the reduced pressure p_R. In Figure 8.6, the function $(s_p^* - s_p)/R$ is plotted against reduced pressure for various reduced temperatures.

In Figure 8.7, the change of entropy of a real gas between two states is shown as the sum of three changes:

$$s_2 - s_1 = (s_2 - s_2^*) + (s_2^* - s_1^*) + (s_1^* - s_1)$$

The term $(s_2 - s_2^*)$ is the entropy difference between the real gas and the gas if it were ideal both at p_2 and T_2. The term $(s_2^* - s_1^*)$ is the entropy difference between states 1 and 2 if the gas were an ideal gas. The term $(s_1^* - s_1)$ is the entropy difference between the real gas and the gas if it were ideal both at p_1 and T_1.

The entropy deviation can also be expressed in terms of the isothermal coefficient μ_T. It was shown that the difference in entropy between the real and ideal gas at constant temperature can be expressed

$$(ds - ds^*)_T = \left[\frac{R}{p} + \left(\frac{\partial s}{\partial p} \right)_T \right] dp_T$$

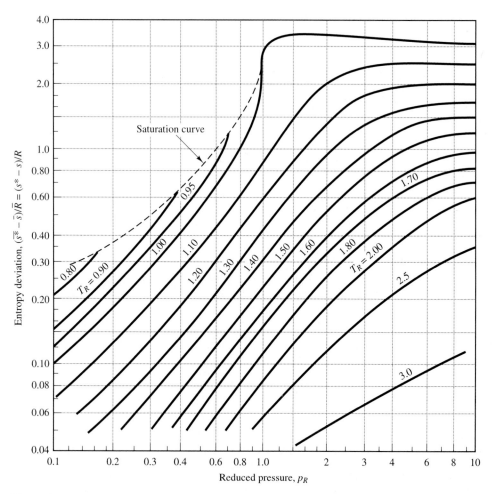

Figure 8.6   Generalized entropy deviation chart. (K. Wark, *Thermodynamics,* 5th ed., McGraw-Hill, Inc., New York, 1988. Based on data from A. L. Lydersen, R. A. Greenkorn, and O. A. Hougen. "Engineering Experiment Station Report No. 4." University of Wisconsin, 1955)

But

$$T\left(\frac{\partial s}{\partial p}\right)_T = \left(\frac{\partial h}{\partial p}\right)_T = \mu_T$$

Therefore,

$$(ds - ds^*)_T = \left[\frac{R}{p} + \frac{\mu_T}{T}\right]dp_T$$

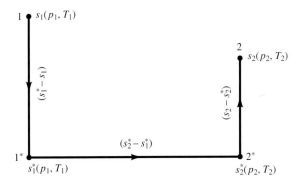

Figure 8.7   Alternative process path to evaluate the entropy change of a real gas.

When integrated, this equation becomes

$$s_{p,T} - s^*_{p,T} = \int_{p \to o}^{p} \left[\frac{R}{p} + \frac{\mu_T}{T} \right] dp_T \qquad (8.89)$$

Equation (8.89) expresses the deviation of entropy in terms of μ_T, p, and T.

Example 8.9

Superheated steam at 35 MPa and 450°C has the following properties: $h = 2672.4$ kJ/kg, $s = 5.1962$ kJ/kg K, and $v = 0.004961$ m³/kg. Determine the values of enthalpy, entropy, and specific volume of steam at 70 MPa and 450°C. The critical temperature, T_c, is 374.14°C; a pseudovalue of p_c for p-V-T computations from the generalized compressibility charts is 22.96 MPa.

SOLUTION

Assume isothermal compression from 35 to 70 MPa. The reduced temperature and pressure at the initial and final states are

$$T_{R_1} = T_{R_2} = \frac{450 + 273.15}{374.14 + 273.15} = 1.1172$$

$$p_{R_1} = \frac{35}{22.96} = 1.524$$

$$p_{R_2} = \frac{70}{22.96} = 3.049$$

At $T_{R_1} = 1.1172$ and $p_{R_1} = 1.524$, according to Figures 8.5 and 8.6,

$$\frac{(\bar{h}^* - \bar{h})_1}{\bar{R}T_c} = 2.11 \qquad \text{and} \qquad \frac{(\bar{s}^* - \bar{s})_1}{\bar{R}} = 1.40$$

Similarly, at $T_{R_2} = 1.1172$ and $p_{R_2} = 3.049$,

$$\frac{(\bar{h}^* - \bar{h})_2}{\bar{R}T_c} = 3.43 \qquad \text{and} \qquad \frac{(\bar{s}^* - \bar{s})_2}{\bar{R}} = 2.27$$

Since $T_1 = T_2$, then $h_1^* = h_2^*$. Thus,

$$\frac{\bar{h}_2 - \bar{h}_1}{\bar{R}T_c} = 2.11 - 3.43 = -1.32$$

or

$$h_2 - h_1 = -1.32 \times \left(\frac{8.3144 \text{ kJ/kg-mol K}}{18.015 \text{ kg/kg-mol}}\right)(647.29 \text{ K}) = -394.3 \text{ kJ/kg}$$

and

$$h_2 = 2672.4 - 394.3 = 2278.1 \text{ kJ/kg}$$

The change of entropy may be calculated as follows:

$$\bar{s}_2 - \bar{s}_1 = (\bar{s}_2 - \bar{s}_2^*) + (\bar{s}_2^* - \bar{s}_1^*) + (\bar{s}_1^* - \bar{s}_1)$$

Since the temperature is constant, $(\bar{s}_2^* - \bar{s}_1^*)$ is a function of pressure alone,

$$\bar{s}_2^* - \bar{s}_1^* = -\bar{R}\ln\frac{p_2}{p_1} = (-8.3144 \text{ kJ/kg-mol K}) \ln 2$$
$$= -5.7628 \text{ kJ/kg-mol K}$$

and

$$\bar{s}_2 - \bar{s}_1 = (-8.3144 \text{ kJ/kg-mol K})(1.40 - 2.27) - 5.7628 \text{ kJ/kg-mol K}$$
$$= -13.996 \text{ kJ/kg-mol K}$$

or

$$s_2 - s_1 = -0.7769 \text{ kJ/kg K}$$

The entropy at state 2 is given by

$$s_2 = 5.1962 - 0.7769 = 4.4193 \text{ kJ/kg K}$$

The specific volume may be determined from the equation

$$v_2 = \frac{Z_2 R T_2}{p_2}$$

At $T_{R_2} = 1.1172$ and $p_{R_2} = 3.049$, the value of Z, obtained from Figure A7, is 0.5. Thus,

$$v_2 = \frac{(0.5)(8.3144 \text{ kJ/kg-mol K})/(18.015 \text{ kg/kg-mol})(723.15 \text{ K})}{70000 \text{ kPa}}$$

$$= 0.0024 \text{ m}^3/\text{kg}$$

Example 8.10

Nitrogen at a pressure of 2 MPa and a temperature of 250 K is compressed in a steady flow process to a pressure of 16.95 MPa and a temperature of 450 K. The mass rate of flow of the nitrogen is 0.1 kg/s. If the heat transfer from the compressor is 4.8 kW, determine the power input to the compressor.

SOLUTION

$$\dot{W} + \dot{Q} = \dot{m}(h_e - h_i)$$

and

$$h_e - h_i = (h_e - h_e^*) + (h_e^* - h_i^*) + (h_i^* - h_i)$$

$$T_{R_i} = \frac{T_i}{T_c} = \frac{250}{126.2} = 1.98 \qquad p_{R_i} = \frac{p_i}{p_c} = \frac{2}{3.39} = 0.59$$

$$T_{R_e} = \frac{T_e}{T_c} = \frac{450}{126.2} = 3.566 \qquad p_{R_e} = \frac{p_e}{p_c} = \frac{16.95}{3.39} = 5.0$$

From the enthalpy deviation chart, $(h_i^* - h_i)/RT_i = 0.1$ so that

$$h_i^* - h_i = \left(\frac{8.3144 \text{ kJ/kg-mol K}}{28.013 \text{ kg/kg-mol}}\right)(126.2 \text{ K})(0.1) = 3.746 \text{ kJ/kg}$$

so that

$$h_e^* - h_e = \left(\frac{8.3144 \text{ kJ/kg-mol K}}{28.013 \text{ kg/kg-mol}}\right)(126.2 \text{ K})(0.25) = 9.364 \text{ kJ/kg}$$

Therefore,

$$h_e - h_i = -9.364 \text{ kJ/kg} + (1.039 \text{ kJ/kg K})(450 - 250) \text{ K} + 3.746 \text{ kJ/kg} = 202.18 \text{ kJ/kg}$$

Substituting in the first law gives

$$\dot{W} = (0.1 \text{ kg/s})(202.18 \text{ kJ/kg}) - (-4.8 \text{ kW}) = 25.018 \text{ kW}$$

Example 8.11

Steam at a pressure of 20 MPa and a temperature of 500°C expands isentropically in a steady flow process to 10 MPa. Determine the change in enthalpy and the final temperature using (a) the generalized enthalpy deviation chart, (b) steam tables.

SOLUTION

(a) Since the flow is isentropic,

$$\bar{s}_e - \bar{s}_i = (\bar{s}_e - \bar{s}_e^*) + (\bar{s}_e^* - \bar{s}_i^*) + (\bar{s}_i^* - \bar{s}_i) = 0$$

At state i,

$$p_{R_i} = \frac{20}{22.09} = 0.9054 \quad \text{and} \quad T_{R_i} = \frac{773.15}{647.29} = 1.1944$$

From Figures 8.5 and 8.6,

$$\frac{\bar{h}_i^* - \bar{h}_i}{\bar{R}T_c} = 0.8 \quad \text{and} \quad \frac{\bar{s}_i^* - \bar{s}_i}{\bar{R}} = 0.52$$

The change of entropy for an ideal gas is

$$\bar{s}_e^* - \bar{s}_i^* = (\bar{s}_e^\circ - \bar{s}_i^\circ) - \bar{R} \ln \frac{p_e}{p_i}$$

At state e,

$$p_{R_e} = \frac{10}{22.09} = 0.4527 \quad \text{and} \quad T_{R_e} = \frac{T_e}{647.29}$$

It is now necessary to assume values of T_e to determine $(\bar{s}_e - \bar{s}_e^*)$, and the correct value of T_e must satisfy the entropy equation.

Problems

8.1 The coefficient of volumetric expansion and the co-efficient of compressibility of a substance are given by

$$\beta = \frac{1}{v}\left(\frac{\partial v}{\partial T}\right)_p = 5 \times 10^{-5} \text{K}^{-1}$$

and

$$\kappa = -\frac{1}{v}\left(\frac{\partial v}{\partial p}\right)_T = 1.2 \times 10^{-11} Pa^{-1}$$

If the substance is kept at a constant volume and the temperature is raised 5°C, determine the final pressure if the initial pressure is 1 atm.

8.2 Assuming *s* to be a function of *v* and *p*, prove that

$$T \, ds = c_p\left(\frac{\partial T}{\partial p}\right)_p dv + c_v\left(\frac{\partial T}{\partial p}\right)_v dp$$

8.3 Using Eqs. (8.4) and the Maxwell relation

$$\left(\frac{\partial T}{\partial v}\right)_{s,n_i} = -\left(\frac{\partial p}{\partial s}\right)_{v,n_i}$$

derive the other three Maxwell relations.

8.4 Check the validity of the Maxwell equation $(\partial s/\partial p)_T = -(\partial v/\partial T)_p$ for steam at a pressure of 0.8 MPa and a temperature of 300°C.

8.5 Using the Clapeyron equation and properties at the triple point, determine the change in the melting temperature of ice due to an increase in pressure to 40 MPa.

8.6 Verify the value of the enthalpy of vaporization of water at 110°C tabulated in the steam tables against the value computed according to the Clapeyron equation.

8.7 Calculate the melting point of ice at a pressure of 20 MPa. The densities of ice and water are 0.917 and 1.000 gm/cm³, respectively, and the enthalpy of melting is 6008 J/gm-mol.

8.8 Determine h_{fg} for water at 100°C from the following data:

T (°C)	p (kPa)
95	84.55
100	101.35
105	120.82

8.9 Using the Clapeyron equation and tabulated properties of refrigerant-12 just above and just below 40°C, determine the enthalpy of vaporization of refrigerant-12 ($M = 120.91$ kg/kg-mol) at 40°C.

8.10 Using the Clapeyron equation, verify the value of enthalpy of vaporization of refrigerant-12 at 400 kPa.

8.11 Using the Clapeyron equation, prove that the equations of the vaporization and sublimation curves are given by

$$p = C_1 e^{-h_{fg}/RT} \quad \text{and} \quad p = C_2 e^{-h_{ig}/RT}$$

where C_1 and C_2 are constants.

8.12 Find expressions for Δh, Δu, and Δs in an isothermal process for a gas, obeying the following equation of state:

$$pv = RT - \frac{a}{v}$$

8.13 Prove, for a simple system having equations of state

$$u = u(T, v) \quad \text{and} \quad p = p(T, v)$$

that

$$\left(\frac{\partial s}{\partial T}\right)_v = \frac{1}{T}\left(\frac{\partial u}{\partial T}\right)_v \quad \text{and} \quad \left(\frac{\partial s}{\partial v}\right)_T = \frac{1}{T}\left(\frac{\partial u}{\partial v}\right)_T + \frac{p}{T}$$

Also noting that entropy is a property, prove that

$$\left(\frac{\partial u}{\partial v}\right)_T = T\left(\frac{\partial p}{\partial T}\right)_v - p$$

8.14   Derive an expression for the change of internal energy of a gas that follows the van der Waal equation of state. Assume constant specific heats.

8.15   Derive an expression for the change of enthalpy of a gas that follows the van der Waals equation of state. Assume constant c_v.

8.16   Using the Maxwell relations, express $(\partial s/\partial v)_T$ and $(\partial s/\partial p)_T$ in terms of properties for a gas obeying the van der Waals equation of state.

8.17   The equation of state for a gas is given by

$$p = \frac{75T}{v} + \frac{150}{T}$$

where p is in kPa, T in K, and v in m³/kg. Determine the change of enthalpy and entropy when the gas is compressed isothermally at 120°C from 10 kPa to 1 MPa.

8.18   Find an expression for $(c_p - c_v)$ for a gas obeying the following equation of state:

$$pv = RT - \frac{a}{v}$$

8.19   Show that

$$c_p = T\left(\frac{\partial p}{\partial T}\right)_s\left(\frac{\partial v}{\partial T}\right)_p$$

and

$$c_v = -T\left(\frac{\partial p}{\partial T}\right)_v\left(\frac{\partial v}{\partial T}\right)_s$$

8.20   The molar specific heat of an ideal gas (molar mass 28.013 kg/kg-mol) at constant pressure is given by

$$\bar{c}_p = 28.90 - 0.1571 \times 10^{-2}T$$
$$+ 0.8081 \times 10^{-5}T^2 - 2.873 \times 10^{-9}T^3$$

where T is in K, and $\bar{c}_p$ in kJ/kg-mol K.
  Find the change in internal energy per kg-mol if the gas is heated from 400 K to 1000 K.

8.21   Determine an expression for $(c_p - c_v)$ for a gas obeying the van der Waals equation of state:

$$\left(p + \frac{a}{v^2}\right)(v - b) = RT$$

8.22   Prove that for a nonideal gas, c_v and c_p are given by the following expressions:

$$c_v = c_v^* + T\int_\infty^v \left(\frac{\partial^2 p}{\partial T^2}\right)_v dv$$

and

$$c_p = c_p^* - T\int_\infty^p \left(\frac{\partial^2 v}{\partial T^2}\right)_p dp$$

where c_v^* and c_p^* are the constant-volume and constant-pressure specific heats for an ideal gas, respectively.

8.23   Using the results of the previous problem, prove that c_v and c_p for a van der Waals gas are given by

$$c_v = c_v^*$$

and for a dilute gas $(a/v^2 \ll p$ and $b \ll v)$

$$c_p = c_p^* + \frac{2a}{RT^2}p = c_p^* + \frac{27}{32}\left[\frac{p/p_c}{(T/T_c)^2}\right]R$$

where c_v^* and c_p^* are the constant-volume and constant-pressure specific heats for an ideal gas, respectively.

8.24   An equation of state for a gas is given by

$$pv = RT + \frac{C}{v}$$

where C is a constant. Determine (a) the coefficient of isothermal compressibility κ, (b) the coefficient of volumetric expansion β, and (c) show that

$$\left(\frac{\partial u}{\partial T}\right)_p = c_p - \frac{Rp\,v^2}{pv^2 + C}$$

8.25 The equation of state of a gas is given by

$$p = \frac{RT}{v - b} - \frac{a}{T(v + c)^2}$$

where a, b, and c are constants. Determine the internal energy u and c_v for that gas. (Hint: Use the expression of $(\partial u / \partial v)_T$ derived in Problem 8.13.)

8.26 The equation of state for a particular vapor is given by the following relationship:

$$p(v - C) = RT + \frac{A}{T}$$

where C and A are constants, and R is the gas constant. Determine an expression for the coefficient of volumetric expansion in terms of R, C, A, T, and p.

8.27 Determine an expression for the Joule–Thomson coefficient for a gas obeying the Clausius equation:

$$p(v - b) = RT$$

What is the expression for the inversion temperature?

8.28 Determine the Joule–Thomson coefficient for steam at 1 MPa and 300°C.

8.29 Steam at a pressure of 5 MPa and a temperature of 400°C is throttled in a steady flow process to 3 MPa. Determine the temperature of the steam after throttling and the average Joule–Thomson coefficient.

8.30 Noting that $(\partial p / \partial v)_T = (\partial^2 p / \partial v^2)_T = 0$ at the critical point, show that the constants in the Dieterici equation are

$$a = \frac{4\bar{R}^2 T_c^2}{e^2 p_c} \qquad \text{and} \qquad b = \frac{\bar{R} T_c}{e^2 p_c}.$$

Find an expression for the coefficient of isothermal compressibility.

8.31 Repeat Problem 8.14 for the Berthelot equation, and find the constants a and b of that equation.

8.32 Determine the equation describing a reversible adiabatic process of a van der Waals gas using the following $T \, ds$ equation:

$$T \, ds = c_v \, dT + T\left(\frac{\partial p}{\partial T}\right)_v dv$$

8.33 Show that

$$\left(\frac{\partial u}{\partial v}\right)_T = \frac{T\beta}{\kappa} - p$$

where β and κ are the coefficients of volumetric expansion and isothermal compressibility.

8.34 The Berthelot equation of state per unit mass is

$$p = \frac{RT}{v - b} - \frac{a}{v^2 T}$$

Show that the change of internal energy and entropy are given by

$$du = c_v \, dT + \frac{2a}{v^2 T} dv$$

and

$$ds = c_v \frac{dT}{T} + \left(\frac{R}{v - b} + \frac{a}{v^2 T^2}\right) dv$$

8.35 Show that $(\partial c_v / \partial v)_T = 0$ for a gas obeying the van der Waals equation of state.

8.36 Prove that

$$(Z - Z_0)_p = -\frac{p}{R} \int_{T_0}^{T} \left(\frac{\mu_T}{T^2}\right)_p dT$$

where Z_0 is the compressibility factor at a pressure p and temperature T_0.

8.37 Determine the specific volume of steam at 30 MPa and 400°C by the use of (a) steam tables, (b) the ideal-gas law, (c) van der Waals equation, (d) compressibility chart.

8.38 Using property deviation charts, evaluate the change in enthalpy and entropy per kg-mol of nitrogen between an initial state defined by 300 K and 100 kPa and a final state 100 K and 7 MPa. Compare your results with values obtained from the nitrogen tables.

8.39 One kilogram of ethane gas in a piston–cylinder assembly is compressed isothermally from initial conditions of 700 kPa and 40°C to a final pressure of 80 MPa. Determine Δh, Δu, and Δs for the process, considering ethane to be a nonideal gas. Assume $c_p = 1.7662$ kJ/kg K for a pressure of 5 kPa and temperatures in the range of 40°C.

8.40 CO_2 enters a compressor operating at steady state at 45°C and 140 kPa at the rate of 50 kg/min, and is compressed reversibly and isothermally to 13.5 MPa. Velocities are negligible. For CO_2, T_c is 304.2 K, and p_c is 7.39 MPa. The average value of c_p can be taken as 0.842 kJ/kg K. Using the enthalpy and entropy deviation charts, find:
(a) the power input to the compressor
(b) the rate of entropy change of CO_2 as it flows through the compressor
(c) the rate of heat interaction
(d) the volumetric flow rate of CO_2 leaving the compressor

8.41 Steam at a pressure of 20 MPa and a temperature of 500°C expands isentropically in a steady flow turbine to a pressure of 4 MPa. Determine the power generated for a mass rate of flow of 1 kg/s using (a) the steam tables, (b) the enthalpy deviation chart. Explain the difference.

8.42 Propane (C_3H_8) is compressed in a reversible adiabatic steady flow process from 1 MPa pressure, 120°C, to 5 MPa pressure. Using property deviation charts, determine the work of compression per kg of propane. ($T = 3700$ K, $p_c = 4.62$ MPa, $M = 44.097$ kg/kg-mol, $c_p = 1.6794$ kJ/kg K.)

8.43 Determine the changes of enthalpy and entropy when CO_2 is compressed from 300 K and 1.0 MPa to 400 K and 10 MPa. Compare your values if CO_2 is assumed an ideal gas.

8.44 Determine the minimum power required to compress 0.1 kg/s of carbon dioxide in an adiabatic steady flow compressor from 100 kPa and 300 K to 10 MPa.

8.45 Ethane (C_2H_6) at 5.5 MPa, 423.2 K is cooled in a heat exchanger operating at steady state to 323.2 K. The volumetric rate of flow entering the heat exchanger is 0.1 m³/s, and the pressure drop through the heat exchanger is negligible. Determine:
(a) the compressibility factor at the initial state
(b) the flow rate of ethane in moles/s
(c) the rate of heat transfer
(d) the heat transfer if ethane were considered an ideal gas
Assume $\bar{c}_p = 53.11$ kJ/kg-mol K, $M = 30.07$ kg/kg-mol, $p_c = 4.88$ MPa, and $T_c = 305.5$ K.

8.46 Propane (C_3H_8) at a temperature of 450 K and a pressure of 5 MPa expands adiabatically in a turbine operating at steady state to a temperature of 350 K and a pressure of 0.5 MPa. Using the enthalpy and entropy deviation charts, determine (a) the work output per kg of propane, (b) the change in entropy per kg of propane. ($T_c = 370$ K, $p_c = 4.62$ MPa, $M = 44.097$ kg/kg-mol, $c_p = 1.6794$ kJ/kg K.)

8.47 Ethane C_2H_6 ($M = 30.03$ kg/kg-mol, $T_c = 305.5$ K, $p_c = 4.88$ MPa, and $c_p = 1.77$ kJ/kg K) at a temperature of 367 K and a pressure of 3.2 MPa expands in a turbine operating at steady state to a pressure of 0.4 MPa and a temperature of 275 K. If the heat transfer from the turbine is equal to 20 percent of the work developed, determine (a) the work output per kg, (b) the change of entropy per kg.

8.48 Nitrogen at 10 MPa and 164 K passes through a throttling valve. Downstream of the valve, the pressure is 0.6 MPa, and the temperature is 140 K. Using the enthalpy deviation chart determine the heat transfer. Neglect changes in kinetic and potential energies, and assume steady state. (For N_2, $p_c = 3.39$ MPa, $T_c = 126.2$ K, and $c_p = 1.0416$ kJ/kg K.)

8.49 Nitrogen initially at 16.95 MPa and 252.4 K expands to a pressure of 10.17 MPa and a temperature of

164 K. Using compressibility and property deviation charts, calculate the changes of internal energy and entropy per kg of nitrogen.

8.50 Nitrogen initially at 300 K and 10 MPa expands adiabatically in a turbine operating at steady state. At the exit, the temperature is 250 K, and the pressure is 4 MPa. Using the enthalpy and entropy deviation charts, determine (a) the work developed, (b) the entropy change, (c) the work developed and the entropy change if nitrogen is considered an ideal gas. (T_c = 126.2 K, p_c = 3.39 MPa, M = 28.013 kg/kg-mol, c_p = 1.0416 kJ/kg K.)

8.51 Superheated steam at 35 MPa and 400°C has the following properties: h = 1987.6 kJ/kg, and s = 4.2126 kJ/kg K. Using the generalized compressibility and the enthalpy and entropy deviation charts, determine the values of enthalpy, entropy, and specific volume at 50 MPa and 400°C. Compare your results with those found in steam tables.

8.52 Methane expands adiabatically in a turbine operating at steady state from 8 MPa and 25°C to 1.5 MPa and −50°C. Determine the work developed by the turbine per kg of methane. What is the entropy change of the methane as it flows through the turbine?

8.53 Saturated water at 277°C expands at constant pressure to 12 times its volume. Using the compressibility and property deviation charts, determine the heat interaction per unit mass. (T_c = 647.3 K, and p_c = 22.09 MPa.)

C8.1 Hydrogen gas at 500 K and 2000 kPa is contained in a rigid tank. Write a computer program to calculate the specific volume of hydrogen using (a) the ideal-gas law, (b) the van der Waals equation of state. Compare the two results. What happens when the gas pressure is lowered? Discuss the results.

C8.2 Two kilograms of helium gas at 500 K and 2000 kPa are contained in a rigid tank. Write a computer program to calculate the specific volume of helium using (a) the ideal-gas law, (b) the virial equation of state. Compare the two results. What happens when the gas pressure is lowered? Discuss the results.

C8.3 Using mathematical software capable of performing symbolic math, determine the specific volumes for Problems C8.1 and C8.2. Compare your result with those obtained using the ideal-gas equation of state.

C8.4 Write a computer program to calculate the enthalpy change of oxygen gas when it changes state from 320 K and 300 kPa to 2200 K and 2000 kPa. Consider a real gas behavior, and use van der Waals equation of state in the analysis. Numerical integration might be needed.

C8.5 Write a computer program to determine the increase in entropy of hydrogen when it changes state from 200 K and 110 kPa to 1900 K and 1800 kPa. Do not assume an ideal-gas behavior, and use the virial equation of state in the analysis. Numerical integration might be needed.

Chapter 9

Nonreactive Gas Mixtures

9.1 Introduction

The thermodynamic behavior of a mixture of gases depends upon the individual properties and the amounts of its constituent gases. Because substances in the gaseous phase are so miscible with each other, there is no limit to the number of different gaseous mixtures that can be formed from a given set of components, and a wide variation is therefore possible in the properties of gaseous mixtures. Furthermore, individual constituents of gaseous mixtures often react chemically with each other, and these reactions introduce another factor that exerts a strong influence on the properties of a gas. This chapter deals only with mixtures of inert gases, in which changes in phase may take place. Properties of reacting gaseous mixtures are discussed in Chapter 10.

A homogeneous gas mixture is frequently treated as if it consisted of a single component rather than many. Properties of the individual constituents of an inert gas tend to be submerged so that the gas behaves, in certain ways, as though it were a single, pure substance. The main constituents of air, for example, are oxygen, nitrogen, argon, and water vapor. But dry air can be treated as a simple gas with a molar mass of 28.97 kg/kg-mol even though this figure represents a composite value based on the molar masses and the proportions of the constituent species.

The first part of this chapter treats mixtures of ideal gases. Equations are derived that express the properties of mixtures in terms of the properties of the constituents. It is assumed that each constituent is unaffected by the presence of the other constituents in the mixture. Mixtures are treated in which no chemical reaction, condensation, or evaporation takes place, and the derived expressions apply, in general, to gas mixtures over a wide range of temperatures and pressures. Although the study centers about ideal gases, mixtures containing real gases often show only negligible deviation from ideal-gas behavior. In the latter part of the chapter, gaseous mixtures in which changes of phase occur are discussed.

9.2 Mixtures of Ideal Gases

Consider a mixture of ideal gases $a, b, c, \ldots$, existing in equilibrium at a pressure p, and occupying a volume V, as shown in Figure 9.1(a). Each constituent occupies the same volume V that the entire mixture occupies, and each constituent is at the same temperature as the mixture. The total mass of the mixture is equal to the sum of the masses of the individual gases, or

$$m = m_a + m_b + m_c + \cdots = \sum_i m_i$$

Similarly, the number of moles of the mixture is the sum of the moles of the individual components:

$$n = n_a + n_b + n_c + \cdots = \sum_i n_i$$

According to *Dalton's model*, the total pressure of a mixture of ideal gases is equal to the sum of the partial pressures of the constituents. The partial pressure of a gas in a mixture is the pressure that it would exert if it alone occupied the

	Constituents	Mixture
(a) V and T are the same for each constituent $(p = p_a + p_b + p_c + \cdots)$	$a:\ m_a, p_a, V, T$ $b:\ m_b, p_b, V, T$ $c:\ m_c, p_c, V, T$	m, p, V, T

	Gas a	Gas b	Gas c
(b) p and T are the same for each constituent $(V = V_a + V_b + V_c + \cdots)$	m_a p V_a T	m_b p V_b T	m_c p V_c T

Figure 9.1   (a) Dalton's and (b) Amagat's models.

whole volume of the mixture at the same temperature. Dalton's model can be written as

$$p = p_a + p_b + p_c + \cdots = \sum_i (p_i)_{V, T} \qquad (9.1)$$

Suppose it were possible to separate gases $a, b, c, \ldots$, and to have each constituent at the pressure and temperature of the mixture, as shown in Figure 9.1(b). According to *Amagat's model*, the sum of the partial volumes of the constituents would be equal to the total volume, or

$$V = V_a + V_b + V_c + \cdots = \sum_i (V_i)_{p, T} \qquad (9.2)$$

It may be remarked that Dalton's model and Amagat's model are consistent with the kinetic theory of gases, which was discussed in Chapter 4. These models assume that no intermolecular forces exist in a mixture of gases and that each constituent acts as if no other constituents were present. Both models are exact for ideal gases but approximate for real gases.

When the mixture is analyzed from the standpoint of Dalton's model, the ideal-gas equation can be applied individually to each constituent:

$$p_a V = n_a \overline{R} T$$
$$p_b V = n_b \overline{R} T$$
$$p_c V = n_c \overline{R} T$$

and so on. Adding the preceding equations gives

$$(p_a + p_b + p_c + \cdots)V = (n_a + n_b + n_c + \cdots)\overline{R} T$$

or

$$pV = n\overline{R} T$$

This equation of state has the same form as the equation of state of a single-component ideal gas so that a mixture of ideal gases also acts like an ideal gas. When the equation of state of each constituent gas is divided by the mixture's equation of state, the following expressions for partial pressures are obtained:

$$p_a = \frac{n_a}{n} p = x_a p$$

$$p_b = \frac{n_b}{n} p = x_b p$$

and so on. Therefore, in general, for the ith component, the partial pressure p is

$$p_i = x_i p \tag{9.3}$$

where x_i is the mole fraction (or molar concentration) of constituent i, and is defined as the ratio of the number of moles of that constituent to the total number of moles in the mixture:

$$x_i = \frac{n_i}{\sum_i n_i} = \frac{n_i}{n} \tag{9.4}$$

The mole fractions of all the components in a gas mixture then must add up to unity:

$$x_a + x_b + \cdots = \sum_i x_i = 1 \tag{9.5}$$

According to Eq. (9.3), the ratio of the partial pressure of any constituent to the total pressure is equal to the mole fraction of that constituent. Since each constituent occupies the same volume and has the same temperature as the mixture, it follows that

$$\frac{p_a}{n_a} = \frac{p_b}{n_b} = \frac{p_c}{n_c} = \cdots = \frac{p_i}{n_i} = \frac{p}{n} \tag{9.6}$$

When the mixture is analyzed from the standpoint of Amagat's model, the ideal-gas law leads to

$$pV_a = n_a \overline{R} T$$
$$pV_b = n_b \overline{R} T$$
$$pV_c = n_c \overline{R} T$$

and so on. Adding the preceding equations gives

$$p(V_a + V_b + V_c + \cdots) = (n_a + n_b + n_c + \cdots)\overline{R} T$$

or

$$pV = n\overline{R} T$$

By dividing the ideal-gas equation for each constituent by the equation for the entire mixture, one obtains the following:

$$\frac{V_a}{V} = \frac{n_a}{n} \qquad \text{or} \qquad V_a = x_a V$$

and, in general, for the ith component, the partial volume V_i is

$$V_i = x_i V \tag{9.7}$$

The ratio of the partial volume of any constituent to the total volume, then, is equal to the mole fraction of that constituent. Since the partial volumes of the constituents are at the same temperature and total pressure, it follows that

$$\frac{V_a}{n_a} = \frac{V_b}{n_b} = \frac{V_c}{n_c} = \cdots = \frac{V_i}{n_i} = \frac{V}{n} \tag{9.8}$$

It can therefore be concluded, from Eqs. (9.6) and (9.7), that the mole fraction of each constituent in a mixture of ideal gases is the same as its volume fraction and, also, the ratio of its partial pressure to the total pressure:

$$x_i = \frac{n_i}{n} = \frac{V_i}{V} = \frac{p_i}{p} \tag{9.9}$$

In the equation of state of a gas mixture, a gas constant involving molar mass appears. The value of this gas constant is determined as follows:
The ideal-gas law for the mixture is

$$pV = mRT \tag{9.10}$$

whereas for each constituent it is

$$p_a V = m_a R_a T$$
$$p_b V = m_b R_b T$$
$$p_c V = m_c R_c T$$

and so on. Since $p = p_a + p_b + p_c + \cdots$,

$$pV = (m_a R_a + m_b R_b + m_c R_c + \cdots)T$$

By combining this equation with Eq. (9.10), one obtains the following:

$$R = \frac{m_a}{m} R_a + \frac{m_b}{m} R_b + \frac{m_c}{m} R_c + \cdots = \sum_i \frac{m_i R_i}{m}$$

or

$$R = m_{f_a} R_a + m_{f_b} R_b + m_{f_c} R_c + \cdots = \sum_i m_{f_i} R_i \qquad (9.11)$$

where m_f, the mass fraction, is the ratio of the mass of a constituent to the total mass of the mixture, or

$$m_{f_i} = \frac{m_i}{\sum_i m_i} \qquad (9.12)$$

Note that the sum of the mass fractions of all the components in the mixture is equal to unity or,

$$m_{f_a} + m_{f_b} + \cdots = \sum_i m_{f_i} = 1 \qquad (9.13)$$

To determine the molar mass of the mixture, the total mass of the mixture is divided by the total number of moles present in the mixture:

$$M = \frac{\sum_i m_i}{\sum_i n_i} = \frac{\sum_i m_i}{\sum_i (m_i/M_i)} = \frac{\sum_i (nM)_i}{\sum_i n_i} \qquad (9.14)$$

The mass of a constituent is equal to the number of moles of that constituent multiplied by its molar mass. Therefore, the masses of the individual components of the mixture are

$$m_a = n_a M_a$$
$$m_b = n_b M_b$$
$$m_c = n_c M_c$$

and so on. Adding the preceding equations yields

$$m_a + m_b + m_c + \cdots = m = n_a M_a + n_b M_b + n_c M_c + \cdots$$

But the total mass of the mixture, m, is equal to

$$m = m_a + m_b + m_c + \cdots = nM$$

where n and M are the total number of moles and the molar mass of the mixture. By combining these two equations, the molar mass of the mixture is obtained:

$$M = \frac{n_a}{n} M_a + \frac{n_b}{n} M_b + \frac{n_c}{n} M_c + \cdots$$

or

$$M = x_a M_a + x_b M_b + x_c M_c + \cdots = \sum_i x_i M_i \qquad (9.15)$$

which expresses M in terms of the mole fraction. By substituting the relation $R_i = \overline{R}/M_i$ in Eq. (9.11), M can be expressed in terms of mass fractions:

$$\frac{\overline{R}}{M} = m_{f_a} \frac{\overline{R}}{M_a} + m_{f_b} \frac{\overline{R}}{M_b} + m_{f_c} \frac{\overline{R}}{M_c} + \cdots$$

from which

$$M = \frac{1}{(m_{f_a}/M_a) + (m_{f_b}/M_b) + (m_{f_c}/M_c) + \cdots} = \frac{1}{\sum_i (m_{f_i}/M_i)} \qquad (9.16)$$

Also, since $\overline{R} = MR$, then

$$M = \frac{\overline{R}}{R} \qquad (9.17)$$

Example 9.1

An ideal-gas mixture consists of 3 kg of nitrogen and 5 kg of carbon dioxide, as shown in Figure 9.2. The pressure and the temperature of the mixture are 300 kPa and 20°C. Find:

(a) the mole fraction of each constituent
(b) the equivalent molar mass of the mixture

3 kg N_2
5 kg CO_2
300 kPa
20°C

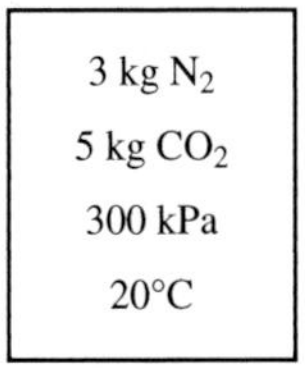

Figure 9.2

(c) the equivalent gas constant of the mixture
(d) the partial pressures and the partial volumes
(e) the volume and density of the mixture

SOLUTION

(a) Since $x_i = n_i / \sum\limits_{i} n_i$,

$$x_{N_2} = \frac{(3 \text{ kg})/(28.013 \text{ kg/kg-mol})}{(3 \text{ kg})/(28.013 \text{ kg/kg-mol}) + (5 \text{ kg})/(44.01 \text{ kg/kg-mol})} = 0.485$$

Similarly,

$$x_{CO_2} = \frac{(5 \text{ kg})/(44.01 \text{ kg/kg-mol})}{(3 \text{ kg})/(28.013 \text{ kg/kg-mol}) + (5 \text{ kg})/(44.01 \text{ kg/kg-mol})} = 0.515$$

(b) Using Eq. (9.15),

$$M = (0.485)(28.013 \text{ kg/kg-mol}) + (0.515)(44.01 \text{ kg/kg-mol})$$
$$= 36.25 \text{ kg/kg-mol}$$

(c) Using Eq. (9.11),

$$R = \frac{3}{8}\left(\frac{8.3144 \text{ kJ/kg-mol K}}{28.013 \text{ kg/kg-mol}}\right) + \frac{5}{8}\left(\frac{8.3144 \text{ kJ/kg-mol K}}{44.01 \text{ kg/kg-mol}}\right)$$
$$= 0.2294 \text{ kJ/kg K}$$

(d) Since the partial pressures are proportional to mole fractions,

$$p_{N_2} = 0.485(300 \text{ kPa}) = 145.5 \text{ kPa}$$
$$p_{CO_2} = 0.515(300 \text{ kPa}) = 154.5 \text{ kPa}$$

Also,

$$V_{N_2} = \frac{m_{N_2} R_{N_2} T}{p} = \frac{(3 \text{ kg})\left(\dfrac{8.3144 \text{ kJ/kg-mol K}}{28.013 \text{ kg/kg-mol}}\right)(293.15 \text{ K})}{300 \text{ kPa}} = 0.870 \text{ m}^3$$

$$V_{CO_2} = \frac{(5 \text{ kg})\left(\dfrac{8.3144 \text{ kJ/kg-mol K}}{44.01 \text{ kg/kg-mol}}\right)(293.15 \text{ K})}{300 \text{ kPa}} = 0.923 \text{ m}^3$$

(e) The volume of the mixture can be obtained by several methods. Noting that each constituent occupies the same volume as the mixture, then

$$V = \frac{mRT}{p} = \frac{m_{N_2}R_{N_2}T}{p_{N_2}} = \frac{m_{CO_2}R_{CO_2}T}{p_{CO_2}}$$

$$= \frac{(8 \text{ kg})(0.2294 \text{ kJ/kg K})(293.15 \text{ K})}{300 \text{ kPa}}$$

$$= \frac{(3 \text{ kg})\left(\dfrac{8.3144 \text{ kJ/kg-mol K}}{28.013 \text{ kg/kg-mol}}\right)(293.15 \text{ K})}{145.5 \text{ kPa}}$$

$$= \frac{(5 \text{ kg})\left(\dfrac{8.3144 \text{ kJ/kg-mol K}}{44.01 \text{ kg/kg-mol}}\right)(293.15 \text{ K})}{154.5 \text{ kPa}} = 1.793 \text{ m}^3$$

The density of the mixture is

$$\rho = \rho_{N_2} + \rho_{CO_2} = \frac{3 \text{ kg}}{1.793 \text{ m}^3} + \frac{5 \text{ kg}}{1.793 \text{ m}^3} = 4.462 \text{ kg/m}^3$$

or

$$\rho = \frac{m}{V} = \frac{8 \text{ kg}}{1.793 \text{ m}^3} = 4.462 \text{ kg/m}^3$$

9.3 Internal Energy, Enthalpy, Specific Heats, and Entropy

According to the Gibbs–Dalton law, an extensive property of a mixture of ideal gases is the sum of the contributions of the individual constituents. Properties of a mixture can be expressed on either a mass or a mole basis and are referred to as per kg or per kg-mol, respectively. Expressions for some of the extensive properties are presented here on a mass basis. The internal energy per unit mass of a mixture is given by

$$u = \frac{m_a}{\sum_i m_i} u_a + \frac{m_b}{\sum_i m_i} u_b + \frac{m_c}{\sum_i m_i} u_c + \cdots$$

$$= \frac{\sum_i m_i u_i}{\sum_i m_i} = \sum_i m_{f_i} u_i \tag{9.18}$$

It was shown previously that

$$pV = m_a R_a T + m_b R_b T + m_c R_c T + \cdots$$

Dividing both sides by $\sum_i m_i$ and substituting $p_i v_i$ for $R_i T$ gives

$$pv = \frac{m_a}{\sum\limits_i m_i}(p_a v_a) + \frac{m_b}{\sum\limits_i m_i}(p_b v_b) + \frac{m_c}{\sum\limits_i m_i}(p_c v_c) + \cdots$$

By adding the foregoing equation to Eq. (9.18), the expression for enthalpy per unit mass is obtained:

$$h = \frac{m_a}{\sum\limits_i m_i} h_a + \frac{m_b}{\sum\limits_i m_i} h_b + \frac{m_c}{\sum\limits_i m_i} h_c + \cdots$$

$$= \frac{\sum\limits_i m_i h_i}{\sum\limits_i m_i} = \sum_i m_{f_i} h_i \tag{9.19}$$

By differentiating the internal energy, Eq. (9.18), and the enthalpy, Eq. (9.19), with respect to T, specific-heat values are obtained for the mixture:

$$c_v = \frac{m_a}{\sum\limits_i m_i} c_{v_a} + \frac{m_b}{\sum\limits_i m_i} c_{v_b} + \frac{m_c}{\sum\limits_i m_i} c_{v_c} + \cdots$$

$$= \frac{\sum\limits_i m_i c_{v_i}}{\sum\limits_i m_i} = \sum_i m_{f_i} c_{v_i} \tag{9.20}$$

$$c_p = \frac{m_a}{\sum\limits_i m_i} c_{p_a} + \frac{m_b}{\sum\limits_i m_i} c_{p_b} + \frac{m_c}{\sum\limits_i m_i} c_{p_c} + \cdots$$

$$= \frac{\sum\limits_i m_i c_{p_i}}{\sum\limits_i m_i} = \sum_i m_{f_i} c_{p_i} \tag{9.21}$$

Similarly, the entropy of the mixture per unit mass is

$$s = \frac{m_a}{\sum\limits_i m_i} s_a + \frac{m_b}{\sum\limits_i m_i} s_b + \frac{m_c}{\sum\limits_i m_i} s_c + \cdots$$

$$= \frac{\sum\limits_i m_i s_i}{\sum\limits_i m_i} = \sum_i m_{f_i} s_i \tag{9.22}$$

Changes in the internal energy and enthalpy of mixtures of ideal gases are

$$u_2 - u_1 = \sum_i m_{f_i}(u_2 - u_1)_i = \int_{T_1}^{T_2} c_v \, dT \tag{9.23}$$

$$h_2 - h_1 = \sum_i m_{f_i}(h_2 - h_1)_i = \int_{T_1}^{T_2} c_p \, dT \tag{9.24}$$

and

$$c_p - c_v = R \tag{9.25}$$

The entropy change of an ideal-gas mixture per unit mass, expressed in terms of temperature and pressure, is

$$s_2 - s_1 = \sum_i m_{f_i}(s_2 - s_1)_i = \sum_i m_{f_i}\left[\int_{T_1}^{T_2} c_{p_i} \, d(\ln T) - R_i \ln \frac{p_{i,2}}{p_{i,1}} \right]$$

$$= \sum_i m_{f_i} \int_{T_1}^{T_2} c_{p_i} \, d(\ln T) - \sum_i m_{f_i} R_i \ln \frac{p_{i,2}}{p_{i,1}}$$

By substituting values obtained from Eqs. (9.11), (9.21), and (9.9), and assuming that the initial temperatures of the gases before mixing are the same, the change of entropy per unit mass of the mixture becomes

$$s_2 - s_1 = \int_{T_1}^{T_2} c_p \frac{dT}{T} - R \ln \frac{p_2}{p_1} \tag{9.26}$$

The foregoing properties can also be expressed on a mole basis. If a superscript $^-$ is used to indicate property that is given on a mole basis, the resultant equations are

$$\bar{u} = \frac{n_a}{\sum_i n_i} \bar{u}_a + \frac{n_b}{\sum_i n_i} \bar{u}_b + \frac{n_c}{\sum_i n_i} \bar{u}_c + \cdots$$

$$= \frac{\sum_i n_i \bar{u}_i}{\sum_i n_i} = \sum_i x_i \bar{u}_i \tag{9.27}$$

$$\bar{h} = \frac{n_a}{\sum_i n_i} \bar{h}_a + \frac{n_b}{\sum_i n_i} \bar{h}_b + \frac{n_c}{\sum_i n_i} \bar{h}_c + \cdots$$

$$= \frac{\sum_i n_i \bar{h}_i}{\sum_i n_i} = \sum_i x_i \bar{h}_i \tag{9.28}$$

$$\bar{c}_v = \frac{n_a}{\sum_i n_i}\bar{c}_{v_a} + \frac{n_b}{\sum_i n_i}\bar{c}_{v_b} + \frac{n_c}{\sum_i n_i}\bar{c}_{v_c} + \cdots$$

$$= \frac{\sum_i n_i \bar{c}_{v_i}}{\sum_i n_i} = \sum_i x_i \bar{c}_{v_i} \tag{9.29}$$

$$\bar{c}_p = \frac{n_a}{\sum_i n_i}\bar{c}_{p_a} + \frac{n_b}{\sum_i n_i}\bar{c}_{p_b} + \frac{n_c}{\sum_i n_i}\bar{c}_{p_c} + \cdots$$

$$= \frac{\sum_i n_i \bar{c}_{p_i}}{\sum_i n_i} = \sum_i x_i \bar{c}_{p_i} \tag{9.30}$$

$$\bar{s} = \frac{n_a}{\sum_i n_i}\bar{s}_a + \frac{n_b}{\sum_i n_i}\bar{s}_b + \frac{n_c}{\sum_i n_i}\bar{s}_c + \cdots$$

$$= \frac{\sum_i n_i \bar{s}_i}{\sum_i n_i} = \sum_i x_i \bar{s}_i \tag{9.31}$$

For ideal gases, changes in the foregoing properties are

$$\bar{u}_2 - \bar{u}_1 = \sum_i x_i(\bar{u}_2 - \bar{u}_1)_i = \int_{T_1}^{T_2} \bar{c}_v \, dT \tag{9.32}$$

$$\bar{h}_2 - \bar{h}_1 = \sum_i x_i(\bar{h}_2 - \bar{h}_1)_i = \int_{T_1}^{T_2} \bar{c}_p \, dT \tag{9.33}$$

$$\bar{c}_p - \bar{c}_v = \bar{R} \tag{9.34}$$

$$\bar{s}_2 - \bar{s}_1 = \sum_i x_i(\bar{s}_2 - \bar{s}_1)_i$$

$$= \sum_i x_i \int_{T_1}^{T_2} \bar{c}_{p_i} \, d(\ln T) - \sum_i x_i \bar{R} \ln \frac{p_{i,2}}{p_{i,1}} \tag{9.35}$$

$$= \int_{T_1}^{T_2} \bar{c}_p \frac{dT}{T} - \bar{R} \ln \frac{p_2}{p_1}$$

Note, from the foregoing equations, that the properties of the mixture per unit mass involve mass fractions, whereas properties per unit mole involve mole fractions.

Example 9.2

For the mixture of Example 9.1, calculate the constant-volume and constant-pressure specific heats. If the mixture is heated at constant volume to 50°C, find the change in internal energy, enthalpy, and entropy of the mixture. Assume average values of specific heats and the ideal-gas law applies.

Also find the foregoing changes if the heating process is performed at constant pressure.

SOLUTION

Using tabulated values of c_v and c_p at an average temperature of 35°C gives,

$$c_v = \sum_i m_{f_i} c_{v_i} = \frac{3}{8}(0.743 \text{ kJ/kg K}) + \frac{5}{8}(0.667 \text{ kJ/kg K})$$

$$= 0.6955 \text{ kJ/kg K}$$

$$c_p = \sum_i m_{f_i} c_{p_i} = \frac{3}{8}(1.039 \text{ kJ/kg K}) + \frac{5}{8}(0.858 \text{ kJ/kg K})$$

$$= 0.9259 \text{ kJ/kg K}$$

The change in internal energy is

$$\Delta u = \int_{T_1}^{T_2} c_v \, dT = (0.6955 \text{ kJ/kg K})[(50 - 20) \text{ K}] = 20.865 \text{ kJ/kg of mixture}$$

and

$$\Delta U = (8 \text{ kg})(20.865 \text{ kJ/kg}) = 166.92 \text{ kJ}$$

The change in enthalpy is

$$\Delta h = \int_{T_1}^{T_2} c_p \, dT = (0.9259 \text{ kJ/kg K})(30 \text{ K}) = 27.777 \text{ kJ/kg of mixture}$$

and

$$\Delta H = (8 \text{ kg})(27.777 \text{ kJ/kg}) = 222.216 \text{ kJ}$$

The change in entropy of the mixture, in terms of changes of temperature and volume, is

$$\Delta s = \int_{T_1}^{T_2} c_v \frac{dT}{T} + R \ln \frac{v_2}{v_1}$$

But since the volume is constant in this problem,

$$\Delta s = (0.6955 \text{ kJ/kg K}) \ln \frac{323.15}{293.15} = 0.6955(0.0974) = 0.0678 \text{ kJ/kg K}$$

and

$$\Delta S = (8 \text{ kg})(0.0678 \text{ kJ/kg K}) = 0.5421 \text{ kJ/K}$$

For the constant-pressure process, the internal energy and enthalpy values are the same as the values calculated in the constant-volume process since they are functions of temperature alone (ideal gas). The change of entropy, as given by Eq. (9.26), is

$$\Delta s = \int_{T_1}^{T_2} c_p \frac{dT}{T} - R \ln \frac{p_2}{p_1}$$

and for a constant-pressure process,

$$\Delta s = (0.9259 \text{ kJ/kg K}) \ln \frac{323.15}{293.15} = 0.9259(0.0974) = 0.0902 \text{ kJ/kg K}$$

and

$$\Delta S = (8 \text{ kg})(0.0902 \text{ kJ/kg K}) = 0.7217 \text{ kJ/K}$$

9.4 Entropy Change Due to Mixing of Ideal Gases

When ideal gases are mixed, a change of entropy occurs as a result of the increase in the disorder of the system. If the initial temperatures of all the constituents are the same, and if the mixing process is adiabatic, then the temperature does not change, but the entropy increases. Let the initial pressure and temperature of the individual gases before mixing be p and T, the same as the total pressure and temperature of the mixture. In the mixing process, each gas may be considered to undergo a free expansion from its initial pressure to its partial pressure in the mixture, with no work interaction. Mixing is an internally irreversible process, and according to the second law of thermodynamics, there is entropy increase so that

$$ds = \sum_i (m_f \, ds)_i \geq 0$$

Since the gases are ideal and the temperature remains constant, both internal energy and enthalpy are unchanged by the mixing process. Applying Eq. (9.26), the change of entropy is

$$\Delta S = -\left(m_a R_a \ln \frac{p_a}{p} + m_b R_b \ln \frac{p_b}{p} + \cdots \right)$$

$$= -\sum_i m_i R_i \ln \frac{p_i}{p} = -\overline{R} \sum_i n_i \ln x_i$$

(9.36)

which on a unit mass basis becomes

$$\Delta s = -\sum_i m_{f_i} R_i \ln \frac{p_{i,\,2}}{p_{i,\,1}}$$

where subscripts 1 and 2 refer to the initial and final states of each component in the mixture. As indicated by Eq. (9.36), the entropy change depends only on the number of moles of the constituent gases and is independent of the nature of these gases. Note that since the mole fraction x_i of each component in the mixture is less than unity, the entropy production due to mixing is positive. The irreversibility in the mixing process is due to the free expansion of each individual gas in order to establish a uniform distribution. Thus different gases may be in thermal and mechanical equilibrium, but they are not in complete thermodynamic equilibrium. When these gases are brought into contact with each other, they diffuse into each other spontaneously, resulting in entropy production. Other factors such as different initial temperatures and pressures of the gases also contribute to the entropy production during a mixing process.

Example 9.3

Find the increase in entropy when 2 kg of oxygen at 70°C are mixed with 6 kg of nitrogen at the same temperature (Figure 9.3). The initial pressure of each constituent is 100 kPa and is the same as that of the mixture. (Assume ideal gases.)

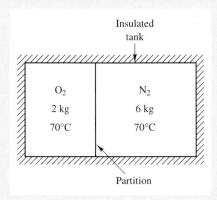

Figure 9.3

SOLUTION

Since the mixing process is isothermal, the change of entropy of the mixture is

$$\Delta S = -m_{O_2} R_{O_2} \ln \frac{p_{O_2}}{p} - m_{N_2} R_{N_2} \ln \frac{p_{N_2}}{p}$$

where

$$\frac{p_{O_2}}{p} = \frac{(2 \text{ kg})/(32 \text{ kg/kg-mol})}{(2 \text{ kg})/(32 \text{ kg/kg-mol}) + (6 \text{ kg})/(28.013 \text{ kg/kg-mol})} = 0.226$$

$$\frac{p_{N_2}}{p} = 1.000 - 0.226 = 0.774$$

Therefore,

$$\Delta S = -(2 \text{ kg})\left(\frac{8.3144 \text{ kJ/kg-mol K}}{32 \text{ kg/kg-mol}}\right) \ln 0.226 - (6 \text{ kg})\left(\frac{8.3144 \text{ kJ/kg-mol K}}{28.013 \text{ kg/kg-mol}}\right) \ln 0.774$$

$$= 0.7728 + 0.4562 = 1.229 \text{ kJ/K}$$

9.5 Mixtures of Ideal Gases at Different Initial Pressures and Temperatures

Three insulated vessels, each containing an ideal gas of mass m_i at temperature T_i and pressure p_i, are shown in Fig. 9.4. The three gases are different from one another, and subscripts a, b, and c identify the particular gas. For simplicity, the procedure is applied to three gases only, but it can be extended to any number of gases. If the interconnecting valves are opened, so that the gases are allowed to diffuse into one another, the final temperature and pressure of the mixture will be at some intermediate value. If no heat or work interactions take place at the boundary of the system, the internal energy of the system, as the first law dictates, is not affected by the mixing process. The internal energy of the mixture per unit mass is

$$u = \frac{m_a}{m} u_a + \frac{m_b}{m} u_b + \frac{m_c}{m} u_c$$

Since u is a function of temperature only for ideal gas and is given by $c_v T$, this equation becomes

$$c_v T = \frac{m_a}{m} c_{v_a} T_a + \frac{m_b}{m} c_{v_b} T_b + \frac{m_c}{m} c_{v_c} T_c = \sum_i \frac{m_i}{m} (c_v)_i T_i$$

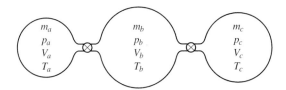

Figure 9.4   Mixing of ideal gases with different initial properties.

If the gases are all monatomic, then $c_v = \frac{3}{2}R$; if they are diatomic, then $c_v = \frac{5}{2}R$; and in either case, the preceding equation becomes

$$mRT = m_a R_a T_a + m_b R_b T_b + m_c R_c T_c$$

from which the temperature of the mixture is calculated:

$$T = \frac{m_a R_a T_a + m_b R_b T_b + m_c R_c T_c}{mR} \tag{9.37}$$

Since $mR = m_a R_b + m_b R_b + m_c R_c$, Eq. (9.37) can be expressed as

$$T = \frac{m_a R_a T_a + m_b R_b T_b + m_c R_c T_c}{m_a R_a + m_b R_b + m_c R_c} = \frac{\sum\limits_i m_i R_i T_i}{\sum\limits_i m_i R_i}$$

From the ideal-gas law, this equation can also be expressed in terms of the pressures, temperatures, and volumes of the individual gases before mixing:

$$T = \frac{p_a V_a + p_b V_b + p_c V_c}{(p_a V_a / T_a) + (p_b V_b / T_b) + (p_c V_c / T_c)} = \frac{\sum\limits_i p_i V_i}{\sum\limits_i p_i V_i / T_i} \tag{9.38}$$

In Eq. (9.37), if $n_i \overline{R} T_i$ is substituted for $m_i R_i T_i$, then

$$n \overline{R} T = n_a \overline{R} T_a + n_b \overline{R} T_b + n_c \overline{R} T_c$$

so that the temperature of the mixture can be calculated from the mole fractions of the constituents and their initial temperatures:

$$T = \frac{n_a}{n} T_a + \frac{n_b}{n} T_b + \frac{n_c}{n} T_c = \sum_i x_i T_i \tag{9.39}$$

Using Eq. (9.37), and by applying the ideal-gas law, an expression for the pressure of the mixture can be obtained:

$$pV = p_aV_a + p_bV_b + p_cV_c$$

which leads to the following:

$$p = \frac{p_aV_a + p_bV_b + p_cV_c}{V} \tag{9.40}$$

Note that the total volume V of the mixture is the sum of the initial volumes V_a, V_b, and V_c. These latter volumes are not the partial volumes of the components of the mixture; they are, instead, the volumes of the vessels that the component gases initially occupied.

Example 9.4

The number of moles, the pressures, and the temperatures of gases *a, b,* and *c* of Figure 9.4 are given in the following table:

Gas	Number of Moles	Pressure (kPa)	Temperature (K)
N_2	1	300	360
CO	3	400	480
O_2	2	600	600

If the containers are connected, and the gases mix freely and adiabatically, find:

(a) the pressure and temperature of the resulting mixture at equilibrium
(b) the change of entropy of each constituent and the total entropy generation due to mixing

SOLUTION

(a) Assuming the gases to have the same c_v, the temperature of the mixture according to Eq. (9.39) is

$$T = \frac{(1)(360 \text{ K}) + (3)(480 \text{ K}) + (2)(600 \text{ K})}{6} = 500 \text{ K}$$

The initial volumes are

$$V_{N_2} = \frac{n_{N_2}\overline{R}T_{N_2}}{p_{N_2}} = \frac{(1 \text{ kg-mol})(8.3144 \text{ kJ/kg-mol K})(360 \text{ K})}{300 \text{ kPa}} = 9.977 \text{ m}^3$$

$$V_{CO} = \frac{(3 \text{ kg-mol})(8.3144 \text{ kJ/kg-mol K})(480 \text{ K})}{400 \text{ kPa}} = 29.932 \text{ m}^3$$

$$V_{O_2} = \frac{(2 \text{ kg-mol})(8.3144 \text{ kJ/kg-mol K})(600 \text{ K})}{600 \text{ kPa}} = 16.629 \text{ m}^3$$

and the total volume is

$$V = 9.977 + 29.932 + 16.629 = 56.538 \text{ m}^3$$

The pressure of the mixture according to the ideal-gas law is

$$p = \frac{n\bar{R}T}{V} = \frac{(6 \text{ kg-mol})(8.3144 \text{ kJ/kg-mol K})(500 \text{ K})}{56.538 \text{ m}^3} = 441.176 \text{ kPa}$$

The same result can also be obtained using Eq. (9.40).

(b) The partial pressures are

$$p_{N_2} = \frac{1}{6}(441.176 \text{ kPa}) = 73.53 \text{ kPa}$$

$$p_{CO} = \frac{3}{6}(441.176 \text{ kPa}) = 220.59 \text{ kPa}$$

$$p_{O_2} = \frac{2}{6}(441.176 \text{ kPa}) = 147.06 \text{ kPa}$$

Therefore,

$$\Delta s_{N_2} = \left(\int_{T_1}^{T_2} c_p \frac{dT}{T} - R \ln \frac{p_2}{p_1} \right)_{N_2}$$

$$= (1.047 \text{ kJ/kg K}) \ln \frac{500}{360} - \frac{8.3144 \text{ kJ/kg-mol K}}{28.013 \text{ kJ/kg-mol}} \ln \frac{73.53}{300}$$

$$= 0.7612 \text{ kJ/kg K} = 21.3245 \text{ kJ/K}$$

$$\Delta s_{CO} = (1.041 \text{ kJ/kg K}) \ln \frac{500}{480} - \frac{8.3144 \text{ kJ/kg-mol K}}{28.010 \text{ kg/kg-mol}} \ln \frac{220.59}{400}$$

$$= 0.2192 \text{ kJ/kg K} = 18.4164 \text{ kJ/K}$$

$$\Delta s_{O_2} = (0.988 \text{ kJ/kg K}) \ln \frac{500}{600} - \frac{8.3144 \text{ kJ/kg-mol K}}{32 \text{ kg/kg-mol}} \ln \frac{147.06}{600}$$

$$= 0.1852 \text{ kJ/kg K} = 11.4847 \text{ kJ/K}$$

and the total change of entropy due to mixing is

$$\Delta S_{mixture} = 21.3245 + 18.4164 + 11.4847 = 51.2256 \text{ kJ/K}$$

In this example, it was assumed that the specific heats were constant at the average temperature of the gas. For greater accuracy, the variation of specific heats with temperature should be taken into consideration. Furthermore, it was assumed that the gases have the same c_v in calculating the temperature of the mixture.

Example 9.5

In a steady-state flow system, 0.1 kg/s of hydrogen at 300 K and nitrogen at 400 K are mixed adiabatically to form a mixed stream at 350 K. The pressure of each stream is atmospheric, and the kinetic and potential energy changes are negligible. Determine:

(a) the mass rate of flow of the nitrogen
(b) the constant-pressure specific heat of the mixture
(c) the rate of irreversibility assuming $T_0 = 25°C$

SOLUTION

(a) Referring to Figure 9.5, the first law applied to the control volume shown is

$$\dot{Q} + \dot{W} + \sum \dot{m}_i \left(h + \frac{V^2}{2} + gz \right)_i - \sum \dot{m}_e \left(h + \frac{V^2}{2} + gz \right)_e = \Delta \dot{E}$$

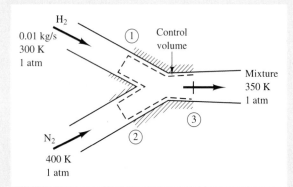

Figure 9.5

For the conditions and assumptions of this problem, it reduces to

$$(\dot{m}_{H_2} h_{H_2})_{T_1} + (\dot{m}_{N_2} h_{N_2})_{T_2} = (\dot{m}_{H_2} h_{H_2} + \dot{m}_{N_2} h_{N_2})_{T_3}$$

Assuming ideal-gas behavior so that $h = c_p T$, then

$$\dot{m}_{H_2}(c_{P_{H_2}} T_{H_2})_{T_1} + \dot{m}_{N_2}(c_{P_{N_2}} T_{N_2})_{T_2} = \dot{m}_{H_2}(c_{P_{H_2}} T_{H_2})_{T_3} + \dot{m}_{N_2}(c_{P_{N_2}} T_{N_2})_{T_3}$$

or

$$\dot{m}_{N_2} = \frac{\dot{m}_{H_2}\left[\left(c_{pH_2}T_{H_2}\right)_{T_3} - \left(c_{pH_2}T_{H_2}\right)_{T_1}\right]}{\left(c_{pN_2}T_{N_2}\right)_{T_2} - \left(c_{pN_2}T_{N_2}\right)_{T_3}}$$

For hydrogen:

At $T_1 = 300$ K, $c_{pH_2} = 14.307$ kJ/kg K
At $T_3 = 350$ K, $c_{pH_2} = 14.427$ kJ/kg K

For nitrogen:

At $T_2 = 400$ K, $c_{pN_2} = 1.044$ kJ/kg K
At $T_3 = 350$ K, $c_{pN_2} = 1.041$ kJ/kg K

Therefore,

$$\dot{m}_{N_2} = \frac{(0.1 \text{ kg/s})[(14.427 \text{ kJ/kg K})(350 \text{ K}) - (14.307 \text{ kJ/kg K})(300 \text{ K})]}{(1.044 \text{ kJ/kg K})(400 \text{ K}) - (1.041 \text{ kJ/kg K})(350 \text{ K})}$$

$$= \frac{(0.1 \text{ kg/s})[(5049.45 - 4292.1) \text{ kJ/kg}]}{(417.6 - 364.35) \text{ kJ/kg}} = 1.4223 \text{ kg/s}$$

(b)

$$c_{p3} = \sum (m_{f_i}c_{p_i})_{T_3} = \left(\frac{0.1 \text{ kg/s}}{1.5223 \text{ kg/s}}\right)(14.427 \text{ kJ/kg K}) + \left(\frac{1.4223 \text{ kg/s}}{1.5223 \text{ kg/s}}\right)(1.041 \text{ kJ/kg K})$$

$$= 0.9477 + 0.9726 = 1.9203 \text{ kJ/kg K}$$

(c) The rate of irreversibility due to mixing is

$$\dot{I} = T_0\left[\dot{m}_{H_2}\left(c_{pH_2}\ln\frac{T_3}{T_1} - R_{H_2}\ln\frac{p_{H_2,m}}{p_{H_2}}\right) + \dot{m}_{N_2}\left(c_{pN_2}\ln\frac{T_3}{T_2} - R_{N_2}\ln\frac{p_{N_2,m}}{p_{N_2}}\right)\right]$$

where c_p values are estimated at average temperatures, and $p_{H_2,m}$ and $p_{N_2,m}$ are the partial pressures of H_2 and N_2 in the mixture. The ratio of the partial pressure to the initial pressure of each species is

$$\frac{p_{H_2,m}}{p_{H_2}} = x_{H_2,m} = \frac{\dfrac{m_{H_2}}{M_{H_2}}}{\dfrac{m_{H_2}}{M_{H_2}} + \dfrac{m_{N_2}}{M_{N_2}}} = \frac{\dfrac{0.1}{2.016}}{\dfrac{0.1}{2.016} + \dfrac{1.4223}{28.013}} = 0.4942$$

and

$$\frac{p_{N_2, m}}{p_{N_2}} = x_{N_2, m} = 1 - 0.4942 = 0.5058$$

Therefore,

$$i = (298.15 \text{ K}) \left\{ (0.1 \text{ kg/s}) \left[14.367 \text{ kJ/kg K ln } \frac{350}{300} - \left(\frac{8.3144 \text{ kJ/kg-mol K}}{2.016 \text{ kg/kg-mol}} \right) \ln 0.4942 \right] \right.$$

$$\left. + (1.4223 \text{ kJ/kg}) \left[(1.0425 \text{ kJ/kg K}) \ln \frac{350}{400} - \left(\frac{8.3144 \text{ kJ/kg-mol K}}{28.013 \text{ kg/kg-mol}} \right) \ln 0.5058 \right] \right\}$$

$$= (298.15 \text{ K})[(0.5122 + 0.0897) \text{ kJ/(Ks)}] = 179.467 \text{ kJ/s}$$

9.6 Real-Gas Mixtures

The properties of a mixture of real gases depend on the behavior of its constituent gases, and property values predicted from ideal-gas behavior can be considerably different. The degree of deviation depends on the temperature and pressure of the mixture.

One way of predicting the p-V-T behavior of a real-gas mixture is to express the compressibility factor of the mixture in terms of the compressibility factors of the individual gases. For a gas mixture of real gases, the pressure of the mixture is

$$p = \frac{Zn\overline{R}T}{V} \tag{9.41}$$

where Z is the compressibility factor for the mixture.

Similarly, the partial pressure of an individual component is

$$p_i = \frac{Z_i n_i \overline{R}T}{V} \tag{9.42}$$

where Z_i is the compressibility factor for component i.

Dalton's model states that the pressure of the mixture is the sum of the pressure exerted by the individual components if they existed at the same temperature and volume of the mixture so that

$$p = \sum_i (p_i)_{T, V} \tag{9.43}$$

Substituting from Eqs. (9.41) and (9.42) into the previous equation gives

$$\frac{Zn\overline{R}T}{V} = \sum_i \frac{Z_i n_i \overline{R}T}{V}$$

or

$$Z = \sum_i (x_i Z_i)_{T,V} \tag{9.44}$$

In the preceding equation, the compressibility factors Z_i are determined at the T and V of the mixture. The equation, however, is based on Dalton's model, which assumes that each component is an ideal rather than a real gas that occupies the whole volume of the mixture with no intermolecular interaction with other components. For this reason, Eq. (9.44) is more valid at low pressures. At high pressures, Amagat's model of additive volumes predicts the behavior of real gases more accurately because it involves the pressure of the mixture that accounts for intermolecular forces. These forces become important at high pressure where the conditions are most favorable for deviation from ideal-gas behavior. Amagat's model states that the volume of the mixture is the sum of the volumes of the individual components if they existed separately at the temperature and pressure of the mixture so that

$$V = \sum_i (V_i)_{T,p} \tag{9.45}$$

Applying the same procedure as before gives the compressibility factor of the mixture as

$$Z = \sum_i (x_i Z_i)_{T,p} \tag{9.46}$$

Another method to evaluate the compressibility factor for a mixture of real gases is to use *Kay's rule*. In this method, the compressibility factor is given by

$$Z = Z(T_R', p_R') \tag{9.47}$$

where $T_R' = T/T_c'$ and $p_R' = p/p_c'$ are the reduced temperature and pressure, and T_c' and p_c' are called the *pseudocritical temperature* and the *pseudocritical pressure* defined as

$$T_c' = \sum_i x_i T_{c_i} \tag{9.48}$$

and

$$p'_c = \sum_i x_i p_{c_i} \qquad (9.49)$$

Properties such as enthalpy and entropy of real-gas mixtures can also be evaluated using the generalized deviation charts developed in Chapter 8. The properties are evaluated using reduced temperature and reduced pressure evaluated at the temperature and pressure of the mixture.

Example 9.6

An insulated rigid tank of volume 0.2 m³ contains 0.25 kg-mol of O_2 and 0.4 kg-mol of CO_2 at 300 K (Figure 9.6). Determine the pressure of the mixture using:

(a) the ideal-gas equation of state
(b) compressibility factors based on Dalton's model
(c) compressibility factors based on Amagat's model
(d) Kay's rule

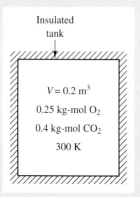

Figure 9.6

SOLUTION

(a) The total number of moles of the mixture is

$$n = 0.25 + 0.4 = 0.65 \text{ kg-mol}$$

The pressure of the mixture, according to the ideal-gas law, is

$$p = \frac{n\overline{R}T}{V} = \frac{(0.65 \text{ kg-mol})(8.3144 \text{ kJ/kg-mol K})(300 \text{ K})}{0.2 \text{ m}^3} = 8106.54 \text{ kPa}$$

(b) The compressibility factor of the mixture is determined assuming that the components exist at the temperature and volume of the mixture.

$$Z = x_{O_2}Z_{O_2} + x_{CO_2}Z_{CO_2}$$

For O_2:

$$T_R = \frac{T}{T_c} = \frac{300}{154.8} = 1.938$$

and the pseudoreduced volume is

$$v_R' = \frac{\overline{v}p_c}{\overline{R}T_c} = \frac{[(0.2 \text{ m}^3)/(0.25 \text{ kg-mol})](5080 \text{ kPa})}{(8.3144 \text{ kJ/kg-mol K})(154.8 \text{ K})} = 3.1576$$

At these values of T_R and v_R', Figure A6 gives $Z_{O_2} = 0.98$.

For CO_2:

$$T_R = \frac{T}{T_c} = \frac{300}{304.2} = 0.9862$$

$$v_R' = \frac{\overline{v}p_c}{\overline{R}T_c} = \frac{[(0.2 \text{ m}^3)/(0.4 \text{ kg-mol})](7390 \text{ kPa})}{(8.3144 \text{ kJ/kg-mol K})(304.2 \text{ K})} = 1.4609$$

At these values of T_R and v_R', Figure A6 gives $Z_{CO_2} = 0.78$.

Therefore,

$$Z = \sum_i x_i Z_i = \left(\frac{0.25}{0.65}\right)(0.98) + \left(\frac{0.4}{0.65}\right)(0.78) = 0.8569$$

The pressure of the mixture is

$$p = \frac{Zn\overline{R}T}{V} = \frac{(0.8569)(0.65 \text{ kg-mol})(8.3144 \text{ kJ/kg-mol K})(300 \text{ K})}{0.2 \text{ m}^3}$$
$$= 6947 \text{ kPa}$$

(c) The compressibility factor of the mixture is determined assuming that the components exist at the temperature and pressure of the mixture.

For O_2:

$$T_R = \frac{300}{154.8} = 1.938$$

To determine p_R, an iterative solution is required since the pressure of the gas mixture is not known. Assuming $p = 7000$ kPa, the value determined in part (b), gives

$$p_R = \frac{7000}{5080} = 1.378$$

so that $Z_{O_2} = 0.97$.

For CO_2:

$$T_R = \frac{300}{304.2} = 0.9862 \qquad p_R = \frac{7000}{7390} = 0.9472$$

so that $Z_{CO_2} = 0.46$.

Therefore,

$$Z = \left(\frac{0.25}{0.65}\right)(0.97) + \left(\frac{0.4}{0.65}\right)(0.46) = 0.3731 + 0.2831 = 0.6562$$

$$p = \frac{(0.6562)(0.65 \text{ kg-mol})(8.3144 \text{ kJ/kg-mol K})(300 \text{ K})}{0.2 \text{ m}^3} = 5319 \text{ kPa}$$

This value is low compared to the assumed value of pressure.

Assume $p = 6000$ kPa.

For O_2:

$$p_R = \frac{6000}{5080} = 1.1811$$

At these values of p_R and $T_R = 1.938$,
$Z_{O_2} = 0.97$.

For CO_2:

$$p_R = \frac{6000}{7390} = 0.8119$$

At these values of p_R and $T_R = 0.9862$,
$Z_{CO_2} = 0.62$.

Therefore,

$$Z = \left(\frac{0.25}{0.65}\right)(0.97) + \left(\frac{0.4}{0.65}\right)(0.62) = 0.7546$$

$$p = \frac{Zn\overline{R}T}{V} = \frac{(0.7546)(0.65)(8.3144)(300)}{0.2} = 6118 \text{ kPa}$$

A third trail indicates that pressure p is 6050 kPa.

(d) To apply Kay's rule, the pseudocritical temperature and pressure must first be determined.

$$T'_c = \sum_i x_i T_{c_i} = \frac{0.25}{0.65}(154.8 \text{ K}) + \frac{0.4}{0.65}(304.2 \text{ K}) = 246.74 \text{ K}$$

$$p'_c = \sum_i x_i p_{c_i} = \frac{0.25}{0.65}(5080 \text{ kPa}) + \frac{0.4}{0.65}(7390 \text{ kPa}) = 6501.54 \text{ kPa}$$

$$T_R = \frac{T}{T'_c} = \frac{300}{2467.4} = 1.2159$$

$$v'_R = \frac{\overline{v}p'_c}{\overline{R}T'_c} = \frac{\left(\frac{0.2}{0.65} \text{ m}^3/\text{kg-mol}\right)(6501.54 \text{ kPa})}{(8.3144 \text{ kJ/kg-mol K})(246.74 \text{ K})} = 0.9751$$

At these values of T_R and v'_R,
$Z \approx 0.80$.

$$p = \frac{Zn\overline{R}T}{V} = \frac{(0.80)(0.65 \text{ kg-mol})(8.3144 \text{ kJ/kg-mol K})(300 \text{ K})}{0.2 \text{ m}^3}$$

$$= 6485 \text{ kPa}$$

Note that the results in this example deviate considerably owing to the high pressure of the mixture. Results based on Amagat's model and Kay's rule are most likely to be accurate.

9.7 Gibbs Phase Rule

This section considers the conditions under which the different phases coexist in equilibrium and determines the minimum number of the properties of a system of several phases that can be varied independently. The discussion is extended to include systems of more than one constituent but excludes chemical reactions, which are discussed in Chapter 10.

Consider a system composed of C components existing in equilibrium in P phases, and assume that the C components exist in all the phases.[1] Suppose the temperature and pressure are the same throughout the system and that both remain constant. In addition to temperature and pressure, the intensive state of each phase is defined if $(C - 1)$ parameters pertaining to composition, such as mole fractions or concentrations, are known. The composition of the system is defined by $(C - 1)$ mole fractions since the last mole fraction can be obtained by noting that the sum of the mole fractions of all components in each phase is equal to unity. The total number of parameters, including the pressure and temperature, that may vary independently is therefore $(C + 1)$ for each phase. For P phases, there are $P(C + 1)$ parameters. Now consider the number of equations available to solve for the $P(C + 1)$ unknowns. These equations express the equilibrium between phases. Thus, $(P - 1)$ equations are obtained by equating the temperature of the phases. An equal number of equations is also obtained by equating the pressure of the phases. Finally, a set of $(P - 1)$ equations may be written for the concentration or chemical potential of each component in the P phases. Adding these equations gives a total number of $(P - 1)(C + 2)$ equations. When this number of equations is subtracted from the number of the unknowns, the number of independent intensive properties that could be varied to define the intensive properties of the phases of the system is obtained. These properties can be chosen arbitrarily and are usually referred to as *degrees of freedom*. If F represents the number of degrees of freedom of a system, then

$$F = P(C + 1) - (P - 1)(C + 2) = C - P + 2 \qquad (9.50)$$

Equation (9.50) is called the *Gibbs phase rule*. It applies to systems that are in a state of equilibrium. Note that the thermodynamic equilibrium of the phases of a system depends on composition and not on the relative amounts of each phase. Applying the phase rule to a single-phase, single-component system, the number of intensive properties to define the state of the system is

$$F = C - P + 2 = 1 - 1 + 2 = 2$$

Thus, any two intensive properties, such as temperature, pressure, or composition, that can be varied independently are sufficient to define the intensive state of a one-component substance. For a one-component substance existing in two phases, $F = 1$; in three phases, $F = 0$.

Figure 4.6 and steam tables or similar property tables may be used to illustrate the foregoing examples for a pure substance. Two properties, such as temperature and pressure, can be varied independently and are sufficient to define the state of a one-component substance in a single phase. For the solid–liquid, solid–

---

[1] It can be shown that the final result is the same whether one or more components is absent from one or more phases.

vapor, or liquid–vapor mixtures, if a property, such as pressure, is chosen, the temperature is fixed since there is only one pressure at which the two phases are in equilibrium. These systems are called *univariant* because one property is sufficient to define the intensive state of the two separate phases. Note that, in order to define the intensive state of the system, it is necessary to define another independent property, such as a mole fraction or specific volume. At the triple point, $F = 0$; that is, the phases of the substance coexist only at unique values of temperature, pressure, and specific volumes of the three phases. This system is called *nonvariant.* Two independent properties, such as two mole fractions, completely define the intensive state of the system. It may be remarked from Eq. (9.50) that the number of phases cannot exceed the number of components by more than two. Another observation is that the maximum number of phases in equilibrium for a single-component substance is three.

For a system comprising ammonia and water in a liquid solution and vapor, the phase rule shows that $F = 2$. This means that specifying the temperature and pressure determines the composition in both the liquid and vapor phases. Alternatively, if the pressure and the composition of one phase are specified, the temperature and the composition of the other phase are also fixed.

9.8 Heterogeneous System of Several Phases

In this section, equilibrium conditions in a system consisting of several multicomponent phases are described. Consider a heterogeneous system of volume V, in which several homogeneous phases $(f = a, b, \ldots r)$ exist in equilibrium. Suppose, also, that each phase consists of $k(i = 1, 2 \ldots k)$ components, and that the number of components in any phase is different from the others.

Within each phase, a change in internal energy is accompanied by a change in entropy, volume, and composition, according to Eq. (6.81):

$$dU_f = T_f \, dS_f - p_f \, dV_f + \sum_{i=1}^{k} (\mu_i \, dn_i)_f \qquad (9.51)$$

A change in the internal energy of the entire system can therefore be expressed as

$$\sum_{f=a}^{r} dU_f = \sum_{f=a}^{r} T_f \, dS_f - \sum_{f=a}^{r} p_f \, dV_f + \sum_{f=a}^{r} \sum_{i=1}^{k} (\mu_i \, dn_i)_f \qquad (9.52)$$

Furthermore, a change in the internal energy of the entire system involves changes in internal energy of the constituent phases:

$$dU = dU_a + dU_b + \cdots + dU_r = \sum_{f=a}^{r} dU_f$$

Likewise, changes in volume, entropy, or chemical composition of the entire system result from contributions from each of the constituent phases:

$$dV = dV_a + dV_b + \cdots + dV_r = \sum_{f=a}^{r} dV_f$$

$$dS = dS_a + dS_b + \cdots + dS_r = \sum_{f=a}^{r} dS_f$$

$$dn = dn_a + dn_b + \cdots + dn_r = \sum_{f=a}^{r} dn_f$$

For a system in equilibrium, however, the internal energy, volume, entropy, and mass[2] are constant so that

$$dU = dV = dS = dn = 0$$

The changes that occur in a heterogeneous system, when the system deviates slightly from the equilibrium state, involve changes in the internal energy, entropy, volume, and mass in each of the phases. Changes in properties of one phase can be expressed in terms of the changes in properties of the remaining phases:

$$\left.\begin{aligned}
dU_a &= -(dU_b + \cdots + dU_r) = -\sum_{j} dU_j \\
V_a &= -\sum_{j} dV_j \\
dS_a &= -\sum_{j} dS_j \qquad \text{and} \\
(dn_i)_a &= -\sum_{j} (dn_i)_j
\end{aligned}\right\} \qquad (9.53)$$

Subscript j in these equations includes all phases except phase a.

In each of these expressions, there are j independent variables and one dependent variable. Equation (9.52) can be rewritten in terms of the dependent variable and the j independent variables:

$$\left(T_a\, dS_a + \sum_{j} T_j\, dS_j \right) - \left(p_a\, dV_a + \sum_{j} p_j\, dV_j \right)$$

$$+ \left[\sum_{i} (\mu_i\, dn_i)_a + \sum_{j}\sum_{i} (\mu_i\, dn_i)_j \right] = 0$$

---

[2] In a chemical reaction, the number of moles may change, but the total mass of the system remains constant. Therefore, the law of conservation of mass $dm = dm_a + dm_b + \cdots + dm_r = 0$ replaces the equation $dn = dn_a + dn_b + \cdots + dn_r$ in the derivation of conditions of equilibrium.

Substituting from Eq. (9.53) into the foregoing equation gives

$$\left(-T_a \sum_j dS_j + \sum_j T_j\, dS_j\right) - \left(-p_a \sum_j dV_j + \sum_j p_j\, dV_j\right)$$

$$+ \left[-\sum_i \sum_j (\mu_{i_a}\, dn_{i_j})_a + \sum_j \sum_i (\mu_i\, dn_i)_j\right] = 0$$

where subscript *ia* refers to component *i* of phase *a*. Rearranging and combining the coefficients of the independent variables dS_j, dV_j, and dn_j, yields

$$\sum_j (T_j - T_a)\, dS_j - \sum_j (p_j - p_a)\, dV_j + \sum_j \sum_i (\mu_{i_j} - \mu_{i_a})\, dn_{i_j} = 0$$

But since dS_j, dV_j, and dn_j are independent, their coefficients must vanish. Consequently,

$$T_j = T_a \qquad p_j = p_a \qquad \mu_{i_j} = \mu_{i_a}$$

These equations represent conditions that exist when the system is in thermal, mechanical, and chemical equilibrium. Evidently, phases of several components must have the same temperature and pressure. In addition, the chemical potential of each of the components *i* within one phase is equal to the chemical potential of the same component in each of the remaining phases. For example, when the liquid phase of a one-component substance is in equilibrium with its vapor, the following is true:

$$T_f = T_g \qquad p_f = p_g \qquad \mu_f = \mu_g$$

where subscripts *f* and *g* refer to liquid and vapor, respectively. Expressing the chemical potential according to Eqs. (6.80) and (6.87) gives

$$\left(\frac{\partial U}{\partial n_f}\right)_{S,\,V} = \left(\frac{\partial U}{\partial n_g}\right)_{S,\,V}$$

and

$$\left(\frac{\partial G}{\partial n_f}\right)_{p,\,T} = \left(\frac{\partial G}{\partial n_g}\right)_{p,\,T}$$

or

$$\bar{g}_f = \bar{g}_g$$

Example 9.7

For a two-phase water–vapor mixture in equilibrium at 100°C, ascertain that the Gibbs function of the liquid phase is equal to the Gibbs function of the vapor phase.

SOLUTION

Using steam tables at 100°C,

$$g_f = h_f - Ts_f = 419.04 \text{ kJ/kg} - (373.15 \text{ K})(1.3069 \text{ kJ/kg K})$$
$$= -68.6297 \text{ kJ/kg}$$

and

$$g_g = h_g - Ts_g = 2676.1 \text{ kJ/kg} - (373.15 \text{ K})(7.3549 \text{ kJ/kg K})$$
$$= -68.3809 \text{ kJ/kg}$$

The value of $g_f \approx g_g$, and if more accurate property values were used, they would be exactly the same, thereby satisfying the criterion of phase equilibrium.

9.9 Chemical Potential and the Ideal-Gas Mixture

The volume and the chemical potential according to Eqs. (6.85) and (6.87) are

$$V = \left(\frac{\partial G}{\partial p}\right)_{T,n_i} \quad \text{and} \quad \mu_i = \left(\frac{\partial G}{\partial n_i}\right)_{p,T,n_j}$$

Since the Gibbs function G is a point function, the following is true:

$$\left(\frac{\partial^2 G}{\partial p \partial n_i}\right)_{T,n_j} = \left(\frac{\partial^2 G}{\partial n_i \partial p}\right)_{T,n_j}$$

and, therefore,

$$\left(\frac{\partial \mu_i}{\partial p}\right)_{T,n_i} = \left(\frac{\partial V}{\partial n_i}\right)_{T,n_j} \tag{9.54}$$

But it can be shown, when the ideal-gas law is applied to the gas mixture, that $(\partial V/\partial n_i)_{T,p,n_j} = \bar{R}T/p = \bar{v}$. The volume V of the gas mixture is also the volume occupied by each component of the mixture, according to Avogadro's law, so that $(\partial V / \partial n_i)_{T,p,n_j} = \bar{v}_i$. The same result can be obtained when the ideal-gas law is applied to component i,

$$\left(\frac{\partial V}{\partial n_i}\right)_{T,p,n_i} = \frac{\bar{R}T}{p_i} = \bar{v}_i$$

This equation can then be substituted in Eq. (9.54) to give

$$\left(\frac{\partial \mu_i}{\partial p}\right)_{T,n_i} = \frac{\bar{R}T}{p_i} = \bar{v}_i \qquad (9.55)$$

Integration of this relationship, based on total derivatives, yields

$$\mu_i = \mu_i^o + \bar{R}T \ln p_i \qquad (9.56)$$

where μ_i^o is the constant of integration and is a function of temperature only. Note that the pressure at the lower limit in the integration either can be unity or can be included in μ_i^o.

Thus, the chemical potential of a component i is a function of the partial pressure of the component and of the temperature of the mixture, which can be expressed as

$$\mu_i = \mu_i(p_i, T) \qquad (9.57)$$

From Dalton's model, the partial pressure, p_i, can be expressed in terms of the mole fraction, x_i, and total pressure p:

$$p_i = x_i p$$

Therefore, at constant temperature, the chemical potential, μ_i, is a function of the total pressure and the mole fraction:

$$\mu_i = \mu_i^o + \bar{R}T \ln (x_i p) \qquad (9.58)$$

The entropy change during the mixing of ideal gases whose original pressure and temperature equal that of the final mixture may be calculated as follows. It will be shown[3] that chemical potential is related to the Gibbs free energy of a mixture in the following way:

$$G = \sum_{i=1}^{k} \mu_i n_i \qquad (9.59)$$

---

[3] See Section 9.10 for the derivation of this equation.

By substituting for μ_i from Eq. (9.58), the Gibbs function of the mixture becomes

$$G = \sum_{i=1}^{k} n_i \mu_i^o + \overline{R}T \sum_{i=1}^{k} n_i \ln (x_i p)$$

The change in free energy of a gaseous mixture, when the system changes from state 1 to state 2, then, is

$$\Delta G = G_2 - G_1 = \overline{R}T \sum_{i=1}^{k} n_i \ln x_i$$

In a constant-pressure process, the free energy change and the entropy change are related as follows:

$$\Delta S = -\frac{\Delta G}{T}$$

Therefore, the entropy change, due to change in composition, is

$$\Delta S = -\overline{R} \sum_{i=1}^{k} n_i \ln x_i \qquad (9.60)$$

This expression is equivalent to Eq. (9.36), which was derived directly from the combined first and second laws. The change of chemical potential with respect to temperature may be derived from the relations

$$S = -\left(\frac{\partial G}{\partial T}\right)_{p,n_i} \qquad \text{and} \qquad \mu_i = \left(\frac{\partial G}{\partial n_i}\right)_{p,T,n_j}$$

to give

$$\left(\frac{\partial \mu_i}{\partial T}\right)_{p,n_i} = -\left(\frac{\partial S}{\partial n_i}\right)_{T,p,n_j} = -\overline{s}_i \qquad (9.61)$$

9.10 The Clausius–Clapeyron and the Gibbs–Duhem Equations

In this section, we derive two important equations governing phase equilibria: The Clausius–Clapeyron equation applies to a single-component system and the Gibbs–Duhem equation applies to a multicomponent system.

Consider the phase equilibrium between a liquid and its vapor in a single-component system. The chemical potentials of the liquid phase and the vapor phase are equal:

$$\mu_f = \mu_g$$

In a single-component system, the chemical potential is a function of only pressure and temperature. The change in chemical potential of the liquid phase and the vapor phases can be expressed in partial differential form:

$$d\mu_f = \left(\frac{\partial \mu_f}{\partial p}\right)_T dp + \left(\frac{\partial \mu_f}{\partial T}\right)_p dT$$

and

$$d\mu_g = \left(\frac{\partial \mu_g}{\partial p}\right)_T dp + \left(\frac{\partial \mu_g}{\partial T}\right)_p dT$$

Substituting for the coefficients of dp and dT from Eqs. (9.55) and (9.61) and noting that $d\mu_f = d\mu_g$, then

$$\bar{v}_f dp - \bar{s}_f dT = \bar{v}_g dp - \bar{s}_g dT$$

Hence,

$$\frac{dp}{dT} = \frac{\bar{s}_g - \bar{s}_f}{\bar{v}_g - \bar{v}_f} = \frac{\bar{s}_{fg}}{\bar{v}_{fg}} \tag{9.62}$$

When equilibrium is established,

$$\bar{s}_g - \bar{s}_f = \frac{\bar{h}_g - \bar{h}_f}{T} = \frac{\bar{h}_{fg}}{T}$$

Equation (9.62) can then be written

$$\frac{dp}{dT} = \frac{\bar{h}_{fg}}{T(\bar{v}_g - \bar{v}_f)} \tag{9.63}$$

This equation, which was also derived in Chapter 8, is the *Clausius–Clapeyron equation*. It relates the pressure and temperature changes required to maintain equilibrium in a single-component system. The same procedure can be followed to derive equations applicable to phase equilibrium between vapor and solid phases or between solid and liquid phases. In the case of solid–vapor and liquid–vapor equilibria, the volume of the solid or the liquid can be neglected since the volume of the vapor is relatively large. As an additional simplification, the ideal-gas law can be assumed to apply to the vapor so that Eq. (9.63) then becomes

$$\frac{d \ln p}{dT} = \frac{\overline{h}_{fg}}{\overline{R}T^2} = \frac{h_{fg}}{RT^2} \tag{9.64}$$

Next, the equilibrium in a multicomponent system will be examined. At constant temperature and pressure, the Gibbs function depends on the chemical potential and the composition:

$$dG = \sum_{i=1}^{k} \mu_i dn_i$$

The mole fraction of each component is

$$x_i = \frac{n_i}{\sum_i n_i}$$

If the quantities, but not proportions, of the constituents change, then

$$dn_i = x_i \, d\sum_i n_i$$

The differential of the Gibbs function then is

$$dG = \sum_{i=1}^{k} \mu_i x_i \, d\sum_{i=1}^{k} n_i$$

Since μ remains constant if the temperature, pressure, and mole fraction do not change, integration leads to

$$G = \sum_{i=1}^{k} \mu_i x_i \int d\left(\sum_{i=1}^{k} n_i\right) = \sum_{i=1}^{k} \mu_i x_i \sum_{i=1}^{k} n_i$$

Therefore,

$$G = \sum_{i=1}^{k} \mu_i n_i \tag{9.65}$$

The differential of Eq. (9.65) is

$$dG = \sum_{i=1}^{k} \mu_i \, dn_i + \sum_{i=1}^{k} n_i \, d\mu_i$$

But

$$dG = -S \, dT + V \, dp + \sum_{i=1}^{k} \mu_i \, dn_i$$

When these two equations are combined, the following results:

$$-S \, dT + V \, dp = \sum_{i=1}^{k} n_i \, d\mu_i \qquad (9.66)$$

This equation, called the *Gibbs–Duhem equation,* shows the relationship between variations in temperature, pressure, and chemical potential of phases of a multi-component system. Note that, according to the Gibbs–Duhem equation, there are $(k + 2)$ variables in a system consisting of k constituents, but only $(k + 1)$ of them can vary independently. The Gibbs–Duhem equation is applied extensively in analyzing equilibrium systems. At constant temperature and pressure, the Gibbs–Duhem equation reduces to

$$\left(\sum_{i=1}^{k} n_i \, d\mu_i \right)_{T,p} = 0 \qquad (9.67)$$

which can also be expressed in mole fractions:

$$\left(\sum_{i=1}^{k} x_i \, d\mu_i \right)_{T,p} = 0 \qquad (9.68)$$

Example 9.8

Using Eq. (9.64), calculate the average enthalpy of vaporization of water at pressures between 20 kPa and 30 kPa. Compare the results with values listed in the steam tables.

SOLUTION

From the Clausius–Clapeyron equation,

$$\int_{p_1}^{p_2} d \ln p = \int_{T_1}^{T_2} \frac{h_{fg}}{RT^2} \, dT$$

Integration gives

$$[\ln p]_{p_1}^{p_2} = \left(\frac{h_{fg}}{R}\right)\left[-\frac{1}{T}\right]_{T_1}^{T_2}$$

so that

$$\ln\frac{p_2}{p_1} = \frac{h_{fg}}{R}\left(\frac{1}{T_1} - \frac{1}{T_2}\right)$$

From steam tables, the saturation temperatures at 20 kPa and 30 kPa are 60.06°C and 69.10°C, respectively. Substituting in the preceding equation gives

$$h_{fg} = \frac{(8.3144 \text{ kJ/kg-mol K})/(18.015 \text{ kg/kg-mol}) \times \ln(30/20)}{1/(333.21 \text{ K}) - 1/(342.25 \text{ K})}$$

$$= 2361 \text{ kJ/kg}$$

The corresponding value from steam tables is 2346.3 kJ/kg at 25 kPa.

9.11 Effect of an Inert Gas on Vapor Pressure

Consider a liquid–vapor mixture of a one-component substance in a constant-volume container. If the mixture is in thermodynamic equilibrium, then

$$T_{l_1} = T_{v_1} \qquad p_{l_1} = p_{v_1} \qquad \text{and} \qquad \mu_{l_1} = \mu_{v_1}$$

where subscript 1 refers to the original state of the one-component system. Now consider the addition of an inert gas to the system, keeping the temperature constant. At equilibrium, the following relations must be satisfied:

$$T_{l_2} = T_{v_2} = T_g \qquad p_{l_2} = p_{v_2} + p_g \qquad \text{and} \qquad \mu_{l_2} = \mu_{v_2}$$

where subscript 2 refers to the final state of the two-component system. The chemical potentials of the liquid and vapor change when the inert gas is added to the system. Furthermore, the change in chemical potential of the liquid phase must be equal in value to the change in the vapor phase, or

$$d\mu_l = d\mu_v \tag{9.69}$$

But

$$\mu = \left(\frac{\partial G}{\partial n}\right)_{p,T} = \bar{g}(p, T)$$

therefore,

$$d\bar{g}_l = d\bar{g}_v \qquad\qquad (9.70)$$

The total differential of $\bar{g}$ is

$$d\bar{g} = \bar{v}\, dp - \bar{s}\, dT$$

which reduces at constant temperature to

$$d\bar{g} = \bar{v}\, dp$$

Substitution into Eq. (9.70) gives

$$\bar{v}_l\, dp_l = \bar{v}_v\, dp_v$$

The volume of the liquid phase can be assumed to remain unchanged. Furthermore, the vapor phase follows the ideal-gas law so that

$$\bar{v}_l\, dp_l = \frac{\bar{R}T}{p_v}\, dp_v$$

Integration gives

$$\frac{p_{v_2}}{p_{v_1}} = e^{(\bar{v}_l/\bar{R}T)(p_{l_2} - p_{l_1})} \qquad\qquad (9.71)$$

Equation (9.71) gives the ratio of the pressures of the vapor phase for states 1 and 2 in terms of the pressure of the liquid in the two states. The exponent in the foregoing equation is very small. As a result, the vapor pressure remains essentially constant, irrespective of the presence of the inert gas. For practical purposes, values of vapor pressure of a one-component substance can be used even when an inert gas is present.

9.12 Human Comfort and Air Conditioning

Air-conditioning systems are used to acclimate homes and workplaces for the comfort of the occupants, or to satisfy the need of operation of equipment in industry. The main factors to be controlled are the air temperature and the relative humidity. Other factors could be air cleanliness, noise, and odor. The comfort condition of people in buildings varies widely, ranging from a temperature of 22 to 26°C and a relative humidity of 40 to 60 percent.

To determine the optimal condition for comfort, it is necessary to know the energy generated by the human body. This in turn, depends on its activity. For an average-size adult, it varies from 70 W with no activity to about 400 W with strenuous physical effort. For the human body to maintain a comfortable temperature, a balance of energy must relate the energy generated to that exchanged with the environment. The energy transfer with the environment takes place by convection, conduction, radiation, and evaporation of body fluids. Air movement around the body enhances the heat transfer by convection and replaces the moist air by fresh air.

In cold weather, the body loses more energy than it generates. To overcome this problem, heavier clothing provides some heat insulation. In addition, the rate of energy generation can be increased by increasing the level of physical activities. In warm or hot weather, the body either may transfer less energy or may receive energy from the surroundings. Light clothing and a reduction of energy generation by reducing the level of physical activity both aid in body comfort. The increase in sweating and its evaporation in hot weather cause a cooling effect and provide the energy balance necessary for human comfort. The rate of evaporation depends on the relative humidity of the surroundings; it is high at low relative humidity, and low at high humidity. This is the reason hot humid weather is most uncomfortable. The human body under these conditions cannot readily dissipate its generated energy either by heat transfer or by evaporative cooling. In dry climate, the body can tolerate high temperatures because cooling is achieved by the increased evaporation of the perspiration.

9.13 Air–Water Vapor Mixtures (Psychrometry)

One common application of thermodynamics centers on control of both temperature and humidity. Atmospheric air is generally the fluid used in these applications. The composition of dry air, which is a mixture of oxygen, nitrogen, argon, carbon dioxide, hydrogen, and inert gases, is shown in Table 9.1 in terms of proportions by mass and proportions by volume. In engineering applications, atmospheric air is considered as a mixture of dry air and water vapor in equilibrium.

The mixture of N_2, Ar, CO_2, H_2, and the inert gases that normally are present in air is usually called *atmospheric nitrogen*. With this definition, the composition of dry air may be written as

	Mass Fraction	*Molar Fraction*
Oxygen	0.23188	0.2099
Atmospheric nitrogen	0.76812	0.7901
	1.00000	1.0000

Table 9.1   Composition of Dry Air ($M = 28.97$ kg/kg-mol; $R = 0.287$ kJ/kg K)

Constituent	Mass Fraction	Molar (Volumetric) Fraction or Partial Pressure
Oxygen, O_2	0.23188	0.2099
Nitrogen, N_2	0.75468	0.7803
Argon, Ar	0.01296	0.0094
Carbon dioxide, CO_2	0.00046	0.0003
Hydrogen, H_2	0.00001	0.0001
Inert gases, such as xenon and krypton	0.00001	—
	1.00000	1.00000

The air–water vapor mixture is a homogeneous two-component system and can be considered to follow the laws of ideal gases for mixtures; both dry air and water vapor individually also tend to behave according to the ideal-gas law. In common applications, air containing water vapor behaves like an ideal gas because the temperature of the dry air is usually very high compared to its critical temperature (for N_2, $T_c = 126.2$ K, and for O_2, $T_c = 154.8$ K) and because the partial pressure of water is relatively low compared to its critical pressure ($p_c = 22.09$ MPa). At low pressure and at temperatures below 50°C, constant-enthalpy lines for water vapor coincide with constant-temperature lines. This means that the enthalpy of water vapor can be assumed to be equal to the enthalpy of saturated vapor at the same temperature with negligible error. It is also assumed that, when the water vapor condenses, there are no dissolved gases in the condensate. Although this section deals with air–water vapor mixtures, the principles and formulas involved can be applied to any mixture of a gas and a condensable vapor. The study of the properties of air and water vapor pertaining to air-conditioning systems is called *psychrometry*. Certain terms used in psychrometry are defined in the following paragraphs.

Saturated air is a mixture of dry air and water vapor in which the partial pressure of the vapor is equal to the saturation pressure of water at the temperature of the mixture. The air in this case cannot hold more moisture, and any additional water vapor introduced will condense. *Unsaturated air* (humid) is a mixture of dry air and superheated water vapor, the partial pressure of the vapor being less than the saturation pressure of water at the temperature of the mixture. *Supersaturated air* is a mixture of dry air and water vapor in which the partial pressure of the water vapor is greater than the saturation pressure of water at the temperature of the mixture.

The three states of unsaturated, saturated, and supersaturated air–water vapor mixtures are represented by points *a, b,* and *c* on the *T-s* diagram of Figure 9.7. Note that as the partial pressure of water–vapor in a mixture increases, at constant temperature, the partial pressure of air decreases accordingly.

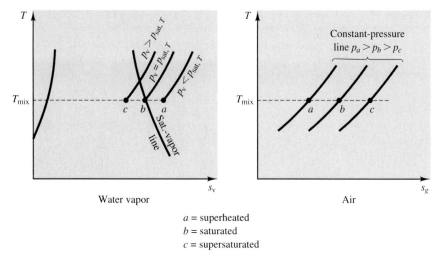

a = superheated
b = saturated
c = supersaturated

Figure 9.7   Temperature–entropy diagrams for water vapor and air.

Absolute humidity, which is also called *specific humidity* or *humidity ratio,* indicates the composition of an air–water vapor mixture. It is defined as the ratio of the mass of vapor to the mass of dry air and is denoted by the symbol ω:

$$\omega = \frac{m_v}{m_a} \tag{9.72}$$

where ω is expressed in kilograms of water vapor per kilogram of dry air. Absolute humidity can be expressed in terms of pressures, rather than masses, by applying the ideal-gas law:

$$
\begin{aligned}
\omega = \frac{m_v}{m_a} &= \frac{p_v V/R_v T}{p_a V/R_a T} = \frac{p_v R_a}{p_a R_v} \\
&= \frac{p_v M_v}{p_a M_a} = \frac{p_v}{p - p_v}\frac{M_v}{M_a} \\
&= 0.622\,\frac{p_v}{p - p_v} \quad \text{(for air–water vapor mixture)}
\end{aligned}
\tag{9.73}
$$

where p is the total pressure.

Relative humidity, which is expressed on a percentage basis, is the ratio of the vapor pressure in the mixture at a given temperature to the saturation pressure of the vapor at the same temperature. The relative humidity indicates the degree to which an air–vapor mixture is saturated. An alternative definition is the ratio of the moisture content in the mixture to the moisture content of an equal quantity of saturated air at the same temperature and total pressure.

Assuming that the ideal-gas equation of state applies to water vapor, relative humidity, φ, can be written as

$$\varphi = \left(\frac{p_v}{p_{v,\,sat}}\right)_T = \frac{R_v T / v_v}{R_v T / v_{v,\,sat}} = \left(\frac{v_{v,\,sat}}{v_v}\right)_T = \left(\frac{\rho_v}{\rho_{v,\,sat}}\right)_T \qquad (9.74)$$

From this equation, then, the relative humidity can also be expressed as the ratio of the density of the water vapor at the given temperature to the density of saturated water vapor at the same temperature.

A relationship between absolute humidity and relative humidity in an air–water vapor mixture can be derived from Eqs. (9.73) and (9.74):

$$\omega = 0.622\frac{p_v}{p_a} = 0.622\left(\frac{p_v}{p_{v,\,sat}}\right)\left(\frac{p_{v,\,sat}}{p_a}\right) = 0.622\varphi\left(\frac{p_{v,\,sat}}{p_a}\right) \qquad (9.75)$$

The *saturation ratio* indicates the degree of saturation and is the ratio of absolute humidity of the air to the absolute humidity of saturated air at the same temperature and total pressure. The saturation ratio, ψ, can be expressed as

$$\psi = \left(\frac{\omega}{\omega_{sat}}\right)_T \qquad (9.76)$$

The saturation ratio ψ can be shown to be related to the relative humidity φ. By substituting for ω and ω_{sat} in Eq. (9.76), one obtains

$$\psi = \left[\left(\frac{p_v}{p_a}\right)\left(\frac{p_{a,\,sat}}{p_{v,\,sat}}\right)\right]_T = \left[\left(\frac{p_v}{p_{v,\,sat}}\right)\left(\frac{p - p_{v,\,sat}}{p - p_v}\right)\right]_T = \varphi\left(\frac{p - p_{v,\,sat}}{p - p_v}\right)_T \qquad (9.77)$$

At room temperature or lower, the vapor pressure of water is small so that the saturation ratio tends to become identical with the relative humidity, φ.

For saturated air, the absolute humidity can be written as

$$\omega_{sat} = 0.622\frac{f(T)}{p - f(T)} \qquad (9.78)$$

where the saturation pressure of water vapor is expressed as a function of temperature $f(T)$, as indicated in steam tables. At 100 percent relative humidity, values of the absolute humidity at various temperatures and at a constant total pressure can be calculated according to Eq. (9.78). The result is shown in Figure 9.8. In

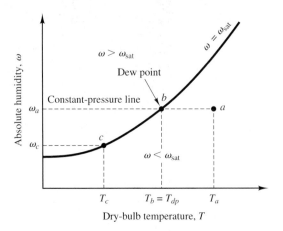

Figure 9.8   Absolute humidity versus dry-bulb temperature.

the region below the saturation curve, $\omega < \omega_{\text{sat}}$ so that the air is unsaturated and the water vapor is superheated. In the region above the saturated curve, $\omega > \omega_{\text{sat}}$, and the air is supersaturated with water vapor. The pressure of the vapor in this case is larger than the saturation pressure of water at that temperature so that sudden and irreversible condensation of water tends to occur, accompanied by a sharp increase in entropy and reestablishment of equilibrium conditions. The resulting mixture of air and water is saturated with water vapor.

In unsaturated air at state a of Figure 9.8 is cooled at constant pressure, its absolute humidity remains constant until saturation is reached, which corresponds to point b. During this phase of cooling, the partial pressure of the water vapor remains constant provided that the total pressure is kept constant. Temperature T_b, which is called the *dew point,* is the temperature at which an unsaturated gas–vapor mixture becomes saturated as a result of isobaric cooling at constant absolute humidity. The dew point is also the saturation temperature corresponding to the initial partial pressure of the vapor ($T_{dp} = T_{\text{sat}}$ at p_v). If cooling proceeds further, condensation begins and both the temperature and the absolute humidity decrease, until point c on the saturation curve is reached. The amount of water condensed per kilogram mass of dry air is the difference in absolute humidity between point a and point c. The same cooling process is represented in Figure 9.9 on the *T-s* diagram. Point a represents the unsaturated state of a mixture. The partial pressure of the vapor remains constant until the dew-point temperature, point b, is reached. Further cooling causes condensation to begin while both the temperature of the mixture and the partial pressure of the vapor are decreased. When a temperature such as T_c is reached, the saturated vapor at state c is in equilibrium with its condensate at state c'.

In a plot of absolute humidity versus temperature, it is possible to indicate points of constant relative humidity. This is shown in Figure 9.10. At a selected total pressure, values of absolute humidity are calculated according to Eq. (9.78), corre-

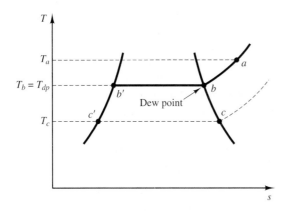

Figure 9.9   Cooling of a gas–vapor mixture at constant pressure.

sponding to saturation conditions ($\varphi = 100$ percent), at various temperatures. At other selected values of φ, absolute humidity is calculated from Eq. (9.75). Since at room temperature (or lower), the relative humidity can be assumed equal to the saturation ratio, and since the saturation ratio is a ratio of two values of absolute humidity at constant temperature, points of constant φ can be determined from

$$\psi_1 = \varphi_1 = \frac{b_1 c_1}{a_1 c_1} = \frac{b_2 c_2}{a_2 c_2} = \frac{b_3 c_3}{a_3 c_3} = \cdots$$

In establishing constant relative humidity lines, the ordinates of the saturation curve are first divided into 10 equal parts, and points at 10 percent increments in relative humidity are established.

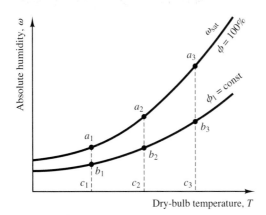

Figure 9.10   Plot of constant relative humidity lines.

The enthalpy of an air–water vapor mixture is the sum of the enthalpies of the dry air and the accompanying water vapor, both taken at the temperature of the mixture. The enthalpy of dry air is based in some cases on a zero value at $-20°C$, whereas enthalpy of water vapor, as found in steam tables, is based on a zero value for the saturated liquid at $0°C$. Although the points of zero enthalpy for air and water vapor are not at the same temperature, the discrepancy can be ignored since interest lies in differences in enthalpy rather than in absolute values of enthalpy. Both the dry air and the water vapor are assumed to follow the ideal-gas law so that their enthalpies are functions of temperature alone. In atmospheric air, the partial pressure of the water vapor is low so that this assumption is valid. In air-conditioning applications, the amount of dry air remains constant, and properties are usually expressed in terms of a unit mass of dry air. The enthalpy, for example, of a mixture of dry air and water vapor per kilogram of dry air is

$$h = h_a + \omega h_v \simeq h_a + \omega h_g \tag{9.79}$$

and relative to an arbitrary reference state ($T = 0$ K),

h_a = enthalpy per unit mass of dry air = $c_{p_a} T_a$
ω = specific humidity
h_v = enthalpy per unit mass of superheated vapor in the mixture
  = $c_{p_f}(T_{dp} - 0) + (h_{fg})_{T_{dp}} + c_{p_v}(T - T_{dp})$
h_g = enthalpy of saturated steam per unit mass at the temperature of the mixture
T_{dp} = dew-point temperature

Because enthalpy is primarily a function of temperature,[4] and not pressure, the enthalpy of saturated steam, h_g, at a given temperature is virtually identical to enthalpy of superheated steam, h_v, at that temperature. Note also that the enthalpy given by Eq. (9.79) expresses the enthalpy of the mixture per kilogram of dry air rather than per kilogram of mixture. The specific volume of a mixture of air and water vapor is given by

$$v = \frac{1}{\rho} = \frac{RT}{p} = \frac{T}{p}\left(\frac{m_a R_a + m_v R_v}{m_a + m_v}\right) = \frac{T}{p}\left(\frac{R_a + \omega R_v}{1 + \omega}\right) \tag{9.80}$$

Since the specific humidity $\omega = \rho_v/\rho_a$, the density of the water vapor is

$$\rho_v = \omega \rho_a = \frac{\omega p_a}{R_a T} \tag{9.81}$$

---

[4] In the range of typical engineering applications of psychrometry. See also the *h-s* diagram for water, and note that constant-enthalpy lines are almost identical to constant-temperature lines in the superheated vapor region at low pressure.

In the case of a gaseous mixture, the density of the mixture is the sum of the densities of the air and water vapor so that

$$\rho = \rho_a + \rho_v = \rho_a + \omega\rho_a = \rho_a(1 + \omega) \qquad (9.82)$$

and the specific volume of the mixture is

$$v = \frac{1}{\rho} = \frac{v_a}{1 + \omega} \qquad (9.83)$$

Example 9.9

An air–water vapor mixture at 20°C and 100 kPa has a relative humidity of 80 percent (Figure 9.11). Determine:

(a) the partial pressures of the vapor and the air
(b) the specific humidity
(c) the saturation ratio
(d) the dew point
(e) the density of the mixture
(f) the amount of water vapor condensed per kilogram of dry air if the mixture is cooled at constant pressure to a temperature of 10°C

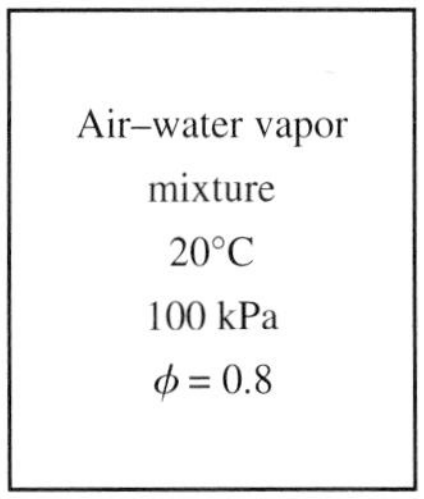

Air–water vapor
mixture
20°C
100 kPa
$\phi = 0.8$

Figure 9.11

SOLUTION

From steam tables at 20°C, $p_{v,\,sat} = 2.339$ kPa. But $\varphi = 0.8 = (p_v/p_{v,\,sat})_{20°C}$. Therefore,

(a)

$$p_v = (0.8)(2.339 \text{ kPa}) = 1.8712 \text{ kPa}$$
$$p_a = p - p_v = 100 - 1.8712 = 98.1288 \text{ kPa}$$

(b)

$$\omega_1 = 0.622 \frac{p_v}{p_a} = 0.622 \frac{1.8712}{98.1288}$$

$$= 0.01186 \text{ kg(water-vapor)/kg(dry-air)}$$

(c)

$$\psi = \left(\frac{\omega}{\omega_{\text{sat}}}\right)_{20°C} = \varphi \frac{p - p_{v,\text{sat}}}{p - p_v} = 0.8 \frac{97.661}{98.1288} = 0.7962$$

(d) From steam tables, at $p_v = 1.8712$ kPa, $T_{dp} = T_{\text{sat at } p_v} = 16.3°C$.

(e)

$$\rho = \rho_a(1 + \omega) = \frac{98.1288 \text{ kPa}}{(0.287 \text{ kJ/kg K})(293.15 \text{ K})}(1 + 0.01186)$$

$$= 1.18017 \text{ kg/m}^3$$

(f) Since 10°C is below the dew point, some of the vapor will condense, leaving a mixture of 100 percent relative humidity.

At 10°C,

$$p_v = p_{v,\text{sat}} = 1.2276 \text{ kPa.}$$

$$\omega_2 = 0.622 \frac{1.2276}{100 - 1.2276} = 0.00773 \text{ kg/kg}$$

The amount of vapor condensed per kilogram of dry air is

$$\omega_1 - \omega_2 = 0.01186 - 0.00773 = 0.00413 \text{ kg/kg}$$

9.14 Adiabatic Saturation Process

In this section, one way of determining the absolute and relative humidities is outlined. Consider a steady stream of air and water vapor flowing through an insulated chamber, as shown in Figure 9.12. The mixture entering the chamber at section 1 is unsaturated at a relative humidity φ_1 less than 100 percent. At the same time, a stream of water is sprayed into the chamber. As the mixture flows through the chamber, some of the liquid water evaporates and is carried

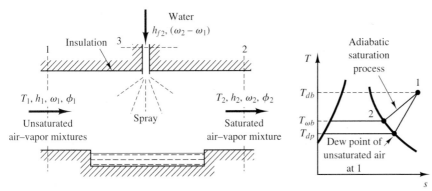

Figure 9.12   The adiabatic saturation process.

away by the air stream. Energy is needed for evaporation; this is reflected in a lowering of the temperature of the air stream. By the time the air reaches section 2, it is assumed that equilibrium between the two streams is established and the air stream is saturated with water vapor. At this point, the chemical potentials of the water in the liquid and vapor states are equal. If the water stream is added adiabatically and the temperature of the water stream does not change during the process, the mixture at section 2 attains a temperature called the *adiabatic saturation temperature*. To maintain steady state, makeup water must be added at a rate equal to that of evaporation, and its temperature is assumed to be equal to adiabatic saturation temperature. The reduction in temperature of the air mixture is a measure of the degree of saturation of the entering mixture.

The adiabatic cooling process is used in air conditioning to lower the temperature and increase the humidity. With no heat or work interaction, the first law of thermodynamics for a steady flow process may then be applied, assuming no changes in kinetic energy or potential energy:

$$\underbrace{h_{a_1} + \omega_1 h_{v_1}}_{\substack{\text{unsaturated} \\ \text{air–vapor mixture}}} + \underbrace{(\omega_2 - \omega_1) h_{f_2}}_{\text{water}} = \underbrace{h_{a_2} + \omega_2 h_{v_2}}_{\substack{\text{saturated} \\ \text{air–vapor mixture}}} \qquad (9.84)$$

From this equation, ω_1 can be determined if the temperature and the pressure at sections 1 and 2 are known. Equation (9.84) can be arranged to take any one of the following forms:

$$h_{a_1} + \omega_1 h_{v_1} - \omega_1 h_{f_2} = h_{a_2} + \omega_2 h_{v_2} - \omega_2 h_{f_2} \qquad (9.85)$$

$$(h_{a_1} + \omega_1 h_{v_1}) - \omega_1 h_{f_2} = (h_{a_2} + \omega_2 h_{f g_2}) \qquad (9.86)$$

$$\omega_1(h_{v_1} - h_{f_2}) = (h_{a_2} - h_{a_1}) + \omega_2 h_{f g_2} \qquad (9.87)$$

Note that in the last two equations, h_v is assumed equal to h_g. Note also that in a real process saturation may not be reached.

9.15 Wet-Bulb Temperature

A simple method of determining the relative humidity of an air–water vapor mixture is by use of dry-bulb and wet-bulb thermometers. The dry-bulb temperature is indicated by a conventional thermometer placed in the mixture. The wet-bulb temperature is measured by a thermometer whose bulb is covered with a wick saturated with water and placed in the air–water vapor stream. If the air surrounding the wet-bulb thermometer is unsaturated, the pressure of the water vapor at the surface of the wick exceeds the partial pressure of the water vapor in the mixture. Evaporation of water from the wick then occurs because of the difference in chemical potential between the liquid and the vapor phases. This evaporation is accompanied by a drop in temperature of the wick, which is reflected in a lowering of the temperature of the wet-bulb thermometer. Heat is transferred from both the wick and the thermometer to sustain the evaporation process. Equilibrium is attained when the chemical potentials of the liquid and vapor phases are identical, and at this point, the wick is at the so-called *wet-bulb temperature*. The wet-bulb temperature is ordinarily less than the dry-bulb temperature; when the dry-bulb thermometer is surrounded by saturated air, both temperatures as well as the dew-point temperature are equal ($T_{db} = T_{wb} = T_{dp}$).

The temperature indicated by a wet-bulb thermometer is very nearly equal to the adiabatic saturation temperature. But some difference does exist. The wet-bulb temperature depends on both mass transfer and heat transfer from the liquid phase; on the other hand, the adiabatic saturation temperature (also called the *thermodynamic wet-bulb temperature*) involves an equilibrium temperature in which no transfer of heat occurs from the liquid phase to the gas phase. Because the rate of exchange of heat and mass affects the readings indicated by the wet-bulb thermometers, the wet-bulb temperature is not a property of the mixture. In practice, however, the wet-bulb temperature is easily determined; on the other hand, the thermodynamic wet-bulb temperature cannot be accurately measured because it is difficult to attain equilibrium between the unsaturated air–vapor mixture and water under adiabatic conditions. Since the difference between the two temperatures is negligible, the wet-bulb temperature is ordinarily used in place of the thermodynamic wet-bulb temperature in humidity calculations. Properties of humid air are readily determined using the wet- and dry-bulb temperatures.

The *evaporative cooling process* is essentially identical to the adiabatic saturation process. When water is sprayed in an air stream, part of the water evaporates by absorbing energy from the air stream, causing an increase in its humidity and a decrease in its temperature. If the system is adiabatic, the evaporative cooling process follows a line of constant wet-bulb temperature ($h \approx$ constant).

Example 9.10

The dry-bulb and wet-bulb temperatures of an air stream are 22°C and 17°C. Determine the specific humidity and relative humidity of the air if the pressure is 101.325 kPa. Assume steady state.

SOLUTION

Referring to the adiabatic saturation process of Figure 9.12, the saturated air at state 2 is at the wet-bulb temperature. The specific humidity, according to Eq. 9.87, is

$$\omega_1 = \frac{(h_{a_2} - h_{a_1}) + \omega_2 h_{fg_2}}{h_{v_1} - h_{f_2}}$$

where ω_2 is the specific humidity at the adiabatic saturation temperature. Assuming $T_{wb} = T_{\text{adia. sat.}}$, then,

$$\omega_2 = 0.622 \frac{p_{v,\text{sat}}}{p - p_{v,\text{sat}}}$$

$$= 0.622 \frac{1.9376}{101.325 - 1.9376}$$

$$= 0.0121 \text{ kg of vapor/kg of dry air}$$

and

$$\omega_1 = \frac{(1.0035 \text{ kJ/kg K})[(17 - 22) \text{ K}] + 0.0121(2461.2 \text{ kJ/kg})}{(2541.7 - 71.38) \text{ kJ/kg}}$$

$$= 0.0100 \text{ kg of vapor/kg of dry air}$$

The specific humidity at state 1 is

$$\omega_1 = 0.622 \frac{p_{v_1}}{p - p_{v_1}}$$

from which

$$p_{v_1} = \frac{\omega_1 p}{\omega_1 + 0.622} = \frac{(0.01)(101.325 \text{ kPa})}{(0.01 + 0.622)} = 1.6032 \text{ kPa}$$

Therefore,

$$\varphi_1 = \left(\frac{p_{v_1}}{p_{v,\text{sat}}}\right) = \frac{1.6032}{2.645} = 0.6061 \quad \text{or} \quad 60.61\%$$

9.16 Psychrometric Chart

A psychrometric chart, shown in Figure 9.13 in skeleton form, provides a graphical representation of several properties of air–water vapor mixtures. (A complete chart is found in the Appendix.) The psychrometric chart is essentially a composition–temperature chart. The specific humidity (identifying composition) and the partial pressure are ordinates; the dry-bulb temperature is the abscissa. On the chart shown in Figure 9.14, lines of constant relative humidity, volume of the mixture (per unit mass of dry air), wet-bulb temperature, and enthalpy of the mixture (per unit mass of dry air) are indicated. Most psychrometric charts are constructed for a constant total pressure of 1 atm (101.325 kPa). Some charts give correction curves for pressures other than atmospheric pressure. For saturated air–water vapor mixture, the dry-bulb, wet-bulb, and dew-point temperatures are identical.

Since the wet-bulb temperature is almost the same as the adiabatic saturation temperature, it is reasonable to expect that the constant enthalpy lines will run almost parallel to the wet-bulb temperatures on the psychrometric chart. Thus, the process of adiabatic humidification (or dehumidification) of air can be considered to be a constant-enthalpy process. Application of the first law of thermodynamics to an adiabatic saturation process shows that

$$h_1 + (\omega_2 - \omega_1)h_f = h_2 \tag{9.88}$$

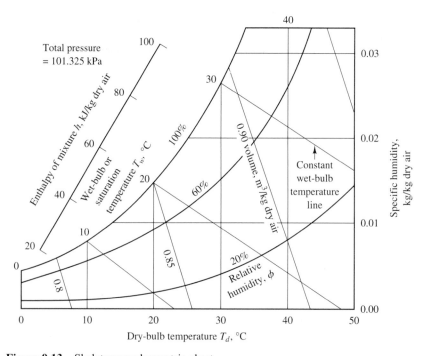

Figure 9.13   Skeleton psychrometric chart.

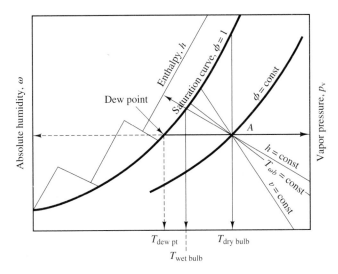

Figure 9.14 Schematic of the psychrometric chart.

where h_f is the enthalpy of the injected water. But clearly, since $h_1 \neq h_2$, the process cannot be considered a constant-enthalpy process. The term $(\omega_2 - \omega_1)h_f$ is, however, small when compared with h_1 or h_2 so that the deviation from a constant-enthalpy process is minor. Some psychrometric charts account for this deviation by including lines of constant "enthalpy deviation."

Air-conditioning processes, such as cooling, heating, humidifying, and dehumidifying, can be represented on the psychrometric chart shown in Figure 9.15.

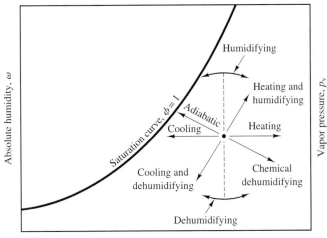

Figure 9.15 Air-conditioning processes.

With this chart, processes can be more readily visualized, and the changes in properties can be determined easily. The use of the chart will be illustrated in Section 9.17.

9.17 Processes Involving Air–Water Vapor Mixtures

Some of the most common processes in air conditioning involving air–water vapor mixtures are mixing, heating and cooling, humidifying, and dehumidifying.

9.17.1 Mixing Process

This process can be illustrated by considering the mixing of two streams of humid air. As shown in Figure 9.16, the original streams are at states 1 and 2, and the final stream is at state 3. If the effects of the kinetic and potential energy are neglected and there is no work other than flow work where matter crosses the boundary, the energy balance for steady flow leads to the following equation:

$$(\dot{m}_{a_1}h_{a_1} + \dot{m}_{v_1}h_{v_1}) + (\dot{m}_{a_2}h_{a_2} + \dot{m}_{v_2}h_{v_2}) + \dot{Q} = (\dot{m}_{a_3}h_{a_3} + \dot{m}_{v_3}h_{v_3})$$

Since $\omega = m_v/m_a$,

$$\dot{m}_{a_1}(h_{a_1} + \omega_1 h_{v_1}) + \dot{m}_{a_2}(h_{a_2} + \omega_2 h_{v_2}) + \dot{Q} = \dot{m}_{a_3}(h_{a_3} + \omega_3 h_{v_3})$$

Denoting $h_a + \omega h_v = h$, the enthalpy of the mixture per kilogram of dry air, the preceding equation becomes

$$\dot{m}_{a_1}h_1 + \dot{m}_{a_2}h_2 + \dot{Q} = \dot{m}_{a_3}h_3 \tag{9.89}$$

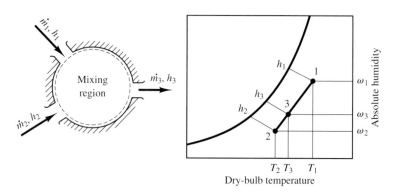

Figure 9.16   Adiabatic mixing process.

The conservation-of-mass equations for the dry air and water vapor are

$$\dot{m}_{a_1} + \dot{m}_{a_2} = \dot{m}_{a_3} \qquad \text{(for dry air)}$$
$$\dot{m}_{v_1} + \dot{m}_{v_2} = \dot{m}_{v_3} \qquad \text{(for water vapor)}$$

The conservation-of-mass equation for water vapor can also be expressed as

$$\omega_1 \dot{m}_{a_1} + \omega_2 \dot{m}_{a_2} = \omega_3 \dot{m}_{a_3}$$

Assuming that mixing occurs adiabatically and substituting for $\dot{m}_{a_3}$, the energy equation becomes

$$\dot{m}_{a_1} h_1 + \dot{m}_{a_2} h_2 = (\dot{m}_{a_1} + \dot{m}_{a_2}) h_3$$

or

$$\frac{\dot{m}_{a_1}}{\dot{m}_{a_2}} = \frac{h_3 - h_2}{h_1 - h_3} \qquad (9.90)$$

Similarly, the conservation-of-mass equation for water vapor is

$$\omega_1 \dot{m}_{a_1} + \omega_2 \dot{m}_{a_2} = \omega_3 (\dot{m}_{a_1} + \dot{m}_{a_2})$$

or

$$\frac{\dot{m}_{a_1}}{\dot{m}_{a_2}} = \frac{\omega_3 - \omega_2}{\omega_1 - \omega_3} \qquad (9.91)$$

It is obvious from Eqs. (9.90) and (9.91) that changes in the enthalpy of the mixture per kilogram of dry air are paralleled by changes in the specific humidity. This is an extremely convenient relationship. As shown in Figure 9.16, if the initial states 1 and 2 are located on the chart, the state of the resulting mixture lies on the line connecting the two states 1 and 2. Point 3 divides the distance 1-2 into two parts, such that the ratio of the lengths 1-3 and 3-2 is equal to the ratio of the masses of dry air in the two streams before they mix.

Example 9.11

The properties of two streams of air are as follows:

	Temperature (°C)	Pressure (kPa)	Relative Humidity (%)	Flow Rate (m³/min)
Stream(1)	15	101.325	20	24
Stream(2)	25	101.325	80	33

If these two streams are mixed adiabatically, and the pressure is maintained at 101.325 kPa, evaluate the following properties of the mixed stream

  (a) relative humidity and specific humidity
  (b) specific volume
Assume steady state.

SOLUTION

Referring to Figure 9.17,

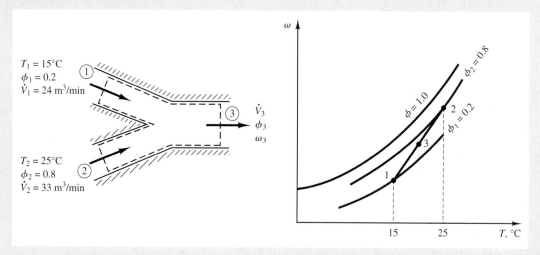

Figure 9.17

(a)

$$\omega_1 = 0.622 \frac{(0.2)(1.7051 \text{ kPa})}{(101.325 - 0.2 \times 1.7051) \text{ kPa}}$$
$$= 0.0021 \text{ kg(water vapor)/kg(dry air)}$$

$$\omega_2 = 0.622 \frac{(0.8)(3.169 \text{ kPa})}{(101.325 - 0.8 \times 3.169) \text{ kPa}}$$
$$= 0.01596 \text{ kg(water vapor)/kg(dry air)}$$

$$\dot{m}_{a_1} = \frac{p_{a_1} \dot{V}_1}{R_a T_1} = \frac{(101.325 - 0.2 \times 1.7051) \text{ kPa}(24 \text{ m}^3/\text{min})}{(0.287 \text{ kJ/kg K})(288.15 \text{ K})}$$
$$= 29.306 \text{ kg of air per minute}$$

$$\dot{m}_{a_2} = \frac{p_{a_2} \dot{V}_2}{R_a T_2} = \frac{(101.325 - 0.8 \times 3.169) \text{ kPa}(33 \text{ m}^3/\text{min})}{(0.287 \text{ kJ/kg K})(298.15 \text{ K})}$$
$$= 38.099 \text{ kg of air per minute}$$

The conservation-of-mass equation for the water vapor is

$$\omega_1 \dot{m}_{a_1} + \omega_2 \dot{m}_{a_2} = \omega_3(\dot{m}_{a_1} + \dot{m}_{a_2})$$

from this equation

$$\omega_3 = \frac{29.306}{67.405} \times 0.0021 + \frac{38.099}{67.405} \times 0.01596$$

$$= 0.00993 \text{ kg(water vapor)/kg(dry air)}$$

The equivalent values from the psychrometric chart are

$$\omega_1 = 0.002, \qquad \omega_2 = 0.016,$$

$$\omega_3 = 0.0099 \text{ kg(water vapor)/kg(dry air)}$$

The vapor pressure, according to Eq. (9.73), is

$$p_{v_2} = \frac{\omega_3 p}{\omega_3 + 0.622} = \frac{(0.00993)(101.325 \text{ kPa})}{0.63193} = 1.5922 \text{ kPa}$$

An energy balance gives

$$\dot{m}_{a_1}(c_{p_a} + \omega_1 c_{p_v})T_1 + \dot{m}_{a_2}(c_{p_a} + \omega_2 c_{p_v})T_2 = \dot{m}_{a_3}(c_{p_a} + \omega_3 c_{p_v})T_3$$

Therefore,

$$(29.306 \text{ kg/min})[(1.0035 \text{ kJ/kg°C}) + (0.0021)(1.8723 \text{ kJ/kg°C})](15°C)$$
$$+ (38.099 \text{ kg/min})[(1.0035 \text{ kJ/kg°C}) + (1.01596)(1.8723 \text{ kJ/kg°C})](25°C)$$
$$= (67.405 \text{ kg/min})[(1.0035 \text{ kJ/kg°C}) + (0.00993)(1.8723 \text{ kJ/kg°C})](T_3)$$

from which $T_3 = 20.7°C$. From the steam tables, at $T_3 = 20.7°C$,

$$p_{v, \text{ sat}} = 2.443 \text{ kPa}$$

Therefore,

$$\varphi_3 = \left(\frac{p_v}{p_{v, \text{ sat}}}\right)_{20.7°C} = \frac{1.5922}{2.443} = 65\%$$

(b)

$$v_3 = \frac{v_{a_1}}{1 + \omega_3}$$

$$= \frac{\left[(0.287 \text{ kJ/kg K})(293.15 \text{ K})/(101.325 - 1.5922) \text{ kPa}\right]}{1.00993}$$

$$= 0.8373 \text{ m}^3/\text{kg}$$

Values obtained directly from the chart are

$$T_3 = 20.7°\text{C} \qquad \varphi_3 = 66 \text{ percent} \qquad v_3 = 0.846 \text{ m}^3/\text{kg}$$

9.17.2 Heating and Cooling Processes

Both the constant-volume process and the constant-pressure process will be considered. In the case of the constant-volume process, the specific volume of each component of the mixture remains constant; also, the heat transfer equals the change in internal energy of the mixture. In the case of the constant-pressure process, the vapor pressure and the specific humidity of the mixture remain constant, provided that no condensation takes place. This process is represented by a horizontal line on the psychrometric chart. If, however, condensation occurs as a result of cooling below the dew point, the specific humidity decreases and the partial pressure of the water vapor becomes equal to the saturation pressure of water at the temperature of the mixture. The following two examples illustrate the constant-volume and the constant-pressure processes.

Example 9.12

A rigid vessel, which has a volume of 1.0 m³, contains dry air at 30°C. If 0.2 kg of water, also at 30°C, is injected into the vessel so that the pressure becomes 100 kPa, find:

(a) the relative humidity of the mixture
(b) the minimum amount of heat that must be transferred to evaporate all the water; also, the resulting temperature and pressure
(c) the relative humidity if the vessel is heated further to 350°C

SOLUTION

(a) At 30°C, $p_{v,\,sat} = 4.246$ kPa. If all the injected water were in the vapor state, then its partial pressure would be

$$p_v = \frac{m_v R_v T}{V} = \frac{(0.2 \text{ kg})(0.46152 \text{ kJ/kg K})(303.15 \text{ K})}{1 \text{ m}^3} = 27.982 \text{ kPa}$$

But the partial pressure of water cannot exceed its saturation vapor pressure. Therefore, the vessel contains air at 100 percent relative humidity and some water in the liquid state.

$$p_a = 100 - 4.246 = 95.754 \text{ kPa}$$

and

$$m_a = \frac{(95.754 \text{ kPa})(1 \text{ m}^3)}{(0.287 \text{ kJ/kg K})(303.15 \text{ K})} = 1.101 \text{ kg}$$

(b) The specific volume of vapor $= 1/0.2 = 5 \text{ m}^3/\text{kg}$, and from steam tables this corresponds to a temperature of 70.242°C and a p_v of 31.72 kPa. The partial pressure of the air is

$$p_a = \frac{(1.101 \text{ kg})(0.287 \text{ kJ/kg K})(343.39 \text{ K})}{1.0 \text{ m}^3} = 108.51 \text{ kPa}$$

The pressure of the mixture $= p_v + p_a = 31.72 + 108.51 = 140.23$ kPa.
   The heat transfer is

$$Q = \Delta U = m_v(\Delta u)_v + m_a(c_v)_a(\Delta T)$$
$$= (0.2 \text{ kg})(2469.9 - 125.78) \text{ kJ/kg} + (1.101 \text{ kg})(0.7165 \text{ kJ/kg K})(70.242 - 30) \text{ K} = 500.57 \text{ kJ}$$

(c) At 350°C and specific volume of 5 m³/kg, the partial pressure of water is

$$p_v = \frac{(0.46152 \text{ kJ/kg K})(623.15 \text{ K})}{5 \text{ m}^3/\text{kg}} = 57.52 \text{ kPa}$$

The vapor pressure p_v can also be obtained from superheated steam tables corresponding to a specific volume of 5 m³/kg and 350°C. The value from the steam tables is $p_v = 57.5$ kPa.

$$\varphi = \left(\frac{p_v}{p_{v,\,sat}}\right)_{350°C} = \frac{57.5}{16513} = 0.348 \text{ percent}$$

Note that this example assumes no temperature change. But there is a slight change due to absorption or release of energy incurred in phase change.

Example 9.13

An air–water vapor mixture at 15°C and 60 percent relative humidity at a pressure of 101.3 kPa is heated at constant pressure to a temperature of 35°C. Determine:

(a) the initial and final specific humidities of the mixture
(b) the final relative humidity, the dew point, and the amount of heat transfer per kilogram of dry air
(c) the amount of water vapor condensed and the heat transfer per kilogram of dry air if the initial mixture is cooled isobarically to 5°C

SOLUTION

(a) At 15°C,

$$p_{v, sat} = 1.7051 \text{ kPa}$$

Hence,

$$p_v = (0.6)(1.7051 \text{ kPa}) = 1.0231 \text{ kPa}$$

The specific humidity and the vapor pressure are unchanged by heating.

$$\omega_1 = \frac{m_v}{m_a} = 0.622 \frac{p_v}{p - p_v} = 0.622 \frac{1.0231 \text{ kPa}}{(101.3 - 1.0231) \text{ kPa}}$$
$$= 0.006346 \text{ kg/kg of dry air}$$

(b) At 35°C,

$$p_{v, sat} = 5.628 \text{ kPa} \qquad \varphi = \frac{1.0231}{5.628} = 18.18 \text{ percent}$$

From steam tables at $p_v = 1.0231$ kPa, $T_{dp} = 7.26$°C.
   The heat transfer is

$$q_{1-2} = h_2 - h_1 = (c_{p_a} + \omega c_{p_v})(T_2 - T_1)$$
$$= [(1.0035 + 0.006346 \times 1.8723) \text{ kJ/kg K}](35 - 15) \text{ K}$$
$$= 20.31 \text{ kJ/kg of dry air}$$

(c) Since 5°C is below the dew point of the mixture, condensation will occur.

At 5°C,

$$p_v = p_{v, sat} = 0.8721 \text{ kPa}$$
$$\omega_3 = 0.622 \frac{0.8721 \text{ kPa}}{(101.325 - 0.8721) \text{ kPa}} = 0.005401 \text{ kg/kg of dry air}$$

$$\text{Mass of vapor condensed} = \omega_1 - \omega_3 = 0.006346 - 0.005401$$
$$= 0.000945 \text{ kg/kg of dry air}$$

Heat transfer per kg of dry air:

$$q_{1-3} = c_{p_a}(T_3 - T_1) - \omega_1 h_{g_1} + \omega_3 h_{g_3} + (\omega_1 - \omega_3)h_{f_3}$$
$$= (1.0035 \text{ kJ/kg K})(5 - 15) \text{ K} - (0.006346)(2528.9 + 0.005401 \times 2510.6) \text{ kJ/kg}$$
$$+ 0.000945(20.98 \text{ kJ/kg})$$

$$= -12.50 \text{ kJ/kg of dry air}$$

For comparison, Figure 9.18 shows the processes of this example on the psychrometric chart.

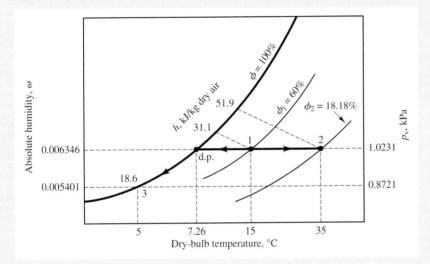

Figure 9.18

9.17.3 Humidifying Processes

Two methods are commonly used to increase the specific humidity of air: One involves adiabatic evaporation; the other involves evaporation with heat interaction. These processes are shown schematically in Figure 9.19. In the adiabatic evaporation process, a relatively dry air stream enters at state 1 and passes through a spray of water. Some of this water evaporates so that the specific humidity of the air increases. The energy for evaporation is provided by the air stream resulting in a drop in the temperature of the air mixture. During this evaporative cooling process, the wet-bulb temperature remains constant until the air mixture becomes saturated (state 2′). In evaporation accompanied by heating, water is sprayed into the air stream as heat is transferred to the system. The cooling due to evaporation opposes the rise in temperature due to heat interaction. The resultant effect may be

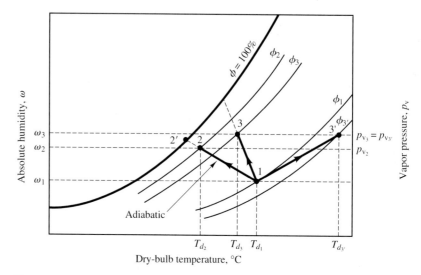

Figure 9.19   Humidification processes with and without heat transfer.

either a net reduction in temperature, as shown by process 1-3 of Figure 9.19, or a net increase in temperature, as shown by 1-3'. In process 1-3, the relative humidity increases; on the other hand, in process 1-3', the relative humidity may increase or decrease depending on $T_{d_{3'}}$.

Example 9.14

Atmospheric air at a temperature of 20°C, pressure 101.325 kPa, and relative humidity 20 percent is to be humidified in a steady-state flow process to a specific humidity of 0.006 kg of vapor/kg of dry air. Utilizing the psychrometric chart, compare the following processes with regard to the final relative humidity, final temperature, and amount of heat transfer per kilogram of dry air:

   (a) adiabatic evaporative
   (b) constant dry-bulb temperature
   (c) constant relative humidity
Assume that the humidifying water is at 20°C.

SOLUTION

The solution is shown in Figure 9.20. The heat interactions are

   (a) $q_{1-2} = 0$
   (b) $q_{1-2'} = 35.5 - 27.4 = 8.1$ kJ/kg of dry air
   (c) $q_{1-2''} = 48.2 - 27.4 = 20.8$ kJ/kg of dry air

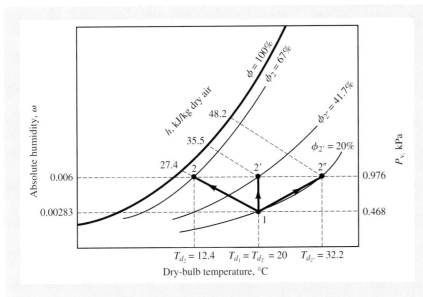

Figure 9.20

The results are tabulated in the following table:

	φ (%)	T_{db} (°C)	q (kJ/kg of dry air)
Initial conditions	20	20	—
Adiabatic process	67	12.4	0
Constant-T_{db} process	41.7	20	8.1
Constant-φ process	20	32.3	20.8

Example 9.15

Air at 1 atm, 30°C, and 60 percent relative humidity enters the dehumidifier shown in Figure 9.21 with a volumetric flow rate of 150 m³/min. The condensate and the saturated mixture leave the dehumidifier at 10°C.

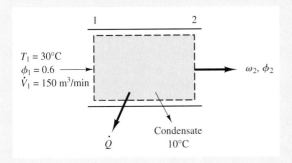

Figure 9.21

Assuming steady-state operation, determine:

(a) the rate at which water condenses
(b) the rate of heat transfer

SOLUTION

(a)

$$\omega_1 = 0.622 \frac{p_{v_1}}{101.325 - p_{v_1}} \quad \text{and} \quad \varphi_1 = 0.6 = \frac{p_{v_1}}{p_{v,\text{sat}}}$$

where $p_{v,\text{sat}} = 4.246$ kPa at 30°C. Therefore,

$$p_{v_1} = 0.6(4.246 \text{ kPa}) = 2.548 \text{ kPa}$$

and

$$\omega_1 = 0.016$$

Also

$$\omega_2 = 0.622 \frac{p_{v_2}}{101.325 - p_{v_2}}$$

where $p_{v_2} = p_{v,\text{sat}} = 1.228$ kPa at 10°C. Therefore,

$$\omega_2 = 0.008$$

and

$$(\omega_1 - \omega_2) = 0.008 \text{ kg vapor/kg dry air}$$

The amount of dry air present is determined from the ideal-gas law:

$$p_a \dot{V} = \dot{m}_{a_1} R_a T_1$$

where

$$p_a = p - p_{v_1} = 101.325 - 2.548 = 98.777 \text{ kPa}$$

Substituting values gives

$$\dot{m}_{a_1} = \frac{p_a \dot{V}}{R_a T_1} = \frac{(98.777 \text{ kPa})(150 \text{ m}^3/\text{min})}{(0.287 \text{ kJ/kg K})(303.15 \text{ K})}$$
$$= 170.30 \text{ kg dry air/min}$$

Therefore, the rate at which water condenses is

$$\dot{m}_a(\omega_1 - \omega_2) = \left(170.30\,\frac{\text{kg dry air}}{\text{min}}\right)\left(0.008\,\frac{\text{kg vapor}}{\text{kg dry air}}\right)$$

$$= 1.362 \text{ kg vapor/min}$$

(b) The rate of heat transfer is given by

$$\dot{Q} + \dot{m}_1 h_1 = \dot{m}_2 h_2 + \dot{m}_{\text{liq}} h_f$$

The enthalpy of the mixture at the inlet is equal to the sum of the enthalpies of the air and vapor components:

$$\dot{m}_1 h_1 = \dot{m}_{v_1} h_{v_1} + \dot{m}_{a_1} h_{a_1}$$

where

$$\dot{m}_{a_1} = 170.30 \text{ kg air/min}$$

and

$$\dot{m}_{v_1} = \omega_1 \dot{m}_{a_1} = (0.016)(170.3 \text{ kg/min}) = 2.725 \text{ kg vapor/min}$$

Also

$$h_{v_1} = c_{p_v} T_1 = (1.8723 \text{ kJ/kg K})(303.15 \text{ K}) = 567.59 \text{ kJ/kg vapor}$$

and

$$h_{a_1} = c_{p_{\text{air}}} T_1 = 304.21 \text{ kJ/kg}$$

Substituting values gives

$$
\begin{aligned}
\dot{m}_1 h_1 &= \dot{m}_{v_1} h_{v_1} + \dot{m}_{a_1} h_{a_1} \\
&= (2.725 \text{ kg/min})(567.59 \text{ kJ/kg}) + (170.30 \text{ kg/min})(304.21 \text{ kJ/kg}) \\
&= 53{,}353.82 \text{ kJ/min}
\end{aligned}
$$

$$\dot{m}_{\text{liq}} h_f = \left(1.362\,\frac{\text{kg vapor}}{\text{min}}\right)(42.01 \text{ kJ/kg}) = 57.22 \text{ kJ/min}$$

As with the inlet, the exit mixture is broken down into its constituents:

$$\dot{m}_2 h_2 = \dot{m}_{v_2} h_{v_2} + \dot{m}_{a_2} h_{a_2}$$

where

$$\dot{m}_{a_2} = \dot{m}_{a_1} = 170.3 \text{ kg dry air/min}$$
$$h_{a_2} = c_{p_a}T_2 = 284.14 \text{ kJ/kg}$$
$$\dot{m}_{v_2} = \dot{m}_{a_2}\omega_2 = (170.3 \text{ kg/min})(0.008) = 1.3624 \text{ kg/min}$$

Noting that

$$h_{v_2} = c_{p_v}T_2 = 530.14 \text{ kJ/kg}$$

therefore,

$$\dot{m}_2 h_2 = (1.3624 \text{ kg/min})(530.14 \text{ kJ/kg}) + (170.3 \text{ kg/min})(284.14 \text{ kJ/kg}) = 49{,}111.3 \text{ kJ/min}$$

The rate of heat interaction is

$$\dot{Q} = \dot{m}_2 h_2 + \dot{m}_{\text{liq}} h_f - \dot{m}_1 h_1$$
$$= 49{,}111.3 + 57.22 - 53{,}353.82 = -4185.3 \text{ kJ/min}$$

A cooling tower tends to operate on an adiabatic-isobaric evaporative cooling process. As shown in Figure 9.22, warm water at the top of the tower is sprayed through nozzles or over baffles to form a large water surface. Unsaturated air is

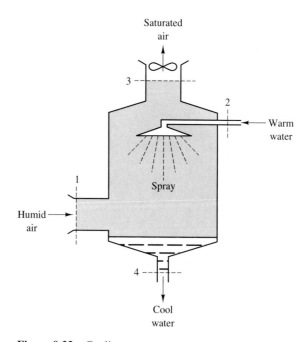

Figure 9.22   Cooling tower.

drawn upward by a fan through the tower and leaves almost saturated. The evaporation process cools the water at constant pressure with negligible heat losses. Since some of the incoming water is evaporated and carried off by the air stream, makeup water is added to balance the mass flow rate of the water.

Example 9.16

It is required to cool 4500 kg of water per hour from 45°C to 30°C in a cooling tower operating at steady state as shown in Figure 9.23. The unsaturated air mixture enters the tower at 20°C and 40 percent relative humidity and leaves at 33°C and 90 percent relative humidity. Assuming the steady-state process to be adiabatic at constant pressure, calculate the rate of air flow and the rate at which water is evaporated.

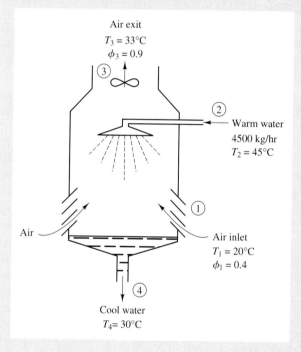

Figure 9.23

SOLUTION

At 20°C,

$$p_{v, \text{sat}_1} = 2.339 \text{ kPa}$$

therefore,

$$p_{v_1} = (0.4)(2.339 \text{ kPa}) = 0.9356 \text{ kPa}$$

$$\omega_1 = 0.622 \frac{p_v}{p - p_v} = 0.622 \frac{0.9356 \text{ kPa}}{(101.325 - 0.9356) \text{ kPa}}$$
$$= 0.0058 \text{ kg/kg of dry air}$$

At 33°C,

$$p_{v, \text{sat}_3} = 5.034 \text{ kPa}$$

therefore,

$$p_{v_3} = (0.9)(5.034 \text{ kPa}) = 4.5306 \text{ kPa}$$
$$\omega_3 = 0.622 \frac{4.5306 \text{ kPa}}{(101.325 - 4.5306) \text{ kPa}}$$
$$= 0.622 \left(\frac{4.5306}{96.7944} \right) = 0.0291 \text{ kg/kg of dry air}$$

$$\text{Mass of water evaporated} = \omega_3 - \omega_1 = 0.0291 - 0.0058$$
$$= 0.0233 \text{ kg/kg of dry air}$$

Combining the conservation-of-mass and energy equations and neglecting changes in kinetic and potential energies, gives

$$\left(h_{a_1} + \omega_1 h_{v_1} \right) + m_{f_2} h_{f_2} = \left(h_{a_3} + \omega_3 h_{v_3} \right) + m_{f_4} h_{f_4}$$

where m_f is the mass of liquid/kg of dry air. This equation can be arranged as

$$c_{p_a}(T_1 - T_3) + \omega_1 h_{g_1} + m_{f_2} h_{f_2} = \omega_3 h_{g_3} + \left[m_{f_2} - (\omega_3 - \omega_1) \right] h_{f_4}$$

Substituting values gives

$$(1.0035 \text{ kJ/kg K})(20 - 33) \text{ K} + (0.0058)(2538.1 \text{ kJ/kg}) + m_{f_2}(188.44 \text{ kJ/kg})$$
$$= (0.0291)(2561.7 \text{ kJ/kg}) + (m_{f_2} - 0.0233)(125.78 \text{ kJ/kg})$$

from which

$$m_{f_2} = 0.7 \text{ kg/kg of dry air}$$

Mass of air–water vapor mixture introduced:

$$\dot{m}_1 = \dot{m}_a(1 + \omega_1) = \left(\frac{\dot{m}_2}{m_{f_2}} \right)(1 + \omega_1) = \left(\frac{4500 \text{ kg of water/hr}}{0.7 \text{ kg of water/kg dry air}} \right) \left(1.0058 \frac{\text{kg of mixture}}{\text{kg of dry air}} \right)$$
$$= 6466.9 \text{ kg/hr}$$

Rate of evaporation of water $= \dot{m}_a(\omega_3 - \omega_1)$

$$= \left[\left(\frac{4500}{0.7}\right) \text{ kg of dry air/hr}\right]\left(0.0233 \; \frac{\text{kg of water evaporated}}{\text{kg of dry air}}\right)$$

$$= 149.79 \text{ kg/hr}$$

9.17.4 Dehumidifying Processes

Air is often subject to dehumidification processes, in which water vapor is re-moved from the air. This may be accomplished chemically by the use of hygro-scopic materials that absorb water vapor. It may also be accomplished by cooling the mixture below its dew point. The water that condenses out is then removed; the remaining mixture may then be heated to the desired temperature.

9.18 Summary

The composition of nonreactive mixtures is described by means of mass or mole fractions of each component in the mixture. Dalton's model states that the total pressure of a mixture of ideal gas is equal to the sum of the partial pressures of its components so that

$$p = \left. \sum_i p_i \right|_{V,T}$$

Amagat's model states that the total volume of the mixture is equal to the sum of the partial volumes of each component of the mixture so that

$$V = \left. \sum_i V_i \right|_{p,T}$$

For a mixture of ideal gas, the mole fraction is equal to the volume or the partial pressure fraction so that

$$x_i = \frac{n_i}{n} = \frac{V_i}{V} = \frac{p_i}{p}$$

The gas constant and the molar mass of the mixture are related to their con-stituents by the relations

$$R = \sum_i m_{f_i} R_i \qquad M = \sum_i x_i M_i$$

and

$$M = \frac{\bar{R}}{R}$$

Extensive properties of a mixture such as internal energy, enthalpy, entropy, and specific heats can be expressed per unit mass as

$$u = \sum_i m_{f_i} u_i \qquad h = \sum_i m_{f_i} h_i \qquad s = \sum_i m_{f_i} s_i$$

$$c_v = \sum_i m_{f_i} c_{v_i} \qquad c_p = \sum_i m_{f_i} c_{p_i}$$

and so on. They can also be expressed per mole as

$$\bar{u} = \sum_i x_i \bar{u}_i \qquad \bar{h} = \sum_i x_i \bar{h}_i \qquad \bar{s} = \sum_i x_i \bar{s}_i$$

$$\bar{c}_v = \sum_i x_i \bar{c}_{v_i} \qquad \bar{c}_p = \sum_i x_i \bar{c}_{p_i}$$

and so on. The entropy change due to mixing depends only on the number of moles of the constituent gases independent of their nature and is given by

$$\Delta S = -\bar{R} \sum_i n_i \ln x_i$$

Properties of real-gas mixtures can be determined by using the generalized compressibility charts. The compressibility factor of the mixture can be expressed in terms of the compressibility factors of the individual components as

$$Z = \sum_i (x_i Z_i)_{T, V} = \sum_i (x_i Z_i)_{T, p}$$

where T, V, and p are properties of the mixture. The compressibility factor of the mixture can also be determined by means of Kay's rule, which gives Z in terms of reduced temperature T/T_c' and reduced pressure p/p_c', where T_c' and p_c' are pseudocritical temperature and pseudocritical pressure, respectively.

The Gibbs phase rule gives the number of properties required to define the equilibrium state of a system in terms of the number of its components and the number of the phases P in which the components exist. It is expressed as

$$F = C - P + 2$$

Complete equilibrium of a thermodynamic system consisting of multicomponent phases requires thermal, mechanical, and chemical equilibrium. The Gibbs–Duhem equation relates these potentials and is given by

$$-S \, dT + V \, dp = \sum_{i=1}^{k} n_i \, d\mu_i$$

Terms applicable to air–water vapor mixtures are specific or absolute humidity, relative humidity, and saturation ratio, which are defined as

Absolute humidity, $\quad \omega = \dfrac{m_v}{m_a} = 0.622 \dfrac{p_v}{p - p_a}$

Relative humidity, $\quad \varphi = \left(\dfrac{p_v}{p_{v,\,sat}}\right)_T$

Saturation ratio, $\quad \psi = \left(\dfrac{\omega}{\omega_{sat}}\right)_T$

The enthalpy of atmospheric air is expressed per unit mass of dry air instead of per unit mass of air–water vapor mixture as

$$h = h_a + \omega h_g$$

The dew point is the temperature at which condensation starts if an air–water vapor mixture is cooled at constant pressure.

The relative humidity and specific humidity can be evaluated by determining the adiabatic saturation temperature. Alternatively, the wet- and dry-bulb temperatures, which can be easily measured, can be used.

Properties and processes of air–water vapor mixture at a specified total pressure can be represented on the psychrometric charts. The specific humidity is plotted versus dry-bulb temperature for different values of relative humidity. During simple heating or cooling, the temperature and the relative humidity change, but the specific humidity remains constant. Humidification and dehumidification imply, respectively, an increase and a decrease in the specific humidity. Lines of constant enthalpy and lines of wet-bulb temperatures are almost parallel on the psychrometric chart.

Problems

9.1 A mixture of ideal gases at a temperature of 20°C has the following composition by volume: $N_2 = 55$ percent, $O_2 = 20$ percent, $CH_4 = 25$ percent. If the partial pressure of CH_4 is 50 kPa, determine:
(a) the partial pressures of N_2 and O_2
(b) the mass proportions of the mixture
(c) R_{mixt} and the volume per mole of the mixture

9.2 Five moles of an ideal-gas mixture has the following mass analysis: $N_2 = 45$ percent, He $= 27$ percent, and $C_6H_6 = 28$ percent. Find:
(a) the analysis by volume and the number of moles of each constituent
(b) the molar mass of the mixture
(c) the volume of the mixture at 300 kPa and 20°C

9.3 The mass analysis of an ideal-gas mixture at 20°C and 140 kPa is 10 percent oxygen, 70 percent nitrogen, 15 percent carbon dioxide, and 5 percent carbon monoxide. Determine:
(a) the partial pressures of each constituent
(b) the gas constant of the mixture
(c) the enthalpy and internal energy of the mixture
(d) the constant-pressure specific heat of the mixture
(e) the entropy of the mixture

9.4 A rigid vessel has a volume of 1 m³ and contains air at 100 kPa and 300 K. Helium at 300 K is added to the vessel until the pressure is raised to 140 kPa. Compute:
(a) the partial pressures of the final components of the mixture comprising oxygen, nitrogen, and helium
(b) the masses of the components of the mixture
(c) the gas constant of the mixture

9.5 Prove that, for a reversible adiabatic process of a mixture of two gases a and b, the individual entropy change of a and b is zero subject to the condition that

$$\bar{c}_{p_a} = \bar{c}_{p_b} = \bar{c}_p$$

Explain.

9.6 Two ideal gases initially have the following properties:

	Volume (m³)	Pressure (kPa)	Temperature (°C)
Gas A	0.02	100	20
Gas B	0.1	30	20

If these two gases are combined in a single vessel of 0.12 m³ volume at temperature of 20°C, what will be the pressure of the mixture?

9.7 Three ideal gases initially have the following properties:

	Volume (m³)	Pressure (kPa)	Temperature (°C)
N_2	0.1	220	20
O_2	0.2	70	50
H_2	0.3	600	−10

If the foregoing gases are mixed adiabatically and the total volume remained constant, find:
(a) the final temperature and pressure of the mixture
(b) the analysis by volume and mass
(c) the partial pressures and partial volumes of the constituents
(d) the change of entropy

9.8 One kilogram of O_2 at 0.1 MPa and 40°C is mixed with 1.5 kg of N_2 at 0.1 MPa and 80°C. If the volume of the mixture is equal to the sum of the volumes of the O_2 and N_2 before mixing, compute:
(a) the temperature and pressure of the mixture
(b) the amount of heat necessary to raise the temperature of the mixture to 100°C
(c) the entropy increase due to the mixing process only
(d) the entropy increase due to the heating process

9.9 Two moles of O_2 are mixed adiabatically with 5 moles of H_2. If both gases are the same temperature and pressure before mixing, what is the change of entropy? What would be the answer if N_2 is substituted for H_2?

9.10 A mixture of gases at 30°C and 12 kPa has the following analysis by mass: $O_2 = 40$ percent, $N_2 = 52$ percent, $H_2 = 8$ percent. Calculate:
(a) the analysis in terms of mole fractions
(b) the gas constant and the molar mass of the mixture
(c) the partial pressures
(d) the density of the mixture

9.11 A mass of 1 kg of an ideal gas at 21 kPa and 550 K is mixed with 1 kg of the same gas at 7 kPa and 320 K. The mixing takes place at constant volume, and during the process the system rejects 300 kJ of heat to the surroundings, which is at 300 K. The molar mass of the gas is 32 kg/kg-mol and $\gamma = 1.33$. Determine (a) the final volume, temperature, and pressure, (b) the entropy production.

9.12 The molar analysis of a gas mixture at 300 K and 100 kPa is 70 percent N_2, 20 percent CO_2, and 10 percent O_2. Determine:
(a) the partial pressure of each component
(b) the analysis in terms of mass fractions
(c) the molar mass of the mixture
(d) the volume occupied by 5 kg of the mixture

9.13 A gas mixture consisting of 2 kg of N_2 and 3 kg of CO_2 is maintained at a pressure of 300 kPa. If the mixture is heated from 25 to 100°C at constant pressure, determine:
(a) the volumetric analysis of the mixture
(b) the heat interaction
(c) the entropy change
Assume constant specific heats.

9.14 A rigid vessel of volume 0.2 m³ contains 1.3 kg of N_2 and 1.8 kg of CO. The initial temperature of the mixture is 25°C. If the mixture is heated to 200°C, determine:
(a) the initial pressure
(b) the final pressure
(c) the heat interaction
(d) the entropy change due to heating

9.15 One kilogram of O_2 is allowed to mix with 3 kg of N_2 until an equilibrium state is attained. Prior to mixing, each gas is at 120°C and 1 atm. If the mixing is adiabatic, determine the entropy production.

9.16 The volumetric analysis of a gas mixture is 15 percent O_2, 10 percent CO_2, and 75 percent N_2. Determine:
(a) the mass analysis
(b) the gas constant and the molar mass of mixture
(c) the change in entropy per unit mass of mixture if each constituent was originally at the same pressure as the mixture, and there is no change in temperature upon mixing

9.17 The volumetric analysis of a gas mixture is 22 percent O_2, 10 percent CO_2, and 68 percent N_2. The initial temperature of the mixture is 700°C. If the mixture expands to five times its initial volume to a final temperature of 500°C, determine the change in entropy per unit mass of mixture.

9.18 The molar analysis of a gas mixture at 25°C and 100 kPa is 60 percent N_2, 25 percent CO_2, and 15 percent O_2. Determine:
(a) the analysis in terms of mass fractions
(b) the partial pressure of each component
(c) the partial volume of each component
(d) the specific volume of the mixture

9.19 Two separate streams of H_2 and CO are mixed adiabatically in a steady-state operation. There are 2 moles of H_2 for each mole of CO in order to simulate the chemical composition of methanol (CH_3OH) as an exiting mixture. If the temperature of the H_2 and CO streams are 100°C and 250°C, respectively, and the pressure is uniform throughout at 200 kPa, determine (a) the temperature of the gas mixture, (b) the entropy production per kg-mol of mixture. Neglect kinetic and potential energy effects.

9.20 The molar analysis of a gas mixture is 60 percent N_2, 22 percent CO_2, and 18 percent water vapor. The mixture enters an adiabatic turbine at 800°C and 650 kPa and expands isentropically to an exit pressure of 100 kPa. If the mass rate of flow is 1.2 kg/s, determine the exit temperature and the power developed by the turbine. Assume steady-state operation, and neglect changes in kinetic and potential energy.

9.21 A gas mixture of the same molar analysis as in the previous problem is compressed adiabatically from an initial state of 25°C and 100 kPa to a final pressure of 325 kPa. If the compression is a steady-state operation and the isentropic compressor efficiency is 70 percent, determine the temperature at the exit and the power required per kilogram of mixture.

9.22 The approximate molar analysis of dry air is N_2 = 0.7803, O_2 = 0.2099, Ar = 0.0094, CO_2 = 0.0003. There are also traces of other gases such as neon and helium. Determine (a) the molar mass and the gas constant for dry air, (b) the composition of dry air in terms of mass fractions.

9.23 A gas mixture enters a turbine operating at steady state at 1200 K and 800 kPa and expands isentropically to 100 kPa. The molar analysis of the gas mixture is 12 percent CO_2, 14 percent H_2O, and 74 percent N_2. If the mass rate of flow is 2.5 kg/s, determine the power generated by the turbine.

9.24 A mixture 0.1 kg of N_2 and 0.2 kg of CO_2, initially at a pressure of 400 kPa and a temperature of 500 K, expands in a reversible process to a pressure of 100 kPa according to the law $pV^{1.2}$ = C. Determine (a) the work interaction and heat interaction, (b) the change in entropy.

9.25   One kg-mol of a gaseous mixture consists of 0.6 mole of CO_2 and 0.4 mole of CH_4. The mixture is at a temperature of 50°C and has a molar specific volume of 0.25 m³/kg-mol. Determine the pressure of the mixture using (a) the ideal-gas equation of state, (b) Kay's rule, (c) the compressibility factor based on Dalton's model.

9.26   Compare the values of dp/dT for saturated water vapor at atmospheric pressure as obtained from the steam tables and the Clausius–Clapeyron equation.

9.27   Determine the pressure of water vapor at 90°C if h_{fg} is 2283.2 kJ/kg at this temperature. Compare your result with the steam tables.

9.28   The vapor pressure of a pure substance UDMH (N_2H_4) is given by the following equation. Assuming that the vapor behaves like an ideal gas, derive an expression of the enthalpy of vaporization. What is the value of h_{fg} at 20°C?

$$\log_{10} p = -\frac{2717.132}{T} - 6.745741 \log_{10} T + 28.000194$$

where T is in K, and p is in mm Hg.

9.29   Determine the enthalpy of vaporization for ammonia if the boiling point at 98.05 kPa is −34°C and its vapor pressure at −20°C is 190.22 kPa.

9.30   A constant-volume vessel of 0.35 m³ capacity contains an air–water vapor mixture at 30°C and 60 percent relative humidity.
(a) If the total pressure is 70 kPa, calculate the specific humidity, R_{mix}, and M_{mix}.
(b) Calculate the amount of water that should be injected into the vessel to obtain a relative humidity of 100 percent at a temperature of 50°C.

9.31   A room with a capacity of 70 m³ contains atmospheric air at 26°C and 101.325 kPa. If the specific humidity of the air is 0.01, calculate:
(a) the mass of water vapor in the room
(b) the relative humidity
(c) the vapor pressure
(d) the dew point
Check the results using the psychrometric chart.

9.32   A 0.3 m³ vessel contains a mixture of air and water vapor at 25°C and 101.3 kPa. If the relative humidity of the mixture is 70 percent, find:
(a) the specific humidity
(b) the vapor pressure
(c) the dew-point and wet-bulb temperature
(d) the mass of vapor and air
(e) the equivalent gas constant
(f) the equivalent molar mass
(g) the amount of water condensed if the mixture is cooled to 10°C in a constant-pressure process

9.33   The dry-bulb and wet-bulb temperatures in a room are 25°C and 17°C, respectively. If the pressure is 1 atm, what is the absolute humidity and relative humidity in the room?

9.34   The temperature of a glass window in a room is 15°C. If the air temperature in the room is 22°C, what is the maximum relative humidity before condensation occurs on the glass?

9.35   A mixture of nitrogen and water vapor at 60°C and 1 atm has a molar analysis of 92 percent N_2 and 8 percent H_2O. The mixture is cooled at constant pressure until condensation begins. Determine the dew-point temperature.

9.36   Atmospheric air at a temperature of 35°C and a pressure of 101.325 kPa has a dew point of 19°C. Determine the specific humidity and the relative humidity of the air. Check your results using the psychrometric chart.

9.37   Air at a temperature of 25°C and a pressure of 101.325 kPa has a relative humidity of 70 percent. Determine the dew-point and the wet-bulb temperature. Check your calculations using the psychrometric chart.

9.38   An air–water vapor mixture at a total pressure of 1 atm flows adiabatically in a duct at steady state. The wet- and dry-bulb temperatures of the mixture are 12°C and 21°C, respectively. Determine the specific humidity, the relative humidity, and the enthalpy of the mixture per kilogram of dry air. Check your calculations using the psychrometric chart.

9.39   An air–water vapor mixture at $T_{db} = 20$°C and 50 percent relative humidity is contained in a piston–

cylinder assembly at a pressure of 29.92 in. Hg abs. It is desired to add steam to the cylinder by throttling from a 140 kPa supply main until the mixture in the cylinder shows a dew-point temperature of 24°C and a φ of 70 percent, still maintaining the pressure of 29.92 in. Hg abs.
(a) How much steam must be added per kilogram of dry air?
(b) What should be the quality of the steam that is supplied from the main?

9.40   A piston–cylinder assembly is filled with an air–water vapor mixture under the following initial conditions:

Initial pressure	= 65 kPa
Initial volume	= 0.28 m³
Initial temperature	= 10°C
Initial relative humidity	= 60 percent

If this mixture is compressed adiabatically and reversibly ($pv^\gamma = C$) until the temperature reaches 50°C, determine:
(a) the mass ratio of vapor to air
(b) the total pressure of the mixture in final state
(c) the relative humidity in final state
(d) the dew point of the mixture at final state

$c_p = 1.0035$ kJ/kg K (air)

$c_p = 1.8723$ kJ/kg K (vapor)

9.41   A mixture of air and water vapor is contained in a piston–cylinder assembly under the following initial conditions:

Initial volume of the cylinder	= 0.28 m³
Initial pressure	= 65 kPa
Initial temperature	= 50°C
Initial relative humidity	= 20 percent

(a) Heat is transferred from the mixture at constant volume until the mixture temperature is reduced to 27°C. Calculate the amount of heat interaction during the process and also the final pressure in the cylinder.
(b) Starting with the initial conditions just given, water vapor at 50°C is added adiabatically to the cylinder while maintaining a constant cylinder pressure of 65 kPa until a final relative humidity of 50 percent is

obtained. What is the mass of water vapor added to the cylinder?

9.42   A closed vessel of volume 0.5 m³ initially contains dry air at a pressure of 1 atm and a temperature of 35°C. Water is introduced in the vessel until the mixture becomes saturated. If the temperature is maintained constant, determine the amount of water vapor introduced in the vessel.

If the vessel is then heated to 70°C and water is introduced to maintain saturation conditions at this temperature, determine the final pressure.

9.43   In a steady flow humidification process, water enters the humidifier at 20°C and air enters at 110 kPa, 20°C, and a specific humidity of 0.01 kg of vapor/kg of dry air. The stream leaving the apparatus has a pressure of 105 kPa, a temperature of 35°C, and a relative humidity of 90 percent. Calculate the amount of heat transfer to the humidifier per kilogram of water evaporated.

9.44   Air at $T_{db} = 55$°C and 26°C dew point, at 200 kPa, is used in a drying process. Under steady-flow equilibrium conditions, the air enters the dryer at the rate of 5000 kg/hr and leaves at 80 percent relative humidity with no change in total pressure.
(a) How much moisture is picked up by the air in kg/hr?
(b) What is the dry-bulb temperature of the leaving air?

9.45   Atmospheric air is cooled and dehumidified as it flows steadily through an air-conditioning unit. The air enters the air conditioner at 27°C and 70 percent relative humidity and leaves at 21°C and 50 percent relative humidity. If the air flow rate at the entrance is 0.5 m³/s, find:
(a) the mass rate of flow of dry air through the system in kg/s
(b) the rate at which water condenses in kg/s
(c) the rate of heat transfer in kJ/s

9.46   Air at 1 atm, 27°C, and 80 percent relative humidity enters a dehumidifier at the rate of 0.5 kg/s. The air flows over cooling coils, where it is cooled and dehumidified. The air is then heated to 27°C and 50 percent relative humidity, and the condensate is removed from the cooling coils at 12°C. Assuming steady state, determine the cooling and heating rates for this process.

9.47   Air enters an evaporative cooler at 1 atm, 35°C, and 30 percent relative humidity. Using the psychrometric chart, determine the temperature of the air at the exit if it leaves saturated.

9.48   Air enters a humidifier-heater at 10°C and 10 percent relative humidity. The air mixture leaves the heater at a temperature of 25°C and a relative humidity of 50 percent. Determine the heat transfer and the amount of water added if the dry air flow rate is 0.75 kg/s. The temperature of the added water is 15°C.

9.49   An air–water vapor mixture enters a humidifier operating at steady state. From the data indicated in Figure 9.24 and assuming all pressures to be 0.1 MPa, calculate (a) φ_3 and ω_3, (b) the rate of heat transfer.

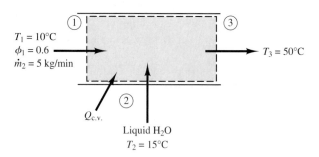

$T_1 = 10°C$
$\phi_1 = 0.6$
$\dot{m}_2 = 5$ kg/min

$T_3 = 50°C$

$Q_{c.v.}$

Liquid H_2O
$T_2 = 15°C$

Figure 9.24

9.50   An air–water vapor mixture at a pressure of 100 kPa flows through a spray chamber operating at steady state. The flow is adiabatic. From the data shown in Figure 9.25, determine ω_3 and φ_3.

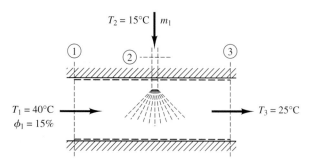

$T_2 = 15°C$ m_1

$T_1 = 40°C$
$\phi_1 = 15\%$

$T_3 = 25°C$

Figure 9.25

9.51   An air–water vapor mixture at 5°C, 100 kPa, and 50 percent relative humidity enters a heater-humidifier operating at steady state. The flow rate of the dry air is 0.1 kg/s. Liquid water at 10°C is sprayed into the mixture at the rate of 0.0022 kg/s. The mixture leaves the unit at 30°C, 100 kPa. Calculate (a) the relative humidity at the outlet, (b) the rate of heat transfer to the unit.

9.52   Air at a relative humidity of 80 percent, a temperature of 26°C, and a pressure of 1 atm enters a dehumidifier operating at steady state. The volumetric flow rate of the air–vapor mixture is 50 m³/min. If the temperature of the saturated air–water vapor mixture and the condensate is 10°C, determine (a) the amount of water condensed, (b) the heat interaction.

9.53   Atmospheric air at 15°C, 1 atm, and 50 percent relative humidity enters a humidifier operating at steady state. Liquid water at 18°C is sprayed into the humidifier. The air–water vapor mixture leaves the humidifier at 28°C, 1 atm, and 60 percent relative humidity. Determine (a) the change in the specific humidity, (b) the amount of heat transfer per kg of dry air.

9.54   Properties of two air–water vapor streams are as follows:

	V (m³/min.)	T (°C)	φ (%)
Stream 1	20	25	70
Stream 2	60	35	60

If the two streams are mixed adiabatically in a steady-state operation to form a single stream, determine the temperature, specific humidity, and relative humidity of the resulting stream. Assume the pressure of all streams is uniform at 1 atm.

9.55   Two air–water vapor streams are mixed steadily and adiabatically at a pressure of 1 atm. The first stream is at 40°C and 45 percent relative humidity. The second stream is at 16°C and 80 percent relative humidity. If the mass rates of flow of the two humid streams are 1 and 1.5 kg/s, respectively, determine the temperature, absolute humidity, and relative humidity of the resulting stream.

9.56 Atmospheric air at a dry-bulb temperature of 10°C enters a heating section of a duct and leaves at 40°C and a relative humidity of 18 percent. If the rate of heat transfer is 2 kW, determine the mass rate of flow and the relative humidity at the inlet. Assume steady-state flow.

9.57 Water flowing at a rate of 5000 gallons per hour is to be cooled from 70°C to 30°C in a cooling tower operating at a steady state. Atmospheric air at 35°C and 25 percent relative humidity is available for the cooling process. Assuming that the air leaves the cooling tower saturated with water vapor at 30°C, find the minimum air circulation rate and the makeup water per hour.

9.58 Air enters a cooling tower operating at steady state at 101.3 kPa, 35°C, 55 percent relative humidity; the air leaves the tower at 30°C and 85 percent relative humidity. Water is cooled from 42°C to 30°C in the tower. The water flow rate to the tower is 6.3 kg/s.
(a) What is the flow rate of the dry air?
(b) What fraction of the water evaporates?
(c) Could the desired cooling have been accomplished by simple heat transfer to the air in a heat exchanger? Explain.

9.59 Water is cooled from 42°C to 25°C in a cooling tower operating at steady state. The mass rate of flow is 2000 kg/hr and is maintained constant at the inlet and exit of the tower. The makeup water is supplied at 20°C. Atmospheric air enters the tower at 30°C and 40 percent relative humidity and leaves at 35°C and 90 percent relative humidity. Determine the mass rate of flow of (a) the dry air, (b) the makeup water. Neglect changes in kinetic and potential energy.

C9.1 Write a computer program to calculate the relative humidity, φ, or the humidity ratio, ω, in an air–ammonia mixture for a given atmospheric pressure, the dry-bulb temperature, T_{db}, and either φ or ω. The saturation pressure of the ammonia can be approximated by

$$\ln p_{sat} = A - \frac{B}{T_{db}} - C \ln T_{db} - DT_{db} + ET_{db}^2$$

where p_{sat} is in psia, T_{db} is in R, and $A = 58.88706$, $B = 7.5873 \times 10^3$, $C = 6.40125$, $D = 9.55176 \times 10^{-4}$, and $E = 3.3986 \times 10^{-6}$.

C9.2 Modify the program in Problem C9.1 to determine the dew-point temperature from the saturation pressure equation evaluated at the partial pressure of the ammonia. You may use any root-finding technique that is convenient.

C9.3 Write a computer program to evaluate the psychrometric properties for an air–water vapor mixture. The input to the program should be dry-bulb temperature, atmospheric pressure, and relative humidity or specific humidity. The program calculates dew-point temperature, enthalpy, specific volume, and either specific humidity or relative humidity. Use the following equations to calculate the properties:

$$p_{sat} = 0.1\exp\left[14.4351 - \frac{53333}{(T_{db} + 273.15)}\right]$$

$$0 \le T_{db} \le 38°C$$

$$T_{db} = \frac{53333}{\left[14.4351 - \ln\left(\frac{p_\omega}{0.1}\right)\right]} - 273.15$$

$$0 \le T_{db} \le 38°C$$

$$h = 1.005T_{db} + \omega(25017 + 1.82T_{db})$$

$$v = (0.286 \times 10^{-8})\frac{(T_{db} + 273.15)}{(p - p_\omega)}$$

The preceding equations are from P. E. Liley, *Mechanical Engineering News* 17(4), 19–20 (1980).

C9.4 Modify the program in Problem C9.3 to determine the properties for any combination of input data. Plot the psychrometric chart, and identify the location of the given input data. You may generate a data file and use any plotting software that is convenient.

C9.5 Using the equations given in Problem C9.3, plot three psychrometric charts covering the dry-bulb temperature ranges of 0–10°C, 10–20°C, and 20–38°C. Use an expanded scale with very finely drawn grid lines to help in the reading of the charts.

Chapter 10

Thermodynamics
of Chemical Reactions

10.1 Introduction

Thermodynamic relationships determine the equilibrium composition of product mixtures resulting from chemical reactions. The nature of the product species, their proportions, their temperature, and their pressure depend upon three major controlling factors: stoichiometry, the first law, and the second law. For one thing, the composition of the products of reaction is affected by material balance considerations. This requires that the total mass of each chemical element in the product mixture be the same as that in the reactants.[1] We shall therefore consider the calculation of elemental composition of various mixtures and examine the laws of stoichiometry. Second, the resultant temperature and pressure of product mixtures depend upon the first law of thermodynamics. This dictates that the total energy be conserved, just as the total mass is. Third, equilibrium composition is determined by relationships that arise indirectly from the second law of thermodynamics. The second law dictates that the equilibrium composition of the products depends on temperature and pressure, whereas the first law indicates that the temperature and pressure of the products depend on their composition. Clearly, then, complex interactions are involved, and problems of calculation of product composition can be handled only by considering both laws simultaneously.

[1] This discussion excludes nuclear reactions.

The energy relations developed in previous chapters are applicable to reacting systems provided that the energy terms include chemical energy to account for the changes in chemical composition. Although the principles outlined in this chapter apply to any chemical reaction, particular emphasis will be given to an important class of chemical reactions, the combustion processes. *Combustion* may be defined as an exothermic reaction between a fuel and an oxidizer in which chemical energy is converted into thermal energy. The topic of chemical kinetics, which predicts reaction rates, is not discussed in this text.

10.2 Stoichiometry and the Chemical Equation

Chemical reactions are indicated by means of chemical equations. The reaction of carbon monoxide and oxygen, for example, may be written as

$$CO(g) + \frac{1}{2}O_2(g) \rightleftharpoons CO_2(g) \tag{10.1}$$

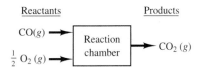

Figure 10.1 Species that enter the reaction chamber are called reactants, and species that exit are called products.

The species at the beginning of the reaction are called *reactants,* whereas those resulting from the reaction are called *products* (Figure 10.1). The arrows in Eq. (10.1) indicate the direction of the reaction. At low temperatures, the reaction proceeds from left to right; at high temperatures, the direction is reversed so that CO_2 tends to dissociate into CO and O_2. In other words, at equilibrium, the tendency of CO and O_2 to form CO_2 is just balanced by the tendency of CO_2 to dissociate to CO and O_2. The letters in parentheses following each species in Eq. (10.1) identify the phases of the various substances, whether solid (s), liquid (ℓ), or gas (g).

The proportions in which molecules of the reactants combine to form the products are also indicated in the chemical equation, and it is more convenient to work with quantities per mole rather than per unit mass. In the preceding equation, 1 mole of CO combines with $1/2$ a mole of O_2 to form 1 mole of CO_2, or 28 kg of CO combine with 16 kg of O_2 to form 44 kg of CO_2. In abbreviated form, the chemical equation may be written

$$\sum_{i=1}^{f} a_i A_i \rightleftharpoons \sum_{j=1}^{k} c_j C_j \tag{10.2}$$

where a_i and c_j are the relative molar proportions in which the various species A_i and C_j interact. The reactants and products in Eq. (10.2) are identified by subscripts $i = 1, 2 \ldots f$ and $j = 1, 2 \ldots k$, respectively. In a chemical reaction, the number of each chemical element and the total mass is conserved although the total number of moles does not necessarily remain unchanged. Mass proportions are readily obtained from molar proportions by multiplying by the corresponding molar masses.

In most combustion processes, air provides the oxygen. The ratio of nitrogen to oxygen in the atmosphere on a mole basis is approximately $0.7901/0.2099 = 3.76$; that is, every mole of oxygen is accompanied by 3.76 moles of nitrogen. On a mass basis, each kilogram of oxygen is accompanied by 3.31 kg $(0.768/0.232)$ of atmospheric nitrogen. Although the nitrogen of the air does not contribute to the oxidation reaction, that is, it is regarded as inert, it leaves the reacting system at the same temperature as the combustion products; consequently, its energy is different from that at entrance. This factor is taken into consideration in the energy balance of the combustion process.

The minimum amount of air required for the complete oxidation of a fuel is called the *theoretical* or *stoichiometric*[2] amount of air. The mass ratio of the fuel to the stoichiometric air is called the stoichiometric fuel–air ratio. The stoichiometric form of Eq. (10.2) is

$$\sum_{i=1}^{f} v_i A_i \rightleftharpoons \sum_{j=1}^{k} v_j C_j \qquad (10.3)$$

where v_i and v_j are called *stoichiometric coefficients* of the species A_i and C_j. As an example, consider the chemical equation for the stoichiometric reaction of benzene with oxygen:

$$C_6H_6(\ell) + 7.5\ O_2(g) \rightarrow 6\ CO_2(g) + 3\ H_2O(g)$$

Since each mole of oxygen, in air, is accompanied by 3.76 moles of nitrogen, the complete equation is

$$C_6H_6(\ell) + 7.5\ O_2(g) + 7.5(3.76)\ N_2(g) \rightarrow$$
$$6\ CO_2(g) + 3\ H_2O(g) + 7.5(3.76)\ N_2(g)$$

or

$$C_6H_6(\ell) + 7.5\ O_2(g) + 28.2\ N_2(g) \rightarrow$$
$$6\ CO_2(g) + 3\ H_2O(g) + 28.2\ N_2(g)$$

The preceding equation indicates that a minimum of 7.5 moles of O_2 accompanied by 28.2 moles of N_2 are necessary to oxidize completely 1 mole of C_6H_6 to form 6 moles of CO_2 and 3 moles of H_2O. Although chemists often balance

---

[2] The term *stoichiometry* comes from Greek and means "the balance of elements."

equations with whole numbers of molecules, the important feature to us is the relative amounts of reactants and products.

The combustion process entails a series of rapid chemical reactions, and the degree of mixing of the fuel and oxidizer is a controlling factor in carrying the process to completion. In an actual combustion process, the amount of air supplied may exceed or be less than the theoretical amount required. If excess air is used, oxygen will appear in the products of combustion; if insufficient air is used, the combustion is incomplete, and products, such as C, CO, OH, H_2, and unburned fuel, may also appear. In the previous example, if 120 percent theoretical air (20 percent excess air) is supplied, the foregoing equation is written

$$C_6H_6(\ell) + 1.2(7.5)\ O_2(g) + 1.2(28.2)\ N_2(g) \rightarrow$$
$$6\ CO_2(g) + 3\ H_2O(g) + 0.2(7.5)\ O_2(g) + 1.2(28.2)\ N_2(g)$$

The excess air in the reactants appears in the products as nitrogen and uncombined oxygen. In practice, some excess air is generally supplied in order to ensure that no unburned fuel is contained in the products or to control the temperature of the combustion chamber. But even with excess air, unless perfect mixing of fuel and oxidizer can be achieved, complete combustion may not occur. On the other hand, a large amount of excess air should also be avoided because the additional oxygen and nitrogen influence the composition of the products and absorb a large proportion of the chemical energy released during combustion. This decreases the temperature of the combustion products and results in a decrease in the rate of heat transfer. In general, inert species influence the equilibrium composition even though they are not involved in the chemical reaction.

The amount of air used in combustion can also be expressed in terms of the *equivalence ratio,* defined as the ratio of the actual air used to the stoichiometric amount of air.

The fuel used in combustion may be solid, liquid, or gas. Coal is a solid fuel whose composition is given in terms of the mass fractions of its chemical elements. Liquid and gaseous fuels in common use consist primarily of hydrocarbons. Liquid hydrocarbons are derived from crude oil through distillation and cracking processes, and their composition is usually given in terms of mass fractions. Gaseous hydrocarbon fuels are obtained from natural gas wells or produced chemically. Their composition is usually given in terms of the mole fractions of their constituents. A fuel often consists of many hydrocarbons, but it is convenient in many cases to express its formula in terms of an equivalent hydrogen–carbon ratio based on atomic quantities. In a typical gasoline, for example, more than 100 different hydrocarbons have been identified. Table 10.1 lists some fuels and their corresponding hydrogen–carbon ratios.

Table 10.1   Hydrogen–Carbon Ratios by Atoms for Some Fuels

Methane	CH_4	4.00	Natural gas	3.5
Ethene	C_2H_4	2.00	Gasoline	2.1
Ethane	C_2H_6	3.00	Kerosene	2.1
Propane	C_3H_8	2.67	Diesel fuel	1.8
Butane	C_4H_{10}	2.50	Heavy fuel oil	1.5
Benzene	C_6H_6	1.00	Bituminous coal	0.6
Hexane	C_6H_{14}	2.33	Crude oils	1.6–1.9
Heptane	C_7H_{16}	2.28		
Octane	C_8H_{18}	2.25		
Decane	$C_{10}H_{22}$	2.20		
Cetane	$C_{16}H_{34}$	2.12		
Ethyl alcohol	C_2H_6O	3.00		

Example 10.1

A natural gas has the following composition by volume:

Gas	Percent
Methane (CH_4)	67.4
Ethane (C_2H_6)	16.8
Propane (C_3H_8)	15.8

Calculate the equivalent hydrogen–carbon ratio for this gas by atoms and mass.

SOLUTION

		Composition by Atom	
Constituent	Mole Fraction	Carbon	Hydrogen
CH_4	0.674	0.674	2.696
C_2H_6	0.168	0.336	1.008
C_3H_8	0.158	0.474	1.264
Totals	1.000	1.484	4.968

The hydrogen–carbon ratio by atoms is $4.968/1.484 = 3.35$ atoms of hydrogen per atom of carbon. The combustion calculations can therefore be based on a hydrocarbon fuel with the chemical formula $C_{1.484}H_{4.968}$. The molar mass of the fuel is

$$(1.484 \text{ kg-mol})(12.01 \text{ kg/kg-mol}) + (4.968 \text{ kg-mol})(1.008 \text{ kg/kg-mol}) = 22.83 \text{ kg/kg-mol of fuel}$$

The hydrogen–carbon ratio by mass is

$$\frac{m_H}{m_C} = \frac{4.968(1.008)}{1.484(12.01)} = 0.281 \text{ kg of hydrogen/kg of carbon}$$

10.3 Stoichiometry Problems

Problems in stoichiometry are divided into two main categories, depending on whether the composition of the fuel or the composition of the reaction products is known.

Example 10.2

Determine the stoichiometric air required for the complete combustion of 1 kg of normal heptane $C_7H_{16}(\ell)$, as shown in Figure 10.2. What is the percentage analysis of the products on a mass and a molar basis?

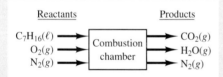

Figure 10.2

SOLUTION

The chemical equation for the combustion process is

$$C_7H_{16}(\ell) + 11\,O_2(g) + 11(3.76)\,N_2(g) \rightarrow 7\,CO_2(g) + 8\,H_2O(g) + 11(3.76)\,N_2(g)$$

The air–fuel ratio on a mass basis is

$$AF = \frac{(11 \text{ kg-mol})\left(32\,\frac{\text{kg}}{\text{kg-mol}}\right) + (11 \times 3.76 \text{ kg-mol})\left(28.013\,\frac{\text{kg}}{\text{kg-mol}}\right)}{(7 \text{ kg-mol})\left(12.01\,\frac{\text{kg}}{\text{kg-mol}}\right) + (8 \text{ kg-mol})\left(2.016\,\frac{\text{kg}}{\text{kg-mol}}\right)}$$

$$= 15.076 \text{ kg air/kg fuel}$$

Alternatively, the air–fuel ratio can be expressed in terms of the molar air–fuel ratio $(\overline{AF})$ by the following relation:

$$AF = \overline{AF}\left(\frac{M_{air}}{M_{fuel}}\right)$$

Since there are $(11 + 11 \times 3.76)$ moles of air, the air–fuel ratio is

$$AF = \frac{[(11 + 11 \times 3.76) \text{ kg-mol}]\left(28.97 \frac{\text{kg}}{\text{kg-mol}}\right)}{(1 \text{ kg-mol})\left(100.2 \frac{\text{kg}}{\text{kg-mol}}\right)} = 15.138 \text{ kg air/kg fuel}$$

The air–fuel ratio on a molar basis is

$$\overline{AF} = \frac{(11 + 11 \times 3.76) \text{ kg-mol}}{1 \text{ kg-mol}} = 52.36 \text{ moles air/mol of fuel}$$

The analysis of the products on a molar and mass basis is as follows:

	By Mole		By Mass	
CO_2	7	12.42%	$7 \times 44.01 = 308.07$	19.13%
H_2O	8	14.20%	$8 \times 18.016 = 144.128$	8.95%
N_2	41.36	73.38%	$41.36 \times 28.013 = 1158.618$	71.92%
Totals	56.36	100.00%	1610.816	100.00%

Example 10.3

Five moles of propane $C_3H_8(g)$ are completely burned at atmospheric pressure with the theoretical amount of air, as shown in Figure 10.3. Determine:
  (a) the volume of air used in the combustion process measured at 101.3 kPa and 25°C
  (b) the partial pressure of each constituent of the products
  (c) the volumetric analysis of the dry products

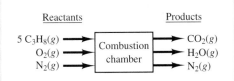

Figure 10.3

SOLUTION

(a) The balanced chemical equation for the reaction is

$$5C_3H_8(g) + 25\,O_2(g) + 25(3.76)\,N_2(g) \rightarrow 15\,CO_2(g) + 20\,H_2O(g) + 25(3.76)\,N_2(g)$$

or

$$5C_3H_8(g) + 25\,O_2(g) + 94.0\,N_2(g) \rightarrow 15\,CO_2(g) + 20\,H_2O(g) + 94.0\,N_2(g)$$

Thus, 25 moles of O_2 are required or $25(1/0.2099) = 119$ moles of air (25 moles of O_2 plus 94 moles of N_2). If air is assumed an ideal gas, then

$$V = \frac{n\overline{R}T}{p} = \frac{(119\text{ kg-mol})(8.3144\text{ kJ/kg-mol K})(25 + 273.15)\text{ K}}{101.3\text{ kPa}}$$
$$= 2912\text{ m}^3$$

(b) The partial pressures of the product constituents are proportional to the mole fractions. Therefore,

$$p_{CO_2} = \frac{15\text{ kg-mol}}{(15 + 20 + 94)\text{ kg-mol}}(101.3\text{ kPa}) = 11.779\text{ kPa}$$

$$p_{H_2O} = \frac{20\text{ kg-mol}}{129\text{ kg-mol}}(101.3\text{ kPa}) \qquad = 15.705\text{ kPa}$$

$$p_{N_2} = \frac{94\text{ kg-mol}}{129\text{ kg-mol}}(101.3\text{ kPa}) \qquad = 73.816\text{ kPa}$$

Total pressure $\qquad\qquad\qquad\qquad\qquad = \overline{101.3\text{ kPa}}$

(c)

$$\frac{V_{CO_2}}{V_{CO_2} + V_{N_2}} = \frac{v_{CO_2}}{v_{CO_2} + v_{N_2}} = \frac{15}{15 + 94} = 13.76\text{ percent}$$

$$\frac{V_{N_2}}{V_{CO_2} + V_{N_2}} = 1 - 0.1376 = 86.24\text{ percent}$$

Example 10.4

The analysis of a sample of coal gives the following values by mass (the remainder is ash):

Carbon	80.7%	Oxygen	5.3%	Sulfur	1.8%
Hydrogen	4.9%	Nitrogen	1.1%		

What is the air–fuel ratio by mass if 20 percent excess air is used in the combustion process?

SOLUTION

For 1 kg of fuel, the following table can be formulated:

	Number of Moles per kg of Fuel	Moles of O_2 Required for Complete Combustion of 1 kg of Fuel	Reaction Equation
C	$\dfrac{0.807}{12.01} = 0.0672$	0.0672	$0.0672\ C(s) + 0.0672\ O_2(g) \rightarrow 0.0672\ CO_2(g)$
H_2	$\dfrac{0.049}{2.016} = 0.0243$	0.01215	$0.0243\ H_2(g) + 0.01215\ O_2(g) \rightarrow 0.0243\ H_2O(g)$
S	$\dfrac{0.018}{32.06} = 0.00056$	0.00056	$0.00056\ S(s) + 0.00056\ O_2(g) \rightarrow 0.00056\ SO_2(g)$
O_2	$\dfrac{0.053}{32} = 0.001656$	0	
N_2	$\dfrac{0.011}{28.013} = 0.000393$	0	

$$O_2 \text{ required} = \overline{0.07991} \text{ mole}$$
$$O_2 \text{ in fuel} = 0.001656 \text{ mole}$$
$$\text{Difference} = \overline{0.07974} \text{ mole } O_2 \text{ per kg of fuel}$$

Stoichiometric air–fuel ratio $= (0.07974 \text{ kg-mol/kg})(100/20.99)(28.97 \text{ kg/kg-mol}) = 11.006 \text{ kg air/kg fuel.}$
With 20 percent excess air, the air–fuel ratio is

$$11.006 \times 1.2 = 13.207 \text{ kg air/kg fuel}$$

Example 10.5

The composition of a hydrocarbon fuel by mass is 85 percent carbon, 13 percent hydrogen, and 2 percent oxygen. Determine:
  (a) the chemically correct mass of air required for the complete combustion of 1 kg of fuel
  (b) the volumetric analysis of the products of combustion and the dew point if the total pressure is 101.3 kPa
  (c) the mass of the water vapor that condenses if the products of combustion are cooled to 15°C

SOLUTION

(a) Consider the combustion of 1 kg of fuel composed of 0.85 kg of carbon, 0.13 kg of hydrogen, and 0.02 kg of oxygen, as shown in Figure 10.4. In the complete combustion process, the carbon oxidizes to CO_2 and the hydrogen to H_2O according to the equations

$$C(s) + O_2(g) \rightarrow CO_2(g)$$

$$H_2(g) + \frac{1}{2} O_2(g) \rightarrow H_2O(g)$$

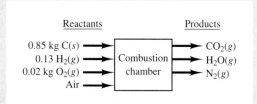

Figure 10.4

The oxygen requirement as dictated by these two equations is, however, reduced by the amount of oxygen already existing in the fuel. The number of moles of O_2 required is

$$\frac{0.85}{12.01} + \frac{0.13}{2.016 \times 2} - \frac{0.02}{32} = 0.07077 + 0.03224 - 0.000625$$

$$= 0.10239 \text{ kg-mol of } O_2/\text{kg of fuel}$$

$$\text{mass of air} = \left[\left(\frac{0.10239}{0.2099} \right) \frac{\text{kg-mol of air}}{\text{kg fuel}} \right] (28.97 \text{ kg/kg-mol air})$$

$$= 14.13 \text{ kg air/kg fuel}$$

(b) The analysis of the products of combustion by mass and volume is given in the following table:

	Moles per kg Fuel	Kilograms per kg Fuel	Molar Analysis
CO_2	$\frac{0.85}{12.01} = 0.07077$	3.11	0.136
H_2O	$\frac{0.13}{2.016} = 0.0645$	1.16	0.124
N_2	$(3.76 \times 0.10239) = 0.385$	10.79	0.740
Totals	0.52027	15.06	1.000

The dew-point temperature is the temperature at which the water vapor in the products starts to condense as the products are cooled. It is the saturation temperature corresponding to the partial pressure of the water vapor in the combustion products:

$$p_v = x_v p = 0.124(101.3 \text{ kPa}) = 12.56 \text{ kPa}$$

and from steam tables, the dew-point temperature is

$$T_{dp} = 50°C$$

(c) Since 15°C is below the dew-point temperature, some water vapor will condense. The vapor pressure is then equal to the saturation vapor pressure corresponding to 15°C.

From steam tables, $p_{v,\,sat}$ at 15°C = 1.7051 kPa and v_g = 77.93 m³/kg. The volume occupied by the water vapor is equal to the volume occupied by the CO_2 and the N_2 at their partial pressure of $(101.3 - 1.7051) = 99.5949$ kPa. The volume according to the ideal-gas law is

$$V = \frac{n\bar{R}T}{p} = \frac{[(0.07077 + 0.385) \text{ kg-mol}](8.3144 \text{ kJ/kg-mol K})(288.15 \text{ K})}{99.5949 \text{ kPa}}$$

$$= 10.963 \text{ m}^3$$

$$\text{Mass of vapor in products} = \frac{V}{v_g} = \frac{10.963 \text{ m}^3}{77.93 \text{ m}^3\text{kg}} = 0.1407 \text{ kg}$$

$$\text{Mass of vapor that condensed} = 1.16 - 0.1407 = 1.0193 \text{ kg}$$

Example 10.6

A hydrocarbon fuel in the vapor state is burned with atmospheric air at 101.3 kPa, 30°C, and 60 percent relative humidity as shown in Figure 10.5. The volumetric analysis of the dry (excluding water vapor) products of combustion is

Product	Percentage	
CO_2	10.00	
O_2	2.37	
CO	0.53	
N_2	87.10	(by difference)
	100.00	

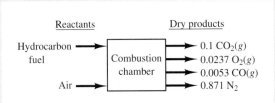

Reactants   Dry products

Hydrocarbon fuel →   Combustion chamber   → 0.1 $CO_2(g)$
                                          → 0.0237 $O_2(g)$
                                          → 0.0053 $CO(g)$
Air →   → 0.871 N_2

Figure 10.5

Calculate:

    (a) the ratio of hydrogen to carbon by mass for the fuel

    (b) the air–fuel ratio by mass

    (c) the air used as a percentage of the stoichiometric value

    (d) the volume of the humid air supplied per kilogram of fuel

SOLUTION

    (a) The following table gives the number of atoms of carbon, oxygen, and nitrogen per mole of dry products:

	Moles (per mole of dry products)	Carbon	Oxygen	Nitrogen
CO_2	0.10	0.10	0.20	—
O_2	0.0237	—	0.0474	—
CO	0.0053	0.0053	0.0053	—
N_2	0.871	—	—	1.742
Totals	1.0000	0.1053	0.2527	1.742

Assuming that H_2O is the only remaining product of combustion, the amount of hydrogen in the fuel may be determined by finding the amount of oxygen used to form the H_2O.

Total moles of O_2 supplied with 0.871 mole of $N_2 = 0.871/3.76 = 0.231$
Moles of O_2 accounted for in the reactions considered $= 0.12635$
Hence the difference used in formation of $H_2O = 0.10465$ moles of O_2
The formation of H_2O can thus be expressed as

$$0.2093\ H_2(g) + 0.10465\ O_2(g) \rightarrow 0.2093\ H_2O(g)$$

which indicates that 0.2093 mole of H_2O is formed per mole of dry products. The hydrogen–carbon ratio in the fuel is

$$\frac{H}{C} = \frac{2 \times 0.2093}{0.1053} = 3.975 \text{ (by atoms)}$$

$$= \frac{(2 \times 0.2093) \times 1.008}{(0.1053) \times 12.01} = 0.339 \text{ (by mass)}$$

    (b)

$$\text{The air–fuel ratio} = \frac{[0.871 \times (1.0/0.79)\text{ kg-mol}](28.97 \text{ kg/kg-mol})}{(0.1053 \times 12.01)\text{ kg} + (0.2093 \times 2.016)\text{ kg}}$$

$$= \frac{31.95}{1.687} = 18.9 \text{ kg air/kg fuel}$$

(c) The number of moles of oxygen required for stoichiometric combustion is

$$0.1053\ C(s) + 0.1053\ O_2(g) \rightarrow 0.1053\ CO_2(g)$$
$$0.2093\ H_2(g) + \underline{0.10465\ O_2(g)} \rightarrow 0.2093\ H_2O(g)$$
$$0.20995 \text{ moles of } O_2 \text{ per mole of dry products}$$

Since the oxygen–air ratio is constant in atmospheric air, the percentage of the air supplied is the same as that of the oxygen. Hence the percentage of the stoichiometric air used is $0.231/0.20995 = 1.1$ or 110 percent.

(d) The number of moles of dry air is $0.871(1/0.79) = 1.102$ moles air/mole dry products, but the mass of fuel $= 1.687$ kg fuel/mole dry products. Therefore, the number of moles of air per kilogram of fuel $= 1.102/1.687 = 0.653$.

At 30°C,

$$p_{v,\,sat} = 4.246 \text{ kPa} \qquad \text{and} \qquad p_v = 0.6(4.246 \text{ kPa}) = 2.5476 \text{ kPa}$$

Using the ideal-gas law and noting that the volume of the humid air is the same as that of dry air, then

$$V = \frac{n_a \overline{R} T}{p_a} = \frac{(0.653 \text{ kg-mol})(8.3144 \text{ kJ/kg-mol K})(303.15 \text{ K})}{(101.3 - 2.5476) \text{ kPa}}$$
$$= 16.667 \text{ m}^3/\text{kg of fuel}$$

Example 10.7

Propane (C_3H_8) reacts with 90 percent of theoretical air. The incomplete combustion produces CO_2, CO, H_2O, and N_2 in the products. Determine:

(a) the balanced reaction equation
(b) the air–fuel ratio on a mass basis
(c) the mass analysis of the dry products
(d) the dew point if the pressure of the products is 1 atm

SOLUTION

(a) The stoichiometric equation is

$$C_3H_8(g) + 5O_2(g) + (3.76)N_2(g) \rightarrow 3CO_2(g) + 4H_2O(g) + 5(3.76)N_2(g)$$

Incomplete combustion gives

$$C_3H_8 + 0.9(5)O_2 + 0.9(5 \times 3.76)N_2 \rightarrow a\ CO_2 + b\ CO + c\ H_2O + 0.9(5 \times 3.76)N_2$$

Applying the conservation of mass (or chemical elements) to the carbon, oxygen, and hydrogen gives

$$C: \qquad 3 = a + b$$

$$O_2: \qquad 0.9(5) = a + \frac{b}{2} + \frac{c}{2}$$

$$H_2: \qquad 4 = c$$

Solving these three equations gives

$$a = 2 \qquad b = 1 \quad \text{and} \quad c = 4$$

The balanced equation is

$$C_3H_8(g) + 4.5O_2(g) + 16.92N_2(g) \rightarrow 2CO_2(g) + CO(g) + 4H_2O(g) + 16.92N_2(g)$$

(b) The air–fuel ratio is

$$\frac{A}{F} = \frac{(4.5)(32) + (16.92)(28)}{(3 \times 12 + 8)}$$

$$= \frac{144 + 473.76}{44} = 14.04 \text{ kg of air/kg of fuel}$$

(c)

	Moles per Mol of Fuel	Mass per Mol of Fuel	Mass Analysis	Mass per Mol of Fuel (dry basis)	Mass Analysis (dry basis)
CO_2	2	$2 \times 44 = 88$	0.1333	88	0.1492
CO	1	$1 \times 28 = 28$	0.0423	28	0.0475
H_2	4	$4 \times 18 = 72$	0.1088	—	—
N_2	16.92	$16.92 \times 28 = 473.76$	0.7159	473.76	0.8033
	23.92	661.76	1.000	589.76	1.000

(d) The mol fraction of H_2O is

$$x_{H_2O} = \frac{p_{H_2O}}{p} = \frac{4}{23.92} = 0.1672$$

Therefore,

$$p_{H_2O} = (0.1672)(101.325) = 16.944 \text{ kPa}$$

From steam tables at $p = 16.944$ kPa, the dew point is

$$T_{dp} \approx 56.34°C$$

10.4 The Chemical System and Conservation of Mass

The number of independent properties that define an equilibrium state of a chemical system depends on the number of parameters necessary to define the ambient medium of the system. Two independent properties, such as temperature and pressure, in addition to parameters indicative of the degree of the reaction are sufficient to define the state of a chemical system. The degree of chemical reaction may be indicated by the number of moles of each component, the chemical potential, or by a parameter λ, called the degree of advance of the chemical reaction. If the last is used, an equation defining the state of a chemical system may be written as

$$\varphi(p, T, \lambda) = 0 \tag{10.4}$$

The advance of the reaction varies from $0(\lambda = 0)$ at the beginning of a stoichiometric reaction to unity ($\lambda = 1.0$) upon termination of the reaction. Thus, the advance of the reaction may be defined according to the following equations:

$$n_1 = n_1^\circ + v_1\lambda$$
$$n_2 = n_2^\circ + v_2\lambda$$
$$.\qquad.\qquad.$$
$$n_r = n_r^\circ + v_r\lambda$$

or, in terms of the ith component ($i = 1, 2 \dots r$),

$$n_i = n_i^\circ + v_i\lambda \tag{10.5}$$

where n_i is the number of moles of component i in the system when the advance of the reaction is λ, n_i° is the number of moles of component i at the beginning of the reaction, and v_i is the number of moles of component i in the stoichiometric reaction.[3] Summation over all components of the reaction gives

$$\sum_{i=1}^{i=r} n_i = \sum_{i=1}^{i=r} n_i^\circ + \lambda\sum_{i=1}^{i=r} v_i$$

or

$$n = n^\circ + \lambda\Delta v \tag{10.6}$$

[3] The value of v_i, as expressed in Eq. (10.5), is positive for products and negative for reactants.

where

$$n = \sum_{i=1}^{i=r} n_i \qquad n^\circ = \sum_{i=1}^{i=r} n_i^\circ \qquad \text{and} \qquad \Delta v = \sum_i v_P - \sum_i v_R$$

where P and R identify products and reactants, respectively. At the beginning of the reaction, $\lambda = 0$ and $n_i = n_i^\circ$. As the reaction progresses, λ increases continuously until one of the reactants disappears and no further reaction takes place. Thus, if component 1 disappears first, the maximum value of λ of the reaction is given by

$$\lambda_{max} = -\frac{n_1^\circ}{v_1} \qquad (10.7)$$

The relation between the number of moles of the reactants can be obtained by differentiating Eq. (10.5). Thus, the change in the number of moles of any component is given by

$$dn_i = v_i \, d\lambda \qquad (10.8)$$

and, accordingly, the proportions of the species created or destroyed have the following relation:

$$\frac{dn_1}{v_1} = \frac{dn_2}{v_2} = \frac{dn_3}{v_3} = \cdots = d\lambda \qquad (10.9)$$

The degree of advance of the reaction λ is a function of composition, which may be expressed in terms of the number of moles $n_1, n_2, \ldots, n_r$ of the components of the system. Therefore, at constant temperature and pressure, the differential of a property $X = X(p, T, \lambda)$ with respect to λ is

$$\left(\frac{\partial X}{\partial \lambda}\right)_{p,T} = \left(\frac{\partial X}{\partial n_1}\right)_{p,T,n_j} \frac{dn_1}{d\lambda} + \left(\frac{\partial X}{\partial n_2}\right)_{p,T,n_j} \frac{dn_2}{d\lambda} + \cdots \left(\frac{\partial X}{\partial n_r}\right)_{p,T,n_j} \frac{dn_r}{d\lambda}$$

But

$$\left(\frac{\partial X}{\partial n_i}\right)_{p,T,n_j} = \bar{x}_i \qquad \text{and} \qquad \frac{dn_i}{d\lambda} = v_i$$

hence,

$$\left(\frac{\partial X}{\partial \lambda}\right)_{p,T} = \bar{x}_1 v_1 + \bar{x}_2 v_2 + \cdots + \bar{x}_r v_r = \sum_{i=1}^{r} \bar{x}_i v_i \qquad (10.10)$$

Equation (10.10) expresses the change of a property X with respect to λ at constant p and T in terms of the molar properties and the stoichiometric number of moles of the components.

Equation (10.5), when multiplied by the molar mass and summed up for all the components, gives the following equation in terms of the masses of the components:

$$\sum_{i=1}^{i=r} n_i M_i = \sum_{i=1}^{i=r} n_i^\circ M_i + \sum_{i=1}^{i=r} v_i M_i \lambda$$

or

$$m = m^\circ + \sum_{i=1}^{i=r} v_i M_i \lambda \qquad (10.11)$$

Since the mass of the system is conserved in a chemical reaction, that is, $m = m^\circ$, it follows that

$$\sum_{i=1}^{i=r} v_i M_i \lambda = 0$$

The foregoing equation is also independent of the value of λ, hence,

$$\sum_{i=1}^{i=r} v_i M_i = 0 \qquad (10.12)$$

which is a mass balance of the stoichiometric equation (Eq. 10.3). Note that Eq. (10.12) substantiates the law of conservation of mass applicable to chemical reactions.

The foregoing relations were derived for a single reaction of one phase. For several reactions and phases, the procedure is to apply the preceding relations to each reaction and to every phase, and to sum up the equations over all the reactions and phases. When phase change occurs as part of the reaction, the energy transfer due to phase change must be taken into consideration.

10.5 Enthalpy of Formation

In a reacting system, the reactants and products are not only different in temperature and pressure but they also have different composition. To evaluate changes in properties of substances involved in a chemical reaction, it is therefore necessary to establish a common base to account for difference in their composition. An enthalpy scale for elements and compounds is established by choosing an arbitrary datum for zero enthalpy. It was conventionally agreed to assign a zero

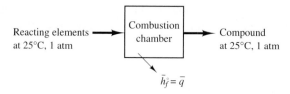

Figure 10.6 The enthalpy of formation of a compound is equal to the energy transfer as the compound is formed from stable elements at the standard state in a steady flow process.

value to the enthalpy of elements in their most stable forms at a *standard reference state* of 1 standard atmospheric pressure and 25°C. If an element has two or more stable forms, the form that is stable at 25°C is chosen as reference. The *enthalpy of formation* of a compound is defined as the enthalpy change that accompanies the formation of the compound at the standard reference state from stable elements at the same state (Figure 10.6). The enthalpy of formation per kg-mol is given the symbol $\bar{h}_f^{\circ}$, where the subscript f refers to the formation of the compound from its elements, and superscript $^{\circ}$ indicates that the property values are at the standard reference state. Thus, if elements A, B, and C at a temperature of 25°C and a pressure of 1 atm combine to form compound $A_a\, B_b\, C_c$ at the same reference state, then the chemical equation can be written

$$aA(g) + bB(g) + cC(g) \rightarrow A_a\, B_b\, C_c(g)$$

An energy balance gives

$$a\bar{h}_A^{\circ} + b\bar{h}_B^{\circ} + c\bar{h}_C^{\circ} + \Delta\bar{h}_{p0, T_0} = \bar{h}_{A_a, B_b, C_c}^{\circ}$$

where $\Delta\bar{h}_{p0, T_0}$ is the enthalpy change at the standard reference state per mole of products. By definition, $\bar{h}_A^{\circ}$, $\bar{h}_B^{\circ}$, and $\bar{h}_C^{\circ}$ are zero, and therefore, the enthalpy of formation is equal to the enthalpy of the compound due to its chemical composition at the standard reference state, or

$$\bar{h}_f^{\circ} = \bar{h}_{A_a, B_b, C_c}^{\circ} = \Delta\bar{h}_{p0, T_0}$$

A positive value of $\bar{h}_f^{\circ}$ means that the enthalpy of formation of the compound is more than the enthalpy of its elements at the reference state (endothermic reaction). Such compounds are often unstable. A negative value of $\bar{h}_f^{\circ}$ indicates that heat interaction is negative when a compound is formed from its elements in a steady flow chemical reaction.

A compound at a temperature T above 25°C will then have an enthalpy equal to the sum of its enthalpy of formation plus a *sensible enthalpy change* (associated with temperature and pressure changes at constant composition) above the standard state, or

$$\overline{h}_{T,p} = \overline{h}_f^\circ + \left(\overline{h}_{T,p} - \overline{h}_{T=25°C,p=1\,\text{atm}}\right)_{\text{const composition}} \quad (10.13)$$

Table 10.2 gives enthalpies of formation for a number of substances. Note that the two values of $\overline{h}_f^\circ$ for water given in the table are for liquid water and water

Table 10.2 Enthalpy of Formation, Gibbs Function of Formation, and Absolute Entropy at 25°C, 1 atm

Substance	Formula	$\overline{h}_f^\circ$ (kJ/kg-mol)	$\overline{g}_f^\circ$ (kJ/kg-mol)	$\overline{s}^\circ$ (kJ/kg-mol K)
Carbon	C (s)	0	0	5.740
Hydrogen	H_2 (g)	0	0	130.680
Nitrogen	N_2 (g)	0	0	191.609
Oxygen	O_2 (g)	0	0	205.147
Carbon monoxide	CO (g)	−110,527	−137,163	197.653
Carbon dioxide	CO_2 (g)	−393,522	−394,389	213.795
Water	H_2O (g)	−241,826	−228,582	188.834
Water	H_2O (ℓ)	−285,830	−237,141	69.950
Hydrogen peroxide	H_2O_2 (g)	−136,106	−105,445	232.991
Ammonia	NH_3 (g)	−45,720	−16,128	192.572
Methane	CH_4 (g)	−74,873	−50,768	186.251
Acetylene	C_2H_2 (g)	+226,731	+209,200	200.958
Ethylene	C_2H_4 (g)	+52,467	+68,421	219.330
Ethane	C_2H_6 (g)	−84,740	−32,885	229.597
Propylene	C_3H_6 (g)	+20,430	+62,825	267.066
Propane	C_3H_8 (g)	−103,900	−23,393	269.917
n-Butane	C_4H_{10} (g)	−126,200	−15,970	306.647
n-Pentane	C_5H_{12} (g)	−146.500	−8,208	348.945
Benzene	C_6H_6 (g)	+82,980	+129,765	269.562
n-Hexane	C_6H_{14} (g)	−167,300	+28	387.979
n-Heptane	C_7H_{16} (g)	−187,900	+8,227	427.805
n-Octane	C_8H_{18} (g)	−208,600	+16,660	466.514
n-Octane	C_8H_{18} (ℓ)	−250,105	+6,741	360.575
n-Dodecane	$C_{12}H_{26}$ (ℓ)	−291,010	+50,150	622.830
Methanol	CH_3OH (g)	−201,300	−162,551	239.709
Ethanol	C_2H_5OH (g)	−235,000	−168,319	282.444
Oxygen	O (g)	+249,170	+231,736	161.058
Hydrogen	H (g)	+217,999	+203,278	114.716
Nitrogen	N (g)	+472,680	+455,540	153.300
Hydroxyl	OH (g)	+38,987	+34,277	183.709

Source: Thermodynamic data in *JANAF Thermochemical Tables,* 3rd ed. Published by the American Chemical Society and the American Institute of Physics, 1985.

vapor, and the difference is equal to $\bar{h}_{fg}$ at 25°C (2442.3 kJ/kg or 44,000 kJ/kg-mol). The table also shows values of Gibbs function of formation and absolute entropy discussed in Section 10.12. Sensible enthalpies of several species are given in the Appendix, Tables A12 to A17.

10.6 Enthalpy and Internal Energy of Reaction

Chemical energy is part of the internal energy of a system and is attributed to the binding energy between the atoms of the substance. In previous chapters, chemical reactions were not considered, and consequently, chemical energy was excluded. In this chapter, however, chemical energy constitutes the major part of the internal energy term in the first law.

In chemical reactions, one important concern is the amount of heat transfer due to the reaction. Two cases will be considered: (1) when the reaction takes place at constant pressure, (2) when it takes place at constant volume. Consider, first, the constant-pressure process.

If no work other than reversible work is done, the first law of thermodynamics applied to a system is

$$\delta Q = dH - V\, dp$$

The differential of the enthalpy $H = H(p, T, \lambda)$ is

$$dH = \left(\frac{\partial H}{\partial p}\right)_{T,\lambda} dp + \left(\frac{\partial H}{\partial T}\right)_{p,\lambda} dT + \left(\frac{\partial H}{\partial \lambda}\right)_{p,T} d\lambda$$

Substituting in the first law and letting $dp = 0$ for a constant-pressure process, yields

$$\delta Q_p = \left(\frac{\partial H}{\partial T}\right)_{p,\lambda} dT + \left(\frac{\partial H}{\partial \lambda}\right)_{p,T} d\lambda$$

The coefficient of dT is

$$\left(\frac{\partial H}{\partial T}\right)_{p,\lambda} = (n\bar{c}_p)_\lambda$$

where $\bar{c}_p$ is the specific heat per mole at constant pressure. The coefficient of $d\lambda$, $(\partial H/\partial \lambda)_{p,T}$, is called the *enthalpy of reaction* and is given the symbol $\Delta H_{p,T}$. It is equal to the amount of heat transfer when the reaction is carried out at constant pressure with no change in temperature. Therefore,

$$\delta Q_p = (n\bar{c}_p)_\lambda\, dT + \Delta H_{p,T}\, d\lambda$$

Integrating the foregoing equation between a reference state (0) and any other state gives

$$Q_p = \int_{T_0}^{T} (n\bar{c}_p)_\lambda \, dT + \int_{\lambda_0}^{\lambda} \Delta H_{p,T} \, d\lambda \qquad (10.14)$$

If the reaction is stoichiometric where λ changes from 0 to 1, and if, further, the temperature at the end of the reaction is equal to the temperature at the beginning of the reaction, then

$$Q_p = \int_{0}^{1} \Delta H_{p,T} \, d\lambda = \Delta H_{p,T} \qquad (10.15)$$

which means that the heat transfer is equal to the enthalpy of reaction subject to the conditions previously stated. According to Eq. (10.10),

$$\Delta H_{p,T} = \left(\frac{\partial H}{\partial \lambda}\right)_{p,T} = \sum_i v_i \bar{h}_i = \left(\sum_i v_i \bar{h}_i\right)_P - \left(\sum_i v_i \bar{h}_i\right)_R \qquad (10.16)$$

where subscripts R and P on the right-hand side of Eq. (10.16) refer to reactants and products, respectively. Note that the enthalpy of reaction is the difference between the enthalpy of the products and the enthalpy of the reactants when complete reaction occurs, and both reactants and products are at the same temperature and pressure. In the case of *exothermic*[4] reactions, $\Delta H_{p,T}$ is negative and

$$\left(\sum_i v_i \bar{h}_i\right)_P < \left(\sum_i v_i \bar{h}_i\right)_R$$

Similarly, if $\Delta H_{p,T}$ is positive, the reaction is *endothermic* and

$$\left(\sum_i v_i \bar{h}_i\right)_P > \left(\sum_i v_i \bar{h}_i\right)_R$$

The second case to be considered is the constant-volume process. The procedure is analogous to the constant-pressure case. The first law applied to a system when only reversible work is involved is

$$\delta Q = dU + p \, dV$$

---

[4] A reaction in which heat interaction is negative is called *exothermic;* a reaction in which heat interaction is positive is called *endothermic.*

The differential of $U = U(V, T, \lambda)$ is

$$dU = \left(\frac{\partial U}{\partial V}\right)_{T,\lambda} dV + \left(\frac{\partial U}{\partial T}\right)_{V,\lambda} dT + \left(\frac{\partial U}{\partial \lambda}\right)_{V,T} d\lambda$$

Substituting for dU in the first law and considering constant volume, then

$$\delta Q_V = \left(\frac{\partial U}{\partial T}\right)_{V,\lambda} dT + \left(\frac{\partial U}{\partial \lambda}\right)_{V,T} d\lambda$$

where $(\partial U/\partial T)_{V,T} = (n\bar{c}_v)_\lambda$, being the specific heat per mole at constant volume, and $(\partial U/\partial \lambda)_{V,T} = \Delta U_{V,T}$ is called the *internal energy of reaction*.
   Therefore,

$$\delta Q_V = (n\bar{c}_v)_\lambda dT + \Delta U_{V,T} \, d\lambda \tag{10.17}$$

where $\Delta U_{V,T}$, according to Eq. (10.10), is given by

$$\Delta U_{V,T} = \left(\frac{\partial U}{\partial \lambda}\right)_{V,T} = \sum_i \nu_i \bar{u}_i = \left(\sum_i \nu_i \bar{u}_i\right)_P - \left(\sum_i \nu_i \bar{u}_i\right)_R \tag{10.18}$$

The relation between $\Delta H_{p,T}$ and $\Delta U_{V,T}$ is obtained as follows:

$$\Delta H_{p,T} = \left[\sum_i \nu_i(\bar{u}_i + p_i\bar{v}_i)\right]_P - \left[\sum_i \nu_i(\bar{u}_i + p_i\bar{v}_i)\right]_R$$

$$= \left[\left(\sum_i \nu_i\bar{u}_i\right)_P - \left(\sum_i \nu_i\bar{u}_i\right)_R\right] + \left[\sum_i \nu_i\bar{p}_i\bar{v}_i\right]_P - \left[\sum_i \nu_i\bar{p}_i\bar{v}_i\right]_R$$

$$= \Delta U_{V,T} + \left[\sum_i \nu_i p_i\bar{v}_i\right]_P - \left[\sum_i \nu_i p_i\bar{v}_i\right]_R$$

In the case of reactions involving liquids and solids, the changes in volume are small and can be neglected in comparison to gaseous substances in the reaction. If, further, the gaseous reactants and products are considered ideal gases, then $p_i\bar{v}_i = \bar{R}T$. Therefore,

$$\Delta H_{p,T} = \Delta U_{V,T} + \bar{R}T\left[\left(\sum_i \nu_i\right)_P - \left(\sum_i \nu_i\right)_R\right] \tag{10.19}$$

Note that the term in brackets is the change in the number of moles due to the reaction. Therefore, $\Delta H_{p,T} = \Delta U_{V,T}$ if there is no change in the number of moles of the gaseous substances due to the reaction.

Example 10.8

The chemical equation for n-octane when burned with oxygen at 25°C is

$$C_8H_{18}(\ell) + 12.5\,O_2(g) \rightarrow 8\,CO_2(g) + 9\,H_2O(g)$$

If the enthalpy of reaction at constant pressure $\Delta h_{p,T} = -44786$ kJ/kg, calculate the internal energy of reaction. Assume all gases to be ideal.

SOLUTION

Neglecting the volume occupied by liquid octane, the change in the number of moles is

$$\sum_i v_i = (9 + 8) - 12.5 = 4.5 \text{ moles/mol of fuel}$$

Using Eq. (10.19),

$$\Delta U_{V,T} = \Delta H_{p,T} - \bar{R}T(4.5)$$

or

$$\Delta u_{V,T} = (-44{,}786 \text{ kJ/kg}) - \frac{(8.3144 \text{ kJ/kg-mol K})(298.15 \text{ K})(4.5)}{114.232 \text{ kg/kg-mol}}$$

$$= -44{,}786 - 97.65 = -44{,}883.65 \text{ kJ/kg of fuel}$$

10.7 Enthalpy of Reaction at the Standard State

The change of enthalpy in a chemical reaction depends only on the initial and final states of the system. The difference between the energy of the reactants and products is equal to the energy transfer during the reaction. As mentioned in Section 10.5, a standard state ($p_0 = 1$ atm, $T_0 = 298.15$ K) is chosen as a basis for comparison. A more appropriate standard state would be at 0 K, but difficulties in measurement at such temperature have excluded this possibility. The *enthalpy of reaction* at the standard state is equal to the difference between the enthalpies of the reactants and products measured at the standard state (Figure 10.7), or

$$\Delta H^\circ = \left(\sum_i v_i \bar{h}_i^\circ\right)_P - \left(\sum_i v_i \bar{h}_i^\circ\right)_R \qquad (10.20)$$

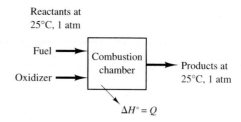

Figure 10.7   The enthalpy of reaction at the standard state is equal to the energy transfer as a fuel is burned with an oxidizer in a steady flow process forming products at the standard state.

Thus, ΔH° is in effect a comparison between the enthalpies of the reactants and products at the standard state. It is a measure of the chemical energy content of the fuel and is equal to the amount of heat transfer when the fuel is burned with oxygen at the standard reference state during a steady flow reaction. A negative value of ΔH° means that the enthalpy of the products is less than that of the reactants (exothermic reaction), which is typical in combustion processes. Table 10.3 gives

Table 10.3   Enthalpy of Combustion and Heating Values at 25°C, 1 atm

Substance	Formula	Molar Mass (kJ/kg-mol)	$\overline{h}_{fg}$ (kJ/kg-mol)	$\Delta\overline{h}^\circ = -HHV$ (kJ/kg-mol)	$\Delta\overline{h}^\circ = -LHV$ (kJ/kg-mol)
Hydrogen	H_2 (g)	2.016		−285,830	−241,826
Carbon monoxide	C (s)	12.011		−393,522	−393,522
Carbon	CO (g)	28.011		−282,995	−282,995
Methane	CH_4 (g)	16.043		−890,309	−802,301
Acetylene	C_2H_2 (g)	26.038		−1,299,605	−1,255,601
Ethylene	C_2H_4 (g)	28.054		−1,411,171	−1,323,163
Ethane	C_2H_6 (g)	30.070		−1,559,794	−1,427,782
Propylene	C_3H_6 (g)	42.081		−2,085,486	−1,926,474
Propane	C_3H_8 (g)	44.094	14,820	−2,219,986	−2,043,970
n-Butane	C_4H_{10} (g)	58.124	21,066	−2,877,038	−2,657,018
n-Pentane	C_5H_{12} (g)	72.151	26,426	−3,536,090	−3,272,066
n-Hexane	C_6H_{14} (g)	86.178	31,552	−4,194,642	−3,886,614
n-Heptane	C_7H_{16} (g)	100.205	36,547	−4,853,394	−4,501,362
n-Octane	C_8H_{18} (g)	114.232	41,484	−5,512,046	−5,116,010
Benzene	C_6H_6 (g)	78.114	33,849	−3,301,602	−3,169,590
Methanol	CH_3OH (g)	32.042	37,430	−764,882	−675,874
Ethanol	C_2H_5OH (g)	46.069	42,309	−1,409,534	−1,277,522

Note: The enthalpies of combustion were calculated using the enthalpies of formation given in Table 10.2.

enthalpies of combustion of several hydrocarbon fuels and oxygen at the standard reference state.

The enthalpy of reaction can, in many cases, be measured by direct calorimetry. It may also be calculated from the enthalpies of formation according to the equation

$$\Delta H^\circ = \left[\sum_i (H_f^\circ)_i \right]_P - \left[\sum_i (H_f^\circ)_i \right]_R \qquad (10.21)$$

which states that the enthalpy of reaction ΔH° is equal to the difference between the enthalpies of formation of all the products and all the reactants.

At any temperature T, other than the standard temperature ($T_0 = 298.15$ K), the enthalpy of reaction ΔH_T depends on the temperature of the reaction. Consider the change of state of a chemical system at atmospheric pressure from state 1 to state 2; first along path a, and second along paths b, c, and d, as shown in Figure 10.8. States 1 and 2 are at $T_0 = 298.15$ K, and states 3 and 4 are at temperature T. The process along path a is the conversion of the reactants to products at a temperature 298.15 K. The enthalpy of reaction is ΔH° and may be computed from Eq. (10.21). Since the change of enthalpy is independent of the path, the same result may be obtained by changing the state of the system from state 1 to state 2 along paths b, c, and d. The changes of enthalpies along these processes are given by

$$\left(\int_{T_0}^{T} \sum_i n_i \bar{c}_{p_i} dT \right)_R \qquad \Delta H_T \qquad \text{and} \qquad \left(\int_{T}^{T_0} \sum_i n_i \bar{c}_{p_i} dT \right)_P$$

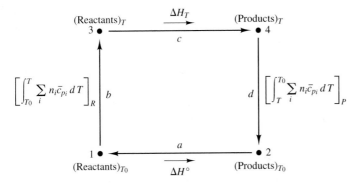

Figure 10.8   Evaluation of ΔH_T at a temperature other than T_0.

Hence

$$\Delta H_T = \left[\sum_i (n_i \bar{h}_{f_i}^\circ)_P - \sum_i (n_i \bar{h}_f^\circ)_R \right]$$

$$- \left(\int_{T_0}^{T} \sum_i n_i \bar{c}_{p_i} dT \right)_R - \left(\int_{T}^{T_0} \sum_i n_i \bar{c}_{p_i} dT \right)_P \qquad (10.22)$$

$$= \sum_i \left(n_i \bar{h}_{f_i}^\circ + \int_{T_0}^{T} n_i \bar{c}_{p_i} dT \right)_P$$

$$- \sum_i \left(n_i \bar{h}_{f_i}^\circ + \int_{T_0}^{T} n_i \bar{c}_{p_i} dT \right)_R$$

A term commonly used in combustion processes is the *heating value* of the fuel. The heating value is numerically equal to the enthalpy of combustion but of opposite sign. There are two types of heating values, depending on the phase of the water formed in the products of combustion. The *higher heating value (HHV)* is obtained when all the water vapor in the products is condensed at the reference temperature T_0. If the water is in the vapor state, the *lower heating value (LHV)* is obtained (Figure 10.9). The relation between these two values is

$$LHV = HHV - m_{H_2O} h_{fg}^\circ = HHV - (2442.3)_{25°C} m_{H_2O} \qquad (10.23)$$

where m_{H_2O} is the mass of the water vapor in the products per kg of fuel. The units of the heating values are usually in kJ/kg of fuel.

Since the change of enthalpy is independent of the number of steps or the nature of the path by which a reaction is carried out, addition or subtraction of reaction equations can be used to interrelate enthalpy changes. As an example, consider the combustion of carbon with oxygen to form carbon monoxide according to the reaction:

$$C(s) + \frac{1}{2} O_2(g) \rightarrow CO(g) \qquad (a)$$

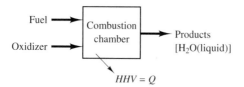

The *HHV* is equal to the heat transfer when H_2O in the products condenses to the liquid state.

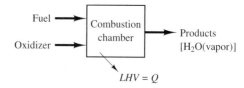

The *LHV* is equal to the heat transfer when H_2O in the products is in the vapor state.

Figure 10.9　Higher and lower heating values.

The enthalpy of combustion for this equation can be calculated from the combustion enthalpies of the following reactions:

$$C(s) + O_2(g) \rightarrow CO_2(g) \qquad \Delta\overline{h}^\circ = -393,522 \text{ kJ/kg-mol} \qquad (b)$$

$$CO(g) + \frac{1}{2}O_2(g) \rightarrow CO_2(g) \qquad \Delta\overline{h}^\circ = -282,995 \text{ kJ/kg-mol} \qquad (c)$$

Subtracting Eq. (c) from Eq. (b) gives Eq. (a), and referring to Figure 10.10, the enthalpy of reaction (a) is

$$\Delta\overline{h}_a^\circ = \Delta\overline{h}_b^\circ - \Delta\overline{h}_c^\circ = -393,522 - (-282,995)$$
$$= -110,527 \text{ kJ/kg-mol}$$

This value is the same as $\overline{h}_{fCO}^\circ$ as found in Table 10.2. As another example, consider the reaction of carbon with water vapor according to the relation

$$C(s) + H_2O(g) \rightarrow H_2(g) + CO(g) \qquad (a)$$

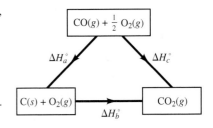

Figure 10.10 Additive property of enthalpy of reaction.

The enthalpy of reaction for this equation can be determined from a knowledge of the enthalpies of combustion of the reactions:

$$C(s) + \frac{1}{2}O_2(g) \rightarrow CO(g) \qquad \Delta\overline{h}^\circ = -110,527 \text{ kJ/kg-mol} \qquad (b)$$

$$H_2(g) + \frac{1}{2}O_2(g) \rightarrow H_2O(g) \qquad \Delta\overline{h}^\circ = -241,826 \text{ kJ/kg-mol} \qquad (c)$$

Noting that subtracting Eq. (c) from Eq. (b) yields Eq. (a), then

$$\Delta\overline{h}_a^\circ = \Delta\overline{h}_b^\circ - \Delta\overline{h}_c^\circ = -110,527 - (-241,826)$$
$$= -131,299 \text{ kJ/kg-mol}$$

A third example is the reaction of methane with oxygen according to the equation

$$CH_4(g) + 2\,O_2(g) \rightarrow CO_2(g) + 2\,H_2O(g) \qquad (a)$$
$$\Delta\overline{h}^\circ = -802,310 \text{ kJ/kg-mol}$$

Both CO_2 and H_2O can also be formed according to the reactions:

$$C(s) + O_2(g) \rightarrow CO_2(g) \qquad \Delta\overline{h}^\circ = -393,522 \text{ kJ/kg-mol} \qquad (b)$$

$$2\,H_2(g) + O_2(g) \rightarrow 2\,H_2O(g) \qquad \Delta\overline{h}^\circ = 2(-241,826) \text{ kJ/kg-mol} \qquad (c)$$

Reaction (a) can now be subtracted from the sum of reactions (b) and (c) to give

$$C(s) + 2\,H_2(g) \rightarrow CH_4(g)$$

The heat interaction in this case is equal to the difference between the enthalpy of formation of the products and reactants, and is given by

$$\Delta \overline{h}^{\circ} = \overline{h}^{\circ}_{f_{CH_4}} = [2(-241,826)] \text{ kJ/kg-mol} + [-393,522] \text{ kJ/kg-mol}$$
$$- (-802,310 \text{ kJ/kg-mol}) = -74,864 \text{ kJ/kg-mol of } CH_4$$

Example 10.9

Find the heat interaction when propane $C_3H_8(g)$ is burned according to the chemical equation

$$C_3H_8(g) + 5\ O_2(g) \rightarrow 3\ CO_2(g) + 4\ H_2O(\ell)$$

Assume both reactants and products are at a pressure of 1 atm and 25°C.

SOLUTION

The enthalpies of formation for the reactants and products are

$$\sum_{R} (H_f^{\circ})_i = (\overline{h}_f^{\circ})_{C_3H_8} + 5(\overline{h}_f^{\circ})_{O_2} = -103,900 + 0 = -103,900 \text{ kJ}$$

$$\sum_{P} (H_f^{\circ})_i = 3(\overline{h}_f^{\circ})_{CO_2} + 4(\overline{h}_f^{\circ})_{H_2O} = (3 \text{ kg-mol})(-393,522 \text{ kJ/kg-mol}) + (4 \text{ kg-mol})(-285,830 \text{ kJ/kg-mol})$$

$$= -2,323,886 \text{ kJ}$$

The heat interaction in this case is equal to the difference between the enthalpy of formation of the products and reactants, and is given by

$$\Delta H^{\circ} = Q_{p_0, T_0} = \sum_{P} (H_f^{\circ})_i - \sum_{R} (H_f^{\circ})_i = -2,323,886 - (-103,900)$$

$$= -2,219,986 \text{ kJ/kg-mol of } C_3H_8$$

Example 10.10

Calculate the enthalpy of reaction at the standard state of benzene $C_6H_6(g)$ when it reacts with oxygen at 25°C to form carbon dioxide and water vapor.

SOLUTION

$$C_6H_6(g) + 7.5\ O_2(g) \rightarrow 6\ CO_2(g) + 3\ H_2O(g)$$

The enthalpies of formation of the different species, according to Table 10.2, are

$$(\bar{h}_f^\circ)_{C_6H_6} = 82{,}980 \text{ kJ/kg-mol}$$
$$(\bar{h}_f^\circ)_{O_2} = 0$$
$$(\bar{h}_f^\circ)_{CO_2} = -393{,}522 \text{ kJ/kg-mol}$$
$$(\bar{h}_f^\circ)_{H_2O} = -241{,}826 \text{ kJ/kg-mol}$$

Therefore,

$$\Delta H^\circ = \sum_P (H_f^\circ)_i - \sum_R (H_f^\circ)_i = [(6 \text{ kg-mol})(-393{,}522 \text{ kJ/kg-mol}) + (3 \text{ kg-mol})(-241{,}826 \text{ kJ/kg-mol})]$$

$$- (82{,}980 \text{ kJ/kg-mol})$$

$$= -3{,}169{,}590 \text{ kJ/kg-mol of } C_6H_6$$

Example 10.11

Calculate the enthalpy of formation of n-heptane if the enthalpy of combustion is $\Delta \bar{h}^\circ = -4{,}853{,}394$ kJ/kg-mol. Assume the water formed in the products to be in the liquid state.

SOLUTION

The combustion equation is

$$C_7H_{16}(g) + 11 \, O_2(g) \rightarrow 7 \, CO_2(g) + 8 \, H_2O(\ell)$$

$$\Delta \bar{h}^\circ = -4{,}853{,}394 \text{ kJ/kg-mol}$$
$$\Delta H^\circ = -4{,}853{,}394 = 7\bar{h}_{f_{CO_2}}^\circ + 8\bar{h}_{f_{H_2O}}^\circ - \bar{h}_{f_{C_7H_{16}}}^\circ$$
$$= (7 \text{ kg-mol})(-393{,}522 \text{ kJ/kg-mol}) + (8 \text{ kg-mol})(-285{,}830 \text{ kJ/kg-mol}) - \bar{h}_{f_{C_7H_{16}}}^\circ$$

From which

$$\bar{h}_{f_{C_7H_{16}}}^\circ = -187{,}900 \text{ kJ/kg-mol of } C_7H_{16}$$

Example 10.12

The enthalpy of combustion of n-octane when the water formed in the products is in the liquid phase is $-48{,}253$ kJ/kg (Table 10.3). What is the value of the enthalpy of combustion if the water in the products remains in the vapor phase?

SOLUTION

The combustion equation is

$$C_8H_{18}(g) + 12.5\ O_2(g) \rightarrow 8\ CO_2(g) + 9\ H_2O(\ell)$$

$$\Delta h° = -48,253 \text{ kJ/kg of fuel}$$

The difference between the two values of enthalpies of combustion per kilogram of fuel is equal to the amount of H_2O formed per kilogram of fuel multiplied by its enthalpy of vaporization.

$$\text{Mass of } H_2O = \frac{(9 \text{ kg-mol})\left(18.016\ \dfrac{\text{kg}}{\text{kg-mol}}\right)}{(8 \text{ kg-mol})\left(12.011\ \dfrac{\text{kg}}{\text{kg-mol}}\right) + (18 \text{ kg-mol})\left(1.008\ \dfrac{\text{kg}}{\text{kg-mol}}\right)}$$

$$= \frac{162.144}{114.232} = 1.4194 \text{ kg/kg of fuel}$$

Therefore,

$$-\Delta h° \text{ with } H_2O(g) = (48,253 \text{ kJ/kg}) - 1.4194(2442.3 \text{ kJ/kg})_{25°C}$$

$$= 48,253 - 3467 = 44,786 \text{ kJ/kg of fuel}$$

10.8 First-Law Analysis for Chemically Reacting Systems

With the preceding definition of the standard state, the enthalpy of formation, and the enthalpy of reaction, the first law for steady state, neglecting kinetic and potential energy changes, may be expressed as

$$W + Q + \left[\sum_i n_i(\bar{h}_f° + \bar{h}_T - \bar{h}°)_i\right]_R = \left[\sum_i n_i(\bar{h}_f° + \bar{h}_T - \bar{h}°)_i\right]_P \quad (10.24)$$

where subscripts R and P refer to reactants and products. Alternatively, this equation can be written in terms of the enthalpy of combustion as

$$W + Q + \left[\sum_i n_i(\bar{h}_T - \bar{h}°)_i\right]_R + (-\Delta H°) = \left[\sum_i n_i(\bar{h}_T - \bar{h}°)_i\right]_P \quad (10.25)$$

Gas tables,[5] which take variations of specific heats into consideration, may be conveniently used to determine the enthalpies of the species, as the following example illustrates.

Example 10.13

One kg-mol of $C_7H_{16}(g)$ at 298.15 K is burned with 100 percent excess air at 400 K in a constant-pressure steady flow process (Figure 10.11). If the products of combustion leave at 700 K, determine the amount of heat transferred.

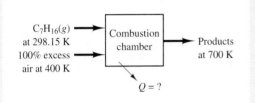

Figure 10.11

SOLUTION

The chemical equation of the reaction is

$$C_7H_{16}(g) + 2(11)\, O_2(g) + 2(11 \times 3.76)\, N_2(g)$$
$$\rightarrow 7\, CO_2(g) + 8\, H_2O(g) + 2(11 \times 3.76)\, N_2(g) + 11\, O_2(g)$$

or

$$C_7H_{16}(g) + 22\, O_2(g) + 82.72\, N_2(g) \rightarrow 7\, CO_2(g) + 8\, H_2O(g) + 82.72\, N_2(g) + 11\, O_2(g)$$

In the absence of shaft work and neglecting changes in kinetic and potential energy, the amount of heat transfer to the chemical system is given by the first law, according to the relation

$$Q = \sum_P n_i \bar{h}_i - \sum_R n_i \bar{h}_i$$

---

[5] Thermodynamic data in *JANAF Thermochemical Tables*, 3rd edition. Published by the American Chemical Society and the American Institute of Physics, 1985. See also Tables A12 to A17 in the Appendix.

$$\sum_R n_i \bar{h}_i = (\bar{h}_f^\circ)_{C_7H_{16}(g)} + 22(\bar{h}_{400} - \bar{h}_{298})_{O_2} + 82.72(\bar{h}_{400} - \bar{h}_{298})_{N_2}$$

$$= (-187{,}900 \text{ kJ}) + (22 \text{ kg-mol})(3025 \text{ kJ/kg-mol}) + (82.72 \text{ kg-mol})(2971 \text{ kJ/kg-mol})$$

$$= -187{,}900 + 66{,}550 + 245{,}761 = 124{,}411 \text{ kJ/kg-mol of fuel}$$

$$\sum_P n_i \bar{h}_i = 7(\bar{h}_f^\circ + \bar{h}_{700} - \bar{h}_{298})_{CO_2} + 8(\bar{h}_f^\circ + \bar{h}_{700} - \bar{h}_{298})_{H_2O} + 82.72(\bar{h}_{700} - \bar{h}_{298})_{N_2} + 11(\bar{h}_{700} - \bar{h}_{298})_{O_2}$$

$$= (7 \text{ kg-mol})[(-393{,}522 + 17{,}754) \text{ kJ/kg-mol}] + (8 \text{ kg-mol})[(-241{,}826 + 14{,}192) \text{ kJ/kg-mol}]$$
$$+ (82.72 \text{ kg-mol})(11{,}937 \text{ kJ/kg-mol}) + (11 \text{ kg-mol})(12{,}499 \text{ kJ/kg-mol})$$

$$= -3{,}326{,}530 \text{ kJ/kg-mol of fuel}$$

Therefore,

$$Q = -3{,}326{,}530 - 124{,}999 = -3{,}451{,}529 \text{ kJ/kg-mol of fuel}$$

The negative sign indicates that the reaction is exothermic.

10.9 Adiabatic Reaction Temperature

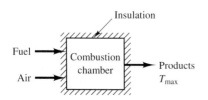

Figure 10.12   The maximum temperature occurs when the combustion is complete with no heat transfer.

Under adiabatic conditions, the temperature attained by the products is called the *adiabatic reaction (or flame) temperature.* The chemical energy released is used internally to raise the temperature of the products. The adiabatic reaction temperature is the maximum temperature that can be achieved in a chemical reaction (Figure 10.12) and, therefore, is a useful parameter in the design of combustion chambers. In actual processes, the combustion temperature is considerably lower than the adiabatic flame temperature owing to the heat interaction with the surroundings and incomplete combustion. Furthermore, dissociation of the products of combustion results in a reduction of the combustion temperature. In combustion processes, excess air is often used as a coolant to control the temperature in combustion chambers. A maximum value of the adiabatic reaction temperature is achieved when the stoichiometric amount of air is used.

Calculation of the adiabatic reaction temperature involves an iterative procedure in order to satisfy the first law which, for steady flow, reduces to

$$\sum_R n_i \bar{h}_i = \sum_P n_i \bar{h}_i \qquad (10.26)$$

The following example illustrates the application of Eq. (10.26).

Example 10.14

Calculate the adiabatic reaction temperature of $C_7H_{16}(g)$ when burned in a steady-state operation with 100 percent excess air at atmospheric pressure (Figure 10.13). Assume no dissociation and the reactants to be at 25°C.

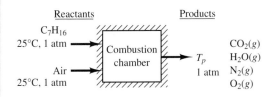

Figure 10.13

SOLUTION

The chemical equation is

$$C_7H_{16}(g) + 22\ O_2(g) + 82.72\ N_2(g) \rightarrow 7\ CO_2(g) + 8\ H_2O(g) + 82.72\ N_2(g) + 11\ O_2(g)$$

Several methods of solving this problem will be presented. The adiabatic reaction temperature will be computed, making use of (a) enthalpy of formation, (b) enthalpy of combustion, (c) specific heat equations.

(a) If enthalpies of formation are used, then

$$\sum_R n_i \bar{h}_i = (\bar{h}_f^\circ)_{C_7H_{16}(g)} = -187,900\ \text{kJ/kg-mol}$$

$$\sum_P n_i \bar{h}_i = 7(\bar{h}_f^\circ + \bar{h}_T - \bar{h}_{298})_{CO_2} + 8(\bar{h}_f^\circ + \bar{h}_T - \bar{h}_{298})_{H_2O} + 82.72(\bar{h}_T - h_{298})_{N_2} + 11(\bar{h}_T - \bar{h}_{298})_{O_2}$$

$$= (7\ \text{kg-mol})[(-393,522 + \Delta\bar{h}_{CO_2})\ \text{kJ/kg-mol}] + (8\ \text{kg-mol})\ [(-241,826 + \Delta\bar{h}_{H_2O})\ \text{kJ/kg-mol}]$$
$$+ (82.72\ \text{kg-mol})(\Delta\bar{h}_{N_2}\ \text{kJ/kg-mol}) + (11\ \text{kg-mol})(\Delta\bar{h}_{O_2}\ \text{kJ/kg-mol})$$

Assuming $T = 1500$ K,

$$\sum_P n_i \bar{h}_i = 7(-393,522 + 61,705) + 8(-241,826 + 48,151) + 82.72(38,405)$$
$$+ 11(40,399) = -250,868\ \text{as compared to } -187,900\ \text{kJ/kg-mol of fuel}$$

Assume $T = 1520$ K,

$$\sum_P n_i \bar{h}_i = 7(-393,522 + 67,878) + 8(-241,826 + 49,102) + 82.72(39,105)$$
$$+ 11(41,172) = -168,639\ \text{as compared to } -187,900\ \text{kJ/kg-mol of fuel}$$

From the preceding trials, a good approximation of T is 1515 K.

(b) If the enthalpy of combustion is used, then

$$-\Delta H^\circ = 44{,}921.53 \text{ kJ/kg of fuel} = 4{,}501{,}362 \text{ kJ/kg-mol of fuel}$$

$$\sum_P n_i \bar{h}_i = 7(\bar{h}_T - \bar{h}_{298})_{CO_2} + 8(\bar{h}_T - \bar{h}_{298})_{H_2O} + 82.72(\bar{h}_T - \bar{h}_{298})_{N_2} + 11(\bar{h}_T - \bar{h}_{298})_{O_2}$$

Assuming a temperature $T = 1515$ K gives

$$\sum_P n_i \bar{h}_i = (7 \text{ kg-mol})(62{,}585 \text{ kJ/kg-mol}) + (8 \text{ kg-mol})(48{,}865 \text{ kJ/kg-mol})$$

$$+ (82.72 \text{ kg-mol})(38{,}930 \text{ kJ/kg-mol}) + (11 \text{ kg-mol})(40{,}979 \text{ kJ/kg-mol})$$

$$= 438{,}095 + 390{,}916 + 3{,}220{,}290 + 450{,}769$$

$$= 4{,}500{,}070 \quad \text{as compared to} \quad 4{,}501{,}362 \text{ kJ/kg-mol of fuel}$$

Note that this method utilizes the enthalpy of combustion, whereas method (a) uses the enthalpies of formation. The relation between these enthalpies, according to Eq. (10.21), is clearly

$$\Delta H^\circ = \sum_P (H_f^\circ)_i - \sum_R (H_f^\circ)_i$$

(c) If the constant-pressure specific heat equations are used, then

$$-\Delta \bar{h}^\circ = 4{,}501{,}362 \text{ kJ/kg-mol of fuel}$$

$$\sum_P n_i \bar{h}_i = 7 \int_{298}^{T} \bar{c}_{pCO_2} \, dT + 8 \int_{298}^{T} \bar{c}_{pH_2O_2} \, dT + 82.72 \int_{298}^{T} \bar{c}_{pN_2} \, dT + 11 \int_{298}^{T} \bar{c}_{pO_2} \, dT$$

Substituting for the values of $\bar{c}_p$ (Table A10 in the Appendix) gives

$$\sum_P n_i \bar{h}_i = 7 \int_{298}^{T} (22.26 + 5.986 \times 10^{-2} T - 3.501 \times 10^{-5} T^2 + 7.469 \times 10^{-9} T^3) dT$$

$$+ 8 \int_{298}^{T} (32.24 + 0.1923 \times 10^{-2} T + 1.055 \times 10^{-5} T^2 - 3.595 \times 10^{-9} T^3) dT$$

$$+ 82.72 \int_{298}^{T} (28.9 - 0.1571 \times 10^{-2} T + 0.8081 \times 10^{-5} T^2 - 2.873 \times 10^{-9} T^3) dT$$

$$+ 11 \int_{298}^{T} (25.48 + 1.52 \times 10^{-2} T - 0.7155 \times 10^{-5} T^2 + 1.312 \times 10^{-9} T^3) dT$$

Assuming values of T, $\sum n_i \bar{h}_i$ for the products can be evaluated and should be equal to $\sum n_i \bar{h}_i$ for the reactants at the correct T.

Assuming $T = 1515$ K,

$$\sum_P n_i \bar{h}_i = 7\left[22.26T + \frac{5.986 \times 10^{-2}}{2}T^2 - \frac{3.501 \times 10^{-5}}{3}T^3 + \frac{7.469 \times 10^{-9}}{4}T^4\right]_{298.15}^{1515}$$

$$+ 8\left[32.24T - \frac{0.1923 \times 10^{-2}}{2}T^2 + \frac{1.055 \times 10^{-5}}{3}T^3 - \frac{3.595 \times 10^{-9}}{4}T^4\right]_{298.15}^{1515}$$

$$+ 82.72\left[28.9T - \frac{0.1571 \times 10^{-2}}{2}T^2 + \frac{0.8081 \times 10^{-5}}{3}T^3 - \frac{2.873 \times 10^{-9}}{4}T^4\right]_{298.15}^{1515}$$

$$+ 11\left[25.48T + \frac{1.52 \times 10^{-2}}{2}T^2 - \frac{0.7155 \times 10^{-5}}{3}T^3 + \frac{1.312 \times 10^{-9}}{4}T^4\right]_{298.15}^{1515}$$

$$= 4{,}504{,}783 \text{ kJ/kg-mol of fuel, which is approximately equal to } -\Delta H°$$

10.10 Second-Law Analysis for Chemically Reacting Systems

In a chemical reaction, the localized energy of the reactants is dispersed, and the products of the reaction have less energy than the reactants, the difference being thermal energy. This dispersal of energy corresponds to an increase in entropy. The second law of thermodynamics dictates that the entropy of an isolated system can never decrease; it may increase or remain unchanged. The equilibrium composition of an isolated system corresponds to the state of maximum entropy subject to the imposed constraints on the variation of the system. Therefore, as explained in Chapter 6, entropy can be used as a criterion of equilibrium for an isolated system, and if entropy generation is positive, the reaction can possibly occur.

Chemical reactions generally take place while the system thermally interacts with a surroundings at constant pressure and temperature. Furthermore, chemical reactions are not restricted to isolated systems. Therefore, entropy is not a readily applicable criterion to determine equilibrium compositions under these conditions. The Gibbs function (free energy) fills this gap and may be used to predict the spontaneity of chemical processes as well as their equilibrium compositions. As explained in Chapter 6, the spontaneous chemical reactions at a fixed temperature and pressure are associated with a decrease in the Gibbs function such that

$$(\Delta G)_{p,T} = (G_2 - G_1)_{p,T} < 0 \qquad (10.27)$$

This means that a reaction can take place spontaneously as long as the Gibbs function decreases, as shown in Figure 10.14. The equilibrium composition corresponds to a minimum value of Gibbs function so that

$$(dG)_{p,T} = 0 \qquad (10.28)$$

Another criterion of equilibrium in a chemical system is given in terms of the chemical potential. At constant temperature and pressure, the differential of the Gibbs function according to Eq. (6.84) is

$$dG_{p,T} = \sum_i \mu_i \, dn_i$$

where μ_i represents the chemical potential, and n_i represents the number of moles of the ith species. But from Eq. (10.9),

$$dn_i = \frac{\nu_i}{\nu_1} dn_1$$

where subscript 1 refers to component 1.
  Therefore,

$$dG_{p,T} = \sum_i \mu_i \frac{\nu_i}{\nu_1} dn_1$$

At equilibrium, since $dG_{p,T} = 0$, and since ν_1 and dn_1 are not zero,

$$\sum_i \nu_i \mu_i = 0 \qquad (10.29)$$

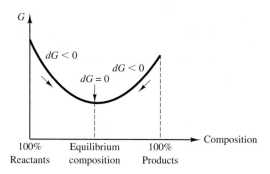

Figure 10.14   Gibbs function as a criterion for chemical equilibrium. At equilibrium, $dG_{p,T} = 0$ and $\Sigma \nu_i \mu_i = 0$.

Equation (10.29) presents the criterion for chemical equilibrium in terms of the chemical potential and the number of moles of the species in the system. The negative of $\sum_i v_i \mu_i$ is called the *affinity* of the reaction. Affinity is discussed in more detail in the following section.

10.11 Chemical Affinity

In a reversible process in which p-V work is the only work involved, the energy relationships, according to the first law of thermodynamics, are as follows:

$$\delta Q = dH - V\,dp$$

Since enthalpy is a function of temperature, pressure, and the extent of the reaction λ, enthalpy may be expressed in the form

$$dH = \left(\frac{\partial H}{\partial T}\right)_{p,\lambda} dT + \left(\frac{\partial H}{\partial p}\right)_{T,\lambda} dp + \left(\frac{\partial H}{\partial \lambda}\right)_{p,T} d\lambda$$

Each of the three coefficients of this equation can be described with equivalent terms:

$$\left(\frac{\partial H}{\partial T}\right)_{p,\lambda} = (n\bar{c}_p)_\lambda = (mc_p)_\lambda$$

$$\left(\frac{\partial H}{\partial p}\right)_{T,\lambda} = V - T\left(\frac{\partial V}{\partial T}\right)_{p,\lambda} \qquad 6$$

$$\left(\frac{\partial H}{\partial \lambda}\right)_{p,T} = \Delta H_{p,T} \qquad \text{(enthalpy of reaction)}$$

---

[6] To obtain this equivalent expression, the first and second laws are first combined in the form
$dH = T\,dS + V\,dp$.
The change of enthalpy as a function of pressure, at constant temperature, is as follows:

$$\left(\frac{\partial H}{\partial p}\right)_{T,\lambda} = T\left(\frac{\partial S}{\partial p}\right)_{T,\lambda} + V$$

But according to the Maxwell relation,

$$\left(\frac{\partial S}{\partial p}\right)_{T,\lambda} = -\left(\frac{\partial V}{\partial T}\right)_{p,\lambda}$$

Consequently,

$$\left(\frac{\partial H}{\partial p}\right)_{T,\lambda} = V - T\left(\frac{\partial V}{\partial T}\right)_{p,\lambda}$$

Therefore, enthalpy can be expressed as

$$dH = (n\bar{c}_p)_\lambda \, dT + \left[V - T\left(\frac{\partial V}{\partial T}\right)_{p,\lambda} \right] dp + \Delta H_{p,T} \, d\lambda$$

By using this expression for enthalpy, the first law then becomes

$$\delta Q = (n\bar{c}_p)_\lambda \, dT - T\left(\frac{\partial V}{\partial T}\right)_{p,\lambda} dp + \Delta H_{p,T} \, d\lambda \qquad (10.30)$$

The rate at which a reaction proceeds depends upon the concentration of the reactants. During the course of a reaction, the concentration of the reactants ordinarily decreases so that the rate of reaction constantly diminishes. Finally, when equilibrium is reached, the rates of the forward and reverse reactions are equal, and the affinity of the reaction in either direction is zero. In an irreversible process, the entropy change and the heat transfer are related through the inequality

$$T \, dS > \delta Q$$

If $\delta Q'$ represents the heat required in excess of that used if the process were reversible, then the preceding inequality can be replaced by an equality,

$$T \, dS = \delta Q + \delta Q' \qquad (10.31)$$

Thus, if $\delta Q / T$ represents the entropy increase due to the reversible portions of a process (external irreversibilities), then $\delta Q'/T$ represents the entropy increase due to the irreversibilities of the process (internal irreversibilities).

The amount of heat $\delta Q'$ in an irreversible reaction is associated with the extent of the reaction. The ratio of the differentials of these terms expresses the potential for chemical change or the affinity

$$z \equiv \frac{\delta Q'}{d\lambda} \qquad (10.32)$$

The entropy change can now be expressed, through Eq. (10.31), in terms of affinity:

$$T \, dS = \delta Q + z \, d\lambda$$

In Eq. (10.30), the heat interaction is expressed as a function of temperature, pressure, and extent of reaction. Entropy and affinity can therefore be introduced into this equation so that

$$dS = \frac{(n\bar{c}_p)_\lambda \, dT}{T} - \left(\frac{\partial V}{\partial T}\right)_{p,\lambda} dp + \left(\frac{\Delta H_{p,T} + z}{T}\right) d\lambda$$

Since entropy is a property that can be determined by temperature, pressure, and λ, it can be expressed as

$$dS = \left(\frac{\partial S}{\partial T}\right)_{p,\lambda} dT + \left(\frac{\partial S}{\partial p}\right)_{T,\lambda} dp + \left(\frac{\partial S}{\partial \lambda}\right)_{p,T} d\lambda$$

The coefficients associated with the extent of the reaction in the preceding two equations can be equated:

$$\left(\frac{\partial S}{\partial \lambda}\right)_{p,T} = \frac{\Delta H_{p,T} + z}{T}$$

From which

$$z = -\Delta H_{p,T} + T\left(\frac{\partial S}{\partial \lambda}\right)_{p,T}$$

But it was shown previously in Eq. (10.10) that entropy and the extent of the reaction λ are related as follows:

$$\left(\frac{\partial S}{\partial \lambda}\right)_{p,T} = \sum_i (v_i \bar{s}_i)_{p,T} = (S_P - S_R)_{p,T} = \Delta S_{p,T}$$

Therefore, affinity can be described by

$$z = -\Delta H_{p,T} + T\Delta S_{p,T} = -\Delta G_{p,T} = -\sum_i v_i \mu_i \qquad (10.33)$$

where $\Delta S_{p,T}$ is the difference in entropy between the products and the reactants when they are at the same pressure and temperature. Evidently, affinity indicates the decrease in Gibbs function that occurs during chemical reaction. When a system is in equilibrium, the change in the Gibbs function is zero, and thus there is no chemical affinity.

10.12 Free Energy and Chemical Equilibrium

During the course of a chemical reaction, many intermediate combinations of species are formed before the final equilibrium composition is attained. In a chemical reaction, the products can react in the reverse direction, forming reactants, just as the reactants form products in the forward direction. At equilibrium, the rates of the forward reaction and the reverse reaction are equal so that the concentration of all chemical species is unchanging. In this section, the factors governing the final composition of a chemical system are discussed, making use of the Gibbs function as a criterion for equilibrium.

Similar to the procedure used in defining the enthalpy of formation, the *Gibbs function of formation* of a compound is equal to the change in the Gibbs function for the reaction in which the compound is formed from its elements.

The Gibbs function at a state other than the standard is given by

$$G = G_f^\circ + (G_{T,p} - G_{25°C, 1 \text{ atm}}) = G_f^\circ + \Delta G \qquad (10.34)$$

Table 10.2 gives the Gibbs function of formation of selected substances at the standard reference state of 25°C and 1 atm. At this state, the stable form of each element is arbitrarily assigned a value of zero free energy of formation.

Thus when a compound is formed from its elements, the resultant free-energy change at the standard state represents the standard free energy of formation G_f° for the compound. The free-energy change of any reaction is given by

$$\Delta G^\circ = \sum_P (G_f^\circ)_i - \sum_R (G_f^\circ)_i \qquad (10.35)$$

where ΔG° is called the *standard Gibbs function of reaction* or the *standard free energy of reaction*. A negative value of ΔG° indicates that reactants at the standard state proceed spontaneously to products at the standard state. Conversely, a positive value of ΔG° indicates the reaction does not take place spontaneously.

The free-energy change of a chemical reaction under isothermal standard-state conditions may be determined according to the equation

$$\Delta G^\circ = \Delta H^\circ - T_0 \Delta S^\circ \qquad (10.36)$$

The change of entropy ΔS° is determined from the absolute entropies of the reactants and products at the standard state according to the equation

$$\Delta S^\circ = \left[\sum_i (S^\circ)_i \right]_P - \left[\sum_i (S^\circ)_i \right]_R \qquad (10.37)$$

Absolute values of entropy of several pure substances at the standard state (25°C and 1 atm pressure) are listed in Table 10.2. According to the third law of thermodynamics, the entropy of pure crystalline substances in thermodynamic equilibrium approaches zero as the temperature of the substance approaches absolute zero. Thus, the third law establishes a datum point for entropy at the absolute of zero temperature relative to which the entropy of each species in a chemical reaction can be evaluated. Note that elements as well as compounds have entropies greater than zero at 25°C and 1 atm pressure. At other temperatures and pressures, the absolute entropy of an ideal gas can be calculated from the Sackur–Tetrode equation, which is developed from statistical mechanics. When the entropy of an ideal gas is known at a certain pressure and temperature, its entropy at another set of temperature and pressure conditions can be calculated from

$$\bar{s}\,(T, p) = \bar{s}^{\circ}(T_0, p_0) + \int_{T_0}^{T} \bar{c}_p \frac{dT}{T} - \bar{R} \ln \frac{p}{p_0}$$

But, according to the third law of thermodynamics, the entropy at the reference state ($T_0 = 0$ K and $p_0 = 1$ atm) is zero. In addition, the integral in the preceding equation reduces to $\bar{s}^{\circ}(T)$ so that the absolute entropy of the ith species in a reacting mixture at a temperature T and a pressure p_i becomes

$$\bar{s}_i\,(T, p_i) = \bar{s}_i^{\circ}(T) - \bar{R} \ln p_i \qquad (10.38)$$

where $p_0 = 1$ atm, and p_i is the partial pressure of component i measured in atmospheres. Values of $\bar{s}^{\circ}(T)$ at the standard reference pressure of 1 atm are given in Tables A11 to A17 for several gases. Note that the entropy of component i is calculated at the temperature and partial pressure of the component.

As in the case of enthalpy of formation, the difference between the entropy of a compound at the standard state and the entropy of its elements at the standard state is the entropy of formation:

$$\Delta S_f^{\circ} = S_{\text{compound}}^{\circ} - \sum S_{\text{elements}}^{\circ} \qquad (10.39)$$

Example 10.15

Determine the change of entropy when a mixture of propane $C_3H_8(g)$ and oxygen $O_2(g)$ at the standard state reacts to form $CO_2(g)$ and $H_2O(g)$ at 1000 K and 2 atm (Figure 10.15). Assume steady flow combustion and both reactants and products are ideal-gas mixtures.

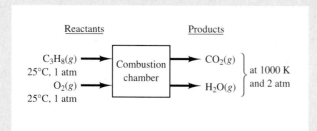

Figure 10.15

SOLUTION

The chemical equation for the complete combustion of propane is

$$C_3H_8(g) + 5\,O_2(g) \rightarrow 3\,CO_2(g) + 4\,H_2O(g)$$

The change of entropy for this reaction is

$$
\begin{aligned}
\Delta S &= \left[\sum_i n_i \bar{s}_i\right]_P - \left[\sum_i n_i \bar{s}_i\right]_R \\
&= \sum_P n_i\left[\bar{s}_i^\circ(T) - \bar{R}\ln\frac{p_i}{p_0}\right] - \sum_R n_i\left[\bar{s}_i^\circ(T) - \bar{R}\ln\frac{p_i}{p_0}\right] \\
&= \left[3\bar{s}_{CO_2}(T_P, p_{CO_2}) + 4\bar{s}_{H_2O}(T_P, p_{H_2O})\right]_P - \left[1\bar{s}_{C_3H_8}(T_R, p_{C_3H_8}) + 5\bar{s}_{O_2}(T_R, p_{O_2})\right]_R
\end{aligned}
$$

where $T_R = 298.15$ K and $T_P = 1000$ K,

$$p_{C_3H_8} = x_{C_3H_8}p = \frac{1}{6}(1\text{ atm}) \qquad p_{CO_2} = x_{CO_2}p = \frac{3}{7}(2\text{ atm})$$

$$p_{O_2} = x_{O_2}p = \frac{5}{6}(1\text{ atm}) \qquad p_{H_2O} = x_{H_2O}p = \frac{4}{7}(2\text{ atm})$$

Using Tables A14 and A15 for the products and Table 10.2 for the reactants, the entropy of each species is

$$\bar{s}_{CO_2}(T_P, p_{CO_2}) = \bar{s}_{CO_2}^\circ(T_P) - \bar{R}\ln\frac{p_{CO_2}}{p_0} = (269.299\text{ kJ/kg-mol K})$$

$$- (8.3144\text{ kJ/kg-mol K})\ln\frac{6}{7} = 270.581\text{ kJ/kg-mol K}$$

$$\bar{s}_{H_2O}(T_P, p_{H_2O}) = \bar{s}_{H_2O}^\circ(T_P) - \bar{R}\ln\frac{p_{H_2O}}{p_0} = (232.738\text{ kJ/kg-mol K})$$

$$- (8.3144\text{ kJ/kg-mol K})\ln\frac{8}{7} = 231.628\text{ kJ/kg-mol K}$$

$$\bar{s}_{C_3H_8}(T_R, p_{C_3H_8}) = \bar{s}^\circ_{C_3H_8}(T_R) - \bar{R} \ln \frac{p_{C_3H_8}}{p_0} = (269.917 \text{ kJ/kg-mol K})$$

$$- (8.3144 \text{ kJ/kg-mol K}) \ln \frac{1}{6} = 284.814 \text{ kJ/kg-mol K}$$

$$\bar{s}_{O_2}(T_R, p_{O_2}) = \bar{s}^\circ_{O_2}(T_R) - \bar{R} \ln \frac{p_{O_2}}{p_0} = (205.147 \text{ kJ/kg-mol K})$$

$$- (8.3144 \text{ kJ/kg-mol K}) \ln \frac{5}{6} = 206.663 \text{ kJ/kg-mol K}$$

The entropy change during the reaction is

$$\Delta S = [(3 \text{ kg-mol})(270.581 \text{ kJ/kg-mol K}) + (4 \text{ kg-mol})(231.628 \text{ kJ/kg-mol K})]$$
$$- [(284.814 \text{ kJ/kg-mol K}) + (5 \text{ kg-mol})(206.663 \text{ kJ/kg-mol K})] = 420.126 \text{ kJ/K per kg-mol of } C_3H_8$$

Example 10.16

Determine the irreversibility of the combustion process of Example 10.15, assuming the pressure to remain constant during the process.

SOLUTION

The irreversibility of the process is

$$I = T_0(\Delta S_{sys} + \Delta S_{surr})$$

The entropy change of the surroundings is

$$\Delta S_{surr} = \frac{Q_{surr}}{T_0} = -\frac{Q_{sys}}{T_0}$$

where T_0 is the temperature of the surroundings. But the heat transfer takes place at constant pressure so that

$$Q_{sys} = H_P - H_R = n_{CO_2}[\bar{h}^\circ_f + (\bar{h}_T - \bar{h}^\circ)]_{CO_2} + n_{H_2O}[\bar{h}^\circ_f + (\bar{h}_T - \bar{h}^\circ)]_{H_2O}$$
$$- n_{C_3H_8}[\bar{h}^\circ_f + (\bar{h}_T - \bar{h}^\circ)]_{C_3H_8} - n_{O_2}(\bar{h}_T - \bar{h}^\circ)_{O_2}$$

$$= (3 \text{ kg-mol})[(-393,522 + 33397) \text{ kJ/kg-mol}] + (4 \text{ kg-mol})[(-241,826 + 26000) \text{ kJ/kg-mol}]$$
$$- (1 \text{ kg-mol})(-103,900 \text{ kJ/kg-mol}) - (5 \text{ kg-mol})(0 \text{ kJ/kg-mol})$$

$$= -1,839,779 \text{ kJ/kg-mol of } C_3H_8$$

$$\Delta S_{surr} = \frac{1839779}{298.15} = 6170.649 \text{ kJ/kg-mol K}$$

and the irreversibility is

$$I = (298.15 \text{ K})(420.126 + 6170.649) \text{ kJ/kg-mol K}$$
$$= 1.965 \times 10^6 \text{ kJ/kg-mol of C}_3\text{H}_8$$

The Gibbs function can be expressed as

$$dG = dH - T \, dS - S \, dT$$

But it was shown that entropy could be expressed in terms of reversible heat interaction and affinity:

$$T \, dS = \delta Q + z \, d\lambda$$

From the first law of thermodynamics, this becomes

$$T \, dS = dH - V \, dp + z \, d\lambda$$

Therefore, the Gibbs function becomes

$$dG = -S \, dT + V \, dp - z \, d\lambda \tag{10.40}$$

Since the Gibbs function is a property determined by temperature, pressure, and extent of reaction,

$$\left(\frac{dG}{d\lambda}\right)_{p,T} = -z \tag{10.41}$$

But it was shown that the Gibbs function could be described by Eq. (10.10):

$$\left(\frac{dG}{d\lambda}\right)_{p,T} = \sum_i v_i \mu_i = \Delta G_P - \Delta G_R$$

These two equations, when combined, lead to

$$z = -\Delta G_{p,T} = (\Delta G_R - \Delta G_P)_{p,T}$$

which is the same as Eq. (10.33).

A relationship can be shown to exist between affinity and enthalpy of reaction. Equation (10.33) can be written as

$$z = -\Delta H_{p,T} + T\Delta S_{p,T}$$

But

$$z = -\Delta G_{p,T}$$

Also, since

$$S = -\left(\frac{\partial G}{\partial T}\right)_{p,\lambda}$$

therefore,

$$\Delta S_{p,T} = -\left(\frac{\partial \Delta G}{\partial T}\right)_{p,\lambda}$$

Substitution from these equations leads to

$$\Delta G_{p,T} = \Delta H_{p,T} + T\left(\frac{\partial \Delta G}{\partial T}\right)_{p,\lambda} \tag{10.42}$$

which is called the *Gibbs–Helmholtz* relation. It shows the effect of changes in temperature on the Gibbs function.

Now Eq. (10.42) may be transformed into an integrable form. First, Eq. (10.33) is expressed as

$$\frac{z}{T} = -\frac{\Delta H_{p,T}}{T} + \Delta S_{p,T}$$

Differentiation with respect to temperature leads to

$$\frac{\partial}{\partial T}\left(\frac{z}{T}\right)_{p,\lambda} = \left(\frac{\Delta H_{p,T}}{T^2}\right)_{p,\lambda} - \frac{1}{T}\left(\frac{\partial \Delta H_{p,T}}{\partial T}\right)_{p,\lambda} + \left(\frac{\partial \Delta S_{p,T}}{\partial T}\right)_{p,\lambda}$$

But from the first and second laws of thermodynamics, it can be shown that

$$\left(\frac{\partial \Delta S_{p,T}}{\partial T}\right)_{p,\lambda} = \frac{1}{T}\left(\frac{\partial \Delta H_{p,T}}{\partial T}\right)_{p,\lambda}$$

Therefore,

$$\frac{\partial}{\partial T}\left(\frac{z}{T}\right)_{p,\lambda} = \left(\frac{\Delta H_{p,T}}{T^2}\right)_{p,\lambda}$$

which, when integrated, gives

$$\frac{z}{T} = \int \frac{\Delta H_{p,T}}{T^2} dT + C \qquad (dp = 0 \text{ and } d\lambda = 0) \qquad (10.43)$$

where C is a constant of integration.

Example 10.17

Determine the Gibbs function of formation of H_2O at the standard state.

SOLUTION

Consider the reaction

$$H_2(g) + \frac{1}{2}O_2(g) \rightarrow H_2O(g)$$

When both reactants and product are at the standard state, the change of the Gibbs function is equal to the reversible work, or

$$\Delta G^\circ = W_{rev} = \Delta H^\circ - T_0 \Delta S^\circ$$
$$= (H_P^\circ - H_R^\circ) - T_0(S_P^\circ - S_R^\circ)$$

where

$$\Delta G^\circ = \sum_P n_i (\bar{g}_f^\circ)_i - \sum_R n_i (\bar{g}_f^\circ)_i$$

Since the Gibbs function and the enthalpy of formation of elements at the standard state is zero,

$$W_{rev} = \Delta G^\circ = \sum_P n_i (\bar{g}_f^\circ)_i = (\bar{g}_f^\circ)_{H_2O}$$

and

$$\Delta H^\circ = \sum_P n_i (\bar{h}_f^\circ)_i - \sum_R n_i (\bar{h}_f^\circ)_i$$
$$= (\bar{h}_f^\circ)_{H_2O} = -241,826 \text{ kJ/kg-mol}$$

The entropy change at the standard state is

$$\Delta S^\circ = \sum_P n_i \bar{s}_i^\circ - \sum_R n_i \bar{s}_i^\circ = (1 \text{ kg-mol})(188.834 \text{ kJ/kg-mol K}) - [(1)(130.68) + (0.5)(205.147) \text{ kJ/K}]$$

$$= -44.42 \text{ kJ/K (for 1 kg-mol of H}_2\text{O)}$$

Therefore,

$$(\bar{g}_f^\circ)_{H_2O} = -241,826 - 298.15(-44.42) = -228,582 \text{ kJ/kg-mol of H}_2\text{O}$$

which agrees with the value given in Table 10.2.

Example 10.18

Ascertain that at an equilibrium pressure of 1 atm the vaporization temperature of water is 100°C.

SOLUTION

Consider the vaporization of water at temperatures near its normal boiling point. At one atmosphere,

$$H_2O(\ell) \rightarrow H_2O(g)$$
$$\Delta h = h_{fg} = 2256.8 \text{ kJ/kg}$$

and

$$\Delta s = s_{fg} = 6.0480 \text{ kJ/kg K}$$

For the system to reach equilibrium, its free energy must decrease toward a minimum value. This means that when Δg is negative, the evaporation process will be spontaneous:

$$\Delta g = \Delta h - T\Delta s < 0 \qquad \text{or} \qquad T\Delta s > \Delta h$$

so that

$$T > \frac{\Delta h}{\Delta s} \qquad \text{or} \qquad T > \frac{2256.8}{6.0480} = 373.15 \text{ K}$$

If, however, Δg is positive, then

$$T\Delta s < \Delta h \qquad \text{and} \qquad T < \frac{\Delta h}{\Delta s} \qquad \text{or} \qquad T < 373.15 \text{ K}$$

This means that as long as the temperature is below the boiling point, the vaporization process at 1 atm pressure is not spontaneous. Note that for the reverse process of condensation,

$$H_2O(g) \longrightarrow H_2O(\ell)$$

the change of the Gibbs function for $T < 373.15$ K is negative, and the condensation of superheated vapor proceeds spontaneously. When $T\Delta s = \Delta h$, $\Delta g = 0$ so that vaporization or condensation at 373.15 K and 1 atm involves no change in the Gibbs function, and the process is therefore reversible. Note also that the difference in the Gibbs function is the driving force for phase change.

10.13 Equilibrium Constant

This section outlines the condition of equilibrium for a reaction given by the equation

$$v_a A + v_b B + \cdots \longrightarrow v_c C + v_d D + \cdots$$

The differential of the Gibbs function is

$$dG = V\,dp - S\,dT$$

Two conditions are now imposed on the preceding expression. The first is that either the reaction is isothermal, that is, the temperature during the reaction remains unchanged, or, alternatively, the temperature is the same at only the beginning and end of the reaction. The second condition is that only ideal gases are involved. The differential of the Gibbs function then becomes

$$dG = V\,dp = \frac{n\bar{R}T}{p}\,dp$$

Integrating the foregoing equation from the standard state to any other state at the same temperature gives

$$G(T, p) - G(T, p_0) = n\bar{R}T \ln \frac{p}{p_0} \tag{10.44}$$

The Gibbs function $G(T, p_0)$ is denoted by $G°(T)$, and the reference pressure p_0 is assigned the value of 1 atm. $G°(T)$ is called the *free energy at the standard state* reference pressure of 1 atm. Equation (10.44) for the ith species can then be written as

$$\bar{g}_i(T, p_i) = \bar{g}_i^\circ(T) + \bar{R}T \ln p_i \qquad (10.45)$$

where p_i is the partial pressure of species i measured in atmospheres.

The change of the Gibbs function, when reactants $A, B, \ldots$, are transformed to products $C, D, \ldots$, in a stoichiometric reaction is

$$\Delta G = \sum_P G_i - \sum_R G_i$$
$$= (v_c \bar{g}_c + v_d \bar{g}_d + \cdots) - (v_a \bar{g}_a + v_b \bar{g}_b + \cdots)$$

where v's are the stoichiometric number of moles of the chemical species. Substitution of the expression for the Gibbs function for each species from Eq. (10.45) gives

$$\Delta G = (v_c \bar{g}_c^\circ + v_d \bar{g}_d^\circ + \cdots) - (v_a \bar{g}_a^\circ + v_b \bar{g}_b^\circ + \cdots) + (v_c \bar{R}T \ln p_c + v_d \bar{R}T \ln p_d + \cdots) - (v_a \bar{R}T \ln p_a + v_b \bar{R}T \ln p_b + \cdots)$$

But the standard-state Gibbs function change is

$$\Delta G^\circ = (v_c \bar{g}_c^\circ + v_d \bar{g}_d^\circ + \cdots) - (v_a \bar{g}_a^\circ + v_b \bar{g}_b^\circ + \cdots)$$

then

$$\Delta G = \Delta G^\circ + \bar{R}T \ln \left[\frac{p_C^{v_c} p_D^{v_d} \cdots}{p_A^{v_a} p_B^{v_b} \cdots} \right] \qquad (10.46)$$

At equilibrium, $dG = 0$ and the change of the Gibbs function at the standard state may then be expressed as

$$-\Delta G^\circ(T) = z^\circ(T) = \bar{R}T \ln \left[\frac{p_C^{v_c} p_D^{v_d} \cdots}{p_A^{v_a} p_B^{v_b} \cdots} \right] = \bar{R}T \ln K_p \qquad (10.47)$$

K_p is called the *equilibrium constant* for the chemical equilibrium of ideal-gas mixtures defined by the equation

$$K_p = e^{-\Delta G^\circ/\bar{R}T} = e^{z^\circ/\bar{R}T} = \frac{p_C^{v_c} p_D^{v_d} \cdots}{p_A^{v_a} p_B^{v_b} \cdots} \qquad (10.48)$$

The subscript p is to emphasize that K_p is defined in terms of the partial pressure of reactants and products. Note that $\Delta G°$ and K_p are independent of the total pressure. They depend only on temperature and the nature of the reactants and products of the reaction. The mixture pressure, however, affects the equilibrium composition of the mixture. Recall also that in choosing the standard state, the unit of pressure for the components was expressed in atmospheres. Therefore, the value of K_p is based on this unit of pressure. In the preceding definition of K_p, the partial pressures of the products are placed in the numerator, and the partial pressures of the reactants are placed in the denominator. When the reaction is reversed, the value of the equilibrium constant becomes the reciprocal of that of the forward reaction. Values of K_p as a function of temperature are usually given in the form of an equation, table, or graph. Table 10.4 gives K_p as a function of temperature for several stoichiometric reactions.

The equilibrium constant may also be expressed in terms of mole fractions. Noting that the mole fraction of a component in an ideal-gas mixture is equal to the ratio of the partial pressure of the component to the total pressure, the equilibrium constant, according to Eq. (10.47), becomes

$$K_p = \frac{n_C^{v_c} n_D^{v_d} \cdots}{n_A^{v_a} n_B^{v_b} \cdots} \left(\frac{p}{\sum_i n_i} \right)^{v_c + v_d + \cdots - v_a - v_b - \cdots} \tag{10.49}$$

or in terms of the mole fraction,

$$K_p = \frac{x_C^{v_c} x_D^{v_d} \cdots}{x_A^{v_a} x_B^{v_b} \cdots} p^{v_c + v_d + \cdots - v_a - v_b - \cdots} \tag{10.50}$$

Note that the equilibrium constant is equal to $\left(x_C^{v_c} x_D^{v_d} \cdots \right) / \left(x_A^{v_a} x_B^{v_b} \cdots \right)$ if $v_c + v_d + \cdots = v_a + v_b + \cdots$, that is, if the number of moles remain unchanged during the reaction. In this case, the pressure has no effect on the equilibrium composition, and, at constant temperature, the volume of the gaseous components remains constant. It will be shown in Section 10.15 that the equilibrium composition of a reacting system at a given temperature and pressure can be determined if the equilibrium constant is known.

The affinity z can be expressed in terms of K_p as

$$z = \bar{R}T \ln K_p - \bar{R}T \sum_i v_i \ln p_i \tag{10.51}$$

But

$$z° = \bar{R}T \ln K_p$$

Table 10.4   Logarithms to the Base e of the Equilibrium Constants, K_p

For the reaction $v_A A + v_B B \rightleftharpoons v_C C + v_D D$, the equilibrium constant K_p is defined as

$$K_p \equiv \frac{p_C^{v_C} p_D^{v_D}}{p_A^{v_A} p_B^{v_B}} \qquad p_o = 0.1 \text{ MPa}$$

Temp. K	$H_2 \rightleftharpoons 2H$	$O_2 \rightleftharpoons 2O$	$N_2 \rightleftharpoons 2N$	$H_2O \rightleftharpoons H_2 + \tfrac{1}{2}O_2$	$H_2O \rightleftharpoons \tfrac{1}{2}H_2 + OH$	$CO_2 \rightleftharpoons CO + \tfrac{1}{2}O_2$	$\tfrac{1}{2}N_2 + \tfrac{1}{2}O_2 \rightleftharpoons NO$
298	−164.003	−186.963	−367.528	−92.210	−106.038	−103.765	−34.934
500	−92.830	−105.623	−213.405	−52.693	−60.166	−57.617	−20.225
1000	−39.810	−45.146	−99.146	−23.161	−25.976	−23.526	−9.355
1200	−30.878	−35.003	−80.025	−18.182	−20.234	−17.868	−7.541
1400	−24.467	−27.741	−66.345	−14.611	−16.122	−13.840	−6.246
1600	−19.638	−22.282	−56.069	−11.925	−13.034	10.828	−5.274
1800	−15.868	−18.028	−48.066	−9.829	−10.629	−8.494	−4.518
2000	−12.841	−14.619	−41.655	−8.145	−8.703	−6.633	−3.913
2200	−10.356	−11.826	−36.404	−6.773	−7.127	−5.116	−3.418
2400	−8.280	−9.495	−32.023	−5.625	−5.813	−3.858	−3.006
2600	−6.519	−7.520	−28.313	−4.652	−4.701	−2.797	−2.658
2800	−5.005	−5.826	−25.129	−3.817	−3.748	−1.891	−2.360
3000	−3.690	−4.356	−22.367	−3.092	−2.923	−1.109	−2.103
3200	−2.538	−3.069	−19.947	−2.458	−2.201	−0.427	−1.878
3400	−1.519	−1.932	−17.810	−1.898	−1.564	0.173	−1.680
3600	−0.611	−0.922	−15.909	−1.400	−0.998	0.702	−1.504
3800	0.201	−0.017	−14.205	−0.953	−0.492	1.178	−1.347
4000	0.934	0.798	−12.671	−0.551	−0.037	1.602	−1.207
4500	2.483	2.520	−9.423	0.301	0.924	2.493	−0.912
5000	3.724	3.898	−6.816	0.986	1.692	3.199	−0.679
5500	4.739	5.027	−4.672	1.549	2.320	3.771	−0.490
6000	5.587	5.969	−2.876	2.020	2.842	4.244	−0.336

Source: Thermodynamic data in *JANAF Thermochemical Tables*, 3rd. ed. Thermal Group, Dow Chemical Company, Midland, Michigan, 1985.

then

$$z = z^\circ - \overline{R}T \sum_i v_i \ln p_i \tag{10.52}$$

At equilibrium $z = 0$ so that

$$\ln K_p = \sum_i \ln p_i^{v_i}$$

or

$$K_p = p_1^{v_1} + p_2^{v_2} + \cdots = \sum_i p_i^{v_i} \tag{10.53}$$

A relation between the equilibrium constant and the enthalpy of reaction may be derived as follows. The equilibrium constant as given by Eq. (10.47) is written

$$\ln K_p = \frac{-\Delta G^\circ}{\overline{R}T}$$

But the free-energy change in an isothermal process when both reactants and products are at the standard state is

$$\Delta G^\circ = \Delta H^\circ - T\Delta S^\circ$$

therefore,

$$\ln K_p = \frac{-(\Delta H^\circ - T\Delta S^\circ)}{\overline{R}T} = -\frac{\Delta H^\circ}{\overline{R}T} + \frac{\Delta S^\circ}{\overline{R}}$$

Differentiation of the preceding equation with respect to T gives

$$\frac{d \ln K_p}{dT} = \frac{\Delta H^\circ}{\overline{R}T^2} - \frac{d(\Delta H^\circ)}{\overline{R}T\, dT} + \frac{d(\Delta S^\circ)}{\overline{R}\, dT}$$

But $\Delta H^\circ / T = \Delta S^\circ$, and the last two terms can be dropped. Thus,

$$\frac{d(\ln K_p)}{dT} = \frac{\Delta H}{\overline{R}T^2}$$

or

$$\frac{d(\ln K_p)}{d(1/T)} = -\frac{\Delta H}{\overline{R}} \tag{10.54}$$

Note that the superscript $^\circ$ has been dropped in ΔH because the enthalpy of an ideal gas is a function of temperature only. ΔH is the enthalpy of reaction at temperature T.

If ΔH is positive (endothermic reaction), K_p increases with increasing temperature, and if ΔH is negative (exothermic reaction), K_p decreases with increasing temperature. Equation (10.54) is called the *van't Hoff equation.* Figure 10.16 shows a plot of $\ln K_p$ versus $1/T$, which is a straight line with a slope of $-\Delta H/R$. The sign of the slope depends on the sign of ΔH, whether positive or negative, corresponding to exothermic or endothermic reactions, respectively. Equation (11.54) may be used to determine the value of ΔH as a function of temperature, provided that the variation of K_p with temperature is known.

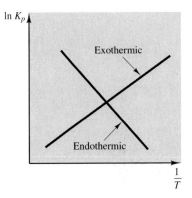

Figure 10.16 $\ln K_p$ versus $1/T$.

Example 10.19

Determine the enthalpy of reaction at 1 atm and 2200 K when carbon monoxide reacts with water vapor (Figure 10.17) according to the reaction

$$CO(g) + H_2O(g) \rightarrow H_2(g) + CO_2(g)$$

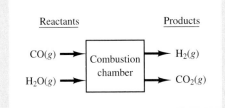

Figure 10.17

SOLUTION

Assuming ΔH to be a constant, integration of the van't Hoff equation gives

$$\ln \frac{(K_p)_2}{(K_p)_1} = -\frac{\Delta H}{R}\left(\frac{1}{T_2} - \frac{1}{T_1}\right)$$

where $(K_p)_1$ and $(K_p)_2$ are the equilibrium constants at two states 1 and 2, and ΔH is the enthalpy of reaction per kg-mol.

To evaluate the equilibrium constant, note that the preceding reaction is the sum of the two following reactions:

$$CO(g) + \frac{1}{2}O_2(g) \rightarrow CO_2(g) \qquad (K_p)_a = \frac{p_{CO_2}}{p_{CO}\,p_{O_2}^{1/2}} \qquad \text{(a)}$$

$$H_2O(g) \rightarrow H_2(g) + \frac{1}{2}O_2(g) \qquad (K_p)_b = \frac{p_{H_2}p_{O_2}^{1/2}}{p_{H_2O}} \qquad (b)$$

Adding,

$$CO(g) + H_2O(g) \rightarrow H_2(g) + CO_2(g) \qquad (c)$$

The equilibrium constant for reaction (c) is

$$(K_p)_c = \frac{p_{H_2}p_{CO_2}}{p_{CO}p_{H_2O}} = (K_p)_a(K_p)_b$$

or

$$\ln (K_p)_c = \ln (K_p)_a + \ln (K_p)_b$$

Thus, the logarithm of the equilibrium constant for reaction (c) is the sum of the logarithms of reactions (a) and (b). This equation is used to evaluate $(K_p)_c$. Choosing two neighboring states (at 2000 and 2400 K) gives:

At 2000 K,

$$\ln (K_p)_1 = +6.635 + (-8.145) = -1.51$$

At 2400 K,

$$\ln (K_p)_2 = +3.860 + (-5.619) = -1.759$$

Substituting values in the van't Hoff equation gives

$$\ln (K_p)_2 - \ln (K_p)_1 = -\frac{\Delta H}{R}\left(\frac{1}{T_2} - \frac{1}{T_1}\right)$$

$$-1.759 - (-1.51) = -\frac{\Delta H}{(8.3144 \text{ kJ/kg-mol K})}\left(\frac{1}{2400 \text{ K}} - \frac{1}{2000 \text{ K}}\right)$$

so that

$$\Delta H = -\frac{(0.249)(8.3144)}{8.333 \times 10^{-5}} = -24,843.43 \text{ kJ/kg-mol}$$

The same result can be obtained by applying the first law to the reaction.

10.14 Fugacity and Activity

In order to determine the equilibrium constant when the gaseous species of a chemical reaction do not follow the ideal-gas law, a new function, *fugacity,* is introduced. When this function is substituted for pressure, the equations for the equilibrium constant become applicable to real gases.

The differential of the Gibbs function of an ideal gas undergoing an isothermal process is

$$dG = V \, dp = \frac{n\bar{R}T}{p} dp = n\bar{R}T d(\ln p) \tag{10.55}$$

At constant temperature, fugacity (pseudopressure) of component i in a mixture is defined as

$$\left.\begin{array}{c} d\bar{g}_i = \bar{R}T d(\ln f_i) \\[2mm] \dfrac{f_i}{p} \to 1 \text{ as } p \to 0 \end{array}\right\} \tag{10.56}$$

where f is called the *fugacity* and has the dimensions of pressure (in atm). Equation (10.56) has the same form as Eq. (10.55) but applies to real gases. The value of fugacity approaches the value of pressure as the latter tends to zero, that is, where ideal-gas conditions apply.

For a real gas with an equation of state $pv = ZRT$, the differential of the Gibbs function at constant temperature is

$$dg_T = v \, dp = \frac{ZRT \, dp_T}{p}$$

but $dg_T = RT d(\ln f)$. Therefore,

$$\frac{Z \, dp}{p} = d(\ln f)$$

When this equation is expressed in terms of the reduced pressure p_R, it becomes

$$Z(d \ln p_R) = d(\ln f)$$

or

$$(Z - 1)d(\ln p_R) = d\left(\ln \frac{f}{p}\right)$$

Integration gives

$$\ln\left(\frac{f}{p}\right) = \int_0^{p_R} (Z - 1)d\ln p_R \tag{10.57}$$

The ratio (f/p) is a function of p_R and Z, which in turn is a function T_R and p_R. This means that $\ln(f/p)$ can be plotted versus p_R for different values of T_R.

Integrating Eq. (10.56) between the standard state ($\,^\circ$) and any other state gives

$$\overline{g}_i - (\overline{g}_i)_{p\to 0} = \overline{R}T \ln\frac{f_i}{f_i^\circ} = \overline{R}T \ln a_i \tag{10.58}$$

where f_i° is the fugacity at very low pressure. The term a is called the *activity*, or *fugacity coefficient*, and is defined as

$$a_i \equiv \frac{f_i}{f_i^\circ} \tag{10.59}$$

Using the same procedure as in Section 10.13, the following expression for a constant-pressure equilibrium constant K_a for a real gas may be obtained:

$$K_a = e^{-\Delta G^\circ/\overline{R}T} = \frac{a_C^{\nu_c} a_D^{\nu_d}\cdots}{a_A^{\nu_a} a_B^{\nu_b}\cdots} \tag{10.60}$$

and

$$\Delta G^\circ = -\overline{R}T \ln K_a \tag{10.61}$$

Note that activities replace partial pressure in the expression for the equilibrium constant.

Example 10.20

Determine the change in the Gibbs function and the equilibrium constant at 298.15 K and 2000 K for the reaction

$$CO_2(g) \rightleftharpoons CO(g) + \frac{1}{2}O_2(g)$$

SOLUTION

(a) At $T = 298.15$ K, the change in the Gibbs function is

$$\Delta G^\circ = \Delta H^\circ - T\Delta S^\circ$$

But

$$\Delta H^\circ = \sum_P n_i \bar{h}_i - \sum_R n_i \bar{h}_i = (\bar{h}_f^\circ)_{CO} - (\bar{h}_f^\circ)_{CO_2}$$
$$= (1 \text{ kg-mol})(-110{,}527 \text{ kJ/kg-mol}) - (1 \text{ kg-mol})(-393{,}522 \text{ kJ/kg-mol}) = 282{,}995 \text{ kJ}$$

The change of entropy at the standard state is

$$\Delta S^\circ = \sum_P n_i \bar{s}_i^\circ - \sum_R n_i \bar{s}_i^\circ = (\bar{s}^\circ)_{CO} + \frac{1}{2}(\bar{s}^\circ)_{O_2} - (\bar{s}^\circ)_{CO_2}$$

$$= (1 \text{ kg-mol})(197.653 \text{ kJ/kg-mol K}) + \left(\frac{1}{2} \text{ kg-mol}\right)(205.147 \text{ kJ/kg-mol K}) - (1 \text{ kg-mol})(213.795 \text{ kJ/kg-mol K})$$

$$= 86.43 \text{ kJ/K}$$

Therefore, the change in Gibbs function is

$$\Delta G^\circ = 282{,}995 \text{ kJ} - (298.15 \text{ K})(86.43 \text{ kJ/K}) = 257{,}225.9 \text{ kJ}$$

Note that the change in the Gibbs function at the standard state can also be determined directly from the molar values of CO and CO_2 listed in Table 10.2.

$$\Delta G^\circ = \nu_{CO}\bar{g}_{CO}^\circ + \nu_{O_2}\bar{g}_{O_2}^\circ - \nu_{CO_2}\bar{g}_{CO_2}^\circ$$
$$= (1 \text{ kg-mol})(-137{,}163 \text{ kJ/kg-mol}) + 0 - (1 \text{ kg-mol})(-394{,}389 \text{ kJ/kg-mol}) = 257{,}226 \text{ kJ}$$

The equilibrium constant is

$$K_{298.15 \text{ K}} = e^{-\Delta G^\circ/\bar{R}T} = e^{-\frac{257{,}226}{8.3144 \times 298.15}} = e^{-103.7646} = 8.6221 \times 10^{-46}$$

which agrees with the value given in Table 10.4.

(b) At $T = 2000$ K,

$$\Delta G^\circ = \Delta H - T\Delta S$$
$$\Delta H = (\bar{h}_f^\circ + \Delta\bar{h})_{CO} + \frac{1}{2}(\Delta\bar{h})_{O_2} - (\bar{h}_f^\circ + \Delta\bar{h})_{CO_2}$$

$$= (1 \text{ kg-mol})[(-110527 + 56{,}744) \text{ kJ/kg-mol}] + \left(\frac{1}{2}\right) \text{ kg-mol}$$

$$\times (59{,}175 \text{ kJ/kg-mol K}) - (1 \text{ kg-mol})(-393{,}522 + 91{,}439) \text{ kJ/kg-mol}$$

$$= 277{,}887.5 \text{ kJ}$$

The change of entropy is

$$\Delta S = (\bar{s})_{CO} + \frac{1}{2}(\bar{s})_{O_2} - (\bar{s})_{CO_2}$$

$$= (1 \text{ kg-mol})(258.714 \text{ kJ/kg-mol K}) + \left(\frac{1}{2}\right) \text{kg-mol}$$

$$\times (268.748 \text{ kJ/kg-mol K}) - (1 \text{ kg-mol})(309.293 \text{ kJ/kg-mol K})$$

$$= 83.795 \text{ kJ/K}$$

Therefore,

$$\Delta G° = (1 \text{ kg-mol})(277,887.5 \text{ kJ/kg-mol}) - (2000 \text{ K})(83.795 \text{ kJ/K})$$
$$= 110,297.5 \text{ kJ}$$

The equilibrium constant is

$$K_{2000 \text{ K}} = e^{-\Delta G°/\bar{R}T} = e^{-\frac{110,297.5}{8.3144 \times 2000}} = e^{-6.6329} = 0.00131$$

which agrees with the value given in Table 10.4.

10.15 Equilibrium Constant and Dissociation

The equilibrium composition of a chemical system at high temperatures can be quite different from that at low temperature owing to possible dissociation and subsequent interactions of the constituents. Although the degree of dissociation increases with temperature, dissociation, in many cases, can be neglected at room temperature and atmospheric pressure. The degree of dissociation of carbon dioxide at atmospheric pressure, for example, is negligible at room temperature but increases to 34.4 percent at 2900 K. Dissociation reactions are endothermic.

Consider the equilibrium composition of a mixture of oxygen and hydrogen at high temperature (Figure 10.18). The following reactions are possible:

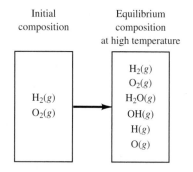

Initial composition

Equilibrium composition at high temperature

$H_2(g)$
$O_2(g)$

$H_2(g)$
$O_2(g)$
$H_2O(g)$
$OH(g)$
$H(g)$
$O(g)$

Figure 10.18 Composition of a mixture of O_2 and H_2 at high temperature.

$$H_2(g) + \frac{1}{2}O_2(g) \rightleftharpoons H_2O$$

$$\frac{1}{2}H_2(g) + \frac{1}{2}O_2(g) \rightleftharpoons OH(g)$$

$$H_2(g) \rightleftharpoons 2 H(g)$$

$$O_2(g) \rightleftharpoons 2 O(g)$$

A chemical system in which the preceding reactions occur contains six species, H_2, O_2, H_2O, OH, H, and O, and therefore six equations are required for the solution. Two are provided by the conservation of the number of atoms (or mass balance) of hydrogen and oxygen such that

for hydrogen: $2n_{H_2} + 2n_{H_2O} + n_{OH} + n_H = N_H$

for oxygen: $n_{H_2O} + 2n_{O_2} + n_{OH} + n_O = N_O$

where N_H and N_O are the total number of atoms of hydrogen and oxygen, respectively. The other four equations can be written in terms of the equilibrium constants. In general, the number of K_p relations needed is equal to the number of species present minus the number of elements in the equilibrium composition. The following example illustrates.

Example 10.21

Compare the degree of dissociation of H_2O and CO_2 at 2000 K. The total pressure in each case is 1.25 atm and 2.25 atm.

SOLUTION

The chemical equation for the dissociation of H_2O is

$$H_2O(g) \rightleftharpoons H_2(g) + \frac{1}{2} O_2(g)$$

Consider 1 mole of H_2O of which a fraction α dissociates to form H_2 and O_2, thus producing α moles of H_2 and $\alpha/2$ moles of O_2 according to the preceding equation (Figure 10.19). The number of moles in the products at equilibrium is, therefore,

$1 - \alpha$	moles of H_2O (undissociated)
α	moles of H_2
$\alpha/2$	moles of O_2

$$\text{Total} = 1 + (\alpha/2)$$

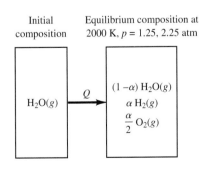

Initial composition   Equilibrium composition at 2000 K, p = 1.25, 2.25 atm

$H_2O(g)$ $\xrightarrow{Q}$ $(1 - \alpha)\, H_2O(g)$
$\alpha\, H_2(g)$
$\frac{\alpha}{2}\, O_2(g)$

Figure 10.19

The partial pressures of the products are

$$p_{H_2O} = \left[\frac{1-\alpha}{1+(\alpha/2)}\right]p_{total} \qquad p_{H_2} = \left[\frac{\alpha}{1+(\alpha/2)}\right]p_{total}$$

$$p_{O_2} = \left[\frac{\alpha/2}{1+(\alpha/2)}\right]p_{total} = \left(\frac{\alpha}{2+\alpha}\right)p_{total}$$

The equilibrium constant for the preceding reaction is

$$K_p = \frac{p_{H_2}p_{O_2}^{1/2}}{p_{H_2O}} = \frac{\left[\dfrac{\alpha}{1+(\alpha/2)}\right]p_{total}\left(\dfrac{\alpha}{2+\alpha}\right)^{1/2}p_{total}^{1/2}}{\left[\dfrac{1-\alpha}{1+(\alpha/2)}\right]p_{total}} = \frac{\alpha^{3/2}p_{total}^{1/2}}{(1-\alpha)(2+\alpha)^{1/2}}$$

where p_{total} is measured in atmospheres. A plot of K_p versus α according to the foregoing equation is shown in Figure 10.20. At 2000 K, Table 10.4 gives K_p (independent of total pressure) as

$$\ln K_p = -8.145 \qquad \text{or} \qquad K_p = 290.2 \times 10^{-6}$$

For the dissociation of CO_2, the chemical equation is

$$CO_2(g) \rightleftharpoons CO(g) + \frac{1}{2}O_2(g)$$

and for a mole fraction α of CO_2 dissociated, the number of moles of the products are

$1-\alpha$	moles of CO_2 (undissociated)
α	moles of CO
$\alpha/2$	moles of O_2

$$\text{Total} = 1 + (\alpha/2)$$

The expression for K_p is the same as in the case of H_2O. At 2000 K,

$$\ln K_p = -6.635 \qquad \text{or} \qquad K_p = 0.001314$$

From Figure 10.20, at 1.25 atm and $(K_p)_{H_2O} = 290.2 \times 10^{-6}$,

$$\alpha_{H_2O} = 0.005113 = 0.5113 \text{ percent}$$

For $(K_p)_{CO_2} = 0.001314$,

$$\alpha_{CO_2} = 0.01393 = 1.393 \text{ percent}$$

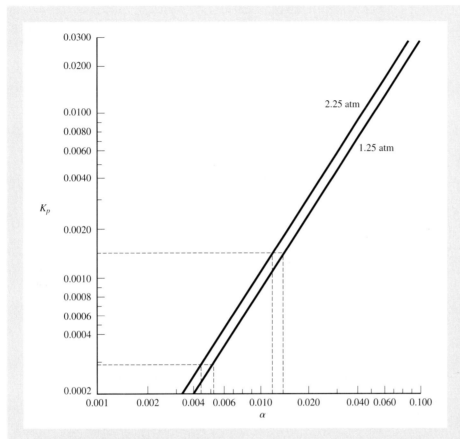

Figure 10.20

At 2.25 atm, the dissociation fractions are

$$\alpha_{H_2O} = 0.004204 = 0.4204 \text{ percent}$$
$$\alpha_{CO_2} = 0.01147 = 1.147 \text{ percent}$$

Note that, although a change in the total pressure does not affect the value of K_p, it affects the equilibrium composition of the mixture unless there is no change in the number of moles during the reaction.

If the temperature of the reaction is unknown, the procedure to determine the composition of the products involves a double trial-and-error solution. At the adiabatic reaction temperature, the reactants and the products are identical in enthalpy value; furthermore, the properties of the product constituents fulfill the equilibrium requirements appropriate for that temperature and pressure. For nonadiabatic conditions, a reaction temperature is first assumed, and the corresponding equilibrium composition is determined using the equations resulting from the assumed dissociation reactions and the balance of the atoms of the elements in the reactants and products. The assumed temperature is ascertained by calculating the enthalpy of both the reactants and products, and equating the difference to the amount of heat transfer in the reaction. If the result does not agree, another reaction temperature is assumed, and the procedure is repeated.

Example 10.22

A system consists initially of 1 mole of normal heptane $C_7H_{16}(g)$ and the stoichiometric amount of air is heated to 2300 K at 1 atm (Figure 10.21). Determine the composition of the equilibrium mixture assuming it is composed of CO_2, H_2O, N_2, O_2, CO, H_2, OH, H, O, and NO. Assume all reactants and products to be ideal gases.

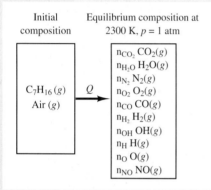

Figure 10.21

SOLUTION

The initial mixture is

$$C_7H_{16}(g) + 11\ O_2(g) + 3.76 \times 11\ N_2(g)$$

Conservation of the atomic species requires the following relations to be satisfied:

$$\frac{n_C}{n_H} = \frac{7}{16} = 0.4375 \tag{1}$$

$$\frac{n_O}{n_H} = \frac{22}{16} = 1.375 \tag{2}$$

$$\frac{n_N}{n_H} = \frac{3.76 \times 22}{16} = 5.16 \tag{3}$$

where n is the number of atoms. Alternatively, conservation of the atomic species can be expressed in terms of the components in the products so that

$$n_C = n_{CO_2} + n_{CO} \tag{4}$$

$$n_H = 2n_{H_2O} + 2n_{H_2} + n_H + n_{OH} \tag{5}$$

$$n_O = 2n_{CO_2} + n_{CO} + n_{H_2O} + 2n_{O_2} + n_{OH} + n_O + n_{NO} \tag{6}$$

$$n_N = 2n_{N_2} + n_{NO} \tag{7}$$

The total pressure of the products is the sum of the partial pressure of the component gases so that

$$p = p_{CO_2} + p_{CO} + p_{H_2O} + p_{O_2} + p_{H_2} + p_{OH} + p_H + p_O + p_{N_2} + p_{NO} = 1 \tag{8}$$

At equilibrium, the following dissociation equations relate the components of the product mixture:

$$CO_2 \rightleftharpoons CO + \frac{1}{2}O_2$$

$$H_2O \rightleftharpoons H_2 + \frac{1}{2}O_2$$

$$H_2O \rightleftharpoons \frac{1}{2}H_2 + OH$$

$$\frac{1}{2}H_2 \rightleftharpoons H$$

$$\frac{1}{2}O_2 \rightleftharpoons O$$

$$\frac{1}{2}N_2 + \frac{1}{2}O_2 \rightleftharpoons NO$$

The equilibrium constants for the preceding reactions at 2300 K are

$$K_1 = \frac{p_{CO}\sqrt{p_{O_2}}}{p_{CO_2}} = 1 \times 10^{-2} \tag{9}$$

$$K_2 = \frac{p_{H_2}\sqrt{p_{O_2}}}{p_{H_2O}} = 2.04 \times 10^{-3} \tag{10}$$

$$K_3 = \frac{p_{OH}\sqrt{p_{H_2}}}{p_{H_2O}} = 1.56 \times 10^{-3} \tag{11}$$

$$K_4 = \frac{p_H}{\sqrt{p_{H_2}}} = 9.66 \times 10^{-3} \tag{12}$$

$$K_5 = \frac{p_O}{\sqrt{p_{O_2}}} = 5.4 \times 10^{-3} \tag{13}$$

$$K_6 = \frac{p_{NO}}{\sqrt{p_{N_2}p_{O_2}}} = 4 \times 10^{-2} \tag{14}$$

There are ten unknown partial pressures and ten equations: six equations for the equilibrium constants, three for the ratios of atomic species, and the total-pressure equation. Note that the number of atoms n of each species is proportional to its partial pressure. The calculation proceeds by first assuming a value for p_{CO_2}/p_{CO} and p_{H_2O}, then calculating n_O/n_H and p to see if they correspond to the values given. Some of the trials are shown in Table 10.5. Column 6 in the table gives the equilibrium composition of the mixture.

Table 10.5   Equilibrium Composition of the Products of Combustion for n-Heptane and Air at 2300 K

Trial No.	1	2	3	4	5	6
p_{CO_2}/p_{CO}	15	10	8	9	8	8.5
p_{H_2O}	0.13	0.135	0.135	0.135	0.136	0.136
p_{O_2}	0.0225	0.01	0.0064	0.0081	0.0064	0.00722
p_O	0.00081	0.00054	0.000432	0.00486	0.000432	0.000458
p_{H_2}	0.00177	0.002755	0.00344	0.003055	0.003465	0.00326
p_H	0.000406	0.000507	0.000566	0.000534	0.000568	0.000551
p_{OH}	0.00482	0.00401	0.00359	0.00381	0.00361	0.00372
n_C	0.1171	0.1225	0.1230	0.1226	0.1238	0.1236
n_H	0.2676	0.2800	0.2811	0.28044	0.2831	0.28259
p_{CO}	0.00732	0.01115	0.01366	0.01266	0.01375	0.0130
p_{CO_2}	0.1097	0.1115	0.1092	0.1103	0.1100	0.1104
n_O	0.40734	0.39370	038382	0.388356	0.386592	0.388418
p_{N_2}	0.690	0.7225	0.725	0.723	0.731	0.730
p_{NO}	small	—	—	—	—	—
n_N	0.138	0.1445	0.145	0.1446	0.1462	0.1460
n_O/n_H	1.52	1.405	1.368	1.385	1.366	1.374
p	0.967326	0.997962	0.997288	0.996545	1.005225	1.004609

10.16 Summary

The energy transfer and the product composition of a chemical reaction depend on the first and second laws of thermodynamics. The reaction must also comply with the law of conservation of mass (or atomic species). The stoichiometric or theoretical air is the minimum amount of air required for the complete oxidation of the fuel. Excess air is usually used to guarantee proper mixing of the fuel and oxidizer, which is essential for complete combustion. The amount of excess air is expressed in terms of the stoichiometric air as a percent excess or percent theoretical air.

In combustion, the chemical energy, in addition to the sensible energy of each species, must be included in applying the first law with all energies referenced to the same standard state (25°C and 1 atm). The enthalpy of formation $\bar{h}_f^\circ$ accounts for the chemical composition of a compound. The enthalpy of formation of all stable elements is assigned a value of zero at the standard reference state. The difference between the enthalpy of the products and the enthalpy of the reactants at the same state is called the enthalpy of reaction.

$$\Delta H_T = \Delta H^\circ + \left[\sum_i n_i(\bar{h}_T - \bar{h}^\circ)_i \right]_P - \left[\sum_i n_i(\bar{h}_T - \bar{h}^\circ)_i \right]_R$$

At the standard state,

$$\Delta H^\circ = \left(\sum_i v_i \bar{h}_{f_i}^\circ \right)_P - \left(\sum_i v_i \Delta \bar{h}_{f_i}^\circ \right)_R$$

The heating value of a fuel is equal to the enthalpy of combustion but of opposite sign. The difference between the higher and lower heating values is the enthalpy of vaporization of the water in the products of combustion.

At steady state, the first law for a chemical reaction in terms of enthalpies of formation is

$$W + Q + \left[\sum_i n_i(\bar{h}_f + \bar{h}_T - \bar{h}^\circ)_i \right]_R = \left[\sum_i n_i(\bar{h}_f^\circ + \bar{h}_T - \bar{h}^\circ)_i \right]_P$$

and in terms of enthalpy of reaction is

$$W + Q + \left[\sum_i n_i(\bar{h}_T - \bar{h}^\circ)_i \right]_R + (-\Delta H^\circ) = \left[\sum_i n_i(\bar{h}_T - \bar{h}^\circ)_i \right]_P$$

Under adiabatic conditions ($Q = 0$) and for $W = 0$, the temperature of the products is a maximum, which is called the *adiabatic flame temperature*. It is determined by equating the enthalpy of the reactants to the enthalpy of the products.

The Gibbs function, rather than entropy, is used as a criterion for equilibrium in combustion processes. The Gibbs function of a compound is

$$G = G_f^\circ + \Delta G = (\Delta H^\circ - T_0 \Delta S^\circ) + \Delta G$$

where G_f° is the Gibbs function of formation at the standard reference state. At equilibrium, the Gibbs function is a minimum so that

$$(dG)_{T,p} = 0$$

The equilibrium constant K_p for an ideal-gas mixture is defined as

$$K_p = e^{-\Delta G^\circ / \overline{R}T} = \frac{p_C^{\nu_c} p_D^{\nu_d \cdots}}{p_A^{\nu_a} p_B^{\nu_b \cdots}} = \frac{n_C^{\nu_c} n_D^{\nu_d \cdots}}{n_A^{\nu_a} n_B^{\nu_b \cdots}} \left(\frac{p}{n_{total}}\right)^{\nu_c + \nu_d + \cdots - \nu_a - \nu_b \cdots}$$

Since ΔG° is a function of temperature only, K_p is also a function of temperature only. The van't Hoff equation indicates the relation of K_p with temperature:

$$\frac{d(\ln K_p)}{dT} = \frac{\Delta H}{\overline{R}T^2}$$

which when integrated becomes

$$\ln \frac{K_{p_2}}{K_{p_1}} = -\frac{\Delta H}{\overline{R}}\left(\frac{1}{T_2} - \frac{1}{T_1}\right)$$

Problems

10.1   The mass analysis of a sample of coal is $0.801, C = H_2 = 0.043$, $O_2 = 0.042$, $N_2 = 0.017$, $S = 0.01$, and the rest is ash. What would be the air–fuel ratio on a mass basis if this coal is burned with 20 percent excess air?

10.2   Calculate the molar analysis of the products of combustion when normal heptane $C_7H_{16}(\ell)$ is burned with 150 percent excess air. What is the dew point of the products if combustion takes place at a pressure of 110 kPa?

10.3   Octane, $C_8H_{18}(\ell)$, is burned with air.
(a) Determine the minimum mass of air required for complete combustion per kilogram of octane.
(b) What is the volume of combustion products per kilogram of fuel at 100 kPa and 480°C?
(c) If the products of combustion are cooled isobarically to a temperature of 20°C, what is the volume per kilogram of fuel?
(d) If 25 percent excess air is used, determine the preceding volumes (at 480°C and 20°C) at 100 kPa.

10.4   The volumetric analysis of natural gas is 21.2 percent C_2H_6 and 78.8 percent CH_4. Find (a) the stoichiometric air–fuel ratio, (b) the percentage by mass of the hydrogen and carbon in the fuel, (c) the mass of CO_2 and H_2O formed per kilogram of fuel.

10.5   The volumetric analysis of the dry products of combustion of a hydrocarbon fuel is as follows:

$CO_2 = 12$ percent
$O_2 = 0$ percent
$CO = 0.8$ percent

If the relative humidity of the atmospheric air at 40°C used for combustion is 50 percent, calculate (a) the air–fuel ratio, (b) the hydrogen–carbon ratio by atoms, (c) the volume of the dry air supplied per mole of dry products.

10.6   The volumetric analysis of the products of combustion of a hydrocarbon fuel on a dry basis is 10 percent CO_2, 1.5 percent CO, 3 percent O_2, and 85.5 percent N_2. Determine the composition of the fuel and the percent excess air used in the combustion process.

10.7   The composition of a hydrocarbon fuel by mass is 84 percent carbon and 16 percent hydrogen. If 20 percent excess air is used to burn this fuel at atmospheric pressure, determine:
(a) the mass of air required per kilogram of fuel (assume complete combustion)
(b) the mass analysis of the products of combustion
(c) the temperature at which moisture appears in the products of combustion

10.8 Atmospheric air at a temperature of 22°C and a relative humidity of 70 percent is used to burn propane $C_3H_8(g)$ in a steady flow combustion chamber. If 100 percent excess air is used, and the combustion process takes place at a constant pressure of 1 atm, determine the dew point of the water vapor in the products. Assume complete combustion and both the reactants and products are ideal gases.

10.9 Hydrogen is burned with 50 percent excess air in a steady flow process. If the pressure of the products is 1 atm, determine the dew-point temperature. What percentage of excess air is necessary to reduce the dew-point temperature by 5°C?

10.10 Calculate the enthalpy of formation per kilogram of propane $C_3H_8(g)$ if the enthalpy of combustion is $\Delta H° = -2,219,986$ kJ/kg-mol. The enthalpies of formation of $H_2O(\ell)$ and CO_2 are

$$\overline{h}^{°}_{f_{H_2O(\ell)}} = -285,830 \text{ kJ/kg-mol}$$
$$\overline{h}^{°}_{f_{CO_2}} = -393,522 \text{ kJ/kg-mol}$$

10.11 Using values of enthalpies of formation from Table 10.2, calculate the enthalpy of combustion of liquid octane at 25°C and 1 atm. (Assume liquid H_2O in the products.)

10.12 (a) Using values of enthalpies of formation for substances other than methane $CH_4(g)$, calculate the enthalpy of formation per kilogram of methane if its enthalpy of combustion $\Delta H° = -50,302$ kJ/kg. (b) Using tabular values of enthalpies of formation, calculate the enthalpy of combustion of propane $C_3H_8(g)$ at the standard state. (Assume gaseous H_2O in the products.)

10.13 Calculate the enthalpy of reaction at 25°C when 1 mole of propane C_3H_8 is burned according to the following equation:

$$C_3H_8(g) + 5\,O_2(g) \rightarrow 3\,CO_3(g) + 4\,H_2O(\ell)$$

Assume steady flow combustion process.

10.14 Ethane $C_2H_6(g)$ at 298 K is burned with 20 percent excess air at 298 K in a steady flow atmospheric

process. If the products of combustion are cooled to 600 K, determine:
(a) the air–fuel ratio by mass
(b) the molar analysis of the products
(c) the enthalpy of reaction at the standard state
(d) the amount of heat transfer

10.15 Calculate the enthalpy of reaction when ammonia $NH_3(g)$ reacts with oxygen to form liquid H_2O and (a) N_2, (b) NO.

10.16 Gaseous propane at 100°C reacts with stoichiometric air at 0°C in a steady flow combustion process. If the products are at 700°C, calculate the enthalpy of reaction at the standard state and the amount of heat transfer.

10.17 Determine the heat interaction when propane is burned with the theoretical amount of air in a steady flow process. The reactants enter the combustion chamber at 25°C and 1 atm, and the products leave at 600 K and 1 atm.

10.18 Propane $C_3H_8(g)$ at 25°C is burned with 30 percent excess air at 298 K in a steady flow atmospheric process. If the products of combustion are cooled to 600 K, determine (a) the air–fuel ratio by mass, (b) the enthalpy of reaction at the standard state, (c) the amount of heat transfer.

10.19 Calculate the enthalpy of combustion of methane $CH_4(g)$ at 500 K. Use Table A10 to determine an average value of $\overline{c}_p$ for methane.

10.20 Propane $C_3H_8(g)$ at 20°C enters a combustion chamber operating at steady state and burns at atmospheric pressure with 150 percent of theoretical air. If the air enters at 40°C and a relative humidity 60 percent, determine:
(a) the air–fuel ratio by mass
(b) the dew point
(c) the amount of the water vapor that condenses in kg/kg of fuel if the products of combustion are cooled to 40°C

10.21 Calculate the enthalpy of combustion of methane $CH_4(g)$ using tabular values for the enthalpy of formation. Assume the reactants and products are at 25°C and 1 atm, and (a) liquid water in the products, (b) water vapor in the products.

10.22   Determine the enthalpy of combustion when gaseous butane at 25°C is burned with (a) atmospheric air, (b) pure oxygen. Assume the water in the products is the liquid phase. What is the lower heating value of butane?

10.23   Using tabulated values of enthalpy of formation, determine the enthalpy of combustion of gaseous butane C_4H_{10} at the standard state assuming (a) water vapor in the products, (b) liquid water in the products.

10.24   One kg-mol of $H_2(g)$ is completely burned to $H_2O(g)$ in a steady flow adiabatic reactor with the stoichiometric amount of air. The hydrogen and air enter at 1 atm and 25°C, and the products leave at 1 atm. Find the temperature of the products.

10.25   Propane is burned with 200 percent theoretical air in a steady flow combustion chamber. If the reactants are at 25°C and 1 atm, determine the adiabatic flame temperature.

10.26   Calculate the adiabatic reaction temperature of methane at 25°C when burned with 200 percent stoichiometric air at 170°C in a steady flow process.

10.27   If the temperature of the products in the previous problem is reduced by 100°C, what is the amount of heat transfer to the surroundings?

10.28   Sulfur at 25°C is burned with 20 percent excess air at a temperature of 400 K in a steady flow process. If the enthalpy of formation of SO_2 is $-296,831$ kJ/kg-mol, determine the adiabatic flame temperature. The specific heat of sulfur dioxide is given by

$$\bar{c} = 32.238 + 0.0219T - 3.527 \times 10^{-6}T^2$$

where $\bar{c}$ is in kJ/kg-mol K, and T is in degrees Kelvin.

10.29   Find the percentage of theoretical oxygen used in burning propane $C_3H_8(g)$ in a steady flow process if the adiabatic reaction temperature is 1400°C. Assume

complete combustion at a constant pressure of 1 atm and the water vapor in the products is in the vapor phase. The reactants are at 25°C.

10.30   The adiabatic reaction temperature when liquid heptane is burned in a steady flow process is 1200 K. The temperature of the fuel is 25°C, and the temperature of the air is 500 K. If the combustion process is complete, what is the percentage of excess air used?

10.31   Carbon monoxide reacts with air in a steady flow process according to the equation (constant pressure)

$$CO(g) + \frac{1}{2}O_2(g) + \frac{3.76}{2}N_2(g) \rightarrow$$
$$CO_2(g) + \frac{3.76}{2}N_2(g)$$

If all reactants are at 25°C and 1 atm, determine (a) the adiabatic flame temperature, (b) the irreversibility of the reaction.

10.32   Liquid octane (C_8H_{18}) is burned with the theoretical amount of air in an internal combustion engine operating at steady state. Both the fuel and the air enter at 25°C and 1 atm, and the products leave the engine at 900 K. If the engine develops 30 kW of power, and the mass rate of fuel is 2×10^{-5} kg-mol/s, determine the rate of heat transfer from the engine.

10.33   A diesel engine uses dodecane, $C_{12}H_{26}(g)$, for fuel. The fuel and the air enter the engine at 25°C, and the products of combustion leave at 600 K. If 200 percent theoretical air is used, and the heat loss from the engine is 200 MJ/kg-mol of fuel, determine the power developed for a fuel rate of 1 kg-mol/hr.

10.34   Calculate the adiabatic reaction temperature when liquid octane at 25°C and 1 atm is burned with the theoretical amount of air in a steady-state combustion process. Assume complete combustion.

10.35   Hydrogen at 25°C and 1 atm is burned with air at the same temperature and pressure in a steady flow process. Plot the adiabatic flame temperature as a function of excess air (0 to 200 percent).

10.36 Calculate the adiabatic reaction temperature when hydrogen is burned with 40 percent excess oxygen in a steady flow process. Assume the reactants are at 25°C and the reaction proceeds to completion.

10.37 Octane, $C_8H_{18}(g)$, is burned with 100 percent excess air in a closed rigid reactor. The initial temperature of the reactants is 25°C, and the initial pressure is 1 atm. If the temperature of the products if 1300 K, determine the heat transfer from the reactor assuming complete combustion. What is the final pressure in the reactor?

10.38 Benzene gas $C_6H_6(g)$ at 25°C is burned in a steady flow process with 20 percent excess air, also at 25°C. Because of inadequate mixing, only 80 percent of the carbon in the fuel burns to CO_2, and the rest is CO. The hydrogen in the fuel burns completely to H_2O. If the temperature of the products of combustion is 900 K, determine (a) the chemical equation of the reaction, (b) the heat transfer.

10.39 A rigid vessel contains a gaseous mixture of 0.006 mole of ethane gas (C_6H_6) and 0.04 moles of O_2 at 25°C and 150 kPa. The mixture undergoes complete combustion, and the final temperature is 1000 K. Determine the final pressure and the heat interaction.

10.40 A rigid vessel contains a mixture of gaseous propane and the stoichiometric amount of air at 25°C and 100 kPa. If the mixture undergoes complete combustion, what is the adiabatic reaction temperature and the maximum pressure in the tank?

10.41 Methane, $CH_4(g)$, and air, both at 25°C and 1 atm, enter an adiabatic reactor operating at steady state. The products of combustion leave the reactor at 1 atm. If 200 percent theoretical air is used, determine the entropy production per kg-mol of methane and the irreversibility of the process.

10.42 Acetylene gas (C_2H_2) at 25°C and 1 atm is burned completely with 20 percent excess air in a steady flow combustion process. The air enters the combustion chamber at 25°C and 1 atm. If the combustion products leave at 900 K and 1 atm, determine the entropy change of the reaction.

10.43 Determine the change of entropy when ethane $C_2H_6(g)$ is burned with $O_2(g)$ in a steady flow process to form $CO_2(g)$ and $H_2O(g)$ at 1000 K. The reactants are at 1 atm and 25°C. Assume the combustion process to occur at a constant pressure of 1 atm and both reactants and products are ideal gases. If the surrounding temperature is 25°C, what is the irreversibility of the process per kg-mol of fuel?

10.44 One mole of propane is burned to completion with 50 percent excess air in a steady flow process. If the reactants are at 25°C and 1 atm, and the products are at 800 K and 1 atm, determine (a) the entropy change of the reaction, (b) the heat transfer to the surroundings, (c) the irreversibility of the process. Assume $T_0 = 25°C$.

10.45 Methane gas (CH_4) at 25°C and 1 atm enters a combustion chamber where it reacts with the theoretical amount of oxygen also at the same temperature and pressure in a steady-state operation. The products leave the combustion chamber at 900 K and 1 atm. Determine:
(a) the amount of heat transfer
(b) the change of entropy of the system
(c) the irreversibility of the process assuming $T_0 = 25°C$

10.46 A gaseous mixture of propane (C_3H_8) and 150 percent theoretical amount of air at 25°C and 1 atm enter a reactor operating at steady state. The products of combustion leave the reactor at 800°C and 1 atm. Assuming complete combustion and an environment temperature $T_0 = 25°C$, determine the irreversibility of the combustion process per kg-mol of fuel.

10.47 Carbon is burned with the theoretical amount of air in a steady flow process. The reactants enter the combustion chamber at 25°C and 1 atm. If the combustion chamber is insulated and the reaction is complete, determine the irreversibility of the combustion process.

10.48 Determine the change of entropy and the Gibbs free energy at the standard state for the reaction

$$CO(g) + \frac{1}{2}O_2(g) \rightarrow CO_2(g)$$

Assume steady flow and all reactants and products are at the standard state.

10.49 Determine the Gibbs free energy of formation of $CO(g)$ using the reaction

$$C(s) + \frac{1}{2}O_2(g) \rightarrow CO(g)$$

Assume all reactants and products are at the standard state.

10.50 Calculate the Gibbs free energy change when propane $C_3H_8(g)$ at 25°C and 1 atm is burned with 400 percent theoretical air. Assume the products are at 25°C and 1 atm.

10.51 Noting that $dG = 0$ at equilibrium, compare the spontaneity of the vaporization and condensation of water at 10 MPa.

10.52 Determine the enthalpy of reaction and the entropy change when CO is burned at 5 atm and 260°C according to the reaction

$$CO(g) + \frac{1}{2}O_2(g) \rightarrow CO_2(g)$$

Assume steady flow process.

10.53 Consider a steady flow chemical reaction in which the reactants and products are in thermal and pressure equilibrium with the surroundings and in which changes in kinetic and potential energies are negligible. Prove that the reversible useful work or the availability of a fuel and an oxidizer is given by

$$W_{\text{rev useful}} = \sum_P n_i \bar{g}_i - \sum_R n_i \bar{g}_i$$

10.54 Carbon monoxide at 1 atm and 25°C reacts with the stoichiometric amount of air also at 1 atm and 25°C to form carbon dioxide and nitrogen. If each constituent in products is at 1 atm and 25°C, evaluate the entropy change of the reaction per kg-mol of CO. What is the total entropy change and the irreversibility of the combustion process?

10.55 Solve the previous problem if the reactants enter as a mixture at 1 atm and 25°C and the products leave as a mixture at 1 atm and 25°C.

10.56 Methane gas (CH_4) at 25°C and 1 atm enters a steady flow combustor and reacts with 150 percent theoretical air entering at 25°C and 1 atm. The reaction equation is

$$CH_4(g) + 3\,O_2(g) + 11.28\,N_2(g) \rightarrow$$
$$CO_2(g) + 2\,H_2O(g) + O_2(g) + 11.28 N_2(g)$$

Heat is transferred from the combustor to a thermal reservoir at 60°C, and the products of combustion leave at 500 K and 1 atm. Make a first-law analysis and a second-law analysis to calculate (a) the heat transfer, (b) the availability of the products of combustion, (c) the irreversibility of the combustion process.

Note that the molar composition of the products will eventually acquire the same molar composition of atmospheric air at 25°C, which is

$$x^{\circ}_{CO_2} = 0.0003 \quad x^{\circ}_{H_2O} = 0.0303 \quad x^{\circ}_{O_2} = 0.2035$$
$$\text{and} \quad x^{\circ}_{N_2} = 0.7567$$

where x° is the molar fraction.

10.57 Using tabular values of the Gibbs function of formation, calculate the equilibrium constant for the reaction $H_2(g) + \frac{1}{2}O_2(g) \rightarrow H_2O(g)$ at 25°C. Compare your result with the values in Table 10.4.

10.58 Methane gas (CH_4) is formed from carbon and hydrogen according to the equation

$$C(s) + 2\,H_2(g) \rightarrow CH_4(g)$$

Calculate the Gibbs function of formation of methane at the standard state, and compare your result with the value of Table 10.2.

10.59 Using values of free energies from Table 10.2, calculate the equilibrium constant at 25°C for the reaction

$$C_2H_6(g) + 3.5\ O_2(g) \rightleftharpoons 2\ CO_2(g) + 3\ H_2O(g)$$

10.60 Compute the equilibrium constant for the reaction

$$CH_4(g) + 2\ O_2(g) \rightleftharpoons CO_2(g) + 2\ H_2O(g)$$

at 25°C and 1 atm.

10.61 What will be the value of the equilibrium constant in the previous problem if the reaction takes place at 450 K?

10.62 What percentage of OH is formed when H_2O is heated to 5000 K at a pressure of 1 atm?

10.63 Compute the equilibrium constant for the reaction

$$NO_2(g) \rightleftharpoons NO(g) + \frac{1}{2}O_2(g)$$

at 2000 K. Note that $\ln K_p = -\Delta G°/RT$ and

$$(\bar{g}_f°)_{NO} = 86456\ \text{kJ/kg-mol}$$
$$(\bar{g}_f°)_{NO_2} = 51392\ \text{kJ/kg-mol}$$

What is the equilibrium composition at 2000 K? Both reactants and products are at atmospheric pressure.

10.64 Nitrogen tetroxide (N_2O_4) dissociates to nitrogen dioxide (NO_2) at relatively low temperatures according to the equation $N_2O_4(g) \rightarrow 2NO_2(g)$. If the degree of dissociation $\alpha = 18.46$ percent at 25°C and 1 atm, what will be the degree of dissociation at 0.5 atm?

10.65 Octane, $C_8H_{18}(\ell)$, is burned with the stoichiometric amount of air at atmospheric pressure in a steady flow process. Determine the equilibrium composition of the products of combustion if the combustion temperature is 1400 K.

10.66 Compute the value of the equilibrium constant for the reaction

$$CO(g) + \frac{1}{2}O_2(g) \rightleftharpoons CO_2(g)$$

at a temperature of 1000 K.

10.67 When oxygen at 1 atm is heated to 3200 K, both O_2 and O are present. Calculate the partial pressures, mole fractions, and mass fractions for the mixture.

10.68 At atmospheric pressure (100 kPa), oxygen begins to dissociate at 2500 K, whereas nitrogen begins to dissociate at 4000 K. Calculate the molar composition and the enthalpy when air is heated to (a) 3000 K, (b) 6000 K.

10.69 Determine the equilibrium composition when 1 kg-mol of carbon monoxide reacts with the theoretical amount of air to form an equilibrium mixture of CO_2, CO, O_2, and N_2 at 2800 K and 1 atm.

10.70 Determine the equilibrium composition when 1 kg-mol of H_2O dissociates to form an equilibrium mixture of H_2O, H_2, and O_2 at 3000 K and 2 atm.

10.71 Using the van't Hoff equation, determine the enthalpy of reaction at 1 atm and 2500 K for the reaction

$$CO_2(g) \rightleftharpoons CO(g) + \frac{1}{2}O_2(g)$$

Compare your result with the value obtained from enthalpy data.

10.72 Using the van't Hoff equation, determine the enthalpy of combustion when hydrogen reacts with oxygen at 2200 K. Check your result using enthalpy data.

10.73 Using the van't Hoff equation, and the tabulated value of the equilibrium constant at 2000 K, ascertain the tabulated value of the equilibrium constant at 2800 K for the reaction

$$CO_2(g) \rightleftharpoons CO(g) + \frac{1}{2}O_2(g)$$

Assume the enthalpy of the reaction does not vary much with temperature.

10.74 Determine the equilibrium composition on a mole basis when hydrogen is heated to a temperature of 4000 K at a pressure of 1 atm.

10.75 One mole of CO_2 is heated in a container to a temperature of 2400 K at a pressure of 1 atm. Determine the molar analysis of the equilibrium composition assuming it consists of CO_2, CO, and O_2.

10.76 Determine the equilibrium composition if one mole of an inert gas such as N_2 is added to the reactants in the previous problem.

10.77 Determine the equilibrium composition when one mole of H_2O is heated to a temperature of 3000 K at a pressure of 3 atm. The species at equilibrium are H_2O, H_2, O_2, and OH.

10.78 One kg-mol of hydrogen reacts with 1 kg-mol of oxygen in an insulated reactor operating at steady state. The hydrogen enters the reactor at 25°C and 1 atm and the oxygen enters at 200°C and 1 atm. Determine the temperature of the products if they leave the reactor at 1 atm and consist of H_2O, H_2, and O_2.

10.79 Using the values of the equilibrium constants for the reactions

$$CO_2(g) \rightarrow CO(g) + \frac{1}{2}O_2(g) \qquad O_2(g) \rightarrow 2\,O(g)$$

determine the equilibrium composition when one mole of CO_2 reacts with half a mole of O_2 at 3400 K and 1 atm to form CO_2, CO, O_2, and O in the products.

10.80 A mixture of one mole of CO_2 and one mole of O_2 is heated at 1 atm to 3000 K in a steady flow process. Determine the equilibrium composition of the products.

10.81 Calculate the molar percentage of NO formed when air is heated at 1 atm in a steady flow process if the temperature of the products is (a) 1000 K, (b) 3000 K. Assume a mixture of O_2, N_2, and NO in the products.

C10.1 Write an interactive computer program to determine the stoichiometric reaction equation for any given hydrocarbon fuel and any desired excess or deficit air. Assume all the hydrogen in the fuel reacts to form water. Print out the equation in an easy-to-read format.

C10.2 Write a computer program to identify the type of hydrocarbon fuel in a stoichiometric reaction. The user can supply the products of combustion in either molar or mass form, and the program returns the corresponding fuel by determining its composition. Since the composition may be in a fractional form, the program should convert the fractional values into whole numbers or, by matching against a table of possible hydrocarbons, determine the most probable fuel.

C10.3 Write a computer program to determine the adiabatic flame temperature for combustion of a user-specified hydrocarbon fuel. Modify the program in Problem C10.1 to allow for the input of temperature of the reactants as well as all the associated enthalpies of formation. Assume constant specific heats for the species. It is suggested to read the input quantities from a data file.

C10.4 Modify the program in Problem C10.3 to include variable specific heats in the analysis.

C10.5 Write a computer program to determine the higher and lower heating values of a hydrocarbon fuel specified by the user. Modify the program in Problem C10.1 to allow for the input of enthalpy of combustion for the reactants and the products. Both the reactants and the products are at the standard reference state. It is suggested to read the input quantities from a data file.

C10.6 Write a computer program to determine the heat transfer in combustion process for a user-defined hydrocarbon fuel. Modify the program in Problem C10.3 to allow for an input of the temperature of the products. Use variable specific heats.

Chapter 11

Advanced Energy Systems

11.1 Introduction

Meeting the electrical power demands of the future will require not only new generating capacity, but also thermodynamically efficient and cost-effective power plants. Optimization of power-generating plants is important to our energy future. This is especially true when the subject is considered in its wider context, including our concern with the conservation of energy resources and with the environmental effects of energy utilization. The combustion of fossil fuels in power plants has led to an increasing concern regarding the potential impact on the environment, and environmental considerations present a continuous challenge to the power industry. In the United States and many other countries worldwide, a "Clean Air Act" establishes limits on emissions in various areas and categories. For these reasons, technological efforts in the commercial exploitation of power generation have focused not only on high efficiency, but also on low emissions.

The meaningful evaluation of the thermodynamic performance of a power plant is the second-law efficiency that compares the power actually produced to the power that could have been produced if the cycle were performed reversibly. Identifying thermal losses is a first step in evaluating any proposed approach for increasing efficiency. But in addition to thermal performance, the criteria for assessment and comparison of power plants includes capital and operating cost and environmental acceptability. Although each of these factors claims priority and competes in importance, they are thermoeconomically coupled to such an extent that it is not uncommon to sacrifice efficiency in favor of lower capital investment. A good example is the use of a simple gas turbine instead of a more efficient steam power plant.

This chapter describes briefly several power systems and methods of improving efficiency by combining different cycles. The latter part of the chapter discusses direct-energy conversion systems.

11.2 Energy Sources

Several energy sources in nature are currently used to generate power. The chemical energy in fossil fuels, nuclear energy, and solar energy are a few examples. Presently, the major source of the world's fuel energy is fossil fuel. *Nuclear energy* is one of the important forms of energy, but its future impact is uncertain, and public acceptance of this type of energy involves controversy. Nuclear power plants are best suited for continuous base-load operation. In most nuclear reactors presently used, the energy released by fissionable fuels such as uranium-235 or plutonium-239 is converted into thermal energy, which in turn is carried by a fluid to the power-generating unit. Figure 11.1 is a schematic of a nuclear power plant. The nuclear reactor replaces the boiler and produces the energy required to generate steam. The primary flow loop operates between the nuclear reactor and the steam generator. In a *pressurized water reactor system,* the primary loop carries water at a high enough pressure to prevent any change of phase. If the water is allowed to boil in the reactor, it is called a *boiling water reactor system.* The primary loop is enclosed in a containment structure to prevent radiation leakage.

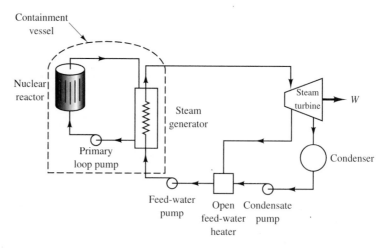

Figure 11.1   Schematic diagram of a nuclear power plant.

Nuclear reactors produce radioactive waste, which presents an environmental problem. The decay period of these wastes may extend to more than a hundred thousand years. In addition, nuclear power plants require large amounts of cooling water to maintain the reactor core at a low temperature, and this results in thermal pollution. Nevertheless, nuclear power plants, if operated safely, represent an important source of power generation.

A renewable unconventional source of energy is *solar energy*. The solar radiation absorbed by the earth is 10^{18} kWh per year, which is 10 times greater than the total fossil resources, including unexplored and nonrecoverable reserves. Solar heat collection devices are combined with a suitable thermodynamic conversion cycle such as the Rankine steam cycle or the gas-turbine cycle. A typical example is shown in Figure 11.2. Tracking mirrors called heliostats are oriented to reflect the direct portion of sunlight into a receiver mounted on top of a tower. The absorbed energy is carried by a working fluid to a turbine coupled to an electric generator, and the exhaust from the turbine is used to preheat the compressed gas in a regenerator. The heat rejection from the cycle is done by cooling water.

Solar energy is nonpolluting, nondepletable, reliable, and free. It is, however, very dilute and is not constant or continuous for terrestrial applications. For these

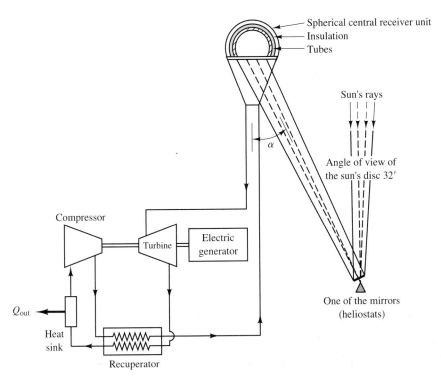

Figure 11.2   Schematic of a central tower receiver with a field of heliostats and a gas turbine.

reasons, solar energy systems must be provided with a standby energy conversion system or energy storage to compensate for the intermittency of sunshine. A hybrid system of solar and fuel-fired boiler can provide continuous energy to the power-generating unit. For summer cooling, solar energy utilization is particularly attractive owing to the coincidence of high insolation and ambient temperature with high cooling load.

In certain locations, *geothermal energy* trapped within the solid crust of the earth may be suitable for commercial exploitation provided that the thermal gradient in the earth is sufficiently steep. Geothermal energy exists in the form of steam, hot water, or hot and molten rock. It is released naturally as geysers, hot springs, or volcanic eruptions. Direct use of geothermal energy has many applications in heating, agriculture, and low-temperature industrial processes. Power-generating plants using geothermal energy have been built, but the available energy is considerably less than from the higher-pressure superheated steam generated in conventional fossil plants. A major problem to be considered in the design of a geothermal power system is the scaling problem, caused by silica precipitation and scaling formation. Figure 11.3 is a schematic of a geothermal power plant.

Another source of energy is ocean thermal gradients. The thermal potential that exists between the ocean-surface warm water and the cold water toward the ocean floor can be used to generate power. The warm surface water boils a working fluid such as ammonia in a Rankine-cycle power plant, whereas the cool deep water serves to condense the vapor leaving the turbine. Figure 11.4 is a schematic of such a plant. Ocean thermal systems have low efficiency owing to the small temperature difference (16°C to 20°C) across which they operate.

Other sources of energy exist in nature such as hydraulic, wind, and tidal energies that are used for power generation.

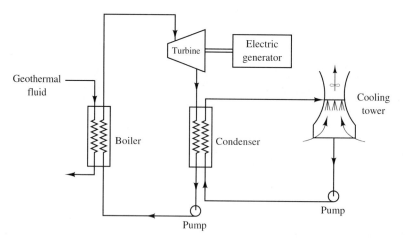

Figure 11.3   Schematic of a binary geothermal power plant.

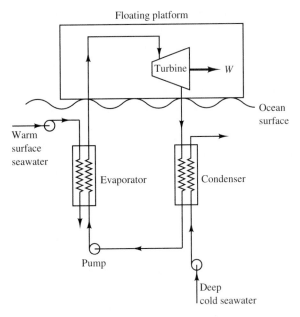

Figure 11.4   Closed-cycle ocean thermal energy conversion system.

11.3 The Combined Cycle

In many power-generation systems, the gas turbine is an excellent candidate for a base-load prime mover. This is mainly due to its high efficiency and its low specific investment cost. The high efficiency of gas turbines is a result of the increase in the firing temperature and the improvement in the design of heat-recovery systems. The main technical difficulties for high-temperature gas-turbine combustors are effective turbine blades and liner cooling and the lower NO_x combustion technology. Development has focused on recuperating the high-temperature energy exhausting the gas turbine. An exhaust-gas-to-compressed-air regenerator transfers the high-temperature exhaust energy to the lower-temperature compressed air entering the combustor. This preheating of the air reduces the fuel required for combustion. The effect of the regenerator, however, decreases as gas-turbine plants operate with higher compression ratios and lower turbine exhaust temperatures. Intercooling between partial compression is used to reduce the energy required for compression and to reestablish the necessary temperature difference between the exhaust gas and the compressed air, thereby rendering the regenerator more effective. Furthermore, the expansion of the hot compressed gases through the turbines is best done in stages. Reheating following a partial turbine expansion is a recognized steam-turbine design resulting in an increase in power output. Reheating increases the available energy (by raising the temperature) to do work. The technical challenges in the analysis and design of advanced gas-turbine studies are directed to three options: (1) intercooling, (2) reheating, and (3) steam cooling of turbine blades.

One way of increasing the efficiency of power generation is to combine two different power cycles using the waste energy from one cycle as the heat source for the other cycle. A common example in commercial use is the gas–steam turbine plant, which combines the Brayton topping cycle with the Rankine steam-turbine bottoming cycle.

The combined-cycle plant comprises one or more gas turbines exhausting into a heat recovery boiler. Figures 11.5 and 11.6 show two possible configurations, with and without supplementary burning between the exit of the power turbine and the steam generator. Supplemental firing increases cycle power output but decreases cycle efficiency owing to the additional fuel required. Figure 11.7 is a *T-s* diagram for the combined cycle shown in Figure 11.5. The combined cycle has achieved a substantially higher efficiency than either the open gas turbine cycle or the steam cycle. The most efficient combined cycle currently available is a multiple turbine–single large steam turbine configuration with an overall efficiency of about 54 percent. This compares favorably with conventional oil or gas power plants with efficiencies of about 38 to 40 percent.

For the combined cycle with a single-pressure steam cycle and no supplementary firing, the minimum temperature difference between the exhaust gases and the saturation temperature corresponding to the steam pressure in the evaporator dictates the mass ratio of the two fluids. This minimum temperature difference is called the *pinch point,* as indicated in Figure 11.8. In multipressure systems, additional energy can be recovered by extracting energy at lower steam pressure levels and correspondingly lower saturation temperatures. Generally, reducing

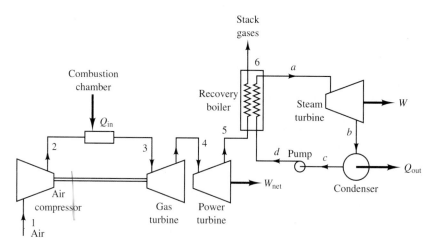

Figure 11.5   Schematic of a combined cycle.

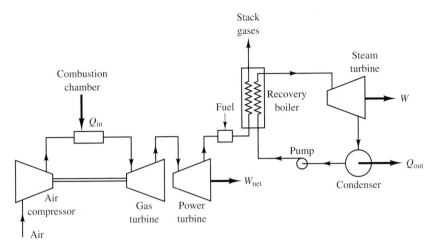

Figure 11.6   Schematic of a combined cycle with supplemental firing.

the pinch point results in an increase in total energy recovery. This, however, decreases the effective temperature difference in the boiler, requiring more surface area. Typical values of temperature differences at the pinch point are in the range of 5°C to 17°C. Note that in a steam cycle, feed-water heating improves the efficiency of the steam cycle but reduces the overall efficiency of the combined cycle because of the controlling influence of the pinch point (see Problems 11.4 and 11.5).

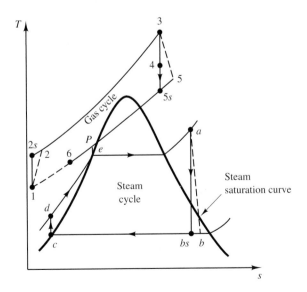

Figure 11.7   *T-s* diagram of the combined cycle.

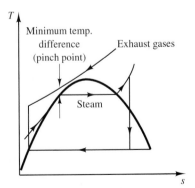

Figure 11.8   *T-s* diagram for a combined cycle indicating the pinch point.

The thermal efficiency of the combined gas–steam power cycle is equal to the total net work output divided by the total heat input:

$$\eta_{comb} = \frac{W_G + W_S}{Q_{input}} = \frac{\eta_G Q_G + \eta_S Q_S}{Q_G + Q_S} \tag{11.1}$$

where subscripts G and S refer to the gas-turbine and steam power plants, respectively. Equation (11.1) can be written as

$$\eta_{comb} = \left(\frac{Q_G}{Q_G + Q_S}\right)\eta_G + \left(\frac{Q_S}{Q_G + Q_S}\right)\eta_S \tag{11.2}$$

where the expressions in parentheses represent the weighted contribution of each cycle to the thermal efficiency of the combined cycle.

Example 11.1

For the gas–steam combined cycle shown in Figure 11.5, the following data apply:
   Gas-turbine cycle:

$$T_1 = 20°C \qquad p_1 = 100 \text{ kPa} \qquad r_p = 14 \qquad \eta_C = 0.84$$
$$T_3 = 1000°C \qquad \eta_T = 0.86$$

Steam cycle:

$$p_a = 5 \text{ MPa} \qquad T_a = 350°C \qquad \eta_T = 0.86$$
$$p_b = 10 \text{ kPa} \qquad w_P \text{ can be neglected}$$

Assume temperature difference at pinch point is 15°C. Basing calculations on a unit mass of air entering the compressor and assuming the fluid properties in the gas-turbine plant to be the same as those of air treated as an ideal gas with constant specific heats ($c_p = 1.01$ kJ/kg K, $\gamma = 1.4$), determine:

   (a) the thermal efficiency of the combined cycle
   (b) the availability utilization of the gas turbine exhaust by the bottoming cycle if the environment temperature $T_0 = 25°C$

SOLUTION

Figure 11.7 shows the T-s diagram of the cycle.
   Topping cycle:

$$\frac{T_{2s}}{T_1} = \left(\frac{p_2}{p_1}\right)^{\gamma-1/\gamma} = (14)^{0.4/1.4} = 2.1254$$

so that

$$T_{2s} = (293.15 \text{ K})(2.1254) = 623.07 \text{ K}$$

To determine T_2,

$$0.84 = \frac{T_{2s} - T_1}{T_2 - T_1} = \frac{623.07 - 293.15}{T_2 - 293.15}$$

from which $T_2 = 685.91$ K

$$w_C = \frac{c_p(T_{2s} - T_1)}{\eta_C} = \frac{(1.01 \text{ kJ/kg K})(623.07 - 293.15) \text{ K}}{0.84}$$
$$= 396.69 \text{ kJ/kg}$$

$$q_{2\text{-}3} = c_p(T_3 - T_2) = (1.01 \text{ kJ/kg K})(1273.15 - 685.91) \text{ K}$$
$$= 593.11 \text{ kJ/kg}$$

$$\frac{T_3}{T_{5s}} = \left(\frac{p_3}{p_5}\right)^{\frac{\gamma-1}{\gamma}} = 2.1254$$

so that

$$T_{5s} = \frac{1273.15}{2.1254} = 599.02 \text{ K}$$

To determine T_5,

$$0.86 = \frac{T_3 - T_5}{T_3 - T_{5s}} = \frac{1273.15 \text{ K} - T_5}{1273.15 \text{ K} - 599.02 \text{ K}}$$

from which $T_5 = 693.4$ K

$$w_T = \eta_T\, c_p(T_{5s} - T_3) = 0.86(1.01 \text{ kJ/kg K})[(599.02 - 1273.15) \text{ K}]$$
$$= -585.55 \text{ kJ/kg}$$

$$w_{\text{net}} = w_C - w_T = 396.69 \text{ kJ/kg} - 585.55 \text{ kJ/kg} = -188.86 \text{ kJ/kg}$$

The thermal efficiency of the topping cycle is

$$\eta = \frac{|w_{\text{net}}|}{q_{2\text{-}3}} = \frac{188.86 \text{ kJ/kg}}{593.11 \text{ kJ/kg}} = 31.84\%$$

Bottoming cycle:

At $p_a = 5$ MPa and $T_a = 350°C$,

$$h_a = 3068.4 \text{ kJ/kg} \qquad \text{and} \qquad s_a = 6.4493 \text{ kJ/kg K}$$

At $p_b = 10$ kPa,

$$h_f = 191.83 \text{ kJ/kg} \qquad h_{fg} = 2392.8 \text{ kJ/kg}$$
$$s_f = 0.6493 \text{ kJ/kg K} \qquad s_{fg} = 7.5009 \text{ kJ/kg K}$$
$$s_a = s_{bs} = s_f + x_{bs}s_{fg}$$

Therefore,

$$6.44493 \text{ kJ/kg K} = 0.6493 \text{ kJ/kg K} + x_{bs}(7.5009 \text{ kJ/kg K})$$

from which $x_{bs} = 0.7732$

$$h_{bs} = h_f + x_{bs}h_{fg} = 191.83 \text{ kJ/kg} + (0.7732)(2392.8 \text{ kJ/kg}) = 2042.04 \text{ kJ/kg}$$
$$w_T = \eta_T(h_{bs} - h_a) = 0.86(2042.04 - 3068.4) \text{ kJ/kg} = -882.67 \text{ kJ/kg}$$
$$q_{d-a} = h_a - h_d = 3068.4 \text{ kJ/kg} - 191.83 \text{ kJ/kg} = 2876.57 \text{ kJ/kg}$$

Neglecting pump work, the thermal efficiency of the bottoming cycle is

$$\eta = \frac{|w_{\text{net}}|}{q_{d-a}} = \frac{882.67 \text{ kJ/kg}}{2876.57 \text{ kJ/kg}} = 30.68\%$$

At the pinch point defined in Figure 11.8, $(T_P - T_e) = 15°C$. But $T_e = T_{\text{sat}}$ at 5 MPa $= 263.99°C$. Therefore,

$$T_P = 263.99°C + 15°C = 278.99°C = 552.14 \text{ K}$$

Applying the first law to the steam generator gives

$$m_{\text{air}}(h_5 - h_P) = m_{\text{steam}}(h_a - h_e)$$

so that

$$\frac{m_s}{m_a} = \frac{c_p(T_5 - T_P)}{h_a - h_e} = \frac{(1.01 \text{ kJ/kg K})[(693.4 - 552.14) \text{ K}]}{(3068.4 - 1154.23) \text{ kJ/kg}} = 0.0745 \text{ kg of steam/kg of air}$$

This ratio is constant and is controlled by the pinch point. The thermal efficiency of the combined cycle is

$$\eta = \frac{w_{\text{net out}}}{q_{\text{input}}} = \frac{188.86 \text{ kJ/kg} + (0.0745)(882.67 \text{ kJ/kg})}{593.11 \text{ kJ/kg}} = 42.93\%$$

Note that the thermal efficiency of the combined cycle is higher than the efficiency of either the gas-turbine or the steam cycle alone.

To determine T_6, energy balance gives

$$m_a c_p(T_5 - T_6) = m_s(h_a - h_d)$$

or

$$T_6 = T_5 - \left(\frac{m_s}{m_a}\right)\frac{h_a - h_d}{c_p}$$

$$= 693.4 - (0.0745)\left(\frac{3068.4 - 191.83}{1.01}\right) = 481.2 \text{ K}$$

The change of availability per unit mass of air flowing through the steam generator is

$$\psi_5 - \psi_6 = (h_5 - h_6) - T_0(s_5 - s_6) = c_p(T_5 - T_6) - T_0(s_5 - s_6)$$

$$= (1.01 \text{ kJ/kg K})[(693.4 - 481.2) \text{ K}] - (298.15 \text{ K})\left[(1.01 \text{ kJ/kg K})\ln\frac{693.4}{481.2}\right]$$

$$= 214.322 \text{ kJ/kg} - 110.011 \text{ kJ/kg} = 104.311 \text{ kJ/kg of air}$$

The work produced by the steam cycle per kilogram of air is

$$|w_{steam}| = (0.0745)(882.67 \text{ kJ/kg}) = 65.7589 \text{ kJ/kg of air}$$

The availability utilization is equal to the work produced by the steam cycle divided by the available energy so that

$$\frac{|w_{steam}|}{\psi_5 - \psi_6} = \frac{65.7589 \text{ kJ/kg}}{104.311 \text{ kJ/kg}} = 63.04\%$$

11.4 The Cheng Dual-Fluid (CDF) Cycle

The conventional combined cycle integrates the Brayton and Rankine cycles in series. The waste energy from the Brayton cycle is used to generate steam that expands in a separate Rankine cycle. The work developed is generated by the gas turbine and the steam turbine separately.

In the combined cycle, superheated steam must be heated to a high enough temperature so that it remains in a superheated state throughout most of the expansion in the steam turbine. This limitation is surmounted in the Cheng dual-fluid cycle because the steam generated in the heat recovery boiler is subsequently heated to a high degree of superheat in the combustor. The steam turbine is eliminated, and a mixture of steam and combustion products expands in the gas turbine.

The CDF cycle combines the gas-turbine cycle and the regenerative Rankine cycles in parallel. As shown in Figure 11.9, the compressed air from the compressors and the steam generated in the heat recovery boiler are combined in the combustor, where further heating takes place. The two fluids expand simultaneously in the turbine to produce power. The mixing of the two fluids in the combustor eliminates the upper temperature limitation required in the Rankine cycle and imposes no operational constraint on the compression ratio of the gas turbine.

In the regenerative gas-turbine cycle, in which the compressed air is heated before combustion, there is a considerable pressure drop on both sides of the regenerator. This means a higher compression ratio is required. At high compression ratio, the temperature of the air leaving the compression is high, and the effectiveness of regeneration is reduced.

In the CDF cycle, the regenerator and the associated mechanical problems in ducting the compressed air to the regenerator and to the combustor inlet are eliminated. Figure 11.10 shows the thermal efficiency plotted as a function of turbine inlet temperature for various compression ratios. Test results have substantiated these values, and efficiencies in excess of 40 percent have been achieved. The output power has also been substantially increased with no increase in turbine inlet temperature.

Water and steam injection in gas-turbine systems has been used in the past in aircraft turbines to increase power at takeoff and also in industrial power plants to suppress NO_x formation. In the CDF cycle, however, the injection of steam is continuous, and although it acts to reduce NO_x, it is also a working fluid that contributes to the power output as it expands with the combustion products in the turbine. However, the losses in the stack gas of a steam-injected gas turbine plant are greater than those from a gas turbine without steam injection. Substituting

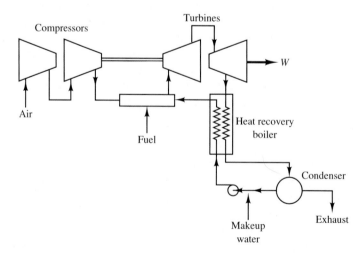

Figure 11.9   Cheng dual-fluid cycle.

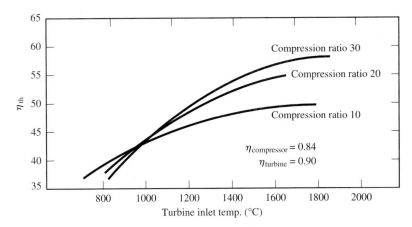

Figure 11.10   Overall cycle efficiency versus turbine inlet temperature.

steam for compressed air is more effective for blade cooling owing to the resulting higher heat transfer coefficient. The amount of steam injected is critical in a
sense that excessive steam injection increases the formation of CO and results in
a decreased efficiency. Hence, a proper balance of the two working fluids is important for maximum efficiency.

11.5 The Humid-Air Turbine (HAT) Cycle

This is a gas-turbine cycle with multistage compression in which water is evaporated and combined with the combustion turbine air stream. In this cycle, the
steam–turbine bottoming cycle is eliminated, and steam is generated using the
low-level energy recovered from the cycle itself. The heating and evaporation of
the water is achieved in the intercoolers between compression stages. Additional
heating is achieved by the energy recovered from the exhaust of the turbine. This
multistep heat transfer takes place with a relatively small thermal potential,
thereby reducing the irreversibilities inherent in the heat transfer process.

A schematic diagram of the HAT cycle is shown in Figure 11.11 with two compressors and a single gas turbine. Several schemes are used to reduce both energy
input and entropy generation. Intercooling, for example, between compression
stages reduces the temperature of the compressed air. This in turn reduces the
power required for compression. At the same time, the energy gained by the circulating water heats and eventually vaporizes the water. The compressed air is
mixed with the generated steam in the saturator. This direct contact is efficient
from a heat transfer point of view. The steam–air mixture passes through a recuperator, where it is heated by the exhaust from the turbine before entering the
combustion chamber. The products of combustion expand through the turbine,
and the exhaust passes through the recuperator and eventually to the stack.

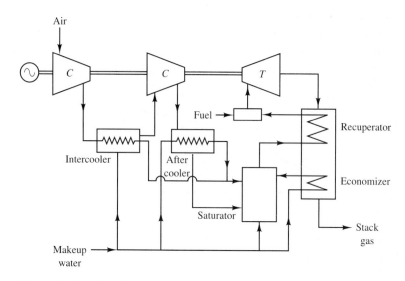

Figure 11.11   Humid-air turbine (HAT) cycle.

One major source of irreversibility in the conventional steam power plant is the large temperature potential between the hot gas and the water as it vaporizes at constant temperature. In the HAT cycle, this effect is reduced because a multistage saturator allows a variable boiling point. The reduction in the temperature difference reduces entropy generation.

The use of steam in the combustor reduces the amount of air that otherwise would be needed to control the temperature of the combustion products to the limits set by the material of the turbine. Because of the large specific enthalpy of vaporization of water, the temperature of the air–steam mixture is reduced. This allows additional recovery of energy from the turbine exhaust.

Preliminary calculations indicate that the thermal efficiency of the HAT cycle compares favorably with coal-based power plants.

11.6 The Chemically Recuperated Gas Turbine (CRGT) Cycle

This cycle incorporates a methane–steam reformer with the gas turbine and uses the energy of the gas turbine exhaust to generate chemical energy in the form of a fuel. The cycle utilizes a catalytic methane–steam reforming process in which methane (CH_4) and steam (H_2O) react at elevated temperature at the surface of a catalyst according to the following reactions:

$$CH_4 + H_2O \rightarrow 3H_2 + CO \qquad + 11{,}150 \text{ kJ/kg of } CH_4$$
$$CO + H_2O \rightarrow CO_2 + H_2 \qquad - 1313 \text{ kJ/kg of } CO$$

The methane–steam reaction is endothermic and is favored by high temperature and low pressure. The subsequent reaction of carbon monoxide and steam is exothermic and is favored at low temperature but is unaffected by pressure. The reactions produce a hydrogen-rich fuel gas that is fired in the gas turbine combustor.

The low-pressure requirement for the reaction has an adverse effect on the gas-turbine cycle whose performance improves with higher pressure. Obviously, an optimum operating pressure must be determined to yield a minimum cost of power generated. Figure 11.12 shows a CRGT cycle in which two turbines are used to drive two compressors. A third turbine is used for power generation. The reformer supplies energy to two combustion chambers.

Hot water from the intercooler is used as feed water to the reformer. Natural gas passes through a bed of activated zinc oxide pellets to remove sulphur compounds. These compounds, if absorbed on the steam-reforming catalyst, cause a loss of catalytic activity. After preliminary heating, the fuel is fed into the reformer. The methane–steam enters the reformer as a two-phase fluid (water and natural gas) and leaves at a temperature approaching the power-turbine exhaust temperature. The mixture is heated by the exhaust gases leaving the power turbine and reacts chemically over the catalyst. Upon leaving the reformer, the high-temperature, hydrogen-rich fuel enters the combustor. The reformer effluent comprises five gases: 40 to 60 percent unreacted steam, 9 to 15 percent unreacted methane, 5 to 6 percent carbon dioxide, 1 to 2 percent carbon monoxide, and 22 to 30 percent hydrogen. The hydrogen contributes less than half the energy input, but owing to its greater flammability, it sustains combustion and causes the residual methane to burn to completion.

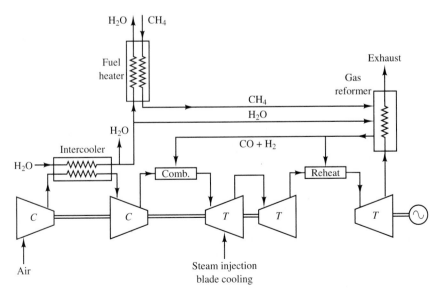

Figure 11.12   Chemically recuperated gas turbine cycle.

The technology in this cycle appears to improve the heat rate, improve emission control (including reduction in NO_x), and increase power output. In addition, because of the absence of sulfur in the flue gas, recovery of heat below the sulfurous acid dew point becomes a significant advantage.

11.7 The Gas Turbine–Ammonia Combined Cycle

In this cycle, ammonia is used both as a fluid to produce power and as a refrigerant to precool the inlet air to the compressor of the gas turbine plant. The increase in the air density due to cooling allows an increase in mass flow resulting in an increased power output. This effect, however, is partially offset by the pressure drop in the air cooler. Figure 11.13 is a schematic of a simple gas-turbine cycle with a "regenerative cooler," and Figure 11.14 is a combined cycle in which the ammonia is used simultaneously as a refrigerant and power fluid. The high-concentration ammonia–vapor mixture leaving the generator expands in an ammonia turbine, and the low-concentration ammonia–water liquid mixture is split into two streams. One stream flows to the steam generator, and the other to a heat exchanger onto the ammonia absorber. The ammonia generator is temperature staged, wherein the pressure is maintained constant in all stages, but the temperature and con-

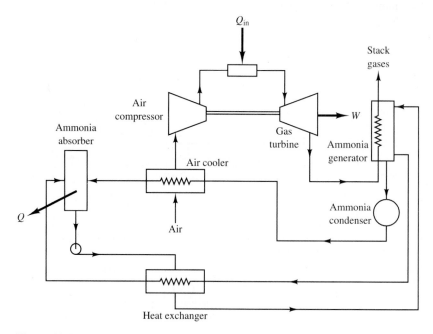

Figure 11.13   Simple ammonia–gas turbine cycle.

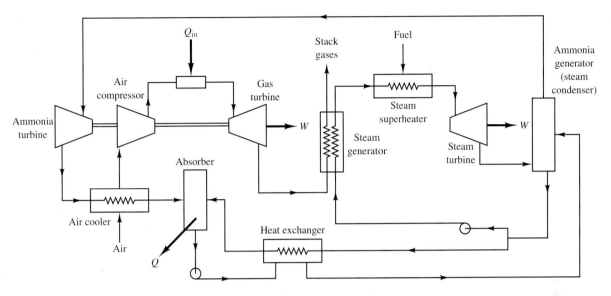

Figure 11.14   Combined ammonia–gas turbine cycle with ammonia used for power and cooling.

centration of the solution change as ammonia is boiled off. The vapor leaving the generator is a mixture of ammonia and steam, and in practice it is rectified in a direct-contact column to yield high-quality ammonia vapor before expanding in the ammonia turbine. Figure 11.15 is a combined gas–steam–ammonia cycle in which a direct-fluid contact in a steam condenser–ammonia generator takes place.

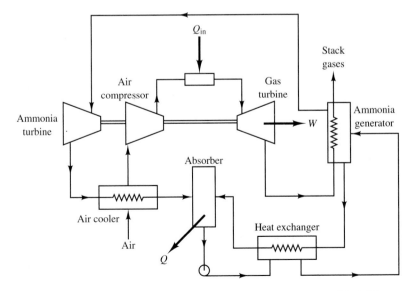

Figure 11.15   Combined gas–steam–ammonia cycle with supplemental firing.

11.8 The Kalina Cycle

In the *Kalina* cycle, a two-component fluid (ammonia and water) allows variation of temperature as the fluid changes phase. Ammonia has a low boiling point compared to water, and by adjusting the proportions of these two components, the temperature at which a change of phase occurs can be varied. This is in contrast to the usual behavior of a one-component substance that remains at constant temperature and pressure as it changes phase.

In order to explain this phenomenon, consider a water–ammonia mixture. Figure 11.16 is a phase diagram for such a mixture at constant pressure. The upper curve is called the *dew-point curve,* whereas the lower curve is called the *bubble-point curve.* As the temperature of a vapor mixture of a certain concentration x is reduced, the dew-point curve indicates the temperature at which condensation starts. Similarly, as the temperature of a liquid mixture of a certain concentration is increased, the bubble-point curve indicates the temperature at which boiling starts. Point a represents the state of a mixture of liquid and vapor in equilibrium at a temperature T and an ammonia concentration x. The concentration of the ammonia in the vapor state, x_G, and the concentration of the ammonia in the liquid state, x_L, are as shown in the figure. Note that for the same composition a temperature change results in a change in both the vapor and liquid concentrations. When a liquid mixture at state b starts to boil, most of the initial vapor is ammonia, whereas the composition of the liquid is equal to the overall composition x. Similarly, when condensation starts at state c, most of the liquid is water, and the vapor composition equals the overall composition. Therefore, by increasing the concentration of the high-boiling point constituent (water) in the mixture, it is possible to increase the temperature of the mixture during vaporization,

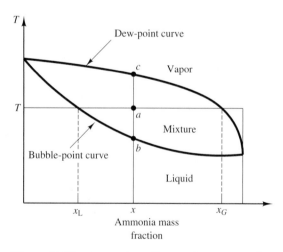

Figure 11.16   A two-phase diagram for an ammonia–water mixture.

thereby reducing the temperature difference between the energy source and the vaporizing fluid. Similarly, by increasing the concentration of the low-boiling point constituent in the condenser, a decrease in the condensing temperature is achieved. The increase in temperature during vaporization and the decrease in temperature during condensation (Figure 11.17) results in a decrease in the temperature difference across which heat is transferred. This in turn results in a decrease in the irreversibility of heat interaction and an improvement in the overall efficiency.

Figure 11.18 shows the Kalina cycle. A mixture of ammonia and water with a high-ammonia concentration is vaporized in the boiler at constant pressure but at continuously increasing temperature. After expanding in the turbine, the mixture is cooled in a heat exchanger before it enters the absorber where the ammonia vapor is absorbed by a low-concentration liquid stream. Upon leaving the absorber, the mixture is split into two streams; one goes to the heat exchanger onto the flash chamber and the other to the condenser. The high-concentration ammonia vapor leaving the flash tank is first mixed with the low-concentration stream in the appropriate proportion to establish the condensing temperature gradient. The resulting mixture is condensed and pumped back to the boiler to complete the cycle. The low-ammonia concentration liquid flows from the flash chamber to the absorber. A practical problem arising from the cycle is the corrosive nature of ammonia, and so there would be the need for an expensive encasement material.

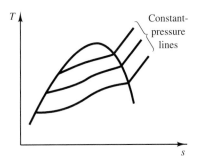

Figure 11.17   *T-s* diagram for an ammonia–water mixture.

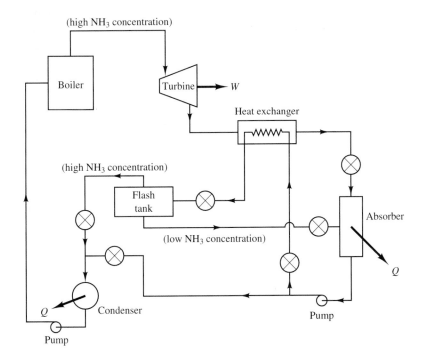

Figure 11.18   Schematic diagram of the Kalina cycle.

11.9 Cogeneration

In many industrial plants, energy transfer as heat, called *process heat,* as well as electric power is needed. The heating load and the power load may vary considerably during the period of a single day and from one industry to another. A balance of these two loads is an important factor in determining the cost of the total energy needs. *Cogeneration* is the simultaneous generation and use of both electric and thermal energies from a single fuel source, and the plant producing these two forms of energy is called a *cogeneration plant.*

Process heating can be accomplished very effectively without significant energy losses. But from the second-law point of view, it is advisable to use high-quality energy to produce work and to use low-quality energy for low-temperature applications such as process heating.

An ideal steam-turbine cogeneration plant utilizes the high-availability steam from the boiler to generate power and the low-availability steam leaving the turbine for process heat. In that respect, there is no waste energy, and there is no need for a condenser. But such a plant allows no variation in the heating and power loads. A balance of these loads is afforded by controlling the amount of the steam supply to the turbine and to the process heating. Figure 11.19 is a schematic of a cogeneration plant with adjustable loads. When the demand for process heating is high, the steam leaving the turbine at 3 is routed to process heating. Additional heating can be supplemented by boiler steam through the expansion valve. If all the generated steam passes through the valve, no power will be generated by the turbine. When the heating load is low, most of the steam will be directed to the turbine to produce power. If there is no demand for process heat, all the steam entering the turbine will pass through the condenser, and the plant will operate as an ordinary steam power plant.

One innovation in cogeneration when a gas turbine is used is to inject part of the steam generated by the turbine exhaust back into the combustion chamber in

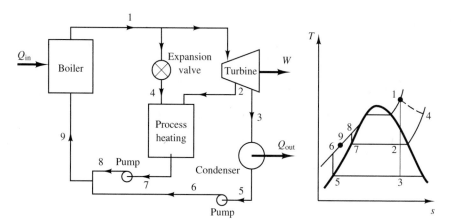

Figure 11.19   Schematic of a cogeneration power plant.

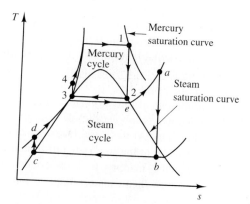

Figure 11.21   Mercury–water binary-vapor cycle.

Figure 11.21 is a binary cycle where a topping Rankine cycle, with mercury as a working fluid, is combined with a steam Rankine cycle. The T-s diagram for this cycle is shown in Figure 11.22. Mercury, after expanding in a turbine, condenses in a heat exchanger. The energy released by the condensing mercury is used to generate steam. The steam is then superheated and expanded through a

Figure 11.22 T-s diagram of a mercury–water binary-vapor cycle.

order to maximize power output. The rest of the steam is used as process steam. Besides increasing the power output and efficiency, steam injection reduces NO_x emissions. But most important, this scheme has excellent flexibility for rapidly changing power and steam loads.

Cogeneration has been found to be economically attractive to many industries. It allows greater control over the cost of meeting total energy needs of industrial plants, providing electric power as well as steam for process use. The performance of a cogeneration plant is usually measured by an *energy utilization factor* or total efficiency defined by

$$\eta_{total} = \frac{W_{\text{net out}} + \text{useful thermal output}}{Q_{input}} \quad (11.3)$$

Utilization factors range from 60 to 80 percent depending upon application. Essentially, more of the fuel input energy is converted into useful output, and the losses due to the use of a condenser are reduced.

Example 11.2

Superheated steam at a pressure of 6 MPa and a temperature of 450°C enters a turbine of a cogeneration plant, as shown in Figure 11.20. Thirty percent of the steam is extracted at a pressure of 0.4 MPa from the turbine for process heating, and the remainder expands to a pressure of 10 kPa. The extracted steam is condensed and leaves the process heater as saturated liquid at 0.4 MPa. It is then mixed with feed water from the condenser, and the mixture is pumped to the steam generator to complete the cycle. Assuming isentropic flow in the turbine and pumps and neglecting changes in kinetic and potential energies, determine the power produced and the rate of process heat supplied if the mass rate of flow of steam entering the turbine is 10 kg/s. What is the utilization factor of the plant?

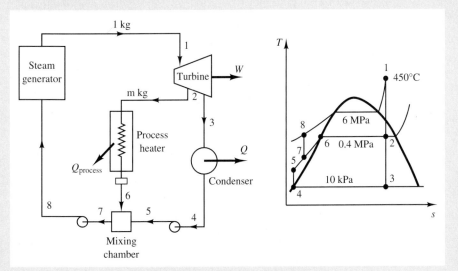

Figure 11.20

SOLUTION

Referring to the *T-s* diagram of the cycle, the following properties indicate energy values per unit mass of steam entering the turbine:

$$h_1 = 3301.8 \text{ kJ/kg} \qquad h_6 = 604.74 \text{ kJ/kg} \qquad h_4 = 191.83 \text{ kJ/kg}$$
$$s_1 = 6.7193 \text{ kJ/kg K}$$

The flow is isentropic in the turbine so that

$$s_1 = s_2 = s_f + x_2 s_{fg}$$
$$6.7193 \text{ kJ/kg K} = 1.7766 \text{ kJ/kg K} + x_2(5.1193 \text{ kJ/kg K})$$
$$\text{from which } x_2 = 0.9655$$
$$h_2 = h_f + x_2 h_{fg} = 604.74 \text{ kJ/kg} + 0.9655(2133.8 \text{ kJ/kg})$$
$$= 2664.93 \text{ kJ/kg}$$
$$s_3 = s_1 = s_f + x_3 s_{fg}$$
$$6.7193 \text{ kJ/kg K} = 0.6493 \text{ kJ/kg K} + x_3(7.5009 \text{ kJ/kg K})$$
$$\text{from which } x_3 = 0.8092$$
$$h_3 = h_f + x_3 h_{fg} = 191.83 \text{ kJ/kg} + 0.8092(2392.8 \text{ kJ/kg})$$
$$= 2128.17 \text{ kJ/kg}$$
$$w_{4\text{-}5} = v_4(p_5 - p_4) = (0.00101 \text{ m}^3/\text{kg})[(400 - 10) \text{ kPa}]$$
$$= 0.39 \text{ kJ/kg}$$
$$h_5 = h_4 + w_{4\text{-}5} = 191.83 \text{ kJ/kg} + 0.39 \text{ kJ/kg} = 192.22 \text{ kJ/kg}$$

Energy balance for the process heater gives the process heat per unit mass of steam entering the turbine:

$$q_{processs} = m(h_6 - h_2) = (0.3 \text{ kg/kg})(604.74 \text{ kJ/kg} - 2664.93 \text{ kJ/kg})$$
$$= -618.057 \text{ kJ/kg}$$

and the rate of process heat is

$$\dot{Q}_{process} = (10 \text{ kg/s})(-618.057 \text{ kJ/kg}) = -6180.57 \text{ kW}$$

Energy balance for the mixing chamber is

$$(1 - m)h_5 + mh_6 = (1)h_7$$

so that

$$h_7 = (0.7 \text{ kg/kg})(192.22 \text{ kJ/kg}) + (0.3 \text{ kg/kg})(604.74 \text{ kJ/kg})$$
$$= 134.554 \text{ kJ/kg} + 181.422 \text{ kJ/kg} = 315.976 \text{ kJ/kg}$$
$$w_{7\text{-}8} = v_7(p_8 - p_7) = (0.001084 \text{ m}^3/\text{kg})[(6000 - 400) \text{ kPa}] = 6.07 \text{ kJ/kg}$$
$$h_8 = h_7 + w_{7\text{-}8} = 315.976 \text{ kJ/kg} + 6.07 \text{ kJ/kg} = 322.046 \text{ kJ/kg}$$

The work developed per unit mass of steam entering the turbine is

$$w_{net} = (h_2 - h_1) + (1 - m)(h_3 - h_2) + (w_{4\text{-}5} + w_{7\text{-}8})$$
$$= (2664.93 - 3301.8) \text{ kJ/kg} + (0.7 \text{ kg/kg})[(2128.17 - 2664.93) \text{ kJ/kg}] + (0.39 + 6.07) \text{ kJ/kg}$$
$$= -636.87 - 375.73 + 6.46 = -1006.14 \text{ kJ/kg}$$

and the power developed is

$$\dot{W} = (10 \text{ kg/s})(-1006.14 \text{ kJ/kg}) = -10061.42 \text{ kW}$$

The energy supplied to the steam in the boiler is utilized for process heating and for power generation. The utilization factor is given by

$$\frac{|\dot{W}_{net}| + |\dot{Q}_{process}|}{\dot{Q}_{input}} = \frac{10061.42 \text{ kW} + 6180.57 \text{ kW}}{(10 \text{ kg/s})[(3301.8 - 322.046) \text{ kJ/kg}]} = \frac{16241.99 \text{ kW}}{29797.54 \text{ kW}}$$
$$= 54.45 \text{ percent}$$

11.10 Binary-Vapor Cycles

In a typical steam power plant, the combustion temperature in the boiler is in the range of 1300°C to 1400°C. This high-temperature energy is used to boil water at a relatively low temperature of about 300°C. The wide range in temperature between the combustion gases and the water or the steam during the heat transfer process is a major source of irreversibility.

Water has a relatively low critical temperature of 374.14°C, and vaporization has to take place below this temperature. This limits the maximum temperature at which most of the energy is absorbed, which takes place during the vaporization process. Water also has a high saturation pressure even at moderate vaporization temperatures. Realizing these inherent limitations at the high-temperature end of the vapor-power cycle, a more suitable fluid with appropriate properties such as mercury can be used in a "topping" cycle superimposed on the steam cycle. Such a combination is called a *binary-vapor cycle*. The mercury–water binary-vapor cycle capitalizes on the properties of mercury in the high range of temperature ($T_{critical}$ about 1500°C, which is well above the metallurgical limit, thus eliminating the need for superheating). At the same time, it makes use of the large enthalpy of vaporization of water at the low range of temperature (reduces the mass flow/kW output). Expanding mercury to a low-condenser temperature results in extremely low pressure, which causes complications due to air leaking into the system. Mercury also has steep saturated-liquid and saturated-vapor lines, which reduce the energy required to bring the condensate to the boiling point and increase the quality after expansion in the turbine.

order to maximize power output. The rest of the steam is used as process steam. Besides increasing the power output and efficiency, steam injection reduces NO_x emissions. But most important, this scheme has excellent flexibility for rapidly changing power and steam loads.

Cogeneration has been found to be economically attractive to many industries. It allows greater control over the cost of meeting total energy needs of industrial plants, providing electric power as well as steam for process use. The performance of a cogeneration plant is usually measured by an *energy utilization factor* or total efficiency defined by

$$\eta_{\text{total}} = \frac{W_{\text{net out}} + \text{useful thermal output}}{Q_{\text{input}}} \tag{11.3}$$

Utilization factors range from 60 to 80 percent depending upon application. Essentially, more of the fuel input energy is converted into useful output, and the losses due to the use of a condenser are reduced.

Example 11.2

Superheated steam at a pressure of 6 MPa and a temperature of 450°C enters a turbine of a cogeneration plant, as shown in Figure 11.20. Thirty percent of the steam is extracted at a pressure of 0.4 MPa from the turbine for process heating, and the remainder expands to a pressure of 10 kPa. The extracted steam is condensed and leaves the process heater as saturated liquid at 0.4 MPa. It is then mixed with feed water from the condenser, and the mixture is pumped to the steam generator to complete the cycle. Assuming isentropic flow in the turbine and pumps and neglecting changes in kinetic and potential energies, determine the power produced and the rate of process heat supplied if the mass rate of flow of steam entering the turbine is 10 kg/s. What is the utilization factor of the plant?

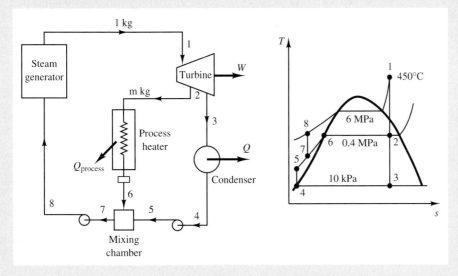

Figure 11.20

SOLUTION

Referring to the *T-s* diagram of the cycle, the following properties indicate energy values per unit mass of steam entering the turbine:

$$h_1 = 3301.8 \text{ kJ/kg} \qquad h_6 = 604.74 \text{ kJ/kg} \qquad h_4 = 191.83 \text{ kJ/kg}$$
$$s_1 = 6.7193 \text{ kJ/kg K}$$

The flow is isentropic in the turbine so that

$$s_1 = s_2 = s_f + x_2 s_{fg}$$
$$6.7193 \text{ kJ/kg K} = 1.7766 \text{ kJ/kg K} + x_2(5.1193 \text{ kJ/kg K})$$
$$\text{from which } x_2 = 0.9655$$
$$h_2 = h_f + x_2 h_{fg} = 604.74 \text{ kJ/kg} + 0.9655(2133.8 \text{ kJ/kg})$$
$$= 2664.93 \text{ kJ/kg}$$
$$s_3 = s_1 = s_f + x_3 s_{fg}$$
$$6.7193 \text{ kJ/kg K} = 0.6493 \text{ kJ/kg K} + x_3(7.5009 \text{ kJ/kg K})$$
$$\text{from which } x_3 = 0.8092$$
$$h_3 = h_f + x_3 h_{fg} = 191.83 \text{ kJ/kg} + 0.8092(2392.8 \text{ kJ/kg})$$
$$= 2128.17 \text{ kJ/kg}$$
$$w_{4\text{-}5} = v_4(p_5 - p_4) = (0.00101 \text{ m}^3/\text{kg})[(400 - 10) \text{ kPa}]$$
$$= 0.39 \text{ kJ/kg}$$
$$h_5 = h_4 + w_{4\text{-}5} = 191.83 \text{ kJ/kg} + 0.39 \text{ kJ/kg} = 192.22 \text{ kJ/kg}$$

Energy balance for the process heater gives the process heat per unit mass of steam entering the turbine:

$$q_{\text{processs}} = m(h_6 - h_2) = (0.3 \text{ kg/kg})(604.74 \text{ kJ/kg} - 2664.93 \text{ kJ/kg})$$
$$= -618.057 \text{ kJ/kg}$$

and the rate of process heat is

$$\dot{Q}_{\text{process}} = (10 \text{ kg/s})(-618.057 \text{ kJ/kg}) = -6180.57 \text{ kW}$$

Energy balance for the mixing chamber is

$$(1 - m)h_5 + mh_6 = (1)h_7$$

so that

$$h_7 = (0.7 \text{ kg/kg})(192.22 \text{ kJ/kg}) + (0.3 \text{ kg/kg})(604.74 \text{ kJ/kg})$$
$$= 134.554 \text{ kJ/kg} + 181.422 \text{ kJ/kg} = 315.976 \text{ kJ/kg}$$
$$w_{7\text{-}8} = v_7(p_8 - p_7) = (0.001084 \text{ m}^3/\text{kg})[(6000 - 400) \text{ kPa}] = 6.07 \text{ kJ/kg}$$
$$h_8 = h_7 + w_{7\text{-}8} = 315.976 \text{ kJ/kg} + 6.07 \text{ kJ/kg} = 322.046 \text{ kJ/kg}$$

The work developed per unit mass of steam entering the turbine is

$$w_{net} = (h_2 - h_1) + (1 - m)(h_3 - h_2) + (w_{4\text{-}5} + w_{7\text{-}8})$$
$$= (2664.93 - 3301.8) \text{ kJ/kg} + (0.7 \text{ kg/kg})[(2128.17 - 2664.93) \text{ kJ/kg}] + (0.39 + 6.07) \text{ kJ/kg}$$
$$= -636.87 - 375.73 + 6.46 = -1006.14 \text{ kJ/kg}$$

and the power developed is

$$\dot{W} = (10 \text{ kg/s})(-1006.14 \text{ kJ/kg}) = -10061.42 \text{ kW}$$

The energy supplied to the steam in the boiler is utilized for process heating and for power generation. The utilization factor is given by

$$\frac{|\dot{W}_{net}| + |\dot{Q}_{process}|}{\dot{Q}_{input}} = \frac{10061.42 \text{ kW} + 6180.57 \text{ kW}}{(10 \text{ kg/s})[(3301.8 - 322.046) \text{ kJ/kg}]} = \frac{16241.99 \text{ kW}}{29797.54 \text{ kW}}$$
$$= 54.45 \text{ percent}$$

11.10 Binary-Vapor Cycles

In a typical steam power plant, the combustion temperature in the boiler is in the range of 1300°C to 1400°C. This high-temperature energy is used to boil water at a relatively low temperature of about 300°C. The wide range in temperature between the combustion gases and the water or the steam during the heat transfer process is a major source of irreversibility.

Water has a relatively low critical temperature of 374.14°C, and vaporization has to take place below this temperature. This limits the maximum temperature at which most of the energy is absorbed, which takes place during the vaporization process. Water also has a high saturation pressure even at moderate vaporization temperatures. Realizing these inherent limitations at the high-temperature end of the vapor-power cycle, a more suitable fluid with appropriate properties such as mercury can be used in a "topping" cycle superimposed on the steam cycle. Such a combination is called a *binary-vapor cycle*. The mercury–water binary-vapor cycle capitalizes on the properties of mercury in the high range of temperature ($T_{critical}$ about 1500°C, which is well above the metallurgical limit, thus eliminating the need for superheating). At the same time, it makes use of the large enthalpy of vaporization of water at the low range of temperature (reduces the mass flow/kW output). Expanding mercury to a low-condenser temperature results in extremely low pressure, which causes complications due to air leaking into the system. Mercury also has steep saturated-liquid and saturated-vapor lines, which reduce the energy required to bring the condensate to the boiling point and increase the quality after expansion in the turbine.

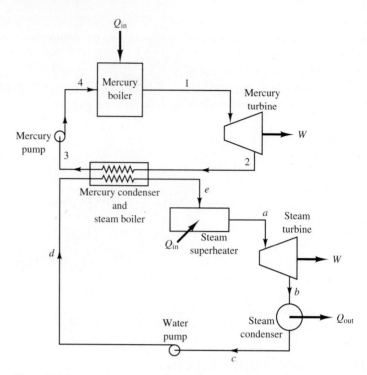

Figure 11.21   Mercury–water binary-vapor cycle.

Figure 11.21 is a binary cycle where a topping Rankine cycle, with mercury as a working fluid, is combined with a steam Rankine cycle. The *T-s* diagram for this cycle is shown in Figure 11.22. Mercury, after expanding in a turbine, condenses in a heat exchanger. The energy released by the condensing mercury is used to generate steam. The steam is then superheated and expanded through a

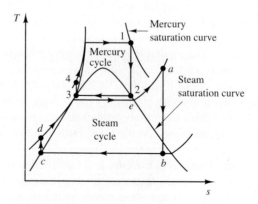

Figure 11.22   *T-s* diagram of a mercury–water binary-vapor cycle.

steam turbine to conventional condenser pressures. A high flow rate of mercury is required because the enthalpy of vaporization of water is about six times that of mercury at normal operating conditions.

Although higher efficiency can be attained by the mercury–water vapor cycle, it has some disadvantages. It has higher fixed costs and complicated special equipment. In addition, mercury is highly toxic.

Example 11.3

In a binary-vapor cycle, dry saturated mercury vapor at a pressure of 689.5 kPa enters a turbine and expands to a pressure of 13.8 kPa. The mercury is condensed in a mercury-condenser steam-boiler where dry saturated steam at a pressure of 4.5 MPa is generated. The steam is then superheated to a temperature of 400°C before expanding in a steam turbine to a condenser pressure of 10 kPa. Assume isentropic flow in the mercury and steam turbines, and neglect pump work. Determine:

(a) the ratio of the mass flow rate of mercury to steam flow
(b) the thermal efficiency of the binary cycle

The properties of mercury are given in the following table.

p (kPa)	T (°C)	h_f (kJ/kg)	h_g (kJ/kg)	s_f (kJ/kg K)	s_g (kJ/kg K)
689.5	486.0	65.48	355.55	0.1377	0.5199
13.8	267.7	36.0	330.26	0.0923	0.6415

SOLUTION

Referring to Figures 11.21 and 11.22, the following are values of properties per unit mass:

For the mercury cycle,

$$h_1 = 355.55 \text{ kJ/kg} \qquad s_1 = 0.5199 \text{ kJ/kg K}$$
$$h_2 = 330.26 \text{ kJ/kg} \qquad h_3 = 36 \text{ kJ/kg}$$

For the steam cycle,

$$h_e = 2797.85 \text{ kJ/kg} \qquad h_a = 3204.7 \text{ kJ/kg}$$
$$s_a = 6.7047 \text{ kJ/kg K} \qquad s_a = s_b$$

so that

$$6.7047 \text{ kJ/kg K} = 0.6493 \text{ kJ/kg K} + x_b(7.5009 \text{ kJ/kg K})$$

from which $x_b = 80.73$ percent

$$h_b = 191.83 \text{ kJ/kg} + (0.8073)(2392.8 \text{ kJ/kg}) = 2123.51 \text{ kJ/kg}$$
$$h_c = 191.83 \text{ kJ/kg}$$

Energy balance for the mercury-condenser steam-boiler gives

$$m_M(h_2 - h_3) = m_S(h_e - h_d)$$

so that

$$\frac{m_M}{m_S} = \frac{h_e - h_d}{h_2 - h_3} = \frac{2797.85 - 191.83}{330.26 - 36} = \frac{2606.02}{294.26}$$
$$= 8.856 \text{ kg of mercury/kg of steam}$$

$$w_M = h_2 - h_1 = 330.26 - 355.55 = -25.29 \text{ kJ/kg of mercury}$$
$$q_M = h_1 - h_3 = 355.55 - 36 = 319.55 \text{ kJ/kg of mercury}$$
$$w_S = h_b - h_a = 2123.51 - 3204.7 = -1081.19 \text{ kJ/kg of steam}$$
$$q_S = h_a - h_e = 3204.7 - 2797.85 = 406.85 \text{ kJ/kg of steam}$$

$$\eta = \frac{W_{\text{net out}}}{Q_{\text{input}}} = \frac{m_M w_M + m_S w_S}{m_M q_M + m_S q_S} = \frac{\left(\dfrac{m_M}{m_S}\right)(w_M) + w_S}{\left(\dfrac{m_M}{m_S}\right)(q_M) + q_S}$$

$$= \frac{(8.856 \text{ kg/kg})(25.29 \text{ kJ/kg}) + 1081.19 \text{ kJ/kg}}{(8.856 \text{ kg/kg})(319.55 \text{ kJ/kg}) + 406.85 \text{ kJ/kg}}$$
$$= 40.32\%$$

The thermal efficiencies of the mercury and steam cycles if operated separately are

$$\eta_M = \frac{|w_M|}{q_M} = \frac{25.29 \text{ kJ/kg}}{319.55 \text{ kJ/kg}} = 7.91\%$$

$$\eta_S = \frac{|w_S|}{q_S} = \frac{1081.19 \text{ kJ/kg}}{3012.87 \text{ kJ/kg}} = 35.89\%$$

Note that the thermal efficiency of the binary cycle is higher than the thermal efficiency of either the mercury cycle or the steam cycle operating alone.

11.11 Direct-Energy Conversion Systems

In conventional power plants, energy is transferred to a fluid from a high-temperature thermal reservoir, the fluid expands, producing mechanical work, which in turn is converted into electrical energy. To complete the cycle, the fluid rejects heat to a low-temperature reservoir. Several losses are incurred in these conversion steps, and the maximum value of efficiency is limited in accordance with the Carnot principle.

Direct-energy conversion systems eliminate some of the intermediate steps, converting the energy of the source directly into electrical energy, thereby circumventing the Carnot principle limitation. Direct-conversion systems for power generation are currently centered on five different approaches, involving thermoelectric converters, thermionic converters, magnetohydrodynamics (MHD), photovoltaic cells, and fuel cells. Thermoelectric converters have been employed in space vehicles, producing from a few watts to several kilowatts of power. As a source of large-scale power, only MHD offers the promise of being feasible. Although these systems are currently either low in efficiency or difficult to operate and control, experimental work continues to show that improved performance characteristics can be attained. We briefly describe some of these energy conversion methods in the following sections and discuss their characteristics from the standpoint of thermodynamics. Certain quantum-physical phenomena associated with the energy bands of solids that are characteristic of these direct-energy conversion techniques are first discussed.

11.12   Energy Bands

Because the distance between adjacent atoms in a solid is small, there is interaction between electrons and the adjacent atoms. As a result of these interactions, the electrons differ among themselves, to some extent, in their allowed energy levels, and they occupy energy bands rather than energy levels. Like the energy levels of the electrons of an isolated atom, these allowed energy bands are discrete, and as shown in Figure 11.23, they are separated by "forbidden bands," which cannot be occupied by an electron.

The distinction between a conductor, an insulator, and a semiconductor may be explained by means of the band theory of solids. When electrons fill an allowed energy band completely, electrons (negative-charge carriers) cannot shift from one atom to another, and so no conduction of electricity is possible. Likewise, if an energy band is completely unoccupied, no electrons are available for conduction. Therefore, only solids with partially occupied energy bands are capable of conducting electricity.

Two energy bands separated by a forbidden band constitute the element that determines the electrical properties of a solid. One energy band is called the *valence* band, and the other is called the *conduction* band. When an electric field is applied to a solid, electrons in the valence band gain small amounts of energy. If this energy enables a large number of electrons to migrate to the conduction band, the solid acts as an electric conductor. Each electron leaving the valence

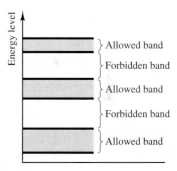

Figure 11.23   Energy bands in a solid.

band creates a void or an unfilled state, called a *hole,* which is equivalent to a positive charge. Both the holes and the electrons contribute to the flow of electric current. A *conductor* is a material in which electrons and holes are plentiful at all temperatures. If the width of the forbidden gap is large, electrons may not acquire enough energy to migrate to the conduction band, even if the valence band is filled. The solid then acts as an *insulator.* The valence band of diamond, for example, is filled, but the forbidden gap is 6 ev wide, and so it is an insulator. A substance in which the valence band is filled, the conduction band is empty, and the forbidden gap is narrow is a *semiconductor.* At high temperatures, large numbers of electrons in a semiconductor can jump across the forbidden gap to the conduction band; at low temperatures, on the other hand, only a small number of electrons can move from the valence band to the conduction band. A material is a semiconductor if the number of electrons in the conduction band depends upon temperature. A semiconductor may act as a conductor or as an insulator, depending upon the energy supplied, thus the name *semiconductor.* The forbidden gaps in silicon and germanium, which are semiconductors, are 1.1 ev and 0.7 ev, respectively.

Two types of semiconductors exist: *intrinsic* semiconductors and *extrinsic* (impurity) semiconductors. In intrinsic semiconductors, thermal excitation is sufficient to produce electrical conduction. In extrinsic semiconductors, impurity atoms are present that donate or accept electrons from the conduction and the valence bands of the main substance, thereby helping conduction due to thermal excitation. Impurity atoms are either donors or acceptors of electrons. The donors are called *n-type* (negative charges); the acceptors are called *p-type* (positive charges). The energy bands of semiconductors resulting from both types are shown in Figure 11.24. The *n*-type impurity atoms introduce discrete energy levels just below the conduction band, and the *p*-type impurity atoms provide energy levels just above the valence band.

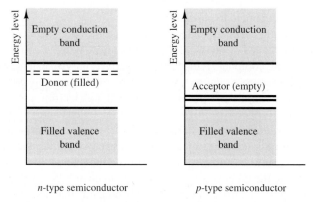

Figure 11.24   *n*- and *p*-types of semiconductors.

11.13 Thermoelectric Effects

Thermoelectric effects refer to phenomena involving the interchange of heat and electrical energy. One familiar example is the heating effect due to the flow of an electric current in a resistor. The electric power dissipation depends on the current flow I and the electric potential V according to the equation

$$P = VI \tag{11.4}$$

If the current flows through a material that obeys Ohm's law, then

$$P = VI = I^2R \tag{11.5}$$

where R is the electric resistance of the material. This dissipation of electrical energy as heat energy is called *ohmic* loss or *Joule heating*. Obviously, this type of energy conversion is irreversible.

Three thermoelectric reversible effects, called the *Seebeck, Peltier*, and *Thomson* effects, are now defined. The Seebeck, Peltier, and Thomson effects occur in both metals and semiconductors, but their effects are more pronounced in semiconductors. The three effects are illustrated in Figure 11.25.

11.13.1 The *Seebeck* Effect

In 1833, Seebeck held the junctions of two dissimilar electrical conductors at different temperatures and noted that an open-circuit voltage was generated that was proportional to the difference in temperature between the two junctions. In equation form, the Seebeck effect is

$$S_{AB} = \lim_{\Delta T \to 0} \frac{\Delta V}{\Delta T} \tag{11.6}$$

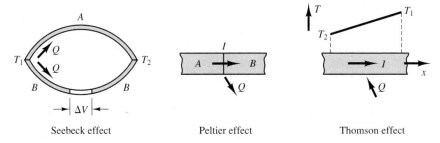

Seebeck effect   Peltier effect   Thomson effect

Figure 11.25   Thermoelectric effects.

where

ΔV = open-circuit potential difference
ΔT = temperature difference between the two junctions
S_{AB} = a coefficient of proportionality dependent on temperature

S_{AB} is called the *Seebeck coefficient* and is usually expressed in units of volts per degree kelvin. The subscript *AB* denotes that *S* depends on the materials constituting the junctions. A familiar example that illustrates the Seebeck effect is the thermocouple. Heat supplied at one of the junctions is converted into electrical energy, which when measured, indicates the temperature.

Two main losses tend to reduce the useful electrical power output provided by the Seebeck effect: heat conduction and electric power loss in the leads. Because a good electrical conductor is also a good heat conductor, much of the heat transferred to a junction is conducted away by the wires so that there are problems in the choice of circuit constants that will lead to optimum power production. At present, even the best metal combinations in a thermocouple achieve an efficiency of only 3 percent in the conversion of heat into electrical energy.

11.13.2 The *Peltier* Effect

In the Peltier effect, which is the inverse of the Seebeck effect, an electric current is sent through a junction of two dissimilar materials *A* and *B*, and this results in either a heating or cooling to the junction. The rate of heating (or cooling) is proportional to the current flow and is given by the equation

$$\dot{Q} = \pi_{AB}I \tag{11.7}$$

where the coefficient π_{AB} is a function of temperature called the *Peltier coefficient*. The subscript *AB* indicates that the Peltier coefficient also depends on the two materials *A* and *B*. A typical unit of π_{AB} is the volt. If current flows in the direction of the potential drop characteristic of the junction, then the temperature of the junction decreases, and a refrigerating effect occurs in the space surrounding the junction. If current flows in the direction of the junction's potential rise, the temperature of the junction increases and heat must be transferred from the junction to maintain constant temperature at the junction.

11.13.3 The *Thomson* Effect

When an electric current flows through a homogeneous material, heat interaction takes place between the material and the surroundings, and the rate of transfer of heat depends on both the quantity of current that flows through the material and the temperature gradient that exists in the material. The *Thomson coefficient* τ is defined as

$$\tau_A = \frac{\delta\dot{Q}/dx}{I(dT/dx)} \tag{11.8}$$

where $\delta\dot{Q}/dx$ is the rate of heat interaction per unit length of the conductor, and dT/dx is the temperature gradient along the length of the conductor. Like the Seebeck and Peltier effects, the Thomson effect is reversible, but it involves only a single substance.

11.14 Thermoelectric Converter

The thermoelectric converter shown in Figure 11.26 consists essentially of two semiconductor blocks connected by a conductor. The conductor receives heat from a thermal source at temperature T_H, whereas the open ends of the blocks reject heat to a low-temperature reservoir at temperature T_L. The sides of the semiconductors are insulated so that heat flow takes place unidimensionally through the semiconductors. When heat is transferred to the high-temperature junction, electrons in n-type semiconductors and holes in p-type semiconductors tend to flow away from the junction. An electric potential is therefore generated, and if the circuit is completed at the low-temperature junction, an electric current flows through the load.

This system may be reversed by supplying electrical energy rather than heat. The current flow then serves to remove heat energy from one junction and to deliver heat energy to the other junction. The process of thermoelectric refrigeration represents an application of the Peltier effect.

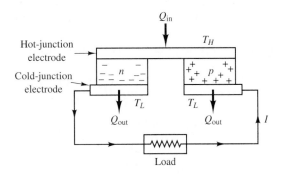

Figure 11.26   Thermoelectric generator.

An index used in rating thermoelectric converters is called the *figure of merit* and is defined as

$$Z = \frac{S}{\rho K_t} \qquad (11.9)$$

where Z is the figure of merit, S is the Seebeck coefficient, ρ is the electrical resistivity, and K_t is the thermal conductivity of the semiconductors. A high figure of merit is achieved by using substances of large Seebeck coefficient, small electrical resistivity, and small thermal conductivity. Maximum values of the figure of merit are obtained when the electric and conduction losses are at a minimum, and when the two losses are approximately equal. Figure 11.27 shows the variations of S and Z as a function of electron density. The maximum value of Z is obtained with semiconductors having an electron density of the order of 10^{19} to 10^{20} electrons/cm^3. Typical values of Z are in the range of 1×10^{-3}/K to 3×10^{-3}/K.

The ideal thermal efficiency of a thermoelectric converter is

$$\text{ideal } \eta_{\text{th}} = \frac{P_{\text{out}}}{\dot{Q}_{\text{in}}} = \left(\frac{T_H - T_L}{T_H} \right) \left[\frac{M - 1}{M + (T_L/T_H)} \right] = \eta_{\text{Carnot}}\, \eta_{\text{rel}} \qquad (11.10)$$

where M is the ratio of the load resistance to the internal resistance of the converter.

$$M = \sqrt{1 + \frac{Z}{2}(T_H + T_L)} \qquad \text{and} \qquad \eta_{\text{rel}} = \frac{M - 1}{M + (T_L/T_H)}$$

The relative efficiency η_{rel} reflects the reduction in efficiency resulting from irreversible losses. The overall efficiencies of actual thermoelectric converters lie between 2 and 6 percent, which are considerably below the efficiencies of the conventional power plants. Special features, such as light weight, small size, and simplicity, make them suitable for space vehicles even though their thermal efficiency is low.

If the thermoelectric converter is operated as a refrigerator, then

$$\text{ideal } \beta_{\text{ref}} = \frac{\dot{Q}_c}{P_{\text{in}}} = \left(\frac{T_L}{T_H - T_L} \right) \left[\frac{M - (T_L/T_H)}{M + 1} \right] \qquad (11.11)$$

$$= (\beta_{\text{ref}})_{\text{Carnot}} (\beta_{\text{ref}})_{\text{rel}}$$

where

$$(\beta_{\text{ref}})_{\text{rel}} = \frac{M - (T_L/T_H)}{M + 1}$$

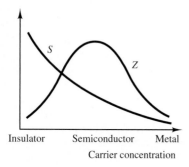

Figure 11.27   *S* and *Z* versus carrier concentration.

If the converter is used as a heat pump, its performance is described by

$$\text{ideal } \beta_{\text{heat pump}} = \frac{\dot{Q}_h}{P_{\text{in}}} = \frac{T_L}{T_H - T_L}\left(1 - 2\frac{M-1}{T_H^2}\right) \qquad (11.12)$$
$$= (\beta_{ht\cdot p})_{\text{Carnot}}(\beta_{ht\cdot p})_{\text{rel}}$$

where

$$(\beta_{ht\cdot p})_{\text{rel}} = \frac{T_L}{T_H}\left(1 - 2\frac{M-1}{T_H^2}\right)$$

11.15 Thermionic Converter

Thermionic converters are high-temperature heat engines in which heat energy is converted directly into electrical energy through thermionic electron emission.

A thermionic converter is a source of low-voltage, high-current electrical energy. The basic elements of the thermionic converter, which is classified as a *diode,* are shown in Figure 11.28. The cathode (emitter) and anode (collector) are separated by either an evacuated or a vapor-filled space. The emitter and collector, coupled to an energy source and sink, respectively, are connected to an electrical load. When heat is transferred to the cathode, electrons in the metal become energetic enough to leave the surface of the metal and to travel toward the anode. An abundance of holes then exists in the cathode, which becomes positive, whereas the anode becomes negative. If a resistive load is placed across the terminals of the electrodes, electric current can be drawn. Thus, the thermionic converter is an engine in which heat transferred to the cathode is converted to electricity.

A major impediment to the flow of electrons is the space-charge barrier. The space charge is formed by electrons and tends to limit the current density.[1] Two solutions to this difficulty are possible. In one method, the space between the electrodes is reduced, thus limiting the emission current of the space charge. In the closed-spaced diode, interelectrode spacing is of the order of 0.001 cm (10×10^{-6} m). Temperatures of 1500 K to 2000 K at the cathode and 1000 K at the anode are common, with a diode output of 2 to 10 W/cm^2.

In the other method, the space charge is suppressed by the introduction of positive ions into the interelectrode space, thus neutralizing the negative space charge (plasma[2] diode). This neutralization is achieved through the presence of a vapor, such as cesium vapor, at low density. Cesium has a low ionization

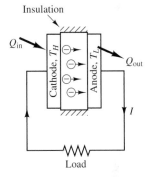

Insulation

Figure 11.28   Thermionic converter.

---

[1] Current density (amp/cm^2) is the product of the velocity of each electron and its charge, summed up for all the electrons emitted by the cathode.
[2] A neutral mixture of ions and electrons is called a *plasma.*

potential (3.88 volts) so that it ionizes readily. Ionization results when cesium atoms collide with the emitted electrons or with the hot cathode wall, according to the equation

$$Cs \rightleftharpoons Cs^+ + e^-$$

In plasma diodes, the interelectrode spacing is of the order of 0.1 cm, and cathode temperatures are in the range of 1600 K to 2600 K. The electrostatic potential diagram of the circuit of a diode in thermodynamic equilibrium is shown in Figure 11.29(a). The potential φ_E is called the *work function* of the emitter (cathode) and is equal to the potential difference between a point in the interior of the cathode and a point on its surface. It represents the barrier potential of the emitter, which is the additional potential an electron must possess in order to be able to leave the surface of the cathode. Similarly, the potential φ_C represents the potential barrier of the collector (anode). The potential φ_g is called the *gap potential* or *contact potential* and is equal to the difference in potential between two points on the surfaces of the electrodes. The *junction potential φ_{EC}* is equal to the potential between the fused junctions of the metals of the emitter and collector.

Figure 11.29(b) shows the simultaneous electronic pressure diagram, which is a mirror image of the potential diagram. Note that the highest pressure occurs in the metals because of their high electron density. The pressure distribution completely balances the force created by the potential distribution at every point of the circuit except when the electrodes are at different temperatures. Any electrons that leave the cathode's surface are accelerated as a result of the potential of the cesium ions adjacent to the cathode surface. The potential of the electrons is thus raised to a value φ', as shown in Figure 11.30. The potential φ_p is the plasma potential drop, and φ_K is potential due to the kinetic energy of the electron. $\varphi_p - \varphi_K$ represents the difference in potential between the emitter surface and the collector surface. The potential φ_C is the collector work function. The output voltage as seen from the diagram is therefore

$$V = \varphi_E + \varphi_K - \varphi_p - \varphi_C \tag{11.13}$$

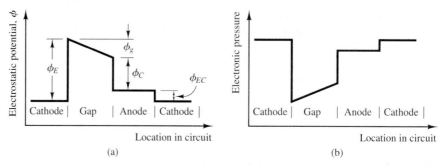

Figure 11.29   Electrostatic potential and electronic pressure diagrams of a diode.

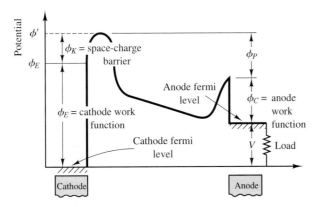

Figure 11.30   Potential diagram of a diode.

The diode current density (current per unit area) is given by the *Richardson–Dushman* equation[3] which applies to both the cathode and the anode:

$$j = AT^2 e^{\varphi'/kT} \tag{11.14}$$

where

j = current density
A = constant equal to 120.1 amp/cm K^2
T = absolute temperature
φ' = work function = $\varphi_E + \varphi_K$
k = Boltzmann's constant

A small current flows in the reverse direction from the anode so that the net diode current is the difference between the current emitted from the cathode and the "back current" from the anode, or

$$j = j_{\text{cathode}} - j_{\text{anode}} \tag{11.15}$$

The current from the anode is small compared to that emitted from the cathode because the anode is much lower in temperature than the cathode. If losses due to heat conduction are not considered, it may be shown that the thermal efficiency of the thermionic converter is

$$\eta_{\text{th}} = \frac{jV}{\dot{q}_e + \dot{q}_r} = \frac{jV}{(\varphi' + 2V_T)j + \dot{q}_r} \tag{11.16}$$

[3] See S. L. Sheldon Chang, *Energy Conversion,* Prentice-Hall, Inc., 1963, for derivation of this equation.

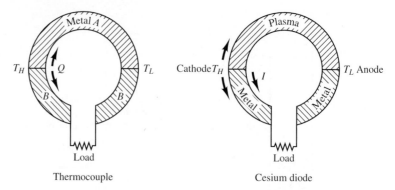

Figure 11.31   Analogy between thermoelectric and thermionic converters.

where $\dot{q}_e$ is the amount of heat energy per unit area and unit time (heat flux) necessary to drive off the electrons from the emitter surface, $\dot{q}_r$ is the radiation heat loss per unit area and unit time from the cathode to the anode, V is the output voltage, V_T is the voltage equivalent of the cathode temperature (kT_{cath}/e), and j is the current density. If the collector work function is reduced, a higher thermal efficiency may be attained; on the other hand, if the work function is very small (~ 0.5 volt), the back current is large.

The thermoelectric converter and the thermionic converter are similar in many respects. As shown in Figure 11.31, metal A in a thermocouple appears as a plasma in a thermionic converter. The junction potential barrier in a thermocouple is the Peltier coefficient; in a thermionic converter, it is the work function.

Thermionic converters are suited for applications that require a large power-to-weight ratio although the maximum efficiency attained at present is 10 percent.

11.16 Fuel Cell

A *fuel cell* is an electrochemical device that converts chemical energy directly into electrical energy. The basic components of a fuel cell that uses hydrogen (or hydrocarbon) as a fuel and oxygen (or air) as an oxidizer are shown in Figure 11.32. Each gas is introduced at high pressure into a chamber; the two chambers are separated from each other by an electrolyte, which may be solid or liquid. Porous ceramic (zirconia oxide), solid polymers, and alkaline solutions (potassium hydroxide) have been successfully used as electrolytes. At elevated temperatures, the electrolyte acts like a sieve, allowing hydrogen ions to migrate through the material. As a result, an ionic potential develops between the two sides of the electrolyte. When a solid electrolyte is used, its surface is first impregnated with a catalyst to facilitate the ion migration process. Impregnation

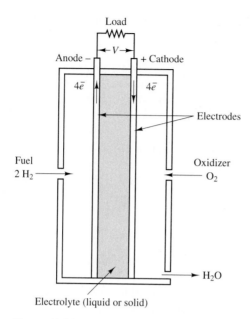

Figure 11.32   Hydrogen–oxygen fuel cell.

also allows the two sides of the electrolyte to be used as electrodes. When a liquid electrolyte is used, separate electrodes are necessary. The electrical load is connected between the anode (hydrogen side) and the cathode (oxygen side). At elevated temperature, solid electrolytes such as polymers are commonly used.

The principle of operation of a fuel cell using hydrogen and oxygen and acidic electrolysis is as follows. At the anode–electrolyte interface, hydrogen molecules ionize, forming two hydrogen ions and two free electrons per molecule. The reaction at the anode can be expressed as

$$2H_2(g) \rightarrow 4H^+ + 4e^-$$

The electrons flow through the external circuit and return to the fuel cell at the cathode, leaving a positive charge at the anode. At the same time, the hydrogen ions diffuse through the electrolyte. When the hydrogen ions reach the cathode, they combine with the electrons and with oxygen molecules to form water. The reaction at the cathode is

$$4e^- + 4H^+ + O_2(g) \rightarrow 2H_2O(\ell)$$

The overall cell reaction is therefore

$$2H_2(g) + O_2(g) \rightarrow 2H_2O(\ell)$$

If an alkaline electrolyte such as potassium hydroxide solution is used, the following reactions take place:

anode: $2H_2(g) + 4(OH)^- \rightarrow 4H_2O(\ell) + 4e^-$
cathode: $O_2(g) + 2H_2O(\ell) + 4e^- \rightarrow 4(OH)^-$
total reaction: $2H_2(g) + O_2(g) \rightarrow 2H_2O(\ell)$

A quantity of energy representing the enthalpy of combustion of the fuel, $\Delta H°$, is released by this chemical reaction. Part of this energy is available for conversion to electrical energy, the maximum amount being equal to the change of the Gibbs function ($-W_e \leq G_1 - G_2$). The electrical energy output per mole of O_2 (or per 2 moles of H_2) is

$$W_e = 4N_a eV \tag{11.17}$$

where N_a is Avogadro's number, e is the electronic charge, and V is the voltage developed.

Under reversible steady-state conditions, the maximum value of W_e is equal to the change in the Gibbs free energy:

$$\begin{aligned} W_e &= G_P - G_R \\ &= \sum_{\text{prod}} n_i (\bar{g}_f° + \bar{g} - \bar{g}°)_i - \sum_{\text{react}} n_i (\bar{g}_f° + \bar{g} - \bar{g}°)_i \end{aligned} \tag{11.18}$$

where $\bar{g}_f°$ is the molar Gibbs function of formation, and n is the number of moles. Therefore, the maximum voltage that can be developed by a fuel cell is

$$V = \frac{G_P - G_R}{n_e N_a e} \tag{11.19}$$

where n_e is the number of electrons released per molecule of oxygen. In forming one kg-mol of liquid water from hydrogen and oxygen, the free-energy change is $-237,141$ kJ/kg-mol of hydrogen. The maximum voltage attainable from the chemical reaction at the standard state is then

$$\begin{aligned} V &= \frac{-2(237141 \times 10^3 \text{ J/kg-mol})}{(4 \text{ electrons})(6.022 \times 10^{26} \text{ molecules/kg-mol})(-1.602 \times 10^{-19} \text{ J/eV})} \\ &= 1.23 \text{ volt} \end{aligned}$$

The maximum efficiency of a fuel cell is

$$\eta = \frac{|W_e|}{(-\Delta H°)} = \frac{n_e N_a eV}{(-\Delta H°)} \tag{11.20}$$

For the hydrogen–oxygen reaction in which liquid water is formed, the enthalpy of combustion is $-285,830$ kJ/kg-mol so that the efficiency is 83 percent.

Unlike heat engines, fuel cells are not subject to Carnot-cycle limitation. A fuel cell can operate isothermally and does not necessarily involve heat interaction with the surroundings. Moreover, the temperature within the fuel cell is low in comparison with devices involving rapid combustion reactions. However, high efficiencies are attained only at relatively light loads and at voltages of about 1 volt. Some advantages of the fuel cell are its simplicity and its high power-to-weight ratio. Its disadvantages include high cost, need for equipment to control the flow of reactants to match reaction rates, problems in selection of component materials that are not attacked by the reactants, and relatively short life, especially at high temperatures.

11.17 Magnetohydrodynamic (MHD) Generator

According to Faraday's law of electromagnetic induction, a changing magnetic field induces an electric field in any conductor located in it. The electric field acts on the free charges in the conductor, causing a current to flow. In a conventional electric generator, electric conductors cross the lines of the magnetic field; in the magnetohydrodynamic (MHD) generator, ionized gas, which acts like an electrical conductor, flows across the lines of a magnetic field. In both cases, a voltage is induced. The simplicity of MHD stems from the fact that the kinetic and enthalpy energies of the ionized gas are converted directly into electric energy without the intermediate mechanical step of rotating the conducting medium across the magnetic field. In the MHD generator, a portion of the energy of the ionized gas is transformed into electrical energy. Figure 11.33 shows the basic components of a MHD generator. Hot ionized gas at a high velocity passes between the poles of an electromagnet. This induces a potential difference between a pair of electrodes placed at right angles to the magnetic field. When the electrodes are connected by a resistive load, a current flows in the circuit.

The energy input to a MHD generator is the ionization energy, which increases with temperature. High electrical conductivity at low temperature is then a prime objective in minimizing energy input. The ionized gas needed for the MHD generator may be obtained by heating a gas to temperatures between 2200 K and 2800 K. A nuclear fuel may serve as the energy source. Alternatively, the ionization can be assisted by seeding the gas with small quantities of a material that ionizes easily, such as cesium or potassium, and the gas may then be used at temperatures as low as 1400 K. After passing through the MHD generator, the ionized gas is usually recirculated through a regenerator, where energy is added to the gas before it reenters the MHD generator.

The electromagnetic force $\vec{F}$ acting on a moving charge is

$$\vec{F} = q(\vec{E} + \vec{V} \times \vec{B}) \tag{11.21}$$

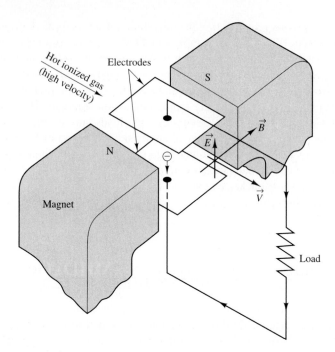

Figure 11.33   Magnetohydrodynamic (MHD) generator.

where

q = charge
$\vec{E}$ = electric field strength
$\vec{V}$ = velocity of the ionized gas
$\vec{B}$ = magnetic field strength
$\times$ indicates vector product

The efficiency of the MHD generator, which is the electric energy produced divided by the energy input, is

$$\eta_{\text{th}} = \frac{VI}{\dot{m}h_{0,\text{in}}} = \frac{h_{0,\text{in}} - h_{0,\text{out}}}{h_{0,\text{in}}} = 1 - \frac{h_{0,\text{out}}}{h_{0,\text{in}}} \qquad (11.22)$$

where

V = voltage produced
I = current flow

$h_{0,\text{in}}$ and $h_{0,\text{out}}$ are total enthalpy[4] per unit mass of the hot gas stream, respectively, entering and leaving the generator.

Much higher temperatures can be tolerated in MHD than in conventional cycles mainly because of the nonexistence of a prime mover. For this reason, MHD generators are considered for use as topping generators for gas turbine and steam turbine power plants. The main problems encountered in MHD systems still center on selection of materials that are capable of withstanding the high temperatures of the ionized gas. Figure 11.34 is a schematic diagram of an MHD generator used as a topping plant. Air is compressed and preheated before entering the combustion chamber upstream of the MHD generator. Seeding is used to ionize the gas as it flows through the magnetic field. At the exit, the energy of the gas serves as an energy source for the bottoming steam cycle.

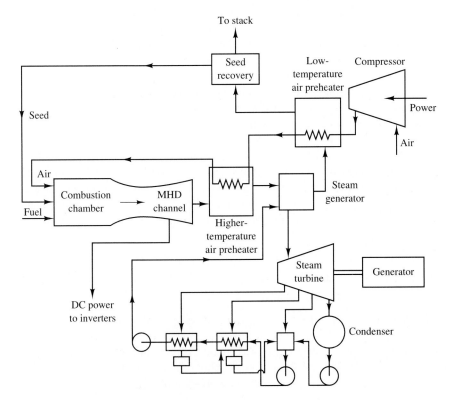

Figure 11.34   Schematic of a combined MHD generator and steam cycle.

---

[4] Total enthalpy h_0 is equal to $h + \dfrac{V^2}{2}$.

11.18 Summary

Fossil and nuclear fuels are the most commonly used sources of energy for power generation. Other sources such as hydraulic, geothermal, and solar are also used but to a much lesser degree. Comparison of an actual cycle with the Carnot cycle reveals that the main irreversibilities in the actual cycle are due to the wide range in temperature difference during the heat interaction processes. In order to improve the performance of power-generating plants, several schemes have been developed to reduce this source of irreversibility.

In the gas–steam combined cycle, the exhaust energy of the gas turbine is used to generate steam, which, upon expanding in a steam turbine, develops additional work. In the Cheng dual-fluid cycle, steam is generated in a heat recovery boiler and injected into the gas-turbine combustion chamber. The humid-air turbine cycle makes use of multistage compression to reduce the power requirement and a multistage saturator that allows a variable boiling point of the steam. The chemically recuperated gas turbine cycle incorporates a methane steam reformer, and in the gas turbine–ammonia combined cycle, ammonia is used as a refrigerant and power fluid. In the Kalina cycle, a mixture of ammonia and water is used as the working fluid. This two-component fluid allows the variation of temperature as the fluid changes phase.

In cogeneration plants, process heat and electric energy are generated simultaneously. The energy developed as work and heat indicates the degree of utilization of the input energy. In the mercury–water binary cycle, mercury is used in a topping cycle. The energy release upon condensing the mercury is used to generate steam, which serves as the working fluid in a steam bottoming cycle.

In direct-energy conversion systems, some intermediate steps in converting energy sources into electric power are bypassed. Further development of fuel cells and MHD generators is subject to material selection to withstand the reactants in the case of fuel cells and the high temperature encountered in MHD.

Problems

11.1 By studying the *T-s* diagram of an ideal regenerative gas-turbine cycle with a regenerator effectiveness of unity show that as the pressure ratio $r_p \rightarrow 1$, the cycle efficiency is given by

$$\eta = 1 - \frac{T_a}{T_b}$$

where T_a and T_b are inlet temperatures to the compressor and turbine, respectively.

11.2 Using a *T-s* diagram, show that the addition of either a single-stage intercooling or a single-stage reheating to a Brayton cycle results in a decrease in the cycle efficiency. Show that the same conclusions also apply to an irreversible simple gas-turbine cycle.

11.3 In a combined gas–steam power plant, air enters the compressor at 300 K and at the rate of 15 kg/s, and experiences a pressure ratio of 10. The air is heated to 1400 K in the combustion chamber. The exhaust gases from the turbine are used to generate steam at 300°C and 8 MPa in the steam boiler. The steam expands in a steam turbine and condenses at 10 kPa. Assuming the flow in the air compressor, the gas turbine, and the steam turbine is isentropic, determine (a) the mass rate of flow of the steam, (b) the net power output, (c) the thermal efficiency of the plant.

11.4 Figure 11.35 shows a combined gas–steam power cycle. Air enters the compressor ($\eta_C = 0.84$) at 25°C and 100 kPa, where it is compressed to 1.5 MPa. The air is then heated to 1000°C before expanding in the gas turbine ($\eta_T = 0.86$). The exhaust of the turbine is used to generate steam at 6 MPa and 300°C. Upon leaving the steam generator, the steam is further superheated at constant pressure to 400°C before entering the steam turbine ($\eta_S = 0.86$). Steam is extracted at a pressure of 1 MPa and mixed with the condensate in an open feed-water heater. The remainder of the steam continues to expand to the condenser pressure of 10 kPa. If the temperature difference at the pinch point is 20°C, represent the cycle on a *T-s* diagram and determine (a) the overall efficiency of the power plant, (b) the availability per unit mass of air at each point of the cycle, (c) the irreversibility in each component. Neglect pump work.

11.5 Determine the improvement in the overall efficiency of the combined cycle of the previous problem if the feed-water heater is eliminated.

11.6 Referring to the gas–steam combined cycle depicted in Problem 11.4, the following data apply:

$T_1 = 25°C$ $p_1 = 100$ kPa $r_p = 15$
$\eta_C = 0.84$ $\eta_T = 0.86$
$T_4 = 1000°C$ $T_5 = 200°C$
$T_7 = 400°C$ $p_7 = 8$ MPa
$p_9 = 0.5$ MPa $p_{10} = 10$ kPa
$| \dot{W}_G + \dot{W}_S | = 500$ MW
No steam superheater and neglect pump work.

Determine (a) the mass ratio of the gas to steam, (b) the mass flow rate of the air, (c) the thermal efficiency of the combined cycle.

11.7 Steam enters the turbine of a cogeneration plant at 6 MPa and 400°C at the rate of 1 kg/s. Saturated steam at a pressure of 0.3 MPa and at a rate of 0.4 kg/s is extracted from the turbine to be used for process heating. The remainder of the steam condenses at 10 kPa. The

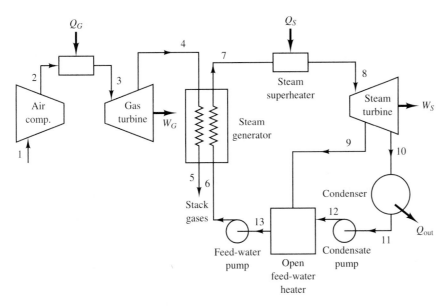

Figure 11.35

steam, leaving the process heater as saturated liquid at 0.3 MPa is mixed with the condensate from the condenser at the same pressure, and the mixture is pumped to the boiler pressure. Assuming isentropic flow in the turbine and pumps, determine (a) the utilization factor of the plant, (b) the net power generated.

11.8   Repeat Problem 11.7 if an open feed-water heater, using 0.1 kg/s of the extracted steam from the turbine, is incorporated in the cycle.

11.9   An industrial plant requires 3 MW of power and 4.5 MW of heat from process steam at 110°C. Steam is generated at a pressure of 2.5 MPa and a temperature of 300°C. It expands in a turbine with an isentropic efficiency of 0.8 to a condenser pressure of 5 kPa. The process steam is bled from the turbine at the required pressure. Determine the heat input to the boiler. Neglect pump work.

11.10   Show that the thermal efficiency of a mercury–steam binary-vapor cycle is given by

$$\eta_{\text{cycle}} = 1 - (1 - \eta_M)(1 - \eta_S)$$

where η_M and η_S are the thermal efficiencies of the mercury and steam cycles, respectively.

11.11   An ideal thermoelectric generator receives heat at the rate of 250 W from an energy source at 200°C and rejects heat to an environment at 25°C. If the figure of merit is 2.1×10^{-3} K^{-1}, determine the thermal efficiency and the power developed.

11.12   An ideal thermoelectric refrigerator removes 200 W of energy from a refrigerated space at -10°C and rejects heat to an environment at 25°C. If the figure of merit is 3×10^{-3} K^{-1}, determine the coefficient of performance and the power required.

11.13   Determine the maximum power output and the figure of merit of a thermoelectric converter operating at steady state between $T_H = 97$°C and $T_L = 32$°C. The average Seebeck coefficient is 2×10^{-3} volts/K, the thermal conductivity is 0.0498 W/mK, and the resistance is 0.07 ohm-m. What is the ideal thermal efficiency of the converter?

11.14   A fuel cell operates steadily at the standard state of 298.15 K and 1 atm and at a cell voltage of 0.7 volt according to the reaction

$$H_2(g) + \frac{1}{2}O_2(g) \rightarrow H_2O(\ell)$$

Calculate the electrical work developed per mole of H_2. What is the electrical work and the heat interaction for reversible operation?

11.15   A bank of oxygen–hydrogen fuel cells operates isothermally at 25°C and atmospheric pressure. The actual voltage of each cell is 70 percent of its ideal voltage when the current is 1.2 A. Assuming that only 80 percent of the hydrogen reacts, determine the power output and the rate of heat transfer of each cell. Assume steady-state operation.

11.16   For a reversible hydrogen–oxygen fuel cell operating isothermally at 25° and 1 atm, determine the flow rate of hydrogen to produce 20 kW. Assume steady-state operation.

11.17   A reversible fuel cell operates isothermally at 25°C and atmospheric pressure. The cell operates at steady state with hydrogen and oxygen gas streams. If the power output is 100 W, determine (a) the volumetric rate of flow of hydrogen at 25°C and 1 atm, (b) the rate of heat transfer from the cell.

Chapter 12

Kinetic Theory and Statistical Thermodynamics

12.1 Introduction

Classical thermodynamics is based on simplified models that cannot explain certain phenomena adequately. For example, it treats a gas as though all its molecules were identical not only in mass but also in dynamic qualities (energy, momentum, velocity). There is evidence that marked differences or fluctuations in dynamic qualities do exist. When these differences are incorporated in the model, and the system is then analyzed, valuable information is derived. In the domain of statistical mechanics, then, substances are considered from the microscopic, not the macroscopic, standpoint; furthermore, differences between molecules in their dynamic qualities are considered.

It is possible, in principle, to apply the laws of mechanics to each individual molecule of a gas and thus determine the behavior of the system. But the behavior of individual particles, such as the molecules of a gas, is complicated, time dependent, and difficult to follow. Furthermore, the macroscopic properties of a substance are determined by the average behavior of a large number of molecules rather than by the behavior of a single molecule at a certain time. A molecule of a gas can show a wide spectrum of velocities ranging, in a short period, from zero to very high values. But the average speed of the molecules in the gas is of more interest in interpreting the thermodynamic properties of the gas than is the speed of each molecule.

In this chapter, several statistical models for thermodynamic analysis are introduced. These models depend on the nature of the particles under consideration and indicate the manner in which a fixed number of particles with a given total energy distribute themselves among the available energy states. When the number of particles is very large, the most probable macrostate of particle distribution is far more common than any other, and the energy distribution of particles of the system can be determined with almost total certainty. Hence, within the framework of statistical mechanics, the most probable macrostate defines thermodynamic equilibrium.

12.2 Distribution of Particle Velocities

The rms velocity[1] of the molecules of a gas has been shown to be related to the temperature of the gas. If all particles at the same temperature are moving at the same speed, then the rms value describes the velocity magnitude of all particles perfectly. But if their speeds differ significantly, then the rms value provides only limited information, and the problem of determining the velocity of all the particles still exists. It then becomes necessary to determine a velocity distribution so that the number of particles having velocities between certain prescribed limits can be determined. In practice, such statistical details do not lend themselves to direct use in subsequent calculations; excessive data in this case are not much more useful than meager data. When, however, the expression for velocity distribution is known, it is possible to derive various velocity parameters that can be used.

In determining the translational velocity distribution, a model is first established that is based on *velocity space*. Consider a volume of a gas at constant temperature whose particles are moving at different velocities. The instantaneous velocity vector of each molecule can be resolved into components v_x, v_y, and v_z. If the three velocity directions serve as axes, the velocity of each molecule may be represented as a vector originating at the origin. The surface of a sphere centered at the origin therefore represents, at some instant of time, all particles of identical speed. Furthermore, all particles of identical speed are assumed to be uniformly distributed since large numbers of particles are involved. Finally, the number of particles in each concentric shell is assumed to bear a simple relationship to the velocity.

Consider an infinitesimal volume $dv_x dv_y dv_z$ in the velocity space shown in Figure 12.1. Let this volume lie within a thin spherical shell of inner and outer radii v and $v + dv$. If N is the total number of molecules of the gas, the spherical shell contains dN_v points, each moving at a velocity between v and $v + dv$ in magnitude. Assume that the number of points in the volume $dv_x dv_y dv_z$ is large so that the volume is of uniform density. Note that the velocity components in the v_x, v_y, and v_z directions of the points in the volume $dv_x dv_y dv_z$ lie between v_x

---

[1] The root-mean-square (rms) velocity is defined by $v_{rms}^2 \equiv \sum_i m_i v_i^2 / \sum_i m_i$, where m_i is the mass of the ith molecule.

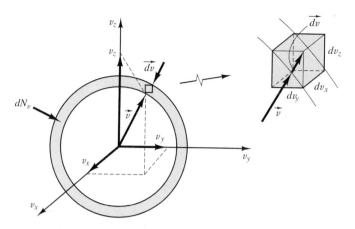

Figure 12.1   Velocity density in a thin spherical shell is constant.

and $v_x + dv_x$, v_y and $v_y + dv_y$, and v_z and $v_z + dv_z$. Consider a narrow zone bounded by two infinite planes parallel to the v_y-v_z plane at a distance v_x from it and dv_x apart. As shown in Figure 12.2(a) the molecules whose velocity vectors terminate between the planes v_x and $v_x + dv_x$ all have their x-component of velocity lying between v_x and $v_x + dv_x$. If dN_{v_x} represents the number of these points, then the fraction dN_{v_x}/N depends on both the distance dv_x between the planes and the location of the planes. Since the location of the planes is a function of v_x, this may be expressed mathematically as

$$\frac{dN_{v_x}}{N} = f(v_x)dv_x \tag{12.1a}$$

where $f(v_x)$ is a function of v_x and is called the *distribution function* for the x-component of velocity. In general, the distribution function indicates the concentration of some quantity per unit "volume" as a function of "position" in an appropriate "space." The product $f(v_x)dv_x$ represents the fraction of molecules with velocities in the x-direction between v_x and $v_x + dv_x$. The term $f(v_x)dv_x$ also expresses the probability that a molecule chosen at random will have an x-component of velocity lying between v_x and $v_x + dv_x$.

The same procedure may be applied to the v_y- and v_z-directions. Since all directions are equally probable, the fraction of molecules having components between v_y and $v_y + dv_y$ and between v_z and $v_z + dv_z$ can be described as

$$\frac{dN_{v_y}}{N} = f(v_y)dv_y \tag{12.1b}$$

and

$$\frac{dN_{v_z}}{N} = f(v_z)dv_z \tag{12.1c}$$

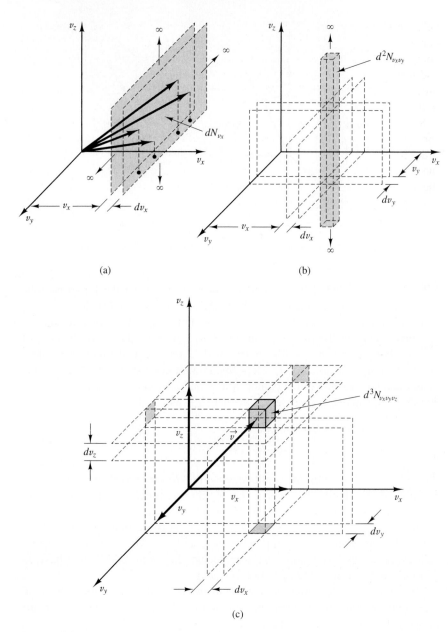

Figure 12.2   Velocity–space diagram.

The foregoing equations imply that the distribution of each component of velocity is independent of the other two.

All the molecules represented by dN_{v_x} have velocities whose v_x-components lie between v_x and $v_x + dv_x$; however, only a small fraction of these points have velocities whose v_y-components lie between v_y and $v_y + dv_y$. If this fraction is denoted by $d^2N_{v_xv_y}/dN_{v_x}$, then $d^2N_{v_xv_y}$ denotes the number of molecules whose components of velocities in the v_x- and v_y-directions lie between v_x and $v_x + dv_x$ and between v_y and $v_y + dv_y$. (The superscript 2 indicates that it is a second-order term.) As shown in Figure 12.2(b), $d^2N_{v_xv_y}$ is equal to the number of points in the vertical cylinder of cross section $dv_x dv_y$ extending between $v_z = \pm\infty$.

Since a gas has a large number of molecules and since there is no preference for any particular direction, the fractions dN_{v_x}/N, $dN_{v_y}/N,\dots$ (and also second-order fractions) can be expected to be the same throughout the entire volume of the gas. Therefore, any small volume of gas contains a representative sample of the velocities of the molecules for the entire volume of the gas. This can be expressed by the following equality:

$$\frac{d^2N_{v_xv_y}}{dN_{v_x}} = \frac{dN_{v_y}}{N}$$

Combining this equation with Eq. (12.1b) gives

$$d^2N_{v_xv_y} = dN_{v_x}f(v_y)dv_y$$

But

$$dN_{v_x} = Nf(v_x)dv_x$$

so that

$$d^2N_{v_xv_y} = Nf(v_x)f(v_y)dv_x\,dv_y \tag{12.2}$$

The foregoing procedure may be extended to the v_z-direction. If the $d^3N_{v_xv_yv_z}$ denotes the number of molecules that have velocity components lying between v_x and $v_x + dv_x$, v_y and $v_y + dv_y$, and v_z and $v_z + dv_z$, then

$$d^3N_{v_xv_yv_z} = Nf(v_x)f(v_y)f(v_z)dv_x\,dv_y\,dv_z \tag{12.3}$$

These points appear in the parallelpiped of volume $dv_x\,dv_y\,dv_z$ of Figure 12.2(c). The density of these velocity points is equal to the number of points divided by the volume, or

$$\rho(v_x,v_y,v_z) = \frac{d^3N_{v_xv_yv_z}}{dv_x\,dv_y\,dv_z} = Nf(v_x)f(v_y)f(v_z) \tag{12.4}$$

Since density in the volume $dv_x\,dv_y\,dv_z$ is a continuous function, the differential of density can be written as

$$d\rho = \left(\frac{\partial\rho}{\partial v_x}\right)dv_x + \left(\frac{\partial\rho}{\partial v_y}\right)dv_y + \left(\frac{\partial\rho}{\partial v_z}\right)dv_z$$

But the density in the volume $dv_x\, dv_y\, dv_z$ is uniform so that $d\rho = 0$ irrespective of the location of the volume $dv_x\, dv_y\, dv_z$ within the shell. Therefore, when Eq. (12.4) is differentiated with respect to the direction variables, the following results:

$$d\rho = 0 = Nf'(v_x)f(v_y)f(v_z)dv_x + Nf'(v_y)f(v_x)f(v_z)dv_y$$
$$+ Nf'(v_z)f(v_x)f(v_y)dv_z$$

where the primes denote differentiation with respect to the argument. This equation reduces to

$$\frac{f'(v_x)}{f(v_x)}dv_x + \frac{f'(v_y)}{f(v_y)}dv_y + \frac{f'(v_z)}{f(v_z)}dv_z = 0 \qquad (12.5)$$

Since all the points in the volume $dv_x\, dv_y\, dv_z$ lie within the spherical shell, the magnitude of the velocity vector v is a constant. This constraint can be expressed as

$$v^2 = v_x^2 + v_y^2 + v_z^2 = \text{constant}$$

Differentiation of the preceding equation gives

$$v_x\, d\, v_x + v_y\, dv_y + v_z\, dv_z = 0 \qquad (12.6)$$

To solve Eq. (12.5) subject to the constraint of Eq. (12.6), *Lagrange's method of undetermined multipliers*, which is outlined in Appendix A18, is applied. Note that there are three unknowns v_x, v_y, and v_z but only two equations. By multiplying Eq. (12.6) by λ and adding the result to Eq. (12.5), the following is obtained:

$$\left(\frac{f'(v_x)}{f(v_x)} + \lambda v_x\right)dv_x + \left(\frac{f'(v_y)}{f(v_y)} + \lambda v_y\right)dv_y + \left(\frac{f'(v_z)}{f(v_z)} + \lambda v_z\right)dv_z = 0$$

where λ is called *Lagrange's undetermined multiplier*.

Since the components of velocity v_x, v_y, and v_z may be considered independent variables, the coefficients of dv_x, dv_y, and dv_z can be individually equated to zero. Therefore,

$$\frac{f'(v_x)}{f(v_x)} + \lambda v_x = 0$$

$$\frac{f'(v_y)}{f(v_y)} + \lambda v_y = 0$$

$$\frac{f'(v_z)}{f(v_z)} + \lambda v_z = 0$$

Integration yields

$$f(v_x) = ae^{-\lambda v_x^2/2}$$
$$f(v_y) = ae^{-\lambda v_y^2/2}$$

and

$$f(v_z) = ae^{-\lambda v_z^2/2}$$

where a is a constant of integration. Note that symmetry indicates that the three foregoing equations have the same integration constant. Although only two equations were available to start with, the proper value of the constant λ makes v_x, v_y, and v_z independent of each other so that four equations rather than only two become available. When the preceding values of $f(v_x)$, $f(v_y)$, and $f(v_z)$ are substituted into Eq. (12.4), the density of velocity points becomes

$$\rho(v) = Na^3 e^{(-\lambda/2)(v_x^2+v_y^2+v_z^2)} = Na^3 e^{(-\lambda/2)(v^2)}$$

Denoting $\lambda/2$ by β^2, then

$$\rho(v) = Na^3 e^{-\beta^2 v^2} \qquad (12.7)$$

Equation (12.7) is the *Maxwell velocity distribution function*. Note that density is a function of the magnitude, but not the direction, of the velocity. This is in accordance with the assumption that the gas is isotropic.

12.3 The Functions a and β

Before Eq. (12.7) can be used, the nature of the two constants that appear in it, a and β, must be established. The total number of molecules is described by

$$N = \int_{v=0}^{\infty} dN_v = \int_0^{\infty} 4\pi v^2 \rho \, dv$$

where dN_v is the number of molecules in a spherical shell of radius v and thickness dv in velocity space. Using Eq. (12.7), then

$$N = \int_0^{\infty} dN_v = 4\pi Na^3 \int_0^{\infty} v^2 e^{-\beta^2 v^2} dv \qquad (12.8)$$

To integrate the preceding expression, first let $x = \beta^2 v^2$, then

$$\int_0^\infty v^2 e^{-\beta^2 v^2}\, dv = \int_0^\infty \left(\frac{x}{\beta^2}\right) e^{-x} \frac{dx}{2x^{1/2}\beta}$$

$$= \frac{1}{2\beta^3}\int_0^\infty x^{(3/2)-1} e^{-x}dx$$

$$= \frac{1}{2\beta^3}\Gamma\left(\frac{3}{2}\right) = \frac{1}{2\beta^3}\frac{\sqrt{\pi}}{2}$$

where $\Gamma(n) = \int_0^\infty x^{n-1}e^{-x}dx$ $(n > 0)$ is called the *gamma function*. Values of the gamma function are tabulated in Table 12.1. The expression for N then becomes

$$N = \left(\frac{2\pi N a^3}{\beta^3}\right)\left(\frac{\sqrt{\pi}}{2}\right) = \pi^{3/2} N\left(\frac{a}{\beta}\right)^3$$

from which

$$a = \frac{\beta}{\sqrt{\pi}} \tag{12.9}$$

The function β still remains to be determined. The translational kinetic energy of the molecules of a gas is described by

$$U = \int_0^\infty \frac{1}{2}mv^2 dN_v$$

Table 12.1 $\Gamma(n) = \int_0^\infty x^{n-1}e^{-x}dx$ $(n > 0)$*

n	$\Gamma(n)$	n	$\Gamma(n)$
$\frac{1}{2}$	$\sqrt{\pi}$	3	2
1	1	$\frac{7}{2}$	$\frac{15}{8}\sqrt{\pi}$
$\frac{3}{2}$	$\frac{\sqrt{\pi}}{2}$	4	6
2	1	$\frac{9}{2}$	$\frac{105}{16}\sqrt{\pi}$
$\frac{5}{2}$	$\frac{3}{4}\sqrt{\pi}$	5	24

*It may be shown that $\Gamma(n + 1) = n\Gamma(n)$.

But from Eq. (12.8) and (12.9),

$$dN_v = \frac{4}{\sqrt{\pi}} v^2 N \beta^3 e^{-\beta^2 v^2} dv$$

Therefore,

$$U = \int_0^\infty \frac{1}{2} mv^2 \frac{4}{\sqrt{\pi}} v^2 N \beta^3 e^{-\beta^2 v^2} dv = \frac{2mN\beta^3}{\sqrt{\pi}} \int_0^\infty v^4 e^{-\beta^2 v^2} dv$$

Again letting $x = \beta^2 v^2$ gives

$$\begin{aligned}
U &= \frac{2mN\beta^3}{\sqrt{\pi}} \int_0^\infty \left(\frac{x}{\beta^2}\right)^2 e^{-x} \frac{dx}{2\beta^2 (x^{1/2}/\beta)} \\
&= \frac{mN}{\sqrt{\pi\beta^2}} \int_0^\infty x^{(5/2)-1} e^{-x} dx \\
&= \left(\frac{mN}{\sqrt{\pi\beta^2}}\right) \Gamma\left(\frac{5}{2}\right) = \frac{3mN}{4\beta^2}
\end{aligned}$$

But the total translational energy U is

$$U = \frac{3}{2} NkT$$

where k is Boltzmann's constant. Equating the two previous equations gives the function β as

$$\beta = \sqrt{\frac{m}{2kT}} \qquad (12.10)$$

From Eq. (12.9), the expression of a is therefore

$$a = \sqrt{\frac{m}{2\pi kT}} \qquad (12.11)$$

Substituting the foregoing expressions of a and β into Eq. (12.8) gives

$$\frac{dN_v}{dv} = \frac{4N}{\sqrt{\pi}} \left(\frac{m}{2kT}\right)^{3/2} v^2 e^{-(m/2kT)v^2} \qquad (12.12)$$

The function dN_v/dv is called the Maxwell–Boltzmann *speed distribution function*. It represents the number of molecules having a speed v per unit range of

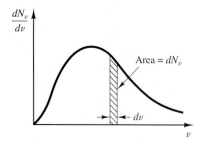

Figure 12.3   Maxwell–Boltzmann speed distribution (Poisson's curve).

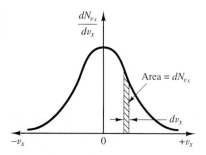

Figure 12.4   Maxwell–Boltzmann distribution for one component of velocity (Gaussian's curve).

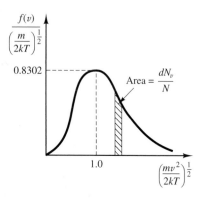

Figure 12.5   Maxwell–Boltzmann speed distribution.

speed. A plot of dN_v/dv versus v is shown in Figure 12.3. An area dv-wide underneath the curve represents the number dN_v, that is, the number of molecules having speeds between v and $v + dv$.

The speed distribution function for each of the three velocity components may now be determined. In the x-direction, for example, Eq. (12.1a) gives

$$\frac{dN_{v_x}}{dv_x} = Nf(v_x) = \frac{N}{\sqrt{\pi}}\left(\frac{m}{2kT}\right)^{1/2} e^{-(m/2kT)v_x^2} \tag{12.13}$$

A plot of dNv_x/dv_x versus v_x is presented in Figure 12.4. The curve is symmetric about the zero-speed axis and has a maximum value at $v_x = 0$ of $N(m/2\pi kT)^{1/2}$. The Maxwell–Boltzmann speed distribution according to Eq. (12.12) may be written in the form

$$\frac{dN_v}{N} = f(v)dv = \frac{4}{\sqrt{\pi}}\left(\frac{m}{2kT}\right)^{3/2} v^2 e^{-mv^2/2kT} dv \tag{12.14}$$

Figure 12.5 shows a plot of the Maxwell–Boltzmann speed distribution law.

An expression for the Maxwell–Boltzmann distribution law for the kinetic energies of the molecules is obtained as follows. In a coordinate space diagram, let the number of molecules having energies between ε and $\varepsilon + d\varepsilon$ be dN_ε. The kinetic energy of a molecule is

$$\varepsilon = \frac{1}{2}mv^2$$

and

$$d\varepsilon = mv\,dv = m\sqrt{\frac{2\varepsilon}{m}}dv = \sqrt{2m\varepsilon}\,dv$$

or

$$dv = \frac{d\varepsilon}{\sqrt{2m\varepsilon}}$$

Substituting for dv from the preceding equation into Eq. (12.12) gives

$$dN_\varepsilon = \frac{4N}{\sqrt{\pi}}\left(\frac{m}{2kT}\right)^{3/2} \frac{2\varepsilon}{m} e^{-\varepsilon/kT} \frac{d\varepsilon}{\sqrt{2m\varepsilon}}$$

or

$$\frac{dN_\varepsilon}{d\varepsilon} = \frac{2N}{\sqrt{\pi}} \frac{\varepsilon^{1/2}}{(kT)^{3/2}} e^{-\varepsilon/kT} \qquad (12.15)$$

A plot of $dN_\varepsilon/d\varepsilon$ versus ε/kT is shown in Figure 12.6. The maximum value of $dN_\varepsilon/d\varepsilon$ corresponds to ε equal to $kT/2$.

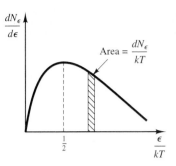

Figure 12.6   Maxwell energy distribution.

12.4 The Distribution Function

The speed distribution function $f(v)$ when multiplied by dv represents the fraction of the molecules with speeds in the interval between v and $v + dv$. Since the number of molecules in that interval is $Nf(v)dv$, the function $f(v)$ can be considered a statistical proportionality factor so that the following equation is satisfied:

$$\int_0^\infty f(v) \, dv = 1 \qquad (12.16)$$

The function f at any instant in time depends on the location of the molecules as well as on their velocities so that

$$f(v) = \varphi(x, y, z, v_x, v_y, v_z, t) \qquad (12.17)$$

When the gas is in statistical equilibrium, the position of the molecules is completely random, and the distribution function can be considered to depend only on v_x, v_y, and v_z; that is,

$$f(v) = \varphi(v_x, v_y, v_z) \qquad (12.18)$$

The previous section outlined the importance of the velocity distribution function as a useful parameter in describing the velocities of the molecules of a gas. The concept may be extended to describe the distributions in other properties, such as density, energy, and square of speeds. In particular, the distribution function can be used to determine three characteristic molecular speeds: the average speed, the root mean square speed, and the most probable speed.

The *average speed* is defined as the sum of the speeds of all molecules divided by the number of molecules. If x is any function of molecular speed, the average value of x is given by the equation

$$x_{\text{avg}} = \int_0^\infty xf(v) \, dv \qquad (12.19)$$

Thus,

$$v_{\text{avg}} = \frac{1}{N} \int_0^\infty v \, dN_v = \int_0^\infty v f(v) \, dv$$

Substituting for $f(v)$ from Eq. (12.14) gives

$$v_{\text{avg}} = \int_0^\infty \frac{4}{\sqrt{\pi}} \left(\frac{m}{2kT} \right)^{3/2} v^3 e^{-(m/2kT)v^2} dv$$

$$= \frac{4}{\sqrt{\pi}} \left(\frac{m}{2kT} \right)^{3/2} \int_0^\infty v^3 e^{-(m/2kT)v^2} dv$$

Let

$$x = \frac{m}{2kT} v^2$$

$$dx = \frac{m}{kT} v \, dv = \sqrt{\frac{2m}{kT}} x^{1/2} dv$$

Substituting for v and dv, the expression for v_{avg} becomes

$$v_{\text{avg}} = \sqrt{\frac{2}{\pi}} \left(\frac{m}{kT} \right)^{3/2} \int_0^\infty \left(\frac{2xkT}{m} \right)^{3/2} e^{-x} \sqrt{\frac{kT}{2m}} \frac{1}{x^{1/2}} dx$$

$$= 2\sqrt{\frac{2kT}{\pi m}} \int_0^\infty x^{2-1} e^{-x} dx \qquad (12.20)$$

$$= 2\sqrt{\frac{2kT}{\pi m}} \Gamma(2) = 2\sqrt{\frac{2kT}{\pi m}}$$

The *root mean square speed* is defined as the square root of the sum of velocity squares of all molecules divided by the number of molecules. Thus

$$v_{\text{rms}}^2 = \frac{1}{N} \int_0^\infty v^2 \, dN_v = \int_0^\infty v^2 f(v) \, dv$$

Substituting from Eq. (12.14) gives

$$v_{\text{rms}}^2 = \int_0^\infty \frac{4}{\sqrt{\pi}} \left(\frac{m}{2kT} \right)^{3/2} v^4 e^{-(m/2kT)v^2} dv$$

$$= \frac{4}{\sqrt{\pi}} \left(\frac{m}{2kT} \right)^{3/2} \int_0^\infty \frac{4k^2T^2}{m^2} e^{-x} x^{3/2} \sqrt{\frac{kT}{2m}} dx$$

$$= \frac{4kT}{\sqrt{\pi m}} \Gamma\left(\frac{5}{2} \right) = \frac{4kT}{\sqrt{\pi m}} \frac{3\sqrt{\pi}}{4} = \frac{3kT}{m}$$

Hence

$$v_{rms} = \sqrt{\frac{3kT}{m}} \tag{12.21}$$

which is similar to Eq. (1.26) of Chapter 1.

The *most probable speed* is the speed at which the largest number of molecules is moving so that it is the speed that occurs most frequently. It is obtained by differentiating the expression of dN_v/dv with respect to v and equating the result to zero.

From Eq. (12.12), we have

$$\frac{4N}{\sqrt{\pi}}\left(\frac{m}{2kT}\right)^{3/2} \frac{d}{dv}(v^2 e^{-(m/2kT)v^2}) = 0$$

or

$$v^2 e^{-(m/2kT)v^2}\left(-\frac{m}{2kT}\right)(2v) + e^{-(m/2kT)v^2}(2v) = 0$$

from which

$$v_{mp} = \sqrt{\frac{2kT}{m}} \tag{12.22}$$

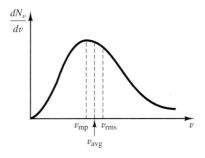

Figure 12.7   Relative magnitudes of v_{mp}, v_{avg}, and v_{rms}.

The relative magnitudes of the three foregoing speeds are shown in Figure 12.7, and their values are in the following proportions:

$$v_{mp} : v_{avg} : v_{rms} = 1 : 1.1284 : 1.2248$$

To calculate the number of molecules with speeds or velocities in a certain range, it is necessary to integrate the speed distribution function between the limits of that range. For example, the number of molecules having speeds between 0 and v is

$$N_{0 \to v} = \int_0^v dN_v$$

But dN_v, according to Eq. (12.12), is

$$dN_v = \frac{4N}{\sqrt{\pi}}\left(\frac{m}{2kT}\right)^{3/2} v^2 e^{-(m/2kT)v^2} dv$$

$$= \frac{4N}{\sqrt{\pi}}\beta^3 v^2 e^{-\beta^2 v^2} dv$$

Letting $x = \beta v = v/v_{\text{mp}}$, then

$$N_{0 \to v} = \frac{4N}{\sqrt{\pi}} \int_0^x x^2 e^{-x^2} dx = -\frac{2N}{\sqrt{\pi}} \int_0^x x \, de^{-x^2}$$

Integration by parts gives

$$N_{0 \to v} = -\frac{2N}{\sqrt{\pi}} \left[x e^{-x^2} - \int_0^x e^{-x^2} dx \right]$$

$$= N \left[\frac{2}{\sqrt{\pi}} \int_0^x e^{-x^2} dx - \frac{2}{\sqrt{\pi}} x e^{-x^2} \right]$$

$$= N \left[erf(x) - \frac{2}{\sqrt{\pi}} x e^{-x^2} \right] \tag{12.23}$$

where $x = (m/2kT)^{1/2} v$.

The function $erf(x)$ appearing is Eq. (12.23) is called the *error function* and is defined as

$$erf(x) \equiv \frac{2}{\sqrt{\pi}} \int_0^x e^{-x^2} dx \tag{12.24}$$

Values of $erf(x)$ as a function of x are given in Table 12.2.

Table 12.2   Values of the Error Function $erf(x) = \dfrac{2}{\sqrt{\pi}} \displaystyle\int_0^x e^{-x^2} dx$

x	$erf(x)$	x	$erf(x)$	x	$erf(x)$
0	0	0.70	0.677801	1.8	0.989091
0.05	0.056372	0.75	0.711156	1.9	0.992790
0.10	0.112463	0.80	0.742101	2.0	0.995322
0.15	0.167996	0.85	0.770668	2.1	0.997021
0.20	0.222703	0.90	0.796908	2.2	0.998137
0.25	0.276326	0.95	0.820891	2.3	0.998857
0.30	0.328627	1.0	0.842701	2.4	0.999311
0.35	0.379382	1.1	0.880205	2.5	0.999593
0.40	0.428392	1.2	0.910314	2.6	0.999764
0.45	0.475482	1.3	0.934008	2.7	0.999866
0.50	0.520500	1.4	0.952285	2.8	0.999925
0.55	0.563323	1.5	0.966105	2.9	0.999959
0.60	0.603856	1.6	0.976348	3.0	0.999978
0.65	0.642029	1.7	0.983790		

Example 12.1

Calculate the average speed, root mean square speed, and most probable speed for oxygen molecules at 0°C.

SOLUTION

The mass of an oxygen molecule is

$$m = \frac{M}{N_a} = \frac{32.0 \text{ kg/kg-mol}}{6.025 \times 10^{26} \text{ molecules/kg-mol}}$$
$$= 5.31 \times 10^{-26} \text{ kg/molecule}$$

and

$$k = 1.38 \times 10^{-23} \text{ J/molecule K}$$

The average speed, according to Eq. (12.20), is

$$v_{avg} = 2\sqrt{\frac{2kT}{\pi m}} = 2\sqrt{\frac{2(1.38 \times 10^{-23} \text{ J/molecule K})(273.15 \text{ K})}{\pi(5.31 \times 10^{-26} \text{ kg/molecule})}}$$
$$= 425 \text{ m/s}$$

The root mean square speed according to Eq. (12.21) is

$$v_{rms} = \sqrt{\frac{3kT}{m}} = \sqrt{\frac{3(1.38 \times 10^{-23} \text{ J/molecule K})(273.15 \text{ K})}{(5.31 \times 10^{-26} \text{ kg/molecule})}}$$
$$= 461 \text{ m/s}$$

The most probable speed according to Eq. (12.22) is

$$v_{mp} = \sqrt{\frac{2kT}{m}} = \sqrt{\frac{2(1.38 \times 10^{-23} \text{ J/molecule K})(273.15 \text{ K})}{(5.31 \times 10^{-26} \text{ kg/molecule})}}$$
$$= 377 \text{ m/s}$$

12.5 Number of Molecules Striking a Unit Surface in a Unit Time

Closely associated with pressure is the number of molecules that strike a surface in a unit time. In order to determine the rate of such collisions, it is first necessary to determine the number of molecules that have a certain speed and are moving in a certain direction. Integration of this expression over the total range

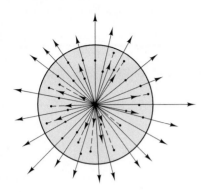

Figure 12.8   Molecular velocity vectors drawn from a common origin.

of speeds and directions leads to an expression of the collision rate. The derivation is done in two steps: first, the number of molecules moving in a certain direction is determined; second, the fraction of these molecules that have a speed v is determined.

Consider a sphere in velocity space, as shown in Figure 12.8. Let the velocity vector of each molecule be drawn from the origin. The radial projection of each vector on the surface of the sphere is represented by a point. The surface of the sphere will then have N points if N particles are considered. Because of the uniform distribution of points in velocity space, the number of points per unit area will be the same[2] irrespective of the position on the surface. If the sphere has a radius r, the number of points per unit area of surface is $N/4\pi r^2$. Consider now a small area dA on the surface of the sphere, as shown in Figure 12.9. This area subtends a solid angle[3] at the origin defined in terms of the small angles $d\theta$ and $d\varphi$ and the radius of the sphere. The elementary area dA expressed in polar coordinates is

$$dA = r^2 \sin \theta \, d\theta \, d\varphi$$

Let the number of points on dA be denoted by $dN_{\theta\varphi}$, where the subscripts indicate the direction of dA with respect to the origin. Since the number of points on any portion of the surface of the sphere is proportional to the area,

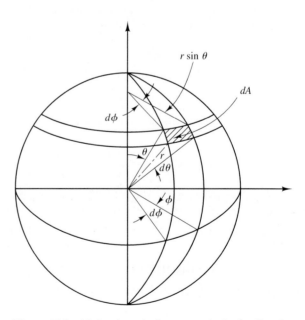

Figure 12.9   Molecular velocity vectors in the $\theta\varphi$ direction.

---

[2] This assumption is valid provided that the unit area is not too small.
[3] The solid angle is defined as the area on the surface of a sphere of unit radius subtended at the center of a sphere.

$$\frac{dN_{\theta\varphi}}{N} = \frac{dA}{4\pi r^2}$$

When the equivalent expression for dA is substituted in this equation, the number of molecules of gas traveling between the directions θ and $\theta + d\theta$ and between φ and $\varphi + d\varphi$ is

$$dN_{\theta\varphi} = \frac{N}{4\pi r^2}(r^2 \sin\theta \, d\theta \, d\varphi) = \frac{N}{4\pi} d\omega \qquad (12.25)$$

where $d\omega(= \sin\theta d\theta d\varphi)$ represents the solid angle at the origin subtended by the area dA. If $dn_{\theta\varphi} = dN_{\theta\varphi}/V$ represents the number of $(\theta\varphi)$ molecules per unit volume (number density), then

$$dn_{\theta\varphi} = \frac{n}{4\pi} d\omega \qquad (12.26)$$

In order to determine the fraction of the $dN_{\theta\varphi}$ molecules that have a speed between v and $v + dv$, consider a sphere of radius v in velocity space. Take the solid angle subtended by a small area dA to be the same for both velocity space and ordinary space. The fraction of the molecules moving with a speed v in the direction bounded by the angles θ and $\theta + d\theta$ and φ and $\varphi + d\varphi$ is equal to the ratio of the area dA_v to the total area of the sphere, or

$$\frac{dN_{\theta\varphi v}}{dN_v} = \frac{dA_v}{4\pi v^2} = \frac{dA}{4\pi r^2} = \frac{dN_{\theta\varphi}}{N}$$

Therefore,

$$dN_{\theta\varphi v} = \frac{dN_{\theta\varphi}}{N} dN_v \qquad (12.27)$$

The foregoing expression could also be obtained from the definition of the distribution function, which, in this case, gives the fraction of the $(\theta\varphi)$ molecules moving at speeds between v and $v + dv$. Substituting for $dN_{\theta\varphi}$ from Eq. (12.25) into Eq. (12.27) gives

$$dN_{\theta\varphi v} = \frac{N}{4\pi} d\omega \frac{dN_v}{N} = \frac{N}{4\pi} d\omega \, f(v) \, dv \qquad (12.28)$$

Dividing both sides of Eq. (12.28) by the volume V gives

$$dn_{\theta\varphi v} = \frac{n}{4\pi} d\omega \, f(v) \, dv \qquad (12.29)$$

where n is the number of molecules per unit volume. Equation (12.29) describes the number of molecules per unit volume that start at the origin of the coordinate system and are confined within a cone of a solid angle $d\omega$ and have a speed between v and $v + dv$.

In an analogous manner, the number of molecules with speeds in the range between v and $v + dv$ starting from a small volume dV in ordinary space and moving in the direction of a small surface dA' at the origin of the coordinate system may be determined. Using Eq. (12.29) and referring to Figure 12.10, the number of these molecules is

$$dn_{\theta\varphi v}\, dV = \left(\frac{n}{4\pi} d\omega'\, f(v)\, dv\right) dV$$

where $d\omega'(= dA'(\cos\theta)/r^2)$ is the solid angle of a cone whose vertex lies in the volume dV and whose base occupies the area dA'. The volume dV is given by

$$dV = r^2 \sin\theta\, d\theta\, d\varphi\, dr \qquad \text{and} \qquad dr = v\, dt$$

These expressions for $d\omega'$ and dV can now be substituted in Eq. (12.29) to determine the number of molecules from dV striking the area dA' in time dt:

$$dC_{\theta\varphi v} = dn_{\theta\varphi v}dV = \left(\frac{n}{4\pi} \frac{dA' \cos\theta}{r^2} f(v)\, dv\right) r^2 \sin\theta\, d\theta\, d\varphi\, v\, dt$$

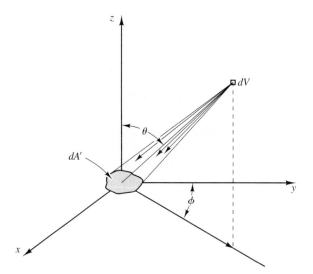

Figure 12.10   Molecular collisions with dA' in the $\theta\varphi$ direction.

Since dt represents the time required for a molecule to travel a distance dr, the number of molecules per unit area and per unit time (molecular flux) that travel this path is

$$\frac{dC_{\theta\varphi v}}{dA'dt} = \frac{n}{4\pi} \sin\theta \cos\theta \; v f(v) \; dv \; d\varphi \; d\theta \tag{12.30}$$

The total number of molecules that strike the wall per unit area per unit time is obtained by integrating the preceding expression in the hemisphere above the *x-y* plane and over all positive velocities, or

$$\frac{C}{dA'dt} = \int_{\varphi=0}^{2\pi} \int_{\theta=0}^{\pi/2} \int_{v=0}^{\infty} \frac{n}{4\pi} \sin\theta \cos\theta \; v f(v) \; dv \; d\theta \; d\varphi$$

The result can be expressed in terms of the average speed:

$$\dot{c} = \frac{C}{dA'dt} = \frac{n}{4} \int_0^{\infty} v f(v) \; dv = \frac{1}{4} n v_{\text{avg}} \tag{12.31}$$

Since v_{avg} can be evaluated from Eq. (12.20), the rate at which the molecules strike a wall per unit area is

$$\dot{c} = n\sqrt{\frac{kT}{2\pi m}} \tag{12.32}$$

By substituting p/kT for n in Eq. (12.32), the frequency of collision with a unit area of wall can be expressed in terms of pressure:

$$\dot{c} = \frac{p}{\sqrt{2\pi mkT}} \tag{12.33}$$

Equation (12.33) may be used to calculate the pressure force on an area. Each collision of a particle with an area A results in a change in momentum of the particle:

$$\text{momentum change due to collision} = 2mv \cos\theta$$

But the total rate of change of momentum per unit area is equal to the pressure. Therefore,

$$dp = \frac{dC_{\theta\varphi v}}{dA'dt}(2mv \cos\theta)$$

$$= \frac{n}{4\pi}v \sin\theta \cos\theta d\theta \, d\varphi(2mv \cos\theta)f(v) \, dv$$

This expression is then integrated over θ from 0 to $\pi/2$, over φ from 0 to 2π, and over v from 0 to ∞ to give

$$p = \frac{1}{3}mn\int_0^\infty v^2 f(v) \, dv = \frac{1}{3}mnv_{rms}^2 = nkT \tag{12.34}$$

This is the same as Eq. (1.18) obtained in Chapter 1.

Example 12.2

Calculate the wall collision frequency per square meter for oxygen at 1 atm and 0°C. Appropriate constants are given in the Appendix.

SOLUTION

$$1 \text{ atm} = 101.325 \text{ kPa}$$

$$m_{O_2} = \frac{32 \text{ kg/kg-mol}}{6.025 \times 10^{26} \text{ molecules/kg-mol}} = 5.31 \times 10^{-26} \text{ kg/molecule}$$

$$k = 1.38 \times 10^{-23} \text{ J/(K} \cdot \text{molecule)}$$

$$T = 273.15 \text{ K}$$

The wall collision frequency according to Eq. (12.33) is

$$\dot{c} = \frac{p}{\sqrt{2\pi mkT}} = \frac{101.325 \times 10^3 \text{ Pa}}{\sqrt{2\pi(5.31 \times 10^{-26} \text{ kg/molecule})(1.38 \times 10^{-23} \text{ J/K} \cdot \text{molecule})(273.15 \text{ K})}}$$

$$= 2.86 \times 10^{27} \text{ molecular collisions/m}^2\text{-s}$$

12.6 Molecular Collisions and Mean Free Path

The velocities at which the molecules of a gas move are described by equations derived previously. But the microscopic view is still not complete since the paths taken by the molecules are yet to be determined. If the particles collide to an appreciable extent with each other, then their diffusional characteristics are different from those when they collide only with the walls of the container. Some indications of the collision tendencies depend on two parameters, the *mean free path* and the *molecular collision* frequency.

The mean free path of a molecule is the average distance it travels between collisions with other molecules. Consider the collision between two molecules of radii r_1 and r_2, as shown in Figure 12.11(a). If the molecules are assumed to be small elastic spheres[4] rather than point masses, the distance between the centers of the molecules is $(r_1 + r_2)$ when they collide. One of the two molecules may therefore be considered to have an effective radius $(r_1 + r_2)$, whereas the other molecule is a point of zero radius, as shown in Figure 12.11(b). All molecules except one may be considered motionless or "frozen" in their respective positions in the gas. Now consider the movement of a single molecule of radius $(r_1 + r_2)$ as it travels in a gas containing stationary molecules of zero radius. The cross-sectional area of the cylindrical volume swept out by the moving molecule is

$$\sigma = \pi(r_1 + r_2)^2$$

where σ is the *collision cross section*. The number of collisions that occur as the molecule travels through this stationary matrix depends on the number of molecules per unit volume that lie within the cylindrical volume. The molecular collision frequency Z, that is, the number of collisions of molecules per unit time irrespective of path deflections, is given by

$$Z = \left(\frac{N}{V}\right)\sigma v_{\mathrm{avg}} = n\sigma v_{\mathrm{avg}}$$

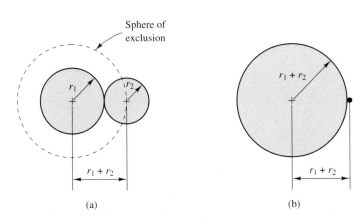

Figure 12.11   Collision of two molecules.

[4] Although it was previously assumed that the molecules are point masses, it may be shown that statistical laws apply to other models and are therefore independent of the assumption of point masses.

where v_{avg} represents the average velocity of the molecules. Assuming identical molecules each of mass m, v_{avg} is

$$v_{avg} = \sqrt{\frac{8kT}{\pi m}}$$

Therefore,

$$Z = \frac{N}{V}\sigma\sqrt{\frac{8kT}{\pi m}}$$

Since the mean free path is the average distance that the molecule travels between collisions with other molecules, the mean free path is given by

$$\Lambda_m = \frac{v_{avg}t}{Zt} = \frac{V}{N\sigma} = \frac{1}{n\sigma}$$

where t represents time interval.

In the preceding derivation, only one molecule was assumed to be moving, and the motion of other molecules was ignored. Because of the motion of other molecules, the average velocity of the original molecule when considered in relation to the other molecules is, in effect, larger than originally considered. The relative average velocity is found to be $(4/3)v_{avg}$. When this correction is taken into account, the foregoing expressions for collision frequency and mean free path become

$$Z = \frac{4}{3}n\sigma\sqrt{\frac{8kT}{\pi m}} = \frac{4}{3}n\sigma v_{avg} \tag{12.35}$$

and

$$\Lambda_m = \frac{3V}{4N\sigma} = \frac{3}{4\sigma n} \tag{12.36}$$

Although the molecular collision frequency depends on the velocity of the molecules, the mean free path is independent of velocity; instead, it depends on the dimensions of the molecules and their number per unit volume. According to Avogadro's law, the concentration of molecules depends only on pressure and temperature. Furthermore, the dimensions of the molecules of all ideal gases are nearly equal, being small compared to the mean free path. Consequently, the mean free path is approximately the same for all gases under the same conditions. For example, the mean path of oxygen at standard conditions is 8×10^{-6} cm. This is approximately 10 times the intermolecular distance and 100 times the molecular diameter.[5]

---

[5] Molecular diameters are approximately 300 Å (3×10^{-8} cm).

Since the mean free path at a given temperature is inversely proportional to pressure, the mean free path of oxygen at 0.01 mm Hg pressure (at 0°C) is of the order of 1 cm.

12.7 Distribution of Free Paths

A molecule, at times, will travel a distance larger than the mean free path before it collides; at other times, it will travel a shorter distance. The magnitude of these free paths shows a typical distribution pattern. An expression describing the distribution of free paths of molecules of a gas, that is, the number of molecules having a free path between x and $x + dx$, is derived as follows. The number of molecules that collide during a time interval dt is proportional to the number of molecules in the gas and to their relative velocities. The change in the number of molecules that do not experience a collision is

$$dN = -PNv\, dt$$

where P is a proportionality factor called the *collision probability*, N is the number of molecules that do not experience a collision in time dt, and v is velocity. This can be expressed in terms of distance as

$$dN = -PN\, dx$$

Integration of the foregoing equation gives

$$N = N_0 e^{-Px}$$

where N_0 is the total number of molecules (Avogadro's number if one mole is considered). Now the mean free path Λ_m represents the average distance traveled by any molecule until it experiences one collision, and it can be evaluated from

$$\Lambda_m = \frac{1}{N_0}\int_0^{N_0} x\, dN = \frac{1}{N_0}\int_0^{\infty} x N_0(-P)e^{-Px}dx \qquad (12.37)$$

It can be shown that $\Lambda_m = 1/P$ so that the collision probability P is the reciprocal of the mean free path. Thus, the number of molecules that travel a path x without collision can be expressed in terms of the mean free path rather than probability:

$$N = N_0 e^{-x/\Lambda_m} \qquad (12.38)$$

Figure 12.12 gives the distribution of free paths, that is, the fraction of the molecules N/N_0 that travel a path equal to x/Λ_m without collision. Note that 37 percent of the molecules travel a distance equal to the mean free path before collision.

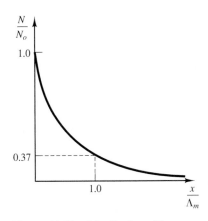

Figure 12.12   Distribution of free paths.

12.8 Thermodynamic Probability

A substance can be considered to consist of a large number of molecules that, although identical in mass, are still not necessarily all alike. The molecules may be considered to be distributed over a number of different states that readily transform from one to another. It is possible to calculate the number of different ways that a system consisting of numerous particles in different states can be arranged. Furthermore, it is possible to predict which of the various arrangements will be the most probable. Such a treatment, in which the most probable distribution is determined from mathematical statistics, can be correlated with certain classical thermodynamic properties.

The approach used in calculating macroscopic properties is based on the theory of probability. A major assumption in the analysis is that all conceivable microstates[6] of a system have the same a priori probability of occurrence. A system in which the energy U, the volume V, and the number of particles N are specified can exist in several different macrostates. The problem lies in determining how many microstates correspond to each macrostate and in determining which particular macrostate has the largest number of microstates. This is the most probable macrostate that determines the equilibrium thermodynamic properties of the system.

Suppose it is required to arrange N *distinguishable*[7] particles into several different groups, with $N_1, N_2, \ldots N_i \ldots$ particles in each group. The groupings may be thought of, more conveniently, in terms of physical compartments, although what is really involved here is a grouping in terms of discrete energy levels. Before examining the number of different ways in which this can be done, the number of ways in which N particles can be arranged in N distinct positions may be first considered. The first particle can occupy any one of the N possible positions. This means that the first particle has a choice of N positions. The second particle has a choice of the remaining $(N - 1)$ positions since one position is already occupied by the first particle. With two positions filled, the third particle has a choice of $(N - 2)$ positions, and so on. This leads to a total of $N(N - 1)(N - 2)\ldots(2)(1) = N!$ possible arrangements of the N particles in the N positions (or energy levels). This is essentially a *permutation process* of the N particles among themselves.

Now suppose that the N particles are arranged so that there are $N_1, N_2, \ldots, N_i, \ldots$ particles in each group. If the number of particles in the first group is N_1, then there are $N_1!$ different arrangements for the particles in that group. Simi-

---

[6] A microstate or a microscopic state of a system is described by the instantaneous particle distribution among the available quantum states. Macroscopic properties are described by the energy distribution of the particles in that microstate that would be observed more frequently than any other distribution.

[7] Quantum statistics considers it possible to distinguish between identical particles even when they are at the same energy level.

larly for the ith group, the number of arrangements is $N_i!$. The total number of ways of arranging all the particles is therefore $(N_1!)(N_2!)...(N_i!)....$

If, however, the particles within a group are indistinguishable from each other, the interchange of particles within a group does not result in a new arrangement. Therefore, the permutations within each group cannot be considered to contribute to the total number of possible arrangements. Since $N!$ represents the number of ways of arranging N particles in N positions, and $(N_1!)(N_2!)...(N_i!)...$ represents the number of ways of distributing N particles in N_i groups when the particles are distinguishable from each other, then the total number of ways of arranging N particles in N_i groups when the particles within each group cannot be distinguished from each other is

$$W = \frac{N!}{N_1!\,N_2!...N_i!...} = \frac{N!}{\prod_i N_i!} \qquad (12.39)$$

The symbol $\prod_i$ denotes continued product over all values of i. W is called the *thermodynamic probability* of a macrostate. It is equal to the number of microstates corresponding to a given macrostate. The larger the value of W, the more probable the corresponding macrodistribution occurs. Thermodynamic probability should not be confused with *mathematical probability*, the latter being the ratio of the number of favorable events to the total number of events. The sum of all mathematical probabilities of the various possibilities of an event to occur is 1. Therefore, mathematical probabilities lie between 0 and 1, the former impossibility and the latter certainty. Note that the ratio of thermodynamic probabilities corresponding to two distributions gives the relative frequency (or probability) with which they occur.

Consider the example of arranging the letters a, b, and c. There are six ways (3!) in which the three letters may be distributed in three separate groups, with only one letter in each group:

A number of arrangements are possible, depending on whether the three letters can be distributed in three groups, two groups, or one group. When only two groups are available for the three letters, the arrangements are shown below.

Note that a change in the order of the letters within a group does not represent another arrangement of the group. The number, rather than the order, of the letters is important in these arrangements, just as the number of particles at each energy level is the significant factor.

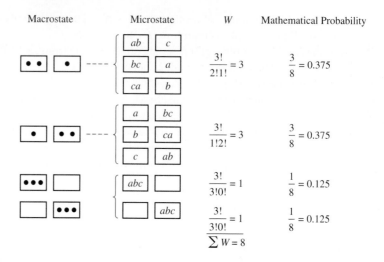

The maximum value of W corresponds to the most probable arrangement of particles. The proof of this statement will be presented subsequently. At the maximum value of W, the rate of change in W must be zero; that is, $dW = 0$. Since W is usually a large number, it is more convenient to maximize the logarithm of W, rather than W itself.

Example 12.3

Consider a system of three particles a, b, and c and four energy levels ε, 2ε, 3ε, and 4ε. If the total energy of the particles is 9ε, find the number of possible arrangements.

SOLUTION

Representing the energy levels horizontally, the possible arrangements are as follows:

Energy	A	B			C					
4ε		ab	ac	bc	a	a	b	b	c	c
3ε	abc				b	c	c	a	a	b
2ε					c	b	a	c	b	a
ε		c	b	a						

with heading: **Macrostate**

In the foregoing distribution, there are three different macrostates and ten microstates. Note that the mathematical probability of each microstate is 0.1, and for the macrostates *A*, *B*, and *C*, they are 0.1, 0.3, and 0.6, as obtained by the following calculation:

Macrostate	W	Mathematical Probability
A	$\dfrac{3!}{3!\,0!} = 1$	$\dfrac{1}{10} = 0.10$
B	$\dfrac{3!}{2!\,1!} = 3$	$\dfrac{3}{10} = 0.30$
C	$\dfrac{3!}{1!\,1!\,1!} = 6$	$\dfrac{6}{10} = 0.60$
	$\sum W = 10$	

If the particles were indistinguishable, only the three macrostates and three microstates exist. These macrodistributions are as follows:

Energy	Macrostate A	B	C
4ε		$\bullet\ \bullet$	$\bullet$
3ε	$\bullet\ \bullet\ \bullet$		$\bullet$
2ε			$\bullet$
ε		$\bullet$	

Example 12.4

Answers to each of four questions may be either right or wrong. Find the distribution of the answers when each choice is as likely to be right as wrong.

SOLUTION

The number of possible arrangements is $2^4 = 16$. If the correct and incorrect arrangements are denoted by + and −, Table 12.3 can be formulated.

Table 12.3   Example 12.4

Number of Correct Answers	Arrangement	Thermodynamic Probability or Relative Frequency of Occurrence	Mathematical Probability
4	+ + + +	$\dfrac{4!}{4!\,0!} = 1$	$\dfrac{1}{16}$
3	$\left\{\begin{array}{cccc} + & + & + & - \\ + & + & - & + \\ + & - & + & + \\ - & + & + & + \end{array}\right\}$	$\dfrac{4!}{3!\,1!} = 4$	$\dfrac{4}{16}$
2	$\left\{\begin{array}{cccc} + & + & - & - \\ + & - & - & + \\ - & - & + & + \\ + & - & + & - \\ - & + & + & - \\ - & + & - & + \end{array}\right\}$	$\dfrac{4!}{2!\,2!} = 6$	$\dfrac{6}{16}$
1	$\left\{\begin{array}{cccc} + & - & - & - \\ - & + & - & - \\ - & - & + & - \\ - & - & - & + \end{array}\right\}$	$\dfrac{4!}{1!\,3!} = 4$	$\dfrac{4}{16}$
0	− − − −	$\dfrac{4!}{0!\,4!} = 1$	$\dfrac{1}{16}$
Total		$\overline{16}$	$\overline{1.000}$

12.9 Maxwell–Boltzmann Statistics

From the mathematical statistics approach, it can be shown that numerous particles, when capable of distributing themselves in several alternative arrangements, will tend to distribute themselves among the various alternatives in the most random way possible. Orderly distributions also occur, but they occur with less probability. The most probable distribution is one that is associated with the least "order."

In real systems, another factor controls the distribution. Different arrangements of molecules are associated with different quantities of energy. In an energy-limited system, each microstate is equally probable, and the system continually passes from one microstate to another. This results in a few number of macrostates that tend to cluster about the same particular value. The clustering effect, owing to the averaging of energy levels, competes with the distributive effect that arises from the disorder associated with probability. The final distribution or the most probable macrostate is a resultant determined by these two competing effects.

In the Maxwell–Boltzmann model of the microstate, particles are distinguishable from each other and there is no limit to the number of particles per quantum state. The collection of the microstates forms the macrostates of the system. The most probable macrostate has the maximum number of microstates, and it is this macrostate that corresponds to thermodynamic equilibrium.

The thermodynamic probability W indicates the number of ways in placing N_1 particles in the first energy level, N_2 particles in the second energy level, and so on. In order to find the most probable macrostate of a system of particles, it is necessary to maximize the number of ways of forming a macrostate by maximizing the expressions describing the particle distribution while imposing applicable conditions of constraint. Since the number of particles is large, it is more convenient, from the arithmetic standpoint, to calculate the logarithm of probability than to calculate probability itself.

The thermodynamic probability W according to Eq. (12.39) is

$$W = \frac{N!}{\prod_i N_i!}$$

The natural logarithm of the foregoing equation is

$$\ln W = \ln N! - \sum_i \ln N_i!$$

Applying Stirling's approximation formula (see the Appendix to this chapter),

$$\ln N! = N \ln N - N \qquad N \gg 1$$

Therefore,

$$\ln W = N \ln N - N - \sum_i (N_i \ln N_i) + \sum_i N_i$$
$$= N \ln N - \sum_i (N_i \ln N_i)$$

Using the symbol δ to denote a small change in $\ln W$ due to particle shifting between energy levels, then

$$\delta(\ln W) = -\delta \sum_i (N_i \ln N_i) \tag{12.40}$$

Equation (12.40) is subject to two conditions of constraint prescribed by the macroscopic state of the system: that the total number of molecules and the internal energy of the system remain unchanged. Thus,

$$\sum_i \delta N_i = dN = 0 \tag{12.41}$$

and

$$\sum_i \varepsilon_i \delta N_i = dU = 0 \tag{12.42}$$

where N_i is the number of molecules in a given energy level ε_i, and the summation is taken over all energy levels. Equation (12.40) can then be written

$$\delta(\ln W) = -\delta \sum_i (N_i \ln N_i)$$

$$= -\sum_i N_i \frac{\delta N_i}{N_i} - \sum_i (\ln N_i)\delta N_i = 0 - \sum_i (\ln N_i)\delta N_i$$

and for the maximum value of ln W (or of W), variation of ln W must vanish for small variations of N_i's,

$$\sum_i (\ln N_i)\delta N_i = 0 \tag{12.43}$$

We apply Lagrange's method of undetermined multipliers to Eq. (12.43) subject to the conditions of constraint given by Eqs. (12.41) and (12.42). First, multiply Eq. (12.41) by $-\ln a$, and Eq. (12.42) by β, and add the result to Eq. (12.43). Then,

$$-\ln a \sum_i \delta N_i + \beta \sum_i \varepsilon_i \delta N_i + \sum_i (\ln N_i)\delta N_i = 0$$

or

$$\sum_i (-\ln a + \beta \varepsilon_i + \ln N_i)\delta N_i = 0$$

But since δN_i's are independent, the terms in parentheses in the summation must vanish for each value of δN_i in order to satisfy the preceding equation. Therefore,

$$-\ln a + \beta \varepsilon_i + \ln N_i = 0$$

or

$$N_i = ae^{-\beta \varepsilon_i} \tag{12.44}$$

N_i in Eq. (12.44) indicates the most probable number of particles having energy ε_i. The total number of particles N is given by

$$N = \sum_i N_i = a \sum_i e^{-\beta \varepsilon_i} \qquad (12.45)$$

The sum in Eq. (12.45) is known as the state-sum or *partition function*[8] of the system and is given the symbol Z. Thus,

$$Z = \sum_i e^{-\beta \varepsilon_i} \qquad (12.46)$$

The summation in the expression of Z is taken over all the states of energy of the system. As will be shown later, all thermodynamic properties may be correlated, either directly or indirectly, to the partition function.

For the most probable macrostate, the number of particles N_i with quantum states in the ith group can now be written in terms of the partition function as

$$N_i^* = N \frac{e^{-\beta \varepsilon_i}}{\sum_i e^{-\beta \varepsilon_i}} = \frac{N}{Z} e^{-\beta \varepsilon_i} \qquad (12.47)$$

The asterisk is to emphasize that N_i^* corresponds to the maximum value of W. Equation (12.47) is called the *Maxwell–Boltzmann distribution law*; it indicates the most probable distribution of particles among the energy levels when the system is in thermodynamic equilibrium.

Example 12.5

Twenty-six particles are arranged in four cells according to the following table:

Cell	Number of Particles	Unit Energy
1	$N_1 = 8$	1.0ε
2	$N_2 = 8$	1.1ε
3	$N_3 = 6$	1.2ε
4	$N_4 = 4$	1.3ε

[8] It is called *partition function* because the particles of a system at equilibrium are partitioned among the different energy states.

(a) If $\delta N_4 = -1$ and $\delta N_3 = 0$, find the new distribution subject to the conditions $\delta N = 0$ and $\delta U = 0$.

(b) If the unit energy for each cell remains the same, which distribution would be most probable?

SOLUTION

(a) The total energy is

$$U = (8 \times 1.0 + 8 \times 1.1 + 6 \times 1.2 + 4 \times 1.3)\varepsilon = 29.2\varepsilon$$

Since $\delta N = 0$ and $\delta U = 0$, we have

$$N'_1 + N'_2 + 6 + 3 = 26 \qquad \text{or} \qquad N'_1 = 17 - N'_2 \tag{a}$$

and

$$(N'_1 \times 1.0 + N'_2 \times 1.1 + 6 \times 1.2 + 3 \times 1.3)\varepsilon = 29.2\varepsilon$$

or

$$N'_1 + 1.1 N'_2 = 18.1 \tag{b}$$

where prime indicates the new distribution. From Eqs. (a) and (b), $N'_1 = 6$ and $N'_2 = 11$.

(b) The probabilities of the two arrangements are

$$W = \frac{26!}{8!\,8!\,6!\,4!} \qquad \text{and} \qquad W' = \frac{26!}{6!\,11!\,6!\,3!}$$

Their ratio is

$$\frac{W}{W'} = 4.42$$

Only 3 of the 26 particles are involved in the shift described here, but the probability that this change of configuration will occur is unfavorable, as indicated by the value of 4.42. It may, however, be shown that, when the particles are arranged in their most probable distribution, a shifting of a few of the particles between energy level changes probability by only a small amount.

12.10 Entropy and Probability

When an isolated system undergoes an irreversible process, such as the free expansion of a gas or the mixing of two gases, the number of possible microstates of the system is increased. In effect, there is an increase in the value of the probability W. In a free expansion, the increase in probability is attributed to the larger volume and to the additional energy states, which are then available to the particles of the system. The molecules of the gas have a larger "choice" of locations or energy levels in the newly acquired volume. At the same time, less is known about the location of a particular molecule after the expansion, for it is then free to move in a larger volume. Similarly, when two gases mix adiabatically, the number of microstates increases, and the molecules are distributed in a greater degree of randomness and disorder. At the same time, the lack of knowledge about the states of the molecules increases.

When a system proceeds from a nonequilibrium state toward an equilibrium state, its entropy increases. This increase in entropy parallels the increase in molecular disorder and the greater probability of attaining the equilibrium state. We further note that both entropy and probability are properties of the system. Therefore, it may be postulated that entropy is related to the probability associated with a state and with the information that may be known about a system. A system reaches a stable equilibrium state when its entropy is at a maximum, and this state is the one of maximum thermodynamic probability and greatest disorder. Boltzmann conceived of this parallelism between entropy and probability and identified the state of maximum probability or least information with maximum entropy. Entropy may therefore be written as

$$S = f(W) \tag{12.48}$$

where f is a function to be determined according to the following reasoning. In calculating the probability W of a system comprising two independent subsystems A and B, we note that each microstate of subsystem A may be chosen in combination with any microstate of subsystem B. The total number of microstates of the combined subsystems A and B is therefore

$$W = W_A W_B \tag{12.49}$$

On the other hand, the entropy S of the system is the sum of the entropies of subsystems A and B:

$$S = S_A + S_B \tag{12.50}$$

The two preceding relations dictate that, in seeking the nature of the function f of Eq. (12.48), one must take into account the respective multiplicative and additive natures of probability and entropy.

The entropy of the foregoing system, according to Eq. (12.48), may then be written in a functional form as

$$S = f(W_A W_B) = f(W_A) + f(W_B)$$

Differentiation of the preceding equation with respect to W_A gives

$$W_B f'(W_A W_B) = f'(W_A)$$

where the primes indicate differentiation with respect to the argument so that

$$f'(W_A W_B) = \frac{\partial f(W_A W_B)}{\partial (W_A W_B)} \quad \text{and} \quad f'(W_A) = \frac{\partial f(W_A)}{\partial W_A}$$

A second differentiation with respect to W_B gives

$$W_A W_B f''(W_A W_B) + f'(W_A W_B) = 0$$

or

$$W f''(W) + f'(W) = 0$$

for which the general solution is

$$f(W) = S = C_1 \ln W + C_2 \qquad (12.51)$$

To determine the constant C_1, consider a free expansion process in which two moles or $2N_a$ molecules of an ideal gas at a pressure p expand from a volume V_1 to $V_2 = 2V_1$. Before expansion, the probability of finding all the $2N_a$ molecules in V_1 is $W_1 = 1$; also, the entropy is given by

$$S_1 = -2\overline{R} \ln p + S_0$$

where p is the pressure measured in units of a standard pressure (say, atmospheres), and S_0 is a constant depending only on temperature. After expansion, the $2N_a$ molecules occupy both the volumes V_1 and $(V_2 - V_1)$, and each volume contains N_a molecules at a pressure equal to $p/2$. The probability and entropy at this state are

$$W_2 = \frac{(2N_a)!}{N_a! N_a!}$$

and

$$S_2 = -2\overline{R} \ln \frac{p}{2} + S_0$$

In evaluating the change in probability, the factorials in W_2 may be calculated, using Stirling's approximation formula:

$$\ln W_2 - \ln W_1 = \ln (2N_a!) - 2 \ln (N_a!) = 2N_a \ln 2$$

But the entropy change is

$$S_2 - S_1 = 2\overline{R} \ln 2$$

Substituting for S and W in Eq. (12.51) gives

$$C_1 = \frac{S_2 - S_1}{\ln W_2 - \ln W_1} = \frac{\overline{R}}{N_a} = k \qquad (12.52)$$

The arbitrary additive constant C_2 can be chosen zero since at $W = 1$, $S = 0$ (completely ordered system). Equation (12.51) can then be written

$$S = k \ln W \qquad (12.53)$$

where k is Boltzmann's constant. Equation (12.53) states that the entropy of an isolated system at a certain state is equal to the natural logarithm of the thermodynamic probability of that state multiplied by Boltzmann's constant. Note that Eq. (12.53) satisfies the additive nature of entropy and the multiplicative nature of probability.

Equation (12.53) provides an interesting interpretation of the second law of thermodynamics. The second law, which, when considered as a consequence of the kinetic theory, indicates that an increase in entropy of a system means that the system changes from a less to a more probable state. This interpretation limits the application of the second law to systems containing large numbers of particles; also, it becomes a statement of a very high probability but not complete certainty.

The foregoing treatment has shown that the molecules distribute themselves among several energy states according to certain patterns, and that in any particular system, a certain distribution is the most probable one even though many distributions are possible. It is then pertinent to consider whether, and how, this most probable configuration differs in some thermodynamic way from the other configurations. Such differences may then indicate the relationship between thermodynamic properties and energy distribution.

12.11 Evaluation of the Parameter β

In the derivation of the Maxwell–Boltzmann distribution law, a Lagrange multiplier β was used. In order to utilize the resultant equation, it is necessary to determine the nature of this constant β. This is done by combining the expression derived by the statistical approach with expressions based on classical thermodynamics. When β is adequately defined, it becomes possible to evaluate all the thermodynamic functions from statistical mechanical relationships (that is, from considerations of energy distribution).

To find an expression for the Lagrange multiplier β, we proceed as follows:
The logarithm of the probability is given by

$$\ln W = N \ln N - \sum_i N_i \ln N_i$$

But according to Eq. (12.47),

$$N_i = \frac{N}{Z} e^{-\beta \varepsilon_i}$$

so that

$$
\begin{aligned}
\ln W &= N \ln N - \sum_i N_i (\ln N - \ln Z - \beta \varepsilon_i) \\
&= N \ln N - \ln N \sum_i N_i + \ln Z \sum_i N_i + \beta \sum_i N_i \varepsilon_i \\
&= N \ln Z + \beta U
\end{aligned}
$$

By introducing entropy, from Eq. (12.53) into this expression, the following expression is obtained:

$$S = Nk \ln Z + k\beta U \tag{12.54}$$

Classical thermodynamics gives the change in the internal energy of a simple system as

$$dU = T \, dS - p \, dV$$

Also,

$$dU = \left(\frac{\partial U}{\partial S}\right)_V dS + \left(\frac{\partial U}{\partial V}\right)_S dV$$

Since S and V are independent, then equating the coefficients of dS gives

$$\left(\frac{\partial U}{\partial S}\right)_V = T \qquad \text{or} \qquad \left(\frac{\partial S}{\partial U}\right)_V = \frac{1}{T} \tag{12.55}$$

Note that this equation may be considered a thermodynamic definition of absolute temperature. The internal energy of a system of N particles is equal to the sum of the energies of the individual particles so that

$$
\begin{aligned}
U &= \sum_i \varepsilon_i N_i \\
&= \sum_i \varepsilon_i \frac{N}{Z} e^{-\beta \varepsilon_i} = \frac{N}{Z} \sum_i \varepsilon_i e^{-\beta \varepsilon_i} = -\frac{N}{Z} \left(\frac{\partial Z}{\partial \beta}\right)_V
\end{aligned}
$$

or

$$\left(\frac{\partial Z}{\partial \beta}\right)_V = -\frac{ZU}{N} \tag{12.56}$$

Differentiating Eq. (12.54) with respect to U at constant volume gives

$$\left(\frac{\partial S}{\partial U}\right)_V = kN\left(\frac{\partial \ln Z}{\partial U}\right)_V + k\beta + kU\left(\frac{\partial \beta}{\partial U}\right)_V$$

Noting that $\partial \ln Z = \partial Z/Z$ and making use of Eq. (12.56), the foregoing equation becomes

$$\left(\frac{\partial S}{\partial U}\right)_V = \frac{kN}{Z}\left(\frac{\partial Z}{\partial \beta}\right)_V\left(\frac{\partial \beta}{\partial U}\right)_V + k\beta + kU\left(\frac{\partial \beta}{\partial U}\right)_V$$

$$= \frac{kN}{Z}\left(-\frac{UZ}{N}\right)\left(\frac{\partial \beta}{\partial U}\right)_V + k\beta + kU\left(\frac{\partial \beta}{\partial U}\right)_V$$

or

$$\left(\frac{\partial S}{\partial U}\right)_V = k\beta \tag{12.57}$$

Equating the right-hand sides of Eqs. (12.55) and (12.57) gives

$$\beta = \frac{1}{kT} \tag{12.58}$$

Note that β is directly related to the absolute temperature T. Replacing the value of β by $1/kT$ in the previous equations in which β appeared gives the following expressions:

$$N_i = \frac{N}{Z}e^{-\varepsilon_i/kT} \tag{12.59}$$

$$Z = \sum_i e^{-\varepsilon_i/kT} \tag{12.60}$$

$$S = Nk \ln Z + \frac{U}{T} \tag{12.61}$$

The internal energy is given by

$$U = \frac{N}{Z}\sum_i \varepsilon_i e^{-\varepsilon_i/kT}$$

But

$$\frac{dZ}{dT} = \frac{1}{kT^2}\sum_i \varepsilon_i e^{-\varepsilon_i/kT}$$

so that

$$U = \frac{N\,kT^2}{Z}\frac{dZ}{dT} = N\,kT^2\left(\frac{\partial \ln Z}{\partial T}\right)_{N,V} \tag{12.62}$$

The Helmholtz function is

$$F = U - TS = -N\,kT \ln Z \tag{12.63}$$

The pressure of the system can be expressed by the relation

$$p = -\left(\frac{\partial F}{\partial V}\right)_{T,N}$$

$$= N\,kT\left(\frac{\partial \ln Z}{\partial V}\right)_{T,N} \tag{12.64}$$

and so on for the other thermodynamic functions.

12.12 Degeneracy of Energy Levels

The previously presented analysis considered situations where particles are distributed among different energy levels. A specialized case occurs frequently where two (or more) energy states differ from each other, from the standpoint of configuration, but their energy values and number of particles are identical. Such cases mean that a greater number of different arrangements are possible. The method of evaluating this effect will now be considered.

If at any given energy level ε_i a number of different states of the system have the same energy, this energy level is said to be *degenerate*. The number of energy states of a given energy level ε_i is called *degeneracy* or *statistical weight* and is denoted by g_i. In the case of a monatomic gas, for example, if N_i is the number of molecules in the ith level, $g_i \gg N_i$ at room temperature (or higher), and the degree of degeneracy increases rapidly with the increase of energy. At low temperatures, the lower-energy states become more populated, and N_i is not much different from g_i. The following analysis will lead to an expression of the Maxwell–Boltzmann distribution law, taking the degeneracy of energy levels into account.

Suppose it is required to arrange N particles into groups with $N_1, N_2,...N_i,...$ particles in the respective groups. Let the energy level of the first group of N_1 particles be ε_1 with a degeneracy g_1, the energy level of the second group of N_2 particles be ε_2 with a degeneracy g_2, and so on. Starting with the first group, since any one of the N particles may occupy any one of the g_1 states of the energy level, there are $g_1 N$ possible arrangements for the first of the N particles. Since any one of the remaining $(N - 1)$ particles may occupy any one of the g_1 states, there are $g_1(N - 1)$ arrangements for the second particle, and so on for the rest of the N_1 particles. The number of possible arrangements in the first group of N_1 particles, taking into account the indistinguishability of particles within the energy level, is then

$$\frac{g_1 N g_1 (N - 1) g_1 (N - 2)...g_1(N - N_1 + 1)}{N_1!} = \frac{g_1^{N_1} N!}{(N - N_1)! N_1!}$$

For the second group of N_2 particles having an energy level ε_2 and a degeneracy g_2, there are $g_2(N - N_1)$ possible arrangements for the first particle, $g_2(N - N_1 - 1)$ for the second, and so on. The number of arrangements in the second group of N_2 particles is then

$$\frac{g_2(N - N_1)g_2(N - N_1 - 1)g_2(N - N_1 - 2)...g_2(N - N_1 - N_2 + 1)}{N_2!}$$

$$= \frac{g_2^{N_2}(N - N_1)!}{(N - N_1 - N_2)! N_2!}$$

In general for the ith group, the number of arrangements is

$$\frac{g_i(N - N_1 - N_2 - ... N_{i-1})g_i(N - N_1 - N_2 - ... N_{i-1} - 1) ... g_i(N - N_1 - N_2 - N_i + 1)}{N_i!}$$

$$= \frac{g_i^{N_i}(N - N_1 - N_2 - ... N_{i-1})!}{(N - N_1 - N_2 - ... - N_i)! N_i!}$$

Multiplying the preceding expressions gives the total number of arrangements as

$$W = \frac{(g_1^{N_1} g_2^{N_2} g_3^{N_3} ...g_i^{N_i}...)N!}{N_1! N_2! N_3!...N_i!...} = N! \prod_i \frac{g_i^{N_i}}{N_i!} \qquad (12.65)$$

Equation (12.65) thus replaces Eq. (12.39) if degeneracy, that is, the number of quantum states in the ith energy level, is taken into consideration.

Applying Stirling's approximation to Eq. (12.65) gives

$$\ln W = \ln N! + \sum_i N_i \ln g_i - \sum_i \ln N_i!$$

$$= N \ln N - N + \sum_i N_i \ln g_i - \sum_i N_i \ln N_i + \sum_i N_i$$

or

$$\ln W = N \ln N + \sum_i N_i \ln \frac{g_i}{N_i} \qquad (12.66)$$

Equation (12.66) is subject to the two following conditions of constraint:

$$\sum_i \delta N_i = dN = 0 \tag{12.67}$$

and

$$\sum_i \varepsilon_i \delta U_i = dU = 0 \tag{12.68}$$

For the maximum value of W and since $dg_i = 0$,

$$\delta \ln W = \sum_i \ln\left(\frac{g_i}{N_i}\right)\delta N_i = 0 \tag{12.69}$$

Multiplying Eqs. (12.67) and (12.68) by the Lagrange multipliers $\ln a$ and $-\beta$ and adding the result to Eq. (12.69) gives

$$\sum_i \left[\ln a - \beta\varepsilon_i + \ln\frac{g_i}{N_i}\right]\delta N_i = 0$$

Since the δN_i's may be treated as independent,

$$\ln\frac{ag_i}{N_i} = \beta\varepsilon_i$$

or

$$N_i^* = ag_i e^{-\beta\varepsilon_i} \tag{12.70}$$

Equation (12.70) gives the most probable particle distribution for the Maxwell–Boltzmann model and is called the *general Maxwell–Boltzmann distribution law*. It differs from Eq. (12.44) by the weighing factor g_i, which accounts for the degeneracy of the energy levels.

The partition function Z now takes the more general form

$$Z = \sum_i g_i e^{-\varepsilon_i/kT} \tag{12.71}$$

In this equation, the partition function is defined in terms of the energy levels and their degeneracies. The summation is taken over all levels of energy accessible to the molecules.

12.13 Other Kinds of Statistics

As important as the Maxwell–Boltzmann distribution law are the Fermi–Dirac and Bose–Einstein statistics. All three statistics apply probability theory to determine the distribution of properties of particles. Each develops a distribution function that gives the average number of particles at a given energy state. At statistical equilibrium, the relative number of molecules at any particular energy level is essentially constant even though the energy states of individual molecules may be constantly changing owing to particle collisions. Which statistical approach to use depends on the nature and the behavior of particles considered. The Maxwell–Boltzmann statistics, for example, describe a classical system comprising identical particles in thermal equilibrium that are distinguishable. An ideal gas in which there is a distribution of particle energies over a wide range is described by a modified Maxwell–Boltzmann statistic, which corrects for the indistinguishability of the molecules. Discrepancies appear, however, when Maxwell–Boltzmann statistics are applied to electrons in a metal. One of these discrepancies is the difference between the observed and the predicted electrical conductivity of a metal. Fermi–Dirac statistics resolve these difficulties.

12.14 Fermi–Dirac Statistics

The essential difference between Fermi–Dirac and Maxwell–Boltzmann statistics arises from the method of defining and counting microstates. Fermi–Dirac statistics are particularly appropriate for a system in which the particles are indistinguishable, such as electrons, protons, and neutrons. Free electrons in a metal, sometimes called *ideal electron gas*, can move quite freely through the entire lattice of positive ions in the metal. Unlike molecules, electrons in a metal show a considerable amount of mobility among themselves because of their high density and because the forces between them are relatively small. As a result, only one particle may occupy a given energy state; that is, no two particles may have the same quantum state, which is a statement of the *Pauli exclusion principle*.[9] Thus, Fermi–Dirac statistics apply to systems of identical indistinguishable particles, such as electrons and protons that obey the Pauli exclusion principle. The following paragraph outlines the derivation of Fermi–Dirac statistics.

Suppose it is required to arrange N indistinguishable particles among energy levels $\varepsilon_1, \varepsilon_2, \ldots$, and so on, having the respective degeneracies $g_1, g_2 \ldots$, and so on. The number of energy states g_i of any energy level ε_i is assumed to be larger than the number of particles N_i in that level. Furthermore, only one particle is allowed in each energy state. Let N_1 be the number of particles assigned to energy level ε_1. The first particle has a choice of g_1 states ($N_i < g_i$); the second

---

[9] The *Pauli exclusion principle* states that no two particles can occupy the same quantum state.

has a choice of $(g_1 - 1)$ states; and so on. The total number of arrangements, taking into account the indistinguishability of the N_1 particles, is

$$\frac{g_1(g_1 - 1)(g_1 - 2)\ldots(g_1 - N_1 + 1)}{N_1!} = \frac{g_1!}{N_1!\,(g_1 - N_1)!}$$

For the ith energy level, the foregoing expression takes the form

$$W_i = \frac{g_i!}{N_i!\,(g_i - N_i)!}$$

The total number of arrangements for all energy states is the product of the individual arrangements of the N_i groups, or

$$W = \prod_i \frac{g_i!}{N_i!\,(g_i - N_i)!} \tag{12.72}$$

Taking the natural logarithm of Eq. (12.72) and making use of Stirling's approximation for $g_i!$ and $(g_i - N_i)!$ gives

$$\begin{aligned}
\ln W &= \sum_i [\ln g_i! - \ln N_i! - \ln (g_i - N_i)!] \\
&= \sum_i [g_i \ln g_i - g_i - N_i \ln N_i + N_i \\
&\qquad\qquad - (g_i - N_i) \ln (g_i - N_i) + (g_i - N_i)] \\
&= \sum_i [g_i \ln g_i - N_i \ln N_i - (g_i - N_i) \ln (g_i - N_i)]
\end{aligned}$$

and

$$\delta(\ln W) = \sum_i \left[-\ln \frac{N_i}{(g_i - N_i)} \right] \delta N_i$$

To determine the most probable distribution, the value of $\ln W$ is maximized:

$$\sum_i \left[\ln \left(\frac{(g_i - N_i)}{N_i} \right) \right] \delta N_i = 0 \tag{12.73}$$

In addition, the following two equations of constraint apply:

$$\sum_i \delta N_i = dN = 0 \tag{12.74}$$

$$\sum_i \varepsilon_i \delta N_i = dU = 0 \tag{12.75}$$

Multiplying Eq. (12.74) by (ln a) and Eq. (12.75) by $(-\beta)$, then adding the result to Eq. (12.73), gives

$$\sum_i \left[\ln a - \beta \varepsilon_i + \ln \frac{g_i - N_i}{N_i} \right] \delta N_i = 0$$

Since each coefficient of δN_i in the summation must be equal to zero,

$$\frac{g_i - N_i}{N_i} = \frac{e^{\beta \varepsilon_i}}{a}$$

or

$$N_i^* = \frac{a g_i e^{-\beta \varepsilon_i}}{1 + a e^{-\beta \varepsilon_i}} \qquad (12.76)$$

Equation (12.76), which describes the distribution of particles according to Fermi–Dirac statistics, was derived by Fermi and was applied by Dirac to electrons in metal.

12.15 Bose–Einstein Statistics

Bose–Einstein statistics are most appropriate for systems consisting of identical indistinguishable particles, such as light quanta or photons. Unlike Fermi–Dirac statistics, the Bose–Einstein distribution law imposes no restriction on the number of particles that can occupy the same energy state, and therefore, the Pauli exclusion principle is not applicable.

In the Maxwell–Boltzmann and Fermi–Dirac statistics, all energy levels were sparsely populated ($N_i \ll g_i$), but in Bose–Einstein statistics, it is assumed all energy levels are densely populated ($N_i \gg g_i$). Proceeding as in Fermi–Dirac statistics, the first particle of the group N_1 has a choice of g_1 states of energy level ε_1. The second particle has a choice of $(g_1 - 1)$ states, in addition to the state occupied by the first particle. This means that there are g_1 arrangements if both particles are at the same state and $[g_1(g_1 - 1)/2]$ arrangements if they are at different states. Therefore, the total number of possible arrangements of the first two particles is

$$W_2 = g_1 + \frac{g_1(g_1 - 1)}{2} = \frac{g_1(g_1 + 1)}{2}$$

The addition of a third particle will give rise to g_1 arrangements when all the particles are in one state, $g_1(g_1 - 1)$ arrangements when two particles are in one state and the third particle is in a different state, and $\{[g_1(g_1 - 1)/2][(g_1 - 2)/3]\}$ arrangements when the three particles are in different states. Thus, the total number of possible arrangements is

$$W_3 = g_1 + g_1(g_1 - 1) + \frac{g_1(g_1 - 1)(g_1 - 2)}{2 \quad 3} = \frac{g_1(g_1 + 1)(g_1 + 2)}{3!}$$

To generalize, then, the number of possible arrangements of N_i particles in the g_i states of the energy level ε_i is

$$W_i = \frac{g_i(g_i + 1)(g_i + 2)...(g_i + N_i - 1)}{N_i!} = \frac{(g_i + N_i - 1)!}{N_i!(g_i - 1)!}$$

The total number of the possible arrangements for all energy levels is equal to the product of all the W_i's, or

$$W = \prod_i W_i = \prod_i \frac{(g_i + N_i - 1)!}{N_i!(g_i - 1)!} \qquad (12.77)$$

Taking the natural logarithm of Eq. (12.77) and neglecting the term 1 compared to $(g_i + N_i)$ or g_i gives

$$\ln W = \sum_i [\ln (g_i + N_i)! - \ln N_i! - \ln g_i!]$$

$$= \sum_i [(g_i + N_i) \ln (g_i + N_i) - N_i \ln N_i - g_i \ln g_i]$$

and

$$\delta(\ln W) = \sum_i \left[\ln \frac{g_i + N_i}{N_i} \right] \delta N_i$$

By maximizing the probability in the preceding equation and by imposing the two conditions of constraint, the following expression for the most probable distribution for the Bose–Einstein model is obtained:

$$N_i^* = \frac{ag_i e^{-\beta \varepsilon_i}}{1 - ae^{-\beta \varepsilon_i}} \qquad (12.78)$$

The Maxwell–Boltzmann, Fermi–Dirac, and Bose–Einstein distributions differ from one another only by the term $ae^{-\beta \varepsilon_i}$ in the denominator. They become essentially identical in value when $ae^{-\beta \varepsilon_i}$ is small compared to unity (or when

$g_i >> N_i$). When $N_i/g_i << 1$, only one particle is found in each quantum state, and the Maxwell–Boltzmann distribution is a good approximation for all cases, particularly at high temperature. This condition, which greatly simplifies mathematical procedure, exists in an ideal gas in which the predominant energies are translational (except at low temperature). The three distributions are compared in Figure 12.13 and in Table 12.4.

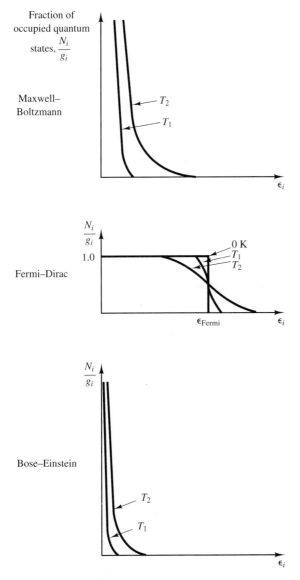

Figure 12.13   Maxwell–Boltzmann, Fermi–Dirac, and Bose–Einstein distribution.

Table 12.4   Comparison Between Maxwell–Boltzmann, Fermi–Dirac, and Bose–Einstein Statistics

	Maxwell–Boltzmann	Fermi–Dirac	Bose–Einstein
Characteristic model	Large number of identical particles that are distinguishable	Large number of identical indistinguishable particles that conform to the Pauli exclusion principle	Large number of identical indistinguishable particles
Thermodynamic probability	$W = N! \prod_i \dfrac{g_i^{N_i}}{N_i!}$	$W = \prod_i \dfrac{g_i!}{N_i!\,(g_i - N_i)!}$	$W = \prod_i \dfrac{(g_i + N_i - 1)!}{N_i!\,(g_i - 1)!}$
Distribution law	$N_i^* = ag_i e^{-\varepsilon_i/kT}$ $= \dfrac{Ng_i e^{-\varepsilon_i/kT}}{Z}$	$N_i^* = \dfrac{ag_i e^{-\varepsilon_i/kT}}{1 + ae^{-\varepsilon_i/kT}}$	$N_i^* = \dfrac{ag_i e^{-\varepsilon_i/kT}}{1 - ae^{-\varepsilon_i/kT}}$
Examples of systems that follow the statistics	Gas molecules* (except near 0 K); electrons at extremely high temperatures	Electrons in metal (except at very high temperatures)	Photons of radiation; gas molecules near 0 K

*Provided that $N_i/g_i \ll 1$ and the value of W is divided by $N!$ to discount the distinguishability of the N particles.

At a temperature of absolute zero, the energy level predicted by the Maxwell–Boltzmann distribution and the Bose–Einstein distribution is zero. On the other hand, the Fermi–Dirac law shows a finite energy level at a temperature of absolute zero.

It is common to write the distribution function for the Fermi–Dirac statistics in the following form:

$$\frac{N_i}{g_i} = \frac{1}{e^{(\varepsilon_i - \varepsilon_f)/kT} + 1} \tag{12.79}$$

where ε_f is the *Fermi level of energy.*

For $(\varepsilon_i - \varepsilon_f)/kT \gg 1$,

$$\frac{N_i}{g_i} = \frac{1}{e^{(\varepsilon_i - \varepsilon_f)/kT}} = \frac{e^{-\varepsilon_i/kT}}{e^{-\varepsilon_f/kT}} = (\text{const.})e^{-\varepsilon_i/kT}$$

so that N_i is proportional to $g_i e^{-\varepsilon_i/kT}$ (Maxwell–Boltzmann statistics).

At $T = 0$ K and for $\varepsilon_i < \varepsilon_f$,

$$\frac{N_i}{g_i} = \frac{1}{e^{-(\varepsilon_f - \varepsilon_i)/kT} + 1} = \frac{1}{e^{-\infty} + 1} = 1$$

At $T = 0$ K and for $\varepsilon_i > \varepsilon_f$,

$$\frac{N_i}{g_i} = \frac{1}{e^{(\varepsilon_i - \varepsilon_f)/kT} + 1} = \frac{1}{e^{\infty} + 1} = 0$$

At $T = 0$ K, the low energy levels become well populated; however, all the electrons cannot be at zero energy level because of the Pauli exclusion principle. This leads to an energy distribution in which all the lowest-energy levels are completely filled, but no more than one electron is contained in any one state. Thus, the number of the occupied energy states starting from zero up to an energy level ε_f is equal to the number of electrons ($N_i = g_i$). The *Fermi energy* ε_f represents the maximum energy at the absolute zero of temperature, and states of higher energies are empty. At higher temperatures, the ratio $N_i/g_i < 1$ approaching zero as ε increases.

Example 12.6

It is required to compare the Fermi–Dirac, Bose–Einstein, and Maxwell–Boltzmann statistics when three particles are arranged in two energy levels. Two particles are at energy level ε_1 having a degeneracy $g_1 = 2$, and one particle at energy level ε_2 having a degeneracy $g_2 = 1$.

SOLUTION

Fermi–Dirac:

$$W = \prod_i \frac{g_i!}{N_i!\,(g_i - N_i)!} = \frac{2!}{2!\,0!} \cdot \frac{1!}{1!\,0!} = 1$$

Bose–Einstein:

$$W = \prod_i \frac{(g_i + N_i - 1)!}{N_i!\,(g_i - 1)!} = \frac{3!}{2!\,1!} \cdot \frac{1!}{1!\,0!} = 3$$

Maxwell–Boltzmann:

$$W = N! \prod_i \frac{g_i^{N_i}}{N_i!} = \frac{3! \cdot (2^2 \cdot 1^1)}{2!\,1!} = 12$$

Here the particles are called a, b, and c since they are distinguishable.

For energy level ε_1,

$$N_1 = 2: \qquad \begin{array}{ccc} n_1 = 2 & n_1 = 1 & n_1 = 0 \\ \text{or} & \text{or} & \\ n_2 = 0 & n_2 = 1 & n_2 = 2 \end{array}$$

$$W = \frac{3!}{2!\,0!\,1!} + \frac{3!}{1!\,1!\,1!} + \frac{3!}{0!\,2!\,1!} = 3 + 6 + 3 = 12 \qquad \text{(as before)}$$

Therefore,

$$W_{FD} : W_{BE} : W_{MB} = 1 : 3 : 12$$

Figure 12.14 illustrates the particle distribution for the three statistics.

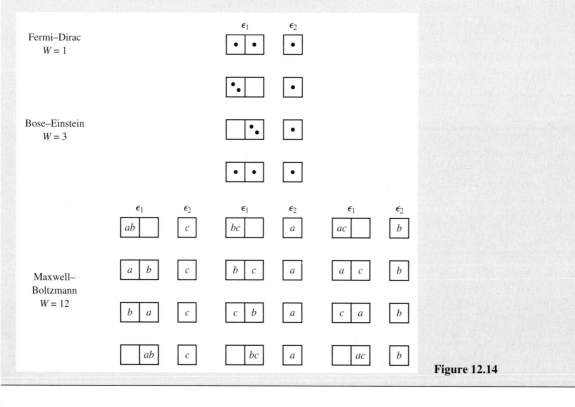

Figure 12.14

12.16 Partition Function

The expressions for the partition function that were derived in previous sections applied to an individual molecule. To describe the properties of a gas consisting of N molecules, an expression of the partition function for a system that consists of N particles is developed. This section outlines the relationship between these two partition functions.

According to Maxwell–Boltzmann statistics, the thermodynamic probability given by Eq. (12.65) is

$$W = N! \prod_i \frac{g_i^{N_i}}{N_i!}$$

Taking the logarithm of the preceding equation and using Stirling's approximation gives

$$\ln W = N \ln N + \sum_i N_i \ln \frac{g_i}{N_i}$$

Substituting from the expression,

$$N_i = \frac{N}{Z} g_i e^{-\varepsilon_i/kT}$$

into the previous equation gives

$$\ln W = N \ln N + \sum_i \left[\left(\frac{N}{Z} g_i e^{-\varepsilon_i/kT} \right) \ln \frac{g_i}{\frac{N}{Z} g_i e^{-\varepsilon_i/kT}} \right]$$

$$= N \ln N + \frac{N}{Z} \sum_i \left(g_i e^{-\varepsilon_i/kT} \right) \ln \frac{Z}{N} - \frac{N}{Z} \sum_i \left(g_i e^{-\varepsilon_i/kT} \ln e^{-\varepsilon_i/kT} \right)$$

$$= N \ln N + \frac{N}{Z} \sum_i \left(g_i e^{-\varepsilon_i/kT} \right) \ln \frac{Z}{N} + \frac{N}{Z} \sum_i \left(g_i \frac{\varepsilon_i}{kT} e^{-\varepsilon_i/kT} \right)$$

But

$$Z = \sum_i g_i e^{-\varepsilon_i/kT} \qquad \text{and} \qquad U = \frac{N \sum_i g_i \varepsilon_i e^{-\varepsilon_i/kT}}{Z}$$

which, when substituted in the preceding equation, gives

$$\ln W = \ln Z^N + \frac{U}{kT} \tag{12.80}$$

Therefore, the system partition function $\mathcal{Z}$ of N distinguishable particles, in terms of the molecular partition function Z, may be written in the form

$$\mathcal{Z} = (Z)^N \tag{12.81}$$

When the particles are indistinguishable, the expression for W given by Eq. (12.65) must be divided by $N!$ so that

$$W = \prod_i \frac{g_i^{N_i}}{N_i!} \tag{12.82}$$

Taking the logarithm of Eq. (12.82) and using Stirling's approximation gives

$$\ln W = \sum_i [N_i \ln g_i - N_i \ln N_i + N_i] = \sum_i N_i \left[\ln \frac{g_i}{N_i} + 1 \right]$$

Substituting for N_i gives

$$\begin{aligned}
\ln W &= \sum_i N_i \left[\ln \frac{g_i}{(N/Z)g_i e^{-\varepsilon_i/kT}} + 1 \right] = \sum_i N_i \left[\ln \frac{Z}{N} + \frac{\varepsilon_i}{kT} + 1 \right] \\
&= N \ln \frac{Z}{N} + \frac{U}{kT} + N = N \ln Z - N \ln N + \frac{U}{kT} + N \tag{12.83} \\
&= N \ln Z - \ln N! + \frac{U}{kT} = \ln \left(\frac{Z^N}{N!} \right) + \frac{U}{kT}
\end{aligned}$$

Therefore, the partition function $\mathcal{Z}$ for a system of indistinguishable particles may be written in terms of the molecular partition function Z as

$$\mathcal{Z} = \frac{(Z)^N}{N!} \tag{12.84}$$

The foregoing result indicates that when Maxwell–Boltzmann statistics are applied to a gas, the expression for W must be divided by $N!$. This division is necessary, since the molecules of a gas are identical, so that from a quantum statistics point of view they are indistinguishable except for their energy levels. Note, however, that the expression of W (when divided by $N!$) does not give the correct number of microstates of indistinguishable particles.

12.17 Statistics of a Monatomic Gas

The Maxwell–Boltzmann velocity distribution law, which was derived in Section 12.3, can also be derived from the foregoing statistics. Consider a volume V containing an ideal monatomic gas of indistinguishable atoms. Let the atoms have different states in any energy level; that is, the gas is degenerate; furthermore, let the number of the allowable quantum states g_i in the ith level be much larger than the number of atoms in that level, that is, $g_i/N_i >> 1$. According to Bose–Einstein statistics, the number of atoms of energy ε_i is

$$N_i = \frac{ag_i e^{-\beta\varepsilon_i}}{1 - ae^{-\beta\varepsilon_i}}$$

But since $g_i/N_i >> 1$, $ae^{-\beta\varepsilon_i} << 1$ and can be neglected in the denominator of the preceding equation. This is a fairly good approximation even at low energies because a is small (except at very low temperature). The Bose–Einstein distribution law then becomes

$$N_i = g_i ae^{-\beta\varepsilon_i} \tag{12.85}$$

which is the Boltzmann distribution law.

Consider next a *momentum space* of coordinates p_x, p_y, and p_z, and let the momentum of each atom in a volume V be represented by a momentum vector starting at the origin. The momentum of the atoms having energies between ε_i and $\varepsilon_i + d\varepsilon_i$ lies in a momentum shell of radius p_i and thickness dp_i. The volume of this shell is

$$dV = (4\pi p_i^2)dp_i$$

Although momentum space is analogous to velocity space, the distribution of the end points of the vectors does not follow the same pattern. In velocity space, changes in magnitude occur in a continuous way, the increments in velocity being infinitesimally small. In momentum space, however, increments in momentum occur in discrete steps, owing to the quantum restrictions imposed on energy. The spacing between the end points of the momentum vectors is still small so that the number of points in a small volume in momentum space can be calculated.

The quantum cell in the six-dimensional phase space of coordinates x, y, z, p_x, p_y, and p_z defines the location of an atom. The Heisenberg uncertainty principle discussed in Section 12.19 imposes the condition that the quantum cell cannot be less than h^3, where h is Planck's constant. An atom whose momentum lies in the spherical shell may be at any location within the volume V, and the number of possible quantum states in the shell (degeneracy) is then

$$g_i = (4\pi p_i^2 dp_i)\left(\frac{V}{h^3}\right)$$

Since $p = mv$,

$$g_i = \frac{4\pi m^3 v_i^2 V \, dv_i}{h^3}$$

Also,[10]

$$a = \frac{N}{Z} = \frac{N}{V}\left(\frac{h^2}{2\pi mkT}\right)^{3/2}$$

Substituting for g_i and a in Eq. (12.85) and noting that $\varepsilon_i = \frac{1}{2}mv_i^2$, then

$$dN = N_i = \left(\frac{4\pi m^3 v_i^2 V \, dv_i}{h^3}\right)\frac{N}{V}\left(\frac{h^2}{2\pi mkT}\right)^{3/2} e^{-m_i v_i^2/2kT}$$

or

$$\frac{dN}{dv_i} = \frac{4N}{\sqrt{\pi}}\left(\frac{m}{2kT}\right)^{3/2} v_i^2 e^{-mv_i^2/2kT} \tag{12.86}$$

Equation (12.86) indicates the speed distribution function and is the same as Eq. (12.12).

12.18 Wave Mechanics and Internal Energy Levels

In the previous sections, formulas were derived for calculating various thermodynamic functions. These are based on the partition function, which describes how a multitude of atoms (or molecules) capable of existing at different energy levels will distribute themselves among the various levels. The remaining problem is to determine these energy values. For this, it becomes necessary to delve into wave mechanics.

Although Newtonian mechanics can describe large-scale or macroscopic systems, it cannot describe the behavior of particles at the atomic scale adequately. Microscopic particles are subject to *quantum restrictions*, which allow the particles to have only discrete values of energy. The behavior of these microscopic particles is described by the Schrödinger wave equation (Section 12.21).

Certain phenomena, such as the diffraction of electrons by a crystal lattice, photoelectric effects,[11] and the radiation emitted from an isothermal enclosure, suggested that these particles can have only definite energy levels. Furthermore, there is evidence that microscopic particles behave in a wavelike manner, operating at definite and discrete values of frequency. Planck proposed that the energy emitted from, or absorbed by, a heated surface takes place in integral

---

[10] See Eq. (13.4) of Chapter 13.

[11] The emission of electrons from a metal due to incident photons is called *photoelectric emission*.

multiples of a unit or a quanta of energy. This allowed him to apply the Maxwell–Boltzmann statistics for a system of photons. The energy of a photon of electromagnetic radiation is proportional to the frequency of radiation according to Planck's equation:

$$\varepsilon = hv \tag{12.87}$$

where h is Planck's constant (equal to 6.626×10^{-34} J-s). Note that the energy of a photon associated with radiation of a given frequency has a single distinct value; that is, it is quantized.

In 1925, De Broglie postulated that a particle of momentum p has associated with it a wave of wavelength λ given by the relation

$$\lambda = \frac{h}{p} \tag{12.88}$$

Equation (12.88) is called the *De Broglie* equation, which is based on Eq. (12.87) and on the classical result that the ratio of energy to momentum in an electromagnetic plane wave equals the velocity of the wave. Equation (12.88) was derived for a photon but applied equally well to material particles. In 1927, Davisson and Germer using electrons, and Stern and Estermann using atoms, confirmed the validity of Eq. (12.88) by diffraction experiments with a crystal lattice. For macroscopic particles, the wavelength is so small that the wavelike properties are very difficult to observe, and thus classical mechanics is sufficient to describe the motion of macroscopic particles to a very high degree of precision.

The foregoing experiments, among many others, confirmed the dual character of atomic particles and suggested that the behavior of matter on an atomic scale is associated with a field having a wavelike character. This field may be described by a function called the *wave function*. A good example is electromagnetic radiation, which behaves like waves, having length and frequency, and also like particles of definite energy and momentum.

Example 12.7

Compare the wavelengths of an electron (mass = 9.1086×10^{-28} gram) when accelerated through a potential of 10,000 volts and a mass of 1 gram moving at a velocity of 5000 cm/s.

SOLUTION

Since 1 electron volt is the energy necessary to accelerate an electron through a potential of 1 volt, the kinetic energy acquired by the electron is

$$\varepsilon = eV$$

where e is the electronic charge ($e = 1.6021 \times 10^{-19}$ coulomb), and V is the voltage drop. The kinetic energy of the electron is

$$\varepsilon = (1.6021 \times 10^{-19} \text{ coul})(10,000 \text{ volts}) = 1.6 \times 10^{-15} \text{ J}$$

but

$$p = \sqrt{2m\varepsilon} = \sqrt{2(9.1086 \times 10^{-28} \text{ gm})(1.6 \times 10^{-8} \text{ gm-cm}^2/\text{s}^2)}$$
$$= 5.4 \times 10^{-18} \text{ gm-cm/s}$$

Therefore, the wavelength of the electron is

$$\lambda = \frac{h}{p} = \frac{6.626 \times 10^{-27} \text{ gm-cm}^2/\text{s}}{5.4 \times 10^{-18} \text{ gm-cm/s}} = 12.27 \times 10^{-10} \text{ cm}$$

In the case of the mass of 1 gram,

$$\lambda = \frac{h}{p} = \frac{6.626 \times 10^{-27} \text{ gm-cm}^2/\text{s}}{(1 \text{ gm})(5000 \text{ cm/s})} = 1.325 \times 10^{-30} \text{ cm}$$

This example demonstrates that, in the case of heavy particles, the wavelengths are so small that classical mechanics is sufficient to describe the motion.

12.19 The Uncertainty Principle

According to quantum mechanics, there is a definite limit to the precision to which the position and momentum of small particles may simultaneously be measured experimentally. The degree of precision is indicated by the *Heisenberg uncertainty principle*. Suppose it is required to measure the momentum and position of an atomic-scale particle, such as an electron. This may be done by observing at least one photon of light of speed c emitted by, or bounced off, the electron. The measuring device might be an ideal optical microscope using light of a short wavelength λ. Since the resolving power of the microscope is inversely proportional to the wavelength of the light used, the degree of accuracy in measuring the position is limited to the wavelength with which the electron is observed. Therefore, shorter wavelengths improve the resolving power and locate the electron's position more accurately. The momentum of the photon, however, changes the momentum of the electron, and consequently its position, so that absolute accuracy of simultaneous measurement of both position and momentum can never be achieved. The energy of a photon is hv, and therefore the error in measuring the momentum would be equal to the momentum of the photon, that is, $hv/c = h/\lambda$, which is inversely proportional to λ. The Heisenberg uncertainty principle shows that the product of the uncertainty in position and the uncertainty in momentum has a minimum value:

$$\Delta p_x \, \Delta x \geq \frac{h}{4\pi} \qquad (12.89)$$

where Δp_x and Δx are the uncertainties in measuring the conjugate momentum and position, respectively, of a particle, and h is Planck's constant. This means that a point in phase space cannot be accurately specified but can be located only within a volume equal to or greater than h^3. Note that at the limiting value indicated by Eq. (12.89) (that is, $\Delta p \Delta x = h/4\pi$), an increased accuracy in measuring momentum can be achieved only at the expense of a decreased knowledge of position. Furthermore, when momentum or position is known precisely, that is, either Δp or Δx is zero, nothing at all is known about the other parameter.

Although this inherent uncertainty exists also in classical mechanics, its effects are unnoticed because of the extremely small magnitude of Planck's constant, and methods used in macroscopic measurements have practically no effect on the measured properties. Certainty in macroscopic measurements in classical mechanics is equivalent to only a high probability in wave mechanics.

Example 12.8

Compare the uncertainties in the velocities of a hydrogen atom and an electron if the position of either particle can be determined with an uncertainty of a 100 μm.

SOLUTION

The uncertainty in momentum, according to Eq. (12.89), is

$$\Delta p \geq \frac{h}{4\pi \Delta x}$$

$$\Delta p \geq \frac{6.626 \times 10^{-27} \text{ gm-cm}^2/\text{s}}{4\pi(100 \times 10^{-8} \text{ cm})} \qquad \text{or}$$

$$\Delta p \geq 0.527 \times 10^{-21} \text{ gm-cm/s}$$

$$\Delta v_H = \frac{\Delta p}{m_H} = \frac{(0.527 \times 10^{-21} \text{ gm-cm/s})}{(1838 \times 9.1086 \times 10^{-28} \text{ gm})} = 0.315 \times 10^3 \text{ cm/s}$$

$$\Delta v_{\text{electron}} = \frac{\Delta p}{m_{\text{electron}}} = \frac{0.527 \times 10^{-21} \text{ gm-cm/s}}{9.1086 \times 10^{-28} \text{ gm}} = 0.579 \times 10^6 \text{ cm/s}$$

Note that for the same uncertainty in position, the uncertainty of velocities is in the ratio of

$$\frac{\Delta v_{electron}}{\Delta v_H} = 1838$$

which is also the ratio of the mass of a hydrogen atom to the mass of an electron.

12.20 The Wave Equation

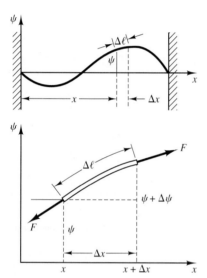

Figure 12.15   Vibration of a string.

Many of the characteristic properties of a substance are associated with the motion of its molecules so that a knowledge of the kinetic behavior of the molecules of a gas should lead to prediction of properties of the gas. But when very large velocities and very small particles are involved, the classical methods of describing the motion do not adequately explain properties. Methods based on a wave interpretation, however, have proved useful because of the analogy between the motion of a particle along a straight line while experiencing elastic collisions and a traveling wave along a string experiencing reflections at the ends of the string. For this reason, the wave equation will now be examined.

Consider the small transverse vibration of a homogeneous string stretched between two points on the x-axis, as shown in Figure 12.15. Let the vibration of the string be confined to the x-ψ plane; that is, there is no lateral vibration. It is assumed that the mass ρ per unit length is uniform and the displacements of the string are small, such that $\Delta\ell \approx \Delta x$. Let F be the uniform tension in the string. The ψ-component of the tension force in the string at (x, Ψ) is given by

$$-F\frac{\partial\Psi}{\partial\ell} = -F\frac{\partial\Psi\partial x}{\partial x\partial\ell} \approx -F\frac{\partial\Psi}{\partial x}$$

whereas at $(x + \Delta x, \psi + \Delta\psi)$, it is given by

$$F\frac{\partial\Psi}{\partial x} + F\frac{\partial^2\Psi}{\partial x^2}\Delta x$$

In the last expression, the remaining terms in Taylor's expansion[12] have been neglected. Equating the net force in the ψ-direction to the mass of element Δx multiplied by its acceleration gives

$$F\frac{\partial^2\Psi}{\partial x^2}\Delta x = \rho\Delta x\frac{\partial^2\Psi}{\partial t^2}$$

---

[12] Taylor's expansion: $f(x + h) = f(x) + hf'(x) + (h^2/2!)f''(x) + (h^3/3!)f'''(x) + \cdots$ (primes indicate differentiation with respect to the argument).

and therefore,

$$\frac{\partial^2 \Psi}{\partial x^2} = \frac{\rho}{F} \frac{\partial^2 \Psi}{\partial t^2}$$

Denoting ρ/F by $1/c^2$ gives the *wave equation* as

$$\frac{\partial^2 \Psi}{\partial x^2} = \frac{1}{c^2} \frac{\partial^2 \Psi}{\partial t^2} \qquad (12.90)$$

where c is called the *wave velocity* of the vibrating string. Equation (12.90) is a homogeneous partial differential equation whose solution may be written as the product of two separate functions, one dependent only on time and the other dependent only on position, so that

$$\Psi(x, t) = f(x)g(t)$$

Taking the partial derivatives of the foregoing expression and substituting into Eq. (12.90) gives

$$\frac{\partial^2[f(x)g(t)]}{\partial x^2} = \frac{1}{c^2} \frac{\partial^2[f(x)g(t)]}{\partial t^2}$$

$$g(t)\frac{d^2f(x)}{dx^2} = \frac{1}{c^2}f(x)\frac{d^2g(t)}{dt^2}$$

$$\frac{1}{f(x)}\frac{d^2f(x)}{dx^2} = \frac{1}{c^2}\left[\frac{1}{g(t)}\right]\frac{d^2g(t)}{dt^2} \qquad (12.91)$$

Through separation of variables, each side of Eq. (12.91) must be equal to the same constant since the two functions are independent of each other. Assuming harmonic waves, the function must repeat itself for each period of vibration or at time intervals equal to $2\pi/\omega$. Therefore, the time-dependent function $g(t)$ may be expressed in the form

$$g(t) = Ae^{-i\omega t}$$

where A is a constant, $i = \sqrt{-1}$, and ω is the angular frequency equal to $2\pi v$. Substituting for $g(t)$ in Eq. (12.91) gives the position-dependent equation as

$$\frac{d^2f(x)}{dx^2} = -\frac{\omega^2}{c^2}f(x) \qquad (12.92)$$

or in terms of wavelength

$$\frac{d^2f(x)}{dx^2} = -\frac{(2\pi)^2}{\lambda^2}f(x) \qquad (12.93)$$

The propagation of longitudinal waves along an elastic homogeneous bar may be described by the wave equation (Eq. 12.90) in which Ψ is interpreted as the longitudinal displacement from the equilibrium position. For longitudinal waves in a fluid such as air, the wave function Ψ represents the density of the medium in which the waves propagate. In such cases, the vibrations occur in three perpendicular directions rather than in only one direction. Three equations analogous to Eq. (12.90) can be formulated, and when they are combined, the three-dimensional wave equation takes the form:

$$\frac{\partial^2\Psi}{\partial x^2} + \frac{\partial^2\Psi}{\partial y^2} + \frac{\partial^2\Psi}{\partial z^2} = \frac{1}{c^2}\frac{\partial^2\Psi}{\partial t^2} \tag{12.94}$$

12.21 Schrödinger's Wave Equation

At this point, the objective lies in determining how the energy of a molecule is distributed. Classically, the energy of a particle is in the form of kinetic and potential energy so that the total energy is

$$\varepsilon = \varepsilon_{\text{kin}} + \varepsilon_{\text{pot}} = \frac{p^2}{2m} + \varepsilon_{\text{pot}}$$

The momentum of a particle is associated with wavelength according to the De Broglie equation

$$p = \frac{h}{\lambda}$$

This expression, relating to wave motion, can be combined with the classical energy equation to give

$$p^2 = \left(\frac{h}{\lambda}\right)^2 = 2m(\varepsilon - \varepsilon_{\text{pot}})$$

Introducing the foregoing result into the one-dimensional wave equation (Eq. 12.93) gives

$$\frac{d^2f(x)}{dx^2} = -\left(\frac{2\pi}{\lambda}\right)^2 f(x) = -(2\pi)^2\left[\frac{2m(\varepsilon - \varepsilon_{\text{pot}})}{h^2}\right]f(x)$$

$$= -\frac{8\pi^2 m}{h^2}(\varepsilon - \varepsilon_{\text{pot}})f(x)$$

For standing waves that correspond to stationary states of the system, the time-dependency of the wave function can be ignored. The function $f(x)$ can then be replaced by Ψ, giving the *Schrödinger's time-independent wave equation* for a single particle in one dimension:

$$\frac{d^2\Psi}{dx^2} + \frac{8\pi^2 m}{h^2}(\varepsilon - \varepsilon_{\text{pot}})\Psi = 0 \qquad (12.95)$$

Equation (12.95) is an ordinary second-order linear differential equation with a variable coefficient (since ε_{pot} is in general a function of x). It gives the allowable values of the discrete energy of a particle of mass m for which solutions exist.

In three dimensions, the Schrödinger equation takes the form:

$$\frac{\partial^2\Psi}{\partial x^2} + \frac{\partial^2\Psi}{\partial y^2} + \frac{\partial^2\Psi}{\partial z^2} + \frac{8\pi^2 m}{h^2}(\varepsilon - \varepsilon_{\text{pot}})\Psi = 0 \qquad (12.96)$$

or

$$\nabla^2\Psi + \frac{8\pi^2 m}{h^2}(\varepsilon - \varepsilon_{\text{pot}})\Psi = 0 \qquad (12.97)$$

where ∇^2 is the Laplacian operator. Note that Eq. (12.97) is time independent, and it relates to the equilibrium state of a system of particles.

The Schrödinger wave equation in spherical polar coordinates may be obtained by making the following substitutions in Eq. (12.96):

$$x = r\sin\theta\cos\varphi \qquad y = r\sin\theta\sin\varphi \qquad z = r\cos\theta$$

and has the form,

$$\frac{1}{r^2}\frac{\partial}{\partial r}\left(r^2\frac{\partial\Psi}{\partial r}\right) + \frac{1}{r^2\sin\theta}\frac{\partial}{\partial\theta}\left(\sin\theta\frac{\partial\Psi}{\partial\theta}\right) + \frac{1}{r^2\sin^2\theta}\frac{\partial^2\Psi}{\partial\varphi^2}$$
$$+ \frac{8\pi^2 m}{h^2}(\varepsilon - \varepsilon_{\text{pot}})\Psi = 0 \qquad (12.98)$$

Now the intensity of a wave is proportional to the square of its amplitude. In the case of particle waves, the square of the amplitude of the wave function is related to particle density, or to the probability of finding a particle in a volume element $dx\, dy\, dz$. But the wave function Ψ may be a complex function; on the other hand, probability must be a positive real value. Consequently, statistical probability density is expressed by means of the product of both the wave function and its conjugate:[13]

$$\Psi\Psi^* = |\Psi|^2 \tag{12.99}$$

The expression $|\Psi|^2 \, dx \, dy \, dz$ can be interpreted as the probability of finding a particle in a given element of volume of the system. In a system of volume V, the probability of finding any particle somewhere in that volume is certain. This criterion serves to determine the probability density according to the normalizing condition,

$$\int_V \Psi\Psi^* dV = 1 \tag{12.100}$$

12.22 Internal Energy Levels

The energy of the molecules of a system may be considered to be associated with translational, rotational, vibrational, electronic, and potential (due to external force fields) modes of motion. If intermolecular forces are considered, a sixth contribution to the total energy is introduced. But in an ideal gas, it is assumed that there is no interaction between the molecules and the total energy of a molecule can be written as

$$\varepsilon = \varepsilon_{trans} + \varepsilon_{rot} + \varepsilon_{vib} + \varepsilon_{elec} + \varepsilon_{pot} \tag{12.101}$$

The Schrödinger equation describes the movement of a wavelike particle as a function of position for different values of the total energy of the particle. Various modes of motion of the molecules may occur, corresponding to these various kinetic and potential energy terms that may be included in the Schrödinger equation. To a first approximation, these different motions may be treated as though they were independent of each other so that each particular motion is then described by its own Schrödinger equation.[14]

According to quantum theory, the internal energies of a system are *quantized;* that is, only discrete values are possible. The numeration of the allowed quantum states constitutes the first step in the statistical analysis of the system.

---

[13] The conjugate of a complex number $(a + ib)$ is $(a - ib)$, where $i = \sqrt{-1}$, and $|a + ib| = |a - ib| = \sqrt{a^2 + b^2}$.

[14] A more refined calculation introduces interaction between, for example, rotational and vibrational modes.

12.23 Translational Energy[15]

This energy is due to the translational kinetic energy of the center of mass of the molecule. Consider the motion of a molecule of a monatomic gas in a box of dimensions a, b, and c. Although Figure 12.16 shows only the dimension a, along the x-axis, it is understood that similar behavior occurs along the y- and z-axes. The molecule is confined between the walls of the box that bound a potential region identified by a uniform potential energy ε_{pot} within the box. Since the motion is unchanged by a change in the zero level of potential energy, we may take $\varepsilon_{pot} = 0$ for convenience inside the box and $\varepsilon_{pot} = \infty$ outside. In order to separate the Schrödinger equation into three differential equations, one for each of the axial directions, assume a solution in which the time-independent wave function Ψ is expressed in the form

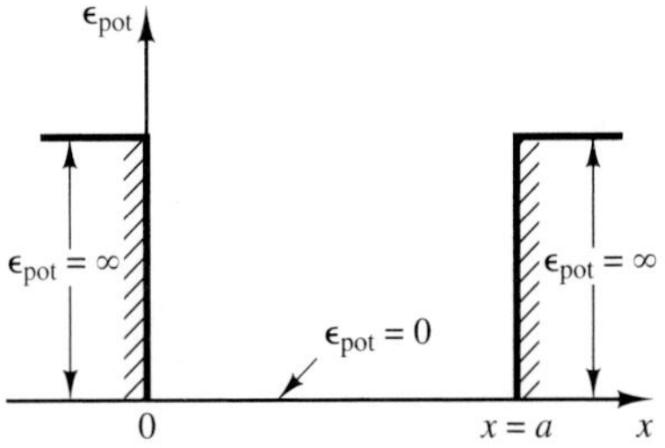

$$\Psi(x, y, z) = X(x)\, Y(y)\, Z(z) \tag{12.102}$$

where $X(x)$ is a function of x only, $Y(y)$ a function of y only, and $Z(z)$ a function of z only. When X, Y, and Z are substituted in the Schrödinger equation (Eq. 12.96), the following is subsequently obtained:

$$\frac{1}{X}\frac{d^2X}{dx^2} + \frac{1}{Y}\frac{d^2Y}{dy^2} + \frac{1}{Z}\frac{d^2Z}{dz^2} + \frac{8\pi^2 m}{h^2}\varepsilon = 0 \tag{12.103}$$

The energy ε can be expressed in terms of its components in each of the three directions:

$$\varepsilon = \varepsilon_x + \varepsilon_y + \varepsilon_z$$

Since translational motion, and energies, in the x-, y-, and z-directions are independent of one another, the wave equation may be expressed separately for each direction:

$$\frac{1}{X}\frac{d^2X}{dx^2} + \frac{8\pi^2 m}{h^2}\varepsilon_x = 0 \tag{12.104}$$

$$\frac{1}{Y}\frac{d^2Y}{dy^2} + \frac{8\pi^2 m}{h^2}\varepsilon_y = 0 \tag{12.105}$$

$$\frac{1}{Z}\frac{d^2Z}{dz^2} + \frac{8\pi^2 m}{h^2}\varepsilon_z = 0 \tag{12.106}$$

Figure 12.16   Particle in a box.

---

[15] Derivation of the expressions of the translational, rotational, and vibrational energies (Sections 12.23 to 12.25) can be skipped without loss of continuity. Emphasis can be placed only on the final results: the expressions of these quantized energies.

Equations (12.104), (12.105), and (12.106) have the same form, and it is sufficient to solve any one of them. The components of potential energy are subject to boundary conditions imposed by the walls of the box since the walls constitute barriers confining the particles to the box. This is equivalent to saying that a particle has no potential energy when inside the box but infinite potential energy when outside. In the region $0 < x < a$, Eq. (12.104) becomes

$$\frac{d^2X}{dx^2} + (M\varepsilon_x)X = 0 \tag{12.107}$$

where

$$M = \frac{8\pi^2 m}{h^2} \tag{12.108}$$

Equation (12.107) is a second-order differential equation in $X(x)$ whose solution is of the form

$$X = A \sin \sqrt{M\varepsilon_x}\, x + B \cos \sqrt{M\varepsilon_x}\, x \tag{12.109}$$

where A and B are constants.

The boundary conditions are introduced into Eq. (12.109):

At $x = 0$, $X = 0$; therefore $B = 0$, and $X = A \sin \sqrt{M\varepsilon_x}\, x$.
At $x = a$, $X = 0$; therefore $\sqrt{M\varepsilon_x}\, a = n_x\pi$, where n_x must be a positive integer.

These integers are termed the *translational quantum numbers*. From this condition, the x-component of the energy may be written

$$\varepsilon_x = \frac{n_x^2 h^2}{8ma^2} \qquad n_x = 1, 2, 3,\ldots \tag{12.110}$$

and

$$X = A \sin \left(\frac{n_x\pi}{a}\right)x$$

Since the translational quantum number n_x is an integer, the energy ε_x can have only certain discrete values, corresponding to specific forms of the wave function. Even though an infinite number of energy values, corresponding to all positive integer values of n_x, give rise to translational motion, these energy values can be at only certain discrete levels. Thus, the energy of a particle in a box is *quantized*.

Similar expressions may be obtained for translational energies in the y- and z-directions. The total translational energy is then

$$\varepsilon = \varepsilon_x + \varepsilon_y + \varepsilon_z = \frac{h^2}{8m}\left[\left(\frac{n_x}{a}\right)^2 + \left(\frac{n_y}{b}\right)^2 + \left(\frac{n_z}{c}\right)^2\right] \quad (12.111)$$

Translational energy is highly degenerate since many combinations of $(n_x/a)^2$, $(n_y/b)^2$, and $(n_z/c)^2$ add to the same value. Such degenerate quantum states have the same energy and can be distinguished only by the set of quantum numbers. Furthermore, the spacing between the translational energy levels is very small. The solution of Eq. (12.96) is given by a set of wave functions as

$$\Psi = A \sin\left(n_x \pi \frac{x}{a}\right) \sin\left(n_y \pi \frac{y}{b}\right) \sin\left(n_z \pi \frac{z}{c}\right) \quad (12.112)$$

which represents a series of standing waves with even and odd half-waves. Figure 12.17 shows Ψ as a function x for $n_x = 1, 2, 3$, and 4 for constant values of y and z equal to $b/2$ and $c/2$, respectively. Figure 12.18 shows the corresponding values of the probability density $|\Psi|^2$, indicating the probability associated with the location of the particle in any particular position.

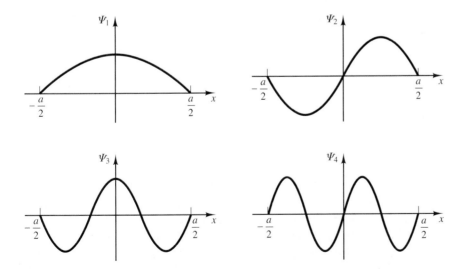

Figure 12.17 Ψ as a function of x for a particle in a box.

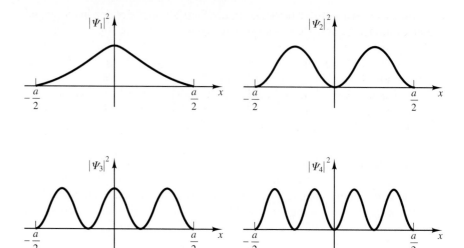

Figure 12.18   Probability density as a function of x for a particle in a box.

12.24 Rotational Energy

Another type of motion of the molecule is rotational motion due to the kinetic energy associated with the molecule rotational velocity and its moment of inertia. To describe rotational motion, it is convenient to express the Schrödinger equation in polar coordinates.

Consider a model of a "dumbbell" diatomic molecule comprising two atoms, as shown in Figure 12.19. The masses of the atoms, m_1 and m_2, may be assumed to be point masses so that the moment of inertia of the molecule about the line joining the two atoms is zero. Perpendicular to the line joining the two atoms,

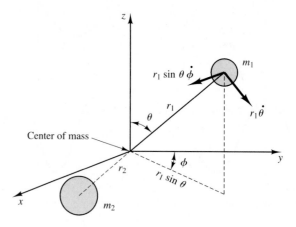

Figure 12.19   Dumbbell molecule.

and along axes that pass through the center of mass, the molecule has two equal moments of inertia. Assume also that the distance between the atoms is constant, and thus treat the molecule as a *rigid rotator*.

Let the origin of the Cartesian coordinates be located at the center of mass of the molecule. In considering the rotation of the molecule about its center of mass, the angles θ and φ determine the orientation of the molecule with respect to the coordinate axes, as shown in Figure 12.19. The velocity of the mass m_1 at a distance r_1 from the origin is given by

$$V_{m_1} = \sqrt{V_\theta^2 + V_\varphi^2} = \sqrt{(r_1\dot{\theta})^2 + (r_1\dot{\varphi}\sin\theta)^2} = r_1\sqrt{\dot{\theta}^2 + \dot{\varphi}^2\sin^2\theta}$$

where $\dot{\theta}$ and $\dot{\varphi}$ are the time rate of change of the angles θ and φ. Similarly, the velocity of the mass m_2 at a distance r_2 from the origin is given by

$$V_{m_2} = r_2\sqrt{\dot{\theta}^2 + \dot{\varphi}^2\sin^2\theta}$$

The kinetic energy of the molecule can thus be written

$$\varepsilon_{\text{rot}} = \frac{1}{2}m_1 V_{m_1}^2 + \frac{1}{2}m_2 V_{m_2}^2$$

$$= \left(\frac{m_1 r_1^2 + m_2 r_2^2}{2}\right)(\dot{\theta}^2 + \dot{\varphi}^2\sin^2\theta)$$

The term $(m_1 r_1^2 + m_2 r_2^2)$ is the moment of inertia, I, of the masses m_1 and m_2 about the center of mass of the molecule. The kinetic energy of the molecule then is

$$\varepsilon_{\text{rot}} = \frac{I}{2}\left(\dot{\theta}^2 + \dot{\varphi}^2\sin^2\theta\right) \tag{12.113}$$

Note that the term $(\dot{\theta}^2 + \dot{\varphi}^2\sin^2\theta)$ is equal to the square of the velocity of a particle that is located at a unit distance from the center of mass. Therefore, the moment of inertia I can be considered to represent the mass of a particle separated from the center of mass by a radial distance of unity.

When the kinetic energy associated with rotational motion, as expressed by Eq. (12.113), is substituted in the Schrödinger equation (Eq. 12.98) expressed in polar coordinates, the following equation is obtained:

$$\frac{1}{\sin\theta}\frac{\partial}{\partial\theta}\left(\sin\theta\frac{\partial\Psi}{\partial\theta}\right) + \frac{1}{\sin^2\theta}\frac{\partial^2\Psi}{\partial\varphi^2} + \frac{8\pi^2 I\varepsilon}{h^2}\Psi = 0 \tag{12.114}$$

Again, the potential energy may be taken to be zero. The solution to Eq. (12.114) is accomplished by assuming a product solution of the form:

$$\Psi(\theta, \varphi) = \Theta(\theta)\Phi(\varphi)$$

where $\Theta(\theta)$ and $\Phi(\varphi)$ are independent functions of θ and φ, respectively. After these values for the wave function are substituted into Eq. (12.114), the variables can be separated in the form

$$\frac{\sin\theta}{\Theta}\frac{\partial}{\partial\theta}\left(\sin\theta\frac{\partial\Theta}{\partial\theta}\right) + \frac{8\pi^2 I\varepsilon}{h^2}\sin^2\theta = -\frac{1}{\Phi}\frac{\partial^2\Phi}{\partial\varphi^2}$$

Since the functions Θ and Φ are independent, each side of the preceding equation can be set equal to a constant, say, m^2. The solution to the equation,

$$-\frac{1}{\Phi}\frac{\partial^2\Phi}{\partial\varphi^2} = m^2 \tag{12.115}$$

has the form

$$\Phi = A\sin m\varphi + B\cos m\varphi$$

The function Φ must be single-valued, and must repeat itself at 2π intervals. Therefore,

$$\sin m\varphi = \sin m(\varphi + 2\pi) = \sin m\varphi\cos 2\pi m + \cos m\varphi\sin 2\pi m$$

and

$$\cos m\varphi = \cos m(\varphi + 2\pi) = \cos m\varphi\cos 2\pi m - \sin m\varphi\sin 2\pi m$$

It can be shown, from these equations, that the conditions $\cos 2\pi m = 1$ and $\sin 2\pi m = 0$ must be simultaneously satisfied. This is possible only if m is an integer.

In order to solve the equation,

$$\frac{\sin\theta}{\Theta}\frac{\partial}{\partial\theta}\left(\sin\theta\frac{\partial\Phi}{\partial\theta}\right) + \frac{8\pi^2 I\varepsilon}{h^2}\sin^2\theta = m^2 \tag{12.116}$$

substitutions are made as follows:

$$D = \frac{8\pi^2 I\varepsilon}{h^2} \tag{12.117}$$

$$x = \cos\theta \tag{12.118}$$

and

$$\Theta(\theta) = (1 - x^2)\frac{m^2}{2}v(x) \tag{12.119}$$

These substitutions lead to

$$(1 - x^2)v'' - 2(m + 1)v' + [D - m(m + 1)]v = 0 \qquad (12.120)$$

where primes denote differentiation with respect to the argument x. The general solution to this second-order equation is a sum of two independent functions multiplied by arbitrary constants. The solution, which is finite for $\theta = 0$, can be expressed as a polynomial in x, and presented in the form of an infinite series:

$$v = \sum_{n=0}^{\infty} a_n x^n \qquad (12.121)$$

Substituting from Eq. (12.121) into Eq. (12.120) ultimately leads to

$$\sum_{n=0}^{\infty} a_n n(n - 1)x^{n-2} - \sum_{n=0}^{\infty} [(n + m)(n + m + 1) - D]a_n x^n = 0$$

If the normalization condition is to be satisfied, the series must terminate with a finite number of terms. At some finite value of n, the coefficient of x must equal zero, and this occurs when

$$D = (n + m)(n + m + 1) = j(j + 1) \qquad (12.122)$$

where $j = m + n$, is an integer since both m and n are integers. This equation for D leads, through Eq. (12.117), to the following expression for rotational energy:

$$\varepsilon_{\text{rot}} = \frac{h^2}{8\pi^2 I}j(j + 1) \qquad j = 0, 1, 2, 3... \qquad (12.123)$$

Note that $j \geq |m|$ and the allowed rotational energy levels are specified by the *rotational quantum number j*. Figure 12.20 shows the allowed rotational energies of diatomic molecules at different values of j. Since the ratio ε_i/kT, as indicated by the Boltzmann law, determines the population of the quantum states, it is readily seen that kT may be used as a reference amount of energy. Figure 12.21 compares the magnitude of rotational energies with kT for CO at 25°C. Consequently, several energy levels will be appreciably occupied at 25°C.

Finally, the solution to Eq. (12.116) may be given:

$$\Psi_{\text{rot}} = (A \sin m\varphi + B \cos m\varphi)CP_j{}^m(\cos \theta) \qquad (12.124)$$

where $P_j{}^m$ is called the *associated Legendre polynomial* of degree j and order m, and C is a constant.

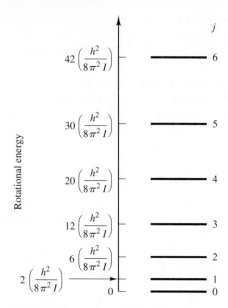

Figure 12.20   Allowed rotational energies of a diatomic molecule.

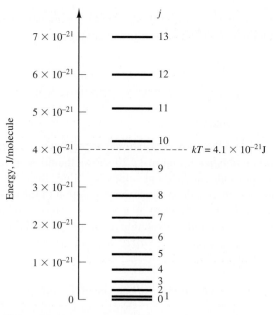

Figure 12.21   Allowed rotational energies for CO at 25°C ($m_e = 11.4 \times 10^{-24}$ gm, $r = 1.13$Å, and $I = 14.5 \times 10^{-40}$ gm-cm^2).

The minimum energy of rotation corresponds to the zero value of j. The quantum number m determines the allowed quantum states for each allowed energy level, and for a given value j, m can assume any quantum state corresponding to any integer value between $-j$ and $+j$ (including zero). Therefore, the degeneracy of each allowed rotational energy level is $(2j + 1)$.

12.25 Harmonic Oscillator

A third type of energy of the molecule is vibrational energy due to the kinetic energy of the linear motion of the atoms as they vibrate and also due to the potential energy associated with intermolecular forces. Particular applications include vibrational motion of diatomic molecules, vibrational motion of atoms in a crystal lattice about their equilibrium positions, and oscillators in the walls of a blackbody radiation cavity that are in equilibrium with radiation present in the cavity of frequency v. The vibration of the atoms of a molecule may be shown to be equivalent to the behavior of a quantum-mechanical system called a *harmonic oscillator*. This model will now be used to determine the quantum-mechanical vibrational energy levels.

Small amplitude sinusoidal oscillation of two atoms of masses m_1 and m_2 along the axis of a diatomic molecule is shown in Figure 12.22. Let r_e be the distance between the atoms at equilibrium, and r_1 and r_2 be the instantaneous displacement of masses m_1 and m_2 from the equilibrium positions measured positively away from the center of mass. The distance r between the atoms at any time is equal to

$$r = r_e \pm (r_2 + r_1) \tag{12.125}$$

Since no external forces are involved in the oscillations and therefore momentum is conserved, the mass of one atom multiplied by its velocity must be equal to the mass of the second multiplied by its corresponding velocity, or

$$m_1 \dot{r}_1 = m_2 \dot{r}_2 \tag{12.126}$$

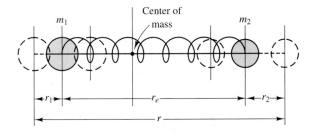

Figure 12.22   Harmonic oscillator.

When Eq. (12.125) is differentiated with respect to time, and the results introduced into Eq. (12.126), the following is obtained:

$$m_1 \dot{r}_1 = m_e \dot{r}$$

where $m_e = (m_1 m_2)/(m_1 + m_2)$ is an equivalent or "reduced" mass. The motion of an oscillating, elastic body of mass m_e is described by

$$m_e \ddot{r} = -Kr \tag{12.127}$$

where K is the spring constant. The solution to this equation is

$$r = A \sin \sqrt{\frac{K}{m_e}} t + B \cos \sqrt{\frac{K}{m_e}} t \tag{12.128}$$

Boundary conditions are as follows:

At $t = 0, r = r_0$.
At $t = 0, \dot{r} = 0$.

The resultant solution is

$$r = r_0 \cos \sqrt{\frac{K}{m_e}} t$$

Apparently, the frequency of oscillation v is given by

$$v = \frac{1}{2\pi} \sqrt{\frac{K}{m_e}} \tag{12.129}$$

These are all results of classical mechanics. But the correct quantum analysis of the harmonic oscillator leads to certain modifications. The Schrödinger equation for simple vibrational motion of a molecule is

$$\frac{d^2 \Psi}{dr^2} + \frac{8\pi^2 m_e}{h^2} (\varepsilon - \varepsilon_{pot}) \Psi = 0$$

The potential energy ε_{pot} is equal to the work done on the oscillator to stretch it through a distance r, or

$$\varepsilon_{pot} = -\int_0^r -Kr \, dr = \frac{1}{2} K r^2$$

Substituting the foregoing value of ε_{pot} into the Schrödinger equation gives

$$\frac{d^2 \Psi}{dr^2} + \frac{8\pi^2 m_e}{h^2} \left(\varepsilon - \frac{1}{2} K r^2 \right) \Psi = 0 \tag{12.130}$$

or

$$\frac{d^2\Psi}{dr^2} + (A - B^2 r^2)\Psi = 0 \tag{12.131}$$

From the following substitution,

$$Br^2 = x^2 \tag{12.132}$$

Eq. (12.130) becomes

$$\frac{d^2\Psi}{dx^2} + (C - x^2)\Psi = 0 \tag{12.133}$$

where $C = A/B$.

   If the solution to Eq. (12.133) is of the form

$$\Psi = e^{-x^2/2} v(x)$$

then substitution into Eq. (12.133) leads to

$$v'' - 2xv' + (C - 1)v = 0 \tag{12.134}$$

Equation (12.134) is called the *Hermite equation*. Its appropriate solution is expressed in the form of the polynomial:

$$v = \sum_{n=0}^{\infty} a_n x^n \tag{12.135}$$

Substituting this polynomial in Eq. (12.134) gives

$$\sum_{n=0}^{\infty} a_n n(n - 1)x^{n-2} - 2\sum_{n=0}^{\infty} a_n n x^n + (C - 1)\sum_{n=0}^{\infty} a_n x^n = 0$$

Letting $k = n - 2$, the following results:

$$2a_2 + \sum_{k=1}^{\infty} \left\{ \left[(k + 2)(k + 1)a_{k+2} x^k \right] \right.$$
$$\left. - [2(k + 2) - (C - 1)]a_k x^{k+2} \right\} + (C - 1)a_0 = 0 \tag{12.136}$$

To satisfy the normalization condition, the series must terminate with a finite number of terms, and this consideration leads to the following condition:

$$C = 2n + 1 = 2\left(n + \frac{1}{2}\right)$$

Since $C = A/B$, and since the values of A and B are given in Eqs. (12.130) and (12.131), substitution leads to the following:

$$\varepsilon = \frac{h}{2\pi} \sqrt{\frac{K}{m_e}} \left(n + \frac{1}{2} \right)$$

But it was shown in Eq. (12.129) that the spring constant K can be related to the classical frequency of vibration, and therefore the vibrational energy can be expressed as

$$\varepsilon_{\text{vib}} = \left(n + \frac{1}{2} \right) h\nu \qquad n = 0, 1, 2,\ldots \qquad (12.137)$$

The integer n is called the *vibrational quantum number*. Equation (12.137) gives the allowable energy levels of a harmonic oscillator. Starting from a minimum value of $\frac{1}{2}h\nu$, the allowable levels are equally spaced, separated from each other by a constant value of $h\nu$. Furthermore, there is a definite quantum energy for each quantum state. Therefore, in one-dimensional problems, there is no degeneracy of the vibrational levels, and there is only one energy state in each level. The allowed vibrational energies of a diatomic molecule are plotted in Figure 12.23. The magnitude of lower-energy states is compared with the value of kT at 25°C for CO in Figure 12.24. Finally, the wave-function solution to Eq. (12.130) is

$$\Psi_{\text{vib}}(r) = D_n e^{-x^2/2} H_n(x) \qquad (12.138)$$

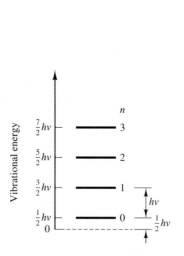

Figure 12.23   Allowed vibrational energies of a diatomic molecule.

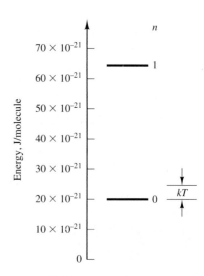

Figure 12.24   Allowed vibrational energies of *CO* at 25°C.

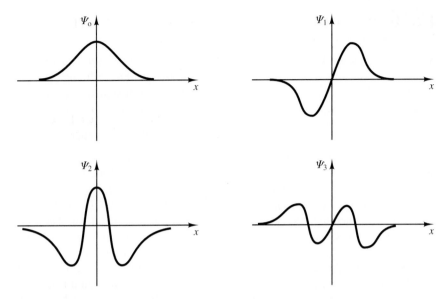

Figure 12.25 Ψ as a function of x for a simple harmonic oscillator.

where $H_n(x)$ is the *Hermite polynomial* of degree n, and D is a normalization constant that may be evaluated by means of the condition (Eq. 12.100). In Figure 12.25 and Figure 12.26, the wave function Ψ and the probability density $|\Psi|^2$ are shown as a function of x.

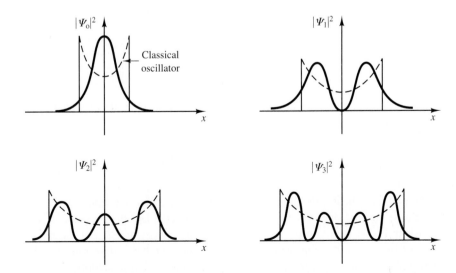

Figure 12.26   Probability density as a function of x for a simple harmonic oscillator.

12.26 Electronic Energy

A fourth energy term involves the motion of the electrons in orbits about the nucleus of each atom of the molecule. The kinetic energy of the orbiting electron and its potential energy due to its existence in the electromagnetic field of the nucleus are the two sources of electronic energy. Here, too, the Schrödinger equation serves as the starting point. Since the motion is of a rotational nature about the nucleus (rather than translational), the Schrödinger equation describing its motion is first expressed in the polar coordinate system. But a multitude of solutions results since both the radial distance from the center of the orbit and the angular displacements along the orbit can vary. In addition, the electron can also rotate about its own center, and discrete energy levels are associated with this motion, also. When the interactions between electrons and nuclei, and between the electrons themselves, are considered, the Schrödinger equation becomes difficult to solve exactly for most atoms and molecules, and various approximation techniques are employed. The spacings between electronic levels are considerably larger than between rotational and vibrational levels, and at moderate temperature, most of the atoms are in the ground electronic level.

12.27 Microscopic Interpretation of Work and Heat

When heat transfer and work transfer to a system are examined from the microscopic point of view, their effects on the particles of the system are quite different. Consider the reversible compression of an ideal monatomic gas. From a macroscopic viewpoint, the reversible work is associated with a change in volume and is given by $\delta W = -p \, dV$ so that the change in the internal energy of the system is

$$dU = -p \, dV + \delta Q \tag{12.139}$$

The total internal energy of the system is the sum of the energies of its particles $\left(U = \sum N_i \varepsilon_i\right)$, and the differential of U is

$$dU = \sum_i N_i \, d\varepsilon_i + \sum_i \varepsilon_i \, dN_i \tag{12.140}$$

This equation indicates that the change in internal energy is the sum of two contributions: the first due to a change in the energy level and the second due to a change in the distribution of particles among the energy levels. Let the molecules of the system be confined in a cube of side length ℓ and volume V. The molecules can move in three dimensions, and their energy according to Eq. (12.111) is given by

$$\varepsilon_i = \frac{h^2}{8M\ell^2}(n_x^2 + n_y^2 + n_z^2) = \frac{h^2 n_i^2}{8MV^{2/3}} \qquad (12.141)$$

where n_x, n_y, and n_z may be regarded as components of a vector $\vec{n}$ of magnitude n. According to this equation $d\varepsilon_i$ is associated with the change in volume; that is, the volume determines the spacing of the energy levels.

As the volume decreases owing to reversible compression, each energy level increases by $d\varepsilon_i$, and the increase in energy of all the particles is

$$\sum N_i d\varepsilon_i$$

For adiabatic reversible compression, the increase in energy is also associated with volume change so that the macroscopic work transfer is equivalent to its counterpart expression on the microscopic scale or

$$\delta W = \sum N_i d\varepsilon_i \qquad (12.142)$$

Note that the particle distribution, that is, the number of particles N_i in each energy level, remains constant, but each particle changes its energy to a different level, as shown in Figure 12.27(a). Note also that Eq. (12.142) is valid only for internally reversible processes; it does not apply to irreversible work interaction such as electrical energy transfer or paddle-wheel work transfer because these are not associated with a volume change.

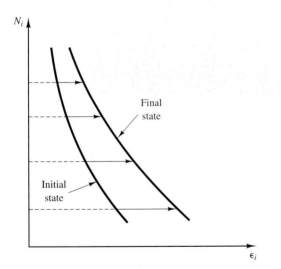

(a) Reversible work transfer in an adiabatic process

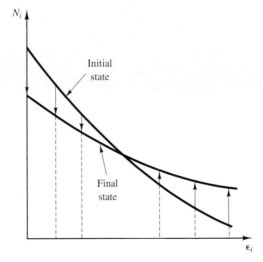

(b) Heat transfer in an internally reversible constant-volume process

Figure 12.27   Microscopic interpretation of heat transfer and work transfer.

Combining Eqs. (12.139) and (12.140) and substituting $\sum N_i d\varepsilon_i$ for $-p\,dV$ gives

$$\delta Q = \sum \varepsilon_i dN_i \tag{12.143}$$

This means that the microscopic interpretation of macroscopic heat transfer to a system for an internally reversible process is a shift of particles from lower- to higher-energy levels, as shown in Figure 12.27(b). Heat transfer in an internally reversible process increases the internal energy of the system by a redistribution of its particles. This redistribution results in an entropy change given by

$$dS = \frac{1}{T_{surr}} \sum \varepsilon_i dN_i \tag{12.144}$$

12.28　Summary

Statistical thermodynamics leads to the evaluation of macroscopic properties based on the molecular behavior of particles. The analysis makes use of the theory of probability in determining the most probable macrostate that is identified with the equilibrium state of the system.

The speed distribution function $f(v)$ indicates the way in which the velocity of particles is distributed in the velocity space. The Maxwell–Boltzmann speed distribution function when multiplied by dv gives the fraction of particles with speeds between v and $v + dv$ so that

$$\frac{dN_v}{N} = f(v)dv = \frac{4}{\sqrt{\pi}} \left(\frac{m}{2kT}\right)^{3/2} v^2 e^{-mv^2/2kT} dv$$

Using the distribution function, the following expressions of the average, root mean square, and most probable velocities can be obtained:

$$v_{avg} = \sqrt{\frac{4kT}{\pi m}}$$

$$v_{rms} = \sqrt{\frac{3kT}{m}}$$

$$v_{mp} = \sqrt{\frac{2kT}{m}}$$

The number of particles with speeds between 0 and v is

$$N_{0 \to v} = N\left[\text{erf}(x) - \frac{2}{\sqrt{\pi}} x e^{-x^2} \right]$$

where $x = v/v_{mp}$.

The number of particles striking a unit surface in a unit time is expressed by the collision frequency,

$$\dot{c} = \frac{p}{\sqrt{2\pi \, mkT}}$$

The mean free path of a molecule is the average distance it travels between collisions. It is given by

$$\Lambda_m = \frac{3}{4\sigma n}$$

where σ is the collision cross section, and n is the number of molecules per unit volume.

The distribution of the free paths gives the fraction of the molecules that travel a path equal to x/Λ_m without collision and is given by

$$N = N_0 e^{-x/\Lambda_m}$$

The distribution of particles among energy levels is based on three models that differ in the method of defining and counting microstates. The thermodynamic probability, which is equal to the number of microstates corresponding to a given macrostate, and the most probable distribution for each of the three statistics are as follows:

Maxwell–Boltzmann: $W = N! \prod_i \dfrac{g_i^{N_i}}{N_i!}$ $N_i^* = ag_i e^{-\varepsilon_i/kT}$

Ferni–Dirac: $W = \prod_i \dfrac{g_i!}{N_i! \, (g_i - N_i)!}$ $N_i^* = \dfrac{ag_i e^{-\varepsilon_i/kT}}{1 + ae^{-\varepsilon_i/kT}}$

Bose–Einstein: $W = \prod_i \dfrac{(g_i! + N_i - 1)!}{N_i! \, (g_i - 1)!}$ $N_i^* = \dfrac{ag_i e^{-\varepsilon_i/kT}}{1 - ae^{-\varepsilon_i/kT}}$

where g_i is the degeneracy of the ith energy level, that is, the number of energy states of the ith energy level.

Entropy is related to thermodynamic probability by the Boltzmann equation

$$S = k \ln W$$

Thermodynamic properties are expressed in terms of the partition function defined by

$$Z = \sum_i g_i e^{-\varepsilon_i/kT}$$

Quantum theory indicates that energy is quantized, and particles of matter can have only discrete energy values. The values of the quantized energy levels are obtained by solutions of the Schrödinger wave equation. The quantized energy for the three modes of motion follows.

For translational energy:

$$\varepsilon = \frac{h^2}{8m}\left[\left(\frac{n_x}{a}\right)^2 + \left(\frac{n_y}{b}\right)^2 + \left(\frac{n_z}{c}\right)^2\right] \qquad n_x, n_y, n_z = 1, 2, 3, \ldots$$

For rotational energy:

$$\varepsilon = \frac{h^2}{8\pi^2 I}j(j + 1) \qquad j = 0, 1, 2, 3, \ldots$$

For vibrational energy:

$$\varepsilon = \left(n + \frac{1}{2}\right)h\nu \qquad n = 0, 1, 2, 3, \ldots$$

Appendix: Stirling's Formula

Stirling's formula provides a useful approximation for the logarithm of a factorial of a large number. The logarithm of $N!$ is given by

$$\ln (N!) = \ln 2 + \ln 3 + \ldots + \ln N$$

Figure 12A shows $\ln N$ plotted versus N. If the abscissa is divided into unit intervals, the rectangular areas defined by the dotted contours are $(1)(\ln 2), (1)(\ln 3), \ldots$ $(1)(\ln N)$.

$$\text{Sum of rectangular-shaped areas} = (1)(\ln 2) + (1)(\ln 3) + \ldots (1)(\ln N)$$
$$= (1)(\ln N!)$$
$$\text{Area under logarithmic curve} = \int_1^N \ln(N)dN$$

If N is large, the area under the logarithmic curve is approximately equal to the steplike series of areas so that

$$\ln N! \approx \int_1^N \ln N \; dN$$

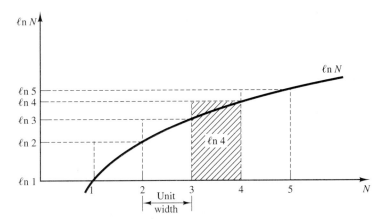

Figure 12A Stirling's approximation.

Performing the integration by parts gives

$$\ln N! \approx N \ln N - \int_1^N dN = N \ln N - N + 1$$

If $N \gg 1$, the preceding equation can be reduced to

$$\ln N! \approx N \ln N - N = N \ln N - N \ln e$$

or

$$N! \approx \left(\frac{N}{e}\right)^N \tag{1}$$

Equation (1) is called *Stirling's approximation formula*. It may be shown that a better approximation for $N!$ is given by

$$N! \approx \sqrt{2\pi N}\left(\frac{N}{e}\right)^N \tag{2}$$

Note that Eq. (2) reduces to Eq. (1) when $N \gg 1$. Table 12A gives a comparison of Eqs. (1) and (2), together with the exact value of $\ln N!$. Note that Stirling's approximation is quite good for large values of N where the relative error approaches zero although the absolute error is not zero.

Table 12A

N	Exact Value	Eq. (1)	Eq. (2)
		ln N!	
5	4.78	3.05	4.77
10	15.10	13.03	15.10
20	42.3	39.9	42.33
30	74.6	72.0	74.66
40	110.3	107.6	110.32
50	148.5	145.6	148.48
100	363.74	360.52	363.74

Problems

12.1 Plot the Maxwell–Boltzmann speed and velocity distributions for 1 kg-mol of nitrogen at temperatures of 100 K and 1000 K.

12.2 The molecules of a gas have a Maxwellian speed distribution given by

$$f(v) = \frac{4}{\sqrt{\pi}} \left(\frac{m}{2kT} \right)^{3/2} v^2 e^{-mv^2/2kT}$$

(a) Check that $\int_0^\infty f(\varepsilon)d\varepsilon = 1$, where $\varepsilon = \frac{1}{2}mv^2$.
(b) Show that the most probable and average kinetic energies per molecule are $1/2kT$ and $3/2kT$.

12.3 Prove that the Maxwell–Boltzmann distribution law is valid for a mixture of ideal gases in thermal equilibrium if the law holds for each gas separately.

12.4 Compute the average speed, root mean square speed, and most probable speed in m/s for hydrogen and carbon monoxide at 0°C and 500°C.

12.5 What is the fractional number of molecules of an ideal gas with speeds between 0 and v_{mp}? Plot the

Maxwell–Boltzmann speed distribution for 1 kg-mol of oxygen at 1000 K.

12.6 Repeat Problem 12.5 if the fractional number of molecules having velocities with x components between $-v_{mp_x}$ and $+v_{mp_x}$ is required. Plot the Maxwell–Boltzmann velocity distribution.

12.7 Calculate the average speed of an ideal gas in the ranges of 0 to v_{mp} and v_{mp} to ∞.

12.8 Determine the fractional number of molecules of an ideal gas with:
(a) speeds between $v = v_{mp}$ and $v = 1.1v_{mp}$; what fraction is moving with a speed greater than v_{mp}?
(b) velocities of x Cartesian components within the same range, that is, between v_{mp} and $1.1v_{mp}$.

12.9 What is the number of oxygen molecules per kg-mol having speeds between 900 and 1000 m/s at 100 K?

12.10 Determine the mass, average kinetic energy, and average speed of an oxygen molecule at 600 K. What fraction of molecules have speeds greater than 500 m/s?

12.11 At what temperature will 60 percent of the atoms in a helium-filled tank have speeds greater than 2000 m/s?

12.12 Determine the percentage of molecules of an ideal gas that have energy in the range of $kT \pm 0.1kT$.

12.13 Assuming Maxwell–Boltzmann energy distribution, determine the fraction of particles of an assembly that have energy between $\varepsilon = 0$ and $\varepsilon = 10^{-20}$ J. Assume the assembly to be at a temperature of 370 K.

12.14 Plot the speed distribution for helium $\left(dN_v/Ndv\right)$ versus v at 250 K for the range of speed from 0 to 2000 m/s. The mass of a helium atom is 6.645×10^{-27} kg.

12.15 Calculate the time rate of molecular impacts on a surface of a square centimeter exposed to air at 300 K if the pressure is (a) atmospheric, (b) 10^{-5} mm Hg.

12.16 Verify that the maximum value in the kinetic energy distribution of a system of particles occurs at $\varepsilon/kT = 1/2$.

12.17 A cubical box, 1 cm on each side, contains hydrogen at 0°C and 1 atm. If the number of molecules in the box is approximately 2.7×10^{19}, calculate the number of collisions made with the walls in each second.

12.18 A closed vessel contains liquid water in equilibrium with its vapor at 100°C and 1 atm. The specific volume of the water vapor at this temperature and pressure is 1673 g/cm^3. Determine:
(a) the number of molecules per cm^3 of vapor
(b) the number of vapor molecules that strike each cm^2 of liquid surface per second
(c) if each molecule that strikes the surface condenses, how many evaporate from each cm^2 per second?
(d) energy required to evaporate one molecule.

12.19 Helium at a pressure of 1 atm and 20°C occupies a sphere 10 cm in diameter. The outside of the sphere is evacuated, and the molecules escape through a pinhole in the sphere of 10^{-7} cm diameter. Assuming isothermal conditions, how long will it take for the pressure to drop to 0.1 atm? (Assume that the wall of the sphere is very thin and is planar in the vicinity of the pinhole.)

12.20 A 10-cm^3 vessel contains an equal number of molecules of helium and oxygen at 300 K and 0.1 atm. One end of the tank contains a porous plug through which the gas can escape. The total hole area is 10^{-4} cm^2. Determine the pressure and N_{O_2}/N_{He} inside the tank as a function of time, assuming the temperature remains constant.

12.21 Air at atmospheric pressure and 300 K leaks into a vacuum system through a small hole. The pressure in the vacuum system is low enough for leakage back to the outside to be negligible. Find the ratio of the number of O_2 to N_2 molecules leaking into the system. If the cross-sectional area of the hole is 10^{-7} cm^2, find the mass of air entering the system per second. Assume the mean molar mass of air is 28.97 kg/kg-mol.

12.22 A spherical tank, 10 cm in diameter, contains oxygen initially at 1 atm and 350 K. The inside walls of the tank are coated with a material that captures $1/10^6$ of the impinging molecules. How long will it take before the pressure decreases to 0.1 atm? Assume constant temperature.

12.23 A 1-liter vacuum tank, initially at a pressure of 10^{-3} mm Hg, is surrounded by air at atmospheric pressure (760 mm Hg). A pinhole of 10^{-4} mm diameter in the wall of the tank allows air molecules to pass through the hole in *both* directions.

Treating the air as a pure substance ($M = 28.97$ kg/kg-mol), develop an expression of the pressure inside the tank as a function of time, and determine the pressure in the tank after 5000 seconds. Assume a constant temperature of 300 K.

12.24 Two vessels containing an ideal gas at different pressures p_1 and p_2 are initially at the same temperature T. It is discovered that gas leaks through a small hole connecting the two constant-pressure vessels. Assuming steady state and adiabatic conditions, calculate:
(a) the rms speed of the leaking gas in each vessel
(b) the rate of leakage per second
(c) the mean energy per molecule of the leaking gas
(d) how you would maintain the vessels at constant temperature

12.25 Calculate the mean free paths for the following gases at 0°C and 100°C. Assume atmospheric pressure.

Gas	Molar Mass (gm/gm-mol)	Molecular Diameter (cm)
A	39.94	3.64×10^{-8}
H_2	2.02	2.18×10^{-8}
He	4.00	2.74×10^{-8}
N_2	28.02	3.75×10^{-8}
CO_2	44.00	4.59×10^{-8}

12.26 Find the number of ways of distributing four distinguishable particles a, b, c, and d in four energy levels. What is the thermodynamic probability of all four particles being in a single energy level? Assume particles to be indistinguishable if they are in a single energy level.

12.27 A box contains three black, four white, and five red balls. What is the mathematical probability of drawing (a) two black balls in succession if the first ball is replaced after drawing, (b) two black balls in succession if the first ball is not replaced after drawing, (c) either a black or a white ball if one ball is drawn?

12.28 It is required to arrange six black and six white particles in two compartments of six positions each. Each particle occupies one of these positions, and each compartment has the same number of particles. Assuming that the black (as well as the white) particles are indistinguishable among themselves, complete the following table, and determine the thermodynamic probabilities of each distribution.

Particles in Compartment 1	Particles in Compartment 2	Number of Arrangements
6B		
5B + 1W		
4B + 2W		
3B + 3W		
2B + 4W		
1B + 5W		
6W		

12.29 Find the number of ways in which ten distinguishable particles can be arranged in three groups such that the groups contain two, three, and five particles. What will be the number of arrangements if the particles in each group are indistinguishable?

12.30 In a cubic cm there are approximately 2.7×10^{19} molecules at standard conditions. What is the mathematical probability if the molecules occupy one-half of the volume leaving the other half empty? What is the probability if the molecules occupy only one-quarter of the volume?

12.31 Calculate the change of entropy when H_2 gas (at 0°C and 1 atm) occupying a volume of 1 cm^3 expands freely and adiabatically to a final volume of 10 cm^3. Check your result by classical thermodynamics.

12.32 A nondegenerate system contains six particles with a total energy of six units. The available energy levels are 0, 1, 2, 3, and 4 units of energy. Determine the possible particle arrangements as a function of energy indicating the most probable distribution.

12.33 A nondegenerate system contains eight particles with a total energy of 4 units. The energy levels have 0, 1, 2, 3, and 4 units of energy. Write down all possible distributions, and determine the most probable distribution.

12.34 Assume that the possible energy levels of an N-particle system are ε, 2ε, 3ε, 4ε,... $n\varepsilon$, where n varies from zero to infinity, and ε is a constant. Assuming the N particles are distributed among the energy levels according to a Boltzmann distribution, determine the expressions for the number of particles in the ith energy level and the partition function. What is the total energy and the constant-volume specific heat of the system?

12.35 Assume an ideal gas obeying Boltzmann's statistics in a force field that acts on each particle of the gas according to the equation $F = -a - bz$, where a and b are constants, and z is the distance above the plane $z = 0$. The particles are all held above the plane $z = 0$. Find the average potential energy of the gas at temperature T in the z-direction.

12.36 The total energy of two identical harmonic oscillators is $7h\nu$. What is the degeneracy of the combined system of the two oscillators?

12.37 Five particles are arranged in two energy levels each of degeneracy 3. What is the thermodynamic probability of each possible macrostate according to Fermi–Dirac and Bose–Einstein statistics?

12.38 Determine the number of ways that three particles having a degeneracy of three can be arranged according to the M–B, B–E, and F–D statistical models.

12.39 Four indistinguishable particles are arranged in two energy levels, each of degeneracy of three. What is the thermodynamic probability of each possible macrostate, and what is the most probable arrangement?

12.40 A system of eight particles has available energy levels of 0, 1ε, 2ε, and 3ε, with degeneracies of 3, 3, 4, and 4, respectively. If the total energy is 6ε, determine the thermodynamic probability of each of the possible macrostates assuming a Maxwell–Boltzmann distribution. Indicate which is the most probable distribution.

12.41 Compare the F–D and B–E statistics when four particles are distributed among two energy-level groups having degeneracies $g_1 = 2$ and $g_2 = 3$. Assume only one particle is allowed in each state.

12.42 Calculate the particle partition function for a system of molecules at 80 K in which the available energies are 0, 2, 6, and 12 units with degeneracies of 1, 3, 5, and 7, respectively. Assume each unit of energy is 5×10^{-22} J.
(a) Which level of energy is the most populated?
(b) What fraction of molecules occupy the most populated level?

12.43 Calculate the translational energy of hydrogen in a system of volume $V = 1$ cm^3 if the translational quantum numbers $n_x = n_y = n_z = 10$. What are the rotational and vibrational energies in the first five energy levels? Assume the frequency of vibration $\nu = 13.2 \times 10^{13}$ cycles/s.

12.44 Prove that the number of translational states with energy less than ε is given by

$$\frac{4\,\pi V}{3\,h^3}(2m\varepsilon)^{3/2}$$

where ε is the average kinetic energy of the particles. For nitrogen molecules in a volume of 1 cm^3 at 25°C, find the number of states with energies less than ε.

12.45 For a particle in a box of side ℓ, determine the degeneracy at each of the following energy levels:

(a) $\varepsilon = 12\left(\dfrac{h^2}{8m\ell^2}\right)$

(b) $\varepsilon = 25\left(\dfrac{h^2}{8m\ell^2}\right)$

12.46 The equation of a translational system,

$$\frac{n_x^2}{a^2} + \frac{n_y^2}{b^2} + \frac{n_z^2}{c^2} = \frac{8m\varepsilon_i}{h^2} \qquad \begin{array}{l} a, b, c \\ n_x, n_y, n_x \end{array} \qquad \begin{array}{l} \text{dimensions} \\ \text{integers} \end{array}$$

is analogous to the equation of an ellipsoid,

$$\frac{x^2}{a^2} + \frac{y^2}{b^2} + \frac{z^2}{c^2} = r^2$$

where r^2 is identified as $8m\varepsilon_i/h^2$. Realizing that only $1/8$ of the volume of the ellipsoid is defined by positive values of the variables necessary to determine the number of states of energies between 0 and ε,
(a) prove that the number of states between ε and $\varepsilon + d\varepsilon$ is

$$g_i = \frac{4\pi m V}{h^3}\sqrt{2m\varepsilon_i}^{1/2}d\varepsilon_i$$

where $V = $ volume of the gas
(b) find an expression for the partition function
(c) using the results of (a) and (b) and the Maxwell–Boltzmann distribution law prove that

$$d\left(\frac{N_i}{N}\right) = \frac{2}{kT\sqrt{\pi}}\frac{(\varepsilon_i/kT)^{1/2}}{e^{\varepsilon_i/kT}}d\varepsilon_i$$

(d) plot $[d(N_i/N)]/[d(\varepsilon_i/kT)]$ versus ε_i/kT, and evaluate the fraction of the molecules between kT and $2kT$

12.47 Calculate the spacing of translational, rotational, and vibrational energy levels for diatomic hydrogen ($r_e = 0.7417 \times 10^{-10}$m) for the first few energy levels. Assume $n_x = n_y = n_z$, and $a = b = c = 1$ cm for translational energy and $v = 13.2 \times 10^{13}$ cycles/s.

12.48 Prove that the average energy of a one-dimensional harmonic oscillator in thermal equilibrium is given by

$$\varepsilon_{avg} = \frac{hv}{e^{hv/kT} - 1}$$

and show that, in classical theory where $kT \gg hv$,

$$\varepsilon_{avg} = kT$$

12.49 Determine the minimum uncertainty in the position of an electron ($m = 9.1096 \times 10^{-31}$ kg) when its speed is measured with an accuracy of 1 percent and found to be 5×10^5 m/s.

C12.1 Write a computer program to evaluate the Stirling's approximation formula:

$$N! = \left(\frac{N}{e}\right)^N \qquad N \gg 1$$

Plot the right- and left-hand side of the equation for a range of N from 1 to 50. Discuss your results. For what value of N is the Stirling approximation within one-tenth of 1 percent of the actual value of the $\ln N!$?

C12.2 Modify the computer program in Problem C12.1 to evaluate the approximation for $N!$:

$$N! = \sqrt{2\pi N} \left(\frac{N}{e}\right)^N$$

C12.3 Write a computer program to generate the data needed to plot the Maxwell–Boltzmann speed distribution for 1 kg-mol of nitrogen at temperatures of 100, 500, 1000, and 2000 K. Use any plotting software that can import a data file.

C12.4 Modify the computer program in Problem C12.3 to plot the speed distribution for nitrogen, helium, carbon dioxide, and highly superheated water vapor. Use a single temperature of 1000 K.

Chapter 13

Applications of Statistical Thermodynamics

13.1 Introduction

The concepts of quantum mechanics and statistical mechanics lend themselves to many applications, and in the field of thermodynamics, they are particularly useful, for they help explain macroscopic properties of systems using statistical averaging techniques.

Thermodynamic properties may be expressed in terms of the partition function and its derivatives. It was previously shown that the explicit form of the partition function indicates how the energy of a system is partitioned among the possible energy levels. Consequently, the first step in evaluating these properties is to compute the value of the partition function. The partition function of a single particle according to Eqs. (12.46) and (12.71) is

$$Z = \sum_{i(\text{states})} e^{-\varepsilon_i/kT} = \sum_{i(\text{levels})} g_i e^{-\varepsilon_i/kT}$$

depending whether the summation is made over either all states or all levels. If the different modes of energy are assumed independent, the total energy of the particle is equal to the sum of the contributing energies or

$$\varepsilon_i = \varepsilon_{\text{trans}} + \varepsilon_{\text{rot}} + \varepsilon_{\text{vib}} + \varepsilon_{\text{elec}}$$

The partition function for one particle then becomes

$$Z = \sum e^{-(\varepsilon_{\text{trans}} + \varepsilon_{\text{rot}} + \varepsilon_{\text{vib}} + \varepsilon_{\text{elec}})/kT}$$
$$= Z_{\text{trans}} Z_{\text{rot}} Z_{\text{vib}} Z_{\text{elec}} \tag{13.1}$$

or the partition function is equal to the product of the contributing partition functions. Equation (13.1) serves as a key equation in determining thermodynamic properties.

In the following sections, thermodynamic properties of monatomic gases, diatomic gases, photons, and simple crystals are determined by application of techniques described in Chapter 12. Particular attention is given to values of specific heat. Cases where only translational, rotational, and vibrational energies contribute to the total energy of the molecules are treated in these applications. The latter part of this chapter discusses the third law of thermodynamics, nonreactive and reactive gas mixtures.

13.2 Ideal Monatomic Gas

The translational energy of the molecules of an ideal monatomic gas constitutes the main energy of the gas. Consider a system of indistinguishable molecules of a monatomic gas in a box of volume V and side dimensions a, b, and c. The energy levels in each of the x-, y-, and z-directions are highly degenerate and are given by Eq. (12.111) as

$$\varepsilon_i = \frac{h^2}{8m}\left[\left(\frac{n_x}{a}\right)^2 + \left(\frac{n_y}{b}\right)^2 + \left(\frac{n_z}{c}\right)^2\right]$$

where the translational quantum numbers n_x, n_y, and n_z are integers (excluding zero) having values 1, 2, 3, ... Since the molecules of the gas are indistinguishable and more than one molecule is allowed in any single quantum state, Bose–Einstein statistics can be applied. The distribution of molecules among the quantum states, for the most probable macrostate, is

$$N_i^* = \frac{a g_i e^{-\varepsilon_i/kT}}{1 - a e^{-\varepsilon_i/kT}} \tag{13.2}$$

where $a = Z/N$. At high temperatures, the molecules of the gas are distributed over many energy states so that $N_i \ll g_i$. For this case, $a e^{-\varepsilon_i/kT} \gg 1$, and this leads to the Boltzmann distribution:

$$N_i^* = N \frac{g_i e^{-\varepsilon_i/kT}}{\sum\limits_i g_i e^{-\varepsilon_i/kT}} \tag{13.3}$$

The translational partition function Z is the product of the particle partition functions Z_x, Z_y, Z_z in the three coordinates axes:

$$Z_{\text{trans}} = \sum_i e^{-\varepsilon_i/kT} = \left[\sum_{n_x=0}^{\infty} e^{-h^2 n_x^2/8mkTa^2} \right] \left[\sum_{n_y=0}^{\infty} e^{-h^2 n_y^2/8mkTb^2} \right]$$
$$\left[\sum_{n_z=0}^{\infty} e^{-h^2 n_z^2/8mkTc^2} \right]$$

Since the coefficients of n_x^2, n_y^2, and n_z^2 in the exponents of the foregoing equation are small for sufficiently high temperatures, the summation signs may be approximated by integration signs. This is also physically justified because translational energy levels are very close. Therefore, plotting the exponent terms versus n_x, n_y, n_z results in almost continuous curves. The preceding equation then takes the form

$$Z_{\text{trans}} = \left[\int_0^{\infty} e^{-h^2 n_x^2/8mkTa^2} \, dn_x \right] \left[\int_0^{\infty} e^{-h^2 n_y^2/8mkTb^2} \, dn_y \right] \left[\int_0^{\infty} e^{-h^2 n_z^2/8mkTc^2} \, dn_z \right]$$

Each of the terms in brackets can be integrated, leading to the following expression for the translational partition function:[1]

$$Z_{\text{trans}} = abc \left(\frac{\sqrt{2\pi mkT}}{h} \right)^3$$

But since the volume $V = abc$, the partition function per molecule becomes

$$Z_{\text{trans}} = V \left(\frac{2\pi mkT}{h^2} \right)^{3/2} \tag{13.4}$$

[1] Note that $\int_0^{\infty} e^{-ax^2} \, dx = \frac{1}{2}(\pi/a)^{1/2}$.

According to Eq. (12.84), the system partition function for N identical indistinguishable molecules is

$$\mathcal{Z}_{trans} = \frac{1}{N!} Z_{trans}^N = \frac{1}{N!} V^N \left(\frac{2\pi mkT}{h^2} \right)^{(3/2)N} \tag{13.5}$$

Since a monatomic gas does not possess vibrational energy and negligible rotational energy, Eqs. (13.4) and (13.5) serve as a basis for the determination of the thermodynamic properties of a monatomic gas at temperatures below electronic excitation.

First, the Helmholtz function can be determined. According to Eq. (12.63), this free energy is

$$F = -kT \ln \mathcal{Z} = -NkT \left[1 + \ln \left(\frac{Z}{N} \right) \right] \tag{13.6}$$

Substituting for $\mathcal{Z}$ (or Z), the Helmholtz function becomes

$$F = -kT \left(-\ln N! + N \ln V + \frac{3}{2} N \ln \frac{2\pi mk}{h^2} + \frac{3}{2} N \ln T \right)$$

and, therefore,

$$F = -N kT \left(-\ln N + 1 + \ln V + \frac{3}{2} \ln \frac{2\pi mk}{h^2} + \frac{3}{2} \ln T \right) \tag{13.7}$$

Since $p = -(\partial F / \partial V)_{T,N}$,

$$p = N kT \left(\frac{\partial \ln Z}{\partial V} \right)_T = N kT \frac{1}{V} \tag{13.8}$$

which is the equation of state of an ideal gas, in agreement with the result obtained from the kinetic theory.

The internal energy according to Eq. (12.62) is given by

$$U = N kT^2 \left(\frac{\partial \ln Z}{\partial T} \right)_V = -N k \left(\frac{\partial \ln Z}{\partial (1/T)} \right)_V \tag{13.9}$$

Substituting for Z from Eq. (13.4), the internal energy of N molecules is

$$U = N kT^2 \left(\frac{3}{2} \frac{1}{T} \right) = \frac{3}{2} N kT = \frac{3}{2} n\bar{R}T \tag{13.10}$$

On a mole basis, the internal energy is

$$\bar{u} = \frac{3}{2} N_a kT = \frac{3}{2} \bar{R} T \qquad (13.11)$$

Enthalpy is given by

$$H = U + pV = N kT \left[T \left(\frac{\partial \ln Z}{\partial T} \right)_V + V \left(\frac{\partial \ln Z}{\partial V} \right)_T \right] \qquad (13.12)$$

Enthalpy per mole can be obtained from internal energy:

$$\bar{h} = \bar{u} + \bar{R} T = \frac{5}{2} \bar{R} T \qquad (13.13)$$

Molar specific heat at constant volume is therefore

$$\bar{c}_v = \left(\frac{\partial \bar{u}}{\partial T} \right)_V = \frac{\bar{R}}{T^2} \left[\frac{\partial^2 \ln Z}{\partial (1/T)^2} \right]_V = \frac{3}{2} \bar{R} \qquad (13.14)$$

whereas molar specific heat at constant pressure is

$$\bar{c}_p = \left(\frac{\partial \bar{h}}{\partial T} \right)_p = \frac{5}{2} \bar{R} \qquad (13.15)$$

and the ratio of specific heats is

$$\gamma = \frac{\bar{c}_p}{\bar{c}_v} = \frac{(5/2)\bar{R}}{(3/2)\bar{R}} = \frac{5}{3} = 1.67 \qquad (13.16)$$

Note that a monatomic gas (possessing only translational and electronic energies) shows no variation of specific heat with temperature. The entropy of the most probable macrostate is obtained by substituting Eq. (13.5) into Eq. (12.54),

$$S = k \ln \mathcal{Z} + \frac{U}{T} = N k \left[1 + \ln \left(\frac{Z}{N} \right) + T \left(\frac{\partial \ln Z}{\partial T} \right)_V \right] \qquad (13.17)$$

which resolves to

$$S = N k \left(-\ln N + \ln V + \frac{3}{2} \ln \frac{2\pi mk}{h^2} + \frac{3}{2} \ln T + \frac{5}{2} \right) \qquad (13.18)$$

Equation (13.18) is called the *Sackur–Tetrode equation* for the absolute entropy of a monatomic gas.

An expression for the chemical potential is

$$\mu_i = \left(\frac{\partial F}{\partial N_i}\right)_{T,V,N_j}$$

where the subscript j indicates that all N_i's are constant except one. From Eq. (13.7), μ becomes

$$\mu = -kT\left(-\ln N + \ln V + \frac{3}{2}\ln\frac{2\pi mk}{h^2} + \frac{3}{2}\ln T\right) \qquad (13.19)$$

The Gibbs free energy can be defined as

$$G = H - TS = F + pV$$

Substituting for H and S in terms of Z gives

$$G = -NkT\left[1 + \ln\left(\frac{Z}{N}\right) - V\left(\frac{\partial \ln Z}{\partial V}\right)_T\right] = -NkT\ln\frac{Z}{N} \qquad (13.20)$$

which resolves to

$$G = -NkT\left(-\ln N + \ln V + \frac{3}{2}\ln\frac{2\pi mk}{h^2} + \frac{3}{2}\ln T\right) \qquad (13.21)$$

Note that the partition function per molecule Z, or the partition function for N molecules $\mathcal{Z}$, can be used to determine thermodynamic properties.

Example 13.1

Determine the change in entropy when an ideal gas expands from an initial state defined by T_1 and V_1 to a final state defined by T_2 and V_2.

SOLUTION:

From Eq. (13.18),

$$S_1 = Nk\left(-\ln N + \ln V_1 + \frac{3}{2}\ln\frac{2\pi mk}{h^2} + \frac{3}{2}\ln T_1 + \frac{5}{2}\right)$$

and

$$S_2 = N k \left(-\ln N + \ln V_2 + \frac{3}{2} \ln \frac{2\pi m k}{h^2} + \frac{3}{2} \ln T_2 + \frac{5}{2} \right)$$

The change in entropy is

$$S_2 - S_1 = N k \left(\ln \frac{V_2}{V_1} + \frac{3}{2} \ln \frac{T_2}{T_1} \right)$$

But $Nk = n\bar{R}$ and $(3/2)N k = n\bar{c}_v$ so that

$$S_2 - S_1 = n \left(\bar{R} \ln \frac{V_2}{V_1} + \bar{c}_v \ln \frac{T_2}{T_1} \right)$$

which agrees with the result obtained by classical thermodynamics.

13.3   Ideal Diatomic Gas

In addition to translational energy, diatomic molecules can have rotational energy and vibrational energy. The total energy of the molecule is the sum of energies from these sources (as well as from electronic sources). The partition function of a diatomic molecule is the product of the partition functions associated with the translation, rotation, and vibration. Effects on specific heat of a diatomic molecule can be ascribed primarily to translational, rotational, and vibrational motion. Since the modes of excitation of an ideal gas can be assumed to be independent of each other, the total constant-volume specific heat is equal to the sum of the translational, rotational, and vibrational specific heats, or

$$\bar{c}_v = \bar{c}_{v_{\text{trans}}} + \bar{c}_{v_{\text{rot}}} + \bar{c}_{v_{\text{vib}}} \tag{13.22}$$

The translational molar specific heat of a diatomic gas, as for a monatomic gas, is

$$\bar{c}_{v_{\text{trans}}} = \frac{3}{2} \bar{R}$$

The translational partition function for a diatomic molecule is the same as that for a monatomic molecule and is given by Eq. (13.4). The rotational partition function per molecule, as shown previously, is

$$Z_{\text{rot}} = \sum_{j=0}^{\infty} (2j + 1)e^{-[j(j+1)/T](h^2/8\pi^2 I k)} \tag{13.23}$$

where the degeneracy, g_i, of each allowed energy level ε_i is indicated by the $(2j + 1)$ term. When various constants in the exponent are lumped together, the result is a constant that is specific for each molecule, θ_r, the *rotational characteristic temperature,* where

$$\theta_r = \frac{h^2}{8\pi^2 Ik} \tag{13.24}$$

θ_r has the dimensions of absolute temperature. Values of θ_r for several gases are given in Table 13.1.

The rotational partition function, when expanded in an infinite series, is

$$Z_{\text{rot}} = \sum_{j=0}^{\infty} (2j + 1)e^{-j(j+1)(\theta_r/T)}$$
$$= 1 + 3e^{-2\theta_r/T} + 5e^{-6\theta_r/T} + 7e^{-12\theta_r/T} + \dots$$

At low temperature (that is, $T \ll \theta_r$), only the first term of the series is significant so that $Z_{\text{rot}} \approx 1$, and, therefore, $\bar{u}_{\text{rot}} = \bar{R} T^2 [\partial(\ln Z_{\text{rot}})/\partial T]_v = 0$. Therefore, only a negligible amount of rotational motion occurs at low temperature. At high temperatures, where $T \gg \theta_r$, molecules acquire rotational energy and occupy quantized energy levels. The rotational energy levels, however, are sufficiently close together, and the summation operation of Eq. (13.23) can be replaced by an integral. The rotational partition function can then be written as

$$Z_{\text{rot}} = \int_0^{\infty} (2j + 1)e^{-j(j+1)(\theta_r/T)} \, dj$$

Table 13.1   Characteristic Temperatures for Rotation and Vibration of Diatomic Molecules

Substance	θ_r (K)	θ_v (K)
H_2	85.5	6140
OH	27.5	5360
HCl	15.3	4300
CH	20.7	4100
N_2	2.86	3340
HI	9.0	3200
CO	2.77	3120
NO	2.47	2740
O_2	2.09	2260
Cl_2	0.347	810
K_2	0.081	140

When the integration is performed, the following result is obtained:

$$Z_{rot} = \frac{T}{\theta_r}$$

One additional factor that affects the rotational partition function must be considered, and this is the symmetry effect. If a molecule is symmetrical, the number of different modes of rotation is only half of that which would occur if it were asymmetrical. This effect is introduced into the partition function through the *symmetry number*, σ, so that the expression of the rotational partition function becomes

$$Z_{rot} = \frac{8\pi^2 I k T}{\sigma h^2} = \frac{T}{\sigma \theta_r} \qquad \text{(high temperature)} \qquad (13.25)$$

The value of σ is equal to 1 for asymmetrical molecules (for example, CO, NO) and is equal to 2 for symmetrical ones (for example, O_2, N_2).

The rotational partition function can now be used to calculate the internal energy due to rotation at high temperature, which is

$$\bar{u}_{rot} = \bar{R}T^2 \frac{\partial(\ln Z_{rot})}{\partial T} = \bar{R}T \qquad (13.26)$$

Equation (13.26) indicates that the internal energy of rotation is a simple function of temperature, and such factors as symmetry or distinguishability have no effect. From the rotational internal energy, the corresponding constant-volume molar specific heat at high temperature can be obtained:

$$\bar{c}_{v_{rot}} = \bar{R} \qquad (13.27)$$

A plot of $\bar{c}_{v_{rot}}/\bar{R}$ against T/θ_r is shown in Figure 13.1. It is seen that the value of $\bar{c}_v$ is zero at low temperature and is equal to $\bar{R}$ at high temperature, but at $(T/\theta_r) = 0.8$, the molar specific heat reaches a maximum of 1.1 $\bar{R}$.

The energy of vibration of a diatomic molecule according to Eq. (12.137) is

$$\varepsilon_{vib} = \left(n + \frac{1}{2}\right)hv \qquad n = 0, 1, 2, 3, \ldots$$

Noting that the vibrational energy levels are nondegenerate ($g_i = 1$), the partition function of vibration then is

$$Z_{vib} = e^{-hv/2Tk} \sum_{n=0}^{\infty} e^{-nhv/Tk} \qquad (13.28)$$

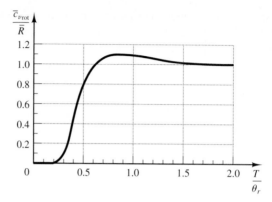

Figure 13.1 $\bar{c}_{v\mathrm{rot}}/\bar{R}$ versus T/θ_r for a diatomic gas.

Here, too, it is convenient to combine constants into a single one, the *vibrational characteristic temperature,* θ_v, so that

$$Z_{\mathrm{vib}} = e^{-\theta_v/2T} \sum_{n=0}^{\infty} e^{-n\theta_v/T} \tag{13.29}$$

where

$$\theta_v = \frac{hv}{k} \tag{13.30}$$

Values of θ_v, expressed in units of temperature, are listed for various molecules in Table 13.1.

The vibrational energy levels are widely spaced, and it is not possible to replace the summation in Eq. (13.29) by an integral. However, the sum of the infinite series in Eq. (13.29) can be reduced to a single fraction, and the vibrational partition function becomes

$$Z_{\mathrm{vib}} = \frac{e^{-\theta_v/2T}}{1 - e^{-\theta_v/T}} \tag{13.31}$$

Based on this expression, internal energy of vibration per mole can then be derived:

$$\bar{u}_{\mathrm{vib}} = \bar{R}T^2 \frac{\partial}{\partial T} (\ln Z_{\mathrm{vib}}) = \bar{R}\theta_v \left[\frac{1}{2} + \frac{1}{e^{\theta_v/T} - 1} \right] \tag{13.32}$$

From this equation, it is evident that even at low temperature, vibrational effects contribute to internal energy, the amount being $\overline{R}\theta_v/2$ per mole. The constant-volume vibrational specific heat can be calculated from internal energy:

$$\overline{c}_{v_{\text{vib}}} = \left(\frac{\partial \overline{u}_{\text{vib}}}{\partial T}\right)_V = \overline{R}\left(\frac{\theta_v}{T}\right)^2 \frac{e^{\theta_v/T}}{(e^{\theta_v/T} - 1)^2} \qquad (13.33)$$

At low temperatures, since $e^{\theta_v/T} \gg 1$, the vibrational molar specific heat reduces to

$$\overline{c}_{v_{\text{vib}}} = \overline{R}\left(\frac{\theta_v}{T}\right)^2 e^{-\theta_v/T} \to 0$$

At high temperatures, on the other hand, since the exponential term can be replaced by an infinite series, the specific heat can be shown to reduce to

$$\overline{c}_{v_{\text{vib}}} = \overline{R}\left(\frac{\theta_v}{T}\right)^2 \frac{[1 + (\theta_v/T) + (\theta_v^2/2T^2) + \cdots]}{[(\theta_v/T) + (\theta_v^2/2T^2) + \cdots]^2} \qquad (13.34)$$

A plot of $\overline{c}_{v_{\text{vib}}}/\overline{R}$ versus T/θ_v is shown in Figure 13.2.

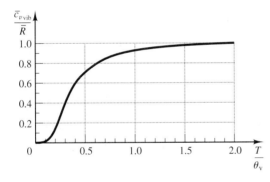

Figure 13.2 $\overline{c}_{v_{\text{vib}}}/\overline{R}$ versus T/θ_v for a diatomic gas.

The contributions to molar specific heat made by the translational, rotational, and vibrational motion of a diatomic molecule may now be added together:

$$\overline{c}_v = \frac{3}{2}\overline{R} + \overline{R}T^2\left(\frac{\partial^2 \ln Z_{\text{rot}}}{\partial T^2}\right)_V + \overline{R}\left(\frac{\theta_v}{T}\right)^2 \frac{e^{\theta_v/T}}{(e^{\theta_v/T} - 1)^2} \qquad (13.35)$$

In Figure 13.3, $\overline{c}_v/\overline{R}$ is plotted against T for a diatomic gas (hydrogen). Because of the individual contributions made by translational, rotational, and vibrational effects, the curve does not show a simple shape.

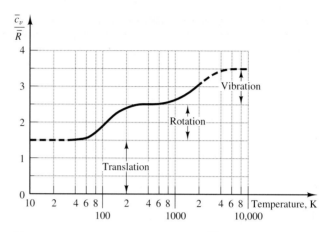

Figure 13.3   Experimental values of $\bar{c}_v/R$ versus T for hydrogen.

The partition function of a diatomic molecule is the product of the partition functions due to translation, rotation, and vibration:

$$Z = V\left(\frac{2\pi mkT}{h^2}\right)^{3/2}\left(\frac{8\pi^2 IkT}{\sigma h^2}\right)\left(\frac{e^{-\theta_v/2T}}{1 - e^{-\theta_v/T}}\right) \tag{13.36}$$

As in the case of monatomic gases, Eq. (13.36) may be used to determine expressions of thermodynamic properties, as will be illustrated in the following examples. Table 13.2 summarizes some formulas of thermodynamic properties expressed in terms of the particle partition function Z and the system partition function $\mathcal{Z}$. In general, evaluation of properties follows the particle partition function.

In Eq. (13.3), ε_i indicates the total energy, including the zero-point energy, of the molecule. The lowest quantum number defines the zero-point energy so that

$$\varepsilon_{0_{\text{trans}}} = \frac{h^2}{8m}\left(\frac{1}{a^2} + \frac{1}{b^2} + \frac{1}{c^2}\right)$$

$$\varepsilon_{0_{\text{rot}}} = 0$$

$$\varepsilon_{0_{\text{vib}}} = \frac{1}{2}hv$$

The ground vibrational state $\varepsilon_0 = \frac{1}{2}hv$ contributes the factor $e^{-\theta_v/2T}$, which appears in the numerator of the partition function Z_v. Since all diatomic molecules possess this ground energy, it is common to express Z_v relative to the ground energy.[2] This results in an additive constant to most of the thermodynamic properties.

---

[2] See Problem 13.27.

Table 13.2   Summary of Formulas of Thermodynamic Properties

$$Z = \sum g_i e^{-\varepsilon_i/kT}$$

$$\frac{N_i^*}{N} = \frac{g_i e^{-\varepsilon_i/kT}}{Z}$$

$$U = \sum \varepsilon_i N_i = N kT^2 \left(\frac{\partial \ln Z}{\partial T}\right)_V$$

$$S = N k \left[1 + \ln\left(\frac{Z}{N}\right) + T\left(\frac{\partial \ln Z}{\partial T}\right)_V\right]$$

$$= N k + N k \ln\left(\frac{Z}{N}\right) + \frac{U}{T}$$

$$F = -N kT \left[1 + \ln\left(\frac{Z}{N}\right)\right]$$

$$G = -N kT \ln\left(\frac{Z}{N}\right)$$

$$H = N kT \left[T\left(\frac{\partial \ln Z}{\partial T}\right)_V + V\left(\frac{\partial \ln Z}{\partial V}\right)_T\right]$$

$$p = N kT \left(\frac{\partial \ln Z}{\partial V}\right)_T$$

$$\mathcal{Z} = \sum g_i e^{-E_i/kT}$$

$$\frac{N_i^*}{N} = \frac{g_i e^{-E_i/kT}}{\mathcal{Z}}$$

$$U = kT^2 \left(\frac{\partial \ln \mathcal{Z}}{\partial T}\right)_{N,V}$$

$$S = k \left[\ln \mathcal{Z} + T\left(\frac{\partial \ln \mathcal{Z}}{\partial T}\right)_V\right]$$

$$= k \ln \mathcal{Z} + \frac{U}{T}$$

$$F = -kT \ln \mathcal{Z}$$

$$G = -kT(\ln \mathcal{Z} - N)$$

$$H = kT \left[T\left(\frac{\partial \ln \mathcal{Z}}{\partial T}\right)_{N,V} + V\left(\frac{\partial \ln \mathcal{Z}}{\partial V}\right)_{N,T}\right]$$

$$p = kT \left(\frac{\partial \ln \mathcal{Z}}{\partial V}\right)_{N,T}$$

Ideal monatomic gas:

$$Z = V \left(\frac{2\pi kTm}{h^2}\right)^{3/2}$$

Ideal diatomic gas:

$$Z_{\text{trans}} = V \left(\frac{2\pi kTm}{h^2}\right)^{3/2}$$

$$Z_{\text{rot}} = \frac{8\pi^2 IkT}{\sigma h^2} \quad \text{when} \quad T \gg \theta_{\text{rot}}$$

$$Z_{\text{vib}} = \frac{e^{-\theta_v/2T}}{1 - e^{-\theta_v/T}}$$

$$S_{\text{trans}} = N k \left(\frac{5}{2} + \ln\frac{Z_{\text{trans}}}{N}\right) = N k \left(\frac{3}{2}\ln M + \frac{5}{2}\ln T - \ln p - 1.155\right)$$

where M is the gram molar mass, T (K) and p in atmospheres.

$$S_{\text{rot}} = N k (1 + \ln Z_{\text{rot}}) = N k (\ln I' + \ln T - \ln \sigma - 2.2)$$

where $I' = I(6.023 \times 10^{23})(10^{16})$ and I is in $(\text{gm/gm-mol})(\text{Å})^2$.

$$S_{\text{vib}} = N k \left[\frac{\theta_v/T}{e^{\theta_v/T} - 1} - \ln\left(1 - e^{-\theta_v/T}\right)\right]$$

$$\theta_r = \frac{h^2}{8\pi^2 Ik}$$

$$\theta_v = \frac{h\nu}{k}$$

The Boltzmann distribution can be expressed in terms of energy measured above the zero point rather than the total energy. In this case, Eq. (13.3) can be written as

$$N_i^* = N \frac{g_i e^{-(\varepsilon_i + \varepsilon_0)/kT}}{\sum_i g_i e^{-(\varepsilon_i + \varepsilon_0)/kT}} = N \frac{g_i e^{-\varepsilon_0/kT} e^{-\varepsilon_i/kT}}{e^{-\varepsilon_0/kT} \sum_i g_i e^{-\varepsilon_i/kT}}$$

$$= N \frac{g_i e^{-\varepsilon_i/kT}}{\sum_i g_i e^{-\varepsilon_i/kT}}$$

where ε_i in the equation indicates energy *above* the zero-point energy.

Example 13.2

Hydrogen is a diatomic molecule comprising two symmetrical atoms and has a rotational characteristic temperature $\theta_r = 85.5$ K. Calculate the most probable distribution of rotational and vibrational energies at 25°C. Plot your results as energy versus particle percentages. Take the vibrational energy level $\varepsilon_i = hv(n)$, that is, relative to the zero-point energy.

SOLUTION:

The Boltzmann distribution law is

$$\frac{N_i^*}{N} = \frac{g_i e^{-\varepsilon_i/kT}}{\sum_i g_i e^{-\varepsilon_i/kT}}$$

For the rotational energy distribution:

$$\frac{N_i^*}{N} = \frac{(2j + 1)e^{-j(j+1)(\theta_r/T)}}{Z_{\text{rot}}}$$

At 25°C,

$$\frac{\theta_r}{T} = \frac{85.5}{298.15} = 0.287$$

Therefore,

$$\frac{N_i^*}{N} = \frac{(2j + 1)e^{-j(j+1)(0.287)}}{Z_{\text{rot}}}$$

To evaluate the foregoing expression, the following table is prepared:

j	$\dfrac{\varepsilon_i}{kT}$	$e^{j(j+1)(0.287)}$	$ge_i^{-\varepsilon_i/kT}$	$\dfrac{N_i^*}{N}$
0	0	1.000	1.000	0.260
1	0.574	0.563	1.690	0.440
2	1.722	0.179	0.895	0.233
3	3.444	0.0319	0.224	0.058
4	5.74	0.0032	0.029	0.0076
5	8.61	0.00018	0.002	0.0005
			$Z_{\text{rot}} = 3.840$	1.000

For the vibrational energy distribution,

$$\frac{N_i^*}{N} = \frac{(1)e^{-[hv(n)]/kT}}{Z_{\text{vib}}} = \frac{e^{-(\theta_v/T)(n)}}{Z_{\text{vib}}}$$

where

$$\frac{\theta_v}{T} = \frac{6140}{298.15} = 20.6$$

The following table can then be prepared:

n	$\left(\dfrac{\theta_v}{T}\right)(n)$	$g_i e^{-\varepsilon_i/kT}$	$\dfrac{N_i^*}{N}$
0	0	1	1
1	20.6	1.13×10^{-9}	0
2	41.2	0	0
		$Z_{\text{vib}} \approx 1.0$	

This means that essentially all the particles are at their lowest vibrational energy level. Figure 13.4 shows the distribution of rotational energy and vibrational energy over the range of energy levels.

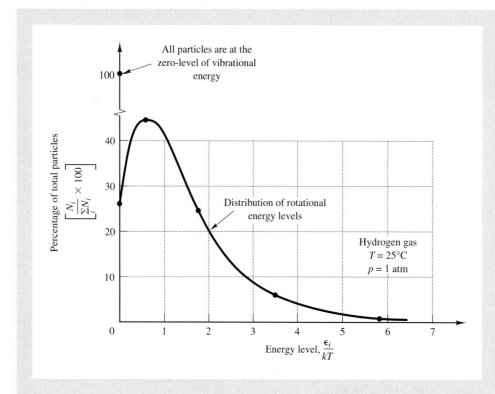

Figure 13.4

Example 13.3

Oxygen molecules have the following physical properties

$$\theta_r = \frac{h^2}{8\pi^2 I k} = 2.09 \text{ K}$$

$$\theta_v = \frac{hv}{k} = 2260 \text{ K}$$

Considering only translational, rotational, and vibrational energies, calculate the following properties per gm-mole at 273 K:

(a) the translational, rotational, and vibrational partition functions
(b) internal energy
(c) enthalpy
(d) Helmholtz and Gibbs functions
(e) entropy

SOLUTION:

(a) The translational partition function for 1 gm-mol is

$$Z_{trans} = \frac{V}{h^3}(2\pi kTm)^{3/2}$$

$$= \frac{22.41 \times 10^{-3} \text{ m}^3/\text{gm-mol}}{(6.626 \times 10^{-34} \text{ J s/molecule})^3}[(2\pi)(1.38 \times 10^{-23} \text{ J/K molecule})(273 \text{ K})(5.31 \times 10^{-26} \text{ kg/molecule})]^{3/2}$$

$$= 34.3 \times 10^{29}$$

The rotational partition function is

$$Z_{rot} = \frac{8\pi^2 IkT}{\sigma h^2} = \frac{T}{\sigma \theta_r} = \frac{273 \text{ K}}{2(2.09 \text{ K})} = 65.3$$

The vibrational partition function is

$$Z_{vib} = \frac{e^{-\theta_v/2T}}{1 - e^{-\theta_v/T}} = 0.0159$$

Therefore,

$$Z = (Z_{trans})(Z_{rot})(Z_{vib}) = 35.6 \times 10^{29}$$

(b) The internal energy is

$$\bar{u} = \frac{3}{2}\bar{R}T + \bar{R}T + \bar{R}\theta_v\left(\frac{1}{2} + \frac{1}{e^{\theta_v/T} - 1}\right)$$

$$= \frac{5}{2}(8.3144 \text{ J/gm-mol K})(273 \text{ K}) + (8.3144 \text{ J/gm-mol K})(2260 \text{ K})\left(\frac{1}{2} + 0\right)$$

$$= 15070 \text{ J/gm-mol}$$

(c) The enthalpy is

$$\bar{h} = \bar{u} + \bar{R}T = 15070 \text{ J/gm-mol K} + (8.3144 \text{ J/gm-mol K})(273 \text{ K})$$
$$= 17340 \text{ J/gm-mol}$$

(d) The Helmholtz and Gibbs functions are

$$\bar{f} = -\bar{R}T\left(1 + \ln\frac{Z}{N_a}\right)$$
$$= -(8.3144 \text{ J/gm-mol K})(273 \text{ K})(1 + 15.59) = -37660 \text{ J/gm-mol}$$

$$\overline{g} = -\overline{R}T \ln \frac{Z}{N_a} = -(8.3144 \text{ J/gm-mol K})(273 \text{ K}) \ln \frac{35.6 \times 10^{29}}{6.025 \times 10^{23}}$$

$$= -35391 \text{ J/gm-mol}$$

(e) The entropy values, according to Table 13.2, are

$$\overline{s}_{trans} = \overline{R}\left(\frac{5}{2} + \ln \frac{Z_{trans}}{N_a}\right) = (8.3144 \text{ J/gm-mol K})\left(\frac{5}{2} + \ln \frac{34.3 \times 10^{29}}{6.025 \times 10^{23}}\right)$$

$$= 150.1 \text{ J/gm-mol K}$$

$$\overline{s}_{rot} = \overline{R}\ln\left(\frac{8\pi^2 IkT}{\sigma h^2}\right) + \overline{R} = \overline{R}(1 + \ln Z_{rot})$$

$$= (8.3144 \text{ J/gm-mol K})(1 + \ln 65.3) = 43.1 \text{ J/gm-mol K}$$

$$\overline{s}_{vib} = \overline{R}\left[\frac{\theta_v/T}{e^{\theta_v/T} - 1} - \ln (1 - e^{-\theta_v/T})\right]$$

$$= (8.3144 \text{ J/gm-mol K})\left[\frac{2260/273}{e^{2260/273} - 1} - \ln (1 - e^{-2260/273})\right]$$

$$= (8.3144 \text{ J/gm-mol K})(2.1 \times 10^{-3} + 3 \times 10^{-4})$$

$$= 0.01996 \text{ J/gm-mol K}$$

The total entropy is

$$\overline{s}_{total} = 150.1 + 43.1 + 0.01996 = 193.22 \text{ J/gm-mol K}$$

13.4 Specific Heat of Ideal Gases

From Sections 13.2 and 13.3, it appears that the average molar specific heat at constant volume can be written as

$$\overline{c}_v = \frac{1}{2}f\overline{R} \tag{13.37}$$

where f is the number of fully *developed degrees of freedom* of the gaseous molecule. For monatomic gases, the value of f is 3, corresponding to the translational kinetic energy of the molecule in three orthogonal directions. The dumbbell model of the diatomic molecule has a nonzero moment of inertia about two of its axes of rotation. Thus, it possesses, in addition to the three degrees of translational freedom, two degrees of rotational freedom. Furthermore, since the atomic bond is not perfectly rigid, the atoms can vibrate along the line joining them. At high temperature, the vibrational movement of two atoms along the line joining them contributes two additional degrees of freedom, one corresponding to the vibrational kinetic energy and the other to the vibrational potential energy of the oscillator. For polyatomic gases, the value of f is 6, three translational and three rotational; the vibrational degrees are again unexcited except at elevated temperatures.

According to the principle of equipartition of energy, the total energy of a system is equally divided among the different degrees of freedom. Each degree of freedom of a gas in thermal equilibrium at temperature T contributes $\frac{1}{2}kT$ per molecule to the kinetic energy; also, each degree of freedom contributes $\frac{1}{2}kT$ per molecule to the vibrational potential energy. Therefore, the average energy associated with the vibration of a diatomic molecule is kT.

The contribution to the constant-volume specific heat is $\frac{1}{2}\overline{R}$ per mole for each degree of freedom. Thus, for a monatomic gas with 3 degrees of freedom, the internal energy per mole is $\frac{3}{2}\overline{R}T$, and $\overline{c}_v$ is $\frac{3}{2}\overline{R}$.

The molar constant-pressure specific heat is

$$\overline{c}_p = \overline{c}_v + \overline{R} = \left(1 + \frac{1}{2}f\right)\overline{R} \qquad (13.38)$$

and the ratio of specific heats is

$$\gamma = \frac{\overline{c}_p}{\overline{c}_v} = \frac{f + 2}{f} \qquad (13.39)$$

Equation (13.39) predicts γ to be 1.67 for monatomic gases, 1.40 for diatomic gases, and 1.33 for polyatomic gases. At high temperature, the vibrational modes become excited, predicting a value of $\overline{c}_v = \frac{7}{2}\overline{R}$ and $\gamma = 1.29$ for diatomic molecules. In general, the more complex the molecule, the higher the values of its specific heat and the smaller the value of γ. The value of γ, however, can never be greater than 1.67 ($f = 3$) or less than 1.0 ($f = \infty$).

The equipartition theorem implies that the specific heat of an ideal gas is independent of temperature. Experiments, however, show that, for a diatomic gas, $\overline{c}_v$ varies with T and approaches the predicted value of $\frac{7}{2}\overline{R}$ only at very high temperature. Therefore, the principle of equipartition of energy does not hold until a sufficiently high temperature is reached where the magnitude of kT is much larger than the separation between the quantized energy levels. Evidently the disagreement arises because classical kinetic theory does not predict the temperature at which the vibrational and rotational degrees of freedom become excited.

According to Eq. (12.137), vibrational oscillations above the ground level require that multiple values of the energy quanta $h\nu$ be supplied. At low temperature, the energy, kT, is small compared to $h\nu$, and since the energy of oscillation can never be less than $h\nu/2$, no change of internal energy is possible, and the vibrational contribution to c_v is zero. Therefore, the diatomic molecule exhibits an average value of c_v of only $\frac{5}{2}k$ per molecule. As the temperature is raised, kT

approaches hv in magnitude. At very high temperature, $kT > hv$, and c_v then agrees with the principle of equipartition of energy.

A similar situation occurs with the rotation of the molecule. The energy of rotation requires multiple values of $k\theta_r$ such that

$$\varepsilon_{rot} = \frac{h^2}{8\pi^2 I} j(j + 1) = k\theta_r j(j + 1) \tag{13.40}$$

For the rotation of a diatomic molecule about an axis perpendicular to the line joining the two atoms, the energy, kT, must equal or exceed a minimum value of rotational energy. At high temperature, the available energy, kT, exceeds this minimum value by a large enough amount to permit rotation at several energy levels so that the principle of equipartition does apply, and $\bar{c}_v$ assumes the classical value of $\bar{R}$. Variation of $\bar{c}_v$ versus temperature for a typical diatomic gas (hydrogen) is shown in Figure 13.3. At low temperature, only translational motion is exhibited, and the value of $\bar{c}_v$ is $\frac{3}{2}\bar{R}$. At 50 K, rotational motion begins to absorb energy so that, when room temperature is reached, rotational motion is fully developed and $\bar{c}_v = \frac{5}{2}\bar{R}$. At about 600 K, vibrational motion begins to occur, and the classical equilibrium value of $\frac{7}{2}\bar{R}$ is attained, ultimately, at still higher temperatures. Additional temperature increase results eventually in ionization of the gas.

Experimentally determined values of specific heat, and specific heat ratios, of various gases are listed in Table 13.3. The agreement between predicted and mea-

Table 13.3   Values of Specific Heats of Various Gases at 1 atm and 15°C (unless otherwise stated)

Gas		$\bar{c}_v/\bar{R}$	$\bar{c}_p/\bar{R}$	γ	$(\bar{c}_p - \bar{c}_v)/\bar{R}$
	He	1.519	2.520	1.659 (18°)	1.001
	Ne			1.64 (19°)	
Monatomic	A	1.509	2.517	1.67	1.008
	Kr			1.68 (19°)	
	Xe			1.66 (19°)	
	H_2	2.438	3.438	1.410	1.000
	N_2	2.448	3.453	1.410	1.005
Diatomic	CO	2.488	3.493	1.404	1.005
	O_2	2.504	3.508	1.401	1.004
	NO	2.512	3.517	1.400	1.005
	Cl_2	3.02	4.11	1.36	1.09
	H_2O	3.3	4.36	1.32 (100°)	1.06
Triatomic	CO_2	3.38	4.407	1.304	1.027
	SO_2	3.79	4.89	1.29	1.10
Polyatomic	NH_3	3.42	4.48	1.31	1.06
	C_2H_4	4.04	5.07	1.26	1.03

sured values is quite good, especially for monatomic gases. The last column of the table gives an indication of the departure of the gas from ideal-gas behavior.

Example 13.4

Using the data of Example 13.2 for hydrogen, calculate the rotational and vibrational constant-volume specific heats at (a) 25°C and atmospheric pressure, (b) 1000°C and atmospheric pressure.

SOLUTION:

(a) The rotational characteristic temperature θ_r, according to Table 13.1, is

$$\theta_r = 85.5 \text{ K}$$

At $T = 25°C$,

$$\frac{\theta_r}{T} = \frac{85.5}{298.15} = 0.287$$

and at $T = 1000°C$,

$$\frac{\theta_r}{T} = \frac{85.5}{1273.15} = 0.0672$$

Internal energy is given by

$$U = \frac{N\sum g_i \varepsilon_i e^{-\varepsilon_i/kT}}{\sum g_i e^{-\varepsilon_i/kT}}$$

Differentiating internal energy with respect to temperature yields the following:

$$(\bar{c}_v)_{\text{rot}} = \bar{R}\left[\frac{\displaystyle\sum_{j=0,1...}^{\infty} g_i r^2 e^{-r}}{\displaystyle\sum_{j=0,1...}^{\infty} g_i e^{-r}} - \left(\frac{\displaystyle\sum_{j=0,1...}^{\infty} g_i r e^{-r}}{\displaystyle\sum_{j=0,1...}^{\infty} g_i e^{-r}}\right)^2\right]$$

where

$$r = \frac{\varepsilon_i}{kT} = \frac{j(j+1)}{T}(\theta_r) \qquad \text{and} \qquad g_i = 2j + 1$$

To calculate the value of $\bar{c}_v$ at 25°C, the following table may be formulated:

j	$j + 1$	$r = j(j + 1)(0.287)$	e^{-r}	$g_i = 2j + 1$	$g_i e^{-r}$	$r(g_i e^{-r})$	$r(rg_i e^{-r})$
0	1	0	1	1	0	0	0
1	2	0.574	0.563	3	1.690	0.970	0.557
2	3	1.722	0.179	5	0.895	1.541	2.65
3	4	3.444	0.0319	7	0.224	0.772	2.66
4	5	5.74	0.0032	9	0.029	0.166	0.955
5	6	8.61	0.00018	11	0.002	0.017	0.148
					3.840	3.466	6.970

At 25°C,

$$(\bar{c}_v)_{\text{rot}} = (8.3144 \text{ J/gm-mol K})\left[\frac{6.970}{3.840} - \left(\frac{3.466}{3.840}\right)^2\right]$$

$$= 8.3178 \text{ J/gm-mol K}$$

At 25°C, $(\bar{c}_v)_{\text{rot}}$ has already reached the classical equilibrium value and, therefore, at 1000°C, $(\bar{c}_v)_{\text{rot}}$ is the classical value.

(b) The vibrational constant-volume specific heat is given by

$$(\bar{c}_v)_{\text{vib}} = (8.3144 \text{ J/gm-mol K})(e^{\theta_v/T})\left[\frac{\theta_v/T}{e^{\theta_v/T} - 1}\right]^2$$

where the vibrational temperature θ_v is equal to 6140 K. At 25°C, $\theta_v/T = 20.6$, and

$$(\bar{c}_v)_{\text{vib}} = (8.3144 \text{ J/gm-mol K})(e^{20.6})\left[\frac{20.6}{e^{20.6} - 1}\right]^2 \approx 0$$

At 1000°C, $\theta_v/T = 4.82$, and

$$(\bar{c}_v)_{\text{vib}} = (8.3144 \text{ J/gm-mol K})(e^{4.82})\left[\frac{4.82}{e^{4.82} - 1}\right]^2 = 1.584 \text{ J/gm-mol K}$$

Example 13.5

The interatomic distance of carbon monoxide is 1.128 Å. What is the constant-volume specific heat at 300 K and atmospheric pressure?

SOLUTION:

$$(\bar{c}_v)_{\text{trans}} = \frac{3}{2}\overline{R}$$

$$I = \left(\frac{m_1 m_2}{m_1 + m_2}\right)r^2 = 1.448 \times 10^{-46} \text{ kg-m}^2/\text{molecule}$$

Therefore (see also Table 13.1),

$$\theta_r = \frac{h^2}{8\pi^2 I k} = \frac{\left(6.626 \times 10^{-34} \dfrac{\text{J s}}{\text{molecule}}\right)^2}{(8\pi^2)\left(1.448 \times 10^{-46} \dfrac{\text{kg-m}^2}{\text{molecule}}\right)\left(1.38049 \times 10^{-23} \dfrac{\text{J}}{\text{K molecule}}\right)}$$

$$= 2.78 \text{ K}$$

and

$$\frac{\theta_r}{T} = 0.00927$$

At 300 K, even very high rotational energy levels are filled (since θ_r corresponds to such a low temperature). Consequently, the classical value of rotational specific heat is reached long before 300 K.

Vibrational effects do not appear for CO at 300 K; much higher temperatures are required, therefore,

$$(\bar{c}_v)_{300 \text{ K}} = \frac{3}{2}\overline{R} + \overline{R} = \frac{5}{2}(8.3144 \text{ J/gm-mol K})$$

$$= 20.786 \text{ J/gm-mol K}$$

13.5 The Photon Gas

Electromagnetic radiation propagates in the form of discrete quanta of energy carried by photons. When the atoms of a system change to different electronic states, they emit or absorb photons of radiation. A photon is a relativistic particle of zero rest mass and has energy equal to $h\nu$, where h is Planck's constant, and ν is the frequency of propagation. Photons have a dualistic nature; they have the properties of a wave and also those of a particle. From a statistical point of view, they can be considered identical particles independent from one another and therefore may be treated as an ideal gas. However, there is no constraint on the total number of photons in each energy state because of the continuous absorption and emittance of electrons. Removing the condition $\sum N_i = N$ in deriving the distribution function and maximizing the thermodynamic probability for the Bose–Einstein statistical model gives the following equilibrium distribution:

$$N_i^* = \frac{g_i}{e^{\,h\nu/kT} - 1} \tag{13.41}$$

where N_i^* is the number of photons in the ith energy level.

When photons are treated as a system of particles in a container of volume V, a wave equation, identical in form to Eq. (12.95), describes the translation of the photons. The solution is a series of standing waves with nodes coinciding with the walls of the container. These boundary conditions require that the wave function equals zero at the walls.

The partition function is given by

$$Z = \sum_{N_1} e^{-N_1\varepsilon_1/kT} \sum_{N_2} e^{-N_2\varepsilon_2/kT} \dots$$

where N_1, N_2, ... are the number of photons at energy levels ε_1, ε_2, ... Each summation can be replaced by a series[3] so that

$$Z = \frac{1}{1 - e^{\varepsilon_1/kT}} \cdot \frac{1}{1 - e^{\varepsilon_2/kT}} \dots$$

and

$$\ln Z = \sum_i \ln \frac{1}{1 - e^{\varepsilon_i/kT}}$$

Summation over the quantum states is now replaced by a sum over wave frequencies so that

$$\ln Z = \sum_\nu g_\nu \ln \frac{1}{1 - e^{h\nu/kT}}$$

The number of quantum states or the degeneracy g_ν in the frequency range ν and $\nu + d\nu$ is given by

$$g_\nu = \frac{8\pi V}{c^3} \nu^2 \, d\nu \tag{13.42}$$

where V is the volume and c is the speed of light.

---

[3] $\displaystyle \sum_N e^{-N\varepsilon/kT} = \sum x^N = 1 + x + x^2 + x^3 + \dots = \frac{1}{1-x}$ $(x = e^{-\varepsilon/kT})$.

Substituting for g_ν gives

$$\ln Z = \frac{8\pi V}{c^3} \int_0^\infty \nu^2 \ln \frac{1}{1 - e^{h\nu/kT}} d\nu$$

Integration by parts,

$$\ln Z = \left[\frac{8\pi V \nu^3}{3c^3} \ln \frac{1}{1 - e^{h\nu/kT}} \right]_0^\infty + \frac{8\pi V h}{3c^3 kT} \int_0^\infty \frac{\nu^3 \, d\nu}{e^{h\nu/kT} - 1}$$

Let $x = h\nu/kT$ so that

$$\ln Z = 0 + \frac{8\pi V}{3\left(\dfrac{hc}{kt}\right)^3} \int_0^\infty \frac{x^3 \, dx}{e^x - 1}$$

Evaluation of the integral in this equation is as follows:

$$\int_0^\infty \frac{x^3}{e^x - 1} \, dx = \int_0^\infty \frac{x^3}{e^x(1 - e^{-x})} \, dx = \int_0^\infty x^3 e^{-x} (1 - e^x)^{-1} \, dx$$

$$= \int_0^\infty x^3 e^{-x} (1 + e^{-x} + e^{-2x} + \ldots + e^{-nx} + \ldots) \, dx$$

$$= \int_0^\infty x^3 \sum_{n=1}^\infty e^{-nx} \, dx = \sum_{n=1}^\infty \int_0^\infty e^{-nx} x^3 dx$$

let $nx = y, dx = dy/n$ so that

$$\int_0^\infty \frac{x^3}{e^x - 1} \, dx = \sum_{n=1}^\infty \frac{1}{n^4} \int_0^\infty e^{-y} y^3 dy$$

$$= \sum_{n=1}^\infty \frac{1}{n^4} \Gamma(4) = 6 \sum_{n=1}^\infty \frac{1}{n^4} = \frac{\pi^4}{15}$$

Therefore,

$$\ln Z = \frac{8\pi V}{3\left(\dfrac{hc}{kT}\right)^3} \int_0^\infty \frac{x^3}{e^x - 1} \, dx = \frac{8\pi^5 V}{45} \left(\frac{kT}{hc}\right)^3 \qquad (13.43)$$

Equation (13.43) is used to determine thermodynamic properties of a photon gas.

Substituting for g_v from Eq. (13.42) into Eq. (13.41) gives the following distribution of photons in the frequency range v to $v + dv$:

$$dN_v = \frac{8\pi V}{c^3} \frac{v^2}{e^{hv/kT} - 1} dv \qquad (13.44)$$

Noting that $\varepsilon_v = hv$, the equilibrium energy density in the frequency range v to $v + dv$ is

$$\frac{\varepsilon_v \, dN_v}{V} = \frac{8\pi h}{c^3} \frac{v^3}{e^{hv/kT} - 1} dv \qquad (13.45)$$

Equation (13.45) is called *Planck's distribution law* for a blackbody radiator. A plot of this equation is shown in Figure 13.5 for two temperatures ($T_2 > T_1$). The peak points of the isotherms shift toward higher frequency as the temperature increases. Differentiating Eq. (13.45) with respect to v and equating the result to zero gives the value of v, which corresponds to the maximum energy density. The result is

$$v_{max} = 1.03 \times 10^{11} T \text{ cycle/s} \qquad (13.46)$$

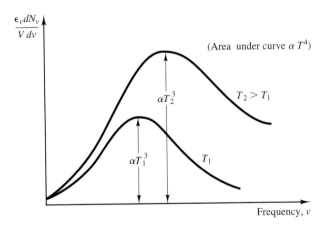

Figure 13.5   Planck's distribution law for blackbody radiation.

Equation (13.46) is called *Wien's displacement law*. Substituting this value of v_{max} into Eq. (13.45) indicates that the maximum energy density is proportional to the third power of absolute temperature.

The total energy per unit volume at any temperature is obtained by integrating Eq. (13.45) over all possible frequency ranges so that

$$\frac{U}{V} = \int_0^\infty \left(\frac{\varepsilon_v \, dN_v}{V \, dv} \right) dv$$

$$= \frac{8\pi h}{c^3} \int_0^\infty \frac{v^3}{e^{hv/kT} - 1} \, dv$$

Substituting x for hv/kT and integrating the preceding equation gives

$$\frac{U}{V} = \frac{8\pi k^4 T^4}{c^3 h^3} \int_0^\infty \frac{x^3}{e^x - 1} \, dx = \left(\frac{8\pi^5 k^4}{15 c^3 h^3} \right) T^4 \qquad (13.47)$$

The same result can be obtained using Eq. (13.43) and the expression of U in terms of the partition function.

Equation (13.47) is called the *Stefan–Boltzmann radiation law*, which states that the energy density at a given temperature is proportional to the fourth power of the absolute temperature. The total energy per unit volume at a temperature T is represented by the area underneath the corresponding curve in Figure 13.5.

A pressure–volume relation for a system of photons undergoing an adiabatic, quasi-static process can be determined as follows. The first law is

$$\delta W = dU$$

so that

$$-p \, dV = dU = d(uV) = u \, dV + V \, du$$

where u is the internal energy per unit volume.

But the radiation pressure $p = \frac{1}{3}u$ (Problem 13.33). Therefore,

$$-\frac{1}{3}u \, dV = u \, dV + V \, du$$

so that

$$V \, du + \frac{4}{3}u \, dV = 0$$

which integrates to

$$uV^{4/3} = \text{constant}$$

or

$$pV^{4/3} = \text{constant} \tag{13.48}$$

Example 13.6

Determine the number of photons in a cavity of 1 cm^3 volume at 298 K.

SOLUTION:

The distribution of the number of photons in the frequency range v to $v + dv$ is given by Eq. (13.44):

$$dN_v = \frac{8\pi V}{c^3} \frac{v^2}{e^{hv/kT} - 1} dv$$

and

$$N = \sum_{v=0}^{\infty} dN_v$$

Assuming that the energy levels are close, the summation sign can be replaced by an integration sign so that

$$N = \int_{v=0}^{v=\infty} \frac{8\pi V}{c^3} \frac{v^2}{e^{hv/kT} - 1} dv$$

$$= \frac{8\pi V}{c^3} \left(\frac{kT}{h}\right)^3 \int_0^\infty \frac{x^2 \, dx}{e^x - 1}$$

where

$$x = hv/kT$$

Using integration tables, the preceding integral has a value of 2.404 so that the number of photons is

$$N = \frac{8\pi V}{c^3} \left(\frac{kT}{h}\right)^3 (2.404) = 8\pi V \left(\frac{kT}{ch}\right)^3 (2.404)$$

$$= 8\pi(1 \text{ cm}^3) \left[\frac{(1.3805 \times 10^{-23} \text{ J/K})(298 \text{ K})}{(2.998 \times 10^{10} \text{ cm/s})(6.626 \times 10^{-34} \text{ J s})}\right]^3 (2.404)$$

$$= 1.447 \times 10^8 \text{ photons}$$

13.6 Simple Crystals

Atoms in the lattice structure of a crystal cannot be treated as independent entities owing to the strong interatomic forces between them. Because of the strong constraints imposed by the lattice, atoms in a simple crystal cannot translate or rotate; they are, however, free to vibrate in three coordinate directions. These vibrations constitute the sole contribution to the crystal internal energy. The oscillations can be treated as small vibrations about equilibrium positions if the temperature is not too high. It is assumed that the vibrations involve both kinetic and potential energy.

Each atom in a crystalline solid has three translational degrees of freedom. An N-particle crystal then has $3N$ independent and equivalent degrees of freedom (only $3N - 6$ degrees of freedom if movement of the entire body is excluded). The partition function is obtained by summation over these degrees of freedom. Although the kinetic energy summation is readily written in terms of the Cartesian coordinate system, the summation of the potential energy of the particles, due to their mutual interactions, becomes a very complex function in Cartesian coordinates. To simplify the form of the potential energy, one may introduce *normal* coordinates.[4] This coordinate transformation enables separation of the Schrödinger equation into simpler differential equations, each of them with a smaller number of variables. Transformation is based on the following assumption: When a system of harmonic oscillators is excited along one normal coordinate, the system vibrates in simple harmonic motion only along that coordinate and does not transfer energy to the other normal coordinates.

The Schrödinger equation may be separated into $3N$ equations, each in terms of an individual normal coordinate, or mode of vibration, corresponding to a particular energy level. Each equation is of the form of a simple harmonic oscillator equation, and the nondegenerate vibrational energy levels obtained from this treatment have the form

$$\varepsilon_i = \left(n + \frac{1}{2} \right) h \nu_i \qquad n = 0, 1, 2, \dots$$

Thus, for small amplitude oscillations, an N-particle crystal may be regarded as an ensemble of $(3N - 6)$ independent harmonic oscillators. Since the oscillators are distinguishable because of their identifications with the lattice points, the statistics follow the Maxwell–Boltzmann model. The partition function is obtained by the usual summation over all independent quantum states. The energy ε_i is summed up over the $3N$ normal modes:

$$\mathcal{Z} = g_0 \sum_{i=1}^{3N} e^{-[(E_0 + \Sigma \varepsilon_i)/kT]} = g_0 \sum_{i=1}^{3N} e^{-E_0/kT} e^{-\Sigma(\varepsilon_i/kT)}$$

[4] Denbigh, K. G., *The Principles of Chemical Equilibrium.* Cambridge: Cambridge University Press, 1957.

But

$$\frac{\varepsilon_i}{kT} = \left(n + \frac{1}{2}\right)\frac{h v_i}{kT} = \left(n + \frac{1}{2}\right)\frac{\theta_v}{T}$$

therefore,

$$\mathcal{Z} = g_0 e^{-E_0/kT} \prod_{i=1}^{3N} e^{-\theta_v/2T}[1 - e^{-\theta_v/T}]^{-1} \tag{13.49}$$

where E_0 is the zero-point energy of the crystal (potential energy), and g_0 is the degeneracy due to different orientation of the atoms in the lattice.

The partition function indicated in Eq. (13.49) is that of the whole crystal consisting of N atoms. No division by $N!$ is required here since the atoms in the immediate vicinity of the fixed lattice points are not free to exchange positions with each other so that they are not indistinguishable even though they are identical. Note also that ε_i's are the energies of the normal modes referenced to the zero-point energy.

The most probable distribution of the atoms according to the Boltzmann distribution is

$$\frac{N_i^*}{3N} = \frac{g_0}{\mathcal{Z}} e^{-[(E_0 + \Sigma \varepsilon_i)/kT]} \tag{13.50}$$

The thermodynamic properties of the crystal may then be obtained, as with a gas, by means of the partition function. The internal energy, Helmholtz function, and entropy are related to the system partition function by

$$U = kT^2 \left(\frac{\partial \ln \mathcal{Z}}{\partial T}\right)_V$$

$$F = -kT \ln \mathcal{Z}$$

$$S = k \ln \mathcal{Z} + kT \left(\frac{\partial \ln \mathcal{Z}}{\partial T}\right)_V$$

If the expression for the partition function, Eq. (13.49), is substituted in these equations, then

$$U = U_0 + h\sum_{i=1}^{3N} v_i(e^{h v_i/kT} - 1)^{-1} \tag{13.51}$$

$$F = -kT \ln g_0 + U_0 + kT\sum_{i=1}^{3N} \ln(1 - e^{-h v_i/kT}) \tag{13.52}$$

$$S = \frac{U - F}{T} = k \ln g_0 - k \sum_{i=1}^{3N} \left[\ln (1 - e^{-hv_i/kT}) - \frac{hv_i}{kT} (e^{hv_i/kT} - 1)^{-1} \right] \quad (13.53)$$

where U_0 represents the minimum internal energy of the crystal and is equal to

$$U_0 = E_0 + \sum_{i=1}^{3N} \frac{hv_i}{2} \qquad (13.54)$$

The specific heat of the crystal at constant volume is

$$c_v = k \sum_{i=1}^{3N} \left(\frac{hv_i}{kT} \right)^2 e^{hv_i/kT} (e^{hv_i/kT} - 1)^{-2} \qquad (13.55)$$

13.7 Einstein Model

The difficulty of evaluating the preceding thermodynamic quantities lies in pre-dicting the $3N$ frequencies of the crystal. In general, some approximations must be made in order to obtain any results. Both Einstein and Debye developed methods of solution. The approximation made at high temperature will be examined first.

At high temperatures, the value of hv/kT becomes much less than unity; al-though the frequency of vibration can increase as the temperature rises, tending to counterbalance the temperature effect, the frequency cannot increase indefi-nitely. The reason is that frequency can increase only as wavelength diminishes, and the wavelength cannot be smaller than the atomic spacing. Therefore, the term $e^{hv/kT}$, which can be expressed as an infinite series, at high temperatures is equivalent to only the first two terms of the series, that is, $[1 + (hv/kT)]$. Then the internal energy and specific heat of the crystal are

$$U = U_0 + h(3N)v \left(1 + \frac{hv}{kT} - 1 \right)^{-1} = U_0 + 3NkT \qquad (13.56)$$

$$\bar{c}_v = 3N_a k = 3\bar{R} \approx 25 \text{ J/gm-mol K} \qquad (13.57)$$

Similarly,

$$F = -kT \ln g_0 + U_0 + 3N_a kT \ln (hv/kT) \qquad (13.58)$$

and

$$S = k \ln g_0 + 3N_a k [1 - \ln (hv/kT)] \qquad (13.59)$$

The results of the foregoing equations agree with those obtained by classical mechanics. According to the principle of equipartition of energy, each of the $3N$ coordinates and moments associated with vibrational motion contributes $\frac{1}{2}kT$ to the internal energy so that the molar specific heat of a crystal should be $3\overline{R}$. According to the law of Dulong and Petit, which was postulated in 1819, the molar constant-volume specific heat of any element in the crystalline solid state is 26.76 J/gm-mol K. Experimental work on atomic and simple ionic crystals shows that specific heat values at room temperatures generally do fall in this range.

But at lower temperatures, specific heats fall below this classical value and, in fact, approach zero. A simple approximation proposed by Einstein in 1907 led to a solution that indicated the correct trend. For a monatomic crystal of identical atoms, Einstein assumed that all $3N$ frequencies of all atoms of the crystal are identical and independent so that Eq. (13.51) becomes

$$U = U_0 + 3Nk\theta_E(e^{\theta_E/T} - 1)^{-1} \tag{13.60}$$

where $\theta_E = h\nu_E/k$ is called the *Einstein characteristic temperature*. Then molar specific heat at constant volume is

$$\overline{c}_v = 3\overline{R}\left(\frac{\theta_E}{T}\right)^2 e^{\theta_E/T}(e^{\theta_E/T} - 1)^{-2} \tag{13.61}$$

By suitable selection of ν_E, calculations based on Eq. (13.61) confirmed specific heat values, measured at temperatures as low as $\theta_E/3$ (about 20 K). At high temperatures, the value of $\overline{c}_v$ calculated by the Einstein method approaches the classical value of $3\overline{R}$. At lower temperatures, however, the deviation is significantly larger because of the assumption that all modes of vibration in the crystal are independent of one another and have the same frequency.

13.8 Debye Model

Debye considered the vibrations of a solid body as a whole, that is as a continuous elastic solid. He analyzed the thermodynamic behavior of crystals by comparing their vibrational characteristics with those of an isotropic (independent of direction) elastic continuum. He was able to derive an expression for the *frequency distribution* of the vibrating atoms in the crystal. In an elastic medium, the frequency density of standing waves is proportional to the square of the frequency:

$$dN = Cv^2\, dv \qquad (13.62)$$

where C is a proportionality constant depending on volume and wave velocity, and dN is the number of modes whose frequencies lie in the range between v and $v + dv$. Debye assumed that this distribution can be employed to describe the crystal vibration frequencies. But because the medium has a discontinuous or atomic structure, it was supposed that the frequencies stop abruptly at an upper limit $v_{\max}$. At this maximum frequency, the total number of normal modes is $3N_a$ for 1 mole. Then,

$$dN = \frac{9N_a}{v_{\max}^3} v^2\, dv \qquad (13.63)$$

The internal energy of the crystal according to Eqs. (13.51) and (13.54) is

$$U = E_0 + h \sum_{i=1}^{3N} \left[\frac{v_i}{2} + v_i (e^{hv_i/kT} - 1)^{-1} \right]$$

$$= E_0 + h \int_0^{3N} \left[\frac{v}{2} + v(e^{hv/kT} - 1)^{-1} \right] dN \qquad (3N \gg 1)$$

Substituting for dN from Eq. (13.63) gives

$$U = E_0 + \frac{9N_a}{v_{\max}^3} h \int_0^{v_{\max}} \left[\frac{v^3}{2} + v^3 (e^{hv/kT} - 1)^{-1} \right] dv$$

which may be rewritten as

$$U = E_0 + \frac{9}{8} N_a k\theta_D + 9N_a kT \left(\frac{T}{\theta_D} \right)^3 \int_0^{\theta_D/T} \frac{x^3}{e^x - 1}\, dx \qquad (13.64)$$

where $\theta_D = hv_D/k$ is the *Debye characteristic temperature*, $v_D = v_{\max}$, and $x = hv/kT$. Table 13.4 gives values of θ_D for several solids.

The molar specific heat at constant volume is

$$\bar{c}_v = 9N_a k \left[4 \left(\frac{T}{\theta_D} \right)^3 \int_0^{\theta_D/T} \frac{x^3}{e^x - 1}\, dx - \frac{\theta_D}{T} (e^{\theta_D/T} - 1)^{-1} \right] \qquad (13.65)$$

Table 13.4   Debye Temperatures for Several Solids

Substance		Temperature Range (K)	θ_D (K)
Lead	Pb	14–573	88
Mercury	Hg	31–232	97
Iodine	I	22–298	106
Cadmium	Cd	50–380	168
Sodium	Na	50–240	172
Silver	Ag	35–873	215
Calcium	Ca	22–62	226
Zinc	Zn	33–673	235
Copper	Cu	14–773	315
Aluminum	Al	19–773	398
Iron	Fe	32–95	453
Diamond	C	30–1169	1860

Equation (13.65) is the Debye equation for the specific heat of a solid. Figure 13.6 shows a plot of $\overline{c}_v/R$ versus T/θ_D based on Eq. (13.65) together with experimental values for several solids.

The integral in Eq. (13.65) cannot be evaluated in a closed form; it has to be evaluated numerically. Equation (13.65) is written as

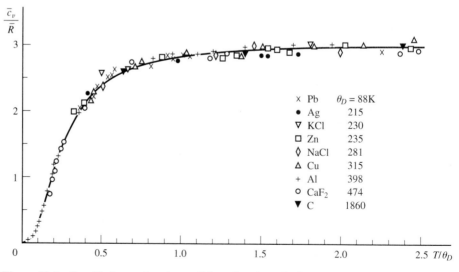

Figure 13.6   Specific heats of various solids as function of T/θ_D.

$$\bar{c}_v = 9N_a k \left[\frac{4}{3}D - \frac{\theta_D}{T}(e^{\theta_D/T} - 1)^{-1} \right] \qquad (13.66)$$

where D is called the *Debye function*, defined as

$$D \equiv 3\left(\frac{T}{\theta_D}\right)^3 \int_0^{\theta_D/T} \frac{x^3}{e^x - 1} \, dx \qquad (13.67)$$

Numerical values of D are given in Table 13.5.

At low temperature, $\theta_D/T \to \infty$ and the second term in the brackets in Eq. (13.65) may be neglected so that

$$\bar{c}_v = 36N_a k \left(\frac{T}{\theta_D}\right)^3 \int_\theta^{\theta_D/T} \frac{x^3}{e^x - 1} \, dx$$

$$= 36N_a k \left(\frac{T}{\theta_D}\right)^3 \int_0^\infty \frac{x^3}{e^x - 1} \, dx \qquad (13.68)$$

$$= 36N_a k \left(\frac{T}{\theta_D}\right)^3 \left(\frac{\pi^4}{15}\right) = 234\bar{R}\left(\frac{T}{\theta_D}\right)^3$$

Table 13.5   Debye Function D

θ_D/T	D	θ_D/T	D
0	1	4	0.1817
0.1	0.9630	4.5	0.1459
0.2	0.9270	5	0.1176
0.3	0.8920	6	0.07758
0.4	0.8580	7	0.05251
0.5	0.8250	8	0.03656
1	0.6744	9	0.02620
1.5	0.5471	10	0.0193
2	0.4411	15	0.00577
2.5	0.3541	20	0.0024
3	0.2836	∞	0
3.5	0.2269		

Equation (13.68) is called the *Debye T^3 law*. It predicts that at low temperature the constant-volume specific heat of a crystal varies as the third power of the absolute temperature. It further predicts that the specific heat is zero at the absolute zero of temperature. Note that in Eq. (13.68) the molar specific heat of the crystal is a function of only the normalized temperature (T/θ_D).

Debye's formula yields results for simple isotropic monatomic lattices, which are in excellent agreement with the experimental data at lower temperatures. At intermediate temperature, the Debye and Einstein formulas do not differ appreciably, and at high normalized temperature, both equations yield the expected classical result ($\bar{c}_v = 3\bar{R}$). Figure 13.7 compares specific heat values according to the Einstein and Debye models as a function of T/θ.

The close agreement between measured data and results calculated by the Debye formula does not necessarily prove the Debye model is correct. In 1937, Blackman made detailed calculations of the frequency distribution for specific crystals and showed that the actual frequency distribution differs significantly from that assumed by Debye. But the statistical averaging process fortuitously led to fair agreement between results.

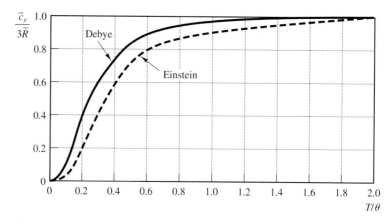

Figure 13.7   Temperature dependence of the constant-volume specific heat of solids according to the Debye theory and the Einstein model with $\theta_E = \theta_D$.

Example 13.7

(a) Calculate the constant-volume specific heat of silver at 40 K; the Debye characteristic temperature of silver is 215 K. (b) If the Einstein $\bar{c}_v$ at 40 K is 3.0 kJ/kg-mol K, estimate the fundamental frequency of oscillation of the atoms according to Einstein analysis.

SOLUTION:

(a) $\dfrac{\theta_D}{T} = \dfrac{215}{40} = 5.375$, and the Debye function $D(5.375) = 0.1004$, thus

$$\bar{c}_v = 9\bar{R}\left[4\left(\frac{T}{\theta_D}\right)^3 \int_0^{\theta_D/T} \frac{x^3}{e^x - 1}\,dx - \frac{\theta_D/T}{e^{\theta_D/T} - 1}\right]$$

$$= 9\bar{R}\left[\frac{4}{3}D(5.375) - \frac{5.375}{e^{5.375} - 1}\right]$$

$$= 9\bar{R}\left[\frac{4}{3}(0.1004) - 0.02501\right]$$

$$= 0.97971\bar{R} = 8.1457 \text{ kJ/kg-mol K}$$

(b) The Einstein constant-volume specific heat is given by

$$(\bar{c}_v)_E = 3\bar{R}\left(\frac{\theta_E/T}{e^{\theta_E/T} - 1}\right)^2 e^{\theta_E/T}$$

Values of $(\bar{c}_v)_E$ corresponding to different values of θ_E/T are calculated in Table 13.6.

Table 13.6

θ_E/T	$e^{\theta_E/T}$	$(\bar{c}_v)_E$ (kJ/kg-mol K)
5	148.413	4.259
6	403.429	2.237
5.7	298.867	2.7298
5.6	270.4264	2.914
5.5	244.6919	3.109
5.55	257.2376	3.01 (close)

From Table 13.6, at $(\bar{c}_v)_E = 3.0$ kJ/kg-mol K,

$$\theta_E/T \approx 5.55$$

From which,

$$\theta_E = (5.55)(40) = 222 \text{ K}$$

but

$$\theta_E = \frac{h\nu_E}{k}$$

therefore,

$$v_E = \frac{(222 \text{ K})(1.38049 \times 10^{-23} \text{ J/K molecule})}{(6.626 \times 10^{-34} \text{ J} \cdot \text{s/molecule})}$$

$$= 4.63 \times 10^{12} \text{ cycle/s}$$

Similarly, if the specific-heat value, $\bar{c}_v$, were 8.1457 kJ/kg-mol K, the fundamental frequency would be calculated, as shown in Table 13.7.

Table 13.7

θ_E/T	$e^{\theta_E/T}$	$(\bar{c}_v)_E$ (kJ/kg-mol K)
4	54.5982	7.5849
3.9	49.4024	8.0001
3.85	46.9931	8.2134
3.86	47.4654	8.1704
3.864	47.6556	8.1533
3.865	47.7033	8.1490 (close)

From Table 13.7,

$$\frac{\theta_E}{T} \approx 3.865 \qquad \text{or} \qquad \theta_E = 154.6 \text{ K}$$

and

$$v_E = 3.22 \times 10^{12} \text{ cycles/s}$$

The fundamental frequency of oscillation of the Debye model is 4.48×10^{12} cycle/s corresponding to a $\bar{c}_v$ value of 8.1457 kJ/kg-mol K. On the other hand, the Einstein equation indicates that, at this $\bar{c}_v$ value, the fundamental frequency is only 3.22×10^{12} cycles/s. At frequencies higher than this, the Einstein equation leads to specific-heat values that are considerably below those of the Debye equation.

13.9 The Third Law of Thermodynamics

Boltzmann's equation defines the entropy of a macrostate of a system in terms of thermodynamic probability: $S \equiv k \ln W$. In order to evaluate entropy by means of this relationship, it is necessary to enumerate the microstates corresponding to a certain macrostate. When a system of indistinguishable particles is at its minimum value of energy, each particle is in its lowest quantum state, and there is only one possible microstate. This microstate corresponds to the minimum possible value of W, which is $W = 1$. Therefore, the minimum value of S is zero. Since the entropy of a system is determined by the mechanical properties of its particles, a system in which the particles contribute nothing to entropy is arbitrarily assigned a zero value of entropy. At low temperatures, perfect crystals tend to form very orderly structures, and their lattice vibrations are at their lowest energies. Therefore, at 0 K, the entropy of such crystals is conceivably zero. Since a temperature of absolute zero is unattainable, measurements cannot be made to prove that the entropy value is zero at zero degrees absolute. It can merely be assumed, on the basis of extrapolation, that this is true.

Nernst postulated in 1906 that the Gibbs free-energy function and the enthalpy of a system approach the same value asymptotically as absolute zero temperature is approached. Furthermore, he indicated that their variation becomes independent of temperature as T approaches zero. This variation is shown in Figure 13.8 and may be expressed mathematically as

$$\lim_{T \to 0} \left(\frac{\partial \Delta G}{\partial T} \right)_p = \lim_{T \to 0} \left(\frac{\partial \Delta H}{\partial T} \right)_p = 0 \qquad (13.69)$$

According to Eq. (8.21), the change in free energy of a constant composition system as a result of temperature changes is related to entropy:

$$-\Delta S = \left(\frac{\partial \Delta G}{\partial T} \right)_p$$

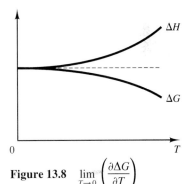

Figure 13.8 $\lim_{T \to 0} \left(\dfrac{\partial \Delta G}{\partial T} \right)_p$

$$= \lim_{T \to 0} \left(\frac{\partial \Delta H}{\partial T} \right)_p = 0.$$

From this information, Eq. (13.69) becomes

$$\lim_{T \to 0} \Delta S = 0 \qquad (13.70)$$

The foregoing equation is called the *third law of thermodynamics*. It indicates that the entropy of a one-component substance in thermodynamic equilibrium approaches zero as the temperature approaches zero. This conclusion is also implicit in the expression relating entropy to the partition function:

$$S = k \left[\ln \mathcal{Z} + T \left(\frac{\partial \ln \mathcal{Z}}{\partial T} \right)_V \right]$$

The determination of the change in entropy of a system undergoing a reversible process is accomplished by evaluating the integral $\int (\delta Q / T)_{\text{rev}}$ since heat interaction takes place reversibly between the system and its surroundings. Experimental data are extrapolated to zero absolute temperature, and the third law is assumed to be valid. The entropy of a system is

$$S = S_0 + \int_0^T \left(\frac{\delta Q}{T}\right)_{\text{rev}} \tag{13.71}$$

where S_0 is the entropy of the given system at 0 K, and δQ is the reversible heat transfer to the system. If the third law does hold, then S_0 equals zero, and Eq. (13.71) can be rewritten as

$$S = \int_0^{T^*} \left(\frac{\delta Q}{T}\right)_{\text{rev}} + \int_{T^*}^T \left(\frac{\delta Q}{T}\right)_{\text{rev}} \tag{13.72}$$

where T^* is the lowest temperature attainable by calorimetric measurements, which is down to a fraction of one degree above 0 K. The first term on the right side of this equation is evaluated by extrapolating from T^* to 0 K. The last term of Eq. (13.72) is evaluated completely from experimental measurements (of specific heat, enthalpy of crystalline transition, enthalpy of fusion, or enthalpy of vaporization, as appropriate). Measurements of specific heat above T^* are made in conjunction with an appropriate theoretical equation, usually the Debye equation.

Above the lowest temperature of measurement, T^*, the molecules of the given substance are distributed among the available quantum states in accordance with the appropriate distribution law. As the temperature is lowered, the distribution of molecules among the available levels of energy changes progressively in such a way that in the limit (at $T = 0$) the molecules have available only one possible state. If at the lowest temperature of measurement, T^*, a crystal of identical atoms can exist in only a single quantum state of energy, its entropy at 0 K is assigned a value of zero, provided that the internal energy corresponding to T^* can be accounted for by the extrapolation from T^* to 0 K.

Note that the third law does not limit the substance to any given crystalline forms. The substance may, at the lowest temperature of measurement, be in any of the several possible crystalline forms, and each one of the different crystalline forms can conform to the requirements of the third law.

In order to confirm the validity of calculations based on the third law, some method of comparison and a criterion for appraisal are required. Both requirements are met by using statistical thermodynamics to determine the absolute entropy of an ideal gas and by comparing this entropy value with the value obtained through calorimetric measurements based on the third law. In the statistical approach, the Sackur–Tetrode equation is used to determine the translational entropy for monatomic gases. In the case of polyatomic gases, the energy levels of allowed particle quantum states are determined spectroscopically, from which the partition function and the entropy are calculated.

In Table 13.8, values of molar entropy calculated by the third law are compared with values determined by statistical methods.

In general, any substance obtainable in pure crystalline form at equilibrium is likely to conform to the requirements of the third law, and absolute value of entropy can be calculated. Nonconformity with the third law occurs because the

Table 13.8[*]   Comparison Between Spectroscopic and Calorimetric Values of Entropy for Several Substances at 1 atm and at the Respective Boiling Points, Except for H_2O, D_2O, H_2, and D_2

Substance	Boiling Point (K)	Entropy (kJ/kg-mol K) Spectroscopically	Calorimetrically
A	87.3	129.0	128.7
O_2	90.13	170.0	170.0
N_2	77.4	153.2	153.6
HCl	188.2	173.2	172.6
Cl_2	238.6	215.4	215.6
CH_4	111.5	152.9	152.7
CO_2	194.7	198.8	198.9
NH_3	239.7	184.4	184.5
CO	83.0	160.1	155.5
NO	121.4	182.8	179.8
N_2O	184.6	202.7	198.0
CH_3D	99.7	165.0	153.5
H_2O	at 298.15	188.5	185.3
D_2O	at 298.15	194.7	191.8
H_2	at 298.15	130.5	124.1
D_2	at 298.15	144.7	141.6

[*]From Socrates, G., *Thermodynamics and Statistical Mechanics.* London, England: Butterworth & Co. Ltd., 1971, Chapter 7.

substance is not in a single quantum state at 0 K, owing to factors such as (a) the presence of two, or more, isotopes; (b) randomness in the structure of the crystal; (c) nonequilibrium distribution of molecules among the quantum states of energy; (d) presence of different molecules in a solution or in a mixture; (e) existence of subcooled (metastable) state rather than crystalline form.

The third law of thermodynamics is useful in calculating thermodynamics properties and in analyzing chemical and phase equilibria. It is also used to explain the behavior of solids at very low temperatures.

Example 13.8

Verify the value of entropy of N_2 as given in Table 13.8. The interatomic distance between the atoms of a nitrogen molecule is 1.0975 Å.

SOLUTION:

The entropy is given by

$$S = k \left[\ln \mathscr{Z} + T \left(\frac{\partial \ln \mathscr{Z}}{\partial T} \right)_V \right]$$

For translation, the partition function per molecule is

$$Z_{\text{trans}} = V \left(\frac{2\pi mkT}{h^2} \right)^{3/2}$$

$$\ln Z_{\text{trans}} = \frac{3}{2} \ln \frac{2\pi mk}{h^2} + \frac{3}{2} \ln T + \ln V$$

But

$$\mathscr{Z}_{\text{trans}} = \frac{Z^N}{N!}$$

Therefore,

$$\ln \mathscr{Z}_{\text{trans}} = N \ln Z - N \ln N + N$$

Also,

$$\left(\frac{\partial \ln \mathscr{Z}_{\text{trans}}}{\partial T} \right)_V = \frac{3}{2} \frac{1}{T}$$

Therefore,

$$\bar{s}_{\text{trans}} = \bar{R} \left(\frac{3}{2} \ln \frac{2\pi mk}{h^2} + \frac{3}{2} \ln T + \ln \bar{v} - \ln N_a + 1 + \frac{3}{2} \right)$$

For 1 mole,

$$\bar{v} = \frac{82.06 \, T}{p} \, \text{cm}^3/\text{gm-mole} \qquad (p \text{ in atm})$$

and

$$m = \frac{M}{6.023 \times 10^{23}} \qquad (M \text{ in gm/gm-mol})$$

Therefore,

$$\bar{s}_{\text{trans}} = \bar{R}\left(\frac{3}{2}\ln M + \frac{5}{2}\ln T - \ln p - 1.155\right)$$

The preceding equation is listed in Table 13.2. At 77.4 K and 1 atm,

$$\bar{s}_{\text{trans}} = 122.35 \text{ kJ/kg-mol K}$$

For rotation,

$$Z_{\text{rot}} = \frac{8\pi^2 IkT}{\sigma h^2}$$

$$\ln Z_{\text{rot}} = \ln \frac{8\pi^2 k}{h^2} + \ln I + \ln T - \ln \sigma$$

Substituting in Eq. (12.56) and considering 1 mole gives (see also Table 13.2)

$$\bar{s}_{\text{rot}} = \bar{R}\left(\ln \frac{8\pi^2 k}{h^2} + \ln I + \ln T - \ln \sigma + 1\right)$$

For a diatomic molecule, $I = [(m_1 m_2)/(m_1 + m_2)]r^2$, where m is the mass of the atom, and r is the bond length. Thus,

$$I_{N_2} = \left[\frac{(14.0067)^2}{28.013}(1.66 \times 10^{-27})\right]\frac{\text{kg}}{\text{molecule}}(1.0975 \times 10^{-10}m)^2$$

$$= 14.003 \times 10^{-47} \text{ kg-m}^2/\text{molecule} \quad ^5$$

Therefore,

$$\bar{s}_{\text{rot}} = (8.3144 \text{ kJ/kg-mol K})\left[\ln \frac{8\pi^2(1.3805 \times 10^{-23})}{(6.626 \times 10^{-34})^2} + \ln (14.003 \times 10^{-47}) + \ln 77.4 - \ln 2 + 1\right]$$

$$= 8.3144(3.4995) = 29.93 \text{ kJ/kg-mol K}$$

For vibration,

$$Z_{\text{vib}} = \frac{1}{1 - e^{-\theta_v/T}}$$

$$\ln Z_{\text{vib}} = -\ln (1 - e^{\theta_v/T})$$

---

[5] Note that $m = m_0 M$, where m is the atomic mass, m_0 is the atomic mass unit equal to 1.66×10^{-27} kg (reciprocal of Avogadro's number), and M is the atomic number.

and

$$\left(\frac{\partial \ln Z_{\text{vib}}}{\partial T}\right)_V = \frac{\theta_v}{T^2} \frac{1}{e^{\theta_v/T} - 1}$$

Therefore,

$$\bar{s}_{\text{vib}} = \bar{R}\left[-\ln\left(1 - e^{\theta_v/T}\right) + \frac{\theta_v}{T}\frac{1}{e^{\theta_v/T} - 1}\right]$$

For N_2, $\theta_v = 3340$ K so that $\theta_v/T = 11.2$ and $\bar{s}_{\text{vib}} \approx 0$. Thus the total entropy is

$$\bar{s} = 122.35 + 29.93 = 155.28 \text{ kJ/kg-mol K}$$

which compares well with the value of Table 13.8.

13.10 Ideal-Gas Mixtures

Methods of determining thermodynamic properties of a mixture of ideal gases can be developed by application of concepts based on statistical techniques. Dalton's model, for example, may be interpreted either at the macroscopic level or at the microscopic level. Dalton's model states that the partial pressure of a constituent is proportional to its mole fraction:

$$p_i = n_i \frac{\overline{R}T}{V}$$

Molecules of any constituent of an ideal-gas mixture may be considered to be unaffected by the presence of molecules of other constituents so that each constituent exerts a pressure independent of the presence of other gases.

Consider a chemically inert system containing gases a, b, c, ... in thermodynamic equilibrium. The particles of each constituent gas may be considered indistinguishable from each other; on the other hand, particles of one constituent may be distinguished from those of the other constituents. If N_a, N_b, N_c, ... represent the number of particles corresponding to each constituent gas, the numbers of microstates corresponding to the most probable arrangements, according to Section 12.12, are

$$W_a = \prod_i \frac{g_{a_i}^{N_{a_i}}}{N_{a_i}!} \qquad W_b = \prod_i \frac{g_{b_i}^{N_{b_i}}}{N_{b_i}!} \qquad W_c = \prod_i \frac{g_{c_i}^{N_{c_i}}}{N_{c_i}!}$$

where g is the degeneracy of the energy levels, and N_{a_i}, N_{b_i}, and N_{c_i}, are the numbers of particles of gases a, b, and c in the ith energy level.

The thermodynamic probability of the mixture is the product of the thermodynamic probabilities of the individual gases, or

$$W = W_a W_b W_c \cdots = \prod_i \frac{g_{a_i}^{N_{a_i}}}{N_{a_i}!} \prod_i \frac{g_{b_i}^{N_{b_i}}}{N_{b_i}!} \prod_i \frac{g_{c_i}^{N_{c_i}}}{N_{c_i}!} \cdots$$

The logarithm of this equation is

$$\ln W = \sum_i (N_{a_i} \ln g_{a_i} - \ln N_{a_i}!) + \sum_i (N_{b_i} \ln g_{b_i} - \ln N_{b_i}!)$$
$$+ \sum_i (N_{c_i} \ln g_{c_i} - \ln N_{c_i}!) + \cdots$$

Applying Stirling's formula ($\ln N_i! \approx N_i \ln N_i - N_i$) gives

$$\ln W = \sum_i (N_{a_i} \ln g_{a_i} - N_{a_i} \ln N_{a_i} + N_{a_i}) + \sum_i (N_{b_i} \ln g_{b_i} - N_{b_i} \ln N_{b_i} + N_{b_i})$$
$$+ \sum_i n(N_{c_i} \ln g_{c_i} - N_{c_i} \ln N_{c_i} + N_{c_i}) + \cdots$$

Taking first derivatives in order to maximize the value of W gives

$$\delta \ln W = \sum_i (\ln g_{a_i} \delta N_{a_i} - \ln N_{a_i} \delta N_{a_i}) + \cdots$$

so that

$$\sum_i \left(\ln \frac{N_{a_i}}{g_{a_i}} \delta N_{a_i} \right) + \sum_i \left(\ln \frac{N_{b_i}}{g_{b_i}} \delta N_{b_i} \right)$$
$$+ \sum_i \left(\ln \frac{N_{c_i}}{g_{c_i}} \delta N_{c_i} \right) + \cdots = 0 \tag{13.73}$$

The following constraints apply:
(a) The number of particles of each gas remains the same so that the net change in the number of particles of each gas is zero,

$$N_a = \sum_i N_{a_i} \qquad \text{or} \qquad \sum_i \delta N_{a_i} = 0 \tag{13.74a}$$

$$N_b = \sum_i N_{b_i} \qquad \text{or} \qquad \sum_i \delta N_{b_i} = 0 \tag{13.74b}$$

$$N_c = \sum_i N_{c_i} \qquad \text{or} \qquad \sum_i \delta N_{c_i} = 0 \tag{13.74c}$$

(b) The total internal energy of the mixture is constant,

$$U = \sum_i \varepsilon_{a_i} N_{a_i} + \sum_i \varepsilon_{b_i} N_{b_i} + \sum_i \varepsilon_{c_i} N_{c_i} + \cdots$$

or

$$dU = \sum_i \left(\varepsilon_{a_i} \delta N_{a_i} + \sum_i \varepsilon_{b_i} \delta N_{b_i} + \sum_i \varepsilon_{c_i} \delta N_{c_i} + \cdots \right) = 0 \qquad (13.75)$$

Lagrange's method of undetermined multipliers is applied by multiplying Eqs. (13.74a), (13.74b), (13.74c), and (13.75), respectively, by $-\ln a_a$, $-\ln a_b$, $-\ln a_c$, and β, and adding the result to Eq. (13.73) yields

$$\sum_i \left(\ln \frac{N_{a_i}}{g_{a_i}} - \ln a_a + \varepsilon_{a_i}\beta \right) \delta N_{a_i} + \sum_i \left(\ln \frac{N_{b_i}}{g_{b_i}} - \ln a_b + \varepsilon_{b_i}\beta \right) \delta N_{b_i}$$

$$+ \sum_i \left(\ln \frac{N_{c_i}}{g_{c_i}} - \ln a_c + \varepsilon_{c_i}\beta \right) \delta N_{c_i} + \cdots = 0$$

Since the values of δN_{a_i}, δN_{b_i}, δN_{c_i}, $\ldots$ may be considered independent, their coefficients in the foregoing equation must vanish separately so that

$$\left(\ln \frac{N_{a_i}}{g_{a_i}} - \ln a_a + \varepsilon_{a_i}\beta \right) = 0 \qquad \text{or} \qquad N_{a_i} = a_a g_{a_i} e^{-\varepsilon_{a_i}\beta}$$

Similarly,

$$N_{b_i} = a_b g_{b_i} e^{-\varepsilon_{b_i}\beta} \qquad \text{and} \qquad N_{c_i} = a_c g_{c_i} e^{-\varepsilon_{c_i}\beta}$$

In Section 12.11, it was shown that the value of β is $1/kT$, whereas a is N/Z, where $Z = \sum_i g_i e^{-\varepsilon_i/kT}$. Substituting these values gives the distribution of the number of molecules as

$$N_{a_i} = \frac{N_a}{Z_a} g_{a_i} e^{-\varepsilon_{a_i}/kT}$$

$$N_{b_i} = \frac{N_b}{Z_b} g_{b_i} e^{-\varepsilon_{b_i}/kT}$$

and so on. The properties of the mixture can now be evaluated by means of expressions outlined in Chapter 12 and by means of the partition function:

$$Z = Z_a Z_b Z_c \cdots = \sum_i g_{a_i} e^{-\varepsilon_{a_i}/kT} \sum_i g_{b_i} e^{-\varepsilon_{b_i}/kT} \sum_i g_{c_i} e^{-\varepsilon_{c_i}/kT} \cdots$$

The internal energy U of the mixture is determined by adding the internal energies of each of the components:

$$U = \sum_i \varepsilon_{a_i} N_{a_i} + \sum_i \varepsilon_{b_i} N_{b_i} + \sum_i \varepsilon_{c_i} N_{c_i} + \cdots$$

$$= \sum_i \varepsilon_{a_i} \frac{N_a}{Z_a} g_{a_i} e^{-\varepsilon_{a_i}/kT} + \sum_i \varepsilon_{b_i} \frac{N_b}{Z_b} g_{b_i} e^{-\varepsilon_{b_i}/kT} + \cdots$$

$$= \frac{N_a \sum_i \varepsilon_{a_i} g_{a_i} e^{-\varepsilon_{a_i}/kT}}{Z_a} + \frac{N_b \sum_i \varepsilon_{b_i} g_{b_i} e^{-\varepsilon_{b_i}/kT}}{Z_b} + \cdots$$

Or

$$U = N_a kT^2 \left(\frac{\partial \ln Z_a}{\partial T}\right)_V + N_b kT^2 \left(\frac{\partial \ln Z_b}{\partial T}\right)_V + \cdots$$

$$= \sum_{x=a,b,c\cdots} N_x kT^2 \left(\frac{\partial \ln Z_x}{\partial T}\right)_V \tag{13.76}$$

The entropy of the mixture is determined from the relation of entropy to maximum probability according to Boltzmann's treatment:

$$S = k \ln W = k\left[\sum_i (N_{a_i} \ln g_{a_i} - N_{a_i} \ln N_{a_i} + N_{a_i}) \right.$$

$$\left. + \sum_i (N_{b_i} \ln g_{b_i} - N_{b_i} \ln N_{b_i} + N_{b_i}) + \cdots \right]$$

$$= k(N_a + N_b + N_c + \cdots) + k\left[\sum_i N_{a_i} \ln \frac{g_{a_i}}{N_{a_i}} N_{b_i} \ln \frac{g_{b_i}}{N_{b_i}} + \cdots \right]$$

$$= kN + k\left[\sum_i N_{a_i}\left(\ln \frac{Z_a}{N_a} + \frac{\varepsilon_{a_i}}{kT} \right) + \sum_i N_{b_i}\left(\ln \frac{Z_b}{N_b} + \frac{\varepsilon_{b_i}}{kT} \right) + \cdots \right]$$

Thus,

$$S = kN + \frac{U}{T} + k\left[\ln \left(\frac{Z_a}{N_a}\right)^{N_a} + \ln \left(\frac{Z_b}{N_b}\right)^{N_b} + \cdots \right]$$

$$= kN + \frac{U}{T} + k \sum_{x=a,b,c,\cdots} \ln \left(\frac{Z_x}{N_x}\right)^{N_x} \tag{13.77}$$

When gases are mixed adiabatically, a change in entropy occurs, and this can be determined from Eq. (13.77):

$$\Delta S = k \left[\left\{ \ln \left(\frac{Z}{N_a} \right)^{N_a} - \ln \left(\frac{Z_a}{N_a} \right)^{N_a} \right\} + \left\{ \ln \left(\frac{Z}{N_b} \right)^{N_b} - \ln \left(\frac{Z_b}{N_b} \right)^{N_b} \right\} + \cdots \right]$$

$$= k \left[\ln \left(\frac{Z}{N_a} \right)^{N_a} + \ln \left(\frac{Z}{N_b} \right)^{N_b} + \ln \left(\frac{Z}{N_c} \right)^{N_c} + \cdots \right]$$

where Z is the partition function after mixing. The partition function of monatomic gases is $Z = V(2\pi mkT/h^2)^{3/2}$; therefore, the partition function is proportional to volume, at constant temperature:

$$\frac{Z}{Z_a} = \frac{V}{V_a}$$

The entropy of mixing then becomes

$$\Delta S = k \left[\ln \left(\frac{V}{V_a} \right)^{N_a} + \ln \left(\frac{V}{V_b} \right)^{N_b} + \ln \left(\frac{V}{V_c} \right)^{N_c} + \cdots \right]$$

At constant temperature, the pressure of each constituent is inversely proportional to its volume so that

$$\Delta S = -kN_a \ln \frac{p_a}{p} - kN_b \ln \frac{p_b}{p} - kN_c \ln \frac{p_c}{p} - \cdots = -k \sum_{x=a,b,c\cdots} N_x \ln \frac{p_x}{p}$$

Since $kN_x = n_x \overline{R} = n_x M_x R_x = m_x R_x$, the entropy change due to mixing is

$$\Delta S = - \sum_{x=a,b,c,\cdots} m_x R_x \ln \frac{p_x}{p}$$

and

$$\Delta s = - \sum_{x=a,b,c,\cdots} m_{f_x} R_x \ln \frac{p_x}{p} \tag{13.78}$$

These equations are equivalent to Eq. (9.36), which was obtained by classical means. The Helmholtz function is

$$F = U - TS$$

$$= -kNT - kT \left[\ln \left(\frac{Z_a}{N_a} \right)^{N_a} + \ln \left(\frac{Z_b}{N_b} \right)^{N_b} + \ln \left(\frac{Z_c}{N_c} \right)^{N_c} + \cdots \right] \tag{13.79}$$

$$= -kNT - kT \sum_{x=a,b,c,\cdots} \ln \left(\frac{Z_x}{N_x} \right)^{N_x}$$

The Gibbs function is

$$G = H - TS = F + pV$$

$$= -kT \sum_{x=a,b,c,\cdots} \ln \left(\frac{Z_x}{N_x}\right)^{N_x} \tag{13.80}$$

The chemical potential μ of each component per molecule can therefore be expressed as

$$\mu_x = \left(\frac{\partial G}{\partial N_x}\right)_{p,T,n_j} = -kT\left(N_x \, d \ln \frac{Z_x}{N_x} + \ln \frac{Z_x}{N_x}\right)$$

$$= -kT\left(\ln \frac{Z_x}{N_x}\right) \tag{13.81}$$

From the equations derived in this section, it is evident that thermodynamic functions indicating properties of nonreactive gas mixtures can be determined by both classical and statistical approaches. In the statistical approach, properties are given in terms of partition functions instead of mole fractions or partial pressures, and in some cases identical expressions are obtained. Evidently, information regarding molecular energies must generally be available if the statistical approach is to be applied. In many current problems, information regarding distribution of molecular energies is often experimentally determined.

13.11 Chemically Reacting System

The equation for the equilibrium constant, as well as expressions for other thermodynamic parameters pertaining to chemical systems, can be deduced from molecular properties. Analogous to the classical approach, the statistical approach leads to the law of mass action (conservation of mass), to the conditions of equilibrium, and to the equilibrium composition.

Although the statistical approach applies to any chemical reaction, a simple reaction involving three chemical species will be considered. The procedure, however, is analogous. Consider the reaction

$$A + B \rightleftharpoons AB$$

in which N_A and N_B molecules of elements A and B react to form N_{AB} molecules of product AB. The molecules of each species in the reactive mixture are considered indistinguishable among themselves, but they are distinguishable from other species. Both reactants and products are assumed to be ideal gases.

Let the energy associated with the particles of each species be denoted by ε_{a_i}, ε_{b_i}, ... , where the subscript $i = 1, 2, 3, \ldots$ denotes the energy level of the particle. The energies of the reactants and product may be measured relative to any arbitrarily chosen datum. The energy of the product AB relative to this datum includes its energy of formation. Table 13.9 indicates the nomenclature describing the distribution of particles of different species and their energies in a chemical system comprising reactants A and B and products AB.

Several constraints are imposed on the behavior of the chemical system. First, the total number of particles of each constituent must not change, or

$$\sum_i \delta N_{a_i} = 0 \tag{13.82a}$$

$$\sum_i \delta N_{b_i} = 0 \tag{13.82b}$$

$$\sum_i \delta N_{ab_i} = 0 \tag{13.82c}$$

The number of any elemental constituents (nuclei), both free and combined, must fulfill the following:

$$N_A = N_a + N_{ab} \tag{13.83a}$$
$$N_B = N_b + N_{ab} \tag{13.83b}$$

Second, the total energy of the system, which is the sum of the energies of its constituents, must not change, or

$$\sum_i \varepsilon_{a_i} \delta N_{a_i} + \sum_i \varepsilon_{b_i} \delta N_{b_i} + \sum_i \varepsilon_{ab_i} \delta N_{ab_i} = 0 \tag{13.84}$$

The total energy is equal to

$$U = U_a + U_b + U_{ab} = \sum_i N_{a_i} \varepsilon_{a_i} + \sum_i N_{b_i} \varepsilon_{b_i} + \sum_i N_{ab_i} \varepsilon_{ab_i}$$

Table 13.9   Number and Energy of Particles of the System $A + B \rightleftharpoons AB$

Constituent	Number of Particles at Different Energy Levels	Total Number of Particles	Energy Levels	Degeneracy of Energy Levels	Energy of Constituents
A	$N_{a_1}, N_{a_2}, \ldots N_{a_i}$	$N_a = \sum_i N_{a_i}$	$\varepsilon_{a_1}, \varepsilon_{a_2}, \ldots \varepsilon_{a_i}$	$g_{a_1}, g_{a_2}, \ldots g_{a_i}$	$U_a = \sum_i N_{a_i} \varepsilon_{a_i}$
B	$N_{b_1}, N_{b_2}, \ldots N_{b_i}$	$N_b = \sum_i N_{b_i}$	$\varepsilon_{b_1}, \varepsilon_{b_2}, \ldots \varepsilon_{b_i}$	$g_{b_1}, g_{b_2}, \ldots g_{b_i}$	$U_b = \sum_i N_{b_i} \varepsilon_{b_i}$
AB	$N_{ab_1}, N_{ab_2}, \ldots N_{ab_i}$	$N_{ab} = \sum_i N_{ab_i}$	$\varepsilon_{ab_1}, \varepsilon_{ab_2}, \ldots \varepsilon_{ab_i}$	$g_{ab_1}, g_{ab_2}, \ldots g_{ab_i}$	$U_{ab} = \sum_i N_{ab_i} \varepsilon_{ab_i}$

Based on the assumption that the particles of each species are indistinguishable from each other, the number of different microstates into which each constituent can be arranged is given by the expressions:

$$W_a = \prod_i \frac{g_{a_i}^{N_{a_i}}}{N_{a_i}!} \qquad W_b = \prod_i \frac{g_{b_i}^{N_{b_i}}}{N_{b_i}!} \qquad \text{and} \qquad W_{ab} = \prod_i \frac{g_{ab_i}^{N_{ab_i}}}{N_{ab_i}!}$$

The total number of microstates for a given macrostate is equal to the product of the preceding expressions of probabilities, or

$$W = W_a W_b W_{ab} = \prod_i \frac{g_{a_i}^{N_{a_i}}}{N_{a_i}!} \prod_i \frac{g_{b_i}^{N_{b_i}}}{N_{b_i}!} \prod_i \frac{g_{ab_i}^{N_{ab_i}}}{N_{ab_i}!} \qquad (13.85)$$

The process of finding the equilibrium distribution of the chemical reaction is performed in two steps. First, the maximum number of microstates for given values of N_a, N_b, and N_{ab} is determined. A second maximization of the probability of this state leads to the identification of those values of N_a, N_b, and N_{ab} that constitute the chemical equilibrium composition. The logarithm of the number of microstates is

$$\ln W = \left(\sum_i N_{a_i} \ln g_{a_i} - \sum_i \ln N_{a_i}! \right) + \left(\sum_i N_{b_i} \ln g_{b_i} - \sum_i \ln N_{b_i}! \right)$$

$$+ \left(\sum_i N_{ab_i} \ln g_{ab_i} - \sum_i \ln N_{ab_i}! \right)$$

Applying Stirling's approximation formula to the last term of each parentheses, the preceding equation becomes

$$\ln W = \left(\sum_i N_{a_i} \ln \frac{g_{a_i}}{N_{a_i}} + N_a \right) + \left(\sum_i N_{b_i} \ln \frac{g_{b_i}}{N_{b_i}} + N_b \right)$$

$$+ \left(\sum_i N_{ab_i} \ln \frac{g_{ab_i}}{N_{ab_i}} + N_{ab} \right) \qquad (13.86)$$

The maximum value of the probability is obtained by equating the differential of Eq. (13.86) to zero:

$$\delta(\ln W) = 0 = -\sum_i \ln \left(\frac{N_{a_i}}{g_{a_i}} \right) \delta N_{a_i}$$

$$- \sum_i \ln \left(\frac{N_{b_i}}{g_{b_i}} \right) \delta N_{b_i} - \sum_i \ln \left(\frac{N_{ab_i}}{g_{ab_i}} \right) \delta N_{ab_i} \qquad (13.87)$$

Lagrange's method of undetermined multipliers can now be applied to Eq. (13.87), when combined, by the use of multipliers $\ln a$ and β, with Eqs. (13.82) and (13.84):

$$\sum_i \left(- \ln \frac{N_{a_i}}{g_{a_i}} + \ln a_a - \beta \varepsilon_{a_i} \right) \delta N_{a_i} + \sum_i \left(- \ln \frac{N_{b_i}}{g_{b_i}} + \ln a_b - \beta \varepsilon_{b_i} \right) \delta N_{b_i}$$

$$+ \sum_i \left(- \ln \frac{N_{ab_i}}{g_{ab_i}} + \ln a_{ab} - \beta \varepsilon_{ab_i} \right) \delta N_{ab_i} = 0$$

Since δN_{a_i}, δN_{b_i}, and δN_{ab_i} can be treated as independent, their coefficients in the preceding equation must vanish separately. This leads to the following expression for the distribution numbers:

$$-\ln \frac{N_{a_i}}{g_{a_i}} + \ln a_a - \beta \varepsilon_{a_i} = 0 \qquad \text{or} \qquad N_{a_i} = a_a g_{a_i} e^{-\beta \varepsilon_{a_i}}$$

Similarly,

$$-\ln \frac{N_{b_i}}{g_{b_i}} + \ln a_b - \beta \varepsilon_{b_i} = 0 \qquad \text{or} \qquad N_{b_i} = a_b g_{b_i} e^{-\beta \varepsilon_{b_i}}$$

and

$$-\ln \frac{N_{ab_i}}{g_{ab_i}} + \ln a_{ab} - \beta \varepsilon_{ab_i} = 0 \qquad \text{or} \qquad N_{ab_i} = a_{ab} g_{ab_i} e^{-\beta \varepsilon_{ab_i}}$$

It can be shown, by methods similar to those outlined in Chapter 12, that the Lagrange multipliers can be expressed as

$$\beta = \frac{1}{kT} \qquad a_a = \frac{N_a}{Z_a} \qquad a_b = \frac{N_b}{Z_b} \qquad \text{and} \qquad a_{ab} = \frac{N_{ab}}{Z_{ab}}$$

where Z is the partition function defined by

$$Z_a = \sum_i g_{a_i} e^{-\varepsilon_i / kT}$$

and so on. Substituting these expressions for a and β, and the distribution numbers (N_{a_i}, N_{b_i}, and N_{ab_i}), into Eq. (13.86) gives the following expression of thermodynamic probability for the most probable distribution:

$$\ln W = \left\{ \sum_i N_{a_i} \ln \left[\frac{g_{a_i}}{(N_a/Z_a)g_{a_i} e^{-\varepsilon_{a_i}/kT}} \right] + N_a \right\}$$

$$+ \left\{ \sum_i N_{b_i} \ln \left[\frac{g_{b_i}}{(N_b/Z_b)g_{a_i} e^{-\varepsilon_{b_i}/kT}} \right] + N_b \right\}$$

$$+ \left\{ \sum_i N_{ab_i} \ln \left[\frac{g_{ab_i}}{(N_{ab}/Z_{ab})g_{ab_i} e^{-\varepsilon_{ab_i}/kT}} \right] + N_{ab} \right\}$$

$$= \left\{ \left[\sum_i N_{a_i} \frac{\varepsilon_{a_i}}{kT} + \sum_i N_{a_i} \ln Z_a - \sum_i N_{a_i} \ln N_a + N_a \right] + \cdots \right\}$$

Thus

$$\ln W = \frac{U}{kT} + (N_a \ln Z_a - N_a \ln N_a + N_a)$$
$$+ (N_b \ln Z_b - N_b \ln N_b + N_b) \qquad (13.88)$$
$$+ (N_{ab} \ln Z_{ab} - N_{ab} \ln N_{ab} + N_{ab})$$

where U is the internal energy of all the species.

The next step is to determine the number of particles N_a, N_b, and N_{ab} that exist in the equilibrium mixture. This is accomplished by finding the maximum value of W in Eq. (13.88). Denoting the maximum probability by W^e, the differential of Eq. (13.88) at its maximum value is

$$\delta(\ln W^e) = 0 = \left(\ln Z_a \delta N_a - N_a \frac{\delta N_a}{N_a} - \ln N_a \delta N_a + \delta N_a \right)$$

$$+ \left(\ln Z_b \delta N_b - N_b \frac{\delta N_b}{N_b} - \ln N_b \delta N_b + \delta N_b \right)$$

$$+ \left(\ln Z_{ab} \delta N_{ab} - N_{ab} \frac{\delta N_{ab}}{N_{ab}} - \ln N_{ab} \delta N_{ab} + \delta N_{ab} \right) \qquad (13.89)$$

$$= (\ln Z_a - \ln N_a)\delta N_a + (\ln Z_b - \ln N_b)\delta N_b + (\ln Z_{ab} - \ln N_{ab})\delta N_{ab}$$

The conditions of constraint imposed on Eq. (13.89), according to Eq. (13.83), are

$$\delta N_a + \delta N_{ab} = 0 \qquad (13.90)$$

and

$$\delta N_b + \delta N_{ab} = 0 \qquad (13.91)$$

Applying Lagrange's method of undetermined multipliers, multiply Eq. (13.90) by a' and Eq. (13.91) by a'' and combine these with Eq. (13.89):

$$(\ln Z_a - \ln N_a + a')\delta N_a + (\ln Z_b - \ln N_b + a'')\delta N_b$$
$$+ (\ln Z_{ab} - \ln N_{ab} + a' + a'')\delta N_{ab} = 0$$

Since N_a, N_b, and N_{ab} can be treated as independent,

$$\ln Z_a - \ln N_a + a' = 0$$
$$\ln Z_b - \ln N_b + a'' = 0$$

and

$$\ln Z_{ab} - \ln N_{ab} + a' + a'' = 0$$

Eliminating a' and a'' from the preceding three equations and denoting the equilibrium distribution numbers by superscript ^{e}, the relation between the species concentration at the equilibrium condition is

$$\ln Z_a Z_b - \ln N_a^e N_b^e - \ln Z_{ab} + \ln N_{ab}^e = 0$$

or

$$K_N = \frac{N_{ab}^e}{N_a^e N_b^e} = \frac{Z_{ab}^e}{Z_a^e Z_b^e} \tag{13.92}$$

where K_N is called the *equilibrium distribution constant*. Therefore, a knowledge of the partition functions of the species of the reaction is sufficient to determine the equilibrium distribution of the different species involved in a chemical reaction. Equation (13.92) is the law of mass action.

The equilibrium constant K_p may be shown to be related to K_N in the following way. The equilibrium constant K_p is

$$K_p = \frac{p_{ab}^e}{p_a^e p_b^e}$$

where p_{ab}^e, p_a^e, and p_b^e are the equilibrium partial pressures of the species AB, A, and B. Their values, according to the ideal-gas law, are given by

$$p_{ab}^e V = N_{ab}^e kT \qquad p_a^e V = N_a^e kT \qquad \text{and} \qquad p_b^e V = N_b^e kT$$

Therefore,

$$K_p = \frac{(N_{ab}^e \, kT/V)}{(N_a^e \, kT/V)(N_b^e \, kT/V)} = \left(\frac{N_{ab}^e}{N_a^e N_b^e}\right)\frac{V}{kT} = \left(\frac{N_{ab}^e}{N_a^e N_b^e}\right)\frac{N}{p} \qquad (13.93)$$

and in terms of the partition function,

$$K_p = \frac{N}{p}\left(\frac{Z_{ab}^e}{Z_a^e Z_b^e}\right) = \frac{N}{p} K_N \qquad (13.94)$$

Expressions for other properties that apply to chemical reaction may similarly be obtained. Of these, entropy, Helmholtz and Gibbs functions, and chemical potential will be considered.

At equilibrium, Boltzmann's law, $S = k \ln W^e$, may be used to obtain the following expression for entropy. Using Eq. (13.88), then

$$S = \frac{U}{T} + kN_a\left(\ln \frac{Z_a}{N_a} + 1\right) + kN_b\left(\ln \frac{Z_b}{N_b} + 1\right) + kN_{ab}\left(\ln \frac{Z_{ab}}{N_{ab}} + 1\right)$$

$$= \frac{U}{T} + k \ln\left[\left(\frac{Z_a}{N_a}\right)^{N_a}\left(\frac{Z_b}{N_b}\right)^{N_b}\left(\frac{Z_{ab}}{N_{ab}}\right)^{N_{ab}}\right] + k(N_a + N_b + N_{ab}) \qquad (13.95)$$

An expression for the Helmholtz function $F = U - TS$ is

$$F = -kT \ln\left[\left(\frac{Z_a}{N_a}\right)^{N_a}\left(\frac{Z_b}{N_b}\right)^{N_b}\left(\frac{Z_{ab}}{N_{ab}}\right)^{N_{ab}}\right] - kT(N_a + N_b + N_{ab}) \qquad (13.96)$$

The Gibbs function $G = H - TS = F + pV$ is given by

$$G = -kT \ln\left[\left(\frac{Z_a}{N_a}\right)^{N_a}\left(\frac{Z_b}{N_b}\right)^{N_b}\left(\frac{Z_{ab}}{N_{ab}}\right)^{N_{ab}}\right] \qquad (13.97)$$

The expressions for the chemical potentials for the species A, B, and AB are

$$\mu_a = \left(\frac{\partial F}{\partial N_a}\right)_{N_b,V,T} = -kT\left[N_a\left(\frac{-Z_a}{N_a^2}\right)\frac{N_a}{Z_a} + \ln \frac{Z_a}{N_a}\right] - kT = -kT \ln \frac{Z_a}{N_a}$$

Similarly,

$$\mu_b = \left(\frac{\partial F}{\partial N_b}\right)_{N_a,V,T} = -kT \ln \frac{Z_b}{N_b}$$

$$\mu_{ab} = \left(\frac{\partial F}{\partial N_{ab}}\right)_{V,T} = -kT \ln \frac{Z_{ab}}{N_{ab}}$$

Adding the first two equations,

$$\mu_a + \mu_b = -kT \ln \frac{Z_a Z_b}{N_a N_b}$$

But at chemical equilibrium, according to Eq. (13.92),

$$\frac{N_a^e N_b^e}{N_{ab}^e} = \frac{Z_a^e Z_b^e}{Z_{ab}^e}$$

Therefore,

$$\mu_a^e + \mu_b^e = \mu_{ab}^e \qquad (13.98)$$

Note that Eq. (13.98) satisfies Eq. (10.29), which is necessary for chemical equilibrium.

The foregoing expressions for the Helmholtz and Gibbs functions can be written in terms of the chemical potential as follows:

$$F = N_a \mu_a + N_b \mu_b + N_{ab} \mu_{ab} - pV \qquad (13.99)$$

$$G = \mu_a(N_a + N_{ab}) + \mu_b(N_b + N_{ab}) = \mu_a N_A + \mu_b N_B \qquad (13.100)$$

The proof of Eqs. (13.99) and (13.100) is left to the reader as an exercise.

The present section, comparable to Section 13.10, indicates that both classical and statistical approaches may be used to derive expressions describing equilibrium conditions in a reactive mixture. In particular, this analogy is illustrated by comparing the equilibrium constant equation (Eq. 13.94), derived from a statistical approach, with Eq. (10.49), derived from a classical approach. For the three-component system consisting of species a, b, and ab,

$$\frac{N_{ab} N}{N_a N_b p} = \frac{n_{ab} n}{n_a n_b p}$$

and the two equations are equivalent.

Example 13.9

Air is heated at a constant pressure of 0.5 atm to 4500 K. The chemical species present are O_2, O, N_2, and N. If $\nu_{O_2} = 4.73 \times 10^{13}$ cycle/s and $\nu_{N_2} = 7.06 \times 10^{13}$ cycles/s, determine the enthalpy of the mixture. Ignore electronic energies and assume air comprises 21 percent O_2 and 79 percent N_2 by mole.

SOLUTION:

The total pressure is the sum of the partial pressures of the species:

$$p = p_{O_2} + p_O + p_{N_2} + p_N = 0.5 \tag{a}$$

At 4500 K,

$$\frac{(p_O)^2}{p_{O_2}} = K_{p_{O_2}} = 12.3419 \tag{b}$$

and

$$\frac{(p_N)^2}{p_{N_2}} = K_{p_{N_2}} = 82 \times 10^{-6} \tag{c}$$

The ratio of the number of moles of N_2 to O_2 is

$$\frac{2\,p_{N_2} + p_N}{2\,p_{O_2} + p_O} = \frac{0.79}{0.21} = 3.76 \tag{d}$$

There are four unknowns, p_O, p_{O_2}, p_N, and p_{N_2}, in the preceding four equations. Solving the equations simultaneously gives

$$p_O = 0.1689 \text{ atm}$$
$$p_{O_2} = 0.0023 \text{ atm}$$
$$p_N = 0.0051 \text{ atm}$$
$$p_{N_2} = 0.3237 \text{ atm}$$

The enthalpy of the mixture is $\bar{h} = \sum_i x_i \bar{h}_i$, where

$$x_O = \frac{p_O}{p} = \frac{0.1689}{0.5} = 0.3378$$

$$x_{O_2} = \frac{p_{O_2}}{p} = \frac{0.0023}{0.5} = 0.0046$$

$$x_N = \frac{p_N}{p} = \frac{0.0051}{0.5} = 0.0102$$

$$x_{N_2} = \frac{p_{N_2}}{p} = \frac{0.3237}{0.5} = 0.6474$$

For O_2:

$$\frac{hv}{kT} = \frac{(6.626 \times 10^{-34} \text{ J s/molecule})(4.73 \times 10^{13} \text{ cycles/s})}{(1.38 \times 10^{-23} \text{ J/K molecule})(4500 \text{ K})} = 0.505$$

Therefore,

$$\bar{h}_{O_2} = \bar{h}_{\text{trans}} + \bar{h}_{\text{rot}} + \bar{h}_{\text{vib}} = \frac{7}{2}\bar{R}T + \frac{hv/kT}{e^{hv/kT} - 1}\bar{R}T$$

$$= \frac{7}{2}(8.3144 \text{ kJ/kg-mol K})(4500 \text{ K}) + \frac{0.505}{e^{0.505} - 1}(8.3144 \text{ kJ/kg-mol K})(4500 \text{ K})$$

$$= 130.952 \times 10^3 + 28.759 \times 10^3 = 159{,}711 \text{ kJ/kg-mol}$$

For O:

$$\bar{h}_O = \frac{5}{2}\bar{R}T + (\bar{h}_f^\circ)_O = \frac{5}{2}(8.3144 \text{ kJ/kg-mol K})(4500 \text{ K}) + 249{,}190 \text{ kJ/kg-mol}$$

$$= 342{,}727 \text{ kJ/kg-mol}$$

where $\bar{h}_f^\circ$ is the enthalpy of formation per kg-mol.
For N_2:

$$\frac{hv}{kT} = \frac{(6.626 \times 10^{-34} \text{ J} \cdot \text{s/molecule})(7.06 \times 10^{13} \text{ cycle/s})}{(1.38 \times 10^{-23} \text{ J/K molecule})(4500 \text{ K})} = 0.753$$

$$\bar{h}_{N_2} = \frac{7}{2}\bar{R}T + \frac{hv/kT}{e^{hv/kT} - 1}\bar{R}T$$

$$= \frac{7}{2}(8.3144 \text{ kJ/kg-mol K})(4500 \text{ K}) + \frac{0.753}{e^{0.753} - 1}(8.3144 \text{ kJ/kg-mol K})(4500 \text{ K})$$

$$= 130.952 \times 10^3 + 25.080 \times 10^3 = 156{,}032 \text{ kJ/kg-mol}$$

For N:

$$\bar{h}_N = \frac{5}{2}\bar{R}T + (\bar{h}_f^\circ) = \frac{5}{2}(8.3144 \text{ kJ/kg-mol K})(4500 \text{ K}) + 472{,}650 \text{ kJ/kg-mol}$$

$$= 566{,}187 \text{ kJ/kg-mol}$$

The enthalpy of the mixture is

$$\bar{h} = \sum_i x_i \bar{h}_i$$

$$= 0.3378(342{,}727 \text{ kj/kg-mol}) + 0.0046(159{,}711 \text{ kJ/kg-mol})$$

$$+ 0.0102(566{,}187 \text{ kJ/kg-mol}) + 0.6474(156{,}032 \text{ kJ/kg-mol})$$

$$= 223{,}298 \text{ kJ/kg-mol of mixture}$$

13.12 Summary

Statistical thermodynamics is used to evaluate macroscopic properties of matter making use of statistical averaging techniques. Properties are generally given in terms of the partition function. For a monatomic gas, the partition function per molecule is

$$Z_{trans} = V \left(\frac{2\pi m k T}{h^2} \right)^{3/2}$$

Other properties are

$$U = \frac{3}{2} N k T \qquad H = \frac{5}{2} N k T$$

$$F = -NkT \left(-\ln N + 1 + \ln V + \frac{3}{2} \ln \frac{2\pi m k}{h^2} + \frac{3}{2} \ln T \right)$$

$$G = -NkT \left(-\ln N + \ln V + \frac{3}{2} \ln \frac{2\pi m k}{h^2} + \frac{3}{2} \ln T \right)$$

$$S = Nk \left(-\ln N + \ln V + \frac{3}{2} \ln \frac{2\pi m k}{h^2} + \frac{3}{2} \ln T + \frac{5}{2} \right)$$

$$c_v = \frac{3}{2} kT \qquad c_p = \frac{5}{2} kT \qquad \gamma = \frac{5}{3}$$

For a diatomic gas, the partition function per molecule is

$$Z = V \left(\frac{2\pi m k T}{h^2} \right)^{3/2} \left(\frac{8\pi^2 I k T}{\sigma h^2} \right) \left(\frac{e^{-\theta_v/2T}}{1 - e^{-\theta_v/T}} \right)$$

For a photon gas, the partition function per photon is

$$\ln Z = \frac{8\pi^5 V}{45} \left(\frac{kT}{hc} \right)^3$$

The radiation density of a blackbody is described by Planck's distribution law:

$$\frac{\varepsilon_v \, dN_v}{V} = \frac{8\pi h}{c^3} \frac{v^3}{e^{nv/kT} - 1} dv$$

The frequency for maximum energy density is given by Wien's displacement law:

$$v_{max} = 1.03 \times 10^{11} T \text{ cycles/s}$$

The energy density is proportional to T^4 and is given by Stefan–Boltzmann radiation law

$$\frac{U}{V} = \left(\frac{8\pi^5 k^4}{15c^3 h^3}\right) T^4$$

The partition function of a simple crystal, treated as a system of N dependent atoms, is

$$Z = g_0 e^{-E_0/kT} \prod_{i=1}^{3N} e^{-\theta_v/2T} [1 - e^{-\theta_v/T}]^{-1}$$

The Einstein model and Debye model predict the value of c_v as a function of temperature. The Einstein model assumes a single frequency of vibration, whereas the Debye model assumes a frequency distribution leading to the Debye T^3 law at low temperature.

The third law of thermodynamics indicates that the entropy of a one-component substance in equilibrium vanishes as the temperature approaches 0 K.

Application of statistical methods to nonreactive ideal-gas mixtures and reactive-gas mixtures yields expressions of properties in terms of the applicable partition functions. For example, the internal energy of a mixture of gases is

$$U = \sum_{x=a,b,c\ldots} N_x k T^2 \left(\frac{\partial \ln Z_x}{\partial T}\right)_V$$

whereas for a single substance

$$U = NkT^2 \left(\frac{\partial \ln Z}{\partial T}\right)_V$$

Other extensive properties follow the same rule.

Problems

13.1 Calculate the molecular translational partition function per cm^3 for H$_2$ at 0°C and N$_2$ at 100°C. Assume standard atmospheric pressure.

13.2 Determine the partition function per molecule of He at standard pressure and 25°C. Calculate the internal energy and entropy based on 1 cm^3 under the same conditions.

13.3 Calculate the rotational and vibrational partition functions per molecule of O$_2$ and H$_2$ at 200 K.

13.4 Calculate the total energy of the first eight energy levels of He molecules in a box of dimensions $1 \times 1 \times 2$ cm. What will be your answer if the box had dimensions $1 \times 1 \times 1$ cm?

13.5   Repeat Problem 13.4 for the translational energy levels of H_2.

13.6   Find the number of the quantized translational energy states of H_2 molecules confined in a box of dimensions $1 \times 1 \times 1$ cm at 300 K, the maximum energy being $\frac{3}{2} kT$ per molecule.

13.7   Calculate the difference between the third and fourth translational energy levels of H_2 molecules confined in a box of dimensions $0.1 \times 0.2 \times 0.3$ cm.

13.8   Determine the most probable rotational energy distribution for a system of unsymmetrical diatomic rigid rotators in which $8\pi^2 IkT/h^2 = 100$. Note that

$$\frac{N_i^*}{N} = \frac{g_i e^{-\varepsilon_i/kT}}{\sum_i g_i e^{-\varepsilon_i/kT}} \qquad g_i = (2j + 1)$$

Plot a curve of N_i/N versus $j(j + 1)$.

13.9   Plot the relative number of H_2 molecules at 1000 K in the first few rotational and vibrational energy states.

$I_{H_2} = 4.64 \times 10^{-48}$ kg-m^2
$v_{H_2} = 1.32 \times 10^{14}$ cycles/s and $\varepsilon_i = hv(n)$ for vibration

What is the enthalpy and c_p of H_2 at 1000 K?

13.10   The quantized vibrational energies of each particle in a system of N particles is given by the expression

$$\varepsilon_{\text{vib}} = nhv$$

where $n = 0, 1, 2, 3 \ldots, \infty$.
  If the particles occupy only the first four energy levels $(n = 0, 1, 2, 3)$, find the partition function and the number of particles in each vibrational energy level. Assume the fundamental frequency of vibration $v = 10^{12}$ cycles/s and $T = 25°$ C.

13.11   Plot the rotational energy distribution for O_2 at 300 K.

13.12   Plot the distribution of CO molecules among the rotational energy levels at 1000 K, and identify the most probable macrostate.

13.13   Plot the vibrational energy distribution for O_2 at 300 and 2260 K, and show the quantized energy levels $(v_{O_2} = 4.65 \times 10^{13}$ cycles/s).

13.14   Compare the rotational energy distributions for N_2 at 60 K and 572 K. Plot your results as percentage of total molecules versus the quantum number j.

13.15   Calculate the fraction of iodine (I_2) molecules that are in their ground state, first excited state, and second excited state at 25°C. The characteristic vibration temperature of I_2 is $\theta_v = 310$ K.

13.16   Calculate and plot the relative distribution of nitrogen in each of the first three vibrational energy levels as a function of temperature in the range of $T = 500$ to 3500 K. The fundamental vibrational frequency of N_2 is $v = 7.06 \times 10^{13}$ cycles/s.

13.17   Prove that the vibrational partition function of a one-dimensional oscillator, when its energy is measured relative to the zero-point energy, is given by

$$Z_{\text{vib}} = \frac{1}{1 - e^{-hv/kT}}$$

Compare this result with Eq. 13.31.

13.18   Prove that the moment of inertia of a diatomic molecule about its center of mass is given by

$$I = m_r r_e^2$$

where $m_r = m_1 m_2/(m_1 + m_2)$ is the reduced mass, m_1 and m_2 are the atomic masses, and r_e is the interatomic distance.

13.19   Calculate the moment of inertia of HI (hydrogen iodide) in gm-cm^2 and the equilibrium interatomic separation in Å ($\theta_r = 9$ K for HI).

13.20 Calculate the vibrational constant-volume specific heat in J/gm-mol K for N_2 and O_2 at 300, 800, and 3000 K. The frequencies are

$$\nu_{N_2} = 6.96 \times 10^{13} \text{ cycles/s}$$

and

$$\nu_{O_2} = 4.65 \times 10^{13} \text{ cycles/s}$$

13.21 Calculate the molar constant-pressure specific heat for O_2 and N_2 at 500, 1000, 1500, and 2000 K. Compare your results with those found in gas tables.

13.22 Consider 1 kg-mol of diatomic oxygen O_2 in thermodynamic equilibrium at 2500 K. Assuming no dissociation, determine:
(a) the percentages of molecules in the vibrational energy level ε_0, ε_2, and ε_5
(b) the enthalpy (including translational, rotational, and vibrational contributions)
(c) the specific heat at constant pressure (assume $\nu_{O_2} = 4.73 \times 10^{13}$ cycles/s)

13.23 Calculate the partition function, enthalpy, entropy, and Gibbs function for 1 mol of CH (molar mass = 13) at 2000 K and 1 atm.

13.24 Calculate the enthalpy, entropy, and Gibbs function of monatomic oxygen at a temperature of 1000 K and a pressure of 1 atm.

13.25 Determine the temperature at which the vibrational specific heat of N_2 reaches 90 percent of its classical value.

13.26 Calculate the rotational contributions to internal energy, entropy, and Gibbs function for 1 mol of diatomic nitrogen at 600 K.

13.27 Show that the vibrational contribution to entropy is independent of the selection of the zero-point energy ε_{v_0}.

13.28 The following table gives properties of CO and N_2:

	Molar Mass	Interatomic Distance (Å)	Classical Value $\overline{C}_v$ at 1 atm and 300 K	Entropy at 298 K (kJ/kg-mol K)
CO	28	1.128	$\frac{5}{2}\overline{R}$	197.653
N_2	28	1.094	$\frac{5}{2}\overline{R}$	191.611

Explain quantitatively the difference in the entropies of the two gases.

13.29 The entropy of carbon monoxide gas CO is 234.531 kJ/kg-mol K at 1000 K and 100 kPa. Check this value by determining the translational, rotational, and vibrational contribution to entropy ($\nu_{CO} = 6.49 \times 10^{13}$ cycles/s).

13.30 Determine the entropy of oxygen gas at 1000 K and 1 atm ($\nu_{O_2} = 4.73 \times 10^{13}$ cycles/s).

13.31 Prove that the entropy and Helmholtz function of a photon gas are given by the following expressions:

$$S = \left(\frac{32\pi^5 k^4 V}{45 c^3 h^3}\right) T^3$$

$$F = -\left(\frac{8\pi^5 k^4 V}{45 c^3 h^3}\right) T^4$$

13.32 Using the thermodynamic relation $p = -(\partial F/\partial V)_T$, calculate the radiation pressure of a photon gas in a vessel at a temperature of 1000 K. What is the relation between p and U/V?

13.33 Prove for a photon gas in equilibrium that the energy, the entropy per unit volume, and the pressure are given by

$$u = \frac{U}{V} = aT^4$$

$$s = \frac{S}{V} = \frac{4}{3}aT^3$$

$$p = \frac{1}{3}aT^4$$

where

$$a = \frac{8\pi^5 k^4}{15 h^3 c^3}$$

13.34 Noting that a photon gas can be treated as a system of molecules, prove that the number of photons striking a unit area per unit time (photon flux) is

$$J_n = \frac{1}{4} n c$$

and the energy flux is

$$J_u = \frac{c}{4} u = \sigma T^4$$

where n and u are the number and energy of photons per unit volume, c is the speed of light, and $\sigma = 2\pi^5 k^4 / 15 h^3 c^2$.

13.35 Prove that the specific heat at constant volume of a photon gas is given by

$$c_v = \frac{(2\pi)^5 k^4 V}{15 (ch)^3} T^3$$

13.36 Determine the radiant energy per unit volume of a blackbody at 1000 K. What fraction of this energy is emitted within the frequency range $0.95 \rightarrow 1.05 \nu_{max}$?

13.37 Calculate the specific heat at constant volume of aluminum at 10 and 900 K ($\theta_D = 398$ K).

13.38 Calculate the specific heat at constant volume of copper at 50 and 500 K.

13.39 Using the Debye theory of specific heats, determine the entropy change of a solid as a function of temperature, and prove that $\Delta s = c_v / 3$.

13.40 Using the Debye model for a solid, prove that

$$\lim_{T \to \infty} \bar{c}_v = 3 k N_a = 3 \bar{R}$$

13.41 Gold has a Debye temperature of 180 K. What is the constant-volume specific heat of gold at 300 K?

Compare your calculated value with that shown in Figure 13.6.

13.42 The Debye temperature of diamond is 1860 K. What is the molar constant-volume specific heat of diamond at 300 K? Compare your calculated value with that shown in Figure 13.6.

13.43 Compare the values of the constant-volume specific heat of diamond at 200 K using the Einstein and Debye equations. The characteristic Einstein temperature is $\theta_E = 1220$ K, and the characteristic Debye temperature is $\theta_D = 1860$ K.

13.44 Noting that c_v of a solid is proportional to the number of excited vibrational modes, show that at the low temperature this number in the frequency range ν to $\nu + d\nu$ is proportional to

$$\left(\frac{\theta_v}{T} \right)^2 e^{-(\theta_v / T)}$$

At what frequency does the maximum of this function occur? What is the limit of this frequency as the temperature approaches absolute zero?

13.45 Nitrogen gas at 300 K is mixed with oxygen gas also at 300 K in the proportions of 79 to 21 by volume to form 1 mol of air. Determine the internal energy and increase in entropy due to mixing. Check your results by classical thermodynamic laws.

C13.1 Write a computer program to determine the specific heat at constant volume of silver using the Einstein statistical model. Plot $\bar{c}_v$ versus T for temperatures ranging from 10 to 2000 K using the frequency ν_E as the parameter for a family of curves. Use the frequency values within the range of $0.1 \nu_{max} < \nu_{max} < 10 \nu_{max}$. The maximum frequency can be calculated from Wien's displacement law. Compare your result with the published value, and discuss any discrepancies.

C13.2 Write a computer program to determine the specific heat at constant volume of silver using the Debye statistical model. Plot the $\bar{c}_v$ versus T for temperatures ranging from 10 to 2000 K using the frequency v_D as the parameter for a family of curves. Use the frequency values within the range of $0.1v_{max} < v_{max} < 10v_{max}$. The maximum frequency can be calculated from Wien's displacement law. Do not use the approximate form of the Debye equation. You may replace the $N_a k$ in the Debye equation by $\overline{R}$. Compare your result with the published value, and discuss any discrepancies. Can you confirm the limiting value of the specific heat, that is, $3\overline{R}$, as the temperature tends to infinity?

C13.3 Write a computer program to calculate the specific heat at constant volume of silver using the Einstein and Debye statistical models. Use the frequencies $v_E = v_D = v_{max}$, where the maximum frequency can be calculated from Wien's displacement law. Plot $c_v/\overline{R}$ versus θ/T, and compare the two models. Do not use the approximate form of the Debye equation. You may replace the $N_a k$ in the Debye equation by $\overline{R}$.

C13.4 Write a computer program to calculate the molar constant-pressure specific heat for O_2 and N_2 at temperatures ranging from 200 to 4000 K. Plot the specific heat values against the temperature. Compare your results with those found in gas tables.

Chapter 14

Introduction
to Irreversible
Thermodynamics

14.1 Introduction

Classical thermodynamics and statistical mechanics treat systems as though they exist only in equilibrium states. They presume that the end states of a system when it interacts with its surroundings is in thermodynamic or statistical equilibrium without concern about the nature of the interaction or the time rates at which mass and energy transfers occur during the course of changing the state. This chapter discusses the nonequilibrium stages that occur when a system changes state in the course of a process.

Classical thermodynamics provides no information about the rate of change or the rate of flow of energy and mass. These effects take place as a result of spatial variations in the intensive properties throughout the system. To determine the rates of flow and the driving forces causing these flows, the extension of the methods of thermodynamics to include irreversible processes is needed.

Irreversible processes may be examined by two approaches. The first is to analyze the microscopic behavior of the particles using statistical averaging methods. Alternatively, the energy flow and the mass flow may be studied macroscopically, considering driving forces and fluxes in conjunction with appropriate transport properties of the system. This latter approach is called *irreversible thermodynamics*. Although the subject is in its development stage, it has proved particularly useful in explaining irreversible transport phenomena. When two or more flows are simultaneously involved, irreversible thermodynamics shows how these flows are related to each other.

Before studying nonequilibrium systems, it is necessary to resolve an ambiguity that exists when the properties of these systems are defined. Suppose heat is flowing through a metal bar, and it is required that the temperature gradient in the bar be measured. The flow of heat may be steady so that no changes take place as a function of time; nevertheless, a nonequilibrium state still exists because there is, in the dimension of space, a thermal gradient along the bar. Since temperature is a meaningful concept only for a system in thermal equilibrium, a temperature-sensing device placed at some point along the bar will give a meaningless reading. However, if a small region is first isolated and its temperature measured when thermal equilibrium is attained, the result then has meaning. If this is repeated at several points along the bar local temperatures can be determined and a spacial distribution of temperature can be established for this nonequilibrium state. In this sense, classical thermodynamics does not acknowledge that heat interaction is inherently a nonequilibrium process. Similar procedures may be applied when other thermodynamic properties are measured, and in all cases the existence of local thermodynamic equilibrium is implied when any property is measured.

Methods of applying irreversible thermodynamics to chemically nonreactive nonequilibrium systems are discussed in this chapter. In addition, laws dealing with energy and mass transport phenomena are formulated.

14.2 Single Phenomenological Flows

Transport or flow phenomena can be described by means of equations that relate *cause* (driving force) with *effect* (flux of energy or mass). The *flux of flow* is defined as the amount of flow per unit area in a unit time. Flow in which only one flux, consisting of either energy or mass, is involved is called *single flow*. When two or more fluxes occur simultaneously, they interfere with each other, and the flow is called *coupled flow*. When coupled flow occurs but one flux predominates, the flow can be treated as though it were a single flow. Temperature gradients in a metal bar, for example, give rise to conductive heat transfer; in addition, they generate voltage and electron concentration gradients in the bar. However, the flow of matter due to these gradients is negligible compared to the primary flow of heat. Similarly, an imposed voltage difference in a conductor causes heat flow resulting in a temperature gradient along the conductor. Fourier noted in 1811 that the heat flow in a homogeneous solid is directly proportional to the temperature gradient. For one-dimensional flow, the heat flux J_Q due to conduction, according to Fourier, is

$$J_Q = -K_t \frac{\partial T}{\partial x} \tag{14.1}$$

where K_t is the thermal conductivity. The negative sign is used to indicate that the heat flow is in the opposite direction to the temperature gradient.

Similarly, when an electrical potential is applied across a resistor, a temperature gradient develops in the resistor, but the flow of electrons caused solely by the temperature gradient is negligible. The current flux or the flux of electrons J_I depends linearly on the potential gradient and is given with sufficient accuracy by *Ohm's law* as

$$J_I = -K_e \frac{\partial \mathcal{E}}{\partial x} \tag{14.2}$$

where K_e is the electrical conductivity.

A third example of single flow is the diffusion process. The diffusion of one species of matter into another is linearly dependent on the concentration gradients. The mass flux due to diffusion J_m, as given by *Fick's law*, is

$$J_m = -D \frac{\partial c}{\partial x} \tag{14.3}$$

where D is the thermal diffusion coefficient. Equation (14.3) assumes that mass flow due to voltage gradients or temperature gradients is negligible.

14.3 Entropy Generation

Differences in intensive properties set up driving forces that cause the flow of matter and energy. As a result, entropy increases until all gradients of intensive properties are zero. In this section, we establish a relation between entropy generation and the mass and energy fluxes. Entropy changes occur in a system due to the interaction of the system with the surroundings and due to irreversible processes that occur internally. These changes may be expressed as

$$dS = dS_{ext} + dS_{int} \tag{14.4}$$

where the subscripts ext and int refer to external and internal effects, respectively. Changes in entropy that arise from external sources may be due to mass transfer or heat interaction with the surroundings. Such changes can be either positive or negative, depending on the direction of the process. The internal contribution of entropy dS_{int} is of particular interest since it indicates the dispersion of energy resulting from internal nonequilibrium processes. According to the second law, dS_{int} is either positive or zero but can never be negative. The change of internal entropy, or the rate of generation of internal entropy, serves as a gauge when comparing irreversible processes. In addition, the rate of entropy generation per unit volume can be related to the driving forces and fluxes of mass and

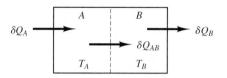

Figure 14.1   Heat interaction between two systems.

energy. Consider, for example, two systems A and B of fixed composition. Let δQ_A and δQ_B represent the heat interaction between systems A and B with the surroundings during an interval of time dt. During the same interval of time, the heat interaction between the two systems is Q_{AB}. As shown in Figure 14.1, if the temperature of each system remains unchanged, an energy balance for system A is

$$\delta Q_{A\,net} = \delta Q_A - \delta Q_{AB}$$

whereas for system B the energy balance is

$$\delta Q_{B\,net} = \delta Q_{AB} - \delta Q_A$$

The change of entropy of system A is

$$dS_A = \frac{\delta Q_A}{T_A} - \frac{\delta Q_{AB}}{T_A}$$

and of system B,

$$dS_B = -\frac{\delta Q_B}{T_B} + \frac{\delta Q_{AB}}{T_B}$$

The change of entropy of the combined systems is

$$dS = dS_A + dS_B = \left(\frac{\delta Q_A}{T_A} - \frac{\delta Q_B}{T_B}\right) + \delta Q_{AB}\left(-\frac{1}{T_A} + \frac{1}{T_B}\right)$$

The entropy change due solely to heat interaction with the surroundings is therefore

$$(dS_{AB})_{ext} = \frac{\delta Q_A}{T_A} - \frac{\delta Q_B}{T_B}$$

whereas the entropy change due to internal effects, resulting from heat interaction between systems A and B is

$$(dS_{AB})_{int} = \delta Q_{AB}\left(-\frac{1}{T_A} + \frac{1}{T_B}\right) = \delta Q_{AB}\left(\frac{T_A - T_B}{T_A T_B}\right) > 0$$

This entropy generation is associated with the irreversible heat interaction between systems A and B. Note that if $T_A > T_B$ and $\delta Q_{AB} > 0$, then $(dS_{AB})_{int} > 0$. On the other hand, if $T_A < T_B$, then $\delta Q_{AB} < 0$, and so $(dS_{AB})_{int}$ is again > 0.

The rate of generation of internal entropy is

$$\frac{(dS_{AB})_{int}}{dt} = \frac{\delta Q_{AB}}{dt}\left(\frac{1}{T_B} - \frac{1}{T_A}\right) > 0$$

If T_A and T_B differ by only a small amount, this equation becomes

$$\frac{dS_{int}}{dt} = \frac{\delta Q}{dt}\frac{dT}{T^2}$$

From this equation, the rate of entropy generation per unit volume, also called the dispersion function σ, may be obtained:

$$\sigma = \frac{\delta Q}{A\,dx\,dt}\frac{dT}{T^2} = -J_Q\frac{1}{T^2}\frac{dT}{dx} = J_Q\frac{d(1/T)}{dx} \tag{14.5}$$

where $A\,dx$ represents volume, and J_Q is the heat flux. The negative sign indicates that heat flow takes place in the direction of the negative temperature gradient. Note that the flow of heat is due to an imbalance in the intensive property $1/T$.

In a similar way, consider the entropy generation due to flow of matter through a semipermeable membrane. As shown in Figure 14.2, a fluid consisting of several species flows steadily and isothermally through a distance dx. A quantity of heat, δQ, must be transferred in order to maintain the isothermal flow. It is assumed that no phase changes occur in the fluid; it is also assumed that such properties as enthalpy and chemical potential vary infinitesimally across the distance dx. The change of entropy then is caused by heat transfer and by internal irreversible processes, which can be expressed as

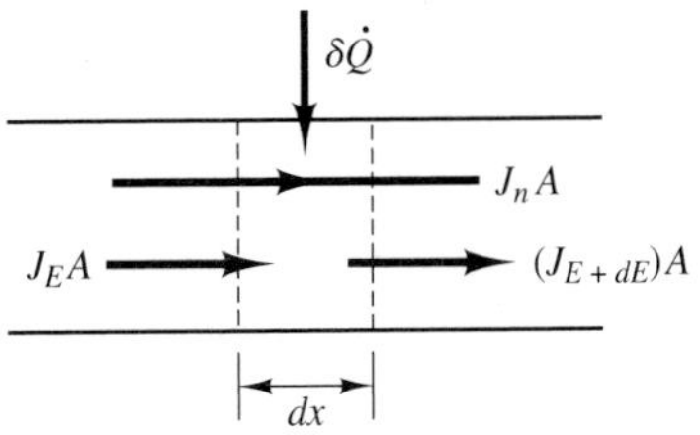

$$dS = dS_{int} + \frac{\delta Q}{T}$$

Rearranging,

$$T\,dS_{int} = T\,dS - \delta Q \tag{14.6}$$

Figure 14.2 Isothermal flow through a semipermeable membrane.

But the first law may be applied to this isothermal, steady-state process to give

$$\delta Q + \sum_i \left[H_i - \left(H_i + \frac{\partial H_i}{\partial x} dx \right) \right] = 0$$

or

$$\delta Q = \sum_i \frac{\partial H_i}{\partial x} dx \tag{14.7}$$

Enthalpy can be expressed, at constant temperature, in terms of the Gibbs free energy as

$$dH = dG + T\ dS$$

But the Gibbs free energy at constant temperature and pressure can be expressed as a function of chemical potential, according to the equation:

$$dG = \sum_i \left(\frac{\partial \mu_i}{\partial x} \right) dx\ dn_i$$

where μ_i and n_i are the chemical potential and the number of moles of the ith species. Therefore,

$$\frac{\partial H_i}{\partial x} dx = \sum_i \left(\frac{\partial \mu_i}{\partial x} \right) dx\ dn_i + T\ dS$$

The first law (Eq. 14.7) can now be expressed as

$$\delta Q = \sum_i \left(\frac{\partial \mu_i}{\partial x} \right) dx\ dn_i + T\ dS$$

Hence,

$$T\ dS_{\text{int}} = T\ dS - \sum_i \left(\frac{\partial \mu_i}{\partial x} \right) dx\ dn_i - T\ dS$$

or

$$T\ dS_{\text{int}} = -\sum_i \left(\frac{\partial \mu_i}{\partial x} \right) dx\ dn_i$$

The rate of internal entropy generation per unit volume can therefore be expressed as

$$\sigma = \frac{dS_{\text{int}}}{A\ dx\ dt} = -\frac{1}{T} \sum_i \left(\frac{\partial \mu_i}{\partial x} \right) \frac{dn_i}{A\ dt}$$

Since $dn_i/A\,dt$ is the molar flux J_{n_i},

$$\sigma = -\frac{1}{T}\sum_i J_{n_i}\frac{\partial \mu_i}{\partial x} = \sum_i J_{n_i}\frac{d(-\mu_i/T)}{\partial x} \geq 0 \qquad (14.8)$$

Note that the flux J_{n_i} is a result of an imbalance in μ_i/T.

It is clear from these two cases that the rate of internal entropy generation per unit volume is the sum of products of the flux and the conjugate driving force. When several fluxes and driving forces are involved, this may be expressed as follows:

$$\sigma = \sum_i J_i X_i \geq 0 \qquad (14.9)$$

where J_i is either the energy flux or the mass flux, and X_i is a function of a potential gradient such as a temperature gradient or a chemical potential gradient.

14.4 Relation Between Flux and Driving Force

In relatively slow processes, the flux of energy or matter J_i is a linear function of the gradient of a potential φ. That is,

$$J_i \propto \frac{\partial \varphi_i}{\partial x}$$

where the subscript i identifies the ith irreversible process. But the gradient of the potential φ also represents the driving force X_i so that

$$X_i = \frac{\partial \varphi_i}{\partial x} \qquad (14.10)$$

Consequently, the flux J_i is proportional to the driving force X_i:

$$J_i = L_i X_i \qquad (14.11)$$

where L_i is a coefficient of proportionality. It is a scalar quantity that depends on the properties of the system.

When two or more flows occur simultaneously in a system, questions arise about relationships between the fluxes and the driving forces. Lars Onsager in 1931 assumed that the flux of each flow bears a linear relationship not only to the conjugate force but also to the other forces to which the system is subjected.

Therefore, in a system consisting of n coupled flows, the ith flux is given by a set of linear equations of the form

$$J_i = L_{i1}X_1 + L_{i2}X_2 + L_{i3}X_3 + \cdots = \sum_{j=1}^{n} L_{ij}X_j \qquad (14.12)$$

Thus, for a three-flow system, the flow rates may be expressed as

$$J_1 = L_{11}X_1 + L_{12}X_2 + L_{13}X_3$$
$$J_2 = L_{21}X_1 + L_{22}X_2 + L_{23}X_3$$
$$J_3 = L_{31}X_1 + L_{32}X_2 + L_{33}X_3$$

In coupled flows, then, the ith flux J_i depends linearly on the conjugate driving force X_i through its *primary* coefficient L_{ii}, and also on the other driving forces, through the coupling coefficients L_{ij}. The coefficients L_{ij} are called the *Onsager phenomenological coefficients*. Onsager assumed that the flux depends on the first power of the driving forces. This is not always true. However, this linear relationship occurs in many phenomenological flows as long as the systems do not deviate markedly from thermodynamic equilibrium.

The coupling of thermomechanical flows can be illustrated by referring to a system in which convective heat transfer occurs. Fluid motion results from potential differences in temperature and in pressure; also, the fluid transports internal energy between source and sink. Under nonequilibrium steady-state flow conditions, all potentials are in dynamic balance, and the transfer of heat takes place at a steady rate. On the other hand, under equilibrium conditions, all thermodynamic potentials are zero and all fluxes vanish. From Eq. (14.12), the mass transfer and energy transfer are

$$J_m = L_{11}X_m + L_{12}X_E$$
$$J_E = L_{21}X_m + L_{22}X_E$$

where J_m represents the mass flux, and J_E represents the energy flux. The thermoelectric effect, which will be discussed in Section 14.6, is an example of a flow in which heat and electrical energy are coupled.

An expression for the rate of entropy generation per unit volume in coupled flows can be obtained by combining Eqs. (14.9) and (14.12):

$$\sigma = \sum_i \sum_j L_{ij}X_iX_j \geq 0 \qquad (14.13)$$

When only two flows are involved, the rate of entropy generation is

$$\sigma = L_{11}X_1^2 + (L_{12} + L_{21})X_1X_2 + L_{22}X_2^2 \geq 0 \qquad (14.14)$$

Since the primary coefficients L_{11} and L_{22} must be positive,[1] it follows that

$$L_{11}L_{22} - \left(\frac{L_{12} + L_{21}}{2}\right)^2 \geq 0 \qquad (14.15)$$

14.5 Onsager Reciprocity Relation

Although the linear relationship between fluxes and flows greatly simplifies the treatment of irreversible transport phenomena, the problem is still not solved. The flux equations require a large number of coefficients. To evaluate these coefficients, a large number of experiments must be performed. Furthermore, these mathematical equations cannot be solved easily. The difficulty is reduced if it can be shown that relationships exist between the various transport coefficients. Onsager postulated that the transport coefficients in coupled flows are symmetrical so that

$$L_{ij} = L_{ji} \qquad (14.16)$$

Equation (14.16) is called the *Onsager reciprocity relation*. It applies in the absence of a magnetic field; it also applies, but in modified form, if a magnetic field is present. When Eq. (14.16) is used, the number of independent constants required to evaluate an irreversible phenomena is reduced. Coefficients that are difficult to measure are calculated through Eq. (14.16) from those that are more readily determined.

Proof of the Onsager reciprocity relation based on statistical-mechanical analysis is available in the literature.[2] The Onsager theorem has been very successful in describing many irreversible phenomena, especially those involving gases. By combining the Onsager reciprocity relation with linear superposition of forces and fluxes, many problems involving irreversible phenomena have been solved. However, the validity of the Onsager relation is still questioned, and further investigation is needed. Furthermore, a more general theory is needed. Like the laws of thermodynamics, the Onsager postulate can be considered a law of nature for it shows the relationships that exist in certain physical phenomena.

---

[1] This may be shown by setting all gradients except one $L_{ij}(i \neq j)$ in Eq. (14.13) to be equal to zero.
[2] See Callen, H. B., *Thermodynamics,* New York: John Wiley & Sons, Inc., 1960, Chapters 16, 17.

14.6 Thermoelectric Phenomena

Methods of irreversible thermodynamics are readily applicable to the analysis of thermoelectric phenomena. In this coupled-flow phenomenon, the simultaneous flows are considered to be linearly superimposed, and the coefficients associated with the superposition are symmetric. As shown in Section 11.13, an electrical potential gradient in a conductor causes heat flow, while a temperature gradient in the conductor causes a flow of electrical current. These two irreversible flows are linearly coupled by the following equations:

$$J_I = -L_{11} \frac{1}{T} \frac{d\mathcal{E}}{dx} - L_{12} \frac{1}{T^2} \frac{dT}{dx} \tag{14.17}$$

$$J_Q = -L_{21} \frac{1}{T} \frac{d\mathcal{E}}{dx} - L_{22} \frac{1}{T^2} \frac{dT}{dx} \tag{14.18}$$

Note that current flow in a conductor can exist when $dT/dx \neq 0$ even if $d\mathcal{E}/dx = 0$. Similarly, heat flux can exist due to current flow even if $dT/dx = 0$. In the absence of a temperature gradient,

$$J_I = -L_{11} \frac{1}{T} \frac{d\mathcal{E}}{dx} \tag{14.19}$$

The coefficients, L_{ij}, of Eqs. (14.17) and (14.18) are related to the transport properties of the conducting solid. According to Ohm's law, the isothermal electrical conductivity K_e is given by

$$J_I = -K_e \frac{d\mathcal{E}}{dx} \tag{14.20}$$

By comparing the two preceding equations, the primary coefficient L_{11} and the electrical conductivity K_e are shown to be related as follows:

$$K_e = \frac{L_{11}}{T} \tag{14.21}$$

Similarly, in the absence of electric current flow, thermal conductivity K_t is related to heat flux through Fourier's equation:

$$J_Q = -K_t \frac{dT}{dx} \tag{14.22}$$

When there is no current flow, J_I is zero, and Eq. (14.17) becomes

$$\left(\frac{d\mathcal{E}/dx}{dT/dx}\right)_{J_I=0} = \left(\frac{d\mathcal{E}}{dT}\right)_{J_I=0} = -\frac{L_{12}}{TL_{11}} \qquad (14.23)$$

From Eq. (14.23), it is clear that an electric potential gradient develops in a body if a temperature gradient exists, even if no electric current flows. This phenomenon is the *Seebeck effect,* discussed in Chapter 11.

The ratio $-d\mathcal{E}/dT$, denoted by ε, is called the *thermoelectric power* of the conductor or the *absolute Seebeck coefficient.* It is a function of temperature. By combining Eqs. (14.21) and (14.23), the coupling coefficient L_{12} may be expressed as follows:

$$L_{12} = \varepsilon T L_{11} = \varepsilon K_e T^2 \qquad (14.24)$$

When Eq. (14.23) is combined with Eq. (14.18), the heat flux can be expressed as

$$J_Q = -\frac{L_{11}L_{22} - L_{12}^2}{L_{11}T^2}\frac{dT}{dx} \qquad (14.25)$$

From the heat flux, Eqs. (14.22) and (14.25), an expression for thermal conductivity is obtained:

$$K_t = \frac{L_{11}L_{22} - L_{12}^2}{L_{11}T^2} \qquad (14.26)$$

This leads to an expression for the L_{22} coefficient as a function of conductivities and thermoelectric power:

$$L_{22} = T^2(K_t + TK_e\varepsilon^2) \qquad (14.27)$$

where both K_e and K_t must, necessarily, have positive values. The L_{ij} coefficients have now been expressed as functions of the current flux; heat flux may also be described in terms of the transport properties:

$$J_I = -K_e\frac{d\mathcal{E}}{dx} - \varepsilon K_e\frac{dT}{dx} \qquad (14.28)$$

and

$$J_Q = -\varepsilon T K_e\frac{d\mathcal{E}}{dx} - (K_t + \varepsilon^2 K_e T)\frac{dT}{dx} \qquad (14.29)$$

By eliminating $d\mathcal{E}/dx$ from Eqs. (14.28) and (14.29), the heat flux becomes

$$J_Q = -K_t \frac{dT}{dx} + \varepsilon T J_I \tag{14.30}$$

Similarly, by eliminating dT/dx, the current flux becomes

$$J_I = \left(\frac{\varepsilon^2 K_e T}{K_t + K_e T} - K_e \right) \frac{d\mathcal{E}}{dx} + \left(\frac{\varepsilon K_e}{K_t + \varepsilon K_e T} \right) J_Q \tag{14.31}$$

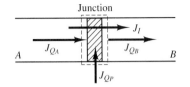

Figure 14.3   Peltier effect.

In addition to the Fourier heat effect and the Seebeck effect, three other phenomena occur when an electric current flows through a junction of two dissimilar metals: (a) the Peltier effect, (b) the Thomson effect, and (c) Joule heat.

When an electric current flows through the junction of two dissimilar metals, the temperature of the junction will change unless heat is transferred to or from the junction. This phenomenon is illustrated in Figure 14.3 and is known as the *Peltier effect*. If isothermal conditions are maintained, Eq. (14.28) becomes

$$J_I = -K_e \frac{d\mathcal{E}}{dx}$$

The heat flux through each metal, according to Eq. (14.29), is

$$J_Q = -\varepsilon T K_e \frac{d\mathcal{E}}{dx}$$

The heat flux through each of the metals forming the junction can be expressed as a function of the current flux:

$$J_{Q_A} = T \varepsilon_A J_I$$
$$J_{Q_B} = T \varepsilon_B J_I$$

The net heat flux to the junction due to the Peltier effect is therefore

$$J_{Q_P} = J_{Q_B} - J_{Q_A} = T(\varepsilon_B - \varepsilon_A) J_I = \pi_{AB} J_I \tag{14.32}$$

The *Peltier coefficient*, π_{AB}, indicates the energy transfer as heat per unit electric current flow and is defined by

$$\pi_{AB} = T(\varepsilon_B - \varepsilon_A) \tag{14.33}$$

Consider next a conductor in which a temperature gradient exists initially. When electric current flows through the conductor, heat must be transferred in order to maintain the original temperature gradient. The direction of heat flow depends on the direction of current flow and on the temperature gradient. This phenomenon is known as the *Thomson effect*. As shown in Figure 14.4, the energy flux J_E is the sum of the heat flux and the electrical power supplied per unit area, or

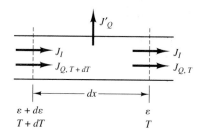

Figure 14.4   Thomson effect.

$$J_E = J_Q + \mathcal{E}J_I$$

Substituting for the heat flux from Eq. (14.30), the energy flux becomes

$$J_E = -K_t \frac{dT}{dx} + (\varepsilon T + \mathcal{E})J_I \qquad (14.34)$$

But the gradient of the energy flux in the x-direction also represents the heat transferred from the conductor:

$$J'_Q = -\frac{\partial}{\partial x}(J_E)$$

By substituting for J_E from Eq. (14.34) into this equation, and noting that there is no change in temperature gradient in the x-direction, the heat transferred from the conductor becomes

$$J'_Q = -\frac{\partial}{\partial x}\left[-K_t \frac{dT}{dx} + (\varepsilon T + \mathcal{E})J_I\right]$$
$$= -\varepsilon J_I \frac{\partial T}{\partial x} - TJ_I \frac{d\varepsilon}{dx} - J_I \frac{d\mathcal{E}}{dx}$$

After substituting for the potential gradient $d\mathcal{E}/dx$ from Eq. (14.28), the preceding equation becomes

$$J'_Q = -TJ_I \frac{d\varepsilon}{dx} + \frac{J_I^2}{K_e} = -TJ_I \frac{d\varepsilon}{dT}\frac{dT}{dx} + \frac{J_I^2}{K_e} \qquad (14.35)$$

The last term in this equation represents *Joule heat* due to electric current flow in the absence of a temperature gradient. This heat is equal to the cooling required to maintain the temperature of the conductor constant. The first term is called *Thomson heat* and represents the additional heating due to current flow through a temperature gradient; it is given by

$$J_{Q\tau} = -TJ_I \frac{d\varepsilon}{dT}\frac{dT}{dx} = \tau J_I \frac{dT}{dx} \qquad (14.36)$$

The *Thomson coefficient*, τ, indicates the energy transfer as heat per unit electric current flow and per unit temperature gradient, and is defined by

$$\tau = T\frac{d\varepsilon}{dT} \tag{14.37}$$

The total heat flux then becomes

$$J_Q' = -\tau J_I\frac{dT}{dx} + \frac{J_I^2}{K_e} \tag{14.38}$$

Differentiating Eq. (14.33) with respect to temperature and using Eq. (14.37) leads to the following expression:

$$\frac{d\pi_{AB}}{dT} = (\varepsilon_B - \varepsilon_A) + (\tau_B - \tau_A) \tag{14.39}$$

Equations (14.33) and (14.39), which are called the *Kelvin relations*, were derived by Kelvin in 1854 from energy considerations alone. Note that the Seebeck, Peltier, and Thomson effects are coupled in Eq. (14.39).

Example 14.1

It is required to determine a temperature T using the thermocouple shown in Figure 14.5. The thermocouple is made of two dissimilar metals A and B. The connectors C and D are of the same materials and are maintained at the same temperature. One junction of the thermocouple is maintained at a *reference temperature T* (0°C).

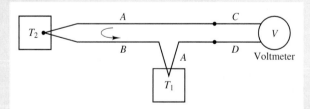

Figure 14.5   Thermocouple.

SOLUTION

Since there is no current flow, Eq. (14.23) indicates that the voltage developed is a function of temperature only so that

$$\left(\frac{d\mathcal{E}}{dT}\right)_{J_l=0} = -\varepsilon(T)$$

where ε is the thermoelectric power.

Integration of this equation for the three conductors A, B, and A gives

$$\mathcal{E}_C - \mathcal{E}_2 = -\int_{T_2}^{T_C} \varepsilon_A(T)dT$$

$$\mathcal{E}_2 - \mathcal{E}_1 = -\int_{T_1}^{T_2} \varepsilon_B(T)dT$$

$$\mathcal{E}_1 - \mathcal{E}_D = -\int_{T_D}^{T_1} \varepsilon_A(T)dT$$

Adding these three equations and noting that $T_C = T_D$, then

$$\mathcal{E}_C - \mathcal{E}_D = \int_{T_1}^{T_2} [\varepsilon_A(T) - \varepsilon_B(T)]dT$$

The voltage developed is a function of the temperature T_1 and T_2, and the materials of the thermocouple. If T_1 is a known (reference) temperature, T_2 can be determined. The output of thermocouples is in the millivolt range and is usually measured by a high-impedance potentiometer or by a dc millivoltmeter. Note that the resistance of the lead wires C and D is of no consequence when a potentiometer is used since no current flows in the thermocouple circuit.

As an example, for the copper–iron thermocouple, $(\mathcal{E}_C - \mathcal{E}_D)$ may be calculated as follows:

$$\varepsilon_{Cu} = 1.34 + 0.0094T \ (\mu V/°C)$$
$$\varepsilon_{Fe} = 17.15 - 0.0482T \ (\mu V/°C)$$

so that

$$\varepsilon_{Cu\text{-}Fe} = -15.81 + 0.0576T$$

If one junction is kept at the ice point (0°C), the voltage generated is

$$\Delta\mathcal{E} = \int_0^T (-15.81 + 0.0576T)dT = \left[-15.81T + 0.0288T^2\right]_0^T$$

If the other junction is at the boiling point of water (100°C),

$$\Delta\mathcal{E} = -1293 \ \mu V$$

Appendix

Table A1   Physical Constants

Speed of light in vacuum	$c = 2.9979 \times 10^8$ m/s
Standard gravitational acceleration	$g = 9.80665$ m/s^2 = 32.174 ft/s^2
Standard atmosphere	1 atm = 101.325 kPa
Avogadro's number	$N_a = 6.022045 \times 10^{26}$ molecules/kg-mol
Planck's constant	$h = 6.626176 \times 10^{-34}$ J s/molecule
Boltzmann's constant	$k = 1.380662 \times 10^{-23}$ J/(K molecule)
Electron rest mass	$m_e = 9.1093 \times 10^{-31}$ kg
Electronic charge	$e = 1.602189 \times 10^{-19}$ coul
Electron volt	$1eV = 1.60209 \times 10^{-19}$ J
Classical electron radius	$r_e = 2.8179 \times 10^{-15}$ m
Universal gas constant	$\overline{R} = 8.3144$ kJ/kg-mol K
	= 0.082056 liter atm/gm-mol K
	= 1545.33 ft lbf/lbm-mol R
	= 1.98586 cal/gm-mol K
	= 1.98586 Btu/lbm-mol R

Table A2   Conversion Factors*

Length	1 m = 10^2 cm = 10^3 mm = 3.281 ft = 39.37 in = 10^{10} Å
	1 cm = 10^{-2} m = 10 mm = 0.03281 ft = 0.3937 in = 10^8 Å
	1 ft = 12 in = 0.3048 m = 30.48 cm = 304.8 mm = 3.048×10^9 Å
	1 in = 0.08333 ft = 0.0254 m = 2.540 cm = 25.40 mm = 2.540×10^8 Å
Mass	1 kg = 1000 gm = 2.2046 lbm
	1 lbm = 453.59 gm = 0.45359 kg
Volume	1 m^3 = 1000 liters = 35.315 ft^3 = 264.17 gal (U.S.)
	1 ft^3 = 0.028317 m^3
	1 gal = 0.13368 ft^3
	1 liter = 1000.0 cm^3 = 0.001 m^3 = 0.035315 ft^3
Density	1 kg/m^3 = 0.062428 lbm/ft^3
	1 lbm/ft^3 = 16.0185 kg/m^3
Force	1 N = 1 kg m/s^2 = 0.224809 lbf = 10^5 dynes
	1 lbf = 4.44822 N = 4.44822×10^5 dynes

(continued)

Table A2 *Continued*

Pressure	$1 \text{ Pa} = 1 \text{ N/m}^2 = 1 \times 10^{-5} \text{ bar} = 1.4504 \times 10^{-4} \text{ psi} = 9.8692 \times 10^{-6} \text{ atm}$
	$1 \text{ bar} = 10^5 \text{ Pa} = 0.98692 \text{ atm} = 14.504 \text{ psi}$
	$1 \text{ lbf/in}^2 \text{ (psi)} = 144 \text{ lbf/ft}^2 = 6894.8 \text{ Pa} = 6.8948 \times 10^{-2} \text{ bar} = 0.068046 \text{ atm}$
	$1 \text{ atm} = 101.325 \text{ kPa} = 14.696 \text{ psi} = 1.01325 \text{ bar}$
Energy	$1 \text{ J} = 1 \text{ N m} = 1 \text{ kg m}^2/\text{s}^2 = 10^7 \text{ ergs}$
	$1 \text{ kJ} = 1 \text{ kW s} = 0.94783 \text{ Btu} = 0.23885 \text{ kcal} = 737.56 \text{ ft lbf}$
	$1 \text{ Btu} = 1.055056 \text{ kJ} = 0.25200 \text{ kcal} = 778.169 \text{ ft lbf}$
	$1 \text{ kcal} = 4.1868 \text{ kJ} = 3.9684 \text{ Btu} = 3088.0 \text{ ft lbf}$
	$1 \text{ kWh} = 3.60 \times 10^3 \text{ kJ} = 2655.2 \times 10^3 \text{ ft lbf} = 3412.2 \text{ Btu} = 859.85 \text{ kcal}$
	$1 \text{ ft lbf} = 1.2851 \times 10^{-3} \text{ Btu} = 1.3558 \times 10^{-3} \text{ kJ}$
Power	$1 \text{ W} = 1 \text{ J/s} = 1 \text{ kg m}^2/\text{s}^3$
	$1 \text{ W} = 3.4122 \text{ Btu/h} = 0.85987 \text{ kcal/h} = 1.34102 \times 10^{-3} \text{ hp} = 0.73756 \text{ ft lbf/s}$
	$1 \text{ Btu/h} = 0.29307 \text{ W} = 0.25200 \text{ kcal/h} = 3.9300 \times 10^{-4} \text{ hp} = 0.21616 \text{ ft lbf/s}$
	$1 \text{ kcal/h} = 1.1630 \text{ W} = 3.9683 \text{ Btu/h} = 1.5595 \times 10^{-3} \text{ hp} = 0.85778 \text{ ft lbf/s}$
	$1 \text{ horsepower (hp)} = 550 \text{ ft lbf/s} = 2544.5 \text{ Btu/h} = 745.70 \text{ W}$
	$1 \text{ ton (cooling capacity)} = 12,000 \text{ Btu/h} = 3.5168 \text{ kW}$
Specific energy	$1 \text{ kJ/kg} = 0.42992 \text{ Btu/lbm} = 0.23885 \text{ kcal/kg} = 334.55 \text{ ft lbf/lbm}$
	$1 \text{ Btu/lbm} = 2.3260 \text{ kJ/kg} = 0.55556 \text{ kcal/kg} = 778.16 \text{ ft lbf/lbm}$
	$1 \text{ kcal/kg} = 4.1868 \text{ kJ/kg} = 1.8000 \text{ Btu/lbm} = 1400.7 \text{ ft lbf/lbm}$
	$1 \text{ ft lbf/lbm} = 2.9891 \times 10^{-3} \text{ Btu/lbm} = 7.1394 \times 10^{-4} \text{ kcal/kg}$
Entropy	$1 \text{ kJ/K} = 0.52657 \text{ Btu/R} = 0.23885 \text{ kcal/K}$
	$1 \text{ Btu/R} = 1.8991 \text{ kJ/K} = 0.45359 \text{ kcal/K}$
	$1 \text{ kcal/K} = 4.1868 \text{ kJ/K} = 2.2047 \text{ Btu/R}$
Specific entropy	$1 \text{ kJ/(kg K)} = 0.23885 \text{ Btu/(lbm R)} = 0.23885 \text{ kcal/(kg K)}$
	$1 \text{ Btu/(lbm R)} = 4.1868 \text{ kJ/(kg K)} = 1.0000 \text{ kcal/(kg K)}$
	$1 \text{ kcal/(kg K)} = 4.1868 \text{ kJ/(kg K)} = 1.0000 \text{ Btu/(lbm R)}$
Temperature	$T \text{ K} = \frac{5}{9} T \text{ R} = \frac{5}{9} (T \text{ °F} + 459.67) = T \text{ °C} + 273.15$
	$T \text{ R} = \frac{9}{5} T \text{ K} = \frac{9}{5} (T \text{ °C} + 273.15) = T \text{ °F} + 459.67$
	$T \text{ °F} = \frac{9}{5} T \text{ °C} + 32$
	$T \text{ °C} = \frac{5}{9} (T \text{ °F} - 32)$
	$(\Delta T) \text{ K} = (\Delta T)° \text{ C}$

*All energy conversions based on International Steam Table values.

Table A3.1 Thermodynamic Properties of Saturated Water and Steam—Temperature Table

T (°C)	p (kPa)	Specific Volume (m³/kg)		Internal Energy (kJ/kg)			Enthalpy (kJ/kg)			Entropy (kJ/kg K)		
		v_f	v_g	u_f	u_{fg}	u_g	h_f	h_{fg}	h_g	s_f	s_{fg}	s_g
0.01	0.6113	0.001000	206.14	.00	2375.3	2375.3	.01	2501.3	2501.4	.0000	9.1562	9.1562
5	0.8721	0.001000	147.12	20.97	2361.3	2382.3	20.98	2489.6	2510.6	.0761	8.9496	9.0257
10	1.2276	0.001000	106.38	42.00	2347.2	2389.2	42.01	2477.7	2519.8	.1510	8.7498	8.9008
15	1.7051	0.001001	77.93	62.99	2333.1	2396.1	62.99	2465.9	2528.9	.2245	8.5569	8.7814
20	2.339	0.001002	57.79	83.95	2319.0	2402.9	83.96	2454.1	2538.1	.2966	8.3706	8.6672
25	3.169	0.001003	43.36	104.88	2304.9	2409.8	104.89	2442.3	2547.2	.3674	8.1905	8.5580
30	4.246	0.001004	32.89	125.78	2290.8	2416.6	125.79	2430.5	2556.3	.4369	8.0164	8.4533
35	5.628	0.001006	25.22	146.67	2276.7	2423.4	146.68	2418.6	2565.3	.5053	7.8478	8.3531
40	7.384	0.001008	19.52	167.56	2262.6	2430.1	167.57	2406.7	2574.3	.5725	7.6845	8.2570
45	9.593	0.001010	15.26	188.44	2248.4	2436.8	188.45	2394.8	2583.2	.6387	7.5261	8.1648
50	12.349	0.001012	12.03	209.32	2234.2	2443.5	209.33	2382.7	2592.1	.7038	7.3725	8.0763
55	15.758	0.001015	9.568	230.21	2219.9	2450.1	230.23	2370.7	2600.9	.7679	7.2234	7.9913
60	19.940	0.001017	7.671	251.11	2205.5	2456.6	251.13	2358.5	2609.6	.8312	7.0784	7.9096
65	25.03	0.001020	6.197	272.02	2191.1	2463.1	272.06	2346.2	2618.3	.8935	6.9375	7.8310
70	31.19	0.001023	5.042	292.95	2176.6	2469.6	292.98	2333.8	2626.8	.9549	6.8004	7.7553
75	38.58	0.001026	4.131	313.90	2162.0	2475.9	313.93	2321.4	2635.3	1.0155	6.6669	7.6824
80	47.39	0.001029	3.407	334.86	2147.4	2482.2	334.91	2308.8	2643.7	1.0753	6.5369	7.6122
85	57.83	0.001033	2.828	355.84	2132.6	2488.4	355.90	2296.0	2651.9	1.1343	6.4102	7.5445
90	70.14	0.001036	2.361	376.85	2117.7	2494.5	376.92	2283.2	2660.1	1.1925	6.2866	7.4791
95	84.55	0.001040	1.982	397.88	2102.7	2500.6	397.96	2270.2	2668.1	1.2500	6.1659	7.4159
100	101.35	0.001044	1.6729	418.94	2087.6	2506.5	419.04	2257.0	2676.1	1.3069	6.0480	7.3549
105	120.82	0.001048	1.4194	440.02	2072.3	2512.4	440.15	2243.7	2683.8	1.3630	5.9328	7.2958
110	143.27	0.001052	1.2102	461.14	2057.0	2518.1	461.30	2230.2	2691.5	1.4185	5.8202	7.2387
115	169.06	0.001056	1.0366	482.30	2041.4	2523.7	482.48	2216.5	2699.0	1.4734	5.7100	7.1833
	MPa											
120	0.19853	0.001060	0.8919	503.50	2025.8	2529.3	503.71	2202.6	2706.3	1.5276	5.6020	7.1296
125	0.2321	0.001065	0.7706	524.74	2009.9	2534.6	524.99	2188.5	2713.5	1.5813	5.4962	7.0775
130	0.2701	0.001070	0.6685	546.02	1993.9	2539.9	546.31	2174.2	2720.5	1.6344	5.3925	7.0269
135	0.3130	0.001075	0.5822	567.35	1977.7	2545.0	567.69	2159.6	2727.3	1.6870	5.2907	6.9777
140	0.3613	0.001080	0.5089	588.74	1961.3	2550.0	589.13	2144.7	2733.9	1.7391	5.1908	6.9299
145	0.4154	0.001085	0.4463	610.18	1944.7	2554.9	610.63	2129.6	2740.3	1.7907	5.0926	6.8833
150	0.4758	0.001091	0.3928	631.68	1927.9	2559.5	632.20	2114.3	2746.5	1.8418	4.9960	6.8379
155	0.5431	0.001096	0.3468	653.24	1910.8	2564.1	653.84	2098.6	2752.4	1.8925	4.9010	6.7935
160	0.6178	0.001102	0.3071	674.87	1893.5	2568.4	675.55	2082.6	2758.1	1.9427	4.8075	6.7502

T												
165	0.7005	0.001108	0.2727	696.56	1876.0	2572.5	697.34	2066.2	2763.5	1.9925	4.7153	6.7078
170	0.7917	0.001114	0.2428	718.33	1858.1	2576.5	719.21	2049.5	2768.7	2.0419	4.6244	6.6663
175	0.8920	0.001121	0.2168	740.17	1840.0	2580.2	741.17	2032.4	2773.6	2.0909	4.5347	6.6256
180	1.0021	0.001127	0.19405	762.09	1821.6	2583.7	763.22	2015.0	2778.2	2.1396	4.4461	6.5857
185	1.1227	0.001134	0.17409	784.10	1802.9	2587.0	785.37	1997.1	2782.4	2.1879	4.3586	6.5465
190	1.2544	0.001141	0.15654	806.19	1783.8	2590.0	807.62	1978.8	2786.4	2.2359	4.2720	6.5079
195	1.3978	0.001149	0.14105	828.37	1764.4	2592.8	829.98	1960.0	2790.0	2.2835	4.1863	6.4698
200	1.5538	0.001157	0.12736	850.65	1744.7	2595.3	852.45	1940.7	2793.2	2.3309	4.1014	6.4323
205	1.7230	0.001164	0.11521	873.04	1724.5	2597.5	875.04	1921.0	2796.0	2.3780	4.0172	6.3952
210	1.9062	0.001173	0.10441	895.53	1703.9	2599.5	897.76	1900.7	2798.5	2.4248	3.9337	6.3585
215	2.104	0.001181	0.09479	918.14	1682.9	2601.1	920.62	1879.9	2800.5	2.4714	3.8507	6.3221
220	2.318	0.001190	0.08619	940.87	1661.5	2602.4	943.62	1858.5	2802.1	2.5178	3.7683	6.2861
225	2.548	0.001199	0.07849	963.73	1639.6	2603.3	966.78	1836.5	2803.3	2.5639	3.6863	6.2503
230	2.795	0.001209	0.07158	986.74	1617.2	2603.9	990.12	1813.8	2804.0	2.6099	3.6047	6.2146
235	3.060	0.001219	0.06537	1009.89	1594.2	2604.1	1013.62	1790.5	2804.2	2.6558	3.5233	6.1791
240	3.344	0.001229	0.05976	1033.21	1570.8	2604.0	1037.32	1766.5	2803.8	2.7015	3.4422	6.1437
245	3.648	0.001240	0.05471	1056.71	1546.7	2603.4	1061.23	1741.7	2803.0	2.7472	3.3612	6.1083
250	3.973	0.001251	0.05013	1080.39	1522.0	2602.4	1085.36	1716.2	2801.5	2.7927	3.2802	6.0730
255	4.319	0.001263	0.04598	1104.28	1496.7	2600.9	1109.73	1689.8	2799.5	2.8383	3.1992	6.0375
260	4.688	0.001276	0.04221	1128.39	1470.6	2599.0	1134.37	1662.5	2796.9	2.8838	3.1181	6.0019
265	5.081	0.001289	0.03877	1152.74	1443.9	2596.6	1159.28	1634.4	2793.6	2.9294	3.0368	5.9662
270	5.499	0.001302	0.03564	1177.36	1416.3	2593.7	1184.51	1605.2	2789.7	2.9751	2.9551	5.9301
275	5.942	0.001317	0.03279	1202.25	1387.9	2590.2	1210.07	1574.9	2785.0	3.0208	2.8730	5.8938
280	6.412	0.001332	0.03017	1227.46	1358.7	2586.1	1235.99	1543.6	2779.6	3.0668	2.7903	5.8571
285	6.909	0.001348	0.02777	1253.00	1328.4	2581.4	1262.31	1511.0	2773.3	3.1130	2.7070	5.8199
290	7.436	0.001366	0.02557	1278.92	1297.1	2576.0	1289.07	1477.1	2766.2	3.1594	2.6227	5.7821
295	7.993	0.001384	0.02354	1305.2	1264.7	2569.9	1316.3	1441.8	2758.1	3.2062	2.5375	5.7437
300	8.581	0.001404	0.02167	1332.0	1231.0	2563.0	1344.0	1404.9	2749.0	3.2534	2.4511	5.7045
305	9.202	0.001425	0.019948	1359.3	1195.9	2555.2	1372.4	1366.4	2738.7	3.3010	2.3633	5.6643
310	9.856	0.001447	0.018350	1387.1	1159.4	2546.4	1401.3	1326.0	2727.3	3.3493	2.2737	5.6230
315	10.547	0.001472	0.016867	1415.5	1121.1	2536.6	1431.0	1283.5	2714.5	3.3982	2.1821	5.5804
320	11.274	0.001499	0.015488	1444.6	1080.9	2525.5	1461.5	1238.6	2700.1	3.4480	2.0882	5.5362
330	12.845	0.001561	0.012996	1505.3	993.7	2498.9	1525.3	1140.6	2665.9	3.5507	1.8909	5.4417
340	14.586	0.001638	0.010797	1570.3	894.3	2464.6	1594.2	1027.9	2622.0	3.6594	1.6763	5.3357
350	16.513	0.001740	0.008813	1641.9	776.6	2418.4	1670.6	893.4	2563.9	3.7777	1.4335	5.2112
360	18.651	0.001893	0.006945	1725.2	626.3	2351.5	1760.5	720.5	2481.0	3.9147	1.1370	5.0526
370	21.03	0.002213	0.004925	1844.0	384.5	2228.5	1890.5	441.6	2332.1	4.1106	.6865	4.7971
374.14	22.09	0.003155	0.003155	2029.6	0	2029.6	2099.3	0	2099.3	4.4298	0	4.4298

Source: Adapted from Joseph H. Keenan, Frederick G. Keyes, Philip G. Hill, and Joan G. Moore, *Steam Tables.* New York: John Wiley & Sons, Inc., 1978.

Table A3.2 Thermodynamic Properties of Saturated Water and Steam—Pressure Table

p (kPa)	T (°C)	Specific Volume (m³/kg)		Internal Energy (kJ/kg)			Enthalpy (kJ/kg)			Entropy (kJ/kg K)		
		v_f	v_g	u_f	u_{fg}	u_g	h_f	h_{fg}	h_g	s_f	s_{fg}	s_g
0.6113	0.01	0.001000	206.14	.00	2375.3	2375.3	.01	2501.3	2501.4	.0000	9.1562	9.1562
1.0	6.98	0.001000	129.21	29.30	2355.7	2385.0	29.30	2484.9	2514.2	.1059	8.8697	8.9756
1.5	13.03	0.001001	87.98	54.71	2338.6	2393.3	54.71	2470.6	2525.3	.1957	8.6322	8.8279
2.0	17.50	0.001001	67.00	73.48	2326.0	2399.5	73.48	2460.0	2533.5	.2607	8.4629	8.7237
2.5	21.08	0.001002	54.25	88.48	2315.9	2404.4	88.49	2451.6	2540.0	.3120	8.3311	8.6432
3.0	24.08	0.001003	45.67	101.04	2307.5	2408.5	101.05	2444.5	2545.5	.3545	8.2231	8.5776
4.0	28.96	0.001004	34.80	121.45	2293.7	2415.2	121.46	2432.9	2554.4	.4226	8.0520	8.4746
5.0	32.88	0.001005	28.19	137.81	2282.7	2420.5	137.82	2423.7	2561.5	.4764	7.9187	8.3951
7.5	40.29	0.001008	19.24	168.78	2261.7	2430.5	168.79	2406.0	2574.8	.5764	7.6750	8.2515
10	45.81	0.001010	14.67	191.82	2246.1	2437.9	191.83	2392.8	2584.7	.6493	7.5009	8.1502
15	53.97	0.001014	10.02	225.92	2222.8	2448.7	225.94	2373.1	2599.1	.7549	7.2536	8.0085
20	60.06	0.001017	7.649	251.38	2205.4	2456.7	251.40	2358.3	2609.7	.8320	7.0766	7.9085
25	64.97	0.001020	6.204	271.90	2191.2	2463.1	271.93	2346.3	2618.2	.8931	6.9383	7.8314
30	69.10	0.001022	5.229	289.20	2179.2	2468.4	289.23	2336.1	2625.3	.9439	6.8247	7.7686
40	75.87	0.001027	3.993	317.53	2159.5	2477.0	317.58	2319.2	2636.8	1.0259	6.6441	7.6700
50	81.33	0.001030	3.240	340.44	2143.4	2483.9	340.49	2305.4	2645.9	1.0910	6.5029	7.5939
75	91.78	0.001037	2.217	384.31	2112.4	2496.7	384.39	2278.6	2663.0	1.2130	6.2434	7.4564
MPa												
.100	99.63	0.001043	1.6940	417.36	2088.7	2506.1	417.46	2258.0	2675.5	1.3026	6.0568	7.3594
0.125	105.99	0.001048	1.3749	444.19	2069.3	2513.5	444.32	2241.0	2685.4	1.3740	5.9104	7.2844
0.150	111.37	0.001053	1.1593	466.94	2052.7	2519.7	467.11	2226.5	2693.6	1.4336	5.7897	7.2233
0.175	116.06	0.001057	1.0036	486.80	2038.1	2524.9	486.99	2213.6	2700.6	1.4849	5.6868	7.1717
0.200	120.23	0.001061	0.8857	504.49	2025.0	2529.5	504.70	2201.9	2706.7	1.5301	5.5970	7.1271
0.225	124.00	0.001064	0.7933	520.47	2013.1	2533.6	520.72	2191.3	2712.1	1.5706	5.5173	7.0878
0.250	127.44	0.001067	0.7187	535.10	2002.1	2537.2	535.37	2181.5	2716.9	1.6072	5.4455	7.0527
0.275	130.60	0.001070	0.6573	548.59	1991.9	2540.5	548.89	2172.4	2721.3	1.6408	5.3801	7.0209
0.300	133.55	0.001073	0.6058	561.15	1982.4	2543.6	561.47	2163.8	2725.3	1.6718	5.3201	6.9919
0.325	136.30	0.001076	0.5620	572.90	1973.5	2546.4	573.25	2155.8	2729.0	1.7006	5.2646	6.9652
0.350	138.88	0.001079	0.5243	583.95	1965.0	2548.9	584.33	2148.1	2732.4	1.7275	5.2130	6.9405
0.375	141.32	0.001081	0.4914	594.40	1956.9	2551.3	594.81	2140.8	2735.6	1.7528	5.1647	6.9175
0.40	143.63	0.001084	0.4625	604.31	1949.3	2553.6	604.74	2133.8	2738.6	1.7766	5.1193	6.8959
0.45	147.93	0.001088	0.4140	622.77	1934.9	2557.6	623.25	2120.7	2743.9	1.8207	5.0359	6.8565
0.50	151.86	0.001093	0.3749	639.68	1921.6	2561.2	640.23	2108.5	2748.7	1.8607	4.9606	6.8213

0.55	155.48	0.001097	0.3427	655.32	1909.2	2564.5	655.93	2097.0	2753.0	1.8973	4.8920	6.7893
0.60	158.85	0.001101	0.3157	669.90	1897.5	2567.4	670.56	2086.3	2756.8	1.9312	4.8288	6.7600
0.65	162.01	0.001104	0.2927	683.56	1886.5	2570.1	684.28	2076.0	2760.3	1.9627	4.7703	6.7331
0.70	64.97	0.001108	0.2729	696.44	1876.1	2572.5	697.22	2066.3	2763.5	1.9922	4.7158	6.7080
0.75	167.78	0.001112	0.2556	708.64	1866.1	2574.7	709.47	2057.0	2766.4	2.0200	4.6647	6.6847
0.80	170.43	0.001115	0.2404	720.22	1856.6	2576.8	721.11	2048.0	2769.1	2.0462	4.6166	6.6628
0.85	172.96	0.001118	0.2270	731.27	1847.4	2578.7	732.22	2039.4	2771.6	2.0710	4.5711	6.6421
0.90	175.38	0.001121	0.2150	741.83	1838.6	2580.5	742.83	2031.1	2773.9	2.0946	4.5280	6.6226
0.95	177.69	0.001124	0.2042	751.95	1830.2	2582.1	753.02	2023.1	2776.1	2.1172	4.4869	6.6041
1.00	179.91	0.001127	0.19444	761.68	1822.0	2583.6	762.81	2015.3	2778.1	2.1387	4.4478	6.5865
1.10	184.09	0.001133	0.17753	780.09	1806.3	2586.4	781.34	2000.4	2781.7	2.1792	4.3744	6.5536
1.20	187.99	0.001139	0.16333	797.29	1791.5	2588.8	798.65	1986.2	2784.8	2.2166	4.3067	6.5233
1.30	191.64	0.001144	0.15125	813.44	1777.5	2591.0	814.93	1972.7	2787.6	2.2515	4.2438	6.4953
1.40	195.07	0.001149	0.14084	828.70	1764.1	2592.8	830.30	1959.7	2790.0	2.2842	4.1850	6.4693
1.50	198.32	0.001154	0.13177	843.16	1751.3	2594.5	844.89	1947.3	2792.2	2.3150	4.1298	6.4448
1.75	205.76	0.001166	0.11349	876.46	1721.4	2597.8	878.50	1917.9	2796.4	2.3851	4.0044	6.3896
2.00	212.42	0.001177	0.09963	906.44	1693.8	2600.3	908.79	1890.7	2799.5	2.4474	3.8935	6.3409
2.25	218.45	0.001187	0.08875	933.83	1668.2	2602.0	936.49	1865.2	2801.7	2.5035	3.7937	6.2972
2.5	223.99	0.001197	0.07998	959.11	1644.0	2603.1	962.11	1841.0	2803.1	2.5547	3.7028	6.2575
3.0	233.90	0.001217	0.06668	1004.78	1599.3	2604.1	1008.42	1795.7	2804.2	2.6457	3.5412	6.1869
3.5	242.60	0.001235	0.05707	1045.43	1558.3	2603.7	1049.75	1753.7	2803.4	2.7253	3.4000	6.1253
4	250.40	0.001252	0.04978	1082.31	1520.0	2602.3	1087.31	1714.1	2801.4	2.7964	3.2737	6.0701
5	263.99	0.001286	0.03944	1147.81	1449.3	2597.1	1154.23	1640.1	2794.3	2.9202	3.0532	5.9734
6	275.64	0.001319	0.03244	1205.44	1384.3	2589.7	1213.35	1571.0	2784.3	3.0267	2.8625	5.8892
7	285.88	0.001351	0.02737	1257.55	1323.0	2580.5	1267.00	1505.1	2772.1	3.1211	2.6922	5.8133
8	295.06	0.001384	0.02352	1305.57	1264.2	2569.8	1316.64	1441.3	2758.0	3.2068	2.5364	5.7432
9	303.40	0.001418	0.02048	1350.51	1207.3	2557.8	1363.26	1378.9	2742.1	3.2858	2.3915	5.6772
10	311.06	0.001452	0.018026	1393.04	1151.4	2544.4	1407.56	1317.1	2724.7	3.3596	2.2544	5.6141
11	318.15	0.001489	0.015987	1433.7	1096.0	2529.8	1450.1	1255.5	2705.6	3.4295	2.1233	5.5527
12	324.75	0.001527	0.014263	1473.0	1040.7	2513.7	1491.3	1193.6	2684.9	3.4962	1.9962	5.4924
13	330.93	0.001567	0.012780	1511.1	985.0	2496.1	1531.5	1130.7	2662.2	3.5606	1.8718	5.4323
14	336.75	0.001611	0.011485	1548.6	928.2	2476.8	1571.1	1066.5	2637.6	3.6232	1.7485	5.3717
15	342.24	0.001658	0.010337	1585.6	869.8	2455.5	1610.5	1000.0	2610.5	3.6848	1.6249	5.3098
16	347.44	0.001711	0.009306	1622.7	809.0	2431.7	1650.1	930.6	2580.6	3.7461	1.4994	5.2455
17	352.37	0.001770	0.008364	1660.2	744.8	2405.0	1690.3	856.9	2547.2	3.8079	1.3698	5.1777
18	357.06	0.001840	0.007489	1698.9	675.4	2374.3	1732.0	777.1	2509.1	3.8715	1.2329	5.1044
19	361.54	0.001924	0.006657	1739.9	598.1	2338.1	1776.5	688.0	2464.5	3.9388	1.0839	5.0228
20	365.81	0.002036	0.005834	1785.6	507.5	2293.0	1826.3	583.4	2409.7	4.0139	.9130	4.9269
21	369.89	0.002207	0.004952	1842.1	388.5	2230.6	1888.4	446.2	2334.6	4.1075	.6938	4.8013
22	373.80	0.002742	0.003568	1961.9	125.2	2087.1	2022.2	143.4	2165.6	4.3110	.2216	4.5327
22.09	374.14	0.003155	0.003155	2029.6	0	2029.6	2099.3	0	2099.3	4.4298	0	4.4298

Table A3.3   Thermodynamic Properties of Superheated Steam

T (°C)	p = .010 MPa (T_{sat} = 45.81°C)				p = .050 MPa (T_{sat} = 81.33°C)			
	v (m³/kg)	u (kJ/kg)	h (kJ/kg)	s (kJ/kg K)	v (m³/kg)	u (kJ/kg)	h (kJ/kg)	s (kJ/kg K)
Sat.	14.674	2437.9	2584.7	8.1502	3.240	2483.9	2645.9	7.5939
50	14.869	2443.9	2592.6	8.1749				
100	17.196	2515.5	2687.5	8.4479	3.418	2511.6	2682.5	7.6947
150	19.512	2587.9	2783.0	8.6882	3.889	2585.6	2780.1	7.9401
200	21.825	2661.3	2879.5	8.9038	4.356	2659.9	2877.7	8.1580
250	24.136	2736.0	2977.3	9.1002	4.820	2735.0	2976.0	8.3556
300	26.445	2812.1	3076.5	9.2813	5.284	2811.3	3075.5	8.5373
400	31.063	2968.9	3279.6	9.6077	6.209	2968.5	3278.9	8.8642
500	35.679	3132.3	3489.1	9.8978	7.134	3132.0	3488.7	9.1546
600	40.295	3302.5	3705.4	10.1608	8.057	3302.2	3705.1	9.4178
700	44.911	3479.6	3928.7	10.4028	8.981	3479.4	3928.5	9.6599
800	49.526	3663.8	4159.0	10.6281	9.904	3663.6	4158.9	9.8852
900	54.141	3855.0	4396.4	10.8396	10.828	3854.9	4396.3	10.0967
1000	58.757	4053.0	4640.6	11.0393	11.751	4052.9	4640.5	10.2964
1100	63.372	4257.5	4891.2	11.2287	12.674	4257.4	4891.1	10.4859
1200	67.987	4467.9	5147.8	11.4091	13.597	4467.8	5147.7	10.6662
1300	72.602	4683.7	5409.7	11.5811	14.521	4683.6	5409.6	10.8382

T (°C)	p = .10 MPa (T_{sat} = 99.63°C)				p = .20 MPa (T_{sat} = 120.23°C)			
	v (m³/kg)	u (kJ/kg)	h (kJ/kg)	s (kJ/kg K)	v (m³/kg)	u (kJ/kg)	h (kJ/kg)	s (kJ/kg K)
Sat.	1.6940	2506.1	2675.5	7.3594	0.8857	2529.5	2706.7	7.1272
100	1.6958	2506.7	2676.2	7.3614				
150	1.9364	2582.8	2776.4	7.6134	.9596	2576.9	2768.8	7.2795
200	2.172	2658.1	2875.3	7.8343	1.0803	2654.4	2870.5	7.5066
250	2.406	2733.7	2974.3	8.0333	1.1988	2731.2	2971.0	7.7086
300	2.639	2810.4	3074.3	8.2158	1.3162	2808.6	3071.8	7.8926
400	3.103	2967.9	3278.2	8.5435	1.5493	2966.7	3276.6	8.2218
500	3.565	3131.6	3488.1	8.8342	1.7814	3130.8	3487.1	8.5133
600	4.028	3301.9	3704.7	9.0976	2.013	3301.4	3704.0	8.7770
700	4.490	3479.2	3928.2	9.3398	2.244	3478.8	3927.6	9.0194
800	4.952	3663.5	4158.6	9.5652	2.475	3663.1	4158.2	9.2449
900	5.414	3854.8	4396.1	9.7767	2.706	3854.5	4395.8	9.4566
1000	5.875	4052.8	4640.3	9.9764	2.937	4052.5	4640.0	9.6563
1100	6.337	4257.3	4891.0	10.1659	3.168	4257.0	4890.7	9.8458
1200	6.799	4467.7	5147.6	10.3463	3.399	4467.5	5147.3	10.0262
1300	7.260	4683.5	5409.5	10.5183	3.630	4683.2	5409.3	10.1982

Table A3.3   *Continued*

T (°C)	p = .30 MPa (T_sat = 133.55°C) v (m³/kg)	u (kJ/kg)	h (kJ/kg)	s (kJ/kg K)	p = .40 MPa (T_sat = 143.63°C) v (m³/kg)	u (kJ/kg)	h (kJ/kg)	s (kJ/kg K)
Sat.	.6058	2543.6	2725.3	6.9919	.4625	2553.6	2738.6	6.8959
150	.6339	2570.8	2761.0	7.0778	.4708	2564.5	2752.8	6.9299
200	.7163	2650.7	2865.6	7.3115	.5342	2646.8	2860.5	7.1706
250	.7964	2728.7	2967.6	7.5166	.5951	2726.1	2964.2	7.3789
300	.8753	2806.7	3069.3	7.7022	.6548	2804.8	3066.8	7.5662
400	1.0315	2965.6	3275.0	8.0330	.7726	2964.4	3273.4	7.8985
500	1.1867	3130.0	3486.0	8.3251	.8893	3129.2	3484.9	8.1913
600	1.3414	3300.8	3703.2	8.5892	1.0055	3300.2	3702.4	8.4558
700	1.4957	3478.4	3927.1	8.8319	1.1215	3477.9	3926.5	8.6987
800	1.6499	3662.9	4157.8	9.0576	1.2372	3662.4	4157.3	8.9244
900	1.8041	3854.2	4395.4	9.2692	1.3529	3853.9	4395.1	9.1362
1000	1.9581	4052.3	4639.7	9.4690	1.4685	4052.0	4639.4	9.3360
1100	2.1121	4256.8	4890.4	9.6585	1.5840	4256.5	4890.2	9.5256
1200	2.2661	4467.2	5147.1	9.8389	1.6996	4467.0	5146.8	9.7060
1300	2.4201	4683.0	5409.0	10.0110	1.8151	4682.8	5408.8	9.8780

T (°C)	p = .50 MPa (T_sat = 151.86°C) v (m³/kg)	u (kJ/kg)	h (kJ/kg)	s (kJ/kg K)	p = .60 MPa (T_sat = 158.85°C) v (m³/kg)	u (kJ/kg)	h (kJ/kg)	s (kJ/kg K)
Sat.	.3749	2561.2	2748.7	6.8213	.3157	2567.4	2756.8	6.7600
200	.4249	2642.9	2855.4	7.0592	.3520	2638.9	2850.1	6.9665
250	.4744	2723.5	2960.7	7.2709	.3938	2720.9	2957.2	7.1816
300	.5226	2802.9	3064.2	7.4599	.4344	2801.0	3061.6	7.3724
350	.5701	2882.6	3167.7	7.6329	.4742	2881.2	3165.7	7.5464
400	.6173	2963.2	3271.9	7.7938	.5137	2962.1	3270.3	7.7079
500	.7109	3128.4	3483.9	8.0873	.5920	3127.6	3482.8	8.0021
600	.8041	3299.6	3701.7	8.3522	.6697	3299.1	3700.9	8.2674
700	.8969	3477.5	3925.9	8.5952	.7472	3477.0	3925.3	8.5107
800	.9896	3662.1	4156.9	8.8211	.8245	3661.8	4156.5	8.7367
900	1.0822	3853.6	4394.7	9.0329	.9017	3853.4	4394.4	8.9486
1000	1.1747	4051.8	4639.1	9.2328	.9788	4051.5	4638.8	9.1485
1100	1.2672	4256.3	4889.9	9.4224	1.0559	4256.1	4889.6	9.3381
1200	1.3596	4466.8	5146.6	9.6029	1.1330	4466.5	5146.3	9.5185
1300	1.4521	4682.5	5408.6	9.7749	1.2101	4682.3	5408.3	9.6906

(continued)

Table A3.3   *Continued*

T (°C)	$p = .80$ MPa ($T_{sat} = 170.43°C$)				$p = 1.00$ MPa ($T_{sat} = 179.91°C$)			
	v (m³/kg)	u (kJ/kg)	h (kJ/kg)	s (kJ/kg K)	v (m³/kg)	u (kJ/kg)	h (kJ/kg)	s (kJ/kg K)
Sat.	.2404	2576.8	2769.1	6.6628	.19444	2583.6	2778.1	6.5865
200	.2608	2630.6	2839.3	6.8158	.2060	2621.9	2827.9	6.6940
250	.2931	2715.5	2950.0	7.0384	.2327	2709.9	2942.6	6.9247
300	.3241	2797.2	3056.5	7.2328	.2579	2793.2	3051.2	7.1229
350	.3544	2878.2	3161.7	7.4089	.2825	2875.2	3157.7	7.3011
400	.3843	2959.7	3267.1	7.5716	.3066	2957.3	3263.9	7.4651
500	.4433	3126.0	3480.6	7.8673	.3541	3124.4	3478.5	7.7622
600	.5018	3297.9	3699.4	8.1333	.4011	3296.8	3697.9	8.0290
700	.5601	3476.2	3924.2	8.3770	.4478	3475.3	3923.1	8.2731
800	.6181	3661.1	4155.6	8.6033	.4943	3660.4	4154.7	8.4996
900	.6761	3852.8	4393.7	8.8153	.5407	3852.2	4392.9	8.7118
1000	.7340	4051.0	4638.2	9.0153	.5871	4050.5	4637.6	8.9119
1100	.7919	4255.6	4889.1	9.2050	.6335	4255.1	4888.6	8.1017
1200	.8497	4466.1	5145.9	9.3855	.6798	4465.6	5145.4	9.2822
1300	.9076	4681.8	5407.9	9.5575	.7261	4681.3	5407.4	9.4543

T (°C)	$p = 1.20$ MPa ($T_{sat} = 187.99°C$)				$p = 1.40$ MPa ($T_{sat} = 195.07°C$)			
	v (m³/kg)	u (kJ/kg)	h (kJ/kg)	s (kJ/kg K)	v (m³/kg)	u (kJ/kg)	h (kJ/kg)	s (kJ/kg K)
Sat.	.16333	2588.8	2784.8	6.5233	.14084	2592.8	2790.0	6.4693
200	.16930	2612.8	2815.9	6.5898	.14302	2603.1	2803.3	6.4975
250	.19234	2704.2	2935.0	6.8294	.16350	2698.3	2927.2	6.7467
300	.2138	2789.2	3045.8	7.0317	.18228	2785.2	3040.4	6.9534
350	.2345	2872.2	3153.6	7.2121	.2003	2869.2	3149.5	7.1360
400	.2548	2954.9	3260.7	7.3774	.2178	2952.5	3257.5	7.3026
500	.2946	3122.8	3476.3	7.6759	.2521	3121.1	3474.1	7.6027
600	.3339	3295.6	3696.3	7.9435	.2860	3294.4	3694.8	7.8710
700	.3729	3474.4	3922.0	8.1881	.3195	3473.6	3920.8	8.1160
800	.4118	3659.7	4153.8	8.4148	.3528	3659.0	4153.0	8.3431
900	.4505	3851.6	4392.2	8.6272	.3861	3851.1	4391.5	8.5556
1000	.4892	4050.0	4637.0	8.8274	.4192	4049.5	4636.4	8.7559
1100	.5278	4254.6	4888.0	9.0172	.4524	4254.1	4887.5	8.9457
1200	.5665	4465.1	5144.9	9.1977	.4855	4464.7	5144.4	9.1262
1300	.6051	4680.9	5407.0	9.3698	.5186	4680.4	5406.5	9.2984

Table A3.3 *Continued*

T (°C)	v (m³/kg)	u (kJ/kg)	h (kJ/kg)	s (kJ/kg K)	v (m³/kg)	u (kJ/kg)	h (kJ/kg)	s (kJ/kg K)
	p = 1.60 MPa (T_{sat} = 201.41°C)				p = 1.80 MPa (T_{sat} = 207.15°C)			
Sat.	.12380	2596.0	2794.0	6.4218	.11042	2598.4	2797.1	6.3794
225	.13287	2644.7	2857.3	6.5518	.11673	2636.6	2846.7	6.4808
250	.14184	2692.3	2919.2	6.6732	.12497	2686.0	2911.0	6.6066
300	.15862	2781.1	3034.8	6.8844	.14021	2776.9	3029.2	6.8226
350	.17456	2866.1	3145.4	7.0694	.15457	2863.0	3141.2	7.0100
400	.19005	2950.1	3254.2	7.2374	.16847	2947.7	3250.9	7.1794
500	.2203	3119.5	3472.0	7.5390	.19550	3117.9	3469.8	7.4825
600	.2500	3293.3	3693.2	7.8080	.2220	3292.1	3691.7	7.7523
700	.2794	3472.7	3919.7	8.0535	.2482	3471.8	3918.5	7.9983
800	.3086	3658.3	4152.1	8.2808	.2742	3657.6	4151.2	8.2258
900	.3377	3850.5	4390.8	8.4935	.3001	3849.9	4390.1	8.4386
1000	.3668	4049.0	4635.8	8.6938	.3260	4048.5	4635.2	8.6391
1100	.3958	4253.7	4887.0	8.8837	.3518	4253.2	4886.4	8.8290
1200	.4248	4464.2	5143.9	9.0643	.3776	4463.7	5143.4	9.0096
1300	.4538	4679.9	5406.0	9.2364	.4034	4679.5	5405.6	9.1818

T (°C)	v (m³/kg)	u (kJ/kg)	h (kJ/kg)	s (kJ/kg K)	v (m³/kg)	u (kJ/kg)	h (kJ/kg)	s (kJ/kg K)
	p = 2.00 MPa (T_{sat} = 212.42°C)				p = 2.50 MPa (T_{sat} = 223.99°C)			
Sat.	.09963	2600.3	2799.5	6.3409	.07998	2603.1	2803.1	6.2575
225	.10377	2628.3	2835.8	6.4147	.08027	2605.6	2806.3	6.2639
250	.11144	2679.6	2902.5	6.5453	.08700	2662.6	2880.1	6.4085
300	.12547	2772.6	3023.5	6.7664	.09890	2761.6	3008.8	6.6438
350	.13857	2859.8	3137.0	6.9563	.10976	2851.9	3126.3	6.8403
400	.15120	2945.2	3247.6	7.1271	.12010	2939.1	3239.3	7.0148
500	.17568	3116.2	3467.6	7.4317	.13998	3112.1	3462.1	7.3234
600	.19960	3290.9	3690.1	7.7024	.15930	3288.0	3686.3	7.5960
700	.2232	3470.9	3917.4	7.9487	.17832	3468.7	3914.5	7.8435
800	.2467	3657.0	4150.3	8.1765	.19716	3655.3	4148.2	8.0720
900	.2700	3849.3	4389.4	8.3895	.21590	3847.9	4387.6	8.2853
1000	.2933	4048.0	4634.6	8.5901	.2346	4046.7	4633.1	8.4861
1100	.3166	4252.7	4885.9	8.7800	.2532	4251.5	4884.6	8.6762
1200	.3398	4463.3	5142.9	8.9607	.2718	4462.1	5141.7	8.8569
1300	.3631	4679.0	5405.1	9.1329	.2905	4677.8	5404.0	9.0291

(continued)

Table A3.3   *Continued*

	p = 3.00 MPa (T_{sat} = 233.90°C)				p = 3.50 MPa (T_{sat} = 242.60°C)			
T (°C)	v (m³/kg)	u (kJ/kg)	h (kJ/kg)	s (kJ/kg K)	v (m³/kg)	u (kJ/kg)	h (kJ/kg)	s (kJ/kg K)
Sat.	.06668	2604.1	2804.2	6.1869	.05707	2603.7	2803.4	6.1253
250	.07058	2644.0	2855.8	6.2872	.05872	2623.7	2829.2	6.1749
300	.08114	2750.1	2993.5	6.5390	.06842	2738.0	2977.5	6.4461
350	.09053	2843.7	3115.3	6.7428	.07678	2835.3	3104.0	6.6579
400	.09936	2932.8	3230.9	6.9212	.08453	2926.4	3222.3	6.8405
450	.10787	3020.4	3344.0	7.0834	.09196	3015.3	3337.2	7.0052
500	.11619	3108.0	3456.5	7.2338	.09918	3103.0	3450.9	7.1572
600	.13243	3285.0	3682.3	7.5085	.11324	3282.1	3678.4	7.4339
700	.14838	3466.5	3911.7	7.7571	.12699	3464.3	3908.8	7.6837
800	.16414	3653.5	4145.9	7.9862	.14056	3651.8	4143.7	7.9134
900	.17980	3846.5	4385.9	8.1999	.15402	3845.0	4384.1	8.1276
1000	.19541	4045.4	4631.6	8.4009	.16743	4044.1	4630.1	8.3288
1100	.21098	4250.3	4883.3	8.5912	.18080	4249.2	4881.9	8.5192
1200	.22652	4460.9	5140.5	8.7720	.19415	4459.8	5139.3	8.7000
1300	.24206	4676.6	5402.8	8.9442	.20749	4675.5	5401.7	8.8723

	p = 4.0 MPa (T_{sat} = 250.40°C)				p = 4.5 MPa (T_{sat} = 257.49°C)			
T (°C)	v (m³/kg)	u (kJ/kg)	h (kJ/kg)	s (kJ/kg K)	v (m³/kg)	u (kJ/kg)	h (kJ/kg)	s (kJ/kg K)
Sat.	.04978	2602.3	2801.4	6.0701	.04406	2600.1	2798.3	6.0198
275	.05457	2667.9	2886.2	6.2285	.04730	2650.3	2863.2	6.1401
300	.05884	2725.3	2960.7	6.3615	.05135	2712.0	2943.1	6.2828
350	.06645	2826.7	3092.5	6.5821	.05840	2817.8	3080.6	6.5131
400	.07341	2919.9	3213.6	6.7690	.06475	2913.3	3204.7	6.7047
450	.08002	3010.2	3330.3	6.9363	.07074	3005.0	3323.3	6.8746
500	.08643	3099.5	3445.3	7.0901	.07651	3095.3	3439.6	7.0301
600	.09885	3279.1	3674.4	7.3688	.08765	3276.0	3670.5	7.3110
700	.11095	3462.1	3905.9	7.6198	.09847	3459.9	3903.0	7.5631
800	.12287	3650.0	4141.5	7.8502	.10911	3648.3	4139.3	7.7942
900	.13469	3843.6	4382.3	8.0647	.11965	3842.2	4380.6	8.0091
1000	.14645	4042.9	4628.7	8.2662	.13013	4041.6	4627.2	8.2108
1100	.15817	4248.0	4880.6	8.4567	.14056	4246.8	4879.3	8.4015
1200	.16987	4458.6	5138.1	8.6376	.15098	4457.5	5136.9	8.5825
1300	.18156	4674.3	5400.5	8.8100	.16139	4673.1	5399.4	8.7549

Table A3.3 *Continued*

T (°C)	$p = 5.0$ MPa ($T_{\text{sat}} = 263.99°C$)				$p = 6.0$ MPa ($T_{\text{sat}} = 275.64°C$)			
	v (m³/kg)	u (kJ/kg)	h (kJ/kg)	s (kJ/kg K)	v (m³/kg)	u (kJ/kg)	h (kJ/kg)	s (kJ/kg K)
Sat.	.03944	2597.1	2794.3	5.9734	.03244	2589.7	2784.3	5.8892
275	.04141	2631.3	2838.3	6.0544				
300	.04532	2698.0	2924.5	6.2084	.03616	2667.2	2884.2	6.0674
350	.05194	2808.7	3068.4	6.4493	.04223	2789.6	3043.0	6.3335
400	.05781	2906.6	3195.7	6.6459	.04739	2892.9	3177.2	6.5408
450	.06330	2999.7	3316.2	6.8186	.05214	2988.9	3301.8	6.7193
500	.06857	3091.0	3433.8	6.9759	.05665	3082.2	3422.2	6.8803
550	.07368	3181.8	3550.3	7.1218	.06101	3174.6	3540.6	7.0288
600	.07869	3273.0	3666.5	7.2589	.06525	3266.9	3658.4	7.1677
700	.08849	3457.6	3900.1	7.5122	.07352	3453.1	3894.2	7.4234
800	.09811	3646.6	4137.1	7.7440	.08160	3643.1	4132.7	7.6566
900	.10762	3840.7	4378.8	7.9593	.08958	3837.8	4375.3	7.8727
1000	.11707	4040.4	4625.7	8.1612	.09749	4037.8	4622.7	8.0751
1100	.12648	4245.6	4878.0	8.3520	.10536	4243.3	4875.4	8.2661
1200	.13587	4456.3	5135.7	8.5331	.11321	4454.0	5133.3	8.4474
1300	.14526	4672.0	5398.2	8.7055	.12106	4669.6	5396.0	8.6199

T (°C)	$p = 7.0$ MPa ($T_{\text{sat}} = 285.88°C$)				$p = 8.0$ MPa ($T_{\text{sat}} = 295.06°C$)			
	v (m³/kg)	u (kJ/kg)	h (kJ/kg)	s (kJ/kg K)	v (m³/kg)	u (kJ/kg)	h (kJ/kg)	s (kJ/kg K)
Sat.	.02737	2580.5	2772.1	5.8133	.02352	2569.8	2758.0	5.7432
300	.02947	2632.2	2838.4	5.9305	.02426	2590.9	2785.0	5.7906
350	.03524	2769.4	3016.0	6.2283	.02995	2747.7	2987.3	6.1301
400	.03993	2878.6	3158.1	6.4478	.03432	2863.8	3138.3	6.3634
450	.04416	2978.0	3287.1	6.6327	.03817	2966.7	3272.0	6.5551
500	.04814	3073.4	3410.3	6.7975	.04175	3064.3	3398.3	6.7240
550	.05195	3167.2	3530.9	6.9486	.04516	3159.8	3521.0	6.8778
600	.05565	3260.7	3650.3	7.0894	.04845	3254.4	3642.0	7.0206
700	.06283	3448.5	3888.3	7.3476	.05481	3443.9	3882.4	7.2812
800	.06981	3639.5	4128.2	7.5822	.06097	3636.0	4123.8	7.5173
900	.07669	3835.0	4371.8	7.7991	.06702	3832.1	4368.3	7.7351
1000	.08350	4035.3	4619.8	8.0020	.07301	4032.8	4616.9	7.9384
1100	.09027	4240.9	4872.8	8.1933	.07896	4238.6	4870.3	8.1300
1200	.09703	4451.7	5130.9	8.3747	.08489	4449.5	5128.5	8.3115
1300	.10377	4667.3	5393.7	8.5473	.09080	4665.0	5391.5	8.4842

(continued)

Table A3.3   *Continued*

T (°C)	v (m³/kg)	u (kJ/kg)	h (kJ/kg)	s (kJ/kg K)	v (m³/kg)	u (kJ/kg)	h (kJ/kg)	s (kJ/kg K)
	$p = 9.0$ MPa ($T_{sat} = 303.40°$C)				$p = 10.0$ MPa ($T_{sat} = 311.06°$C)			
Sat.	.02048	2557.8	2742.1	5.6772	.018026	2544.4	2724.7	5.6141
325	.02327	2646.6	2856.0	5.8712	.019861	2610.4	2809.1	5.7568
350	.02580	2724.4	2956.6	6.0361	.02242	2699.2	2923.4	5.9443
400	.02993	2848.4	3117.8	6.2854	.02641	2832.4	3096.5	6.2120
450	.03350	2955.2	3256.6	6.4844	.02975	2943.4	3240.9	6.4190
500	.03677	3055.2	3386.1	6.6576	.03279	3045.8	3373.7	6.5966
550	.03987	3152.2	3511.0	6.8142	.03564	3144.6	3500.9	6.7561
600	.04285	3248.1	3633.7	6.9589	.03837	3241.7	3625.3	6.9029
650	.04574	3343.6	3755.3	7.0943	.04101	3338.2	3748.2	7.0398
700	.04857	3439.3	3876.5	7.2221	.04358	3434.7	3870.5	7.1687
800	.05409	3632.5	4119.3	7.4596	.04859	3628.9	4114.8	7.4077
900	.05950	3829.2	4364.8	7.6783	.05349	3826.3	4361.2	7.6272
1000	.06485	4030.3	4614.0	7.8821	.05832	4027.8	4611.0	7.8315
1100	.07016	4236.3	4867.7	8.0740	.06312	4234.0	4865.1	8.0237
1200	.07544	4447.2	5126.2	8.2556	.06789	4444.9	5123.8	8.2055
1300	.08072	4462.7	5389.2	8.4284	.07265	4460.5	5387.0	8.3783
	$p = 12.5$ MPa ($T_{sat} = 327.89°$C)				$p = 15.0$ MPa ($T_{sat} = 342.24°$C)			
Sat.	.013495	2505.1	2673.8	5.4624	.010337	2455.5	2610.5	5.3098
350	.016126	2624.6	2826.2	5.7118	.011470	2520.4	2692.4	5.4421
400	.02000	2789.3	3039.3	6.0417	.015649	2740.7	2975.5	5.8811
450	.02299	2912.5	3199.8	6.2719	.018445	2879.5	3156.2	6.1404
500	.02560	3021.7	3341.8	6.4618	.02080	2996.6	3308.6	6.3443
550	.02801	3125.0	3475.2	6.6290	.02293	3104.7	3448.6	6.5199
600	.03029	3225.4	3604.0	6.7810	.02491	3208.6	3582.3	6.6776
650	.03248	3324.4	3730.4	6.9218	.02680	3310.3	3712.3	6.8224
700	.03460	3422.9	3855.3	7.0536	.02861	3410.9	3840.1	6.9572
800	.03869	3620.0	4103.6	7.2965	.03210	3610.9	4092.4	7.2040
900	.04267	3819.1	4352.5	7.5182	.03546	3811.9	4343.8	7.4279
1000	.04658	4021.6	4603.8	7.7237	.03875	4015.4	4596.6	7.6348
1100	.05045	4228.2	4858.8	7.9165	.04200	4222.6	4852.6	7.8283
1200	.05430	4439.3	5118.0	8.0987	.04523	4433.8	5112.3	8.0108
1300	.05813	4654.8	5381.4	8.2717	.04845	4649.1	5376.0	8.1840

Table A3.3 *Continued*

T (°C)	p = 17.5 MPa (T_sat = 354.75°C)				p = 20.0 MPa (T_sat = 365.81°C)			
	v (m³/kg)	u (kJ/kg)	h (kJ/kg)	s (kJ/kg K)	v (m³/kg)	u (kJ/kg)	h (kJ/kg)	s (kJ/kg K)
Sat.	.007920	2390.2	2528.8	5.1419	.005834	2293.0	2409.7	4.9269
400	.012447	2685.0	2902.9	5.7213	.009942	2619.3	2818.1	5.5540
450	.015174	2844.2	3109.7	6.0184	.012695	2806.2	3060.1	5.9017
500	.017358	2970.3	3274.1	6.2383	.014768	2942.9	3238.2	6.1401
550	.019288	3083.9	3421.4	6.4230	.016555	3062.4	3393.5	6.3348
600	.02106	3191.5	3560.1	6.5866	.018178	3174.0	3537.6	6.5048
650	.02274	3296.0	3693.9	6.7357	.019693	3281.4	3675.3	6.6582
700	.02434	3398.7	3824.6	6.8736	.02113	3386.4	3809.0	6.7993
800	.02738	3601.8	4081.1	7.1244	.02385	3592.7	4069.7	7.0544
900	.03031	3804.7	4335.1	7.3507	.02645	3797.5	4326.4	7.2830
1000	.03316	4009.3	4589.5	7.5589	.02897	4003.1	4582.5	7.4925
1100	.03597	4216.9	4846.4	7.7531	.03145	4211.3	4840.2	7.6874
1200	.03876	4428.3	5106.6	7.9360	.03391	4422.8	5101.0	7.8707
1300	.04154	4643.5	5370.5	8.1093	.03636	4638.0	5365.1	8.0442

T (°C)	p = 25.0 MPa				p = 30.0 MPa			
	v (m³/kg)	u (kJ/kg)	h (kJ/kg)	s (kJ/kg K)	v (m³/kg)	u (kJ/kg)	h (kJ/kg)	s (kJ/kg K)
375	.0019731	1798.7	1848.0	4.0320	.0017892	1737.8	1791.5	3.9305
400	.006004	2430.1	2580.2	5.1418	.002790	2067.4	2151.1	4.4728
425	.007881	2609.2	2806.3	5.4723	.005303	2455.1	2614.2	5.1504
450	.009162	2720.7	2949.7	5.6744	.006735	2619.3	2821.4	5.4424
500	.011123	2884.3	3162.4	5.9592	.008678	2820.7	3081.1	5.7905
550	.012724	3017.5	3335.6	6.1765	.010168	2970.3	3275.4	6.0342
600	.014137	3137.9	3491.4	6.3602	.011446	3100.5	3443.9	6.2331
650	.015433	3251.6	3637.4	6.5229	.012596	3221.0	3598.9	6.4058
700	.016646	3361.3	3777.5	6.6707	.013661	3335.8	3745.6	6.5606
800	.018912	3574.3	4047.1	6.9345	.015623	3555.5	4024.2	6.8332
900	.021045	3783.0	4309.1	7.1680	.017448	3768.5	4291.9	7.0718
1000	.02310	3990.9	4568.5	7.3802	.019196	3978.8	4554.7	7.2867
1100	.02512	4200.2	4828.2	7.5765	.020903	4189.2	4816.3	7.4845
1200	.02711	4412.0	5089.9	7.7605	.022589	4401.3	5079.0	7.6692
1300	.02910	4626.9	5354.4	7.9342	.024266	4616.0	5344.0	7.8432

(continued)

Table A3.3   *Continued*

T (°C)	p = 35.0 MPa				p = 40.0 MPa			
	v (m³/kg)	u (kJ/kg)	h (kJ/kg)	s (kJ/kg K)	v (m³/kg)	u (kJ/kg)	h (kJ/kg)	s (kJ/kg K)
375	.0017003	1702.9	1762.4	3.8722	.0016407	1677.1	1742.8	3.8290
400	.002100	1914.1	1987.6	4.2126	.0019077	1854.6	1930.9	4.1135
425	.003428	2253.4	2373.4	4.7747	.002532	2096.9	2198.1	4.5029
450	.004961	2498.7	2672.4	5.1962	.003693	2365.1	2512.8	4.9459
500	.006927	2751.9	2994.4	5.6282	.005622	2678.4	2903.3	5.4700
550	.008345	2921.0	3213.0	5.9026	.006984	2869.7	3149.1	5.7785
600	.009527	3062.0	3395.5	6.1179	.008094	3022.6	3346.4	6.0114
650	.010575	3189.8	3559.9	6.3010	.009063	3158.0	3520.6	6.2054
700	.011533	3309.8	3713.5	6.4631	.009941	3283.6	3681.2	6.3750
800	.013278	3536.7	4001.5	6.7450	.011523	3517.8	3978.7	6.6662
900	.014883	3754.0	4274.9	6.9886	.012962	3739.4	4257.9	6.9150
1000	.016410	3966.7	4541.1	7.2064	.014324	3954.6	4527.6	7.1356
1100	.017895	4178.3	4804.6	7.4057	.015642	4167.4	4793.1	7.3364
1200	.019360	4390.7	5068.3	7.5910	.016940	4380.1	5057.7	7.5224
1300	.020815	4605.1	5333.6	7.7653	.018229	4594.3	5323.5	7.6969

T (°C)	p = 50.0 MPa				p = 60.0 MPa			
	v (m³/kg)	u (kJ/kg)	h (kJ/kg)	s (kJ/kg K)	v (m³/kg)	u (kJ/kg)	h (kJ/kg)	s (kJ/kg K)
375	.0015594	1638.6	1716.6	3.7639	.0015028	1609.4	1699.5	3.7141
400	.0017309	1788.1	1874.6	4.0031	.0016335	1745.4	1843.4	3.9318
425	.002007	1959.7	2060.0	4.2734	.0018165	1892.7	2001.7	4.1626
450	.002486	2159.6	2284.0	4.5884	.002085	2053.9	2179.0	4.4121
500	.003892	2525.5	2720.1	5.1726	.002956	2390.6	2567.9	4.9321
550	.005118	2763.6	3019.5	5.5485	.003956	2658.8	2896.2	5.3441
600	.006112	2942.0	3247.6	5.8178	.004834	2861.1	3151.2	5.6452
650	.006966	3093.5	3441.8	6.0342	.005595	3028.8	3364.5	5.8829
700	.007727	3230.5	3616.8	6.2189	.006272	3177.2	3553.5	6.0824
800	.009078	3479.8	3933.6	6.5290	.007459	3441.5	3889.1	6.4109
900	.010283	3710.3	4224.4	6.7882	.008508	3681.0	4191.5	6.6805
1000	.011411	3930.5	4501.1	7.0146	.009480	3906.4	4475.2	6.9127
1100	.012496	4145.7	4770.5	7.2184	.010409	4124.1	4748.6	7.1195
1200	.013561	4359.1	5037.2	7.4058	.011317	4338.2	5017.2	7.3083
1300	.014616	4572.8	5303.6	7.5808	.012215	4551.4	5284.3	7.4837

Table A3.4   Thermodynamic Properties of Compressed Liquid Water

	$p = 5$ MPa ($T_{sat} = 263.99°C$)				$p = 10$ MPa ($T_{sat} = 311.06°C$)			
T (°C)	v (m³/kg)	u (kJ/kg)	h (kJ/kg)	s (kJ/kg K)	v (m³/kg)	u (kJ/kg)	h (kJ/kg)	s (kJ/kg K)
Sat.	.0012859	1147.8	1154.2	2.9202	.0014524	1393.0	1407.6	3.3596
0	.0009977	.04	5.04	.0001	.0009952	.09	10.04	.0002
20	.0009995	83.65	88.65	.2956	.0009972	83.36	93.33	.2945
40	.0010056	166.95	171.97	.5705	.0010034	166.35	176.38	.5686
60	.0010149	250.23	255.30	.8285	.0010127	249.36	259.49	.8258
80	.0010268	333.72	338.85	1.0720	.0010245	332.59	342.83	1.0688
100	.0010410	417.52	422.72	1.3030	.0010385	416.12	426.50	1.2992
120	.0010576	501.80	507.09	1.5233	.0010549	500.08	510.64	1.5189
140	.0010768	586.76	592.15	1.7343	.0010737	584.68	595.42	1.7292
160	.0010988	672.62	678.12	1.9375	.0010953	670.13	681.08	1.9317
180	.0011240	759.63	765.25	2.1341	.0011199	756.65	767.84	2.1275
200	.0011530	848.1	853.9	2.3255	.0011480	844.5	856.0	2.3178
220	.0011866	938.4	944.4	2.5128	.0011805	934.1	945.9	2.5039
240	.0012264	1031.4	1037.5	2.6979	.0012187	1026.0	1038.1	2.6872
260	.0012749	1127.9	1134.3	2.8830	.0012645	1121.1	1133.7	2.8699
280					.0013216	1220.9	1234.1	3.0548
300					.0013972	1328.4	1342.3	3.2469

	$p = 15$ MPa ($T_{sat} = 342.24°C$)				$p = 20$ MPa ($T_{sat} = 365.81°C$)			
T (°C)	v (m³/kg)	u (kJ/kg)	h (kJ/kg)	s (kJ/kg K)	v (m³/kg)	u (kJ/kg)	h (kJ/kg)	s (kJ/kg K)
Sat.	.0016581	1585.6	1610.5	3.6848	.002036	1785.6	1826.3	4.0139
0	.0009928	.15	15.05	.0004	.0009904	.19	20.01	.0004
20	.0009950	83.06	97.99	.2934	.0009928	82.77	102.62	.2923
40	.0010013	165.76	180.78	.5666	.0009992	165.17	185.16	.5646
60	.0010105	248.51	263.67	.8232	.0010084	247.68	267.85	.8206
80	.0010222	331.48	346.81	1.0656	.0010199	330.40	350.80	1.0624
100	.0010361	414.74	430.28	1.2955	.0010337	413.39	434.06	1.2917
120	.0010522	498.40	514.19	1.5145	.0010496	496.76	517.76	1.5102
140	.0010707	582.66	598.72	1.7242	.0010678	580.69	602.04	1.7193
160	.0010918	667.71	684.09	1.9260	.0010885	665.35	687.12	1.9204
180	.0011159	753.76	770.50	2.1210	.0011120	750.95	773.20	2.1147
200	.0011433	841.0	858.2	2.3104	.0011388	837.7	860.5	2.3031
220	.0011748	929.9	947.5	2.4953	.0011693	925.9	949.3	2.4870
240	.0012114	1020.8	1039.0	2.6771	.0012046	1016.0	1040.0	2.6674
260	.0012550	1114.6	1133.4	2.8576	.0012462	1108.6	1133.5	2.8459
280	.0013084	1212.5	1232.1	3.0393	.0012965	1204.7	1230.6	3.0248
300	.0013770	1316.6	1337.3	3.2260	.0013596	1306.1	1333.3	3.2071
320	.0014724	1431.1	1453.2	3.4247	.0014437	1415.7	1444.6	3.3979
340	.0016311	1567.5	1591.9	3.6546	.0015684	1539.7	1571.0	3.6075
360					.0018226	1702.8	1739.3	3.8772

(continued)

Table A3.4   *Continued*

T (°C)	p = 30 MPa				p = 50 MPa			
	v (m³/kg)	u (kJ/kg)	h (kJ/kg)	s (kJ/kg K)	v (m³/kg)	u (kJ/kg)	h (kJ/kg)	s (kJ/kg K)
Sat.								
0	.0009856	.25	29.82	.0001	.0009766	.20	49.03	−.0014
20	.0009886	82.17	111.84	.2899	.0009804	81.00	130.02	.2848
40	.0009951	164.04	193.89	.5607	.0009872	161.86	211.21	.5527
60	.0010042	246.06	276.19	.8154	.0009962	242.98	292.79	.8052
80	.0010156	328.30	358.77	1.0561	.0010073	324.34	374.70	1.0440
100	.0010290	410.78	441.66	1.2844	.0010201	405.88	456.89	1.2703
120	.0010445	493.59	524.93	1.5018	.0010348	487.65	539.39	1.4857
140	.0010621	576.88	608.75	1.7098	.0010515	569.77	622.35	1.6915
160	.0010821	660.82	693.28	1.9096	.0010703	652.41	705.92	1.8891
180	.0011047	745.59	778.73	2.1024	.0010912	735.69	790.25	2.0794
200	.0011302	831.4	865.3	2.2893	.0011146	819.7	875.5	2.2634
220	.0011590	918.3	953.1	2.4711	.0011408	904.7	961.7	2.4419
240	.0011920	1006.9	1042.6	2.6490	.0011702	990.7	1049.2	2.6158
260	.0012303	1097.4	1134.3	2.8243	.0012034	1078.1	1138.2	2.7860
280	.0012755	1190.7	1229.0	2.9986	.0012415	1167.2	1229.3	2.9537
300	.0013304	1287.9	1327.8	3.1741	.0012860	1258.7	1323.0	3.1200
320	.0013997	1390.7	1432.7	3.3539	.0013388	1353.3	1420.2	3.2868
340	.0014920	1501.7	1546.5	3.5426	.0014032	1452.0	1522.1	3.4557
360	.0016265	1626.6	1675.4	3.7494	.0014838	1556.0	1630.2	3.6291
380	.0018691	1781.4	1837.5	4.0012	.0015884	1667.2	1746.6	3.8101

Table A3.5 Thermodynamic Properties of Saturated Ice Water–Vapor

T (°C)	p (kPa)	Specific Volume (m³/kg)		Internal Energy (kJ/kg)			Enthalpy (kJ/kg)			Entropy (kJ/kg K)		
		$v_i \times 10^3$	v_g	u_i	u_{ig}	u_g	h_i	h_{ig}	h_g	s_i	s_{ig}	s_g
.01	.6113	1.0908	206.1	-333.40	2708.7	2375.3	-333.40	2834.8	2501.4	-1.221	10.378	9.156
0	.6108	1.0908	206.3	-333.43	2708.8	2375.3	-333.43	2834.8	2501.3	-1.221	10.378	9.157
-2	.5176	1.0904	241.7	-337.62	2710.2	2372.6	-337.62	2835.3	2497.7	-1.237	10.456	9.219
-4	.4375	1.0901	283.8	-341.78	2711.6	2369.8	-341.78	2835.7	2494.0	-1.253	10.536	9.283
-6	.3689	1.0898	334.2	-345.91	2712.9	2367.0	-345.91	2836.2	2490.3	-1.268	10.616	9.348
-8	.3102	1.0894	394.4	-350.02	2714.2	2364.2	-350.02	2836.6	2486.6	-1.284	10.698	9.414
-10	.2602	1.0891	466.7	-354.09	2715.5	2361.4	-354.09	2837.0	2482.9	-1.299	10.781	9.481
-12	.2176	1.0888	553.7	-358.14	2716.8	2358.7	-358.14	2837.3	2479.2	-1.315	10.865	9.550
-14	.1815	1.0884	658.8	-362.15	2718.0	2355.9	-362.15	2837.6	2475.5	-1.331	10.950	9.619
-16	.1510	1.0881	786.0	-366.14	2719.2	2353.1	-366.14	2837.9	2471.8	-1.346	11.036	9.690
-18	.1252	1.0878	940.5	-370.10	2720.4	2350.3	-370.10	2838.2	2468.1	-1.362	11.123	9.762
-20	.1035	1.0874	1128.6	-374.03	2721.6	2347.5	-374.03	2838.4	2464.3	-1.377	11.212	9.835
-22	.0853	1.0871	1358.4	-377.93	2722.7	2344.7	-377.93	2838.6	2460.6	-1.393	11.302	9.909
-24	.0701	1.0868	1640.1	-381.80	2723.7	2342.0	-381.80	2838.7	2456.9	-1.408	11.394	9.985
-26	.0574	1.0864	1986.4	-385.64	2724.8	2339.2	-385.64	2838.9	2453.2	-1.424	11.486	10.062
-28	.0469	1.0861	2413.7	-389.45	2725.8	2336.4	-389.45	2839.0	2449.5	-1.439	11.580	10.141
-30	.0381	1.0858	2943	-393.23	2726.8	2333.6	-393.23	2839.0	2445.8	-1.455	11.676	10.221
-32	.0309	1.0854	3600	-396.98	2727.8	2330.8	-396.98	2839.1	2442.1	-1.471	11.773	10.303
-34	.0250	1.0851	4419	-400.71	2728.7	2328.0	-400.71	2839.1	2438.4	-1.486	11.872	10.386
-36	.0201	1.0848	5444	-404.40	2729.6	2325.2	-404.40	2839.1	2434.7	-1.501	11.972	10.470
-38	.0161	1.0844	6731	-408.06	2730.5	2322.4	-408.06	2839.0	2430.9	-1.517	12.073	10.556
-40	.0129	1.0841	8354	-411.70	2731.3	2319.6	-411.70	2838.9	2427.2	-1.532	12.176	10.644

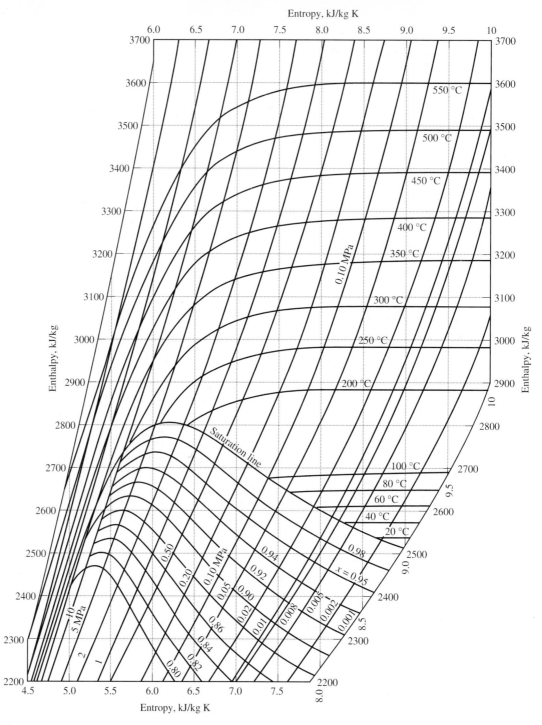

Figure A1 Mollier diagram for water. (*Source:* Adapted from Joseph H. Keenan, Frederick G. Keyes, Philip G. Hill, and Joan G. Moore, *Steam Tables, SI Units,* John Wiley & Sons, Inc. New York, 1978.)

Table A4.1 Thermodynamic Properties of Saturated Ammonia—Temperature Table

T (°C)	p (kPa)	v_f	v_{fg}	v_g	h_f	h_{fg}	h_g	s_f	s_{fg}	s_g
		Specific Volume (m³/kg)			*Enthalpy (kJ/kg)*			*Entropy (kJ/kg K)*		
−50	40.88	0.001424	2.6239	2.6254	−44.3	1416.7	1372.4	−0.1942	6.3502	6.1561
−48	45.96	0.001429	2.3518	2.3533	−35.5	1411.3	1375.8	−0.1547	6.2696	6.1149
−46	51.55	0.001434	2.1126	2.1140	−26.6	1405.8	1379.2	−0.1156	6.1902	6.0746
−44	57.69	0.001439	1.9018	1.9032	−17.8	1400.3	1382.5	−0.0768	6.1120	6.0352
−42	64.42	0.001444	1.7155	1.7170	−8.9	1394.7	1385.8	−0.0382	6.0349	5.9967
−40	71.77	0.001449	1.5506	1.5521	0.0	1389.0	1389.0	0.0000	5.9589	5.9589
−38	79.80	0.001454	1.4043	1.4058	8.9	1383.3	1392.2	0.0380	5.8840	5.9220
−36	88.54	0.001460	1.2742	1.2757	17.8	1377.6	1395.4	0.0757	5.8101	5.8858
−34	98.05	0.001465	1.1582	1.1597	26.8	1371.8	1398.5	0.1132	5.7372	5.8504
−32	108.37	0.001470	1.0547	1.0562	35.7	1365.9	1401.6	0.1504	5.6652	5.8156
−30	119.55	0.001476	0.9621	0.9365	44.7	1360.0	1404.6	0.1873	5.5942	5.7815
−28	131.64	0.001481	0.8790	0.8805	53.6	1354.0	1407.6	0.2240	5.5241	5.7481
−26	144.70	0.001487	0.8044	0.8059	62.6	1347.9	1410.5	0.2605	5.4548	5.7153
−24	158.78	0.001492	0.7373	0.7388	71.6	1341.8	1413.4	0.2967	5.3864	5.6831
−22	173.93	0.001498	0.6768	0.6783	80.7	1335.6	1416.2	0.3327	5.3188	5.6515
−20	190.22	0.001504	0.6222	0.6237	89.7	1329.3	1419.0	0.3648	5.2520	5.6205
−18	207.71	0.001510	0.5728	0.5743	98.8	1322.9	1421.7	0.4040	5.1860	5.5900
−16	226.45	0.001515	0.5280	0.5296	107.8	1316.5	1424.4	0.4393	5.1207	5.5600
−14	246.51	0.001521	0.4874	0.4889	116.9	1310.0	1427.0	0.4744	5.0561	5.5305
−12	267.95	0.001528	0.4505	0.4520	126.0	1303.5	1429.5	0.5093	4.9922	5.5015
−10	290.85	0.001534	0.4169	0.4185	135.2	1296.8	1432.0	0.5440	4.9290	5.4730
−8	315.25	0.001540	0.3863	0.3878	144.3	1290.1	1434.4	0.5785	4.8664	5.4449
−6	341.25	0.001546	0.3583	0.3599	153.5	1283.3	1436.8	0.6128	4.8045	5.4173
−4	368.90	0.001553	0.3328	0.3343	162.7	1276.4	1439.1	0.6469	4.7432	5.3901
−2	398.27	0.001559	0.3094	0.3109	171.9	1269.4	1441.3	0.6808	4.6825	5.3633
0	429.44	0.001566	0.2879	0.2895	181.1	1262.4	1443.5	0.7145	4.6223	5.3369
2	462.49	0.001573	0.2683	0.2698	190.4	1255.2	1445.6	0.7481	4.5627	5.3108
4	497.49	0.001580	0.2502	0.2517	199.6	1248.0	1447.6	0.7815	4.5037	5.2852
6	534.51	0.001587	0.2335	0.2351	208.9	1240.6	1449.6	0.8148	4.4451	5.2599
8	573.64	0.001594	0.2182	0.2198	218.3	1233.2	1451.5	0.8479	4.3871	5.2350
10	614.95	0.001601	0.2040	0.2056	227.6	1225.7	1453.3	0.8808	4.3295	5.2104
12	658.52	0.001608	0.1910	0.1926	237.0	1218.1	1455.1	0.9136	4.2725	5.1861
14	704.44	0.001616	0.1789	0.1805	246.4	1210.4	1456.8	0.9463	4.2159	5.1621
16	752.79	0.001623	0.1677	0.1693	255.9	1202.6	1458.5	0.9788	4.1597	5.1385
18	803.66	0.001631	0.1574	0.1590	265.4	1194.7	1460.0	1.0112	4.1039	5.1151
20	857.12	0.001639	0.1477	0.1494	274.9	1186.7	1461.5	1.0434	4.0486	5.0920
22	913.27	0.001647	0.1388	0.1405	284.4	1178.5	1462.9	1.0755	3.9937	5.0692
24	972.19	0.001655	0.1305	0.1322	294.0	1170.3	1464.3	1.1075	3.9392	5.0467
26	1033.97	0.001663	0.1228	0.1245	303.6	1162.0	1465.6	1.1394	3.8850	5.0244

(continued)

Source: Adapted from National Bureau of Standards Circular No. 142, *Tables of Thermodynamic Properties of Ammonia.*

Table A4.1   *Continued*

T (°C)	p (kPa)	Specific Volume (m³/kg)			Enthalpy (kJ/kg)			Entropy (kJ/kg K)		
		v_f	v_{fg}	v_g	h_f	h_{fg}	h_g	s_f	s_{fg}	s_g
28	1098.71	0.001671	0.1156	0.1173	313.2	1153.6	1466.8	1.1711	3.8312	5.0023
30	1166.49	0.001680	0.1089	0.1106	322.9	1145.0	1467.9	1.2028	3.7777	4.9805
32	1237.41	0.001689	0.1027	0.1044	332.6	1136.4	1469.0	1.2343	3.7246	4.9589
34	1311.55	0.001698	0.0969	0.0986	342.3	1127.6	1469.9	1.2656	3.6718	4.9374
36	1389.03	0.001707	0.0914	0.0931	352.1	1118.7	1470.8	1.2969	3.6192	4.9161
38	1469.92	0.001716	0.0863	0.0880	361.9	1109.7	1471.5	1.3281	3.5669	4.8950
40	1554.33	0.001726	0.0815	0.0833	371.7	1100.5	1472.2	1.3591	3.5148	4.8740
42	1642.35	0.001735	0.0771	0.0788	381.6	1091.2	1472.8	1.3901	3.4630	4.8530
44	1734.09	0.001745	0.0728	0.0746	391.5	1081.7	1473.2	1.4209	3.4112	4.8322
46	1829.65	0.001756	0.0689	0.0707	401.5	1072.0	1473.5	1.4518	3.3595	4.8113
48	1929.13	0.001766	0.0652	0.0669	411.5	1062.2	1473.7	1.4826	3.3079	4.7905
50	2032.62	0.001777	0.0617	0.0635	421.7	1052.0	1473.7	1.5135	3.2561	4.7696

Table A4.2   Thermodynamic Properties of Saturated Ammonia—Pressure Table

p (kPa)	T (°C)	Specific Volume (m³/kg)			Enthalpy (kJ/kg)			Entropy (kJ/kg K)		
		v_f	v_{fg}	v_g	h_f	h_{fg}	h_g	s_f	s_{fg}	s_g
40	−50.4	.001421	2.683	2.684	−45.9	1417.8	1371.9	−0.201	6.370	6.169
60	−43.3	.001442	1.834	1.836	−14.6	1398.3	1383.6	−0.063	6.063	6.023
80	−37.9	.001456	1.401	1.402	9.1	1383.2	1392.3	0.039	5.883	5.922
100	−33.6	.001467	1.136	1.138	28.4	1370.7	1399.1	0.120	5.725	5.845
150	−25.3	.001489	0.777	0.778	66.1	1345.6	1411.7	0.274	5.432	5.706
200	−18.9	.001507	0.593	0.594	94.9	1325.7	1420.6	0.389	5.217	5.605
300	−9.2	.001535	0.405	0.406	138.7	1294.2	1432.9	0.557	4.906	5.463
400	−1.9	.001559	0.309	0.310	172.4	1269.0	1441.4	0.683	4.681	5.364
500	4.1	.001579	0.249	0.251	200.3	1247.4	1447.7	0.784	4.501	5.285
600	9.3	.001598	0.209	0.210	224.3	1228.3	1452.6	0.869	4.351	5.220
800	17.9	.001631	0.157	0.159	264.7	1195.3	1460.0	1.009	4.109	5.118
1000	24.9	.001660	0.126	0.128	298.3	1166.7	1465.0	1.122	3.916	5.038
1200	31.0	.001686	0.105	0.107	327.5	1141.0	1468.4	1.218	3.754	4.971
1400	36.3	.001710	0.090	0.092	353.4	1117.4	1470.8	1.301	3.613	4.914
1600	41.1	.001732	0.079	0.081	376.9	1095.5	1472.4	1.375	3.488	4.864
1800	45.4	.001752	0.070	0.072	398.5	1074.9	1473.4	1.442	3.376	4.819
2000	49.4	.001771	0.633	0.065	418.4	1055.4	1473.9	1.504	3.274	4.778

Table A4.3 Thermodynamic Properties of Superheated Ammonia

Abs. Press. (kPa) Sat. Temp. (°C)		Temperature (°C)											
		−20	−10	0	10	20	30	40	50	60	70	80	100
50 (−46.54)	v	2.4474	2.5481	2.6482	2.7479	2.8473	2.9464	3.0453	3.1441	3.2427	3.3413	3.4397	
	h	1435.8	1457.0	1478.1	1499.2	1520.4	1541.7	1563.0	1584.5	1606.1	1627.8	1649.7	
	s	6.3256	6.4077	6.4865	6.5625	6.6360	6.7073	6.7766	6.8441	6.9099	6.9743	7.0372	
75 (39.18)	v	1.6233	1.6915	1.7591	1.8263	1.8932	1.9597	2.0261	2.0923	2.1584	2.2244	2.2903	
	h	1433.0	1454.7	1476.1	1497.5	1518.9	1540.3	1561.8	1583.4	1605.1	1626.9	1648.9	
	s	6.1190	6.2028	6.2828	6.3597	6.4339	6.5058	6.5756	6.6434	6.7096	6.7742	6.8373	
100 (−33.61)	v	1.2110	1.2631	1.3145	1.3654	1.4160	1.4664	1.5165	1.5664	1.6163	1.6659	1.7155	1.8145
	h	1430.1	1452.2	1474.1	1495.7	1517.3	1538.9	1560.5	1582.2	1604.1	1626.0	1648.0	1692.6
	s	5.9695	6.0552	6.1366	6.2144	6.2894	6.3618	6.4321	6.5003	6.5668	6.6316	6.6950	6.8177
125 (−29.08)	v	0.9635	1.0059	1.0476	1.0889	1.1297	1.1703	1.2107	1.2509	1.2909	1.3309	1.3707	1.4501
	h	1427.2	1449.8	1472.0	1493.9	1515.7	1537.5	1559.3	1581.1	1603.0	1625.0	1647.2	1691.8
	s	5.8512	5.9389	6.0217	6.1006	6.1763	6.2494	6.3201	6.3887	6.4555	6.5206	6.5842	6.7072
150 (−25.23)	v	0.7984	0.8344	0.8697	0.9045	0.9388	0.9729	1.0068	1.0405	1.0740	1.1074	1.1408	1.2072
	h	1424.1	1447.3	1469.8	1492.1	1514.1	1536.1	1558.0	1580.0	1602.0	1624.1	1646.3	1691.1
	s	5.7526	5.8424	5.9266	6.0066	6.0831	6.1568	6.2280	6.2970	6.3641	6.4295	6.4933	6.6167
200 (−18.86)	v		0.6199	0.6471	0.6738	0.7001	0.7261	0.7519	0.7774	0.8629	6.8282	0.8533	0.9035
	h		1442.0	1465.5	1488.4	1510.9	1533.2	1555.5	1577.7	1599.9	1622.2	1644.6	1689.6
	s		5.6863	5.7737	5.8559	5.9342	6.0091	6.0813	6.1512	6.2189	6.2849	6.3491	6.4732
250 (−13.67)	v		0.4910	0.5135	0.5354	0.5568	0.5780	0.5989	0.6196	0.6401	0.6605	0.6809	0.7212
	h		1436.6	1461.0	1484.5	1507.6	1530.3	1552.9	1575.4	1597.8	1620.3	1642.8	1688.2
	s		5.5609	5.6517	5.7365	5.8165	5.8928	5.9661	6.0368	6.1052	6.1717	6.2365	6.3613
300 (−9.23)	v			0.4243	0.4430	0.4613	0.4792	0.4968	0.5143	0.5316	0.5488	0.5658	0.5997
	h			1456.3	1480.6	1504.2	1527.4	1550.3	1573.0	1595.7	1618.4	1641.1	1686.7
	s			5.5493	5.6366	5.7186	5.7963	5.8707	5.9423	6.0114	6.0785	6.1437	6.2693
350 (−5.35)	v			0.3605	0.3770	0.3929	0.4086	0.4239	0.4391	0.4541	0.4689	0.4837	0.5129
	h			1451.5	1476.5	1500.7	1524.4	1547.6	1570.7	1593.6	1616.5	1639.3	1685.2
	s			5.4600	5.5502	5.6342	5.7135	5.7890	5.8615	5.9314	5.9990	6.0647	6.1910
400 (−1.89)	v			0.3125	0.3274	0.3417	0.3556	0.3692	0.3826	0.3959	0.4090	0.4220	0.4478
	h			1446.5	1472.4	1497.2	1521.3	1544.9	1568.3	1591.5	1614.5	1637.6	1683.7
	s			5.3803	5.4735	5.5597	5.6405	5.7173	5.7907	5.8613	5.9296	5.9957	6.1228
450 (1.26)	v			0.2752	0.2887	0.3017	0.3143	0.3266	0.3387	0.3506	0.3624	0.3740	0.3971
	h			1441.3	1468.1	1493.6	1518.2	1542.2	1565.9	1589.3	1612.6	1635.8	1682.2
	s			5.3078	5.4042	5.4926	5.5752	5.6532	5.7275	5.7989	5.8678	5.9345	6.0623

(continued)

Table A4.3   *Continued*

Abs. Press. (kPa) Sat. Temp. (°C)		Temperature (°C)											
		20	30	40	50	60	70	80	100	120	140	160	180
500 (4.14)	v	0.2698	0.2813	0.2926	0.3036	0.3144	0.3251	0.3357	0.3565	0.3771	0.3975		
	h	1489.9	1515.0	1539.5	1563.4	1587.1	1610.6	1634.0	1680.7	1727.5	1774.7		
	s	5.4314	5.5157	5.5950	5.6704	5.7425	5.8120	5.8793	6.0079	6.1301	6.2472		
600 (9.29)	v	0.2217	0.2317	0.2414	0.2508	0.2600	0.2691	0.2781	0.2957	0.3130	0.3302		
	h	1482.4	1508.6	1533.8	1558.5	1582.7	1606.6	1630.4	1677.7	1724.9	1772.4		
	s	5.3222	5.4102	5.4923	5.5697	5.6436	5.7144	5.7826	5.9129	6.0363	6.1541		
700 (13.81)	v	0.1874	0.1963	0.2048	0.2131	0.2212	0.2291	0.2369	0.2522	0.2672	0.2821		
	h	1474.5	1501.9	1528.1	1553.4	1578.2	1602.6	1626.8	1674.6	1722.4	1770.2		
	s	5.2259	5.3179	5.4029	5.4826	5.5582	5.6303	5.6997	5.8316	5.9562	6.0749		
800 (17.86)	v	0.1615	0.1696	0.1773	0.1848	0.1920	0.1991	0.2060	0.2196	0.2329	0.2459	0.2589	
	h	1466.3	1495.0	1522.2	1548.3	1573.7	1598.6	1623.1	1671.6	1719.8	1768.0	1816.4	
	s	5.1387	5.2351	5.3232	5.4053	5.4827	5.5562	5.6268	5.7603	5.8861	6.0057	6.1202	
900 (21.54)	v		0.1488	0.1559	0.1627	0.1693	0.1757	0.1820	0.1942	0.2061	0.2178	0.2294	
	h		1488.0	1516.2	1543.0	1569.1	1594.4	1619.4	1668.5	1717.1	1765.7	1814.4	
	s		5.1593	5.2508	5.3354	5.4147	5.4897	5.5614	5.6968	5.8237	5.9442	6.0594	
1000 (24.91)	v		0.1321	0.1388	0.1450	0.1511	0.1570	0.1627	0.1739	0.1847	0.1954	0.2058	0.2162
	h		1480.6	1510.0	1537.7	1564.4	1590.3	1615.6	1665.4	1714.5	1763.4	1812.4	1861.7
	s		5.0889	5.1840	5.2713	5.3525	5.4299	5.5021	5.6392	5.7674	5.8888	6.0047	6.1159
1200 (30.96)	v			0.1129	0.1185	0.1238	0.1289	0.1338	0.1434	0.1526	0.1616	0.1705	0.1792
	h			1497.1	1526.6	1554.7	1581.7	1608.0	1659.2	1709.2	1758.9	1808.5	1858.2
	s			5.0629	5.1560	5.2416	5.3215	5.3970	5.5379	5.6687	5.7919	5.9091	6.0214
1400 (36.28)	v			0.0944	0.0995	0.1042	0.1088	0.1132	0.1216	0.1297	0.1376	0.1452	0.1528
	h			1483.4	1515.1	1544.7	1573.0	1600.2	1652.8	1703.9	1754.3	1804.5	1854.7
	s			4.9534	5.0530	5.1434	5.2270	5.3053	5.4501	5.5836	5.7087	5.8273	5.9406
1600 (41.05)	v				0.0851	0.0895	0.0937	0.0977	0.1053	0.1125	0.1195	0.1263	0.1330
	h				1502.9	1534.4	1564.0	1592.3	1646.4	1698.5	1749.7	1800.5	1851.2
	s				4.9584	5.0543	5.1419	5.2232	5.3722	5.5084	5.6355	5.7555	5.8699
1800 (45.39)	v				0.0739	0.0781	0.0820	0.0856	0.0926	0.0992	0.1055	0.1116	0.1177
	h				1490.0	1523.5	1554.6	1584.1	1639.8	1693.1	1745.1	1796.5	1847.7
	s				4.8693	4.9715	5.0635	5.1482	5.3018	5.4409	5.5699	5.6914	5.8069
2000 (49.38)	v				0.0648	0.0688	0.0725	0.0760	0.0824	0.0885	0.0943	0.0999	0.1054
	h				1476.1	1512.0	1544.9	1575.6	1633.2	1687.6	1740.4	1792.4	1844.1
	s				4.7834	4.8930	4.9902	5.0786	5.2371	5.3793	5.5104	5.6333	5.7499

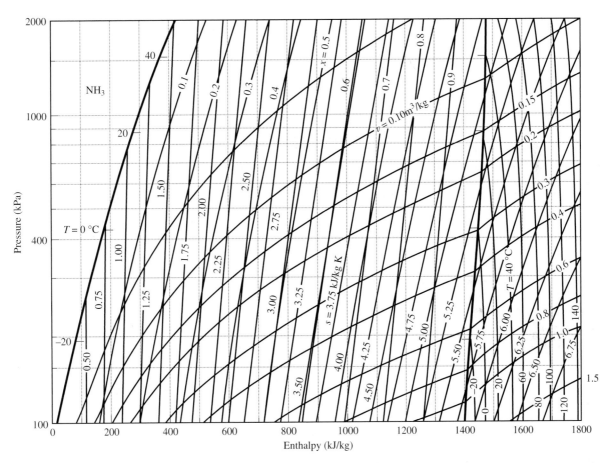

Figure A2 Pressure–enthalpy diagram for ammonia. (*Source:* J. B. Jones and G. A. Hawkins, *Engineering Thermodynamics,* 2nd ed. John Wiley & Sons, Inc., New York, 1986.)

Table A5.1 Thermodynamic Properties of Saturated Refrigerant-12 (Dichlorodifluoromethane)—Temperature Table

T (°C)	p (MPa)	Specific Volume (m³/kg)			Enthalpy (kJ/kg)			Entropy (kJ/kg K)		
		v_f	v_{fg}	v_g	h_f	h_{fg}	h_g	s_f	s_{fg}	s_g
−90	0.0028	0.000608	4.414937	4.415545	−43.243	189.618	146.375	−0.2084	1.0352	0.8268
−85	0.0042	0.000612	3.036704	3.037316	−38.968	187.608	148.640	−0.1854	0.9970	0.8116
−80	0.0062	0.000617	2.137728	2.138345	−34.688	185.612	150.924	−0.1630	0.9609	0.7979
−75	0.0088	0.000622	1.537030	1.537651	−30.401	183.625	153.224	−0.1411	0.9266	0.7855
−70	0.0123	0.000627	1.126654	1.127280	−26.103	181.640	155.536	−0.1197	0.8940	0.7744
−65	0.0168	0.000632	0.840534	0.841166	−21.793	179.651	157.857	−0.0987	0.8630	0.7643
−60	0.0226	0.000637	0.637274	0.637910	−17.469	177.653	160.184	−0.0782	0.8334	0.7552
−55	0.0300	0.000642	0.490358	0.491000	−13.129	175.641	162.512	−0.0581	0.8051	0.7470
−50	0.0391	0.000648	0.382457	0.383105	−8.772	173.611	164.840	−0.0384	0.7779	0.7396
−45	0.0504	0.000654	0.302029	0.302682	−4.396	171.558	167.163	−0.0190	0.7519	0.7329
−40	0.0642	0.000659	0.241251	0.241910	−0.000	169.479	169.479	−0.0000	0.7269	0.7269
−35	0.0807	0.000666	0.194732	0.195398	4.416	167.368	171.784	0.0187	0.7027	0.7214
−30	0.1004	0.000672	0.158703	0.159375	8.854	165.222	174.076	0.0371	0.6795	0.7165
−25	0.1237	0.000679	0.130487	0.131166	13.315	163.037	176.352	0.0552	0.6570	0.7121
−20	0.1509	0.000685	0.108162	0.108847	17.800	160.810	178.610	0.0730	0.6352	0.7082
−15	0.1826	0.000693	0.090326	0.091018	22.312	158.534	180.846	0.0906	0.6141	0.7046
−10	0.2191	0.000700	0.075946	0.076646	26.851	156.207	183.058	0.1079	0.5936	0.7014
−5	0.2610	0.000708	0.064255	0.064963	31.420	153.823	185.243	0.1250	0.5736	0.6986
0	0.3086	0.000716	0.054673	0.055389	36.022	151.376	187.397	0.1418	0.5542	0.6960
5	0.3626	0.000724	0.046761	0.047485	40.659	148.859	189.518	0.1585	0.5351	0.6937
10	0.4233	0.000733	0.040180	0.040914	45.337	146.265	191.602	0.1750	0.5165	0.6916
15	0.4914	0.000743	0.034671	0.035413	50.058	143.586	193.644	0.1914	0.4983	0.6897
20	0.5673	0.000752	0.030028	0.030780	54.828	140.812	195.641	0.2076	0.4803	0.6879
25	0.6516	0.000763	0.026091	0.026854	59.653	137.933	197.586	0.2237	0.4626	0.6863
30	0.7449	0.000774	0.022734	0.023508	64.539	134.936	199.475	0.2397	0.4451	0.6848
35	0.8477	0.000786	0.019855	0.020641	69.494	131.805	201.299	0.2557	0.4277	0.6834
40	0.9607	0.000798	0.017373	0.018171	74.527	128.525	203.051	0.2716	0.4104	0.6820
45	1.0843	0.000811	0.015220	0.016032	79.647	125.074	204.722	0.2875	0.3931	0.6806
50	1.2193	0.000826	0.013344	0.014170	84.868	121.430	206.298	0.3034	0.3758	0.6792
55	1.3663	0.000841	0.011701	0.012542	90.201	117.565	207.766	0.3194	0.3582	0.6777
60	1.5259	0.000858	0.010253	0.011111	95.665	113.443	209.109	0.3355	0.3405	0.6760
65	1.6988	0.000877	0.008971	0.009847	101.279	109.024	210.303	0.3518	0.3224	0.6742
70	1.8858	0.000897	0.007828	0.008725	107.067	104.255	211.321	0.3683	0.3038	0.6721
75	2.0874	0.000920	0.006802	0.007723	113.058	99.068	212.126	0.3851	0.2845	0.6697
80	2.3046	0.000946	0.005875	0.006821	119.291	93.373	212.665	0.4023	0.2644	0.6667
85	2.5380	0.000976	0.005029	0.006005	125.818	87.047	212.865	0.4201	0.2430	0.6631
90	2.7885	0.001012	0.004246	0.005258	132.708	79.907	212.614	0.4385	0.2200	0.6585
95	3.0569	0.001056	0.003508	0.004563	140.068	71.658	211.726	0.4579	0.1946	0.6526
100	3.3440	0.001113	0.002790	0.003903	148.076	61.768	209.843	0.4788	0.1655	0.6444
105	3.6509	0.001197	0.002045	0.003242	157.085	49.014	206.099	0.5023	0.1296	0.6319
110	3.9784	0.001364	0.001098	0.002462	168.059	28.425	196.484	0.5322	0.0742	0.6064
112	4.1155	0.001792	0.000005	0.001797	174.920	0.151	175.071	0.5651	0.0004	0.5655

Source: Copyright 1955 and 1956, E. I. du Pont de Nemours & Company, Inc. Reprinted by permission. Adapted from English units.

Table A5.2 Thermodynamic Properties of Saturated Refrigerant-12—Pressure Table

p (MPa)	T (°C)	Specific Volume (m³/kg)		Enthalpy (kJ/kg)			Entropy (kJ/kg K)	
		v_f	v_g	h_f	h_{fg}	h_g	s_f	s_g
0.06	−41.42	0.0006578	0.2575	−1.25	170.19	168.94	−0.0054	0.7290
0.10	−30.10	0.0006719	0.1600	8.78	165.37	174.15	0.0368	0.7171
0.12	−25.74	0.0006776	0.1349	12.66	163.48	176.14	0.0526	0.7133
0.14	−21.91	0.0006828	0.1168	16.09	161.78	177.87	0.0663	0.7102
0.16	−18.49	0.0006876	0.1031	19.18	160.23	179.41	0.0784	0.7076
0.18	−15.38	0.0006921	0.09225	21.98	158.82	180.80	0.0893	0.7054
0.20	−12.53	0.0006862	0.08354	24.57	157.50	182.07	0.0992	0.7035
0.24	−7.42	0.0007040	0.07033	29.23	155.09	184.32	0.1168	0.7004
0.28	−2.93	0.0007111	0.06076	33.35	152.92	186.27	0.1321	0.6980
0.32	1.11	0.0007177	0.05351	37.08	150.92	188.00	0.1457	0.6960
0.40	8.15	0.0007299	0.04321	43.64	147.33	190.97	0.1691	0.6928
0.50	15.60	0.0007438	0.03482	50.67	143.35	194.02	0.1935	0.6899
0.60	22.00	0.0007566	0.02913	56.80	139.77	196.57	0.2142	0.6878
0.70	27.65	0.0007686	0.02501	62.29	136.45	198.74	0.2324	0.6860
0.80	32.74	0.0007802	0.02188	67.30	133.33	200.63	0.2487	0.6845
0.90	37.37	0.0007914	0.01942	71.93	130.36	202.29	0.2634	0.6832
1.0	41.64	0.0008023	0.01744	76.26	127.50	203.76	0.2770	0.6820
1.2	49.31	0.0008237	0.01441	84.21	122.03	206.24	0.3015	0.6799
1.4	56.09	0.0008448	0.01222	91.46	116.76	208.22	0.3232	0.6778
1.6	62.19	0.0008660	0.01054	98.19	111.62	209.81	0.3329	0.6758

Table A5.3 Thermodynamic Properties of Superheated Refrigerant-12

T (°C)	p = 0.05 MPa (T_sat = −45.2°C)			p = 0.10 MPa (T_sat = −30.1°C)		
	v (m³/kg)	h (kJ/kg)	s (kJ/kg K)	v (m³/kg)	h (kJ/kg)	s (kJ/kg K)
−20	0.341857	181.042	0.7912	0.167701	179.861	0.7401
−10	0.356227	186.757	0.8133	0.175222	185.707	0.7628
0	0.370508	192.567	0.8350	0.182647	191.628	0.7849
10	0.384716	198.471	0.8562	0.189994	197.628	0.8064
20	0.398863	204.469	0.8770	0.197277	203.707	0.8275
30	0.412959	210.557	0.8974	0.204506	209.866	0.8482
40	0.427012	216.733	0.9175	0.211691	216.104	0.8684
50	0.441030	222.997	0.9372	0.218839	222.421	0.8883
60	0.455017	229.344	0.9565	0.225955	228.815	0.9078
70	0.468978	235.774	0.9755	0.233044	235.285	0.9269
80	0.482917	242.282	0.9942	0.240111	241.829	0.9457
90	0.496838	248.868	1.0126	0.247159	248.446	0.9642

(continued)

Table A5.3   *Continued*

T (°C)	p = 0.15 MPa (T_sat = −20.2°C)			p = 0.20 MPa (T_sat = −12.5°C)		
	v (m³/kg)	h (kJ/kg)	s (kJ/kg K)	v (m³/kg)	h (kJ/kg)	s (kJ/kg K)
−10	0.114716	184.619	0.7318			
0	0.119866	190.660	0.7543	0.088608	189.669	0.7320
10	0.124932	196.762	0.7763	0.092550	195.878	0.7543
20	0.129930	202.927	0.7977	0.096418	202.135	0.7760
30	0.134873	209.160	0.8186	0.100228	208.446	0.7972
40	0.139768	215.463	0.8390	0.103989	214.814	0.8178
50	0.144625	221.835	0.8591	0.107710	221.243	0.8381
60	0.149450	228.277	0.8787	0.111397	227.735	0.8578
70	0.154247	234.789	0.8980	0.115055	234.291	0.8772
80	0.159020	241.371	0.9169	0.118690	240.910	0.8962
90	0.163774	248.020	0.9354	0.122304	247.593	0.9149
100				0.125901	254.339	0.9332
110				0.129483	261.147	0.9512

T (°C)	p = 0.25 MPa (T_sat = −6.3°C)			p = 0.30 MPa (T_sat = −0.8°C)		
	v (m³/kg)	h (kJ/kg)	s (kJ/kg K)	v (m³/kg)	h (kJ/kg)	s (kJ/kg K)
0	0.069752	188.644	0.7139	0.057150	187.583	0.6984
10	0.073024	194.969	0.7366	0.059984	194.034	0.7216
20	0.076218	201.322	0.7587	0.062734	200.490	0.7440
30	0.079350	207.715	0.7801	0.065418	206.969	0.7658
40	0.082431	214.153	0.8010	0.068049	213.480	0.7869
50	0.085470	220.642	0.8214	0.070635	220.030	0.8075
60	0.088474	227.185	0.8413	0.073185	226.627	0.8276
70	0.091449	233.785	0.8608	0.075705	233.273	0.8473
80	0.094398	240.443	0.8800	0.078200	239.971	0.8665
90	0.097327	247.160	0.8987	0.080673	246.723	0.8853
100	0.100238	253.936	0.9171	0.083127	253.530	0.9038
110	0.103134	260.770	0.9352	0.085566	260.391	0.9220

Table A5.3   *Continued*

T (°C)	p = 0.40 MPa (T_sat = 8.2°C)			p = 0.50 MPa (T_sat = 15.6°C)		
	v (m³/kg)	h (kJ/kg)	s (kJ/kg K)	v (m³/kg)	h (kJ/kg)	s (kJ/kg K)
20	0.045836	198.762	0.7199	0.035646	196.935	0.6999
30	0.047971	205.428	0.7423	0.037464	203.814	0.7230
40	0.050046	212.095	0.7639	0.039214	210.656	0.7452
50	0.052072	218.779	0.7849	0.040911	217.484	0.7667
60	0.054059	225.488	0.8054	0.042565	224.315	0.7875
70	0.056014	232.230	0.8253	0.044184	231.161	0.8077
80	0.057941	239.012	0.8448	0.045774	238.031	0.8275
90	0.059846	245.837	0.8638	0.047340	244.932	0.8467
100	0.061731	252.707	0.8825	0.048886	251.869	0.8656
110	0.063600	259.624	0.9008	0.050415	258.845	0.8840
120	0.065455	266.590	0.9187	0.051929	265.862	0.9021
130	0.067298	273.605	0.9364	0.053430	272.923	0.9198

T (°C)	p = 0.60 MPa (T_sat = 22.0°C)			p = 0.70 MPa (T_sat = 27.7°C)		
	v (m³/kg)	h (kJ/kg)	s (kJ/kg K)	v (m³/kg)	h (kJ/kg)	s (kJ/kg K)
30	0.030422	202.116	0.7063			
40	0.031966	209.154	0.7291	0.026761	207.580	0.7148
50	0.033450	216.141	0.7511	0.028100	214.745	0.7373
60	0.034887	223.104	0.7723	0.029387	221.854	0.7590
70	0.036285	230.062	0.7929	0.030632	228.931	0.7799
80	0.037653	237.027	0.8129	0.031843	235.997	0.8002
90	0.038995	244.009	0.8324	0.033027	243.066	0.8199
100	0.040316	251.016	0.8514	0.034189	250.146	0.8392
110	0.041619	258.053	0.8700	0.035332	257.247	0.8579
120	0.042907	265.124	0.8882	0.036458	264.374	0.8763
130	0.044181	272.231	0.9061	0.037572	271.531	0.8943
140				0.038673	278.720	0.9119
150				0.039764	285.946	0.9292

(continued)

Table A5.3   *Continued*

T (°C)	$p = 0.80$ MPa ($T_{sat} = 32.8°C$)			$p = 0.90$ MPa ($T_{sat} = 37.4°C$)		
	v (m³/kg)	h (kJ/kg)	s (kJ/kg K)	v (m³/kg)	h (kJ/kg)	s (kJ/kg K)
40	0.022830	205.924	0.7016	0.019744	204.170	0.6982
50	0.024068	213.290	0.7248	0.020912	211.765	0.7131
60	0.025247	220.558	0.7469	0.022012	219.212	0.7358
70	0.026380	227.766	0.7682	0.023062	226.564	0.7575
80	0.027477	234.941	0.7888	0.024072	233.856	0.7785
90	0.028545	242.101	0.8088	0.025051	241.113	0.7987
100	0.029588	249.260	0.8283	0.026005	248.355	0.8184
110	0.030612	256.428	0.8472	0.026937	255.593	0.8376
120	0.031619	263.613	0.8657	0.027851	262.839	0.8562
130	0.032612	270.820	0.8838	0.028751	270.100	0.8745
140	0.033592	278.055	0.9016	0.029639	277.381	0.8923
150	0.034563	285.320	0.9189	0.030515	284.687	0.9098

T (°C)	$p = 1.00$ MPa ($T_{sat} = 41.7°C$)			$p = 1.20$ MPa ($T_{sat} = 49.3°C$)		
	v (m³/kg)	h (kJ/kg)	s (kJ/kg K)	v (m³/kg)	h (kJ/kg)	s (kJ/kg K)
50	0.018366	210.162	0.7021	0.014483	206.661	0.6812
60	0.019410	217.810	0.7254	0.015463	214.805	0.7060
70	0.020397	225.319	0.7476	0.016368	222.687	0.7293
80	0.021341	232.739	0.7689	0.017221	230.398	0.7514
90	0.022251	240.101	0.7895	0.018032	237.995	0.7727
100	0.023133	247.430	0.8094	0.018812	245.518	0.7931
110	0.023993	254.743	0.8287	0.019567	252.993	0.8129
120	0.024835	262.053	0.8475	0.020301	260.441	0.8320
130	0.025661	269.369	0.8659	0.021018	267.875	0.8507
140	0.026474	276.699	0.8839	0.021721	275.307	0.8689
150	0.027275	284.047	0.9015	0.022412	282.745	0.8867
160	0.028068	291.419	0.9187	0.023093	290.195	0.9041

Table A5.3   *Continued*

T (°C)	p = 1.40 MPa (T_sat = 56.1°C)			p = 1.60 MPa (T_sat = 62.2°C)		
	v (m³/kg)	h (kJ/kg)	s (kJ/kg K)	v (m³/kg)	h (kJ/kg)	s (kJ/kg K)
60	0.012579	211.457	0.6876			
70	0.013448	219.822	0.7123	0.011208	216.650	0.6959
80	0.014247	227.891	0.7355	0.011984	225.177	0.7204
90	0.014997	235.766	0.7575	0.012698	233.390	0.7433
100	0.015710	243.512	0.7785	0.013366	241.397	0.7651
110	0.016393	251.170	0.7988	0.014000	249.264	0.7859
120	0.017053	258.770	0.8183	0.014608	257.035	0.8059
130	0.017695	266.334	0.8373	0.015195	264.742	0.8253
140	0.018321	273.877	0.8558	0.015765	272.406	0.8440
150	0.018934	281.411	0.8738	0.016320	280.044	0.8623
160	0.019535	288.946	0.8914	0.016864	287.669	0.8801
170				0.017398	295.290	0.8975
180				0.017923	302.914	0.9145

T (°C)	p = 1.80 MPa (T_sat = 67.9°C)			p = 2.00 MPa (T_sat = 72.9°C)		
	v (m³/kg)	h (kJ/kg)	s (kJ/kg K)	v (m³/kg)	h (kJ/kg)	s (kJ/kg K)
70	0.009406	213.049	0.6794			
80	0.010187	222.198	0.7057	0.008704	218.859	0.6909
90	0.010884	230.835	0.7298	0.009406	228.056	0.7166
100	0.011526	239.155	0.7524	0.010035	236.760	0.7402
110	0.012126	247.264	0.7739	0.010615	245.154	0.7624
120	0.012697	255.228	0.7944	0.011159	253.341	0.7835
130	0.013244	263.094	0.8141	0.011676	261.384	0.8037
140	0.013772	270.891	0.8332	0.012172	269.327	0.8232
150	0.014284	278.642	0.8518	0.012651	277.201	0.8420
160	0.014784	286.364	0.8698	0.013116	285.027	0.8603
170	0.015272	294.069	0.8874	0.013570	292.822	0.8781
180	0.015752	301.767	0.9046	0.014013	300.598	0.8955

(continued)

Table A5.3   *Continued*

T (°C)	p = 2.50 MPa (T_{sat} = 84.2°C)			p = 3.00 MPa (T_{sat} = 94.0°C)		
	v (m³/kg)	h (kJ/kg)	s (kJ/kg K)	v (m³/kg)	h (kJ/kg)	s (kJ/kg K)
90	0.006595	219.562	0.6823			
100	0.007264	229.852	0.7103	0.005231	220.529	0.6770
110	0.007837	239.271	0.7352	0.005886	232.068	0.7075
120	0.008351	248.192	0.7582	0.006419	242.208	0.7336
130	0.008827	256.794	0.7798	0.006887	251.632	0.7573
140	0.009273	265.180	0.8003	0.007313	260.620	0.7793
150	0.009697	273.414	0.8200	0.007709	269.319	0.8001
160	0.010104	281.540	0.8390	0.008083	277.817	0.8200
170	0.010497	289.589	0.8574	0.008439	286.171	0.8391
180	0.011879	297.583	0.8752	0.008782	294.422	0.8575
190	0.011250	305.540	0.8926	0.009114	302.597	0.8753
200	0.011614	313.472	0.9095	0.009436	310.718	0.8927

T (°C)	p = 3.50 MPa (T_{sat} = 102.6°C)			p = 4.00 MPa (T_{sat} = 110.3°C)		
	v (m³/kg)	h (kJ/kg)	s (kJ/kg K)	v (m³/kg)	h (kJ/kg)	s (kJ/kg K)
110	0.004324	222.121	0.6750			
120	0.004959	234.875	0.7078	0.003736	224.863	0.6771
130	0.005456	245.661	0.7349	0.004325	238.443	0.7111
140	0.005884	255.524	0.7591	0.004781	249.703	0.7386
150	0.006270	264.846	0.7814	0.005172	259.904	0.7630
160	0.006626	273.817	0.8023	0.005522	269.492	0.7854
170	0.006961	282.545	0.8222	0.005845	278.684	0.8063
180	0.007279	291.100	0.8413	0.006147	287.602	0.8262
190	0.007584	299.528	0.8597	0.006434	296.326	0.8453
200	0.007878	307.864	0.8775	0.006708	304.906	0.8636
210				0.006972	313.380	0.8813
220				0.007228	321.774	0.8985
230				0.007477	330.108	0.9152

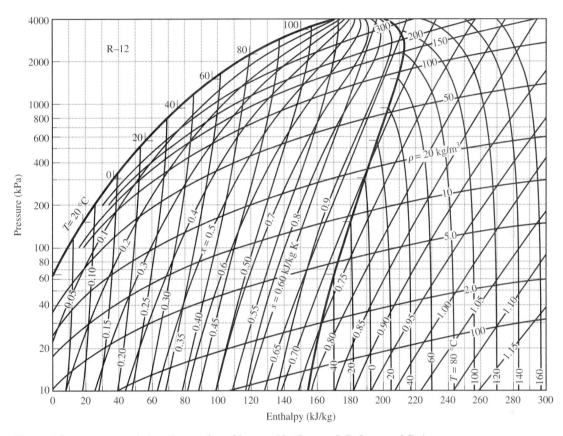

Figure A3 Pressure–enthalpy diagram for refrigerant-12. (*Source:* J. B. Jones and G. A. Hawkins, *Engineering Thermodynamics,* 2nd ed. John Wiley & Sons, Inc., New York, 1986.)

Table A6.1 Thermodynamic Properties of Saturated Refrigerant-134a (1,1,1,2-Tetrafluoroethane)—Temperature Table

T (°C)	p (kPa)	Specific Volume (m³/kg)		Enthalpy (kJ/kg)			Entropy (kJ/kg K)	
		v_f	v_g	h_f	h_{fg}	h_g	s_f	s_g
−40	51.14	0.0007	0.3614	148.4	225.9	374.3	0.7967	1.7655
−35	66.07	0.0007	0.2843	154.6	222.8	377.4	0.8231	1.7586
−30	84.29	0.0007	0.2260	160.9	219.6	380.6	0.8492	1.7525
−25	106.32	0.0007	0.1817	167.3	216.4	383.7	0.8750	1.747
−20	132.67	0.0007	0.1474	173.7	213.1	386.8	0.9005	1.7422
−15	163.90	0.0007	0.1207	180.2	209.7	389.8	0.9257	1.7379
−10	200.60	0.0008	0.0996	186.7	206.2	392.9	0.9507	1.7341
−5	243.39	0.0008	0.0828	193.3	202.5	395.9	0.9755	1.7308
0	292.93	0.0008	0.0693	200.0	198.8	398.8	1.0000	1.7278
5	349.87	0.0008	0.0583	206.8	194.9	401.7	1.0244	1.7252
10	414.92	0.0008	0.0494	213.6	190.9	404.5	1.0485	1.7229
15	488.78	0.0008	0.0421	220.5	186.8	407.3	1.0726	1.7208
20	572.25	0.0008	0.0360	227.5	182.5	410.0	1.0964	1.7189
25	666.06	0.0008	0.0309	234.6	178.0	412.6	1.1202	1.7171
30	771.02	0.0008	0.0266	241.8	173.3	415.1	1.1439	1.7155
35	887.91	0.0009	0.0230	249.2	168.3	417.5	1.1676	1.7138
40	1017.61	0.0009	0.0200	256.6	163.2	419.8	1.1912	1.7122
45	1161.01	0.0009	0.0174	264.2	157.7	421.9	1.2148	1.7105
50	1319.00	0.0009	0.0151	271.9	151.9	423.8	1.2384	1.7086
55	1492.59	0.0009	0.0132	279.8	145.8	425.6	1.2622	1.7064
60	1682.76	0.0010	0.0115	287.9	139.2	427.1	1.2861	1.7039
65	1890.54	0.0010	0.0100	296.2	132.1	428.3	1.3102	1.7009
70	2117.34	0.0010	0.0087	304.8	124.4	429.1	1.3347	1.6971
75	2364.31	0.0010	0.0075	313.7	115.8	429.5	1.3597	1.6924
80	2632.97	0.0011	0.0065	322.9	106.3	429.2	1.3854	1.6863
85	2925.11	0.0011	0.0055	332.8	95.3	428.1	1.4121	1.6782
90	3242.87	0.0012	0.0046	343.4	82.1	425.5	1.4406	1.6668
95	3589.44	0.0013	0.0037	355.6	64.9	420.5	1.4727	1.6489
100	3969.94	0.0015	0.0027	373.2	33.8	407.0	1.5187	1.6092
101	4051.35	0.0018	0.0022	383.0	13.0	396.0	1.5447	1.5794

Source: Adapted from *Thermodynamic Properties of HFC-134a (1,1,1,2-tetrafluoroethane),* Dupont Company, Wilmington, Delaware, 1993, with permission.

Table A6.2 Thermodynamic Properties of Saturated Refrigerant 134a—Pressure Table

p (MPa)	T (°C)	Specific Volume (m³/kg)		Enthalpy (kJ/kg)			Entropy (kJ/kg K)	
		v_f	v_g	h_f	h_{fg}	h_g	s_f	s_g
0.06	−36.91	0.00071	0.31114	152.2	224.0	376.2	0.8130	1.7611
0.08	−31.09	0.00072	0.23747	159.5	220.4	379.9	0.8435	1.7538
0.10	−26.34	0.00073	0.19246	165.6	217.2	382.8	0.8681	1.7484
0.12	−22.29	0.00073	0.16207	170.7	214.7	385.4	0.8888	1.7443
0.14	−18.75	0.00074	0.14010	175.3	212.2	387.5	0.9068	1.7411
0.16	−15.58	0.00074	0.12344	179.4	210.1	389.5	0.9228	1.7384
0.18	−12.71	0.00075	0.11039	183.2	208.0	391.2	0.9372	1.7361
0.20	−10.08	0.00075	0.09985	186.6	206.2	392.8	0.9503	1.7342
0.22	−7.64	0.00076	0.09117	189.8	204.5	394.3	0.9624	1.7325
0.24	−5.37	0.00076	0.08389	192.8	202.8	395.6	0.9736	1.7310
0.26	−3.24	0.00077	0.07769	195.7	201.2	396.9	0.9841	1.7297
0.28	−1.24	0.00077	0.07235	198.3	199.8	398.1	0.9939	1.7285
0.30	0.66	0.00077	0.06770	200.9	198.3	399.2	1.0032	1.7275
0.32	2.46	0.00078	0.06361	203.3	196.9	400.2	1.0120	1.7265
0.34	4.18	0.00078	0.05998	205.6	195.6	401.2	1.0204	1.7256
0.36	5.82	0.00078	0.05675	207.9	194.3	402.2	1.0283	1.7248
0.38	7.4	0.00079	0.05384	210.0	193.1	403.1	1.0360	1.7240
0.40	8.91	0.00079	0.05122	212.1	191.8	403.9	1.0433	1.7234
0.50	15.71	0.00081	0.04114	221.5	186.2	407.7	1.0759	1.7205
0.60	21.54	0.00082	0.03432	229.7	181.1	410.8	1.1038	1.7183
0.70	26.68	0.00083	0.02939	237.0	176.5	413.5	1.1282	1.7166
0.80	31.29	0.00085	0.02565	243.7	172.0	415.7	1.1500	1.7150
0.90	35.49	0.00086	0.02271	249.9	167.8	417.7	1.1699	1.7137
1.0	39.35	0.00087	0.02034	255.6	163.9	419.5	1.1881	1.7124
1.2	46.28	0.00089	0.01674	266.2	156.2	422.4	1.2208	1.7100
1.4	52.39	0.00092	0.01413	275.7	149.0	424.7	1.2498	1.7076
1.6	57.88	0.00094	0.01215	284.5	142.0	426.5	1.2759	1.7050
1.8	62.87	0.00096	0.01058	292.6	135.2	427.8	1.2999	1.7022
2.0	67.47	0.00099	0.00931	300.4	128.4	428.8	1.3223	1.6991
2.2	71.72	0.00102	0.00825	307.8	121.5	429.3	1.3433	1.6956
2.4	75.69	0.00104	0.00735	314.9	114.6	429.5	1.3632	1.6917
2.6	79.41	0.00107	0.00657	321.8	107.5	429.3	1.3823	1.6872
2.8	82.91	0.00111	0.00589	328.6	100.1	428.7	1.4007	1.6819
3.0	86.22	0.00114	0.00528	335.3	92.3	427.6	1.4188	1.6758

Table A6.3   Thermodynamic Properties of Superheated Refrigerant-134a

T (°C)	$p = 0.05$ MPa ($T_{sat} = -40.43°C$)			$p = 0.10$ MPa ($T_{sat} = -26.34°C$)		
	v (m³/kg)	h (kJ/kg)	s (kJ/kg K)	v (m³/kg)	h (kJ/kg)	s (kJ/kg K)
−40	0.36982	374.3	1.7675	—	—	—
−30	0.38745	381.8	1.7990	—	—	—
−25	0.39604	385.6	1.8144	0.19372	383.9	1.7527
−20	0.40469	389.5	1.8297	0.19829	387.9	1.7685
−10	0.42194	397.3	1.8599	0.20734	395.8	1.7994
0	0.43898	405.2	1.8895	0.21626	404.0	1.8297
10	0.45579	413.3	1.9186	0.22502	412.2	1.8593
20	0.47259	421.5	1.9473	0.23370	420.6	1.8883
30	0.48948	430.0	1.9755	0.24231	429.1	1.9169
40	0.50607	438.5	2.0034	0.25088	437.7	1.9450
50	0.52274	447.3	2.0309	0.25940	446.6	1.9727
60	0.53937	456.2	2.0580	0.26788	455.5	2.0001
70	0.55586	465.3	2.0848	0.27624	464.7	2.0270
80	0.57241	474.5	2.1113	0.28466	473.9	2.0537
90	0.58893	483.9	2.1375	0.29300	483.4	2.0800
100	0.60533	493.4	2.1635	0.30139	492.9	2.1060

T (°C)	$p = 0.15$ MPa ($T_{sat} = -17.12°C$)			$p = 0.20$ MPa ($T_{sat} = -10.08°C$)		
	v (m³/kg)	h (kJ/kg)	s (kJ/kg K)	v (m³/kg)	h (kJ/kg)	s (kJ/kg K)
−15	0.13259	390.3	1.7464	—	—	—
−10	0.13576	394.4	1.7622	0.09989	392.9	1.7344
0	0.14196	402.7	1.7931	0.10478	401.4	1.7661
10	0.14806	411.1	1.8233	0.10953	409.9	1.7968
20	0.15404	419.6	1.8528	0.11417	418.5	1.8267
30	0.15995	428.2	1.8817	0.11874	427.3	1.8560
40	0.16581	436.9	1.9101	0.12324	436.1	1.8847
50	0.17156	445.8	1.9380	0.12767	445.1	1.9129
60	0.17734	454.9	1.9656	0.13207	454.2	1.9406
70	0.18305	464.0	1.9927	0.13643	463.4	1.9679
80	0.18875	473.4	2.0195	0.14075	472.8	1.9948
90	0.19440	482.8	2.0459	0.14505	482.3	2.0214
100	0.20000	492.4	2.0720	0.14932	491.9	2.0476

Table A6.3 *Continued*

T (°C)	p = 0.25 MPa (T_sat = −4.29°C)			p = 0.30 MPa (T_sat = 0.66°C)		
	v (m³/kg)	h (kJ/kg)	s (kJ/kg K)	v (m³/kg)	h (kJ/kg)	s (kJ/kg K)
0	0.08241	400.0	1.7441	—	—	—
5	0.08442	404.4	1.7599	0.06921	403.1	1.7415
10	0.08638	408.7	1.7754	0.07092	407.5	1.7573
20	0.09023	417.5	1.8059	0.07424	416.4	1.7883
30	0.09399	426.3	1.8355	0.07748	425.4	1.8183
40	0.09768	435.3	1.8645	0.08063	434.4	1.8477
50	0.10131	444.3	1.8930	0.08372	443.6	1.8764
60	0.10489	453.5	1.9209	0.08678	452.8	1.9045
70	0.10845	462.8	1.9484	0.08978	462.1	1.9322
80	0.11197	472.2	1.9755	0.09276	471.6	1.9594
90	0.11545	481.8	2.0021	0.09571	481.2	1.9862
100	0.11891	491.4	2.0285	0.09863	490.9	2.0126
110	0.12235	501.3	2.0544	0.10153	500.8	2.0387
120	0.12579	511.2	2.0801	0.10444	510.8	2.0645

T (°C)	p = 0.40 MPa (T_sat = 8.91°C)			p = 0.50 MPa (T_sat = 15.71°C)		
	v (m³/kg)	h (kJ/kg)	s (kJ/kg K)	v (m³/kg)	h (kJ/kg)	s (kJ/kg K)
10	0.05152	404.9	1.7269	—	—	—
20	0.05421	414.2	1.7590	0.04213	411.8	1.7347
30	0.05680	423.4	1.7900	0.04434	421.3	1.7667
40	0.05930	432.7	1.8201	0.04646	430.8	1.7975
50	0.06172	442.0	1.8493	0.04851	440.4	1.8274
60	0.06411	451.4	1.8779	0.05050	449.9	1.8565
70	0.06645	460.8	1.9059	0.05244	459.5	1.8849
80	0.06876	470.4	1.9335	0.05434	469.2	1.9128
90	0.07102	480.1	1.9605	0.05621	479.0	1.9401
100	0.07328	489.9	1.9872	0.05806	488.9	1.9670
110	0.07551	499.9	2.0134	0.05989	498.9	1.9934
120	0.07772	509.9	2.0394	0.06169	509.0	2.0195
130	0.07991	520.1	2.0649	0.06348	519.3	2.0452
140	0.08210	530.4	2.0902	0.06526	529.6	2.0707

(continued)

Table A6.3 *Continued*

	$p = 0.60$ MPa ($T_{sat} = 21.54°C$)			$p = 0.70$ MPa ($T_{sat} = 26.68°C$)		
T (°C)	v (m³/kg)	h (kJ/kg)	s (kJ/kg K)	v (m³/kg)	h (kJ/kg)	s (kJ/kg K)
25	0.03502	414.2	1.7299	—	—	—
30	0.03600	419.2	1.7463	0.02999	416.8	1.7278
40	0.03787	428.9	1.7780	0.03171	426.9	1.7606
50	0.03967	438.7	1.8086	0.03333	436.9	1.7919
60	0.04140	448.4	1.8382	0.03488	446.8	1.8221
70	0.04308	458.1	1.8671	0.03638	456.7	1.8515
80	0.04472	468.0	1.8953	0.03784	466.7	1.8801
90	0.04633	477.9	1.9229	0.03925	476.7	1.9080
100	0.04790	487.8	1.9500	0.04064	486.8	1.9354
110	0.04945	497.9	1.9767	0.04201	496.9	1.9622
120	0.05100	508.1	2.0030	0.04336	507.2	1.9887
130	0.05252	518.4	2.0288	0.04468	517.6	2.0147
140	0.05402	528.8	2.0544	0.04600	528.0	2.0404

	$p = 0.80$ MPa ($T_{sat} = 31.29°C$)			$p = 0.90$ MPa ($T_{sat} = 35.49°C$)		
T (°C)	v (m³/kg)	h (kJ/kg)	s (kJ/kg K)	v (m³/kg)	h (kJ/kg)	s (kJ/kg K)
40	0.02705	424.8	1.7445	0.02340	422.6	1.7293
50	0.02856	435.1	1.7767	0.02482	433.2	1.7625
60	0.02998	445.2	1.8076	0.02616	443.5	1.7942
70	0.03135	455.3	1.8374	0.02743	453.8	1.8246
80	0.03267	465.4	1.8664	0.02864	464.1	1.8540
90	0.03395	475.5	1.8947	0.02981	474.3	1.8826
100	0.03520	485.7	1.9223	0.03096	484.6	1.9106
110	0.03642	495.9	1.9494	0.03207	494.9	1.9379
120	0.03763	506.3	1.9761	0.03316	505.3	1.9647
130	0.03881	516.7	2.0023	0.03423	515.8	1.9911
140	0.03998	527.2	2.0281	0.03530	526.4	2.0170
150	0.04114	537.9	2.0535	0.03634	537.1	2.0426

Table A6.3 *Continued*

	$p = 1.0$ MPa ($T_{\text{sat}} = 39.35°C$)			$p = 1.20$ MPa ($T_{\text{sat}} = 46.28°C$)		
T (°C)	v (m³/kg)	h (kJ/kg)	s (kJ/kg K)	v (m³/kg)	h (kJ/kg)	s (kJ/kg K)
40	0.02044	420.2	1.7147	—	—	—
50	0.02181	431.2	1.7491	0.01722	426.8	1.7235
60	0.02308	441.8	1.7816	0.01842	438.1	1.7580
70	0.02427	452.3	1.8126	0.01951	449.1	1.7905
80	0.02541	462.7	1.8425	0.02054	459.8	1.8215
90	0.02650	473.1	1.8715	0.02151	470.5	1.8513
100	0.02756	483.5	1.8997	0.02245	481.2	1.8802
110	0.02859	493.9	1.9273	0.02335	491.8	1.9083
120	0.02959	504.4	1.9543	0.02423	502.5	1.9358
130	0.03058	515.0	1.9809	0.02508	513.2	1.9627
140	0.03155	525.6	2.0070	0.02592	523.9	1.9891
150	0.03250	536.3	2.0326	0.02675	534.8	2.0150
160	0.03345	547.2	2.0579	0.02756	545.7	2.0406
170	0.03438	558.1	2.0829	0.02836	556.8	2.0657

	$p = 1.40$ MPa ($T_{\text{sat}} = 52.39°C$)			$p = 1.60$ MPa ($T_{\text{sat}} = 57.88°C$)		
T (°C)	v (m³/kg)	h (kJ/kg)	s (kJ/kg K)	v (m³/kg)	h (kJ/kg)	s (kJ/kg K)
55	0.01445	427.9	1.7174	—	—	—
60	0.01502	434.0	1.7357	0.01239	429.3	1.7134
70	0.01607	445.6	1.7700	0.01344	441.7	1.7503
80	0.01703	456.8	1.8023	0.01437	453.6	1.7842
90	0.01793	467.8	1.8331	0.01522	465.0	1.8162
100	0.01878	478.8	1.8628	0.01602	476.2	1.8467
110	0.01960	489.6	1.8915	0.01677	487.4	1.8761
120	0.02038	500.5	1.9194	0.01750	498.4	1.9046
130	0.02115	511.3	1.9467	0.01820	509.5	1.9323
140	0.02189	522.2	1.9734	0.01887	520.5	1.9594
150	0.02263	533.2	1.9997	0.01954	531.6	1.9860
160	0.02334	544.2	2.0255	0.02018	542.7	2.0120
170	0.02405	555.4	2.0508	0.02082	554.0	2.0376

(continued)

Table A6.3   *Continued*

T (°C)	p = 1.80 MPa (T_sat = 62.87°C)			p = 2.0 MPa (T_sat = 67.47°C)		
	v (m³/kg)	h (kJ/kg)	s (kJ/kg K)	v (m³/kg)	h (kJ/kg)	s (kJ/kg K)
65	0.01082	430.8	1.7110	—	—	—
70	0.01134	437.4	1.7306	0.00959	432.5	1.7101
80	0.01227	450.0	1.7667	0.01055	446.1	1.7493
90	0.01309	462.0	1.8001	0.01137	458.8	1.7845
100	0.01386	473.6	1.8317	0.01211	470.8	1.8173
110	0.01457	485.0	1.8618	0.01279	482.6	1.8483
120	0.01524	496.3	1.8909	0.01344	494.1	1.8781
130	0.01589	507.5	1.9191	0.01405	505.5	1.9068
140	0.01652	518.7	1.9466	0.01463	516.9	1.9347
150	0.01713	530.0	1.9734	0.01520	528.3	1.9619
160	0.01772	541.2	1.9997	0.01575	539.7	1.9884
170	0.01830	552.5	2.0256	0.01629	551.1	2.0145

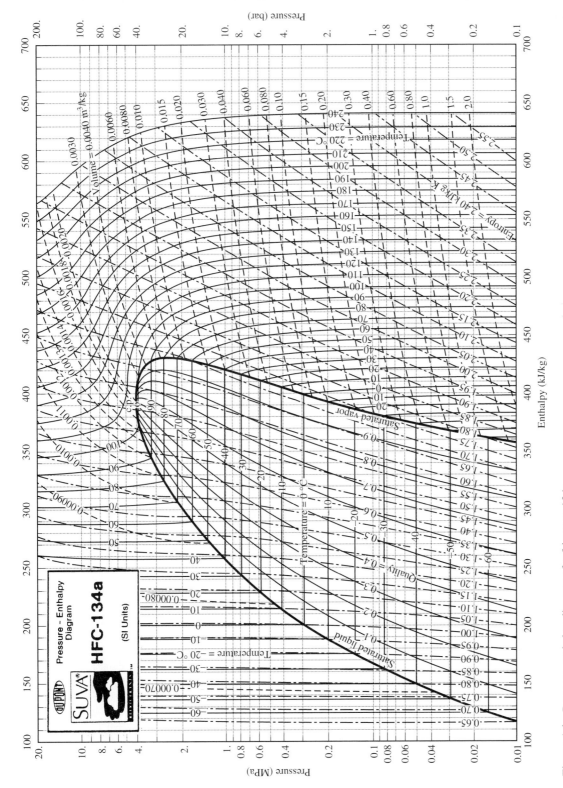

Figure A4 Pressure–enthalpy diagram of refrigerant-134a.

907

Table A7   Critical Constants

Substance	Formula	Molar Mass	Temp. (K)	Pressure (MPa)	Volume (m³/kg-mol)
Ammonia	NH_3	17.031	405.5	11.35	0.0725
Argon	Ar	39.948	150.8	4.87	0.0749
Bromine	Br_2	159.808	588.0	10.30	0.1272
Carbon dioxide	CO_2	44.01	304.1	7.38	0.0939
Carbon monoxide	CO	28.01	132.9	3.50	0.0932
Chlorine	CL_2	70.906	416.9	7.98	0.1238
Deuterium (normal)	D_2	4.032	38.4	1.66	—
Fluorine	F_2	37.997	144.3	5.22	0.0663
Helium	He	4.003	5.19	0.227	0.0574
Helium³	He	3.017	3.31	0.114	0.0729
Hydrogen (normal)	H_2	2.016	33.2	1.30	0.0651
Krypton	Kr	83.80	209.4	5.50	0.0912
Neon	Ne	20.183	44.4	2.76	0.0416
Nitric oxide	NO	30.006	180.0	6.48	0.0577
Nitrogen	N_2	28.013	126.2	3.39	0.0898
Nitrogen dioxide	NO_2	46.006	431.0	10.1	0.1678
Nitrous oxide	N_2O	44.013	309.6	7.24	0.0974
Oxygen	O_2	31.999	154.6	5.04	0.0734
Sulfur dioxide	SO_2	64.063	430.8	7.88	0.1222
Water	H_2O	18.015	647.3	22.09	0.0568
Xenon	Xe	131.30	289.7	5.84	0.1184
Acetylene	C_2H_2	26.038	308.3	6.14	0.1127
Benzene	C_6H_6	78.114	562.2	4.89	0.2590
n-Butane	C_3H_{10}	58.124	425.2	3.80	0.2550
Carbon tetrachloride	CCL_4	153.823	556.4	4.56	0.2759
Chlorodifluoroethane (142)	CH_3CCLF_2	100.495	410.3	4.25	0.2310
Chlorodifluoromethane (22)	$CHCLF_2$	86.469	369.3	4.97	0.1656
Chloroform	$CHCL_3$	119.378	536.4	5.37	0.2389
Dichlorodifluoromethane (12)	CCL_2F_2	120.914	385.0	4.14	0.2167
Dichlorofluoroethane (141)	CH_3CCL_2F	116.95	481.5	4.54	0.2520
Dichlorofluoromethane (21)	$CHCL_2F$	102.923	451.6	5.18	0.1964
Difluoroethane (152)	CHF_2CH_3	66.05	386.4	4.52	0.1795
Ethane	C_2H_6	30.070	305.4	4.88	0.1483
Ethyl alcohol	C_2H_5OH	46.069	513.6	6.14	0.1671
Ethylene	C_2H_4	28.054	282.4	5.04	0.1304
n-Heptane	C_7H_{16}	100.205	540.3	2.74	0.4320
n-Hexane	C_6H_{14}	86.178	507.5	3.01	0.3700
Methane	CH_4	16.043	190.4	4.60	0.0992
Methyl alcohol	CH_3OH	32.042	512.6	8.09	0.1180
Methyl chloride	CH_3CL	50.488	416.3	6.70	0.1389
n-Octane	C_8H_{18}	114.232	568.8	2.49	0.4920
n-Pentane	C_5H_{16}	72.151	469.7	3.37	0.3040
Propane	C_3H_8	44.094	369.8	4.25	0.2030
Propene	C_3H_6	42.081	364.9	4.60	0.1810

Table A7 *Continued*

Propyne	C$_3$H$_4$	40.065	402.4	5.63	0.1640
Trichlorofluoromethane	CCL$_3$F	137.37	471.2	4.38	0.2478

Source: R. C. Reid, J. M. Prausnitz, and B. E. Poling, *The Properties of Gases and Liquids*, 4th ed., (McGraw-Hill Book, New York, 1987, and NIST Thermophysics Division, 1989.

Pseudocritical Properties for Air: Air is a mixture of gases and therefore does not have true critical properties. However, pseudocritical properties can be defined using Kay's rule for mixtures. Values for pseudocritical temperature and pressure obtained in this manner for air (M = 28.97) are: T_c = 133 K; p_c = 3.76 MPa.

Table A8 Properties of Various Solids and Liquids at 25°C

Solid	ρ (kg/m^3)	c_p (kJ/kg K)	Liquid	ρ (kg/m^3)	c_p (kJ/kg K)
Aluminum	2700	0.9	Ammonia	606	4.8
Concrete	2300	0.65	Benzene	879	1.72
Copper	8900	0.386	Butane	556	2.469
Glass	2300	0.8	Ethanol	783	2.456
Granite	2700	1.017	Glycerine	1200	2.40
Graphite	2500	0.711	Iso-octane	692	2.1
Iron	7840	0.450	Mercury	13560	0.139
Lead	11310	0.128	Methanol	787	2.55
Rubber (soft)	1100	1.84	Oil (light)	910	1.8
Sand (dry)	1450–1750	0.8	Propane	510	2.54
Silver	10470	0.235	R-12	1310	0.971
Steel (AISI302)	8050	0.48	R-134a	1206	1.43
Tin	5730	0.217	Water	997	4.184
Wood (most)	350–700	1.76			

Table A9 Properties of Various Ideal Gases at 300 K

Gas	Chemical Formula	Molar Mass	R (kJ/kg K)	c_p (kJ/kg K)	c_v (kJ/kg K)	γ
Acetylene	C$_2$H$_2$	26.038	0.3193	1.6986	1.3793	1.231
Air		28.97	0.2870	1.0035	0.7165	1.400
Ammonia	NH$_3$	17.031	0.48819	2.1300	1.6418	1.297
Argon	Ar	39.948	0.20813	0.5203	0.3122	1.667
Butane	C$_4$H$_{10}$	58.124	0.14304	1.7164	1.5734	1.091
Carbon dioxide	CO$_2$	44.01	0.18892	0.8418	0.6529	1.289
Carbon monoxide	CO	28.01	0.29683	1.0413	0.7445	1.400
Ethane	C$_2$H$_6$	30.07	0.27650	1.7662	1.4897	1.186
Ethanol	C$_2$H$_5$OH	46.069	0.18048	1.427	1.246	1.145
Ethylene	C$_2$H$_4$	28.054	0.29637	1.5482	1.2518	1.237
Helium	He	4.003	2.07703	5.1926	3.1156	1.667
Hydrogen	H$_2$	2.016	4.12418	14.2091	10.0849	1.409
Methane	CH$_4$	16.04	0.51835	2.2537	1.7354	1.299

continued

Table A9   *Continued*

Gas	Chemical Formula	Molar Mass	R (kJ/kg K)	c_p (kJ/kg K)	c_v (kJ/kg K)	γ
Methanol	CH_3OH	32.042	0.25948	1.4050	1.1455	1.227
Neon	Ne	20.183	0.41195	1.0299	0.6179	1.667
Nitrogen	N_2	28.013	0.29680	1.0416	0.7448	1.400
Nitrous oxide	N_2O	44.013	0.18891	0.8793	0.6904	1.274
n-Octane	C_8H_{18}	114.23	0.07279	1.7113	1.6385	1.044
Oxygen	O_2	31.999	0.25983	0.9216	0.6618	1.393
Propane	C_3H_8	44.097	0.18855	1.6794	1.4909	1.126
Steam	H_2O	18.015	0.46152	1.8723	1.4108	1.327
Sulfur dioxide	SO_2	64.059	0.12979	0.6236	0.4938	1.263
Sulfur trioxide	SO_3	80.058	0.10386	0.6346	0.5307	1.196

Table A10   Expressions for Constant-Pressure Specific Heat as a Function of Temperature for Various Ideal Gases

$$\bar{c}_p = a + bT + cT^2 + dT^3 \qquad (T \text{ in K}, \bar{c}_p \text{ in kJ/kg-mol K})$$

Substance	Formula	a	$b \times 10^2$	$c \times 10^5$	$d \times 10^9$	Temperature range (K)
Nitrogen	N_2	28.90	−0.1571	0.8081	−2.873	273–1800
Oxygen	O_2	25.48	1.520	−0.7155	1.312	273–1800
Air		28.11	0.1967	0.4802	−1.966	273–1800
Hydrogen	H_2	29.11	−0.1916	0.4003	−0.8704	273–1800
Carbon monoxide	CO	28.16	0.1675	0.5372	−2.222	273–1800
Carbon dioxide	CO_2	22.26	5.986	−3.501	7.469	273–1800
Water vapor	H_2O	32.24	0.1923	1.055	−3.595	273–1800
Nitric oxide	NO	29.34	−0.09395	0.9747	−4.187	273–1500
Nitrous oxide	N_2O	24.11	5.8632	−3.562	10.58	273–1500
Nitrogen dioxide	NO_2	22.9	5.715	−3.52	7.87	273–1500
Ammonia	NH_3	27.568	2.5630	0.99072	−6.6909	273–1500
Sulfur	S_2	27.21	2.218	−1.628	3.986	273–1800
Sulfur dioxide	SO_2	25.78	5.795	−3.812	8.612	273–1800
Sulfur trioxide	SO_3	16.40	14.58	−11.20	32.42	273–1300
Acetylene	C_2H_2	21.8	9.2143	−6.527	18.21	273–1500
Benzene	C_6H_6	−36.22	48.475	−31.57	77.62	273–1500
Methanol	CH_4O	19.0	9.152	−1.22	−8.039	273–1000
Ethanol	C_2H_6O	19.9	20.96	−10.38	20.05	273–1500
Hydrogen chloride	HCl	30.33	−0.7620	1.327	−4.338	273–1500
Methane	CH_4	19.89	5.024	1.269	−11.01	273–1500
Ethane	C_2H_6	6.900	17.27	−6.406	7.285	273–1500
Propane	C_3H_8	−4.04	30.48	−15.72	31.74	273–1500
n-Butane	C_4H_{10}	3.96	37.15	−18.34	35.00	273–1500
i-Butane	C_4H_{10}	−7.913	41.60	−23.01	49.91	273–1500
n-Pentane	C_5H_{12}	6.774	45.43	−22.46	42.29	273–1500
n-Hexane	C_6H_{14}	6.938	55.22	−28.65	57.69	273–1500
Ethylene	C_2H_4	3.95	15.64	−8.344	17.67	273–1500
Propylene	C_3H_6	3.15	23.83	−12.18	24.62	273–1500

Source: B. G. Kyle, *Chemical and Process Thermodynamics,* Prentice-Hall, Englewood Cliffs, NJ, 1984. Used with permission.

Table A11   Ideal-Gas Properties of Air (Standard Entropy at 0.1 MPa)

T (K)	u (kJ/kg)	h (kJ/kg)	$s°$ (kJ/kg K)	p_r	v_r
200	142.768	200.174	6.46260	0.27027	493.4661
220	157.071	220.218	6.58812	0.37700	389.1494
240	171.379	240.267	6.64535	0.51088	313.2735
260	185.695	260.323	6.72562	0.67573	256.5840
280	200.022	280.390	6.79998	0.87556	213.2566
300	214.364	300.473	6.86926	1.11458	179.4906
320	228.726	320.576	6.93413	1.39722	152.7277
340	243.113	340.704	6.99515	1.72814	131.1994
360	257.532	360.863	7.05276	2.11226	113.6542
380	271.988	381.060	7.10735	2.55479	99.18824
400	286.487	401.299	7.15926	3.06119	87.13665
420	301.035	421.589	7.20875	3.63727	77.00250
440	315.640	441.934	7.25607	4.28916	68.40881
460	330.306	462.340	7.30142	5.02333	61.06577
480	345.039	482.814	7.34499	5.84663	54.74787
500	359.844	503.360	7.38692	6.76629	49.27771
520	374.726	523.982	7.42736	7.78997	44.51426
540	389.689	544.686	7.46642	8.92569	40.34439
560	404.736	565.474	7.50422	10.18197	36.67649
580	419.871	586.350	7.54084	11.56771	33.43581
600	435.097	607.316	7.57638	13.09232	30.56089
620	450.415	628.375	7.61090	14.76564	28.00082
640	465.828	649.528	7.64448	16.59801	25.71315
660	481.335	670.776	7.67717	18.60025	23.66228
680	496.939	692.120	7.70903	20.78367	21.81816
700	512.639	713.561	7.74010	23.16010	20.15529
720	528.435	735.098	7.77044	25.74188	18.65192
740	544.328	756.731	7.80008	25.54188	17.28943
760	560.316	778.460	7.82905	31.57347	16.05476
780	576.400	800.284	7.85740	34.85061	14.92504
800	592.577	822.202	7.88514	38.38777	13.89724
820	608.847	844.212	7.91232	42.19998	12.95785
840	625.208	866.314	7.93895	46.30283	12.09771
860	641.659	888.506	7.96506	50.71250	11.30876
880	658.198	910.786	7.99068	55.44570	10.58391
900	674.824	933.152	8.01581	60.51977	9.91692
920	691.534	955.603	8.04048	65.95261	9.30223
940	708.328	978.138	8.06472	71.76272	8.73495
960	725.203	1000.753	8.08853	77.96920	8.21069
980	742.157	1023.448	8.11193	84.59178	7.72555
1000	759.189	1046.221	8.13493	91.65077	7.27604
1020	776.297	1069.069	8.15756	99.16715	6.85905
1040	793.478	1091.991	8.17981	107.1625	6.47175
1060	810.731	1114.984	8.20171	115.6590	6.11164

continued

Table A11   *Continued*

T (K)	u (kJ/kg)	h (kJ/kg)	s° (kJ/kg K)	p_r	v_r
1080	828.054	1138.048	8.22327	124.6796	5.77643
1100	845.445	1161.180	8.24449	134.2478	5.46408
1120	862.903	1184.379	8.26539	144.3878	5.17272
1140	880.426	1207.642	8.28598	155.1244	4.90068
1160	898.012	1230.969	8.30626	166.4833	4.64642
1180	915.660	1254.357	8.32625	178.4907	4.40857
1200	933.367	1277.805	8.34596	191.1735	4.18586
1250	977.888	1336.677	8.39402	226.0192	3.68804
1300	1022.751	1395.892	8.44046	265.7144	3.26257
1350	1067.936	1455.429	8.48539	310.7426	2.89711
1400	1113.426	1515.270	8.52891	361.6191	2.58171
1450	1159.202	1575.398	8.57111	418.8942	2.30831
1500	1205.253	1635.800	8.61208	483.1554	2.07031
1550	1251.547	1696.446	8.65185	554.9577	1.86253
1600	1298.079	1757.329	8.69051	634.9670	1.68035
1650	1344.834	1818.436	8.72811	723.8560	1.52007
1700	1391.801	1879.755	8.76472	822.3320	1.37858
1750	i438.970	1941.275	8.80039	931.1376	1.25330
1800	1486.331	2002.987	8.83516	1051.051	1.14230
1850	1533.873	2064.882	8.86908	1182.887	1.04204
1900	1581.591	2126.951	8.90219	1327.498	0.95445
1950	1629.474	2189.186	8.93452	1485.772	0.87521
2000	1677.518	2251.581	8.96611	1658.635	0.80410
2050	1725.714	2314.128	8.99699	1847.076	0.74012
2100	1774.057	2376.823	9.02721	2052.108	0.68242
2150	1822.541	2439.659	9.05678	2274.788	0.63027
2200	1871.161	2502.630	9.08573	2516.216	0.58305
2250	1919.912	2565.733	9.11409	2777.537	0.54020
2300	1968.790	2628.962	9.14189	3059.939	0.50124
2350	2017.789	2692.313	9.16913	3364.657	0.46576
2400	2066.907	2755.782	9.19586	3692.973	0.43338
2450	2116.138	2819.366	9.22208	4046.215	0.40378
2500	2165.480	2883.059	9.24781	4425.759	0.37669
2550	2214.929	2946.859	9.27308	4833.030	0.35185
2600	2264.481	3010.763	9.29790	5269.504	0.32903
2650	2314.133	3074.767	9.32228	5736.707	0.30805
2700	2363.883	3138.868	9.34625	6236.214	0.28872
2750	2413.727	3203.064	9.36980	6769.656	0.27089
2800	2463.663	3267.351	9.39297	7338.714	0.25443
2850	2513.687	3331.726	9.41576	7945.124	0.23921
2900	2563.797	3396.188	9.43818	8590.676	0.22511
2950	2613.990	3460.733	9.46025	9227.216	0.21205
3000	2664.265	3525.359	9.48198	10006.645	0.19992

Source: Thermodynamic data in *JANAF Thermochemical Tables,* 3rd ed., Thermal Group, Dow Chemical, Midland, MI, 1985.

Table A12 Ideal-Gas Enthalpy and Absolute Entropy for N_2 and N (Entropies at 0.1 MPa)

Temp. (K)	Nitrogen, Diatomic (N_2) $\overline{h}_f^\circ = 0$ kJ/kg-mol $M = 28.013$		Nitrogen, Monatomic (N) $\overline{h}_f^\circ = 472\ 680$ kJ/kg-mol $M = 14.007$	
	$(\overline{h}_T^\circ - \overline{h}^\circ)$ (kJ/kg-mol)	$\overline{s}^\circ$ (kJ/kg-mol K)	$(\overline{h}_T^\circ - \overline{h}^\circ)$ (kJ/kg-mol)	$\overline{s}^\circ$ (kJ/kg-mol K)
0	−8670	0	−6197	0
100	−5768	159.811	−4119	130.593
200	−2857	179.985	−2040	145.001
298.15	0	191.609	0	153.300
300	54	191.789	38	153.429
400	2971	200.181	2117	159.408
500	5911	206.739	4196	164.047
600	8894	212.176	6274	167.836
700	11 937	216.866	8353	171.041
800	15 046	221.017	10 431	173.816
900	18 223	224.757	12 510	176.264
1000	21 463	228.170	14 589	178.454
1100	24 760	231.313	16 667	180.436
1200	28 109	234.226	18 746	182.244
1300	31 503	236.943	20 824	183.908
1400	34 936	239.487	22 903	185.448
1500	38 405	241.880	24 982	186.882
1600	41 904	244.138	27 060	188.224
1700	45 529	246.275	29 139	189.484
1800	48 978	248.304	31 218	190.672
1900	52 548	250.234	33 296	191.796
2000	56 137	252.074	35 375	192.863
2200	63 361	255.517	39 534	194.844
2400	70 640	258.684	43 695	196.655
2600	77 963	261.614	47 860	198.322
2800	85 323	264.341	52 033	199.868
3000	92 715	266.891	56 218	201.311
3200	100 134	269.285	60 420	202.667
3400	107 577	271.541	64 646	203.948
3600	115 041	273.675	68 902	205.164
3800	122 525	275.698	73 194	206.325
4000	130 027	277.622	77 532	207.437
4200	137 545	279.456	81 920	208.508
4400	145 078	281.208	86 367	209.542
4600	152 625	282.885	90 877	210.544
4800	160 187	284.494	95 457	211.519
5000	167 763	286.041	100 111	212.469
5200	175 352	287.529	104 843	213.397
5400	182 955	288.964	109 655	214.305
5600	190 571	290.348	114 550	215.195
5800	198 201	291.687	119 528	216.068
6000	205 848	292.984	124 590	216.926

Source: Thermodynamic data in *JANAF Thermochemical Tables,* 3rd ed. Published by the American Chemical Society and the American Institute of Physics, 1985.

Table A13   Ideal-Gas Enthalpy and Absolute Entropy for O_2 and O (Entropies at 0.1 MPa)

Temp. (K)	Oxygen, Diatomic (O_2) $\bar{h}_f^\circ = 0$ kJ/kg-mol $M = 31.999$		Oxygen, Monatomic (O) $\bar{h}_f^\circ = 249\ 170$ kJ/kg-mol $M = 16.00$	
	$(\bar{h}_T^\circ - \bar{h}^\circ)$ (kJ/kg-mol)	$\bar{s}^\circ$ (kJ/kg-mol K)	$(\bar{h}_T^\circ - \bar{h}^\circ)$ (kJ/kg-mol)	$\bar{s}^\circ$ (kJ/kg-mol K)
0	−8683	0	−6725	0
100	−5779	173.307	−4518	135.947
200	−2868	193.485	−2186	152.153
298.15	0	205.147	0	161.058
300	54	205.329	41	161.194
400	3025	213.871	2207	167.430
500	6084	220.693	4343	172.197
600	9244	226.451	6462	176.060
700	12 499	231.466	8570	179.310
800	15 835	235.921	10 671	182.116
900	19 241	239.931	12 767	184.585
1000	22 703	243.578	14 860	186.790
1100	26 212	246.922	16 950	188.782
1200	29 761	250.010	19 039	190.599
1300	33 344	252.878	21 126	192.270
1400	36 957	255.556	23 212	193.816
1500	40 399	258.068	25 296	195.254
1600	44 266	260.434	27 381	196.599
1700	47 958	262.672	29 464	197.862
1800	51 673	264.796	31 547	199.053
1900	55 413	266.818	33 630	200.179
2000	59 175	268.748	35 713	201.247
2200	66 769	272.366	39 878	203.232
2400	74 453	275.709	44 045	205.045
2600	82 224	278.819	48 216	206.714
2800	90 079	281.729	52 391	208.261
3000	98 013	248.466	56 574	209.704
3200	106 023	287.050	60 767	211.057
3400	114 102	289.499	64 971	212.332
3600	122 245	291.826	69 190	213.537
3800	130 447	294.044	73 424	214.682
4000	138 705	296.162	77 675	215.772
4200	147 015	298.189	81 945	216.814
4400	155 374	300.133	86 234	217.812
4600	163 783	302.002	90 543	218.769
4800	172 240	303.801	94 872	219.690
5000	180 749	305.538	99 222	220.578
5200	189 311	307.217	103 592	221.435
5400	197 923	308.844	107 982	222.264
5600	206 618	310.424	112 391	223.065
5800	215 375	311.960	116 818	223.842
6000	224 210	313.457	121 264	224.596

Table A14 Ideal-Gas Enthalpy and Absolute Entropy for CO_2 and CO (Entropies at 0.1 MPa)

Temp. (K)	Carbon Dioxide (CO_2) $\bar{h}_f^\circ = -393\ 522$ kJ/kg-mol $M = 41.01$		Carbon Monoxide (CO) $\bar{h}_f^\circ = -110\ 527$ kJ/kg-mol $M = 28.01$	
	$(\bar{h}_T^\circ - \bar{h}^\circ)$ (kJ/kg-mol)	$\bar{s}^\circ$ (kJ/kg-mol K)	$(\bar{h}_T^\circ - \bar{h}^\circ)$ (kJ/kg-mol)	$\bar{s}^\circ$ (kJ/kg-mol K)
0	−9364	0	−8671	0
100	−6458	179.009	−5769	165.850
200	−3414	199.975	−2858	186.025
298.15	0	213.795	0	197.653
300	69	214.025	54	197.833
400	4003	225.314	2976	206.238
500	8305	234.901	5931	212.831
600	12 907	243.283	8942	218.319
700	17 754	250.750	10 023	223.066
800	22 806	257.494	15 177	227.277
900	28 030	263.645	18 401	231.074
1000	33 397	269.299	21 690	234.538
1100	38 884	274.528	25 035	237.726
1200	44 473	279.390	28 430	240.679
1300	50 148	283.932	31 868	243.431
1400	55 896	288.191	35 343	246.006
1500	61 705	292.199	38 850	248.426
1600	67 569	295.983	42 385	250 707
1700	73 480	299.566	45 945	252.865
1800	79 431	302.968	49 526	254.912
1900	85 419	306.205	53 126	256.859
2000	91 439	309.293	56 744	258.714
2200	103 562	315.070	64 021	262.182
2400	115 779	320.385	71 324	265.359
2600	128 073	325.305	78 673	268.300
2800	140 433	329.885	86 074	271.042
3000	152 852	334.169	93 504	273.605
3200	165 321	338.192	100 960	276.011
3400	177 836	341.986	108 438	278.278
3600	190 393	345.574	115 937	280.421
3800	202 989	348.979	123 454	282.453
4000	215 622	352.219	130 989	284.386
4200	228 290	355.310	138 540	286.228
4400	240 991	358.264	146 106	287.988
4600	253 725	361.094	153 687	289.673
4800	266 489	363.810	161 282	291.289
5000	279 283	366.422	168 890	292.842
5200	292 109	368.937	176 511	294.336
5400	304 971	371.364	184 146	295.777
5600	317 870	373.709	191 775	297.164
5800	330 806	375.979	199 434	298.508
6000	343 779	378.178	207 106	299.808

Table A15   Ideal-Gas Enthalpy and Absolute Entropy for H_2O and OH (Entropies at 0.1 MPa)

Temp. (K)	Water (H_2O) $\overline{h}_f^\circ = -241\ 826$ kJ/kg-mol $M = 18.015$		Hydroxyl (OH) $\overline{h}_f^\circ = -38\ 987$ kJ/kg-mol $M = 17.007$	
	$(\overline{h}_T^\circ - \overline{h}^\circ)$ (kJ/kg-mol)	$\overline{s}^\circ$ (kJ/kg-mol K)	$(\overline{h}_T^\circ - \overline{h}^\circ)$ (kJ/kg-mol)	$\overline{s}^\circ$ (kJ/kg-mol K)
0	−9904	0	−9172	0
100	−6615	152.388	−6139	149.590
200	−3282	175.485	−2976	171.592
298.15	0	188.834	0	183.709
300	62	189.042	55	183.894
400	3452	198.788	3035	192.466
500	6925	206.534	5992	199.066
600	10 501	213.052	8943	204.447
700	14 192	218.739	11 902	209.007
800	18 002	223.825	14 880	212.983
900	21 938	228.459	17 888	216.526
1000	26 000	232.738	20 935	219.736
1100	30 191	236.731	24 024	222.680
1200	34 506	240.485	27 160	225.408
1300	38 942	244.035	30 342	227.955
1400	43 493	247.407	33 569	230.346
1500	48 151	250.620	36 839	232.602
1600	52 908	253.690	40 151	234.740
1700	57 758	256.630	43 502	236.771
1800	62 693	259.451	46 889	238.707
1900	67 706	262.161	50 310	240.557
2000	72 790	264.769	53 762	242.327
2200	83 153	269.706	60 752	245.658
2400	93 741	274.312	67 841	248.741
2600	104 520	278.625	75 017	251.613
2800	115 464	282.680	82 267	254.300
3000	126 549	286.504	89 584	256.824
3200	137 757	290.120	96 960	259.203
3400	149 073	293.550	104 387	261.455
3600	160 485	296.812	111 863	263.591
3800	171 980	299.919	119 381	265.624
4000	183 552	302.887	126 939	267.562
4200	195 191	305.726	134 534	269.415
4400	206 892	308.448	142 164	271.189
4600	218 650	311.061	149 827	272.893
4800	230 458	313.574	157 521	274.530
5000	242 313	315.993	165 246	276.107
5200	254 215	318.327	173 001	277.627
5400	266 164	320.582	180 784	279.096
5600	278 161	322.764	188 597	280.517
5800	290 204	324.877	196 438	281.892
6000	302 295	326.926	204 308	283.226

Table A16 Ideal-Gas Enthalpy and Absolute Entropy for H_2 and H (Entropies at 0.1 MPa)

Temp. (K)	Hydrogen, Diatomic (H_2) $\overline{h}_f^\circ = 0$ kJ/kg-mol $M = 2.016$		Hydrogen, Monatomic (H) $\overline{h}_f^\circ = 217\ 999$ kJ/kg-mol $M = 1.008$	
	$(\overline{h}_T^\circ - \overline{h}^\circ)$ (kJ/kg-mol)	$\overline{s}^\circ$ (kJ/kg-mol K)	$(\overline{h}_T^\circ - \overline{h}^\circ)$ (kJ/kg-mol)	$\overline{s}^\circ$ (kJ/kg-mol K)
0	−8467	0	−6197	0
100	−5468	100.727	−4119	92.009
200	−2774	119.412	−2040	106.417
298.15	0	130.680	0	114.716
300	53	130.858	38	114.845
400	2959	139.216	2117	120.825
500	5882	145.737	4196	125.463
600	8811	151.077	6274	129.253
700	11 749	155.606	8353	132.457
800	14 702	159.548	10 431	135.232
900	17 676	163.051	12 510	137.681
1000	20 680	166.216	14 589	139.871
1100	23 719	169.112	16 667	141.852
1200	26 797	171.790	18 746	143.660
1300	29 918	174.288	20 824	145.324
1400	33 082	176.633	22 903	146.865
1500	36 290	178.846	24 982	148.299
1600	39 541	180.944	27 060	149.640
1700	42 835	182.940	29 139	150.900
1800	46 169	184.846	31 217	152.088
1900	49 541	186.669	33 296	153.212
2000	52 951	188.418	35 375	154.278
2200	59 876	191.718	39 532	156.260
2400	66 928	194.785	43 689	158.068
2600	74 096	197.654	47 846	159.732
2800	81 369	200.349	52 004	161.272
3000	88 740	202.891	56 161	162.706
3200	96 202	205.299	60 318	164.048
3400	103 750	207.587	64 475	165.308
3600	111 380	209.767	68 632	166.496
3800	119 089	211.851	72 790	167.620
4000	126 874	213.848	76 947	168.686
4200	134 734	215.765	81 104	169.700
4400	142 667	217.610	85 261	170.667
4600	150 670	219.389	89 418	171.591
4800	158 741	221.106	93 576	172.476
5000	166 876	222.767	97 733	173.325
5200	175 071	224.374	101 890	174.140
5400	183 322	225.931	106 047	174.924
5600	191 621	227.440	110 204	175.680
5800	199 963	288.903	114 362	176.410
6000	208 341	230.323	118 519	177.114

Table A17   Ideal-Gas Enthalpy and Absolute Entropy for NO and NO$_2$ (Entropies at 0.1 MPa)

Temp. (K)	Nitric Oxide (NO) $\overline{h}_f^\circ = 90291$ kJ/kg-mol $M = 30.006$		Nitrogen Dioxide (NO$_2$) $\overline{h}_f^\circ = 33100$ kJ/kg-mol $M = 46.005$	
	$(\overline{h}_T^\circ - \overline{h}^\circ)$ (kJ/kg-mol)	$\overline{s}^\circ$ (kJ/kg-mol K)	$(\overline{h}_T^\circ - \overline{h}^\circ)$ (kJ/kg-mol)	$\overline{s}^\circ$ (kJ/kg-mol K)
0	−9192	0	−10 186	0
100	−6073	177.031	−6861	202.563
200	−2951	198.747	−3495	225.852
298.15	0	210.758	0	240.034
300	55	210.943	68	240.262
400	3040	219.529	3927	251.342
500	6059	226.263	8099	260.638
600	9144	231.886	12 555	268.755
700	12 307	236.761	17 250	275.988
800	15 548	241.087	22 138	282.512
900	18 858	244.985	27 179	288.449
1000	22 229	248.536	32 344	293.889
1100	25 653	251.799	37 605	298.903
1200	29 120	254.816	42 946	303.550
1300	32 626	257.621	48 351	307.876
1400	36 164	260.243	53 808	311.920
1500	39 729	262.703	59 309	315.715
1600	43 319	265.019	64 846	319.288
1700	46 929	267.208	70 414	322.663
1800	50 557	269.282	76 007	325.861
1900	54 201	271.252	81 624	328.897
2000	57 859	273.128	87 259	331.788
2200	65 212	276.632	98 577	337.181
2400	72 606	279.849	109 947	342.128
2600	80 034	282.822	121 357	346.694
2800	87 491	285.585	132 799	350.934
3000	94 973	288.165	144 267	354.889
3200	102 477	290.587	155 756	358.597
3400	110 000	292.867	167 262	362.084
3600	117 541	295.022	178 783	365.377
3800	125 098	297.065	190 316	368.495
4000	132 671	299.008	201 859	371.455
4200	140 257	300.858	213 412	374.274
4400	147 857	302.626	224 973	376.963
4600	155 469	304.318	236 540	379.534
4800	163 094	305.940	248 114	381.996
5000	170 730	307.499	259 692	384.360
5200	178 377	308.998	271 276	386.631
5400	186 034	310.443	282 863	388.818
5600	193 703	311.838	294 455	390.926
5800	201 381	313.185	306 049	392.960
6000	209 070	314.488	317 647	394.926

A.18 Lagrange's Method of Undetermined Multipliers

Lagrange's method of undetermined multipliers is a mathematical tool for finding stationary values of a function, subject to conditions of constraint. Stationary values of a function occur at a maximum, a minimum, or a point of inflection. The function is stable only when it is stationary at a minimum value. Consider the problem of finding a stationary value of a function of several variables given by

$$f(x_1, x_2, \ldots, x_n) = 0 \tag{1}$$

The variables $x_1, x_2, \ldots, x_n$, n in number, are related according to the equations of constraint given by

$$\begin{aligned}
\varphi_1(x_1, x_2, \ldots, x_n) &= 0 \\
\varphi_2(x_1, x_2, \ldots, x_n) &= 0 \\
&\vdots \\
\varphi_m(x_1, x_2, \ldots, x_n) &= 0
\end{aligned} \tag{2}$$

where m is the number of equations of constraint. Note that $m \leq n$.

When there are as many conditions of constraint as there are variables (that is, $m = n$), then the value of the function f is unique and can readily be determined by solving simultaneously n equations in n unknowns. But when $m < n$, the equations must be combined in such a way that f is maximized (or minimized). Note that $(n - m)$ equations are then needed; when added to the existing m constraint equations, they give a total of n equations.

If the function f and its derivatives are continuous, then the differential of f is given by

$$df = \frac{\partial f}{\partial x_1} dx_1 + \frac{\partial f}{\partial x_2} dx_2 + \cdots + \frac{\partial f}{\partial x_n} dx_n = \sum_{i=1}^{n} \frac{\partial f}{\partial x_i} dx_i \tag{3}$$

In the vicinity of a stationary point of the function f, the change of f must be equal to zero, or

$$df = 0 = \sum_{i=1}^{n} \frac{\partial f}{\partial x_i} dx_i \tag{4}$$

The differentials of Eq. (2) are

$$d\varphi_j = 0 = \sum_{i=1}^{n} \frac{\partial \varphi_j}{\partial x_i} dx_i \qquad j = 1, 2, \ldots, m \tag{5}$$

Lagrange's method consists of multiplying $d\varphi_1$ by λ_1, $d\varphi_2$ by λ_2, ... , and so on, where $\lambda_1, \lambda_2, \ldots , \lambda_m$ are arbitrary functions called *undetermined multipliers*, and adding these expressions to Eq. (4). A single equation is obtained, in which all the variables (rather than $n - 1$) may be treated as independent.

$$\frac{df}{\partial x_1} dx_1 + \frac{\partial f}{\partial x_2} dx_2 + \cdots + \frac{\partial f}{\partial x_n} dx_n$$

$$+ \lambda_1 \left(\frac{\partial \varphi_1}{\partial x_1} dx_1 + \frac{\partial \varphi_1}{\partial x_2} dx_2 + \cdots + \frac{\partial \varphi_1}{\partial x_n} dx_n \right)$$

$$+ \lambda_2 \left(\frac{\partial \varphi_2}{\partial x_1} dx_1 + \frac{\partial \varphi_2}{\partial x_2} dx_2 + \cdots + \frac{\partial \varphi_2}{\partial x_n} dx_n \right)$$

$$\vdots$$

$$+ \lambda_m \left(\frac{\partial \varphi_m}{\partial x_1} dx_1 + \frac{\partial \varphi_m}{\partial x_2} dx_2 + \cdots + \frac{\partial \varphi_m}{\partial x_n} dx_n \right) = 0$$

or

$$\left(\frac{\partial f}{\partial x_1} + \lambda_1 \frac{\partial \varphi_1}{\partial x_1} + \lambda_2 \frac{\partial \varphi_2}{\partial x_1} + \cdots + \lambda_m \frac{\partial \varphi_m}{\partial x_1} \right) dx_1$$

$$+ \left(\frac{\partial f}{\partial x_2} + \lambda_1 \frac{\partial \varphi_1}{\partial x_2} + \lambda_2 \frac{\partial \varphi_2}{\partial \varphi_2} + \cdots + \lambda_m \frac{\partial \varphi_m}{\partial x_2} \right) dx_2$$

$$\vdots$$

$$+ \left(\frac{\partial f}{\partial x_n} + \lambda_1 \frac{\partial \varphi_1}{\partial x_n} + \lambda_2 \frac{\partial \varphi_2}{\partial x_n} + \cdots + \lambda_m \frac{\partial \varphi_m}{\partial x_n} \right) dx_n = 0$$

Only those values of $\lambda_1, \lambda_2, \ldots , \lambda_m$ are acceptable that make the coefficients of $dx_1, dx_2, \ldots , dx_n$ in Eq. (6) vanish separately so that n separate equations are obtained from Eq. (6). This is essentially equivalent to treating each x as independent. After the m equations are solved for the λ's, the remaining $(n - m)$ equations are solved for the unknowns $x_1, x_2, \ldots , x_n$.

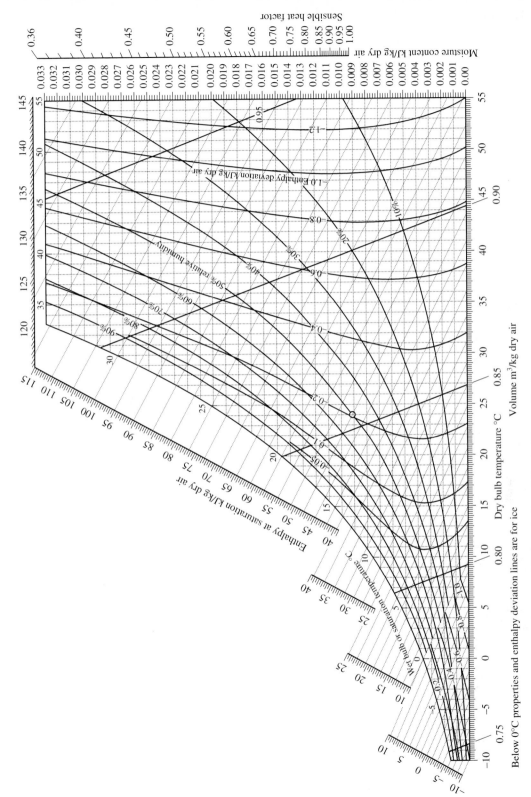

Figure A5 Psychrometric chart. (Reproduced by permission of Carrier Corporation.)

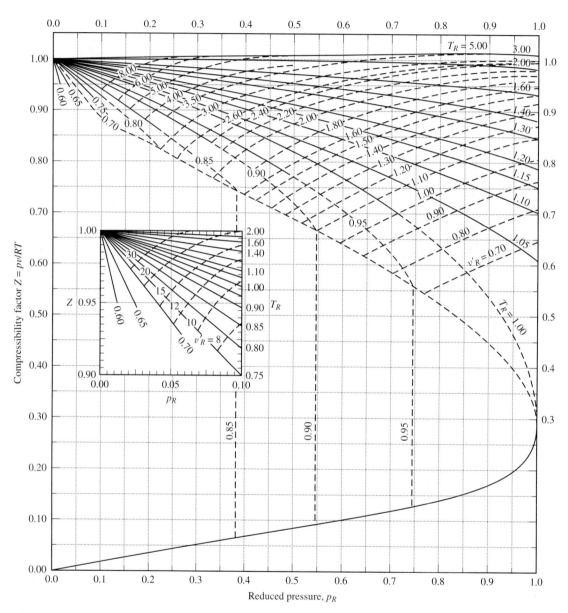

Figure A6   Generalized compressibility chart, $p_R \leq 1$. (Modified by Peter E. Liley, *Chemical Engineering*, 94 [1987]. *Original source:* E. E. Obert, *Concepts of Thermodynamics,* McGraw-Hill, Inc., 1960.)

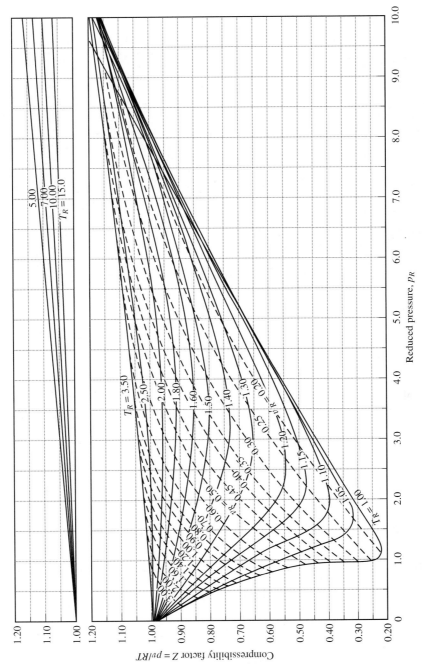

Figure A7 Generalized compressibility chart, $p_R \leq 10$. (*Source:* E. E. Obert, *Concepts of Thermodynamics,* McGraw-Hill, Inc., 1960)

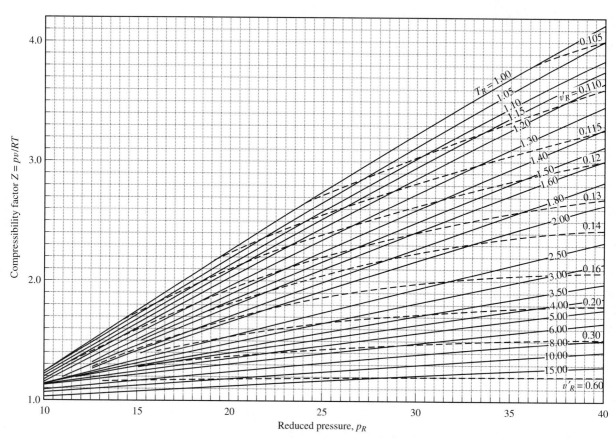

Figure A8   Generalized compressibility chart, $10 \leq p_R \leq 40$. (*Source:* E. E. Obert, *Concepts of Thermodynamics,* McGraw-Hill, Inc., 1960.)

Selected References

Balmer, R., *Thermodynamics,* West, St. Paul, Minn., 1990

Bejan, A., *Advanced Engineering Thermodynamics,* Wiley, New York, 1982.

Bejan, A., *Entropy Generation Through Heat and Fluid Flow,* Wiley-Interscience, New York, 1982.

Black, W. Z., and Hartley, J. G., *Thermodynamics,* Harper & Row, New York, 1985.

Burghardt, M. D., *Engineering Thermodynamics with Applications,* 3rd ed., Harper & Row, New York, 1986.

Callen, H. B., *Thermodynamics,* Wiley, New York, 1960.

Cengel, Y. A., and Boles, M. A., *Thermodynamics,* 2nd ed., McGraw-Hill, New York, 1994.

Crawford, F. H., *Heat, Thermodynamics and Statistical Physics,* Harcourt, Brace, Jovanovich, New York, 1963.

Denbigh, K., *The Principles of Chemical Equilibrium,* Cambridge University Press, 1954

Dixon, J. R., *Thermodynamics I. An Introduction to Energy,* Prentice-Hall, Englewood Cliffs, N.J., 1976.

Doolittle, J. S., and Hale, J. H., *Thermodynamics for Engineers,* Wiley, New York, 1984.

Faires, V. M., *Thermodynamics,* 4th ed., Macmillan, New York, 1962.

Hatsopoulous, G. N., and Keenan, J. H., *Principles of General Thermodynamics,* Wiley, New York, 1965, 1981.

Holman, J. P., *Thermodynamics,* 4th ed., McGraw-Hill, New York, 1988.

Howell, J. R., and Buckins, R. O., *Fundamentals of Engineering Thermodynamics,* McGraw-Hill, New York, 1987, 1992.

Huang, F. F., *Engineering Thermodynamics Fundamentals and Applications,* Macmillan, New York, 1976.

Jones, J. B., and Hawkins, G. A., *Engineering Thermodynamics,* 2nd ed., Wiley, New York, 1986.

Karlekar, B. V., *Thermodynamics for Engineers,* Prentice-Hall, Englewood Cliffs, N.J., 1982.

Keenan, J. H., *Thermodynamics,* MIT Press, Cambridge, Mass., 1970.

Kestin, J., *A Course in Thermodynamics, vol. 1,* Blaisdell, Waltham, Mass., 1966.

King, A. L., *Thermophysics,* W. H. Freeman and Co., San Francisco, 1962.

Lee, J. F., Sears, F. W., and Turcotte, D. L., *Statistical Thermodynamics,* 2nd ed., Addison-Wesley, Reading, Mass., 1973.

Lewis, G. N., and Randall, M., *Thermodynamics,* 2nd ed. (revised by K. S. Pitzer and L. Brewer), McGraw-Hill, New York, 1961.

Look, D. C., Jr. and Sauer, Jr., H. S., *Engineering Thermodynamics,* PWS Engineering, Boston, 1986.

Moran, M. J., *Availability Analysis: A Guide to Efficient Energy Use,* Prentice-Hall, Englewood Cliffs, N.J., 1982.

Moran, M. J., and Shapiro, H. N., *Fundamentals of Engineering Thermodynamics,* Wiley, New York, 1988.

Penner, S. S., *Thermodynamics for Scientists and Engineers,* Addison-Wesley, Reading, Mass., 1968.

Reynolds, W. C., and Perkins, H. C., *Engineering Thermodynamics,* 2nd ed., McGraw-Hill, New York, 1977.

Sears, F. W., *An Introduction to Thermodynamics, The Kinetic Theory of Gases, and Statistical Mechanics,* Addison-Wesley, Reading, Mass., 1950.

Sears, F. W., *Thermodynamics,* 2nd ed., Addison-Wesley, Reading, Mass., 1953.

Sonntag, R. E., and Van Wylen, G. J., *Introduction to Thermodynamics, Classical & Statistical,* 2nd ed., Wiley, New York, 1971, 1982.

Soo, S. L., and Tan, C. W., *Analytical Thermodynamics,* Prentice-Hall, Englewood Cliffs, N.J., 1962

Van Wylen, G. J., and Sonntag, R. E., *Fundamentals of Thermodynamics,* 3rd ed., Wiley, New York, 1985.

Wark, K., *Thermodynamics,* 5th ed., McGraw-Hill, New York, 1988.

Zemansky, M. W., *Heat and Thermodynamics,* 5th ed., McGraw-Hill, New York, 1968.

Answers to Selected Problems

4.6 12.875%, 87.125%

4.9 0.162 kg, 0.018 kg

4.10 99.49%, 0.51%

4.17 (a) 0.4448 kg, 3.1446 kg; (b) −6909.86 kJ

4.18 164.5 kJ/kg

4.20 (a) 457.72 kJ/kg, (b) 547.4 kJ/kg

4.22 22.5155 kJ/kg, −187.973 kJ/kg

4.23 136.55 kJ/kg

4.25 1192.6 K, −85.56 kJ, 405.997 kJ/kg

4.27 −179.55 kJ/kg, 64.46 kJ/kg

4.28 12.7084 kJ/kg

4.31 −715.69 kJ/kg

4.32 8096.26 kJ, 1.095 MPa

4.37 (a) 2607.75 kJ/kg, (b) 0.6043

4.44 949.66 kJ

4.46 (a) −59.16 kJ/kg, (b) 475.98 kJ/kg

4.49 (a) 0.221, (b) −149.6 kJ/kg, (c) 519.09 kJ/kg

4.54 (a) 201.5 kJ/kg, (b) 200.25 kJ/kg, (c) 198.253 kJ/kg

4.58 0.0386 kg/s

4.59 0.234

4.61 0.2987

4.66 833.7 kJ

4.67 11281.7 kJ

5.5 (a) 22.34 kW, (b) 6.17 kW

5.10 481.52 K, 1500 kJ

5.12 392.68 kJ

5.13 (a) 475.6 K, (b) 656.08 kJ, (c) 156.1 kJ, (d) 39.03%

5.21 (a) −0.50724 kJ/min K, (b) 0.84761 kJ/min K, (c) 0.34037 kJ/min K

5.22 68.1705 kJ/K

5.23 (a) 0.46325 kJ/kg K, (b) 0.25424 kJ/kg K

5.26 0.0896 kJ/K

5.29 0.343 kJ/K

5.37 0.696 kJ/kg K

5.39 319.79 K

5.40 3.924 kJ/K

5.46 (a) −22.52 kJ, (b) 22.6%

5.48 (a) 17.97%, (b) 0.263 kJ/kg K

5.50 (a) 0.0126 kg/s, (b) 9.69 kPa

5.51 0.612

5.53 (a) −88.285 kJ/kg, (b) −173.415 kJ/kg

5.60 (a) 35.736%, (b) −479.96 kJ/kg

5.62 599.32 K, 299.54 K

5.65 −305.49 kJ/kg, 1.78565 kJ/kg K, −747.73 kJ/kg

5.66 0.1022 kJ/K, 0.1126 kJ/K

5.67 452.41 K, 0.2238 kJ/K

5.72 (a) 0.0605 kJ/kg K, (b) 0.315 kJ/kg K

5.78 (a) 301.62 K, (b) 0.2096 kJ/kg K

5.84 46.05 kW, −0.524 kJ/kg K, 60.4 kW

5.90 (a) 521 K, (b) 174.53 kW, (c) 0.076 kJ/K.s, (d) 0.2316 m³/s

5.92 0.6019 kg/s

5.93 (a) 928 m/s, (b) 894 m/s, (c) 87 J/kg K

5.94 134.25 kJ/hr

5.95 0.0754 kJ/kg K

6.1 (a) −80 kJ, (b) −60 kJ

6.2 −140.804 kJ

6.4 (a) −10.64 kJ, (b) 10.64 kJ

6.5 354.45 K, 52.788 kJ

6.8 −336.6 kJ/kg

6.9 42.618 kJ, 167.175 kJ

6.10 5.047 kW

6.12 0.929, 54.08 kJ/kg, −941.554 kJ/kg

6.13 (a) 0.858, (b) 293.31 K, (c) 1.16 kJ/s, (d) 4.21 cm²

6.16 143.62 kJ/kg

6.18 (a) −156.24 kJ/kg, (b) −211.71 kJ/kg,

(c) −207.99 kJ/kg, (d) 51.75 kJ/kg

6.19 345.62 kJ/kg, 320.13 kJ/kg, 25.59 kJ/kg

6.20 (a) −100.35 kJ/kg, (b) −177.68 kJ/kg, (c) −214.52 kJ/kg, (d) 77.33 kJ/kg

6.23 584.11 m/s, 32.74 kJ/kg

6.25 376 kJ

6.26 (a) 397.53 kPa, (b) 120.09 kJ/kg, (c) −66.88 kJ/kg, (d) 66.88 kJ/kg

6.30 −148 kJ/kg, 185.7 kJ/kg

6.33 279.42 kJ, −115.32 kJ

6.34 128.11 kJ/kg

6.37 (a) 135.912 kJ/kg, (b) 64.61°C, (c) 17.264 kJ/kg

6.39 0.3061

6.40 (a) 551.925 kJ/kg, (b) 306.932 kJ/kg, (c) 244.993 kJ/kg, (d) 0.307, (e) 16.957 kW

6.41 288.4 kJ/kg

6.43 (a) 7.0185 kJ, 2.5463 kJ; (b) 4.4722 kJ

6.47 (a) 7.71816 kJ/kg K, (b) −3.12974 kJ/kg K, (c) 1253.33 kJ/kg

6.50 7.849 kJ/kg

6.51 (a) 69.97 kJ, (b) 30.15 kJ

6.53 −40.06 kW

6.54 (a) 50.25 kW, (b) 128.22 kW

6.55 −479 kJ/kg, −99.75 kJ/kg, 0.259 kJ/kg K

6.56 150.95 kW

6.57 (a) 186.4 kJ/kg, (b) 221.05 kJ/kg, (c) 23.85 kJ/kg

6.58 (a) 532.237 kJ/kg, (b) 135.14 kJ/kg, (c) 181.493 kJ/kg, (d) 0, (e) 33.18 kJ/kg

6.59 (a) 49.44 kW, (b) 46.92 kW

6.60 (a) 168.724, 106.864 kJ/kg;
         (b) 381.97 kW; (c) 380.87 K
6.63 (a) 285.875 kJ/kg,
         (b) 235.94 kJ/kg,
         (c) 49.94 kJ/kg,
         (d) 235.94 kJ/kg
6.64 (a), (b) 519.9 kW;
         (c) 62.13 kW; (d) 518.2 kW
6.65 (a) 761.8 kJ/kg, 195.81 kJ/kg;
         (b) −521.82 kJ/kg,
            −566 kJ/kg;
         (c) −628.84 kJ/kg
6.67 (a) 211.26 kJ/kg, 35.45 kJ/kg;
         (b) −80.28 kJ/kg,
            −175.81 kJ/kg;
         (c) −168 kJ/kg;
         (d) 95.53 kJ/kg
6.68 (a) 95.546 kJ/kg,
         (b) −91.82 kJ/kg,
         (c) 91.82 kJ/kg
6.69 (a) 2015.3 kJ/kg,
         (b) −193.313 kJ/kg,
         (c) −470.918 kJ/kg,
         (d) 277.605 kJ/kg
6.70 (a) 58.58 kJ/kg;
         (b) 1152.49 kJ/kg,
            586.405 kJ/kg; (c) 0.897
6.71 56.48%
6.72 89.34%
6.73 (a) 209.812 kJ/kg, (b) 0.7081
6.74 (a) 1.8437 kW, (b) 91.98%
7.1   150.16 kJ/kg, 18.46 kJ/kg
7.3   17.56 kW, −8.78 kW,
         20.125 kW
7.5   (a) 0.927 kJ/kg K,
         (b) 5.31 kg/kW-hr,
         (c) 276.385 kJ/kg
7.9   (a) 33.84%, (b) 34.09%
7.11 37.67%
7.13 38.37%
7.16 40.19%, 37.82%
7.17 (a) 312 kJ/kg, (b) 56.47%,
         (c) 23.61 kJ/kg
7.20 (a) 2.338 kg, (b) −360.57 kJ,
         (c) 54.93%, (d) 222.65 kPa

7.22 44.13 kW, 840.53 kPa
7.24 60.91%, 657.49 kPa
7.28 61.396%, 56.473%
7.33 58.4%, 0, 103.27 kJ/kg,
         21.2 kJ/kg, 0, 626.4 kJ/kg
7.38 0.268
7.41 (a) 49.3%, (b) 4.91 kg/s
7.44 0.2528
7.48 (a) 19.006%, (b) 28.793%
7.51 (a) −392.261 kJ/kg,
         (b) 0.4821, (c) 245.86 kJ/kg
7.56 (a) 5.453, 0.002592 kg/s;
         (b) 6.429
7.62 (a) 8.1 kW, (b) 1.54 kW,
         (c) 5.263
7.64 1.713, 71.677 kJ/kg
7.66 (a) 15.0646 kW,
         (b) 13.4329 kW, (c) 0.892
8.1   206.609 atm
8.5   2.96 K
8.7   271.67 K
8.17 1158.37 kJ/kg,
         −345.35 kJ/kg K
8.20 13.51 MJ/kg-mol
8.28 0.0125 K/kPa
8.29 384.78°C,
         15.225×10^{-3} °C/kPa

8.32 $\left(p + \dfrac{a}{v^2} \right)(v - b)^\gamma = \text{constant}$

8.39 −342.11 kJ/kg,
         −419.86 kJ/kg, 2.085 kJ/kg K
8.40 (a) 197.62 kW,
         (b) −1.16 kW/K,
         (c) −369.05 kW,
         (d) 0.00115 m³/s
8.42 115.75 kJ/kg
8.43 1816.44 kJ/kg-mol,
         −12.076 kJ/kg-mol K
8.44 45.8 kW
8.46 (a) −108.641 kJ/kg,
         (b) 0.1169 kJ/kg K
8.47 (a) 117.4 kJ/kg,
         (b) 0.087 kJ/kg K
8.48 49.164 kJ/kg

8.50 (a) −42.716 kJ/kg;
         (b) 0.10489 kJ/kg K;
         (c) −52.08 kJ/kg,
            0.08204 kJ/kg K
8.52 −119.5 kJ/kg, 0.3436 kJ/kg K
8.53 1047.2 kJ/kg
9.6   41.67 kPa
9.8   (a) 65.26°C, 100 kPa;
         (b) 61.577 kJ;
         (c) 0.46878 kJ/K;
         (d) 0.17321 kJ/K
9.9   34.82 kJ/K, same
9.11 (a) 18.683 m³, 244.5K,
            6.8 kPa;
         (b) 0.53 kJ/K
9.13 (a) 0.512, 0.488;
         (b) 346.2 kJ;
         (c) 0.1079 kJ/K
9.15 0.6145 kJ/K
9.17 0.2416 kJ/kg K
9.19 (a) 423.48 K,
         (b) 5.6727 kJ/kg-mol K
9.20 717.61 K, 568.25 kW
9.21 444.69 K, 163.37 kW
9.23 −1388.3 kW
9.28 591.86 kJ/kg
9.33 0.0088, 44.6%
9.34 63.84%
9.35 42°C
9.38 0.005, 0.3195, 33.77 kJ/kg of
         dry air
9.39 0.0057 kg/kg, superheated
         steam
9.42 0.0198 kg, 144 kPa
9.43 3194.1 kJ/kg water
         evaporated
9.46 −14.5651 kW, 5.8944 kW
9.50 0.01311 kg/kg, 65.11%
9.53 (a) 0.0091 kg vapor/kg dry
         air, (b) 35.81 kJ/kg dry air
9.54 32.5°C, 0.01952, 63%
9.59 2368.3 kg/hr, 52.36 kg/hr
10.1 12.662 kg air/kg fuel
10.3 (a) 15.12 kg air/ kg fuel;
         (b) 35.15 m³/kg fuel;

(c) 12.247 m^3/kg fuel;
(d) 15.558 m^3/kg fuel,
    43.326 m^3/kg fuel

10.5 (a) 16.21 kg air/kg fuel,
    (b) 3.37,
    (c) 29.438 m^3/kg-mol of dry
      products

10.8 45.416°C

10.9 94%

10.12 −74,850 kJ/kg-mol,
    −2,043,990 kJ/kg-mol

10.15 (a) −382,555 kJ/kg-mol,
    (b) −292,115 kJ/kg-mol

10.16 −2,043,990 kJ/kg-mol fuel,
    −1,449,840 kJ/kg mol fuel

10.22 (a) −2,877,080 kJ/kg-mol,
    (b) −2,877,080 kJ/kg-mol,
    (c) 2,657,089.8 kJ/kg-mol

10.25 1510 K

10.26 1590 K

10.27 −79362 kJ/kg-mol fuel

10.29 172%

10.30 256.8%

10.31 2670 K, 47730.8 kJ/kg-mol
    of CO

10.33 −2467.6 kW

10.37 −1,923,276 kJ/kg-mol fuel,
    4.465 atm

10.41 940.815 kJ/kg-mol K,
    280,504 kJ/kg-mol

10.42 467 kJ/kg-mol K

10.44 (a) 1349.792 kJ/kg-mol K,
    (b) −1,439,682 kJ/kg-mol,
    (c) 1,842,122 kJ/kg-mol

10.47 109.307.5 kJ/kg-mol

10.50 −2,108,310 kJ/kg-mol

10.56 (a) −707,386 kJ/kg-mol,
    (b) 39,663 kJ/kg-mol,
    (c) 717,838 kJ/kg-mol

10.62 0.9065

10.64 0.257

10.73 −1.896

10.78 2520 K

10.81 (a) 0, (b) 0.2178

11.3 (a) 2.2684 kg/s,
    (b) 8.5537 MW,
    (c) 60.97%

11.4 (a) 42.21%

11.5 44.15% (1.94%
  improvement)

11.6 (a) 12.0176, (b) 181.15 kg/s,
    (c) 42.42%

11.7 (a) 0.601, (b) −911.33 kW

11.8 (a) 0.5716, (b) −911.45 kW

11.9 12.985 MW

11.13 0.242 W, 1.245 × 10^{-3}/K,
    1.66%

11.14 −135.06 MJ/kg-mol H$_2$,
    −48.689 MJ/kg-mol H$_2$

11.15 1.0332 W, −0.7452 W

11.17 0.0376 m^3/hr, 74.67 kW

12.7 $0.7v_{mp}$, $1.45v_{mp}$

12.8 (a) 8.38%, 57.73%;
    (b) 1.69%

12.9 5.784 × 10^{25} molecules

12.11 1027 K

12.12 0.081

12.13 72.98%

12.15 (a) 2.856 × 10^{23}
      collisions/cm^2 s,
    (b) 3.77 × 10^{15}
      collisions/cm^2 s

12.19 1.55 × 10^5 years

12.21 0.249, 1.38 × 10^{-9} kg/s

12.22 5.32 min

12.27 (a) $\dfrac{1}{16}$, (b) $\dfrac{1}{22}$, (c) $\dfrac{7}{12}$

12.29 (a) 3,628,800; (b) 2520

12.36 7

12.42 2.574, (a) ε = 2 units,
    (b) 0.471

12.49 1.1573 × 10^{-8} m

13.1 2.426 × 10^{24},
    2.02 × 10^{26}

13.2 7.725 × 10^{24},
    151.988 × 10^{-3} J,
    20 × 10^{-3} J/K

13.6 3.86 × 10^{24} states

13.7 2.278 × 10^{-38} J/molecule

13.19 4.475 × 10^{-40} gm-cm^2,
    1.65 Å

13.23 (a) 129.173 × 10^{33} /kg-mol,
    (b) 80,273.3 kJ/kg-mol,
    (c) 230.27 kJ/kg-mol K,
    (d) −38,027.8 kJ/kg-mol

13.24 20,786 kJ/kg-mol,
    119.375 kJ/kg-mol K,
    −98,589.7 kJ/kg-mol

13.25 2969 K

13.30 145.3857 kJ/kg-mol K

13.32 2.52 × 10^{-4} Pa, $p = \dfrac{1}{3}\left(\dfrac{U}{T}\right)$

13.36 7.5635 × 10^{-10}
    J/cm^3-molecule, 6.6%

13.37 0.03086 kJ/kg-mol K,
    24.68 kJ/kg K

13.41 24.692 kJ/kg-mol K

Index